Oxytocin, Vasopressin, and Related Peptides in the Regulation of Behavior

Edited by

Elena Choleris
University of Guelph, Ontario, Canada

Donald W. Pfaff
Rockefeller University, New York, USA

Martin Kavaliers
University of Western Ontario, London, Canada

CAMBRIDGE
UNIVERSITY PRESS

CAMBRIDGE UNIVERSITY PRESS
Cambridge, New York, Melbourne, Madrid, Cape Town,
Singapore, São Paulo, Delhi, Mexico City

Cambridge University Press
The Edinburgh Building, Cambridge CB2 8RU, UK

Published in the United States of America by Cambridge University Press, New York

www.cambridge.org
Information on this title: www.cambridge.org/9780521190350

First published 2013

Printed and bound in the United Kingdom by the MPG Books Group

A catalog record for this publication is available from the British Library

Library of Congress Cataloging in Publication data
Oxytocin, vasopressin, and related peptides in the regulation of behavior /
edited by Elena Choleris, Donald W. Pfaff, Martin Kavaliers.
p. ; cm.
Includes bibliographical references and index.
ISBN 978-0-521-19035-0 (hardback)
I. Choleris, Elena. II. Pfaff, Donald W., 1939– III. Kavaliers, Martin.
[DNLM: 1. Behavior – drug effects. 2. Oxytocin – pharmacology.
3. Behavior Control – methods. 4. Behavior, Animal – drug effects.
5. Mental Disorders – drug therapy. 6. Vasopressins – pharmacology. QV 173]
615.7′042 – dc23 2012039023

ISBN 978-0-521-19035-0 Hardback

Oxytocin, Vasopressin, and Related Peptides in the Regulation of Behavior

The mammalian neurohypophyseal peptide hormones oxytocin and vasopressin act to mediate human social behavior – they affect trust and social relationships and have an influence on avoidance responses. Describing the evolutionary roots of the effects that these neuropeptides have on behavior, this book examines remarkable parallel findings in both humans and non-human animals.

The chapters are structured around three key issues: the molecular and neurohormonal mechanisms of peptides; phylogenetic considerations of their role in vertebrates; and their related effects on human behavior, social cognition, and clinical applications involving psychiatric disorders such as autism. A final chapter summarizes current research perspectives and reflects on the outlook for future developments.

Providing a comparative overview and featuring contributions from leading researchers, this is a valuable resource for graduate students, researchers, and clinicians in this rapidly developing field.

Elena Choleris is Professor of Psychology and Neuroscience at the University of Guelph, Ontario, Canada. Her main field of expertise is the neurobiology of social behavior in rodents.

Donald W. Pfaff is Professor and Head of the Laboratory of Neurobiology and Behavior at the Rockefeller University, New York, USA. A Member of the National Academy of Sciences, he was awarded the 2011 Lehrman Lifetime Achievement Award by the Society for Behavioral Neuroendocrinology.

Martin Kavaliers is Professor of Psychology and Neuroscience at the University of Western Ontario, London, Canada. His main field of expertise is the neurobiology of biobehavioral responses to naturalistic stressors in rodents.

Contents

Part II Behavioral studies – Comparative approach

Part III Human studies

Color plates will fall between pages 210 and 211

Preface

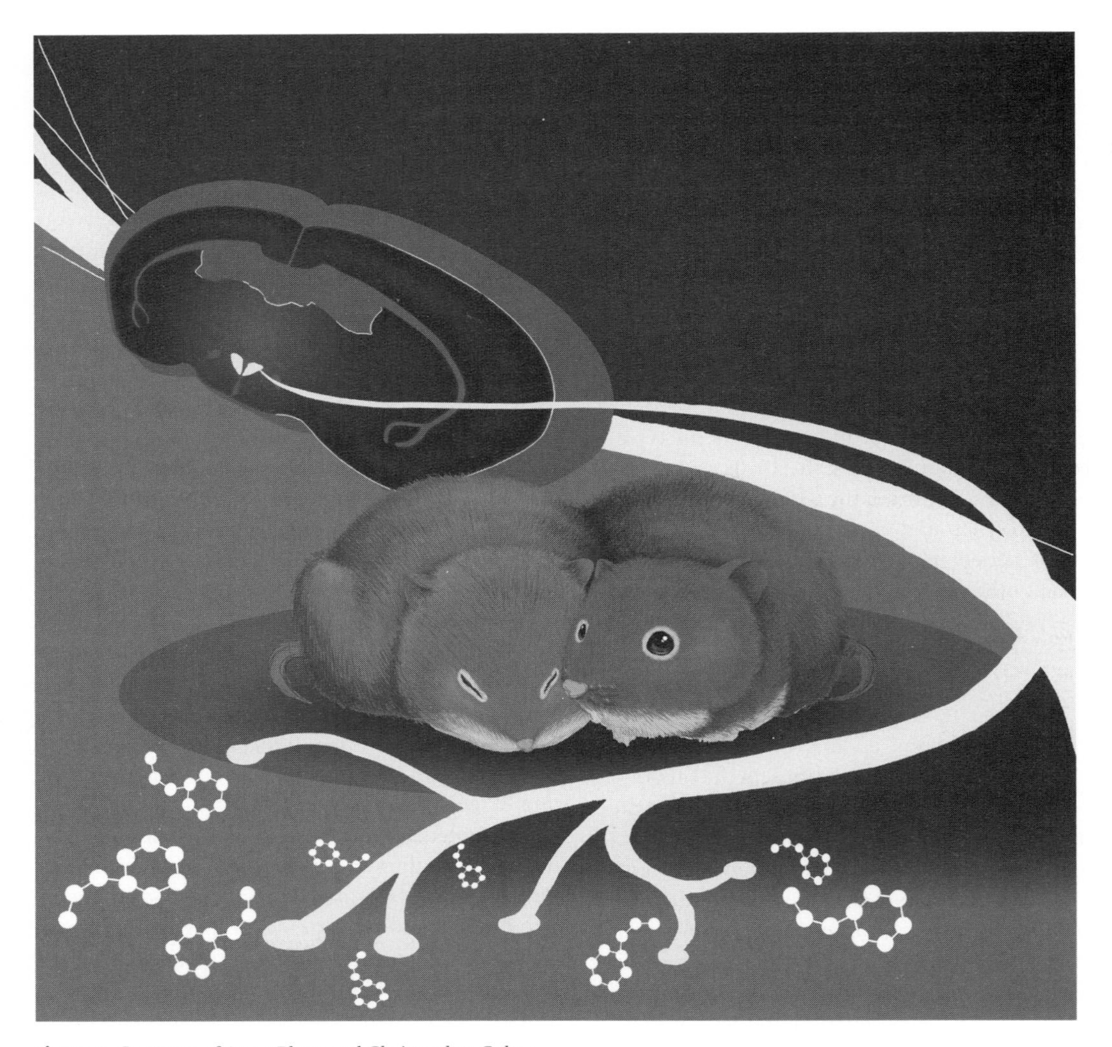

Figure 1 Courtesy of Anna Phan and Christopher Gabor.

Comparative approaches to oxytocin, vasopressin, and vertebrate behavior

This text is intended by the three of us to serve upper-level undergraduate students and beginning graduate students who are interested in how relatively well understood neurochemical systems regulate natural behaviors in animals, including humans. Some of the strongest causal links discovered, to date, between molecular biological phenomena and behavioral regulation have to do with hormones. This is especially true for hormones whose chemistry is relatively simple. Classically, those causal links have involved steroid hormones, produced in peripheral organs, telling the brain what is going on in the rest of the body, and thus allowing the brain to regulate behavior in a manner consonant with the state of the body. In this text, chapters explicate molecular/behavioral regulation in the opposite direction: hormones that are produced in the vertebrate brain, by specific groups of nerve cells in the basal forebrain, not only enter the circulation but also act as neuromodulators within the central nervous system. Oxytocin and arginine vasopressin, whose chemical structures in the vertebrates were elucidated during the 1950s and whose genes were cloned during the 1980s, each has only nine amino acids and each peptide has its structure constrained by a disulfide bridge. Differing from each other by only two animo acids, the two neuropeptides or "nonapeptides" have a fascinating role across the vertebrates.

As described, you will see in this text that oxytocin, vasopressin, and related neuropetides have a variety of behavioral actions in vertebrate animals ranging from fishes to humans. In the broadest sense the two hormones produced in the brain are "telling" the body what behavioral and physiological function these particular basal forebrain cell groups need to have accomplished. A series of foundational chapters lay the basis for understanding the regulation and expression of oxytocin and vasopressin systems. This is followed by a number of chapters that utilize a phylogenetic/comparative approach to describe the behavioral roles of oxytocin and vasopressin and related neuropetides across vertebrate species. Finally, a number of chapters consider the roles of oxytocin and vasopressin in the modulation of human behavior.

Evolutionary foundations and the roles of oxytocin/vasopressin-related neuropetides in invertebrates

Although this text is designed to provide a comparative behavioral approach to oxytocin and vasopressin and related peptides in the vertebrates, to more fully appreciate the roles of oxytocin and vasopressin it is useful to understand their evolutionary history and invertebrate foundations. Although oxytocin and vasopressin are only found in mammals, members of the two neuropeptide systems constitute one of the most ancient and evolutionarily conserved neuropetide systems. OT and AVP belong to a large superfamily found in a wide range of vertebrate and invertebrate (e.g., hydra, worms and some insect) species (for reviews see Archer, 1972; Donaldson and Young, 2008; Goodson 2008). In the jawed vertebrates oxytocin-like and vasopressin-like neuropetide lineages arose from a common ancestral gene by local duplication in a gawed vertebrate ancestor (Goodson, 1998). Invertebrates, with a few exceptions (e.g., cephalopods), have only one oxytocin/vasopressin gene family homolog (e.g., annetocin (annelid worms), conopressin (snails, sea hare, leeches), inotocin (some insects)) (Donaldson and Young, 2008). Interestingly, in the insects oxytocin/vasopressin-like peptides were found in flies, mosquitoes, some beetles but not in the more advanced eusocial honey bee (Stafflinger et al., 2008).

The molecular structure and behavioral actions mediated by these neuropeptides and their receptors in the invertebrates are in many respects comparable to those of vertebrates. For example, just as oxytocin and vasopressin are produced in the neurosecretory magnocellular neurons in the vertebrate hypothalamus so the oxytocin/vasopressin homolog, annetocin, is expressed in, and released

from, the sensory neurosecretory "brain" counterpart of annelid worms. Indeed, the annelid neurons express the same micro-RNAs and transcription factors as do the neurosecretory magnocellular neurons of vertebrates (Tesmair-Raible et al., 2007). At a functional level oxytocin/vasopressin-like neuropetide involvement in osomoregulation and fluid balance is also evident across the animal phyla (Goodson, 1998). It is tempting to speculate that this early involvement in the regulation of responses to osmotic stress may lay the foundation for the evolution of neuropetide mechanisms that modulate interactions with the environment and stress responses.

Oxytocin and vasopressin's association with reproduction, parental and socio-sexual behaviors and responses are also evolutionarily conserved, even though the specific behaviors affected can be species and taxa specific. For example, several members of the oxytocin/vasopressin family evoke response related to reproduction in annelids and leeches (Fujino et al., 1999; Wagenaar et al., 2010). Similarly, conopressin, a molluscan (snail) homolog of oxytocin/vasopressin, modulates ejaculation in males and egg-laying in females. (Oumi et al., 1996). These early reproductive roles may have set the stage for the evolution of the involvement of these neuropetides in various socio-sexual functions described in this book for the vertebrates. Snails present another particularly fascinating example of the evolutionarily flexibility of the oxytocin/vasopressin system. The venom of cone snails contains an endogenous vasopressin analog, conporessin-T, that functions as a vasopressin antagonist. These venoms, which are injected through specialized mouth parts of the cone snail and are used to catch prey or for protection against predators, may in part exert their actions thorough modifications in the effects of vasopressin-like neuropetides (Dutertre et al., 2008). Finally, in the most advanced of the molluscs, the cephalopods (octopus, cuttlefish), there are two superfamilies of oxytocin/vasopressin-like peptides members (octopressin and cephaloctocin (Minakat, 2010)) that exert effects on cuttlefish learning and memory similar to those of OT/AVP in mammals (Bardou et al., 2010). As described, you will see in this text a range of behavioral roles of oxytocin and vasopressin the vertebrates that build upon these invertebrate foundations.

Comments to us by students and other readers will be welcome, because shortcomings of the current effort could be remedied in a second edition of this text.

Finally, we want to thank our editors at the Cambridge University Press, Chris Curcio and Martin Griffiths, for shepherding this project through the publication process.

E. C., D. W. P. and M. K.
July 2012

REFERENCES

Archer, R. and Cauvet, J. (1995). The neurohypophysial endocrine regulatory cascade: precursors, mediators, receptors and effectors. *Fontiers in Neuroendocrinology*, 16, 237–289.

Bardou, I., Leprince, J., Chichery, R., Vaudry, H., and Agin, V. (2010). Vasopressin/oxytocin-related peptides influence long-term memory of a passive avoidance task in the cuttlefish, *Sepia officinalis. Neurobiology of Learning and Memory*, 93, 240–247.

Donaldson, Z. R. and Young, J. L. (2008). Oxytocin, vasopressin, and the neurogenetics of sociality. *Science*, 322, 900–904.

Dutertre, S., Croker, D., Daly, N. L., Andersson, A., Muttenthaler, M., Lumsden, N. G., Craikm D. J., Alewood, P. F., Guillon, G., and Lewis, R. J. (2008). Conopressin-T from *Conus tulipa* reveals an antagonist switch in vasopressin-like peptides. *Journal of Biological Chemistry*, 283, 7100–7108.

Fujino, Y., Nagahama, T., Oumi, T., Uken, K., Morishita, F., Furukawa, Y., Matasushima, O., Ando, M., Takahama, H., Sataks, H., et al. (1999). Possible functions of oxytocin/vasopressin-superfamily peptides in annelids with special reference to reproduction and osmoregulation. *Journal of Experimental Zoology*, 284, 401–406.

Goodson, J. L. (2008) Neuropeptides and the evolutionary patterning of sociality. *Progress in Brain Research*, 170, 3–15

Hoyle, C. H. V. (1998) Neuropetide families: evolutionary perspectives. *Regulatory Peptides*, 73, 1–33.

Minakata, H. (2010). Oxytocin/vasopressin and gonadotropin-releasing hormone from cephalopods to vertebrates. *Annals of the New York Academy of Science*, 1200, 3–42.

Oumi, T., Ukena, K., Matasushima, O., Ikeda, T., Fujita, T., Minakata, H., and Nomoto, K. (1996). Annetocin, an anelid oxytocin-related peptide, induces egg-laying behavior in the earthworm, *Eisenia foetida*. *Journal of Experimental Zoology*, 276, 151–156.

Stafflinger, E., Hansen, K. K., Hauser, F., Scneider, M., Cazzamali, G., Williamson, M., and Gimmelikhuijzen, C. J. P. (2008) Cloning and identification of an oxytocin/vasopressin-like receptor and its ligand from insects. *Proceedings of the National Academy of Sciences USA*, 105, 3262–3267.

Tessmar-Raible, K., Raible, F., Christodoulou, F., Guy, K., Rembold, M., Hausen, H., and Arendt, D. (2007). Conserved sensory-neurosecretory cell types in annelid and fish forebrain: insights into hypothalamus evolution. *Cell*, 129, 1389–1400.

Wagennar, D. A., Sarhas Hamilton, M., Huang, T., Kristan, W. B., and French, K. A. (2010) A hormone-activated central pattern generator for courtship. *Current Biology*, 20, 487–495.

Contributors

H. Elliott Albers
Center for Behavioral Neuroscience, Neuroscience Institute, Georgia State University, Atlanta, GA, USA

Reut Avinun
Neurobiology, Hebrew University, Jerusalem, Israel

Karen L. Bales
Department of Psychology, University of California, Davis; and California National Primate Research Center, CA, USA

Jorge A. Barraza
Center for Neuroeconomics Studies, Claremont Graduate University, Claremont, CA, USA

Michael T. Bowen
School of Psychology, University of Sydney, Australia

Sunny K. Boyd
Department of Biological Sciences, University of Notre Dame, Notre Dame, IN, USA

Heather K. Caldwell
Laboratory of Neuroendocrinology and Behavior, Department of Biological Sciences and School of Biomedical Sciences, Kent State University, Kent, OH, USA

Elena Choleris
Department of Psychology and Neuroscience Program, University of Guelph, Guelph, Ontario, Canada

Amy E. Clipperton-Allen
Department of Psychology and Neuroscience Program, University of Guelph, Guelph, Ontario, Canada

Bruce S. Cushing
Department of Biology and Integrated Bioscience Program, University of Akron, Akron, OH, USA

Monica B. Dhakar
Laboratory of Neuroendocrinology and Behavior, Department of Biological Sciences and School of Biomedical Sciences, Kent State University, Kent, OH, USA

Riccardo Dore
Department of Psychology and Neuroscience Program, University of Guelph, Guelph, Ontario, Canada

Richard P. Ebstein
Psychology Department, National University of Singapore, Singapore, Psychology Department, Hebrew University, Jerusalem, Israel

Craig F. Ferris
Department of Psychology, Northeastern University, Boston, MA, USA

Sara M. Freeman
Center for Translational Social Neuroscience, Division of Behavioral Neuroscience and Psychiatric Disorders, Yerkes National Primate Research Center, Emory University, Atlanta, GA, USA

James L. Goodson
Department of Biology, Indiana University, Bloomington, IN, USA

Joshua J. Green
Albert Einstein College of Medicine and Montefiore Medical Center, New York, NY, USA

Haruhiro Higashida
Department of Biophysical Genetics, Kanazawa University Graduate School of Medicine, Kanazawa, Japan

Eric Hollander
Albert Einstein College of Medicine and Montefiore Medical Center, New York, NY, USA

Salomon Israel
Psychology Department, Hebrew University, Jerusalem, Israel

Martin Kavaliers
Department of Psychology, University of Western Ontario, London, Ontario, Canada

Keith M. Kendrick
School of Life Sciences and Technology, University of Electronic Science and Technology of China, Chengdu, PR China

Ariel Knafo
Psychology Department, Hebrew University, Jerusalem, Israel

Yoav Litvin
Laboratory of Neurobiology and Behavior, Rockefeller University, New York, NY, USA

Olga Lopatina
Department of Biophysical Genetics, Kanazawa University Graduate School of Medicine, Kanazawa, Japan and Department of Biochemistry, Krasnoyarsk State Medical University, Russia

David Mankuta
Department of Obstretics and Gynecology, Hadassah Medical Center, Hebrew University, Jerusalem, Israel

Iain S. McGregor
School of Psychology, University of Sydney, Australia

Richard H. Melloni, Jr.
Department of Psychology, Northeastern University, Boston, MA, USA

Inga D. Neumann
Department of Neurobiology and Animal Physiology, University of Regensburg, Regensburg, Germany

Jerome H. Pagani
Section on Neural Gene Expression, National Institute of Mental Health, NIH, DHHS. Bethesda, MD, USA

Cort A. Pedersen
Department of Psychiatry, University of North Carolina at Chapel Hill, Chapel Hill, NC, USA

Donald W. Pfaff
Laboratory of Neurobiology and Behavior, Rockefeller University, New York, NY, USA

Anna Phan
Department of Psychology and Neuroscience Program, University of Guelph, Guelph, Ontario, Canada

Benjamin J. Ragen
Department of Psychology, University of California, Davis; and California National Primate Research Center, CA, USA

Amina Sarwat
Department of Biophysical Genetics, Kanazawa University Graduate School of Medicine, Kanazawa, Japan

Idan Shalev
Neurobiology, Hebrew University, Jerusalem, Israel

Erica L. Stevenson
Laboratory of Neuroendocrinology and Behavior, Department of Biological Sciences and School of Biomedical Sciences, Kent State University, Kent, OH, USA

Bonnie Taylor
Albert Einstein College of Medicine and Montefiore Medical Center, New York, NY, USA

Richmond R. Thompson
Department of Psychology, Neuroscience Program, Bowdoin College, Brunswick, ME, USA

Florina Uzefovsky
Psychology Department, Hebrew University, Jerusalem, Israel

Erwin H. van den Burg
Department of Neurobiology and Animal Physiology, University of Regensburg, Regensburg, Germany

James C. Walton
Department of Neuroscience, Ohio State University, Wexner Medical Center, Columbus, OH, USA

Scott R. Wersinger
Department of Psychology, University at Buffalo, SUNY, Buffalo, NY, USA

Nurit Yirmiya
Psychology Department, Hebrew University, Jerusalem, Israel

Larry J. Young
Center for Translational Social Neuroscience, Division of Behavioral Neuroscience and Psychiatric Disorders, Yerkes National Primate Research Center, Emory University, Atlanta, GA.

W. Scott Young, III
Section on Neural Gene Expression, National Institute of Mental Health, NIH, DHHS. Bethesda, MD, USA

Paul J. Zak
Center for Neuroeconomics Studies, Claremont Graduate University, Claremont, CA, USA

PART I

Oxytocin and vasopressin systems

Anatomy, function, and development

Oxytocin, vasopressin, and their interplay with gonadal steroids

Monica B. Dhakar, Erica L. Stevenson, and Heather K. Caldwell

1.1 Overview

The neuropeptides oxytocin (OT) and vasopressin (AVP) are two evolutionarily ancient neurohormones known to influence mammalian sex-specific and species-specific behaviors. The gonadal steroids are also important modulators of many mammalian behaviors. Thus, it is not surprising that there are profound and complex interactions between these two systems. This chapter will provide an overview of the OT and AVP systems, including their interactions with gonadal steroids.

1.2 Oxytocin and vasopressin

OT and AVP are composed of nine amino acids and differ from one another by only two amino acid residues, specifically those in the third and eighth positions (as reviewed in Hara, 1990; Burbach et al., 2001; Young and Gainer, 2003; Caldwell et al., 2008; Donaldson and Young, 2008; Lee et al., 2009). Their gene structures are also similar as they are the result of a gene duplication of the ancestral vasotocin gene, which occurred approximately 700 million years ago (Acher and Chauvet, 1995). While OT and AVP are prominent only in mammals, they are a part of a peptide family that is conserved across phyla (as reviewed in Caldwell, 2008; Lee, 2009). Both OT and AVP amino acids sequences are largely conserved across mammalian species, with a notable exception in OT in some species of New World primates. This novel OT was dubbed [P8] OT due to the substitution of a proline for a leucine in the eighth position (Lee, 2011). Across species, OT and AVP are important to the regulation of social interactions, with OT being mostly identified with bonding between individuals, and AVP being mostly identified with the regulation of aggression and male parental care; though, their roles are not nearly so restricted as these generalizations.

1.2.1 Oxytocin and vasopressin gene and protein structures

Within a species, the *OT* and *AVP* genes are located on the same chromosome (i.e., chromosome 2 in mice, 20 in humans, and 3 in rats) and contain three exons and two introns. The genes are oriented in opposing transcriptional direction on the chromosome and are separated by a region of DNA referred to as the intergenic region (IGR). The IGR is highly variable across species, being approximately 11 kbp in rat (Mohr et al., 1988) and human (Gainer et al., 2001) and approximately 3.6 kbp in mouse (Hara et al., 1990). The significance of the IGR is not completely understood, but in the hypothalamus, portions of the IGR appear to be critical for the normal expression of OT and AVP (Fields et al., 2003; Young and Gainer, 2003).

Oxytocin, Vasopressin, and Related Peptides in the Regulation of Behavior, ed. E. Choleris, D. W. Pfaff, and M. Kavaliers.

OT and AVP are each synthesized as part of a larger preprohormone, which contains a signal peptide, the biologically active peptide, a neurophysin, and a glycoprotein. The first exon encodes the 5′ non-coding promoter region, the nonapeptide, the tripeptide processing signal, and the first nine residues of the neurophysin. The second exon encodes for the bulk of the neurophysin molecule (residues 10–76), and the third exon encodes the remainder of the neurophysin (77–93/95 residues), including the COOH terminal, as well as the glycopeptide of the AVP preprohormone (Gainer et al., 2001; Gimpl and Fahrenholz, 2001). While neurophysin does not possess biological activity, it is thought to play a role in protecting OT and AVP from enzymatic degradation (de Bree, 2000). Also, as AVP is considered less biologically stable than OT, it has been proposed that the glycopeptide portion of the AVP preprohormone may be important for folding of the AVP precursor (Barat et al., 2004).

Based on the sequence analysis of *OT* and *AVP* in several species, including rat (Ivell and Richter, 1984), human (Sausville et al., 1985), cow (Ruppert et al., 1984), sheep (Ivell et al., 1990), and mouse (Hara et al., 1990), the transcriptional start site of the genes are found downstream of a TATA-like sequence in the 5′ flanking region (Hara, 1990; Gainer et al., 1994). Upstream of this start site, within the putative promoter, there are several regulatory elements that provide an opportunity for gonadal steroids to affect the transcription of *OT* and *AVP*. There are estrogen response elements (EREs), which allow estrogens bound to estrogen receptor alpha (ERα) or estrogen receptor beta (ERβ) to directly affect transcription, as well as a highly conserved DNA segment called the multiple hormone response element (HRE), which binds to multiple members of the retinoic acid and thyroid hormone receptor superfamily (Mohr et al., 1988; Adan et al., 1991; Richard and Zingg, 1991). There are also identified promoter regions that allow for the more "indirect" action of gonadal steroids, including activator protein-1 (AP-1) and activator protein-2 (AP-2) sites. The details of these regulatory elements will be detailed below (Section 1.4).

1.2.2 Distribution of oxytocin and vasopressin neurons and fibers

OT and AVP are primarily synthesized in the magnocellular neurons of the paraventricular (PVN) and supraoptic nuclei (SON) of the hypothalamus. The axons of these neurons project to the posterior pituitary, ultimately releasing OT and AVP into the bloodstream where their peripheral effects can be exerted. OT and AVP fibers are widely distributed within the central nervous system (CNS) and originate from other neurons, either within the PVN, SON, or elsewhere. It is the actions of OT and AVP in many of the subcortical regions, described below, that are involved in the regulation of aspects of social and sexual behavior in mammals.

1.2.2.1 Oxytocin

In most species, central OT production is limited to the PVN and SON. However, in mice, smaller quantities of OT appear to be produced by neurons in the bed nucleus of stria terminalis (BNST), the medial preoptic area (MPOA), and the amygdala (Castel and Morris, 1988; Jirikowski et al., 1990; Wang et al., 1996). There are also subtle species differences found in the extrahypothalamic distribution of OT-immunoreactive (OT-ir) neurons. In rats and humans, the parvocellular neurons of the PVN provide robust projections to the olfactory bulb, the dorsal and ventral hippocampus, the amygdala, the substantia nigra and substantia grisea, the nucleus of solitary tract and the nucleus ambiguous of the brainstem, and to the substantia gelatinosa of the spinal cord (Buijs, 1978; Sofroniew, 1980; Swanson and Kuypers, 1980; Rinaman, 1998). In contrast to the rat, the guinea pig (*Cavia porcellus*) shows prominent OT-ir neurons in the visual pathway, the retrochiasmatic and subchiasmatic areas, and in the medial preoptic nucleus (MPN) (Dubois-Dauphin et al., 1989b), and in the garden mouse (*Eliomys quercinus L.*), there are OT-ir neurons in the prefrontal cortex, the claustrum, and the septum (Hermes et al., 1988). Species differences have also been

detected in the distribution of OT-ir cells within four closely related species of voles (Wang et al., 1996).

1.2.2.2 Vasopressin

Unlike OT, central AVP is produced in a variety of brain areas other than the PVN and SON. In all mammals studied to date, there are AVP-producing neurons found in the preoptic area (POA) and anterior hypothalamus (AH), and in many species there are AVP-immunoreactive (AVP-ir) cells in the BNST and medial amygdala (MeA). These latter two sites send projections to the lateral septum (LS) and lateral habenular nucleus (LHb). The suprachiasmatic nucleus (SCN), the primary mammalian circadian clock, is also an area rich in AVP-producing neurons (Buijs et al., 1978; Castel et al., 1990).

The AVP system has been examined in many species, including: marmosets (*Callithrix jacchus*) (Wang et al., 1997), golden hamsters (*Mesocricetus auratus*) (Delville et al., 1994), prairie voles (*Microtus ochrogaster*) (Wang and De Vries, 1993; Wang, 1995), meadow voles (*Microtus pennsylvanicus*) (Wang, 1995), Djungarian hamsters (*Phosdopus sungorus*) (Dubois-Dauphin et al., 1994), European hamsters (*Cricetus cricetus*) (Buijs et al., 1986), garden dormouse (*Eliomys quercinus*) (Hermes et al., 1990), and a variety of laboratory rat and mouse strains (Sofroniew, 1983; De Vries and al-Shamma, 1990). In most mammalian species, AVP-ir fibers can be found in the POA, anterior and lateral hypothalamic areas, midbrain tegmentum, periaquaductal grey, locus coeruleus (LC), LS, LHb, nucleus of the solitary tract, and area postrema (Moore and Lowry, 1998).

1.3 Oxytocin and vasopressin receptors

To date, only one receptor subtype has been identified for OT, the oxytocin receptor (OTR). Whereas three receptor subtypes have been identified for AVP: the vasopressin 1a receptor (AVPR1a), the vasopressin 1b receptor (AVPR1b), and the vasopressin 2 receptor (AVPR2). As the AVPR2 is not found centrally it will not be discussed in this chapter, but is reviewed in (Barberis et al., 1998; Foletta et al., 2002; Bankir et al., 2010). While there are instances of dramatic gonadal steroid-dependent species and sex differences in OT and AVP, gonadal steroid-dependent changes in distributions of the OTR and the AVPR1a seem to contribute greatly to behavioral differences within and between species.

1.3.1 Oxytocin receptor structure and distribution

The OTR was first sequenced from human myometrium by Kimura and colleagues (1992). Subsequently, the OTR has been cloned and sequenced in a variety of species, including rat (Rozen et al., 1995), vole (Young et al., 1996), mouse (Kubota et al., 1996), rhesus monkey (Salvatore et al., 1998), cow (Bathgate et al., 1995), and pig (Gorbulev et al., 1993). The OTR is a member of the G protein-coupled receptor family and signals via activation of $G_{q/11\alpha}$ class GTP binding proteins and generation of inositol triphosphate and 1,2-diacylglycerol (for review, see Gimpl and Fahrenholz, 2001; Young and Gainer, 2003).

In humans, the chromosomal location of the OTR has been mapped to gene locus 3p25–3p26.2, using fluorescence *in situ* hybridization, (Inoue et al., 1994; Michelini et al., 1995; Simmons et al., 1995). The OTR gene contains 4 exons and 3 introns and spans approximately 17 kbp and encodes approximately 389 amino acids. Exon 1 and 2 encode the 5′ non-coding region and exons 3 and 4 encode the receptor protein. The transcription start site lays 618–621bp upstream of the methionine initiation codon. Twenty-eight to 31bp upstream of the transcription start site is a TATA-like motif and 65bp upstream is a potential specific protein-1 (SP-1) binding site. The 5′ flanking region, while lacking a classic ERE, has binding sites for other transcription regulating factors such as AP-1, AP-2, GATA-1, and c-Myb. It also contains two half-palindromic 5′-GGTCA-3′ and one half-palindromic 5′-TGACC-3′ ERE motifs (Inoue et al., 1994).

The amino acid sequence of the mouse and the rat OTR have 91% and 93% homology with the human OTR, respectively (Rozen et al., 1995; Kubota et al., 1996). The gene structure in mouse is similar to that in humans, except that the promoter region lacks an apparent TATA box, but does contain putative interleukin-response elements and an ERE (Kubota et al., 1996). The rat OTR, on the other hand, spans more than 20 kbp and contains 3 exons. The promoter region contains multiple putative interleukin-response elements but lacks an apparent TATA or CCATT box (Rozen et al., 1995). The 5′-flanking region has been shown to contain a palindromic ERE within 4 kbp of the translational start site (Bale and Dorsa, 1997).

The OTR is widely distributed throughout the central nervous system, however regional distribution shows marked species and sex differences. The distribution of the OTR has been extensively studied in the rat brain using two radiolabeled ligands: 1) a tritiated OT ($[^3H]$OT) and 2) an iodinated OT antagonist ($[^{125}I]$OTA) (Tribollet et al., 1992). Specific OT binding sites are found in numerous areas including anterior olfactory nucleus, cell groups of the olfactory tubercle, LS, BNST, hypothalamic ventromedial nucleus (VMH), PVN, central amygdala (CeA), MeA, shell of the nucleus accumbens, ventral subiculum of the hippocampus, and the caudoputamen region (Brinton et al., 1984; De Kloet et al., 1985; Tribollet et al., 1988; Tribollet et al., 1990). In the rat brain, the expression of OTR mRNA, as detected by *in situ* hybridization, corresponds to the location of OT binding sites (Yoshimura et al., 1993). This suggests that the site of synthesis of protein is the same as the location of the OTR in the central nervous system.

1.3.2 Vasopressin receptor structure and distribution

The AVPR1 was initially characterized by Jard and colleagues (1983) and was later broken down into subtypes, AVPR1a and AVPR1b (Antoni, 1984; Jard et al., 1986). Both receptor subtypes are G protein-coupled receptors that activate $G_{\alpha q/11}$ GTP binding proteins, which in turn activate phospholipase C with the help of $G_{\beta\lambda}$ (Michell et al., 1979; Jard et al., 1987). The *AVPR1a* gene is made up of two exons divided by one intron (~1.8 kbp) and spans 3.8 kbp total. Suggestive of a common ancestry, most of the AVP/OT receptor family has six of the seven transmembrane domains encoded by a single exon, and the seventh transmembrane domain encoded by a separate exon. The cDNA is made up of 1354 nucleotides that produce 394 amino acids (Morel et al., 1992); with the rat and human AVPR1a sharing 72% of their amino acid sequence (Thibonnier et al., 1994). The AVPR1a gene promoter contains three transcriptional initiation sites at –405, –243 and –236bp upstream of the start codon, the major sites are at –243 and –236bp (Murasawa et al., 1995). The promoter for the *AVPR1a* contains no TATA or CCAAT promoter elements and has a high G and C content (~62%) (Murasawa et al., 1995). The AVPR1a gene promoter also contains several regulatory elements, including AP-1, AP-2, and SP-1 binding sequences (Murasawa et al., 1995).

Unlike the *AVPR1a*, the *AVPR1b* is made up of three exons and two introns in mouse and rat strains. Exon 2 encodes six of the seven transmembrane domains, while exon 3 encodes for the seventh domain. There are two transcription start sites at –861 and –830bp relative to the start codon. The *AVPR1b*, like the *AVPR1a*, does not contain a proximal TATA box. However, there is a CACA box and an inverted GAGA box present in the promoter, which is unusual for G protein-coupled receptors. There is no sequence homology between the *AVPR1b* promoter regions of the mouse and rat except for the location of the CACA and inverted GAGA boxes, suggestive that these are the sequences conserved in the *AVPR1b*. The promoter also contains several regulatory elements, including three AP-1, five AP-2, three SP-1, and two CCAAT-enhancer-binding-protein (C/EBP) sites as well as a glucocorticoid response element (GRE) (Rabadan-Diehl et al., 2000).

The AVPR1a is widely distributed throughout the central nervous system, but its localization

is fairly conserved across mouse and rat strains (Johnson et al., 1993; Tribollet et al., 1997; Young et al., 2000). Radiolabeled receptor autoradiography reveals AVPR1a binding in the LS, neocortical layer IV, hippocampal formation, amygdalostriatal area, BNST, hypothalamus, ventral tegmental area (VTA), substantia nigra, superior colliculus, dorsal raphe, nucleus of the solitary tract, and superior olive (Johnson et al., 1993). *In situ* hybridization histochemistry shows prominent AVPR1a mRNA in the olfactory bulb, hippocampal formation, LS, SCN, PVN, AH, arcuate nucleus, LHb, VTA, substantia nigra, LC, inferior olive, area postrema, and nucleus of the solitary tract (Ostrowski et al., 1994; Szot et al., 1994).

The AVPR1b appears not to be as widely distributed as the AVPR1a. It is highly abundant in the anterior pituitary, where it is found on the corticotrophes (Antoni, 1984), In the brain, *in situ* hybridization histochemistry and immunocytochemistry have localized the AVPR1b to the olfactory bulb, piriform cortical layer II, septum, cerebral cortex, hippocampus, PVN, SCN, cerebellum, and red nucleus (Lolait et al., 1995; Saito et al., 1995; Vaccari et al., 1998; Hernando et al., 2001; Stemmelin et al., 2005; Young et al., 2006). However, a later study found that the distribution of the AVPR1b in rat, mouse, and human was limited, with prominence in the anterior pituitary, the CA2 region of hippocampus, and a few cells in anterior amygdala (Young et al., 2006). This latter study may better reflect the distribution of the AVPR1b, as the earlier studies by Lolait et al. (1995) and Vaccari et al. (1998) used sequences that had significant identity with the AVPR1a and the OTR.

1.4 The gonadal steroids

The two primary classes of gonadal steroids are the androgens and the estrogens. While predominately synthesized in the testes and ovaries, they can also be generated in other tissues such as adrenal glands, liver, and fat, or locally synthesized in the brain. Like all steroids, the gonadal steroids are derived from cholesterol through the process of steroidogenesis. The common precursor hormone for all androgens and estrogens is dehydroepiandrosterone (DHEA), which is a “weak androgen” produced primarily in the adrenal cortex. For a summary of gonadal steroid synthesis, starting with DHEA, see Figure 1.1 (Torn et al., 2003; Steckelbroeck et al., 2004; Bauman et al., 2006). Interestingly, all of the enzymes necessary to metabolize testosterone into other androgens, or estradiol, can be found in the brain (Guennoun et al., 1995). Thus, in the brain, depending on what enzymes are locally available, the presence of testosterone can ultimately result in either, or both, androgenic and estrogenic effects. As many behaviors are associated with an animal’s reproductive status, it is not surprising that gonadal steroids are important regulators of the OT and AVP systems.

1.4.1 Possible interactions among oxytocin, vasopressin, and gonadal steroids

So, how do the gonadal steroids interact with the OT and AVP systems? There are numerous ways for this to occur. The most direct “classical” action is through the binding of gonadal steroids to their respective intracellular receptors such as ERα and ERβ for the estrogens, and the androgen receptor (AR) for the androgens. These hormone-receptor complexes dimerize and are then translocated from the cytosol to the nucleus where they can directly interact with the DNA through response elements (i.e., an ERE or an androgen response element (ARE) in the promoter regions of the *OT* and *AVP* genes or the genes for their receptors. While there can be crosstalk between the estrogens and androgens and their respective receptors, this is limited due to differences in their affinity for the receptors. Even within a given class of gonadal steroids, different estrogens or androgens may have more or less affinity for their ERs or AR, respectively. For example, dihydrotestosterone (DHT) has a higher affinity for the AR compared to testosterone (3–5 times), but does not bind well with ERα or ERβ (Handa et al., 1987; Kuiper et al., 1998). Similarly, estradiol binds to ERα and ERβ with equal affinity (Kuiper et al., 1998),

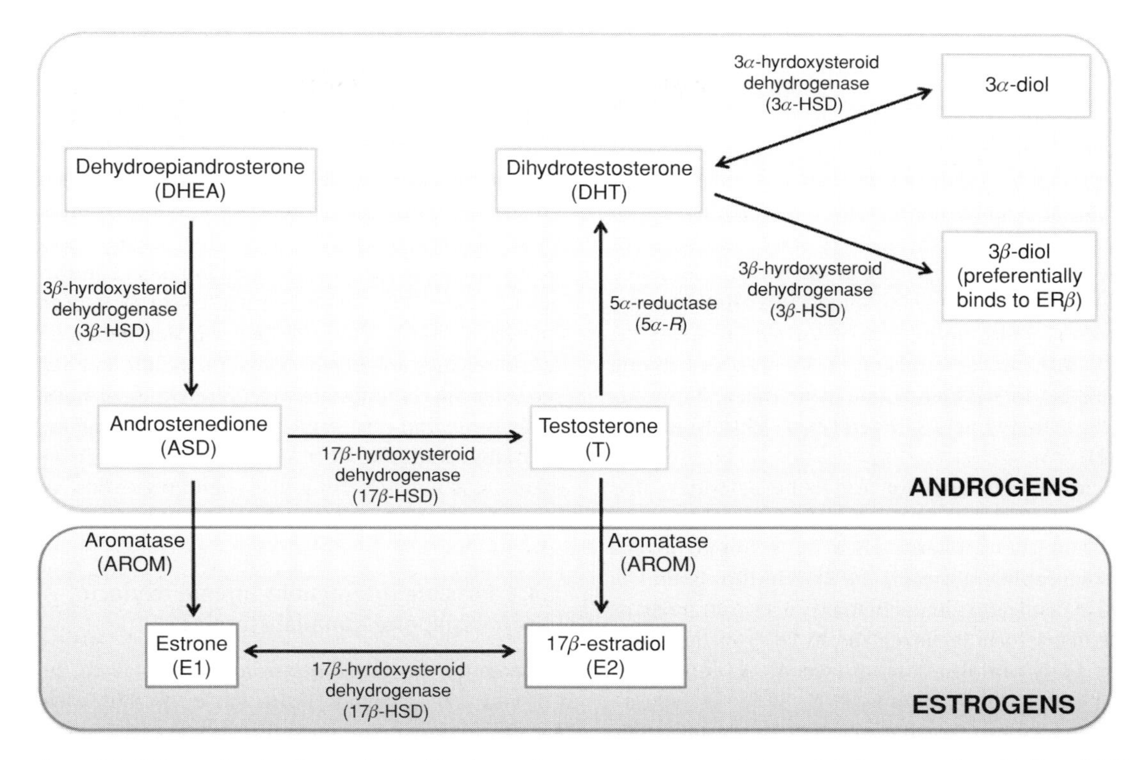

Figure 1.1 Pathway showing the synthesis of the key androgens and estrogens that regulate the expression of OXT and AVP peptides and receptors. Adapted and modified from Handa et al., 2009. See color version in plates section.

but has a very low affinity for the AR (Handa et al., 1987).

It is also important to note that there are numerous subtypes of intracellular receptors. The originally described ERβ would now be better termed ERβ1 due to the discovery of at least five different splice variants of the receptor. These variants include one lacking the third exon (δ3), one lacking the fourth exon (δ4), one with an insert between exons 5 and 6 (β2), and combinations of the three (Peterson, 1998; Hanstein et al., 1999; Price et al., 2000; Price et al., 2001). There are also interactions between estrogenic and androgenic pathways. For example, 5α-androstane-3β,17β-diol (3β-diol), an androgen metabolite, has estrogen-like effects since it preferentially binds and activates ERβ but has a low binding affinity for the AR (Weihua et al., 2002). 5α-androstane-3α,17β-diol (3α-diol) on the other hand has little to no affinity for either ER but preferentially binds to the AR (Kuiper et al., 1998).

Estrogens and androgens bound to receptors can also exert effects by interacting with other regulatory sites on the DNA, such as AP-2 sites, which stimulate transcription via protein kinase A and C pathways (Imagawa, 1987); AP-1-like sequences, the sole function of which is to regulate gene transcription in response to immediate early gene induction (e.g., Fos and Jun); and a SP-1 transcription factor binding site, which directly binds to DNA to enhance gene expression during early development (Webb et al., 1995; Marin et al., 1997; Price et al., 2001; Kim et al., 2005). The presence of one or more of these regulatory elements in the promoters of the

OT, AVP, or their receptor genes allows for myriad types of transcriptional regulation.

1.5 Gonadal steroid regulation of oxytocin and vasopressin

1.5.1 Oxytocin

There is ample evidence suggesting that gonadal steroids regulate various aspects of the OT system in both males and females. In rats, increased expression of OT mRNA in certain areas of the hypothalamus are coincident with the onset of puberty and vary across the estrous cycle; both of these events are associated with increased concentrations of circulating estrogens (Van Tol et al., 1988; Zingg and Lefebvre, 1988; Miller et al., 1989a). Gonadectomy of prepubertal and adult female rats decreases OT mRNA, and estradiol replacement during puberty can increase OT mRNA (Miller et al., 1989a; Chibbar et al., 1990). Similarly, the expression of OT, as measured by mRNA levels in the SON and PVN, mirrors the fluctuations in estradiol and progesterone concentrations that are found during pregnancy and lactation (Van Tol et al., 1988; Miller et al., 1989a; Crowley et al., 1993). In a study by Crowley and colleagues (1995), which mimicked the hormone levels of pregnancy in rats, treatment with estradiol and progesterone followed by progesterone withdrawal increases OT mRNA in the hypothalamus (Crowley et al., 1995). Apart from regulating the transcription of OT, estrogens also affect serum and pituitary levels of OT, axonal and dendritic release of OT, as well as the electrical activity of OT neurons (Yamaguchi et al., 1979; Akaishi and Sakuma, 1985; Skowronski et al., 1987; Van Tol et al., 1988; Wang and De Vries, 1995).

The actions of estrogens on OT are primarily mediated via ERα and ERβ. These receptors are often overlapping in their CNS distribution; though, the PVN is a pronounced exception. Within the PVN ERα is found at low or undetectable levels (Shughrue et al., 1997; Simonian and Herbison, 1997; Alves et al., 1998; Hrabovszky et al., 1998; Greco et al., 2001), suggesting that any effects of estrogens on PVN-derived OT may be mediated solely by ERβ. Support for a modulatory role of ERβ on OT neurons comes from immunocytochemical and *in situ* hybridization studies. These studies have found that ERβ is expressed in OT producing neurons of the SON and PVN of rats, being most abundant in the caudal regions of the PVN (Alves et al., 1998; Hrabovszky et al., 1998; Shughrue et al., 2002; Hrabovszky et al., 2004), in the SON of mice (Sar and Stumpf, 1980), and in the PVN of guinea pigs (Warembourg and Poulain, 1991). Further, studies conducted in ERβ knockout (ERβKO) mice have found that treatment with estradiol has no effect on OT mRNA expression compared to wild-type (WT) mice, suggesting that the estrogenic regulation of OT is likely via ERβ (Nomura et al., 2002; Patisaul et al., 2003).

The aforementioned *in vivo* effects of estradiol on OT expression prompted a series of *in vitro* studies designed to examine whether estrogen-dependent activation of the OT system could be attributed to the *direct* interaction of ERβ with the OT promoter. As the OT promoter in human and rat contains a highly conserved DNA binding site approximately –160 nucleotides from the transcription start site, which is homologous to the classic ERE palindromic sequence, there is the potential for direct genomic effects by estradiol on *OT* transcription (Mohr and Schmitz, 1991). Co-transfection studies in a heterologous cell culture system using promoter-reporter constructs confirm that the human and rat OT promoters are activated by estradiol (Peter et al., 1990; Richard and Zingg, 1990; Adan et al., 1993). Though, based on *in vitro* work in bovine and human this -160 HRE does not appear to directly interact with estradiol bound to ERβ, indicating that estrogens might regulate the *OT* gene via some indirect mechanism (Stedronsky et al., 2002). One possible mechanism for the *indirect* effects of estradiol on the *OT* gene is through the interaction of the –160 HRE with nuclear orphan receptors. Recently, the human OT –160 HRE has been found to have a strong affinity for the nuclear orphan receptors, steroidogenic

factor-1 (SF-1), and chicken ovalbumin upstream promoter transcription factor I (COUP-TF I) (Wehrenberg et al., 1994a; Wehrenberg et al., 1994b; Stedronsky et al., 2002). Ligand-activated estrogen receptors likely bind to these nuclear orphan receptors, thus facilitating their binding to the OT promoter, ultimately causing estrogen-dependent upregulation of OT gene expression (Giguere et al., 1988; Giguere, 2002; Sanchez et al., 2002; Koohi et al., 2005).

Yet another way that estrogens can indirectly alter OT gene transcription is through a non-genomic mechanism. Recently, a membrane-bound estrogen receptor, the GPCR-30 (recently renamed GPER), was characterized. The GPER is a member of the 7 transmembrane G-protein coupled superfamily of receptors. It is found in OT producing neurons of the SON and PVN, providing a means by which estrogens could affect oxytocinergic cells. It is speculated that the GPER mediates the rapid actions of estrogens on the OT system through activation of adenylyl cyclase, intracellular calcium mobilization, and generation of phosphatidylinositol-3,4,5-triphosphate (Revankar et al., 2005; Sakamoto et al., 2007).

1.5.1.1 Sexual dimorphisms

Sexual dimorphisms in the distribution of OT are not as common as with AVP (detailed below), with differences being species specific (Buijs et al., 1978; Haussler et al., 1990; Wang et al., 1996; Rosen et al., 2008). In studies where sex differences have been found, the number of OT-ir neurons, as well as the amount of OT as measured by immunoassay, tends to be higher in females compared to males (Haussler et al., 1990). A summary of changes in OT-ir that are thought to be gonadal steroid-dependent can be found in Table 1.1.

1.5.2 Vasopressin

In vivo, the AR and ERβ have been localized to brain regions rich in AVP or AVP receptors, especially the PVN, SON, and BNST (Laflamme et al., 1998; Suzuki and Handa, 2005). Further, the AVP promoter contains at least one ERE and several AP-1 and HRE sites (Shapiro et al., 2000). Work *in vitro* has found that gonadal steroids can have either facilitatory or inhibitory effects on AVP gene transcription, depending on what hormone/receptor complex binds to the DNA. Facilitation of AVP gene transcription occurs through the activation of ERα and ERβ (Pak et al., 2005; Pak et al., 2007). Estradiol and DHT, both metabolites of testosterone (Naftolin et al., 1975; Selmanoff et al., 1977), are known to affect AVP expression. Castrated rats treated with estradiol and DHT have a full restoration of AVP-ir and mRNA expression within the BNST and MeA. However, estradiol administered alone only partially restores AVP-ir and mRNA expression in these brain areas (de Vries et al., 1986; De Vries et al., 1994; Wang and de Vries, 1995). Interestingly, there is also evidence that AVP transcription can be inhibited by DHT when it is bound to the AR (Pak et al., 2007). However, the mechanism by which this occurs is not yet understood. With a lack of an apparent ARE in the AVP promoter it is unclear how DHT is influencing the transcription of the *AVP* gene.

There is evidence that a metabolite of DHT, 3β-diol, interacts with the AVP promoter. While 3β-diol is derived from an androgen, it preferentially binds to ERβ and has been found to stimulate AVP promoter activity though ERβ1 and ERβ2 (Pak et al., 2007). In culture, 3β-diol significantly increases ERE-mediated promoter activity to levels greater than that achieved by estradiol. This effect is specific to the presence of the ERE, as there is no effect on AVP promoter activity in the presence of only an AP-1 site (Pak et al., 2005). When ERβ splice variants are cotransfected with a firefly luciferase reporter construct containing the AVP promoter in the presence of 3β-diol, it was determined that 3β-diol increases AVP promoter activity in the presence of the ERβ1 and ERβ2 splice variants, but not the ERβ1δ3 splice variant (Pak et al., 2007). (This latter finding was not surprising since the ERβ1δ3 splice variant lacks exon 3, which encodes for the second finger of the

Table 1.1 Oxytocin immunoreactivity in specific neuroanatomical areas in different species. Gonadal steroid/castration sensitivity is for both males and females. Sexual dimorphisms shown in plain text indicate male > female unless depicted by italicized text (*italics* = females > males). Abbreviations: MPOA-AH, medial preoptic area-anterior hypothalamus; PVN, paraventricular nucleus; SON, supraoptic nucleus.

Species	Gonadal Steroid/ Castration Sensitivity	Sexual Dimorphisms	References
Rat *(Rattus rattus)*	MPOA-AH, PVN, SON		Caldwell et al., 1994; 1987; 1988; Jirikowski et al., 1988
Mouse (species not specified)		*PVN*	Haussler et al., 1990
Brandt's vole *(Lasipodomys brandtii)*		*PVN*	Xu et al., 2010
Greater long-tailed hamster *(Tscherskia triton)*		*PVN*	Xu et al., 2010

DNA binding domain (Peterson, 1998), and thus is prevented from binding directly to the DNA). Interestingly, when the ERE in the AVP promoter is mutated using site-directed mutagenesis, luciferase activity is still increased in the presence of ERβ1 and ERβ2 when 3β-diol is added. Taken together these studies confirm that 3β-diol bound to ERβ interacts with the AVP promoter, but its action is not through a direct interaction with the ERE (Pak et al., 2007).

One possible explanation for how 3β-diol acts through ERβ independent of an ERE, is ERβ activation of coregulatory proteins. ERs are known to interact with a variety of coregulatory proteins that aid in gene transcription. The glucocorticoid receptor interacting protein 1 (GRIP1), a steroid receptor coactivator, has been shown to be an important regulator of ER signaling (Hong et al., 1996; Norris et al., 1998). GRIP1 can be recruited by ERβ, but it is dependent upon a ligand being bound (Kraichely et al., 2000). *In vitro* work has demonstrated that GRIP1 is required for 3β-diol/ERβ1-induced AVP promoter activity (Pak et al., 2007). Thus, it is through this *indirect* mechanism that 3β-diol is able to stimulate AVP promoter activity.

In vivo studies in mice with or without specific ERs or the AR have found brain region-specific effects of gonadal steroids on AVP expression. In a hormone replacement study using gonadectomized AR knockout (ARKO) and ERα knockout (ERαKO) mice, estradiol treatment increases AVP-ir in the LS of WT males and ARKO females as compared to controls. ARKO and ERαKO males on the other hand have reduced AVP-ir in the MeA compared to the other genotypes, suggesting a sex- and receptor-specific downregulation of AVP in the MeA in the presence of estradiol (Scordalakes and Rissman, 2004). Also, in the PVN of WT males, estradiol treatment decreases AVP-ir (Nomura et al., 2002). In ERβKO mice, this decrease in AVP-ir within the PVN is prevented when estradiol is administered, suggesting that ERβ is necessary for normal AVP expression within the PVN (Nomura et al., 2002). These data, along with the work implicating DHT and 3β-diol in the facilitation of AVP promoter activity via ERβ (Pak et al., 2007), confirm that regulation of AVP expression by ERβ is ligand dependent. Taken together, these findings suggest that AR and ERα increase AVP expression in the LS and MeA, while ERβ increases AVP expression in the PVN. Further, while the mechanism remains unknown, the ARKO studies provide proof of regulation of AVP expression by androgen activation of AR, as opposed to aromatization of testosterone into estradiol. Further studies must be done in order to better understand the functions of the gonadal steroid receptors in the regulation of AVP transcription in other regions of the brain.

Thus, depending on which gonadal steroids and gonadal steroid receptors are found in AVP neurons, there can be differential effects on AVP transcription. The significance of this type of regulation is that a hormone like DHT, and its conversion into 3β-diol can result in opposing effects on AVP transcription, depending on what receptors are available in a specific brain area. This site-specific modulation of AVP transcription may be particularly important in the development and maintenance of sexually dimorphic brain structures, which will be discussed in the next section.

1.5.2.1 Sexual dimorphisms

As AVP is linked to the modulation of a variety of sex-specific behaviors, differences in the distribution of AVP and/or its receptors between the sexes are often influenced by differences in circulating gonadal steroids. For instance, there is evidence that androgens and estrogens can differentially alter AVP expression and receptor binding (de Vries et al., 1992; de Vries et al., 1994). In a number of mammalian species, there is an elevation in the number of AVP-ir neurons in the MeA, BNST, LHb, and LS in males as compared to females (de Vries and Buijs, 1983; de Vries et al., 1984; Miller et al., 1989c; de Vries and al-Shamma, 1990; Wang et al., 1993). The AVP projections from the BNST to the LS and LHb are also sensitive to changes in circulating gonadal hormones (de Vries et al., 1985; Miller et al., 1992). In fact, the AVP neurons of the BNST have become one of the most recognizable steroid-dependent regions of the brain, with males having more AVP-ir neurons than females. There are, however, exceptions to the rule that males express more AVP than females. For example, female guinea pigs have higher AVP-ir in the inferior colliculus, ventral trapezoid body, and dorsal cochlear nucleus than males (Dubois-Dauphin et al., 1987; 1989a). In mice, the MPOA and mammillary nuclei have higher AVP receptor binding in females than in males (Dubois-Dauphin et al., 1996). Further, in deer mice, the centromedial thalamus is sexually dimorphic in AVP receptor binding, with females having higher levels than males (Insel et al., 1991).

Gonadectomized males and females have reduced AVP density in regions of the brain that receive innervation from the BNST and MeA, which contain parvocellular neurons, but not the PVN and SON, which contain mainly magnocellular AVP-ir neurons (de Vries and Buijs, 1983; de Vries et al., 1984; de Vries et al., 1985; van Leeuwen et al., 1985; Miller et al., 1989b). Testosterone administered to female and gonadectomized male rats increases the density of AVP neurons in the BNST, LS, and MeA (de Vries and Buijs, 1983; Wang et al., 1993; de Vries et al., 1994). Also, while gonadectomy in Syrian hamsters has no effect on AVP-ir (Albers et al., 1991), there is an effect of gonadectomy on AVP-ir in Siberian hamsters (Dubois-Dauphin et al., 1994). In primates, such as marmosets, macaques and humans, the BNST has similar AVP-ir levels in both sexes, while male rodents typically have more AVP-ir than females. However, in contrast to rodents, primates have little to no AVP-ir in the LS (Fliers et al., 1986; Caffe et al., 1989; Wang et al., 1997). Taken together, these data indicate sex- and species-specific differences in the steroid-dependent regulation of AVP expression. A summary of changes in AVP-ir that are thought to be gonadal steroid-dependent can be found in Table 1.2.

1.6 Gonadal steroid regulation of oxytocin and vasopressin receptors

1.6.1 The oxytocin receptor

Like OT, the expression of the OTR is also sensitive to gonadal steroids. While there are a variety of effects in the periphery (Soloff, 1975; 1982; Adachi and Oku, 1995; Bale and Dorsa, 1995a; b; Ostrowski et al., 1995; Breton et al., 1996; Wathes et al., 1996; Young et al., 1997), this section will focus on the brain. In rats, treatment with estradiol or testosterone increases OT binding and OTR mRNA within the VMH (an

Table 1.2 Vasopressin-immunoreactivity in specific neuroanatomical areas between different species. Gonadal steroid/castration sensitivity and seasonal variation is for males unless otherwise denoted by asterisks (** males and females). Sexual dimorphisms shown in plain text depicts males > females unless otherwise noted by italicized text (*italics* = females > males). For a more detailed table refer to (Goodson and Bass, 2001). Abbreviations: BNST, bed nucleus of the stria terminalis; DB, diagonal band; HP, hippocampus; IC, inferior colliculus; LC, locus coeruleus; LH, lateral hypothalamus; LS, lateral septum; MeA, medial amygdala; PAG, periaquaductal grey; SCN, suprachiasmatic nucleus; SON, supraoptic nucleus; VTA, ventral tegmental area; VTB, ventral trapezoid body.

Species	Gonadal Steroid/ Castration Sensitivity	Seasonal/ Photoperiodic Variation	Sexual Dimorphisms	References
Marmoset *(Callithrix jacchus)*			BNST	Wang et al., 1997
Guinea pig *(Caviaporcellus)*			*IC, VTB*	Dubois-Dauphin et al., 1987; 1989a
European hamster *(Cricetus cricetus)*	DB, LH, LS, MeA, PAG, HP, VTA	DB**, LH**, LS**, MeA**, PAG**, HP** VTA**	LH, LS, MeA	Buijs et al., 1986
Garden dormouse *(Eliomys quercinus)*		DB**, LC**, LH**, LS**, MeA**, PAG**, HP**, VTA**	LC, LH, LS, MeA, PAG, DB, HP, VTA	Hermes et al., 1990
Greater Egyptian jerboa *(Jaculus orientalis)*	BNST, MeA	BNST, MeA	BNST, MeA	Lakhdar-Ghazal et al., 1995
Golden hamster *(Mesocrietus auratus)*			SON	Delville et al., 1994
Prairie vole *(Microtus orchrogaster)*	BNST, MeA		BNST, MeA	Wang et al., 1993, 1995
Djungarian hamster *(Phodopus sungorus)*	BNST, MeA			Dubois-Dauphin et al., 1994
Rat *(Rattus rattus)*	BNST, MeA, SCN		*BNST, MeA*	al-Shamma et al., 1996; de Vries et al., 1990; Miller et al., 1992; Planas et al., 1995; Swaab et al., 1995; van Leeuwen et al., 1985; Wang et al., 1993

area of the brain known to contain a high density of estrogen-containing cells), which decreases following gonadectomy and is restored following hormone replacement (Coirini et al., 1989; Johnson et al., 1989a; Johnson et al., 1989b; Johnson et al., 1991; Bale and Dorsa, 1995a; 1995b). Other regions of the brain that show changes in the expression of the OTR in response to gonadal steroids include the

BNST, LS, CeA, and MPOA (Johnson et al., 1989a; Patchev et al., 1993).

How estrogens induce the expression of the OTR is still unclear. As mentioned previously, the rat and mouse *OTR* promoter contains a classic ERE, and the rat, mouse, and human *OTR* promoter contains several half-palindromic ERE motifs (Inoue et al., 1994; Rozen et al., 1995; Kubota et al., 1996; Bale and Dorsa, 1997). However, both DNA binding and transfection studies using the rat promoter indicate only a weak transcriptional response to estradiol (Bale and Dorsa, 1997). Thus, it is likely that the half-palindromic ERE motifs may act synergistically to mediate the responses to estrogens (Kato et al., 1992; Sanchez et al., 2002).

There is also evidence for non-genomic regulation of the OTR by estrogens. In a study conducted by Bale and colleagues (2001) using female rats, it was determined that inhibition of protein kinase C (PKC) decreases estradiol-mediated increases in OT binding in VMH. Conversely, administration of PKC increases OT binding in absence of estrogens, indicating that estrogens might regulate the *OTR* through a PKC-dependent pathway. There are other responsive elements on the DNA, such as the interleukin response element, a cAMP response element, AP-1, AP-2, AP-3, and AP-4 sites, as well as data demonstrating that administration of interleukin-6 along with estrogens significantly increases the expression of OTR mRNA. Collectively, these data provide strong evidence that estrogens can regulate OTR expression via these indirect mechanisms (Young et al., 1997; Gimpl and Fahrenholz, 2001).

In males, the effects of testosterone on the OTR protein and mRNA expression levels seem to be mediated by its metabolites estradiol and DHT, though a direct action of testosterone on OTR regulation cannot be excluded (Coirini et al., 1989; Johnson et al., 1991; Bale and Dorsa, 1995b). Changes in the expression of the OTR in response to androgens differ according to the species. In contrast to rats, castrated mice have two-fold increases in OT binding compared to gonadally intact and testosterone treated-castrated mice (Insel et al., 1993). In mice, ERα does not appear to be required for basal expression of the OTR, but is necessary for inducing the expression of OTR in response to treatment with estrogens (Young et al., 1998). So far no ARE has been found in the *OTR* promoter. Hence, further studies are required to elucidate the exact mechanism by which androgens regulate OTR expression as well as to explain the observed species differences.

1.6.1.1 Sexual dimorphisms

There are species-specific and neuroanatomical-specific sexual dimorphisms in OTR distribution. In rats, sexual dimorphisms are found in the distribution of the OTR in areas such as VMH and the dorsal horn of the spinal cord, with males having higher expression than females (Uhl-Bronner et al., 2005). In prairie voles, females have a higher density of OT binding than males in the medial prefrontal cortex (Smeltzer et al., 2006). In two different species of mice, sexual dimorphisms are more prominent in the polygamous species, deer mice, as compared to monogamous California mice (Insel et al., 1991). In deer mice, the males tend to have higher levels of [^{125}I]OTA binding in most of the forebrain areas, except the CA1 region of the hippocampus. A summary of changes in the OTR that are thought to be gonadal steroid dependent can be found in Table 1.3.

1.6.2 Vasopressin receptors

The AVPR1a is also sensitive to changes in circulating concentrations of gonadal steroids. In hamsters, gonadectomy and lactation decrease AVPR1a mRNA and receptor binding in the lateral aspects of the MPOA and posterior hypothalamus (Delville et al., 1995; Johnson et al., 1995; Young et al., 2000). In adult rats, AVPR binding is unaffected by changes in gonadal steroid levels (Tribollet et al., 1990), whereas estrogens given to adolescent female rats increase AVPR1a transcription in the POA of the hypothalamus (Funabashi et al., 2000). AVPR1a density is also greater in males than in females in the

Table 1.3 Oxytocin receptor binding in specific neuroanatomical areas in different species. Gonadal steroid/castration sensitivity is for both males and females unless otherwise denoted by asterisks (*male only). Sexual dimorphisms shown in plain text indicate male > female unless depicted by italicized text (*italics* = females > males). Abbreviations: BNST, bed nucleus of stria terminalis; CeA, central amygdala; LS, lateral septum; mPFC, medial prefrontal cortex; VMH, ventromedial hypothalamus.

Species	Gonadal Steroid/ Castration Sensitivity	Sexual Dimorphisms	References
Rat *(Rattus rattus)*	BNST, VMH, Islands of Calleja, CeA, Olfactory tubercle	*VMH*, dorsal horn of the spinal cord	Coirini et al., 1989; De Kloet et al., 1985; Johnson et al., 1989a; 1989b; Ulh-Bronner et al., 2005
Prairie vole *(Microtus orchrogaster)*		*mPFC*	Smeltzer et al., 2006
Mouse *(Mus musculus)**	LS, Me A, VMH, hippocampus		Insel et al., 1993
Singing Mouse *(Scotinomys xerampelimus)*	MeA, CAl-hippocampus		Campbell et al., 2009
Deer Mouse *(Peromyscus maniculatus)*		Dorsal LS, BNST, *CA1-hippocampus*	Insel et al., 1991

premammillary nuclei of Syrian and Siberian hamsters (Dubois-Dauphin et al., 1991; Delville et al., 1995).

There is also evidence that androgens regulate the AVPR1a, though it is likely brain area specific (Clancy et al., 1994; Young et al., 2000). Since the aromatase enzyme is found throughout the brain, the effects of testosterone could be mediated through estrogen receptors, by its conversion to estradiol. Evidence for this is found in hamsters where testosterone is known to decrease AVPR1a mRNA in the MPN, which is rich in both the AR and ERs, but has no effect in the LS, which is rich only in the AR (Romeo et al., 1999). To date, the exact mechanism behind the gonadal steroid regulation of the AVPR1a is unknown.

Evidence of gonadal steroid modulation of AVPR1a also comes from photoperiod studies in Siberian and Syrian hamsters. In these species, exposure to short "winter-like" photoperiods results in gonadal regression, which leads to low levels of circulating gonadal steroids that are concurrent with changes in social behavior. These gonadally regressed hamsters have reduced AVPR1a receptor binding within the MPOA (Dubois-Dauphin et al., 1994; Caldwell and Albers, 2003; 2004; Caldwell et al., 2008). As of yet there is no evidence that the AVPR1b is regulated by gonadal steroids.

1.6.2.1 Sexual dimorphisms

In species in which the AVP system is commonly studied, sexual dimorphisms in receptor distributions have been found, with males typically having higher AVPR1a expression. Siberian male hamsters have increased AVPR1a in their premammillary nuclei compared to females (Dubois-Dauphin et al., 1991; Dubois-Dauphin et al., 1994; Delville et al., 1995). Syrian hamsters show sexual dimorphisms in AVP receptor binding in the ventrolateral

Table 1.4 Vasopressin receptor 1a binding in specific neuroanatomical areas in different species. Gonadal steroid/castration sensitivity and seasonal variation is for males unless otherwise denoted by asterisks (** males and females). Sexual dimorphisms shown in plain text depicts males > females unless otherwise noted by italicized text (*italics* = females > males). For a more detailed table refer to (Goodson and Bass, 2001). Abbreviations: BNST, bed nucleus of the stria terminalis; CMT, centromedial thalamus; M, mammilary nuclei; MPN, medial preoptic nucleus; mPOA, medial preoptic area; MTu medial tuberal nucleus; premammilary nucleus; PM, premammilary nucleus; POA-LH, preoptic area-lateral hypothalamus; VLH, ventrolateral hypothalamus; VMN, vetromedial nucleus of the hypothalamus.

Species	Gonadal Steroid/ Castration Sensitivity	Seasonal/ Photoperiodic Variation	Sexual Dimorphisms	References
Golden hamster *(Mesocrietus auratus)*	BNST, MPN, POA-LH, VLH**		VLH	Delville et al., 1995; 1996; Johnson et al., 1995; Young et al., 2000
Mouse *(Mus musculus)*			*MPOA, M*	Dubois-Dauphin et al., 1996
Deer mouse *(Peromyscus)*			*CMT*	Insel, et al., 1991
Djungarian hamster *(Phodopus sungorus)*	VMN	VMN**	MTu, PM	Dubois-Dauphin et al., 1991; 1994

hypothalamus (Delville et al., 1995). In rats, there is a sexually dimorphic distribution of AVPR1a receptors in the POA and the sexually dimorphic nucleus (Bloch and Gorski, 1988). Mice also show sex-related differences in AVP binding sites in the POA as well as the hypothalamic mammillary nuclei (Dubois-Dauphin et al., 1996). A summary of changes in AVPR1a that are thought to be gonadal steroid-dependent can be found in Table 1.4.

1.7 Conclusion

Through the interplay of OT, AVP, and gonadal steroids, dynamic changes in behavior within a species can be achieved. Given the diversity of species-specific behaviors it is perhaps not surprising that the interactions of gonadal steroids with the OT and AVP systems is complicated and in many ways not well understood. With many different gonadal steroids regulating the expression of OT and AVP, as well as their receptors, there are countless ways in which OT and AVP can be modulated within specific brain regions. By continuing to examine species differences in the gonadal steroid-dependent gene regulation of the OT and AVP systems, important insights into what mechanisms are conserved across species and what mechanisms are species specific will be gained.

Acknowledgments

The authors would like to express their gratitude to Drs. Colleen Novak and W. Scott Young, III for taking the time to make suggestions for the improvement of this manuscript. Research was supported in part by NIH MH083963 awarded to HKC.

Abbreviations

3α-diol, 5α-androstane-3α,17β-diol; 3β-diol, 5α-androstane-3β,17β-diol; 3α-HSD, 3α-hydroxysteroid oxidoreductase; 3β-HSD, 3β-hydroxysteroid oxidoreductase; 3β-HSD, 3β-hyrdroxysteroid dehydrogenase; AH, anterior hypothalamus; AP-1, activator protein-1; AP-2, activator protein-2;

AR; androgen receptor; ARE, androgen response element; ARKO, androgen receptor knockout; ASD, androstenedione; AVP, vasopressin; AVP-ir, vasopressin-immunoreactive; BNST, bed nucleus of stria terminalis; CeA, central amygdala; CMT, centromedial thalamus; CNS, central nervous system; COUP-TF I, chicken ovalbumin upstream promoter transcription factor I; DB, diagonal band; DHEA, dehydroepiandrosterone; DHT, dihydrotestosterone; E1, estrone; E2, estradiol; ERE, estrogen response element; ERα, estrogen receptor alpha; ERαKO, estrogen receptor α knockout; ERβ, estrogen receptor β; ERβKO, estrogen receptor beta knockout; GRIP1, glucocorticoid receptor interacting protein 1; HP, hippocampus; HRE, hormone response element; IC, inferior colliculus; IGR, intergenic region; KO, knockout; LC, locus coeruleus; LHb, lateral habenula; LH, lateral hypothalamus; LS, lateral septum; M, mammilary nuclei; MeA, medial amygdala; mPFC, medial prefrontal cortex; MPN, medial preoptic nucleus; MPOA, medial preoptic area; MPOA-AH, medial preoptic area-anterior hypothalamus; MTu, medial tuberal nucleus; OT, oxytocin; OT-ir, OT-immunoreactive; PAG, periaquaductal grey; PKC, protein kinase C; PM, premammilary nucleus; POA, preoptic area; POA-LH preoptic area-lateral hypothalamus; PVN, paraventricular nucleus; SCN, suprachiasmatic nucleus; SF-1, steroidogenic factor-1; SON, supraoptic nucleus; SP-1, specific protein-1; VLH, ventrolateral hypothalamus; VMH, ventromedial hypothalamus; VMN, ventromedial nucleus of the hypothalamus; VTA, ventral tegmental area; VTB, ventral trapezoid body; WT, wild-type.

REFERENCES

Acher, R. and Chauvet, J. (1995). The neurohypophysial endocrine regulatory cascade: precursors, mediators, receptors, and effectors. *Front Neuroendocrinol*, 16, 237–289.

Adachi, S. and Oku, M. (1995). The regulation of oxytocin receptor expression in human myometrial monolayer culture. *J Smooth Muscle Res*, 31, 175–187.

Adan, R. A., Cox, J. J., Beischlag, T. V., and Burbach, J. P. (1993). A composite hormone response element mediates the transactivation of the rat oxytocin gene by different classes of nuclear hormone receptors. *Mol Endocrinol*, 7, 47–57.

Adan, R. A., Walther, N., Cox, J. J., Ivell, R., and Burbach, J. P. (1991). Comparison of the estrogen responsiveness of the rat and bovine oxytocin gene promoters. *Biochem Biophys Res Commun*, 175, 117–122.

Akaishi, T. and Sakuma, Y. (1985). Estrogen excites oxytocinergic, but not vasopressinergic cells in the paraventricular nucleus of female rat hypothalamus. *Brain Res*, 335, 302–305.

Albers, H. E., Rowland, C. M., and Ferris, C. F. (1991). Arginine-vasopressin immunoreactivity is not altered by photoperiod or gonadal hormones in the Syrian hamster (Mesocricetus auratus). *Brain Res*, 539, 137–142.

al-Shamma, H. A. and De Vries, G. J. (1996). Neurogenesis of the sexually dimorphic vasopressin cells of the bed nucleus of the stria terminalis and amygdala of rats. *J Neurobiol*, 29, 91–98.

Alves, S. E., Lopez, V., McEwen, B. S., and Weiland, N. G. (1998). Differential colocalization of estrogen receptor beta (ERbeta) with oxytocin and vasopressin in the paraventricular and supraoptic nuclei of the female rat brain: an immunocytochemical study. *Proc Natl Acad Sci USA*, 95, 3281–3286.

Antoni, F. A. (1984). Novel ligand specificity of pituitary vasopressin receptors in the rat. *Neuroendocrinology*, 39, 186–188.

Bale, T. L., Davis, A. M., Auger, A. P., Dorsa, D. M., and McCarthy, M. M. (2001). CNS region-specific oxytocin receptor expression: importance in regulation of anxiety and sex behavior. *J Neurosci*, 21, 2546–2552.

Bale, T. L. and Dorsa, D. M. (1995a). Regulation of oxytocin receptor messenger ribonucleic acid in the ventromedial hypothalamus by testosterone and its metabolites. *Endocrinology*, 136, 5135–5138.

Bale, T. L. and Dorsa, D. M. (1995b). Sex differences in and effects of estrogen on oxytocin receptor messenger ribonucleic acid expression in the ventromedial hypothalamus. *Endocrinology*, 136, 27–32.

Bale, T. L. and Dorsa, D. M. (1997). Cloning, novel promoter sequence, and estrogen regulation of a rat oxytocin receptor gene. *Endocrinology*, 138, 1151–1158.

Bankir, L., Bichet, D. G., and Bouby, N. (2010). Vasopressin V2 receptors, ENaC, and sodium reabsorption: a risk factor for hypertension? *Am J Physiol Renal Physiol*, 299, F917–928.

Barat, C., Simpson, L., and Breslow, E. (2004). Properties of human vasopressin precursor constructs: inefficient monomer folding in the absence of copeptin as a potential contributor to diabetes insipidus. *Biochemistry*, 43, 8191–8203.

Barberis, C., Mouillac, B., and Durroux, T. (1998). Structural bases of vasopressin/oxytocin receptor function. *J Endocrinol*, 156, 223–229.

Bathgate, R., Rust, W., Balvers, M., Hartung, S., Morley, S., and Ivell, R. (1995). Structure and expression of the bovine oxytocin receptor gene. *DNA Cell Biol*, 14, 1037–1048.

Bauman, D. R., Steckelbroeck, S., Williams, M. V., Peehl, D. M., and Penning, T. M. (2006). Identification of the major oxidative 3alpha-hydroxysteroid dehydrogenase in human prostate that converts 5alpha-androstane-3alpha,17beta-diol to 5alpha-dihydrotestosterone: a potential therapeutic target for androgen-dependent disease. *Mol Endocrinol*, 20, 444–458.

Bloch, G. J. and Gorski, R. A. (1988). Cytoarchitectonic analysis of the SDN-POA of the intact and gonadectomized rat. *J Comp Neurol*, 275, 604–612.

Breton, C., Neculcea, J., and Zingg, H. H. (1996). Renal oxytocin receptor messenger ribonucleic acid: characterization and regulation during pregnancy and in response to ovarian steroid treatment. *Endocrinology*, 137, 2711–2717.

Brinton, R. E., Wamsley, J. K., Gee, K. W., Wan, Y. P., and Yamamura, H. I. (1984). [3H]oxytocin binding sites in the rat brain demonstrated by quantitative light microscopic autoradiography. *Eur J Pharmacol*, 102, 365–367.

Buijs, R. M. (1978). Intra- and extra-hypothalamic vasopressin and oxytocin pathways in the rat. Pathways to the limbic system, medulla oblongata and spinal cord. *Cell Tissue Res*, 192, 423–435.

Buijs, R. M., Pevet, P., Masson-Pevet, M., Pool, C. W., de Vries, G. J., Canguilhem, B., and Vivien-Roels, B. (1986). Seasonal variation in vasopressin innervation in the brain of the European hamster (Cricetus cricetus). *Brain Res*, 371, 193–196.

Buijs, R. M., Swaab, D. F., Dogterom, J., and van Leeuwen, F. W. (1978). Intra- and extrahypothalamic vasopressin and oxytocin pathways in the rat. *Cell Tissue Res*, 186, 423–433.

Burbach, J. P., Luckman, S. M., Murphy, D., and Gainer, H. (2001). Gene regulation in the magnocellular hypothalamo-neurohypophysial system. *Physiol Rev*, 81, 1197–1267.

Caffe, A. R., Van Ryen, P. C., Van der Woude, T. P., and Van Leeuwen, F. W. (1989). Vasopressin and oxytocin systems in the brain and upper spinal cord of Macaca fascicularis. *J Comp Neurol*, 287, 302–325.

Caldwell, H. K. and Albers, H. E. (2003). Short-photoperiod exposure reduces vasopressin (V1a) receptor binding but not arginine-vasopressin-induced flank marking in male Syrian hamsters. *J Neuroendocrinol*, 15, 971–977.

Caldwell, H. K. and Albers, H. E. (2004). Photoperiodic regulation of vasopressin receptor binding in female Syrian hamsters. *Brain Res*, 1002, 136–141.

Caldwell, H. K., Lee, H. J., Macbeth, A. H., and Young, W. S., III. (2008). Vasopressin: behavioral roles of an "original" neuropeptide. *Prog Neurobiol*, 84, 1–24.

Caldwell, J. D., Greer, E. R., Johnson, M. F., Prange, A. J., Jr., and Pedersen, C. A. (1987). Oxytocin and vasopressin immunoreactivity in hypothalamic and extrahypothalamic sites in late pregnant and postpartum rats. *Neuroendocrinology*, 46, 39–47.

Caldwell, J. D., Walker, C. H., Pedersen, C. A., Barakat, A. S., and Mason, G. A. (1994). Estrogen increases affinity of oxytocin receptors in the medial preoptic area-anterior hypothalamus. *Peptides*, 15, 1079–1084.

Campbell, P., Ophir, A. G., and Phelps, S. M. (2009). Central vasopressin and oxytocin receptor distributions in two species of singing mice. *J Comp Neurol*, 516, 321–333.

Castel, M., Feinstein, N., Cohen, S., and Harari, N. (1990). Vasopressinergic innervation of the mouse suprachiasmatic nucleus: an immuno-electron microscopic analysis. *J Comp Neurol*, 298, 172–187.

Castel, M. and Morris, J. F. (1988). The neurophysin-containing innervation of the forebrain of the mouse. *Neuroscience*, 24, 937–966.

Chibbar, R., Toma, J. G., Mitchell, B. F., and Miller, F. D. (1990). Regulation of neural oxytocin gene expression by gonadal steroids in pubertal rats. *Mol Endocrinol*, 4, 2030–2038.

Clancy, A. N., Whitman, C., Michael, R. P., and Albers, H. E. (1994). Distribution of androgen receptor-like immunoreactivity in the brains of intact and castrated male hamsters. *Brain Res Bull*, 33, 325–332.

Coirini, H., Johnson, A. E., and McEwen, B. S. (1989). Estradiol modulation of oxytocin binding in the ventromedial hypothalamic nucleus of male and female rats. *Neuroendocrinology*, 50, 193–198.

Crowley, R. S., Insel, T. R., O'Keefe, J. A., and Amico, J. A. (1993). Cytoplasmic oxytocin and vasopressin gene transcripts decline postpartum in the hypothalamus of the lactating rat. *Endocrinology*, 133, 2704–2710.

Crowley, R. S., Insel, T. R., O'Keefe, J. A., Kim, N. B., and Amico, J. A. (1995). Increased accumulation of oxytocin messenger ribonucleic acid in the hypothalamus of the female rat: induction by long term estradiol and progesterone administration and subsequent progesterone withdrawal. *Endocrinology*, 136, 224–231.

de Bree, F. M. (2000). Trafficking of the vasopressin and oxytocin prohormone through the regulated secretory pathway. *J Neuroendocrinol*, 12, 589–594.

De Kloet, E. R., Voorhuis, T. A., and Elands, J. (1985). Estradiol induces oxytocin binding sites in rat hypothalamic ventromedial nucleus. *Eur J Pharmacol*, 118, 185–186.

Delville, Y., Conklin, L. S., and Ferris, C. F. (1995). Differential expression of vasopressin receptor binding in the hypothalamus during lactation in golden hamsters. *Brain Res*, 689, 147–150.

Delville, Y., Koh, E. T., and Ferris, C. F. (1994). Sexual differences in the magnocellular vasopressinergic system in golden hamsters. *Brain Res Bull*, 33, 535–540.

Delville, Y., Mansour, K. M., and Ferris, C. F. (1996). Testosterone facilitates aggression by modulating vasopressin receptors in the hypothalamus. *Physiol Behav*, 60, 25–29.

de Vries, G. J. and al-Shamma, H. A. (1990). Sex differences in hormonal responses of vasopressin pathways in the rat brain. *J Neurobiol*, 21, 686–693.

de Vries, G. J. and Buijs, R. M. (1983). The origin of the vasopressinergic and oxytocinergic innervation of the rat brain with special reference to the lateral septum. *Brain Res*, 273, 307–317.

de Vries, G. J., Buijs, R. M., and Sluiter, A. A. (1984). Gonadal hormone actions on the morphology of the vasopressinergic innervation of the adult rat brain. *Brain Res*, 298, 141–145.

de Vries, G. J., Buijs, R. M., Van Leeuwen, F. W., Caffe, A. R., and Swaab, D. F. (1985). The vasopressinergic innervation of the brain in normal and castrated rats. *J Comp Neurol*, 233, 236–254.

de Vries, G. J., Crenshaw, B. J., and al-Shamma, H. A. (1992). Gonadal steroid modulation of vasopressin pathways. *Ann N Y Acad Sci*, 652, 387–396.

de Vries, G. J., Duetz, W., Buijs, R. M., van Heerikhuize, J., and Vreeburg, J. T. (1986). Effects of androgens and estrogens on the vasopressin and oxytocin innervation of the adult rat brain. *Brain Res*, 399, 296–302.

de Vries, G. J., Wang, Z., Bullock, N. A., and Numan, S. (1994). Sex differences in the effects of testosterone and its metabolites on vasopressin messenger RNA levels in the bed nucleus of the stria terminalis of rats. *J Neurosci*, 14, 1789–1794.

Donaldson, Z. R. and Young, L. J. (2008). Oxytocin, vasopressin, and the neurogenetics of sociality. *Science*, 322, 900–904.

Dubois-Dauphin, M., Barberis, C., and de Bilbao, F. (1996). Vasopressin receptors in the mouse (Mus musculus) brain: sex-related expression in the medial preoptic area and hypothalamus. *Brain Res*, 743, 32–39.

Dubois-Dauphin, M., Theler, J. M., Ouarour, A., Pevet, P., Barberis, C., and Dreifuss, J. J. (1994). Regional differences in testosterone effects on vasopressin receptors and on vasopressin immunoreactivity in intact and castrated Siberian hamsters. *Brain Res*, 638, 267–276.

Dubois-Dauphin, M., Theler, J. M., Zaganidis, N., Dominik, W., Tribollet, E., Pevet, P., Charpak, G., and Dreifuss, J. J. (1991). Expression of vasopressin receptors in hamster hypothalamus is sexually dimorphic and dependent upon photoperiod. *Proc Natl Acad Sci USA*, 88, 11163–11167.

Dubois-Dauphin, M., Tribollet, E., and Dreifuss, J. J. (1987). A sexually dimorphic vasopressin innervation of auditory pathways in the guinea pig brain. *Brain Res*, 437, 151–156.

Dubois-Dauphin, M., Tribollet, E., and Dreifuss, J. J. (1989a). Distribution of neurohypophysial peptides in the guinea pig brain. I. An immunocytochemical study of the vasopressin-related glycopeptide. *Brain Res*, 496, 45–65.

Dubois-Dauphin, M., Tribollet, E., and Dreifuss, J. J. (1989b). Distribution of neurohypophysial peptides in the guinea pig brain. II. An immunocytochemical study of oxytocin. *Brain Res*, 496, 66–81.

Fields, R. L., House, S. B., and Gainer, H. (2003). Regulatory domains in the intergenic region of the oxytocin and vasopressin genes that control their hypothalamus-specific expression in vitro. *J Neurosci*, 23, 7801–7809.

Fliers, E., Guldenaar, S. E., van de Wal, N., and Swaab, D. F. (1986). Extrahypothalamic vasopressin and oxytocin in the human brain; presence of vasopressin cells in the bed nucleus of the stria terminalis. *Brain Res*, 375, 363–367.

Foletta, V. C., Brown, F. D., and Young, W. S., III. (2002). Cloning of rat ARHGAP4/C1, a RhoGAP family member expressed in the nervous system that colocalizes with the Golgi complex and microtubules. *Brain Res Mol Brain Res*, 107, 65–79.

Funabashi, T., Shinohara, K., Mitsushima, D., and Kimura, F. (2000). Estrogen increases arginine-vasopressin V1a receptor mRNA in the preoptic area of young but not of middle-aged female rats. *Neurosci Lett*, 285, 205–208.

Gainer, H., Fields, R. L., and House, S. B. (2001). Vasopressin gene expression: experimental models and strategies. *Exp Neurol*, 171, 190–199.

Gainer, H., Pant, H. C., and Cohen, R. S. (1994). Squid optic lobe synaptosomes: what can they tell us about presynaptic protein synthesis? *J Neurochem*, 63, 387–389.

Giguere, V. (2002). To ERR in the estrogen pathway. *Trends Endocrinol Metab*, 13, 220–225.

Giguere, V., Yang, N., Segui, P., and Evans, R. M. (1988). Identification of a new class of steroid hormone receptors. *Nature*, 331, 91–94.

Gimpl, G. and Fahrenholz, F. (2001). The oxytocin receptor system: structure, function, and regulation. *Physiol Rev*, 81, 629–683.

Goodson, J. L. and Bass, A. H. (2001). Social behavior functions and related anatomical characteristics of vasotocin/vasopressin systems in vertebrates. *Brain Res Brain Res Rev*, 35, 246–265.

Gorbulev, V., Buchner, H., Akhundova, A., and Fahrenholz, F. (1993). Molecular cloning and functional characterization of V2 [8-lysine] vasopressin and oxytocin receptors from a pig kidney cell line. *Eur J Biochem*, 215, 1–7.

Greco, B., Allegretto, E. A., Tetel, M. J., and Blaustein, J. D. (2001). Coexpression of ER beta with ER alpha and progestin receptor proteins in the female rat forebrain: effects of estradiol treatment. *Endocrinology*, 142, 5172–5181.

Guennoun, R., Fiddes, R. J., Gouezou, M., Lombes, M., and Baulieu, E. E. (1995). A key enzyme in the biosynthesis of neurosteroids, 3 beta-hydroxysteroid dehydrogenase/delta 5-delta 4-isomerase (3 beta-HSD), is expressed in rat brain. *Brain Res Mol Brain Res*, 30, 287–300.

Handa, R. J., Weiser, M. J., and Zuloaga, D. G. (2009). A role for the androgen metabolite, 5 alpha-androstane-3beta-diol, in modulating oestrogen receptor beta-mediated regulation of hormonal stress reactivity. *J. Neuroendocrinol.*, 21, 351–358.

Handa, R. J., Stadelman, H. L., and Resko, J. A. (1987). Effect of estrogen on androgen receptor dynamics in female rat pituitary. *Endocrinology*, 121, 84–89.

Hanstein, B., Liu, H., Yancisin, M. C., and Brown, M. (1999). Functional analysis of a novel estrogen receptor-beta isoform. *Mol Endocrinol*, 13, 129–137.

Hara, Y., Battey, J., and Gainer, H. (1990). Structure of mouse vasopressin and oxytocin genes. *Brain Res Mol Brain Res*, 8, 319–324.

Haussler, H. U., Jirikowski, G. F., and Caldwell, J. D. (1990). Sex differences among oxytocin-immunoreactive neuronal systems in the mouse hypothalamus. *J Chem Neuroanat*, 3, 271–276.

Hermes, M. L., Buijs, R. M., Masson-Pevet, M., and Pevet, P. (1988). Oxytocinergic innervation of the brain of the garden dormouse (Eliomys quercinus L.). *J Comp Neurol*, 273, 252–262.

Hermes, M. L., Buijs, R. M., Masson-Pevet, M., and Pevet, P. (1990). Seasonal changes in vasopressin in the brain of the garden dormouse (Eliomys quercinus L.). *J Comp Neurol*, 293, 340–346.

Hernando, F., Schoots, O., Lolait, S. J., and Burbach, J. P. (2001). Immunohistochemical localization of the vasopressin V1b receptor in the rat brain and pituitary gland: anatomical support for its involvement in the central effects of vasopressin. *Endocrinology*, 142, 1659–1668.

Hong, H., Kohli, K., Trivedi, A., Johnson, D. L., and Stallcup, M. R. (1996). GRIP1, a novel mouse protein that serves as a transcriptional coactivator in yeast for the hormone binding domains of steroid receptors. *Proc Natl Acad Sci USA*, 93, 4948–4952.

Hrabovszky, E., Kallo, I., Hajszan, T., Shughrue, P. J., Merchenthaler, I., and Liposits, Z. (1998). Expression of estrogen receptor-beta messenger ribonucleic acid in oxytocin and vasopressin neurons of the rat supraoptic and paraventricular nuclei. *Endocrinology*, 139, 2600–2604.

Hrabovszky, E., Kallo, I., Steinhauser, A., Merchenthaler, I., Coen, C. W., Petersen, S. L., and Liposits, Z. (2004). Estrogen receptor-beta in oxytocin and vasopressin neurons of the rat and human hypothalamus: Immunocytochemical and in situ hybridization studies. *J Comp Neurol*, 473, 315–333.

Imagawa, M. (1987). Transcription factor AP-2 mediates induction by two different signal transduction pathways: protein kinase C and cAMP. *Cell*, 51, 251.

Inoue, T., Kimura, T., Azuma, C., Inazawa, J., Takemura, M., Kikuchi, T., Kubota, Y., Ogita, K., and Saji, F. (1994). Structural organization of the human oxytocin receptor gene. *J Biol Chem*, 269, 32451–32456.

Insel, T. R., Gelhard, R., and Shapiro, L. E. (1991). The comparative distribution of forebrain receptors for neurohypophyseal peptides in monogamous and polygamous mice. *Neuroscience*, 43, 623–630.

Insel, T. R., Young, L., Witt, D. M., and Crews, D. (1993). Gonadal steroids have paradoxical effects on brain oxytocin receptors. *J Neuroendocrinol*, 5, 619–628.

Ivell, R., Hunt, N., Abend, N., Brackman, B., Nollmeyer, D., Lamsa, J. C., and McCracken, J. A. (1990). Structure and ovarian expression of the oxytocin gene in sheep. *Reprod Fertil Dev*, 2, 703–711.

Ivell, R. and Richter, D. (1984). Structure and comparison of the oxytocin and vasopressin genes from rat. *Proc Natl Acad Sci USA*, 81, 2006–2010.

Jard, S. (1983). Vasopressin: mechanisms of receptor activation. *Prog Brain Res*, 60, 383–394.

Jard, S., Barberis, C., Audigier, S., and Tribollet, E. (1987). Neurohypophyseal hormone receptor systems in brain and periphery. *Prog Brain Res*, 72, 173–187.

Jard, S., Gaillard, R. C., Guillon, G., Marie, J., Schoenenberg, P., Muller, A. F., Manning, M., and Sawyer, W. H. (1986). Vasopressin antagonists allow demonstration of a novel type of vasopressin receptor in the rat adenohypophysis. *Mol Pharmacol*, 30, 171–177.

Jirikowski, G. F., Caldwell, J. D., Pedersen, C. A., and Stumpf, W. E. (1988). Estradiol influences oxytocin-immunoreactive brain systems. *Neuroscience*, 25, 237–248.

Jirikowski, G. F., Sanna, P. P., and Bloom, F. E. (1990). mRNA coding for oxytocin is present in axons of the hypothalamo-neurohypophysial tract. *Proc Natl Acad Sci USA*, 87, 7400–7404.

Johnson, A. E., Audigier, S., Rossi, F., Jard, S., Tribollet, E., and Barberis, C. (1993). Localization and characterization of vasopressin binding sites in the rat brain using an iodinated linear AVP antagonist. *Brain Res*, 622, 9–16.

Johnson, A. E., Barberis, C., and Albers, H. E. (1995). Castration reduces vasopressin receptor binding in the hamster hypothalamus. *Brain Res*, 674, 153–158.

Johnson, A. E., Coirini, H., Ball, G. F., and McEwen, B. S. (1989a). Anatomical localization of the effects of 17 beta-estradiol on oxytocin receptor binding in the ventromedial hypothalamic nucleus. *Endocrinology*, 124, 207–211.

Johnson, A. E., Coirini, H., McEwen, B. S., and Insel, T. R. (1989b). Testosterone modulates oxytocin binding in the hypothalamus of castrated male rats. *Neuroendocrinology*, 50, 199–203.

Johnson, A. E., Coirini, H., Insel, T. R., and McEwen, B. S. (1991). The regulation of oxytocin receptor binding in the ventromedial hypothalamic nucleus by testosterone and its metabolites. *Endocrinology*, 128, 891–896.

Kato, S., Tora, L., Yamauchi, J., Masushige, S., Bellard, M., and Chambon, P. (1992). A far upstream estrogen response element of the ovalbumin gene contains several half-palindromic 5′-TGACC-3′ motifs acting synergistically. *Cell*, 68, 731–742.

Kim, K., Barhoumi, R., Burghardt, R., and Safe, S. (2005). Analysis of estrogen receptor alpha-Sp1 interactions in breast cancer cells by fluorescence resonance energy transfer. *Mol Endocrinol*, 19, 843–854.

Kimura, T., Tanizawa, O., Mori, K., Brownstein, M. J., and Okayama, H. (1992). Structure and expression of a human oxytocin receptor. *Nature*, 356, 526–529.

Koohi, M. K., Ivell, R., and Walther, N. (2005). Transcriptional activation of the oxytocin promoter by oestrogens uses a novel non-classical mechanism of oestrogen receptor action. *J Neuroendocrinol*, 17, 197–207.

Kraichely, D. M., Sun, J., Katzenellenbogen, J. A., and Katzenellenbogen, B. S. (2000). Conformational changes and coactivator recruitment by novel ligands for estrogen receptor-alpha and estrogen receptor-beta: correlations with biological character and distinct differences among SRC coactivator family members. *Endocrinology*, 141, 3534–3545.

Kubota, Y., Kimura, T., Hashimoto, K., Tokugawa, Y., Nobunaga, K., Azuma, C., Saji, F., and Murata, Y. (1996). Structure and expression of the mouse oxytocin receptor gene. *Mol Cell Endocrinol*, 124, 25–32.

Kuiper, G. G., Lemmen, J. G., Carlsson, B., Corton, J. C., Safe, S. H., van der Saag, P. T., van der Burg, B., and Gustafsson, J. A. (1998). Interaction of estrogenic chemicals and phytoestrogens with estrogen receptor beta. *Endocrinology*, 139, 4252–4263.

Laflamme, N., Nappi, R. E., Drolet, G., Labrie, C., and Rivest, S. (1998). Expression and neuropeptidergic characterization of estrogen receptors (ERalpha and ERbeta) throughout the rat brain: anatomical evidence of distinct roles of each subtype. *J Neurobiol*, 36, 357–378.

Lakhdar-Ghazal, N., Dubois-Dauphin, M., Hermes, M. L., Buijs, R. M., Bengelloun, W. A., and Pevet, P. (1995). Vasopressin in the brain of a desert hibernator, the jerboa (Jaculus orientalis): presence of sexual dimorphism and seasonal variation. *J Comp Neurol*, 358, 499–517.

Lee, A. G., Cool, D. R., Grunwald, W. C., Jr., Neal, D. E., Buckmaster, C. L., Cheng, M. Y., Hyde, S. A., Lyons, D. M., and Parker, K. J. (2011). A novel form of oxytocin in New World monkeys. *Biol Lett*, 7, 584–587.

Lee, H. J., Macbeth, A. H., Pagani, J. H., and Young, W. S., III. (2009). Oxytocin: the great facilitator of life. *Prog Neurobiol*, 88, 127–151.

Lolait, S. J., O'Carroll, A. M., Mahan, L. C., Felder, C. C., Button, D. C., Young, W. S., III, Mezey, E., and Brownstein, M. J. (1995). Extrapituitary expression of the rat V1b vasopressin receptor gene. *Proc Natl Acad Sci USA*, 92, 6783–6787.

Marin, M., Karis, A., Visser, P., Grosveld, F., and Philipsen, S. (1997). Transcription factor Sp1 is essential for early embryonic development but dispensable for cell growth and differentiation. *Cell*, 89, 619–628.

Michelini, S., Urbanek, M., Dean, M., and Goldman, D. (1995). Polymorphism and genetic mapping of the human oxytocin receptor gene on chromosome 3. *Am J Med Genet*, 60, 183–187.

Michell, R. H., Kirk, C. J., and Billah, M. M. (1979). Hormonal stimulation of phosphatidylinositol breakdown with particular reference to the hepatic effects of vasopressin. *Biochem Soc Trans*, 7, 861–865.

Miller, F. D., Ozimek, G., Milner, R. J., and Bloom, F. E. (1989a). Regulation of neuronal oxytocin mRNA by ovarian steroids in the mature and developing hypothalamus. *Proc Natl Acad Sci USA*, 86, 2468–2472.

Miller, M. A., Urban, J. H., and Dorsa, D. M. (1989b). Steroid dependency of vasopressin neurons in the bed nucleus of the stria terminalis by in situ hybridization. *Endocrinology*, 125, 2335–2340.

Miller, M. A., Vician, L., Clifton, D. K., and Dorsa, D. M. (1989c). Sex differences in vasopressin neurons in the bed nucleus of the stria terminalis by in situ hybridization. *Peptides*, 10, 615–619.

Miller, M. A., DeVries, G. J., al-Shamma, H. A., and Dorsa, D. M. (1992). Decline of vasopressin immunoreactivity and mRNA levels in the bed nucleus of the stria terminalis following castration. *J Neurosci*, 12, 2881–2887.

Mohr, E. and Schmitz, E. (1991). Functional characterization of estrogen and glucocorticoid responsive elements in the rat oxytocin gene. *Brain Res Mol Brain Res*, 9, 293–298.

Mohr, E., Schmitz, E., and Richter, D. (1988). A single rat genomic DNA fragment encodes both the oxytocin and vasopressin genes separated by 11 kilobases and oriented in opposite transcriptional directions. *Biochimie*, 70, 649–654.

Moore, F. L. and Lowry, C. A. (1998). Comparative neuroanatomy of vasotocin and vasopressin in amphibians and other vertebrates. *Comp Biochem Physiol C Pharmacol Toxicol Endocrinol*, 119, 251–260.

Morel, A., O'Carroll, A. M., Brownstein, M. J., and Lolait, S. J. (1992). Molecular cloning and expression of a rat V1a arginine vasopressin receptor. *Nature*, 356, 523–526.

Murasawa, S., Matsubara, H., Kijima, K., Maruyama, K., Mori, Y., and Inada, M. (1995). Structure of the rat V1a vasopressin receptor gene and characterization of its promoter region and complete cDNA sequence of the 3′-end. *J Biol Chem*, 270, 20042–20050.

Naftolin, F., Ryan, K. J., Davies, I. J., Reddy, V. V., Flores, F., Petro, Z., Kuhn, M., White, R. J., Takaoka, Y., and Wolin, L. (1975). The formation of estrogens by central neuroendocrine tissues. *Recent Prog Horm Res*, 31, 295–319.

Nomura, M., McKenna, E., Korach, K. S., Pfaff, D. W., and Ogawa, S. (2002). Estrogen receptor-beta regulates transcript levels for oxytocin and arginine vasopressin in the hypothalamic paraventricular nucleus of male mice. *Brain Res Mol Brain Res*, 109, 84–94.

Norris, J. D., Fan, D., Stallcup, M. R., and McDonnell, D. P. (1998). Enhancement of estrogen receptor transcriptional activity by the coactivator GRIP-1 highlights the role of activation function 2 in determining estrogen receptor pharmacology. *J Biol Chem*, 273, 6679–6688.

Ostrowski, N. L., Lolait, S. J., and Young, W. S., III. (1994). Cellular localization of vasopressin V1a receptor messenger ribonucleic acid in adult male rat brain, pineal, and brain vasculature. *Endocrinology*, 135, 1511–1528.

Ostrowski, N. L., Young, W. S., III, and Lolait, S. J. (1995). Estrogen increases renal oxytocin receptor gene expression. *Endocrinology*, 136, 1801–1804.

Pak, T. R., Chung, W. C., Hinds, L. R., and Handa, R. J. (2007). Estrogen receptor-beta mediates dihydrotestosterone-induced stimulation of the arginine vasopressin promoter in neuronal cells. *Endocrinology*, 148, 3371–3382.

Pak, T. R., Chung, W. C., Lund, T. D., Hinds, L. R., Clay, C. M., and Handa, R. J. (2005). The androgen metabolite, 5alpha-androstane-3beta, 17beta-diol, is a potent modulator of estrogen receptor-beta1-mediated gene transcription in neuronal cells. *Endocrinology*, 146, 147–155.

Patchev, V. K., Schlosser, S. F., Hassan, A. H., and Almeida, O. F. (1993). Oxytocin binding sites in rat limbic and hypothalamic structures: site-specific modulation by adrenal and gonadal steroids. *Neuroscience*, 57, 537–543.

Patisaul, H. B., Scordalakes, E. M., Young, L. J., and Rissman, E. F. (2003). Oxytocin, but not oxytocin receptor, is Regulated by oestrogen receptor beta in the female mouse hypothalamus. *J Neuroendocrinol*, 15, 787–793.

Peter, J., Burbach, H., Adan, R. A., Tol, H. H., Verbeeck, M. A., Axelson, J. F., Leeuwen, F. W., Beekman, J. M., and Ab, G. (1990). Regulation of the rat oxytocin gene by estradiol. *J Neuroendocrinol*, 2, 633–639.

Peterson, D. N. (1998). Identification of estrogen receptor beta2, a functional variant of estrogen receptor beta

expressed in normal rat tissue. *Endocrinology*, 139, 1082–1092.

Planas, B., Kolb, P. E., Raskind, M. A., and Miller, M. A. (1995). Sex difference in coexpression by galanin neurons accounts for sexual dimorphism of vasopressin in the bed nucleus of the stria terminalis. *Endocrinology*, 136, 727–733.

Price, R. H., Jr., Butler, C. A., Webb, P., Uht, R., Kushner, P., and Handa, R. J. (2001). A splice variant of estrogen receptor beta missing exon 3 displays altered subnuclear localization and capacity for transcriptional activation. *Endocrinology*, 142, 2039–2049.

Price, R. H., Jr., Lorenzon, N., and Handa, R. J. (2000). Differential expression of estrogen receptor beta splice variants in rat brain: identification and characterization of a novel variant missing exon 4. *Brain Res Mol Brain Res*, 80, 260–268.

Rabadan-Diehl, C., Lolait, S., and Aguilera, G. (2000). Isolation and characterization of the promoter region of the rat vasopressin V1b receptor gene. *J Neuroendocrinol*, 12, 437–444.

Revankar, C. M., Cimino, D. F., Sklar, L. A., Arterburn, J. B., and Prossnitz, E. R. (2005). A transmembrane intracellular estrogen receptor mediates rapid cell signaling. *Science*, 307, 1625–1630.

Richard, S. and Zingg, H. H. (1990). The human oxytocin gene promoter is regulated by estrogens. *J Biol Chem*, 265, 6098–6103.

Richard, S. and Zingg, H. H. (1991). Identification of cis-acting regulatory elements in the human oxytocin gene promoter. *Mol Cell Neurosci*, 2, 501–510.

Rinaman, L. (1998). Oxytocinergic inputs to the nucleus of the solitary tract and dorsal motor nucleus of the vagus in neonatal rats. *J Comp Neurol*, 399, 101–109.

Romeo, R. D., Diedrich, S. L., and Sisk, C. L. (1999). Estrogen receptor immunoreactivity in prepubertal and adult male Syrian hamsters. *Neurosci Lett*, 265, 167–170.

Rosen, G. J., de Vries, G. J., Goldman, S. L., Goldman, B. D., and Forger, N. G. (2008). Distribution of oxytocin in the brain of a eusocial rodent. *Neuroscience*, 155, 809–817.

Rozen, F., Russo, C., Banville, D., and Zingg, H. H. (1995). Structure, characterization, and expression of the rat oxytocin receptor gene. *Proc Natl Acad Sci USA*, 92, 200–204.

Ruppert, S., Scherer, G., and Schutz, G. (1984). Recent gene conversion involving bovine vasopressin and oxytocin precursor genes suggested by nucleotide sequence. *Nature*, 308, 554–557.

Saito, M., Sugimoto, T., Tahara, A., and Kawashima, H. (1995). Molecular cloning and characterization of rat V1b vasopressin receptor: evidence for its expression in extra-pituitary tissues. *Biochem Biophys Res Commun*, 212, 751–757.

Sakamoto, H., Matsuda, K., Hosokawa, K., Nishi, M., Morris, J. F., Prossnitz, E. R., and Kawata, M. (2007). Expression of G protein-coupled receptor-30, a G protein-coupled membrane estrogen receptor, in oxytocin neurons of the rat paraventricular and supraoptic nuclei. *Endocrinology*, 148, 5842–5850.

Salvatore, C. A., Woyden, C. J., Guidotti, M. T., Pettibone, D. J., and Jacobson, M. A. (1998). Cloning and expression of the rhesus monkey oxytocin receptor. *J Recept Signal Transduct Res*, 18, 15–24.

Sanchez, R., Nguyen, D., Rocha, W., White, J. H., and Mader, S. (2002). Diversity in the mechanisms of gene regulation by estrogen receptors. *Bioessays*, 24, 244–254.

Sar, M. and Stumpf, W. E. (1980). Simultaneous localization of [3H]estradiol and neurophysin I or arginine vasopressin in hypothalamic neurons demonstrated by a combined technique of dry-mount autoradiography and immunohistochemistry. *Neurosci Lett*, 17, 179–184.

Sausville, E., Carney, D., and Battey, J. (1985). The human vasopressin gene is linked to the oxytocin gene and is selectively expressed in a cultured lung cancer cell line. *J Biol Chem*, 260, 10236–10241.

Scordalakes, E. M. and Rissman, E. F. (2004). Aggression and arginine vasopressin immunoreactivity regulation by androgen receptor and estrogen receptor alpha. *Genes Brain Behav*, 3, 20–26.

Selmanoff, M. K., Brodkin, L. D., Weiner, R. I., and Siiteri, P. K. (1977). Aromatization and 5alpha-reduction of androgens in discrete hypothalamic and limbic regions of the male and female rat. *Endocrinology*, 101, 841–848.

Shapiro, R. A., Xu, C., and Dorsa, D. M. (2000). Differential transcriptional regulation of rat vasopressin gene expression by estrogen receptor alpha and beta. *Endocrinology*, 141, 4056–4064.

Shughrue, P., Scrimo, P., Lane, M., Askew, R., and Merchenthaler, I. (1997). The distribution of estrogen receptor-beta mRNA in forebrain regions of the estrogen receptor-alpha knockout mouse. *Endocrinology*, 138, 5649–5652.

Shughrue, P. J., Dellovade, T. L., and Merchenthaler, I. (2002). Estrogen modulates oxytocin gene expression in regions of the rat supraoptic and paraventricular nuclei that contain estrogen receptor-beta. *Prog Brain Res*, 139, 15–29.

Simmons, C. F., Jr., Clancy, T. E., Quan, R., and Knoll, J. H. (1995). The oxytocin receptor gene (OTR) localizes to human chromosome 3p25 by fluorescence in situ hybridization and PCR analysis of somatic cell hybrids. *Genomics*, 26, 623–625.

Simonian, S. X. and Herbison, A. E. (1997). Differential expression of estrogen receptor alpha and beta immunoreactivity by oxytocin neurons of rat paraventricular nucleus. *J Neuroendocrinol*, 9, 803–806.

Skowronski, R., Beaumont, K., and Fanestil, D. D. (1987). Modification of the peripheral-type benzodiazepine receptor by arachidonate, diethylpyrocarbonate and thiol reagents. *Eur J Pharmacol*, 143, 305–314.

Smeltzer, M. D., Curtis, J. T., Aragona, B. J., and Wang, Z. (2006). Dopamine, oxytocin, and vasopressin receptor binding in the medial prefrontal cortex of monogamous and promiscuous voles. *Neurosci Lett*, 394, 146–151.

Sofroniew, M. V. (1980). Projections from vasopressin, oxytocin, and neurophysin neurons to neural targets in the rat and human. *J Histochem Cytochem*, 28, 475–478.

Sofroniew, M. V. (1983). Morphology of vasopressin and oxytocin neurones and their central and vascular projections. *Prog Brain Res*, 60, 101–114.

Soloff, M. S. (1975). Uterine receptor for oxytocin: effects of estrogen. *Biochem Biophys Res Commun*, 65, 205–212.

Soloff, M. S. (1982). Oxytocin receptors and mammary myoepithelial cells. *J Dairy Sci*, 65, 326–337.

Steckelbroeck, S., Jin, Y., Gopishetty, S., Oyesanmi, B., and Penning, T. M. (2004). Human cytosolic 3alpha-hydroxysteroid dehydrogenases of the aldo-keto reductase superfamily display significant 3beta-hydroxysteroid dehydrogenase activity: implications for steroid hormone metabolism and action. *J Biol Chem*, 279, 10784–10795.

Stedronsky, K., Telgmann, R., Tillmann, G., Walther, N., and Ivell, R. (2002). The affinity and activity of the multiple hormone response element in the proximal promoter of the human oxytocin gene. *J Neuroendocrinol*, 14, 472–485.

Stemmelin, J., Lukovic, L., Salome, N., and Griebel, G. (2005). Evidence that the lateral septum is involved in the antidepressant-like effects of the vasopressin V1b receptor antagonist, SSR149415. *Neuropsychopharmacology*, 30, 35–42.

Suzuki, S. and Handa, R. J. (2005). Estrogen receptor-beta, but not estrogen receptor-alpha, is expressed in prolactin neurons of the female rat paraventricular and supraoptic nuclei: comparison with other neuropeptides. *J Comp Neurol*, 484, 28–42.

Swaab, D. F., Slob, A. K., Houtsmuller, E. J., Brand, T., and Zhou, J. N. (1995). Increased number of vasopressin neurons in the suprachiasmatic nucleus (SCN) of 'bisexual' adult male rats following perinatal treatment with the aromatase blocker ATD. *Brain Res Dev Brain Res*, 85, 273–279.

Swanson, L. W. and Kuypers, H. G. (1980). The paraventricular nucleus of the hypothalamus: cytoarchitectonic subdivisions and organization of projections to the pituitary, dorsal vagal complex, and spinal cord as demonstrated by retrograde fluorescence double-labeling methods. *J Comp Neurol*, 194, 555–570.

Szot, P., Bale, T. L., and Dorsa, D. M. (1994). Distribution of messenger RNA for the vasopressin V1a receptor in the CNS of male and female rats. *Brain Res Mol Brain Res*, 24, 1–10.

Thibonnier, M., Auzan, C., Madhun, Z., Wilkins, P., Berti-Mattera, L., and Clauser, E. (1994). Molecular cloning, sequencing, and functional expression of a cDNA encoding the human V1a vasopressin receptor. *J Biol Chem*, 269, 3304–3310.

Torn, S., Nokelainen, P., Kurkela, R., Pulkka, A., Menjivar, M., Ghosh, S., Coca-Prados, M., Peltoketo, H., Isomaa, V., and Vihko, P. (2003). Production, purification, and functional analysis of recombinant human and mouse 17beta-hydroxysteroid dehydrogenase type 7. *Biochem Biophys Res Commun*, 305, 37–45.

Tribollet, E., Audigier, S., Dubois-Dauphin, M., and Dreifuss, J. J. (1990). Gonadal steroids regulate oxytocin receptors but not vasopressin receptors in the brain of male and female rats. An autoradiographical study. *Brain Res*, 511, 129–140.

Tribollet, E., Barberis, C., and Arsenijevic, Y. (1997). Distribution of vasopressin and oxytocin receptors in the rat spinal cord: sex-related differences and effect of castration in pudendal motor nuclei. *Neuroscience*, 78, 499–509.

Tribollet, E., Barberis, C., Jard, S., Dubois-Dauphin, M., and Dreifuss, J. J. (1988). Localization and pharmacological characterization of high affinity binding sites for vasopressin and oxytocin in the rat brain by light microscopic autoradiography. *Brain Res*, 442, 105–118.

Tribollet, E., Dubois-Dauphin, M., Dreifuss, J. J., Barberis, C., and Jard, S. (1992). Oxytocin receptors in the central

nervous system. Distribution, development, and species differences. *Ann NY Acad Sci*, 652, 29–38.

Uhl-Bronner, S., Waltisperger, E., Martinez-Lorenzana, G., Condes Lara, M., and Freund-Mercier, M. J. (2005). Sexually dimorphic expression of oxytocin binding sites in forebrain and spinal cord of the rat. *Neuroscience*, 135, 147–154.

Vaccari, C., Lolait, S. J., and Ostrowski, N. L. (1998). Comparative distribution of vasopressin V1b and oxytocin receptor messenger ribonucleic acids in brain. *Endocrinology*, 139, 5015–5033.

van Leeuwen, F. W., Caffe, A. R., and de Vries, G. J. (1985). Vasopressin cells in the bed nucleus of the stria terminalis of the rat: sex differences and the influence of androgens. *Brain Res*, 325, 391–394.

Van Tol, H. H., Bolwerk, E. L., Liu, B., and Burbach, J. P. (1988). Oxytocin and vasopressin gene expression in the hypothalamo-neurohypophyseal system of the rat during the estrous cycle, pregnancy, and lactation. *Endocrinology*, 122, 945–951.

Wang, Z. (1995). Species differences in the vasopressin-immunoreactive pathways in the bed nucleus of the stria terminalis and medial amygdaloid nucleus in prairie voles (Microtus ochrogaster) and meadow voles (Microtus pennsylvanicus). *Behav Neurosci*, 109, 305–311.

Wang, Z., Bullock, N. A., and de Vries, G. J. (1993). Sexual differentiation of vasopressin projections of the bed nucleus of the stria terminals and medial amygdaloid nucleus in rats. *Endocrinology*, 132, 2299–2306.

Wang, Z. and de Vries, G. J. (1993). Testosterone effects on paternal behavior and vasopressin immunoreactive projections in prairie voles (Microtus ochrogaster). *Brain Res*, 631, 156–160.

Wang, Z. and de Vries, G. J. (1995). Androgen and estrogen effects on vasopressin messenger RNA expression in the medial amygdaloid nucleus in male and female rats. *J Neuroendocrinol*, 7, 827–831.

Wang, Z., Moody, K., Newman, J. D., and Insel, T. R. (1997). Vasopressin and oxytocin immunoreactive neurons and fibers in the forebrain of male and female common marmosets (Callithrix jacchus). *Synapse*, 27, 14–25.

Wang, Z., Zhou, L., Hulihan, T. J., and Insel, T. R. (1996). Immunoreactivity of central vasopressin and oxytocin pathways in microtine rodents: a quantitative comparative study. *J Comp Neurol*, 366, 726–737.

Warembourg, M. and Poulain, P. (1991). Presence of estrogen receptor immunoreactivity in the oxytocin-containing magnocellular neurons projecting to the neurohypophysis in the guinea-pig. *Neuroscience*, 40, 41–53.

Wathes, D. C., Mann, G. E., Payne, J. H., Riley, P. R., Stevenson, K. R., and Lamming, G. E. (1996). Regulation of oxytocin, oestradiol and progesterone receptor concentrations in different uterine regions by oestradiol, progesterone and oxytocin in ovariectomized ewes. *J Endocrinol*, 151, 375–393.

Webb, P., Lopez, G. N., Uht, R. M., and Kushner, P. J. (1995). Tamoxifen activation of the estrogen receptor/AP-1 pathway: potential origin for the cell-specific estrogen-like effects of antiestrogens. *Mol Endocrinol*, 9, 443–456.

Wehrenberg, U., Ivell, R., Jansen, M., von Goedecke, S., and Walther, N. (1994a). Two orphan receptors binding to a common site are involved in the regulation of the oxytocin gene in the bovine ovary. *Proc Natl Acad Sci USA*, 91, 1440–1444.

Wehrenberg, U., von Goedecke, S., Ivell, R., and Walther, N. (1994b). The orphan receptor SF-1 binds to the COUP-like element in the promoter of the actively transcribed oxytocin gene. *J Neuroendocrinol*, 6, 1–4.

Weihua, Z., Lathe, R., Warner, M., and Gustafsson, J. A. (2002). An endocrine pathway in the prostate, ERbeta, AR, 5alpha-androstane-3beta,17beta-diol, and CYP7B1, regulates prostate growth. *Proc Natl Acad Sci USA*, 99, 13589–13594.

Xu, L., Pan, Y., Young, K. A., Wang, Z., and Zhang, Z. (2010). Oxytocin and vasopressin immunoreactive staining in the brains of Brandt's voles (Lasiopodomys brandtii) and greater long-tailed hamsters (Tscherskia triton). *Neuroscience*, 169, 1235–1247.

Yamaguchi, K., Akaishi, T., and Negoro, H. (1979). Effect of estrogen treatment on plasma oxytocin and vasopressin in ovariectomized rats. *Endocrinol Jpn*, 26, 197–205.

Yoshimura, R., Kiyama, H., Kimura, T., Araki, T., Maeno, H., Tanizawa, O., and Tohyama, M. (1993). Localization of oxytocin receptor messenger ribonucleic acid in the rat brain. *Endocrinology*, 133, 1239–1246.

Young, L. J., Huot, B., Nilsen, R., Wang, Z., and Insel, T. R. (1996). Species differences in central oxytocin receptor gene expression: comparative analysis of promoter sequences. *J Neuroendocrinol*, 8, 777–783.

Young, L. J., Muns, S., Wang, Z., and Insel, T. R. (1997). Changes in oxytocin receptor mRNA in rat brain during pregnancy and the effects of estrogen and interleukin-6. *J Neuroendocrinol*, 9, 859–865.

Young, L. J., Wang, Z., Cooper, T. T., and Albers, H. E. (2000). Vasopressin (V1a) receptor binding, mRNA expression and transcriptional regulation by androgen in the Syrian hamster brain. *J Neuroendocrinol*, 12, 1179–1185.

Young, L. J., Wang, Z., Donaldson, R., and Rissman, E. F. (1998). Estrogen receptor alpha is essential for induction of oxytocin receptor by estrogen. *Neuroreport*, 9, 933–936.

Young, W. S., III and Gainer, H. (2003). Transgenesis and the study of expression, cellular targeting and function of oxytocin, vasopressin, and their receptors. *Neuroendocrinology*, 78, 185–203.

Young, W. S., Li, J., Wersinger, S. R., and Palkovits, M. (2006). The vasopressin 1b receptor is prominent in the hippocampal area CA2 where it is unaffected by restraint stress or adrenalectomy. *Neuroscience*, 143, 1031–1039.

Zingg, H. H. and Lefebvre, D. L. (1988). Oxytocin and vasopressin gene expression during gestation and lactation. *Brain Res*, 464, 1–6.

2

Oxytocin and vasopressin release and their receptor-mediated intracellular pathways that determine their behavioral effects

Inga D. Neumann and Erwin H. van den Burg

2.1 Introduction

The nonapeptides oxytocin (OT) and arginine vasopressin (AVP), which differ only by two amino acids, form an essential part of the hypothalamo-neurohypophysial system (HNS) that regulates physiological and behavioral functions via peripheral and central actions. Key to the HNS are magnocellular neurons within the hypothalamus that have served as one of the most valuable model systems in neuroendocrinology and neuroscience ever since the original description of the HNS by the German biologist Ernst Scharrer in 1928. Outstanding discoveries have been made utilizing OT and AVP neurons. These include neuropeptide synthesis, axonal transport and neurohypophysial secretion, stimulus-secretion coupling, as well as intracerebral neuropeptide release including central release from their dendrites and perikarya. Moreover, multiple neuronal morphological, neuroendocrine and behavioral actions of OT and AVP, therefore classified as neuropeptides of the brain (Landgraf and Neumann, 2004), were revealed, which are mediated by their specific receptor proteins (Thibonnier et al., 1998; Gimpl and Fahrenholz, 2001).

Due to dendritic and/or somatic release, OT and AVP from the magnocellular neurons contribute to the central effects elicited by these neuropeptides. In addition, OT and AVP released from hypothalamic parvocellular neuron axons control behavior in a variety of brain regions.

OT and AVP currently attract intense attention due to their striking effects in the context of social interactions and modulation of stress responses. With respect to social behaviors, to date, these neuropeptides are involved in the regulation of maternal care and aggression (both), intermale aggression (AVP), pair bonding (both), sexual behavior (OT), social memory (both), social support (OT), and human trust (OT). Moreover, both brain OT and AVP are important regulators of the stress response, for example, OT downregulates neuronal, hormonal and anxiety responses, whereas AVP exerts anxiogenic effects. These discoveries make these neuromodulator/neurotransmitter systems of the brain promising candidates for psychotherapeutic intervention and treatment of anxiety- and depression-related diseases, social phobia, autism, and postpartum depression (Hollander et al., 2003; Slattery and Neumann, 2010a;b).

Although detailed mechanisms of neuronal synthesis (Altstein and Gainer, 1988; Chu and Zingg, 1999; Shughrue et al., 2002; Itoi et al., 2004; Mohr and Richter, 2004), release (Landgraf and Neumann, 2004) and regional receptor distribution (see below) are well established, our knowledge regarding OT receptor-(OTR) or AVP receptor-(AVP-R) mediated intracellular signaling is limited. Moreover, little is

Oxytocin, Vasopressin, and Related Peptides in the Regulation of Behavior, ed. E. Choleris, D. W. Pfaff, and M. Kavaliers.
Published by Cambridge University Press.

known about the involvement of distinct neuronal signaling cascades that mediate the behavioral or physiological processes regulated by OT and AVP release. Therefore, this chapter will summarize (i) selected social and stressful stimuli, which trigger the release of OT / AVP within distinct brain regions, and (ii) the behavioral significance of such locally released neuropeptide. Further, (iii) we will review existing knowledge on OTR- and AVP-R-mediated intracellular signaling cascades that may underlie the behavioral effects.

2.2 Social stimuli triggering intracerebral release of OT and AVP

Local release patterns of a given neuropeptide determine its concentration in the extracellular fluid of that brain area and the cascade of events from receptor binding to intracellular signaling and subsequent neurophysiological changes. The method of choice for monitoring stimulus-dependent fluctuations in the release of neuropeptides in freely behaving animals is microdialysis (for review, see (Landgraf and Neumann, 2004; Neumann, 2007; Veenema and Neumann, 2008)). In this way, OT and AVP release have been quantified in limbic, hypothalamic and cortical brain areas including the septum (Demotes-Mainard et al., 1986; Landgraf et al., 1988; Neumann and Landgraf, 1989), the hippocampus (Landgraf et al., 1988; Neumann and Landgraf, 1989), the central amygdala, (Bosch et al., 2005) the olfactory bulb (Kendrick et al., 1988; Levy et al., 1995; Larrazolo-Lopez et al., 2008), the suprachiasmatic nucleus (Kalsbeek et al., 1995), and their nuclei of origin, that is, the hypothalamic supraoptic (SON) and paraventricular (PVN) nuclei (Moos et al., 1989; Pow and Morris, 1989; Landgraf and Ludwig, 1991; Neumann et al., 1993a; Neumann et al., 1993b; Ludwig et al., 1994). Within the SON and PVN, such release is likely to occur from neuronal dendrites and somata (Landgraf and Neumann, 2004; Ludwig and Leng, 2006).

A variety of physiological, emotional and pharmacological stimuli were described to trigger intracerebral OT and AVP release in a neuropeptide-specific, stimulus-dependent and locally restricted manner and are reviewed in detail elsewhere (for reviews see citations above). It is important to mention that the central release of OT and AVP may occur independently of peripheral secretion into blood from magnocellular neurohypophysial terminals, but under many physiological circumstances the release of OT or AVP into the different target compartments, that is, brain and blood, respectively, was found to be coupled (Neumann et al., 1993a; Neumann et al., 1994; Neumann et al., 1995; Engelmann et al., 1998; Engelmann et al., 1999).

Social stimuli seem to be of particular relevance for the OT and AVP system; although differentiation between reproductive/physiological and social stimuli is not always easy. For example, OT release is triggered by parturition-related stimuli as well as by suckling in the lactating animal within several brain areas including the PVN (Neumann et al., 1993a; Neumann et al., 1993b). Also, AVP is released within hypothalamic regions during psychosocial stress (social defeat) and involved in behavioral and neuroendocrine regulation (Wotjak et al., 1996).

2.2.1 Central OT release during reproductive functions: Context to stress regulation

During parturition and suckling, hypothalamic OT neurons are highly active and synchronized as a prerequisite for high amounts of peripheral secretion. OT is released within the SON and PVN both during parturition and suckling in the lactating animal. Such locally released OT within the SON was found to autoregulate OT neurons and to facilitate synchronization of bursting activity of OT neurons during the milk-ejection reflex (Moos et al., 1984; Neumann et al., 1993b; Neumann et al., 1994; Neumann et al., 1996). OT is also released in other brain regions including the bed nucleus of the stria terminalis (BNST) (Ingram and Moos, 1992), the septum, hippocampus and olfactory bulb during female reproduction (Kendrick et al., 1988; Neumann and Landgraf, 1989; Landgraf et al., 1991), where it is relevant for the onset (Pedersen and

Prange, 1979; van Leengoed et al., 1987; Neumann et al., 1996), and fine-tuned maintenance (Pedersen and Boccia, 2003), of maternal behavior as well as offspring recognition (as shown in sheep; (Kendrick et al., 1997)), which are equally important for offspring survival. Moreover, OT is released within the PVN and central amygdala during maternal aggression and defense of the offspring ((Bosch et al., 2005)) – a behavioral phenomenon found in the peripartum period of almost all species. Thus, central release of OT as a result of an activated brain OT system significantly contributes to complex physiological and behavioral adaptations described in the peripartum period (for reviews see (Brunton and Russell, 2008; Slattery and Neumann, 2008)). Importantly, the high activity of the brain OT system plays a crucial role in the attenuation of emotional and physiological stress responses in the peripartum period of the female including a state of anxiolysis (Slattery and Neumann, 2008).

However, is there a comparable physiological situation that results in such an activated brain OT system in males and concurrent attenuated stress responsiveness? In male rats, we could show that hypothalamic OT release within the PVN is triggered by sexual activity (Waldherr and Neumann, 2007), where it is critically involved in the regulation of male sexual functions (Argiolas and Melis, 2004). Such centrally released OT was found to mediate the reduced level of anxiety-related behavior revealed 30 min to 4 h after a 30-min mating period (Waldherr and Neumann, 2007). Thus, in the context of both female and male reproduction, social stimuli trigger the activation of central OT release, which in turn modulates physiological and behavioral functions and the accompanying attenuation of anxiety and physiological stress responses.

2.2.2 Central AVP release during reproductive functions

The brain AVP system is also activated during female reproduction (Walker et al., 2001; Bosch and Neumann, 2008; Bosch et al., 2010; Kessler et al., 2010), and has only recently been described to play a central role in the regulation of maternal behavior and aggression (Nephew and Bridges, 2008; Bosch, 2010; Nephew et al., 2010). We could show that the display of maternal behavior including maternal aggression is largely dependent on the substantial release of AVP, acting on AVPR in the central amygdala and PVN (Bosch and Neumann, 2010). Thus, it seems that OT and AVP act in concert to mediate maternal aggression: OT reduces anxiety and, hence, makes the animals more courageous to approach a potentially threatening intruder; while AVP elicits aggressive behavior of the dam (see also (Bosch, 2010)). In contrast to OT, brain AVP exerts anxiogenic effects, which have been localized within the medio-lateral septum (Landgraf et al., 1995).

2.2.3 Central OT release during social stressors in females: Social defeat of virgins or maternal defense of lactating dams

Central OT release has also been monitored in response to stressful social experiences, both in males (resident–intruder test) and females (maternal defeat/defense). Placing a virgin female intruder rat into the home cage of a lactating resident dam in the presence of her litter is a stressful experience for both (Neumann et al., 2001). During lactation, dams are highly aggressive while protecting their offspring against potentially dangerous conspecifics. In virgin intruders, exposure to maternal defeat stimulates a significant rise in OT release within the PVN, but not within the amygdala or the lateral septum (Bosch et al., 2004) indicating a strict region-dependent pattern of OT release.

Also, exposure of the lactating resident dam to the virgin intruder (maternal defense) triggers OT release within the PVN and the central amygdala as revealed by microdialysis (Bosch et al., 2004; Bosch et al., 2005). However, an increased local release of OT was only found in rats that displayed a particularly high level of maternal aggression, that is, in rats selectively bred for high anxiety-related behavior (HAB). In contrast, OT release was unaltered (PVN), or even decreased (amygdala), in dams that were less aggressive. Importantly, the amount of locally

released OT directly correlated with the amount of aggression displayed by the dam during the maternal defense test (Bosch et al., 2005).

2.2.4 Central OT and AVP release during social stressors in males

Intracerebral OT and AVP release have also been monitored in male rats in response to social stressors such as during exposure to a larger and aggressive conspecific (social defeat) or during contact to an intruder rat, that is, the resident–intruder test – a test for intermale aggression. In male rats, exposure to social defeat selectively stimulates OT release within the SON, but not the PVN (Engelmann et al., 1999). In contrast, OT secretion into blood remains unchanged in response to this emotional stressor, indicating independent central and peripheral release patterns. Interestingly, a different scenario was found for AVP, which, while also being contained within the central compartment, was released within the PVN, but not the SON, in response to the same stimulus (Wotjak et al., 1996).

2.2.5 Central AVP release during intermale aggression

Brain AVP has been linked to the regulation of intermale aggression and territorial defense. The AVP circuit involved in male aggressive and territorial behaviors comprises AVP neurons located in the medial amygdala and the BNST, which project to the lateral septum (De Vries and Buijs, 1983), and of AVP projections to hypothalamic regions, especially the anterior hypothalamus (Ferris et al., 1997; Gobrogge et al., 2007). The septal area is of particular interest in the context of aggression regulation, as pharmacological manipulation of the local AVP system in castrated hamsters, or rats, altered intermale aggression (Koolhaas et al., 1991; Ferris and Delville, 1994). However, both a positive (Bester-Meredith et al., 1999) as well as a negative (Compaan et al., 1993; Everts et al., 1997) correlation between septal AVP and intermale aggression have been hypothesized. Here, detailed knowledge regarding the dynamics of local AVP patters seems essential.

In male Wistar rats bred for low (LAB) versus high (HAB) anxiety-related behavior (Neumann et al., 2010) we found line-dependent differences in AVP release within the septum during intermale aggression. LAB rats generally display a higher level of aggressive behavior than HAB and non-selected males during the resident–intruder test (Beiderbeck et al., 2007; Veenema et al., 2007), which is associated with a significant decrease in septal AVP release, whereas local AVP release increases in HAB resident males. These findings favor a negative correlation between septal AVP and intermale aggression in these rat lines. However, synthetic AVP-administration in the septum of high-aggressive male LAB rats or a specific vasopressin 1A receptor (AVPR1a) antagonist [d$(CH_2)_5$Tyr(Me)AVP] in that of low-aggressive male HAB rats via retrodialysis did not alter the level of aggression in either (Beiderbeck et al., 2007). In contrast, local AVP increased anxiety-related behavior in LAB males, and the AVPR1a antagonist reduced social investigation in HABs (Beiderbeck et al., 2007). Thus, aggression-induced changes in septal AVP release might alter other related behaviors, which are indirectly beneficial, and relevant, in the context of aggression such as anxiety.

Further, dysregulation in the AVP and/or OT system might significantly influence male aggressive and other social behaviors, and contribute to social behavioral dysfunctions. Indeed, excessive aggression in personality-disordered subjects was found to correlate with high AVP concentrations in the cerebrospinal fluid (Coccaro et al., 1998), which likely reflects – at least globally – central release patterns (Landgraf and Neumann, 2004).

2.3 Anxiolytic effect of centrally released OT

Brain OT has generally been described as an important regulator of neuroendocrine and behavioral stress-coping strategies. This includes its anxiolytic

actions and the inhibitory effects on HPA-axis responses (Windle et al., 1997; Neumann et al., 2000a; Neumann et al., 2000b; Neumann et al., 2000c). Specifically, OTR-mediated effects on basal and/or stimulated HPA-axis responses were found in the PVN and SON (Neumann et al., 2000a; Neumann et al., 2006). Whereas OT inhibits the basal level of HPA axis activity, OT may also contribute to the adequate HPA axis response to an acute challenge, for example, during forced swimming (Neumann, 2002).

Recently, we demonstrated that bilateral OT infusion into the PVN significantly lowered anxiety-related behavior in male (Blume et al., 2008) and female (Hillerer, Van den Burg, Slattery, and Neumann, unpublished) rats. Moreover, in support of the local anxiolytic effects of synthetic OT, OT release within the PVN following successful mating was found to mediate mating-induced anxiolysis ((Waldherr and Neumann, 2007); and see above). Thus, dendritically released OT within the hypothalamus is likely to contribute not only to specific physiological demands (e.g., sexual functions, milk ejection), but also to the simultaneous adaptation in behavioral and neuroendocrine stress responsiveness such as reduced anxiety. Interestingly, a close link between anxiety levels and the display of social behavior such as intermale aggression, maternal behavior, maternal aggression, and social interaction has been repeatedly described (Beiderbeck et al., 2007; Veenema et al., 2007; Bosch and Neumann, 2008).

In addition to the PVN, the central amygdala has also been identified as a target region of OT and local anxiolytic effects have been described (Bale et al., 2001; Neumann, 2001).

In summary, when released from various neuronal structures within distinct brain regions in response to a physiological, social or stressful stimulus, OT and AVP significantly contribute to neuroendocrine and behavioral adaptations of an individual. The biological and evolutionary advantage of somatodendritic or somatic, that is, of non-synaptic, release of neuropeptides like OT and AVP are likely to be the liberation from the constraints of point-to-point signaling (wired transmission (Fuxe et al., 1985)). In turn, such a mechanism enhances the significance of neuromodulatory functions, in a rather paracrine manner, after diffusion from the original sites of neuropeptide release. Thus, the neuromodulatory functions of OT and AVP contribute to an overwhelming complexity of subsequent information processing. This complexity is further amplified by the variety of intracellular responses that can occur following OT or AVP binding to their respective receptors, which are discussed in the following sections.

2.4 Neuronal responses underlying OT and AVP central effects

OTR and AVPR are members of the seven-transmembrane G protein-coupled receptor (GPCR) family. Three AVPRs have been found, of which the AVPR1a and AVPR1b are expressed in the brain. Only one OTR appears to exist that can occur in a high and a low affinity state, depending on high or low cholesterol levels in the cellular membrane, respectively (Gimpl and Fahrenholz, 2001). The neuronal responses to OT or AVP stimulation not only depend on the type and affinity state of the receptor expressed by the neuron, but also on the brain region, the cell type and the intracellular pathways that are activated by the liganded receptor. Current research concerning the activities of OT or AVP aims to unravel the mechanisms of action at the neuronal level underlying their behavioral and physiological responses described above. However, most of the intracellular molecular responses have been studied in various somatic rather than neuronal cells and cell lines. It appears that several uniform responses are found in essentially every cell type studied, whereas others are rather cell-type specific. Some of the intracellular responses to OTR and AVPR activation in the brain have directly been linked to cognitive processes, which will be discussed at the end of this chapter. First, we will briefly summarize the distribution of OTR and AVPR, as this is important for the behaviors

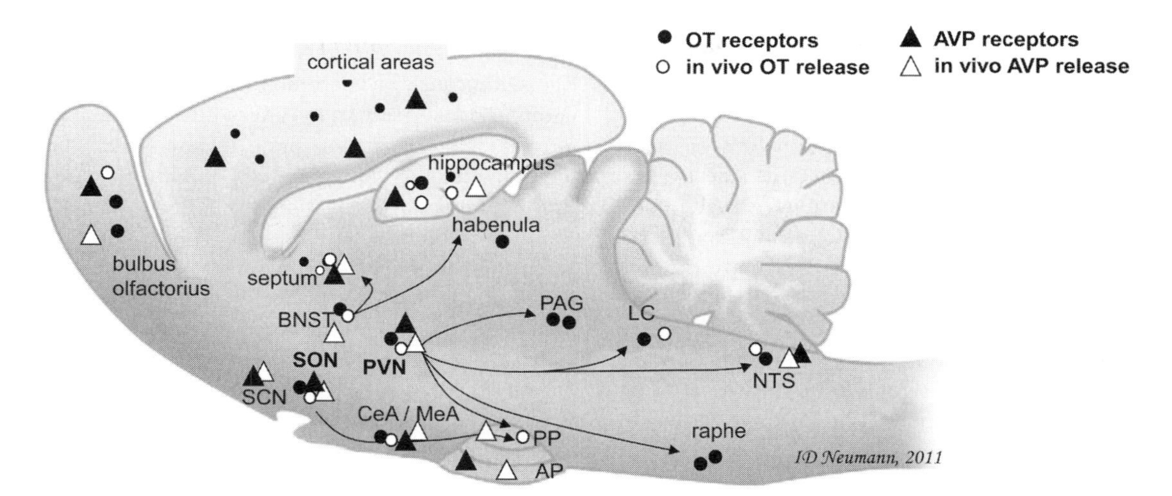

Figure 2.1 OTR and AVPR localization and overlap with OT and AVP release sites in the rat brain. AP, anterior pituitary; BNST, bed nucleus of the stria terminalis; CeA / MeA, central and medial nuclei of amygdala; LC, locus coeruleus; NTS, nucleus of the solitary tract; PAG, periaqueductal grey; PP, posterior pituitary; PVN, paraventricular nucleus; SCN, suprachiasmatic nucleus; SON, supraoptic nucleus.

controlled by OT and AVP, which will be followed by a general picture of intracellular consequences of OT and AVP binding to their respective receptors.

2.4.1 Oxytocin and vasopressin receptor distribution

The physiological relevance of OT or AVP release has been emphasized by the good correlation between the sites of release and the location of the receptors. The OTR has a wide distribution that is clearly more extended than the sites where OT release has been monitored (Figure 2.1). Using OTR venus knock-in mice, where the fluorescent protein *venus* is under the control of the OTR promoter, Yoshida and colleagues demonstrated apparent OTR localization in many cortical, limbic, hypothalamic, and brain stem regions (Yoshida et al., 2009). However, the distributions of OTR as well as AVPR1a vary among species. These differences in OTR and AVPR distribution correlate well with distinct behavioral phenotypes of even closely related species. For example, in two species of singing mice, which have a highly developed vocal system for communication, AVPR1a are highly expressed in two regions important for the generation of vocal signals, the periaqueductal gray and anterior hypothalamus, in the more vocal species. Conversely, the AVPR1a is more expressed in the anterior and laterodorsal hypothalamus, as is the OTR in the hippocampus and medial amygdala, of the other species, which lives in groups of lower density. Together, a pattern emerges in which the differential OTR and AVPR distribution is important for vocal communication, and associated social and spatial memory related to group density (Campbell et al., 2009). Similar differences related to social behavior have been found in the polygamous montane versus the monogamous prairie voles (see Chapter 8), or the social versus solitary tuco-tucos (Wang et al., 1997; Beery et al., 2008). In rats and mice, AVP binding in the brain is partially gender specific, with more apparent binding in the medial preoptic area / anterior hypothalamus and the mammillary nucleus in females, which might be important for reproductive functions (Dubois-Dauphin et al., 1996).

The AVPR1b has a more restricted expression pattern than the AVPR1a. The highest AVPR1b density

is found in corticotroph cells of the adenohypophysis, where they modulate the stress response of the HPA axis, and the median eminence. In rats, lower levels of AVPR1b are found in the hippocampus, the hypothalamus (including the PVN), the amygdala, the cerebellum, and the circumventricular organs (Hernando et al., 2001). However, using highly specific riboprobes in mice (as opposed to the use of an antibody in the rat study) a relatively narrow distribution was revealed, which was limited to the hippocampus, PVN, and amygdala (Young et al., 2006). It should be noted here that autoradiography studies to delineate AVPR1b distribution have not been performed to date, because of a lack of specific radiolabeled AVPR1b agonists (Roper et al., 2010).

2.4.2 General signaling properties of OTR and AVPR

Proper ligand binding and subsequent receptor activation depend on four aromatic amino acid residues (Phe, Trp, Phe, and Tyr; amino acid numbers based on the human OTR) in the second extracellular domain that are found in all vertebrate OTR and AVPR sequenced to date (Conner et al., 2007). As described in Figure 2.2, the OTR and AVPR1a / AVPR1b are coupled to $G_{q/11}$ α class GTP-binding proteins that stimulate the activity of phospholipase C and subsequent hydrolysis of phosphatidylinositol (Jard et al., 1987; Thibonnier et al., 2001). For the AVPR1a it has been shown that these effects depend on the second intracellular loop of the receptor (Liu and Wess, 1996), whereas in the OTR, G-protein binding occurs at the receptor´s C-terminal cytoplasmic tail (Hoare et al., 1999). Phosphatidylinositol hydrolysis leads to the formation of inositol-1,4,5-trisphosphate (IP3) and diacylglycerol (DAG). IP3 stimulates the release of Ca from intracellular stores, whereas DAG activates protein kinase C (PKC) (Figure 2.2). Together, these universal responses coupled to the activation of $G_{q/11}$ α class GTP-binding proteins initiate a series of intracellular events that enable adequate cellular responses to OT and AVP (Gimpl and Fahrenholz, 2001). However, in addition to these classical responses, there are other neuronal responses to OTR and AVPR activation that depend on transactivation of the epidermal growth factor receptor (EGFR). Examples of this will be given below.

2.4.3 OTR-mediated intracellular signaling in somatic cells

OTR have been shown not only to couple to $G_{q/11}$ α, but also to G_i α class GTP-binding proteins, which activate the MAPK p38 in transfected Chinese hamster ovary cells (Hoare et al., 1999) (Figure 2.2). Also, G_i α proteins appear to be important for both stimulation and inhibition of cell proliferation, depending on whether the OTR they are coupled to are localized within or outside caveolin-enriched microdomains (Guzzi et al., 2002). Caveolins are structural proteins that interact with signaling molecules at the membrane, and form particular membrane microdomains to facilitate proper intracellular signaling (reviewed in (Stern and Mermelstein, 2010)). Pharmacological and molecular analyses in HEK293 cells stably transfected with OTR showed that OTR located outside the caveolin-enriched microdomains inhibit cell proliferation by activating a G_i α – PLC – c-Src – PI3K–dependent pathway, whereas OTR located within the caveolin-enriched microdomains signal via a G_i α – PLC – c-Src – PI3K–independent pathway (Rimoldi et al., 2003). c-Src is known to transactivate the EGFR, which subsequently promotes PI3K activity (Soltoff and Cantley, 1996; Wetzker and Bohmer, 2003) (see also Figure 2.2). Therefore, one might predict that blocking the EGFR prevents the antiproliferative effect of OT, leaving the mitogenic effect of OTR localized within caveolin-enriched domains intact. However, it appeared that AG1478 (an inhibitor of EGFR kinase activity) abolished both the antiproliferative and mitogenic effects, indicating that both depend on EGFR transactivation. In support of this, in both cases, transactivation is achieved via a metalloprotease, as indicated by the absence of effects in the presence of a specific metalloprotease blocker (CRM197) (Rimoldi et al.,

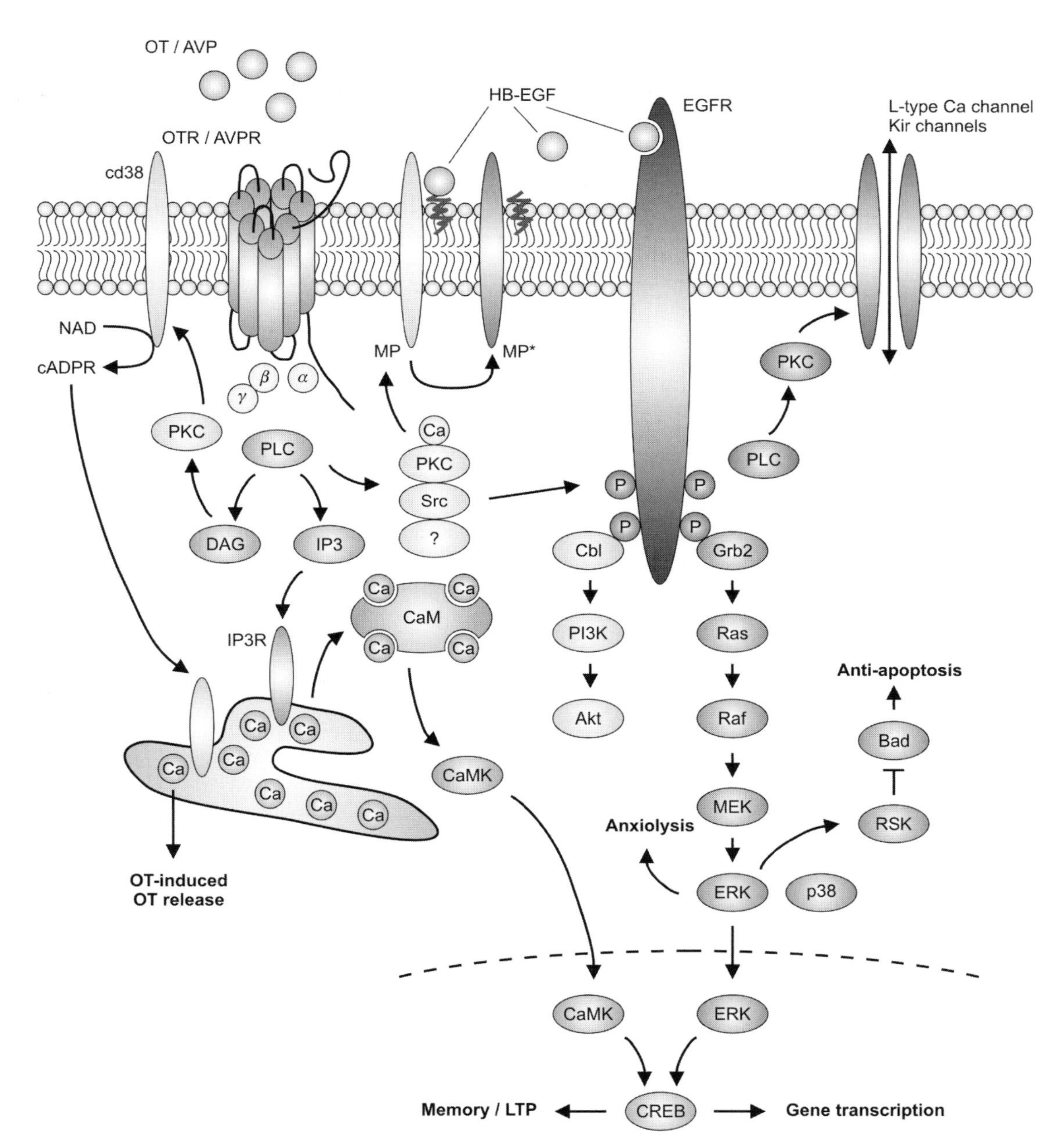

Figure 2.2 Pathways and associated processes controlled by OT and AVP. This scheme has been constructed on the basis of the publications cited in the text. Please note that these concern studies in different neurons in several brain regions, and that not all the processes depicted in the figure necessarily occur in every cell type, or are implicated in any type of behavior.

Blue module: OT- / AVP-binding to their receptors (OTR / AVPR) activates phospholipase C (PLC), which mobilizes inositol trisphosphate (IP3) and diacylglycerol (DAG). IP3 recruits Ca^{2+} from intracellular stores by binding to

2003). This is an additional mechanism of EGFR transactivation employed by other GPCR, where PLC recruits a factor (in some cases this factor is thought to be Ca^{2+}, Src, or PKC) that activates a metalloprotease. This, in turn, induces shedding of membrane-bound EGF, which subsequently activates its receptor (Wetzker and Bohmer, 2003) (Figure 2.2).

Interestingly, the OTR excluded from the caveolin-enriched microdomains persistently transactivate the EGFR and one of its downstream effectors, extracellular signal-regulated kinase (ERK), whereas the receptors included in the caveolin-enriched domain do so only transiently. The long-term activation of EGFR and ERK is required to increase $p21^{WAF/CIP1}$ protein levels, necessary to inhibit proliferation of the transfected HEK293 cells (Rimoldi et al., 2003). Thus, signaling via the OTR not only depends on its affinity state dictated by plasma membrane cholesterol levels as mentioned earlier, but also on cell membrane microdomains. The duration of intracellular signaling events is one of the determinants of the outcome of OTR activation as well.

In other cells, two intracellular signaling pathways rather operate in concert to bring about particular responses: in the bovine endometrial epithelial cell line, OT-induced prostaglandin production (to induce luteolysis at the end of an infertile period) and cyclooxygenase 2-expression depend on two pathways, which both rely on transactivation of the EGFR. The first pathway is a mitogen-activated protein kinase (MAPK) pathway leading to the phosphorylation of ERK, whereas the second concerns a PI3K – Akt sequence (Krishnaswamy et al., 2010) (see also Figure 2.2).

2.4.4 OTR-mediated intracellular signaling in the brain

Many of the intracellular signaling pathways that are recruited by liganded OTR observed in somatic cells have also been reported in neurons. These include the recruitment of intracellular Ca^{2+}, activation of PKC, transactivation of the EGFR and subsequent activation of ERK. In neurons, these effectors of receptor activation may be involved in the control of, for example, neurotransmitter and neuromodulator release, cellular excitability, plasticity, and gene expression that shape cognitive functions, behavior and physiology in response to external or internal stimuli. Some examples will be presented below.

A particular neuronal signaling system that seems to be specific for autoregulation of OT release

(*cont.*) the IP3 receptor (IP3R). The resulting increase of intracellular Ca^{2+} activates calmodulin (CaM) and Ca^{2+}/calmodulin-dependent kinases (CaMK); part of activated CaMK translocates to the nucleus.
Light green module: Although DAG activates protein kinase C (PKC) following both OTR / AVPR receptor binding, PKC activates cd38 only in the case of OTR activation, leading to the formation of cADPR. cADPR stimulates the release of intracellular Ca^{2+} independently from IP3, resulting in autostimulation of OT release in the SON and PVN.
Purple module: OTR and AVPR activation transactivate the epidermal growth factor receptor (EGFR) via two pathways. First, PLC may recruit Ca^{2+}, PKC, Src, or an unknown factor (cell-type dependent), which then activate a metalloproteinase (MP). Activated MP (MP*) liberates membrane-bound EGF (or heparin-binding EGF, HB-EGF), which binds to and activates the EGFR. Second, Src phosphorylates the EGFR directly.
Orange module: Activated EGFR binds the adaptor molecule Cbl, which binds PI3K. PI3K activates Akt, which is a central mediator of many intracellular processes.
Brown module: Activated EGFR binds the adaptor molecule complex Shc/Grb2/Sos (depicted as Grb2), which activates Ras. This activates the MAPK cascade Raf – MEK1/2 – ERK1/2, as well as p38. Activated MEK1/2 is necessary for the anxiolytic activity of OT. It also inhibits apoptosis when phosphorylated following AVPR activation via the phosphorylation of RSK, and subsequent phosphorylation deactivation of Bad. ERK1/2 and CaMK in the nucleus phosphorylate CREB to induce LTP and improvement of spatial memory (OT-mediated), and to control gene expression.
Dark green module: OT and AVP exert fast effects on ion channels, including L-type Ca^{2+} and inwardly rectifying K^+ channels (Kir channels), which might depend on a PLC – PKC sequence.
See color version in plates section.

concerns the CD38/Cyclic ADP-ribose system. CD38 is a transmembrane glycoprotein that catalyses the formation of cyclic ADP-ribose (cADPR) from the dinucleotide nicotinamide adenine dinucleotide (NAD^+). cADPR can mobilize Ca^{2+} from intracellular Ca^{2+} stores, independently from IP3 activity (reviewed in (Lee, 2001); Figure 2.2). Recruitment of Ca^{2+} from intracellular stores by CD38 is important for activation of OT neurons and central OT release and, therefore, the cognitive effects mediated by OT (Jin et al., 2007). In the hypothalamic PVN and SON, local OT release activates OTR on neighboring OT neurons (for review see (Landgraf and Neumann, 2004; Ludwig and Leng, 2006)). In crude membrane preparations of hypothalamic tissue of male mice, OTR binding activates CD38 activity and increases intracellular Ca^{2+} levels (Lopatina et al., 2010). This strongly suggests that the OTR can be coupled to CD38 signaling, and that this is important for OT-induced OT release (Lopatina et al., 2010). This could be a mechanism underlying the synchronization of OT release during the milk ejection reflex in lactating rats, for example (see above).

Other neuronal effects of OT include the rapid modulation of inwardly rectifying K^+ channels, and hence neuronal excitability (Gravati et al., 2010). In murine immortalized GnRH-positive undifferentiated migrating neurons, OTR activation exerts a dual action on K^+ channels via two different pathways. First, OT inhibits an inwardly rectifying current that is mediated by pertussis toxin (PTX)-resistant G protein (probably G_q), and depends on PLC activation. Second, OT stimulates an inwardly rectifying current probably via a direct effect of a PTX-sensitive G_i protein. The latter current is stimulated 0.2 s following the onset of OT exposure, in contrast to the 1.7 s that is necessary for the current inhibition. The physiological relevance of these opposing effects is not known, but it has been speculated that they contribute to neuronal growth and differentiation, as these experiments were done in cells derived from immature, migrating neurons (Gravati et al., 2010).

2.4.5 Involvement of OTR-activated neuronal signaling in cognition and anxiety

In a few instances, a direct link between the activation of OTR-dependent intracellular pathways and OT-mediated cognitive and emotional behaviors could be shown. OT promotes memory functions and exerts anxiolytic effects – both effects display a slower onset than those described in the previous paragraph, have a relatively long duration, and require the activation of ERK.

In the hippocampus of female mice during motherhood, ERK phosphorylated following OTR activation induces CREB phosphorylation and long-term potentiation (LTP). This improves spatial memory, which might serve to better remember the extended locations of water and food (Tomizawa et al., 2003).

Also, neuronal hypothalamic ERK phosphorylation seems to be importantly involved in OT-mediated anxiolysis. In the PVN of male rats, bilateral infusion of synthetic OT reduces anxiety-related behavior in several independent tests for anxiety (Blume et al., 2008). Similarly, stimulation of endogenous OT release within the hypothalamic PVN during successful mating in male rats is involved in mating-induced anxiolysis (Waldherr and Neumann, 2007). We found that exogenous OT transactivates the EGFR in hypothalamic neurons within the PVN and SON, and then activates a MAPK pathway leading to the phosphorylation of ERK within 10 min (Blume et al., 2008) (Figure 2.2). This timing coincides with the anxiolytic effect of OT (Blume et al., 2008). Importantly, when phosphorylation of ERK is blocked by the MAPK kinase (MEK) inhibitor U0126, the anxiolytic effect of OT is eliminated (Blume et al., 2008). ERK phosphorylation decreases from 10 min onward, and returns to basal levels shortly after 30 min, yet the anxiolytic activity of OT persists for 4 h following OT infusion (unpublished) or successful mating (Waldherr and Neumann, 2007). One could thus assume that phosphorylated ERK activates one or more substrates that are more stably activated to maintain OT-induced anxiolytic activity over time.

2.4.6 AVPR-mediated intracellular signaling in somatic cells

The intracellular effects elicited by AVPR activation in somatic cells are quite similar to those observed following OTR activation, and will therefore be discussed only briefly. Most of the described effects of AVPR activation concern those that involve AVPR1a, which is because of a lack of reliable pharmacological tools to study AVPR1b. Also, most of the somatic AVP research has focused on the AVPR2, which signals via $G\alpha_s$ class GTP-binding proteins that stimulate the activity of adenylyl cyclase and subsequent accumulation of cAMP and protein kinase A (PKA) activity. However, as the AVPR2 is only expressed in the kidney and not in the brain, this will not be discussed further.

It is of interest to note that AVPR1a activation transactivates the EGFR via both metalloproteases/shedding of EGF and directly via c-Src in the rat vascular smooth muscle A-10 cell line, and that the route of transactivation determines which immediate early gene (IEG) is transcribed (Fuentes et al., 2008). When the metalloproteinases are inhibited by MMPII and GM6001, the AVP-induced upregulation of the IEG c-fos is abolished, whereas that of Egr-1 is not. In contrast, the c-Src inhibitor PPI blocks the upregulation of Egr-1 expression, without affecting c-fos expression. Additional experiments with c-Src dominant negative mutant cells reveal similar gene expression changes. Also, transactivation of the EGFR via c-Src activates the raf – MEK – ERK pathway, which was found to control Egr-1, but not c-fos, gene expression. Thus, it appears that transactivation of the EGFR by AVPR1a leads to two different functional intracellular pathways with different physiological outcome in smooth muscle cells. Finally, in the same cells, AVP induced the heterodimerization of the AVPR1a and the EGFR, as shown by immunoprecipitation assays (Fuentes et al., 2008). This supports the hypothesis that a multimeric protein complex is an important prerequisite for EGFR transactivation (Wetzker and Bohmer, 2003).

2.4.7 AVPR-mediated intracellular signaling in the brain

Generally, cellular responses to OT and AVP were mainly studied in somatic cell lines, whereas specific neuronal responses are only studied occasionally, for example, in the hypothalamic neuronal cell line H32, which expresses AVPR1a, AVPR1b, and OTR. Here, the transactivation of the EGFR and the subsequent activation of ERK are necessary for GAGA binding activity of nuclear proteins to the AVPR1b promoter to induce AVPR1b gene expression (Volpi et al., 2006). In the same cells, it was additionally found that activation of ERK, Ca^{2+}/calmodulin-dependent kinase (CaMK), and PKC are necessary for the antiapoptotic actions of AVP (Chen et al., 2008). The complex intracellular signaling response leads to phosphorylation – inactivation of the proapoptotic protein Bad (see also Figure 2.2), and a reduction of cytosolic cytochrome c and caspase-3 activation. It has therefore been suggested that AVP might have neuroprotective properties in the brain by activating ERK, CaMK, and PKC (Chen et al., 2008).

In cortical neurons of E18 foetal rats, AVPR1a activation by means of a receptor agonist induces the activation of the phosphatidylinositol pathway, followed by activation of PKC, and phosphorylation and subsequent opening of L-type calcium channels (see also Figure 2.2). This sequence of events activates the recruitment of intracellular calcium to fully activate the neuronal responses to AVP and might underlie memory-enhancing effects of AVP (Son and Brinton, 2001).

Surprisingly, activation of the AVPR1a has been reported to activate adenylyl cyclase in vasopressinergic neurons in the SON themselves (Sabatier et al., 1998), and in facial motoneurons in the brainstem (Wrobel et al., 2011). In the latter, AVP promotes the formation of cAMP, which directly gates a cation channel that generates a sustained inward current (Wrobel et al., 2011). To date, the mechanisms underlying AVP-mediated activation of the adenylyl cyclase – cAMP – PKA pathway are not known but

could be the consequence of Ca^{2+} and calmodulin recruitment (Sunahara and Taussig, 2002).

In general, the intracellular mechanisms involved in complex behavioral functions of brain AVP, such as its anxiogenic effect or its involvement in various social behaviors, are completely unknown. Such AVPR-mediated effects are likely to be dependent on the neuronal subtypes and brain regions involved, and on the specific stimuli activating not only the AVP but also other brain transmitter systems. Also, neuronal responses are influenced by the functioning of other cell types in the brain, such as astrocytes that express AVPR. For example, in cortical astrocytes AVP induces an increase in intracellular Ca^{2+} concentrations, and activates CaMKII, PKC, and ERK (Zhao and Brinton, 2004). Activated CaMKII and ERK translocate to the nucleus, where they stimulate the phosphorylation of the transcription factor CREB. It is of interest to note that both CaMKII and ERK are required for phosphorylation of CREB, which makes CREB a node of crosstalk between two activated intracellular pathways (Zhao and Brinton, 2004). The phosphorylation of CREB by CaMKII and ERK leads to suppression of the proinflammatory cytokines interleukin 1β and tumor necrosis factor α, and these may influence neuronal functioning (Zhao and Brinton, 2004).

In some other cases, the intracellular pathway involved in a certain behavior is known, but this has not been linked to intracerebral OT and AVP release or the activation of their receptors. For instance, it is known that AVP is an important mediator of aggressive behavior (see above), and that aggressive behavior is correlated with the phosphorylation of ERK in brain regions that govern this sort of behavior and express AVPR (Trainor et al., 2010). However, it has not been shown that the sequence of local AVP release, subsequent receptor activation, and phosphorylation of ERK regulates aggressive behavior.

2.5 Conclusions and perspectives

Microdialysis coupled with pharmacological experiments has revealed that OT and AVP release within various hypothalamic and limbic brain regions are important for social and stress-related behaviors. This has revived interest in the underlying biochemical and molecular receptor-mediated mechanisms of these behaviors. Although in some instances intracellular signaling pathways that mediate a particular behavior are at least partially known (OT-mediated anxiolysis and spatial memory enhancement), our knowledge in general on this topic is rather fragmentary. As OT- and AVP-driven behaviors have been well characterized, the systematic study of the molecular mediators of this behavior will prove to be a valuable tool to better understand how behavior in general is modulated by neuropeptides, in health and disease.

Acknowledgment

The authors are grateful to Dr. D. A. Slattery for critical reading of the manuscript.

REFERENCES

Altstein, M. and Gainer, H. (1988). Differential biosynthesis and posttranslational processing of vasopressin and oxytocin in rat brain during embryonic and postnatal development. *Journal of Neuroscience*, 8, 3967–3977.

Argiolas, A. and Melis, M. R. (2004). The role of oxytocin and the paraventricular nucleus in the sexual behaviour of male mammals. *Physiology and Behavior*, 83, 309–317.

Bale, T. L., Davis, A. M., Auger, A. P., Dorsa, D. M., and McCarthy, M. M. (2001). CNS region-specific oxytocin receptor expression: importance in regulation of anxiety and sex behavior. *Journal of Neuroscience*, 21, 2546–2552.

Beery, A. K., Lacey, E. A., and Francis, D. D. (2008). Oxytocin and vasopressin receptor distributions in a solitary and a social species of tuco-tuco (Ctenomys haigi and Ctenomys sociabilis). *Journal of Comparative Neurology*, 507, 1847–1859.

Beiderbeck, D. I., Neumann, I. D., and Veenema, A. H. (2007). Differences in intermale aggression are accompanied by opposite vasopressin release patterns within the septum in rats bred for low and high anxiety. *European Journal of Neuroscience*, 26, 3597–3605.

Bester-Meredith, J. K., Young, L. J., and Marler, C. A. (1999). Species differences in paternal behavior and aggression in peromyscus and their associations with vasopressin immunoreactivity and receptors. *Hormones and Behavior,* 36, 25–38.

Blume, A., Bosch, O. J., Miklos, S., et al. (2008). Oxytocin reduces anxiety via ERK1/2 activation: local effect within the rat hypothalamic paraventricular nucleus. *European Journal of Neuroscience,* 27, 1947–1956.

Bosch, O. J. (2010). Maternal nurturing is dependent on her innate anxiety: The behavioral roles of brain oxytocin and vasopressin. *Hormones and Behavior,* 59, 202–212.

Bosch, O. J., Kromer, S. A., Brunton, P. J., and Neumann, I. D. (2004). Release of oxytocin in the hypothalamic paraventricular nucleus, but not central amygdala or lateral septum in lactating residents and virgin intruders during maternal defence. *Neuroscience,* 124, 439–448.

Bosch, O. J., Meddle, S. L., Beiderbeck, D. I., Douglas, A. J., and Neumann, I. D. (2005). Brain oxytocin correlates with maternal aggression: link to anxiety. *Journal of Neuroscience,* 25, 6807–6815.

Bosch, O. J. and Neumann, I. D. (2008). Brain vasopressin is an important regulator of maternal behavior independent of dams' trait anxiety. *Proceedings of the National Academy of Sciences of the United States of America,* 105, 17139–17144.

Bosch, O. J. and Neumann, I. D. (2010). Vasopressin released within the central amygdala promotes maternal aggression. *European Journal of Neuroscience,* 31, 883–891.

Bosch, O. J., Pfortsch, J., Beiderbeck, D. I., Landgraf, R., and Neumann, I. D. (2010). Maternal behaviour is associated with vasopressin release in the medial preoptic area and bed nucleus of the stria terminalis in the rat. *Journal of Neuroendocrinology,* 22, 420–429.

Brunton, P. J. and Russell, J. A. (2008). The expectant brain: adapting for motherhood. *Nature Reviews,* 9, 11–25.

Campbell, P., Ophir, A. G., and Phelps, S. M. (2009). Central vasopressin and oxytocin receptor distributions in two species of singing mice. *Journal of Comparative Neurology,* 516, 321–333.

Chen, J., Volpi, S., and Aguilera, G. (2008). Anti-apoptotic actions of vasopressin in H32 neurons involve MAP kinase transactivation and Bad phosphorylation. *Experimental Neurology,* 211, 529–538.

Chu, K. and Zingg, H. H. (1999). Activation of the mouse oxytocin promoter by the orphan receptor RORα. *Journal of Molecular Endocrinology,* 23, 337–346.

Coccaro, E. F., Kavoussi, R. J., Hauger, R. L., Cooper, T. B., and Ferris, C. F. (1998). Cerebrospinal fluid vasopressin levels: correlates with aggression and serotonin function in personality-disordered subjects. *Archives of General Psychiatry,* 55, 708–714.

Compaan, J. C., Buijs, R. M., Pool, C. W., De Ruiter, A. J., and Koolhaas, J. M. (1993). Differential lateral septal vasopressin innervation in aggressive and nonaggressive male mice. *Brain Research Bulletin,* 30, 1–6.

Conner, M., Hawtin, S. R., Simms, J., et al. (2007). Systematic analysis of the entire second extracellular loop of the V(1a) vasopressin receptor: key residues, conserved throughout a G-protein-coupled receptor family, identified. *Journal of Biological Chemistry,* 282, 17405–17412.

De Vries, G. J. and Buijs, R. M. (1983). The origin of the vasopressinergic and oxytocinergic innervation of the rat brain with special reference to the lateral septum. *Brain Research,* 273, 307–317.

Demotes-Mainard, J., Chauveau, J., Rodriguez, F., Vincent, J. D., and Poulain, D. A. (1986). Septal release of vasopressin in response to osmotic, hypovolemic and electrical stimulation in rats. *Brain Research,* 381, 314–321.

Dubois-Dauphin, M., Barberis, C., and de Bilbao, F. (1996). Vasopressin receptors in the mouse (Mus musculus) brain: sex-related expression in the medial preoptic area and hypothalamus. *Brain Research,* 743, 32–39.

Engelmann, M., Ebner, K., Landgraf, R., Holsboer, F., and Wotjak, C. T. (1999). Emotional stress triggers intrahypothalamic but not peripheral release of oxytocin in male rats. *Journal of Neuroendocrinology,* 11, 867–872.

Engelmann, M., Wotjak, C. T., and Landgraf, R. (1998). Differential central and peripheral release of vasopressin and oxytocin in response to swim stress in rats. *Advances in Experimental Medicine and Biology,* 449, 175–177.

Everts, H. G., De Ruiter, A. J., and Koolhaas, J. M. (1997). Differential lateral septal vasopressin in wild-type rats: correlation with aggression. *Hormones and Behavior,* 31, 136–144.

Ferris, C. F. and Delville, Y. (1994). Vasopressin and serotonin interactions in the control of agonistic behavior. *Psychoneuroendocrinology,* 19, 593–601.

Ferris, C. F., Melloni, R. H., Jr., Koppel, G., Perry, K. W., Fuller, R. W., and Delville, Y. (1997). Vasopressin/serotonin interactions in the anterior hypothalamus control aggressive behavior in golden hamsters. *Journal of Neuroscience,* 17, 4331–4340.

Fuentes, L. Q., Reyes, C. E., Sarmiento, J. M., et al. (2008). Vasopressin upregulates the expression of growth-related immediate-early genes via two distinct EGF receptor transactivation pathways. *Cellular Signalling*, 20, 1642–1650.

Fuxe, K., Wikstrom, A. C., Okret, S., et al. (1985). Mapping of glucocorticoid receptor immunoreactive neurons in the rat tel- and di-encephalon using a monoclonal antibody against rat liver glucocorticoid receptor. *Endocrinology*, 117, 1803–1812.

Gimpl, G. and Fahrenholz, F. (2001). The oxytocin receptor system: structure, function, and regulation. *Physiological Reviews*, 81, 629–683.

Gobrogge, K. L., Liu, Y., Jia, X., and Wang, Z. (2007). Anterior hypothalamic neural activation and neurochemical associations with aggression in pair-bonded male prairie voles. *Journal of Comparative Neurology*, 502, 1109–1122.

Gravati, M., Busnelli, M., Bulgheroni, E., et al. (2010). Dual modulation of inward rectifier potassium currents in olfactory neuronal cells by promiscuous G protein coupling of the oxytocin receptor. *Journal of Neurochemistry*, 114, 1424–1435.

Guzzi, F., Zanchetta, D., Cassoni, P., et al. (2002). Localization of the human oxytocin receptor in caveolin-1 enriched domains turns the receptor-mediated inhibition of cell growth into a proliferative response. *Oncogene*, 21, 1658–1667.

Hernando, F., Schoots, O., Lolait, S. J., and Burbach, J. P. (2001). Immunohistochemical localization of the vasopressin V1b receptor in the rat brain and pituitary gland: anatomical support for its involvement in the central effects of vasopressin. *Endocrinology*, 142, 1659–1668.

Hoare, S., Copland, J. A., Strakova, Z., et al. (1999). The proximal portion of the COOH terminus of the oxytocin receptor is required for coupling to g(q), but not g(i). Independent mechanisms for elevating intracellular calcium concentrations from intracellular stores. *Journal of Biological Chemistry*, 274, 28682–28689.

Hollander, E., Novotny, S., Hanratty, M., et al. (2003). Oxytocin infusion reduces repetitive behaviors in adults with autistic and Asperger's disorders. *Neuropsychopharmacology*, 28, 193–198.

Ingram, C. D. and Moos, F. (1992). Oxytocin-containing pathway to the bed nuclei of the stria terminalis of the lactating rat brain: immunocytochemical and in vitro electrophysiological evidence. *Neuroscience*, 47, 439–452.

Itoi, K., Jiang, Y. Q., Iwasaki, Y., and Watson, S. J. (2004). Regulatory mechanisms of corticotropin-releasing hormone and vasopressin gene expression in the hypothalamus. *Journal of Neuroendocrinology*, 16, 348–355.

Jard, S., Barberis, C., Audigier, S., and Tribollet, E. (1987). Neurohypophyseal hormone receptor systems in brain and periphery. *Progress in Brain Research*, 72, 173–187.

Jin, D., Liu, H. X., Hirai, H., et al. (2007). CD38 is critical for social behaviour by regulating oxytocin secretion. *Nature*, 446, 41–45.

Kalsbeek, A., Buijs, R. M., Engelmann, M., Wotjak, C. T., and Landgraf, R. (1995). In vivo measurement of a diurnal variation in vasopressin release in the rat suprachiasmatic nucleus. *Brain Research*, 682, 75–82.

Kendrick, K. M., Da Costa, A. P., Broad, K. D., et al. (1997). Neural control of maternal behaviour and olfactory recognition of offspring. *Brain Research Bulletin*, 44, 383–395.

Kendrick, K. M., Keverne, E. B., Chapman, C., and Baldwin, B. A. (1988). Intracranial dialysis measurement of oxytocin, monoamine and uric acid release from the olfactory bulb and substantia nigra of sheep during parturition, suckling, separation from lambs and eating. *Brain Research*, 439, 1–10.

Kessler, M. S., Bosch, O. J., Bunck, M., Landgraf, R., and Neumann, I. D. (2010). Maternal care differs in mice bred for high vs. low trait anxiety: Impact of brain vasopressin and cross-fostering. *Social Neuroscience*, 1–13.

Koolhaas, J. M., Moor, E., Hiemstra, Y., and Bohus, B. (1991). The testosterone-dependent vasopressinergic neurons in the medial amygdala and lateral septum: involvement in social behaviour of male rats. In Jard, S., Jamison, R. (eds.) *Vasopressin*. INSERM/Libbey, Paris/London, pp. 213–219.

Krishnaswamy, N., Lacroix-Pepin, N., Chapdelaine, P., Taniguchi, H., Kauffenstein, G., Chakravarti, A., Danyod, G., and Fortier, M. A. (2010). Epidermal growth factor receptor is an obligatory intermediate for oxytocin-induced cyclooxygenase 2 expression and prostaglandin F2 alpha production in bovine endometrial epithelial cells. *Endocrinology*, 151, 1367–1374.

Landgraf, R., Gerstberger, R., Montkowski, A., et al. (1995). V1 vasopressin receptor antisense oligodeoxynucleotide into septum reduces vasopressin binding, social discrimination abilities, and anxiety-related behavior in rats. *Journal of Neuroscience*, 15, 4250–4258.

Landgraf, R. and Ludwig, M. (1991). Vasopressin release within the supraoptic and paraventricular nuclei of the

rat brain: osmotic stimulation via microdialysis. *Brain Research*, 558, 191–196.

Landgraf, R., Neumann, I., and Pittman, Q. J. (1991). Septal and hippocampal release of vasopressin and oxytocin during late pregnancy and parturition in the rat. *Neuroendocrinology*, 54, 378–383.

Landgraf, R., Neumann, I., and Schwarzberg, H. (1988). Central and peripheral release of vasopressin and oxytocin in the conscious rat after osmotic stimulation. *Brain Research*, 457, 219.

Landgraf, R. and Neumann, I. D. (2004). Vasopressin and oxytocin release within the brain: a dynamic concept of multiple and variable modes of neuropeptide communication. *Frontiers in Neuroendocrinology*, 25, 150–176.

Larrazolo-Lopez, A., Kendrick, K. M., Aburto-Arciniega, M., et al. (2008). Vaginocervical stimulation enhances social recognition memory in rats via oxytocin release in the olfactory bulb. *Neuroscience*, 152, 585–593.

Lee, H. C. (2001). Physiological functions of cyclic ADP-ribose and NAADP as calcium messengers. *Annual Review of Pharmacology and Toxicology*, 41, 317–345.

Levy, F., Kendrick, K. M., and Goode, J. A., Guevara-Guzman, R., and Keverne, E. B. (1995). Oxytocin and vasopressin release in the olfactory bulb of parturient ewes: changes with maternal experience and effects on acetylcholine, gamma-aminobutyric acid, glutamate and noradrenaline release. *Brain Research*, 669, 197–206.

Liu, J. and Wess, J. (1996). Different single receptor domains determine the distinct G protein coupling profiles of members of the vasopressin receptor family. *Journal of Biological Chemistry*, 271, 8772–8778.

Lopatina, O., Liu, H. X., Amina, S., Hashii, M., and Higashida, H. (2010). Oxytocin-induced elevation of ADP-ribosyl cyclase activity, cyclic ADP-ribose or Ca(2+) concentrations is involved in autoregulation of oxytocin secretion in the hypothalamus and posterior pituitary in male mice. *Neuropharmacology*, 58, 50–55.

Ludwig, M., Callahan, M. F., Neumann, I., Landgraf, R., and Morris, M. (1994). Systemic osmotic stimulation increases vasopressin and oxytocin release within the supraoptic nucleus. *Journal of Neuroendocrinology*, 6, 369–373.

Ludwig, M. and Leng, G. (2006). Dendritic peptide release and peptide-dependent behaviours. *Nature Reviews*, 7, 126–136.

Mohr, E. and Richter, D. (2004). Subcellular vasopressin mRNA trafficking and local translation in dendrites. *Journal of Neuroendocrinology*, 16, 333–339.

Moos, F., Freund-Mercier, M. J., Guerne, Y., Guerne, J. M., Stoeckel, M. E., and Richard, P. (1984). Release of oxytocin and vasopressin by magnocellular nuclei in vitro: specific facilitatory effect of oxytocin on its own release. *Journal of Endocrinology*, 102, 63–72.

Moos, F., Poulain, D. A., Rodriguez, F., Guerne, Y., Vincent, J. D., and Richard, P. (1989). Release of oxytocin within the supraoptic nucleus during the milk ejection reflex in rats. *Experimental Brain Research*, 76, 593–602.

Nephew, B. C. and Bridges, R. S. (2008). Central actions of arginine vasopressin and a V1a receptor antagonist on maternal aggression, maternal behavior, and grooming in lactating rats. *Pharmacology, Biochemistry, and Behavior*, 91, 77–83.

Nephew, B. C., Byrnes, E. M., and Bridges, R. S. (2010). Vasopressin mediates enhanced offspring protection in multiparous rats. *Neuropharmacology*, 58, 102–106.

Neumann, I., Douglas, A. J., Pittman, Q. J., Russell, J. A., and Landgraf, R. (1996). Oxytocin released within the supraoptic nucleus of the rat brain by positive feedback action is involved in parturition-related events. *Journal of Neuroendocrinology*, 8, 227–233.

Neumann, I., Koehler, E., Landgraf, R., and Summy-Long, J. (1994). An oxytocin receptor antagonist infused into the supraoptic nucleus attenuates intranuclear and peripheral release of oxytocin during suckling in conscious rats. *Endocrinology*, 134, 141–148.

Neumann, I. and Landgraf, R. (1989). Septal and hippocampal release of oxytocin, but not vasopressin, in the conscious lactating rat during suckling. *Journal of Neuroendocrinology*, 1, 305.

Neumann, I., Landgraf, R., Bauce, L., and Pittman, Q. J. (1995). Osmotic responsiveness and cross talk involving oxytocin, but not vasopressin or amino acids, between the supraoptic nuclei in virgin and lactating rats. *Journal of Neuroscience*, 15, 3408–3417.

Neumann, I., Ludwig, M., Engelmann, M., Pittman, Q. J., and Landgraf, R. (1993a). Simultaneous microdialysis in blood and brain: oxytocin and vasopressin release in response to central and peripheral osmotic stimulation and suckling in the rat. *Neuroendocrinology*, 58, 637–645.

Neumann, I., Russell, J. A., and Landgraf, R. (1993b). Oxytocin and vasopressin release within the supraoptic and paraventricular nuclei of pregnant, parturient and lactating rats: a microdialysis study. *Neuroscience*, 53, 65–75.

Neumann, I. D. (2001). Alterations in behavioral and neuroendocrine stress coping strategies in pregnant,

parturient and lactating rats. *Progress in Brain Research*, 133, 143–152.

Neumann, I. D. (2002). Involvement of the brain oxytocin system in stress coping: interactions with the hypothalamo-pituitary-adrenal axis. *Progress in Brain Research*, 139, 147–162.

Neumann, I. D. (2007). Stimuli and consequences of dendritic release of oxytocin within the brain. *Biochemical Society Transactions*, 35, 1252–1257.

Neumann, I. D., Kromer, S. A., Toschi, N., and Ebner, K. (2000a). Brain oxytocin inhibits the (re)activity of the hypothalamo-pituitary-adrenal axis in male rats: involvement of hypothalamic and limbic brain regions. *Regulatory Peptides*, 96, 31–38.

Neumann, I. D., Torner, L., Toschi, N., and Veenema, A. H. (2006). Oxytocin actions within the supraoptic and paraventricular nuclei: differential effects on peripheral and intranuclear vasopressin release. *American Journal of Physiology. Regulatory, Integrative and Comparative Physiology*, 291, R29–36.

Neumann, I. D., Torner, L., and Wigger, A. (2000b). Brain oxytocin: differential inhibition of neuroendocrine stress responses and anxiety-related behaviour in virgin, pregnant and lactating rats. *Neuroscience*, 95, 567–575.

Neumann, I. D., Toschi, N., Ohl, F., Torner, L., and Kromer, S. A. (2001). Maternal defence as an emotional stressor in female rats: correlation of neuroendocrine and behavioural parameters and involvement of brain oxytocin. *European Journal of Neuroscience*, 13, 1016–1024.

Neumann, I. D., Veenema, A. H., and Beiderbeck, D. I. (2010). Aggression and anxiety: social context and neurobiological links. *Frontiers in Behavioral Neuroscience*, 4, 12.

Neumann, I. D., Wigger, A., Torner, L., Holsboer, F., and Landgraf, R. (2000c). Brain oxytocin inhibits basal and stress-induced activity of the hypothalamo-pituitary-adrenal axis in male and female rats: partial action within the paraventricular nucleus. *Journal of Neuroendocrinology*, 12, 235–243.

Pedersen, C. A. and Boccia, M. L. (2003). Oxytocin antagonism alters rat dams' oral grooming and upright posturing over pups. *Physiology and Behavior*, 80, 233–241.

Pedersen, C. A. and Prange, A. J., Jr. (1979). Induction of maternal behavior in virgin rats after intracerebroventricular administration of oxytocin. *Proceedings of the National Academy of Sciences of the United States of America*, 76, 6661–6665.

Pow, D. V. and Morris, J. F. (1989). Dendrites of hypothalamic magnocellular neurons release neurohypophysial peptides by exocytosis. *Neuroscience*, 32, 435–439.

Rimoldi, V., Reversi, A., Taverna, E., et al. (2003). Oxytocin receptor elicits different EGFR/MAPK activation patterns depending on its localization in caveolin-1 enriched domains. *Oncogene*, 22, 6054–6060.

Roper, J., O'Carroll, A. M., Young, W., III, and Lolait, S. (2010). The vasopressin Avpr1b receptor: molecular and pharmacological studies. *Stress*, 14, 98–115.

Sabatier, N., Richard, P., and Dayanithi, G. (1998). Activation of multiple intracellular transduction signals by vasopressin in vasopressin-sensitive neurones of the rat supraoptic nucleus. *Journal of Physiology*, 513 (Pt 3), 699–710.

Shughrue, P. J., Dellovade, T. L., and Merchenthaler, I. (2002). Estrogen modulates oxytocin gene expression in regions of the rat supraoptic and paraventricular nuclei that contain estrogen receptor-beta. *Progress in Brain Research*, 139, 15–29.

Slattery, D. A. and Neumann, I. D. (2008). No stress please! Mechanisms of stress hyporesponsiveness of the maternal brain. *Journal of Physiology*, 586, 377–385.

Slattery, D. A. and Neumann, I. D. (2010a). Chronic icv oxytocin attenuates the pathological high anxiety state of selectively bred Wistar rats. *Neuropharmacology*, 58, 56–61.

Slattery, D. A. and Neumann, I. D. (2010b). Oxytocin and major depressive disorder: experimental and clinical evidence for links to aetiology and possible treatment. *Pharmaceuticals*, 3, 702–724.

Soltoff, S. P. and Cantley, L. C. (1996). p120cbl is a cytosolic adapter protein that associates with phosphoinositide 3-kinase in response to epidermal growth factor in PC12 and other cells. *Journal of Biological Chemistry*, 271, 563–567.

Son, M. C. and Brinton, R. D. (2001). Regulation and mechanism of L-type calcium channel activation via V1a vasopressin receptor activation in cultured cortical neurons. *Neurobiology of Learning and Memory*, 76, 388–402.

Stern, C. M. and Mermelstein, P. G. (2010). Caveolin regulation of neuronal intracellular signaling. *Cellular and Molecular Life Sciences*, 67, 3785–3795.

Sunahara, R. K. and Taussig, R. (2002). Isoforms of mammalian adenylyl cyclase: multiplicities of signaling. *Molecular Interventions*, 2, 168–184.

Thibonnier, M., Berti-Mattera, L. N., Dulin, N., Conarty, D. M., and Mattera, R. (1998). Signal transduction pathways of the human V1-vascular, V2-renal, V3-pituitary

vasopressin and oxytocin receptors. *Progress in Brain Research*, 119, 147–161.

Thibonnier, M., Coles, P., Thibonnier, A., and Shoham, M. (2001). The basic and clinical pharmacology of nonpeptide vasopressin receptor antagonists. *Annual Review of Pharmacology and Toxicology*, 41, 175–202.

Tomizawa, K., Iga, N., Lu, Y. F., et al. (2003). Oxytocin improves long-lasting spatial memory during motherhood through MAP kinase cascade. *Nature Neuroscience*, 6, 384–390.

Trainor, B. C., Crean, K. K., Fry, W. H., and Sweeney, C. (2010). Activation of extracellular signal-regulated kinases in social behavior circuits during resident-intruder aggression tests. *Neuroscience*, 165, 325–336.

van Leengoed, E., Kerker, E., and Swanson, H. H. (1987). Inhibition of post-partum maternal behaviour in the rat by injecting an oxytocin antagonist into the cerebral ventricles. *Journal of Endocrinology*, 112, 275–282.

Veenema, A. H. and Neumann, I. D. (2008). Central vasopressin and oxytocin release: regulation of complex social behaviours. *Progress in Brain Research*, 170, 261–276.

Veenema, A. H., Torner, L., Blume, A., Beiderbeck, D. I., and Neumann, I. D. (2007). Low inborn anxiety correlates with high intermale aggression: link to ACTH response and neuronal activation of the hypothalamic paraventricular nucleus. *Hormones and Behavior*, 51, 11–19.

Volpi, S., Liu, Y., and Aguilera, G. (2006). Vasopressin increases GAGA binding activity to the V1b receptor promoter through transactivation of the MAP kinase pathway. *Journal of Molecular Endocrinology*, 36, 581–590.

Waldherr, M. and Neumann, I. D. (2007). Centrally released oxytocin mediates mating-induced anxiolysis in male rats. *Proceedings of the National Academy of Sciences of the United States of America*, 104, 16681–16684.

Walker, C. D., Toufexis, D. J., and Burlet, A. (2001). Hypothalamic and limbic expression of CRF and vasopressin during lactation: implications for the control of ACTH secretion and stress hyporesponsiveness. *Progress in Brain Research*, 133, 99–110.

Wang, Z., Young, L. J., Liu, Y., and Insel, T. R. (1997). Species differences in vasopressin receptor binding are evident early in development: comparative anatomic studies in prairie and montane voles. *Journal of Comparative Neurology*, 378, 535–546.

Wetzker, R. and Bohmer, F. D. (2003). Transactivation joins multiple tracks to the ERK/MAPK cascade. *Nature Reviews. Molecular and Cellular Biology*, 4, 651–657.

Windle, R. J., Shanks, N., Lightman, S. L., and Ingram, C. D. (1997). Central oxytocin administration reduces stress-induced corticosterone release and anxiety behavior in rats. *Endocrinology*, 138, 2829–2834.

Wotjak, C. T., Kubota, M., Liebsch, G., et al. (1996). Release of vasopressin within the rat paraventricular nucleus in response to emotional stress: a novel mechanism of regulating adrenocorticotropic hormone secretion? *Journal of Neuroscience*, 16, 7725–7732.

Wrobel, L. J., Dupre, A., and Raggenbass, M. (2011). Excitatory action of vasopressin in the brain of the rat: role of cAMP signaling. *Neuroscience*, 172, 177–186.

Yoshida, M., Takayanagi, Y., Inoue, K., et al. (2009). Evidence that oxytocin exerts anxiolytic effects via oxytocin receptor expressed in serotonergic neurons in mice. *Journal of Neuroscience*, 29, 2259–2271.

Young, W. S., Li, J., Wersinger, S. R., and Palkovits, M. (2006). The vasopressin 1b receptor is prominent in the hippocampal area CA2 where it is unaffected by restraint stress or adrenalectomy. *Neuroscience*, 143, 1031–1039.

Zhao, L. and Brinton, R. D. (2004). Suppression of proinflammatory cytokines interleukin-1beta and tumor necrosis factor-alpha in astrocytes by a V1 vasopressin receptor agonist: a cAMP response element-binding protein-dependent mechanism. *Journal of Neuroscience*, 24, 2226–2235.

3

Regulation of oxytocin and vasopressin secretion

Involvement of the calcium amplification pathway through cyclic ADP-ribose and CD38

Haruhiro Higashida, Olga Lopatina, and Amina Sarwat

3.1 Introduction

Oxytocin (OT) and arginine-vasopressin (AVP) play a critical role in social recognition and behavior in mammals (Carter, 2003; Kosfeld et al., 2005; Zak, 2007; Donaldson and Young, 2008; Neumann, 2008; Insel, 2010; Higashida et al., 2010). OT is secreted dendritically from neurons in the paraventricular nucleus (PVN) and supraoptic nucleus (SON) of the hypothalamus and spread to other brain areas (McGregor et al., 2008; Skuse and Gallgher, 2009). Adrenaline in the SON neurons stimulates oxytocinergic neurons, which results in the local release of OT in the brain (Tobin et al., 2010). Thus, locally released OT causes excitation of OT neurons by activating OT receptors expressed in the neurons of both the PVN and SON (Adan et al., 1995; Freund-Mercier et al., 1994; Young et al., 1997). This excitation leads to a facilitative OT release, known as autoregulation (Moos et al., 1984; Neumann et al., 1996). The autoregulation, OT-induced OT release, occurs during uterine contraction in labor and milk ejection in lactation (Richard et al., 1991). This chapter focuses on the release and autoregulation of OT and secretion of AVP, in relation to the regulation of social recognition in male mice and its implications for human autism.

OT receptors, seven-transmembrane proteins, couple with $G_{q/11}$ and stimulate the production of inositol 1,4,5-trisphosphate (IP_3) and diacylglycerol (DAG) through the activation of phospholipase C (PLC) (Gimpl and Fahrenholz, 2001), resulting in activation of Ca^{2+} signals and protein kinase C (PKC). This PLC- and IP_3-dependent Ca^{2+} signaling may function in autoregulation (Lambert et al., 1994; see also Figure 3.4). Another Ca^{2+} signalling pathway of cyclic ADP-ribose (cADPR) has recently been shown to be present in many tissues, including the nervous system (Lee, 2001; Higashida et al., 2007). It is known that intracellular cADPR concentrations are regulated in many different ways such as the activation of ADP-ribosyl cyclase or CD38, via heterotrimeric GTP-binding proteins or phosphorylation downstream of the G-protein-coupled receptor signaling pathway (Boittin et al., 2003; Higashida et al., 1997 and 1999; Sternfeld et al., 2003). Specifically, the activation of ADP-ribosyl cyclase or CD38 by cyclic GMP- or cyclic AMP-dependent protein kinases has been reported in *Aplysia califonica* and liver cells (Graeff et al., 1998; Kim et al., 2010), LAK cells (Rah et al., 2005) and arterial smooth muscle cells (Boittin et al., 2003). However, before our report was published in 2007 (Jin et al., 2007), the mechanism of how ADP-ribosyl cyclase or CD38 are activated after OT receptor stimulation in the hypothalamus, which may lead to secretion of OT, had not been examined. cADPR is a product of internal ADP-ribosyl cyclase or ectopic CD38 and acts primarily as an intracellular second messenger in the nervous system (Lee, 2001; Higashida et al., 2010).

Oxytocin, Vasopressin, and Related Peptides in the Regulation of Behavior, ed. E. Choleris, D. W. Pfaff, and M. Kavaliers.
Published by Cambridge University Press.

It has been reported that cADPR applied extracellularly stimulates intracellular ryanodine receptors after internalization by the nucleotide-transporting capacity of CD38 in fibroblasts and astrocytes (De Flora et al., 2004). However, to date there have been no reports regarding whether extracellular cADPR is active in neurons. Therefore, the second question that will be addressed here is whether ADP-ribosyl cyclase/CD38-dependent cADPR and $[Ca^{2+}]_i$ signaling are involved in the autoregulatory positive feedback of OT release in the hypothalamus and posterior pituitary in adult male mice in a PKC-sensitive fashion downstream of OT receptors.

3.2 Involvement of cyclic ADP-ribose

3.2.1 Effects of oxytocin on the activity of membrane-bound form of ADP-ribosyl cyclase and cyclic ADP-ribose levels

Application of OT stimulates ADP-ribosyl cyclase activity, when measured fluorimetrically by cyclic GDP-ribose (cGDPR) production (Graeff et al., 1994), in crude membranes prepared from the hypothalamus and posterior pituitary of adult male mice. cGDPR production increased in a concentration-dependent manner upon exposure to OT for 5 min (Figures 3.1A and B). Maximum increase in ADP-ribosyl cyclase activity by 10 nM OT was observed in the hypothalamus (158 $\pm$ 7% ($n = 5$) of pre-exposure levels) and in the pituitary (278 $\pm$ 57% ($n = 5$)) (Lopatina et al., 2010). Simultaneous application of vasotocin, an OT receptor antagonist (Lopatina et al., 2010), inhibits the OT-induced increase of ADP-ribosyl cyclase activity.

Tissue cADPR concentrations ($[cADPR]_i$) in the hypothalamus of male mice was 227.9 $\pm$ 52.6 nmol/mg protein ($n = 4$) before stimulation. $[cADPR]_i$ dose-dependently increased during incubation with OT for 5 min. The maximal activation of ADP-ribosyl cyclase to 1050 $\pm$ 72 nmol/min/m protein ($n = 4$) (4.6-fold of the control level) was obtained by 10 nM OT (Lopatina et al., 2010).

3.2.2 Inhibition of oxytocin-induced activation of the ADP-ribosyl cyclase

To examine the possibility that ADP-ribosyl cyclase is also activated by protein kinases via the OT-signaling pathway, staurosporine (PKI-STSP), one of the most potent and widely used cell permeable inhibitors of protein kinases, was used to measure ADP-ribosyl cyclase activity after OT stimulation for 5 min in the hypothalamus and pituitary (Lopatina et al., 2010). As shown in Figures 3.1C and D, a significant inhibition was obtained by 5 nM PKI-STSP, with no greater inhibition evident at the higher dose (25 nM) in both tissues, suggesting that PKI-STSP inhibits selectively PKC. In addition, a more specific PKC inhibitor, Calphostin C, also inhibited in both tissues. Taken together, these results indicate that PKC may play a role in the activation of ADP-ribosyl cyclase by OT.

3.2.3 Effect of oxytocin on $[Ca^{2+}]_i$ increases in isolated nerve endings

In isolated hypothalamic neurons, it is hard to identify cell types. But, in isolated nerve endings of neurohypophysis, it is easy to judge oxytocinergic nerve endings from those of vasopressin due to the larger size of the former (OuYoung et al., 2004). Application of 100 pM OT to such nerve endings resulted in two phases of $[Ca^{2+}]_i$increases in its time course: a rapid initial increase and a sustained elevation lasting for 5 min (Figures 3.1E–G). OT elicited elevations of the maximum $[Ca^{2+}]_i$ (145 $\pm$ 21% ($n = 4$)) at 20 s after application (Figure 3.1F). An IP_3 receptor antagonist, Xestospongin C, slowed down the initial increase with little or no effect on the sustained level in $[Ca^{2+}]_i$ increases (Figure 3.1G). Pretreatment with 10 μM 8-bromo-cADPR, an antagonistic cADPR analog, inhibited the OT-mediated sustained $[Ca^{2+}]_i$increases along with a decreased initial increase, that resembles the effect of the OT receptor antagonist. Calphostin C had little effect on the initial phase, but did elicit a significant decrease in the sustained level. In the Ca^{2+} free condition, OT-mediated increase of $[Ca^{2+}]_i$ showed small but

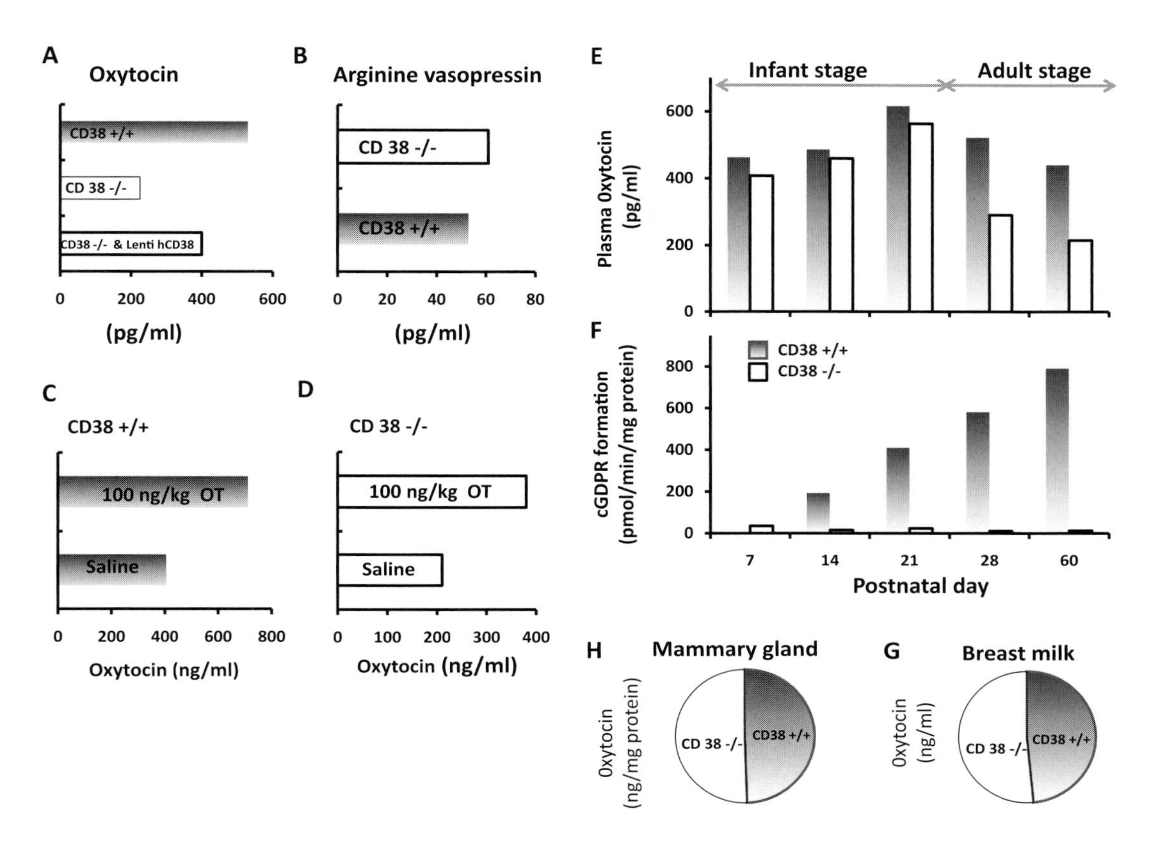

Figure 3.1 Oxytocin and vasopressin concentrations. A, B, Levels of plasma OT (A) and AVP (B) in adult $CD38^{+/+}$ and $CD38^{-/-}$ mice. C,D, Concentrations of OT in CSF of adult $CD38^{+/+}$ (C) and $CD38^{-/-}$ (D) mice after subcutaneous injection of 100 ng OT per kg body weight. E, F, Time course of plasma OT levels (E) and ADP-ribosyl cyclase activity (measured as the rate of cyclic GDP-ribose formation by whole-cell homogenates isolated from the hypothalamus). F, during development in $CD38^{+/+}$ and $CD38^{-/-}$ mice. G, H, Concentration of OT in the mammary gland tissue (G) and milk (H) in lactating $CD38^{+/+}$ and $CD38^{-/-}$ dams. Full bars – $CD38^{+/+}$ mice, open bars – $CD38^{-/-}$ mice.

not significant changes in both phases. Therefore, IP_3 and cADPR inhibitors act differentially on the two phases of $[Ca^{2+}]_i$ elevations (Lopatina et al., 2010).

3.2.4 Effect of oxytocin on promoting oxytocin release from isolated nerve endings

OT-induced OT release from isolated nerve endings over 5 min was examined under various conditions. Application of 100 pM OT, resulted in 2.1-fold increase in OT release for 5 min (1.4 ratio ($n = 4$)) from the control level with vasotocin as a control null level (0.67 ratio, $n = 4$; Figure 3.1H). This effect of OT-mediated OT release was significantly inhibited to control levels by addition of Calphostin C and 8-bromo-cADPR. Xestospongin C slightly inhibited the release, reducing it to the same level as obtained under the extracellular Ca^{2+}-free condition (Lopatina et al., 2010). The sensitivity of antagonists on autoregulation of OT release (Figure 3.1H) resembles the inhibitory pattern obtained on $[Ca^{2+}]_i$ levels (Figures 3.1F and G).

3.2.5 Oxytocin release by extracellular application of cyclic ADP-ribose

High potassium-induced depolarization produced a 2- or 8-fold increase in OT secretion from isolated mouse hypothalamic neurons or their axon terminals in the posterior pituitary gland, respectively, (data not shown). OT release was enhanced approximately 4-fold by application of extracellular 100 μM β-NAD^+, a precursor of cADPR (Refer to figure 3.4 in Jin et al., 2007). The increase was completely blocked by 8-bromo-cADPR. To further confirm the involvement of cADPR, we tested the effects of extracellular application of several β-NAD^+ metabolites or a synthetic analog under identical conditions (Lopatina et al., 2010). Only cADPR showed the potentiation effect, indicating that OT release utilizes the cADPR/ryanodine calcium amplification system as intracellular signaling.

3.2.6 Involvement of TRPM2 channels in oxytocin release

The melastatin-related transient receptor potential channel 2 (TRPM2, previously named TRPC7 or LTRPC2) with ADPR hydrolase activity is a Ca^{2+}-permeable cation channel, and can be activated by either β-NAD^+, ADP-ribose or cADPR (Perraud et al., 2001). TRPM2 activation by cADPR is promoted at body temperature (>35 °C) and is involved in insulin secretion in pancreatic β cells (Togashi et al., 2006). In addition, TRPM2 channels are expected to couple to receptor functions through cADPR formation (Beck et al., 2006). Therefore, we addressed the question of whether extracellular cADPR can activate $[Ca^{2+}]_i$ signalling via CD38 or TRPM2 channels downstream of OT receptors. We used NG108–15 mouse neuroblastoma × rat glioma hybrid cells that possess CD38 (Higashida et al., 2007) but not OT receptors (Nirenberg et al., 1983). These are useful model neurons for investigating cADPR signaling, as we have already reported that intracellular application of cADPR causes membrane voltage-dependent increases in $[Ca^{2+}]_i$ (Hashii et al., 2000).

We measured $[Ca^{2+}]_i$ in wild-type NG108–15 cells before and after stimulation with extracellularly applied 50 μM cADPR. As shown in Figure 3.2I, NG108–15 cells showed significant increases in $[Ca^{2+}]_i$ upon extracellular challenge with 50 μM cADPR and heating to 40 °C. The average peak level was 169 ± 12% ($n = 5$; $P < 0.01$), which was significantly larger than the basal levels at 35 °C ($P < 0.001$). Under a temperature shift to 40 °C with no agonist, $[Ca^{2+}]_i$ level was 111 ± 2% of that at 35 °C ($n = 6$; Figure 3.2J). These $[Ca^{2+}]_i$ amplifications in response to the combination of heat and cADPR are very similar to those observed in HEK-239 cells expressing TRPM2 channels (Togashi et al., 2006). The $[Ca^{2+}]_i$ increases elicited by cADPR and heat (69%) are equivalent to the increased levels obtained by application of 50 mM KCl, which depolarize these cells to –10 mV from the resting membrane potentials of –50 to –60 mV (Higashida et al., 2007). It has been reported that stimulation by 50 mM KCl elevates $[Ca^{2+}]_i$ levels to 200 nM from the resting concentration of 70 nM in NG108–15 cells (Hashii et al., 2000). In addition, we reported that the 50-mM depolarizing stimulation elicits increases of OT release by 1.5-fold in primary tissues isolated from of the mouse hypothalamus and pituitary (Jin et al., 2007).

Little or no cADPR-mediated $[Ca^{2+}]_i$ elevation was observed at 40 °C in the absence of extracellular Ca^{2+}. This suggests that Ca^{2+} influx resulting in elevated $[Ca^{2+}]_i$, probably occurs through non-selective cation TRPM2 channels and was triggered in response to the combination of cADPR and heat. To further confirm this, we used a TRPM2 channel inhibitor, 2-aminoethoxydiphenyl borate (2-APB) (Togashi et al., 2008). The $[Ca^{2+}]_i$ increases elicited by extracellularly applied cADPR (50 μM) plus heat were completely inhibited by pretreatment with 2-APB, at a concentration (30 μM), capable of inhibiting human TRPM2 channels.

8-bromo-cADPR (50 μM) significantly inhibited the cADPR/heat-induced increase in $[Ca^{2+}]_i$: with the peak level at 121% ± 9% of the prestimulation level ($n = 6$; $P < 0.01$; Figure 3.2L). Similarly, an inhibitory effect of 8-bromo-cADPR was observed

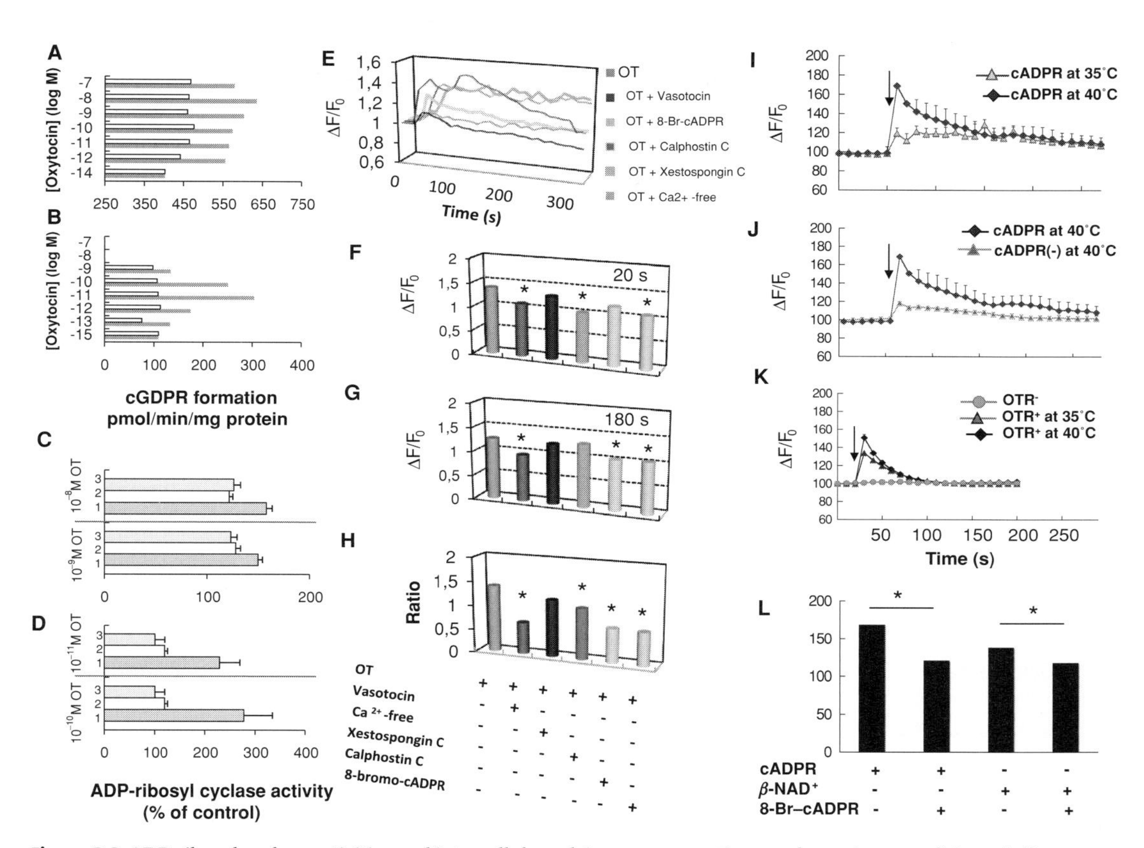

Figure 3.2 ADP-ribosyl cyclase activities and intracellular calcium concentrations under various conditions. A, B, ADP-ribosyl cyclase activities were measured as the rate of cyclic GDP-ribose formation by homogenates isolated from mouse hypothalamus (A) and posterior pituitary (B) for 5 min under various concentrations of OT with (open bars) or without 1 mM of the OT receptor antagonist, vasotocin (filled bars). C, D, ADP-ribosyl cyclase activity, presented as percentages of control cyclic GDP-ribose formation activity, were measured in the presence or absence of OT (4 different concentrations indicated), PKI-STSP (5 nM) or PKC-inhibitor calphostin C (100 nM) in the hypothalamus (C) and pituitary (D). 1 – OT present, 2 – OT and PKI-STSP, 3 – OT and calphostin C. E, Average time courses of changes in $[Ca^{2+}]_i$ elicited with 100 pM OT with or without 1 mM OT receptor antagonist vasotocin, 100 nM PKC-inhibitor calphostin C, 2 mM IP_3-inhibitor Xestospongin C, 100 μM 8-bromo-cADPR, and extracellular Ca^{2+}. F, G, Average increases in $[Ca^{2+}]_i$ measured at 20 s (F) and 180 s (G) after OT stimulation. Data are shown as changes in fluorescence divided by resting fluorescence, i.e. $\Delta F/F_0$. H, OT concentrations are presented as OT release ratio (arbitrary unit) in an isolated nerve endings under 100 pM OT stimulation (5 min) with or without 1 mM of the oxytocin receptor antagonist, vasotocin, 100 nM PKC-inhibitor calphostin C, 2 mM IP_3-Inhibitor Xestospongin C, 100 mM 8-bromo-cADPR, and extracellular Ca^{2+}, as indicated. I, Time courses of changes in $[Ca^{2+}]_i$ drawn from fluorescence imaging comparing stimulation at 35 and 40 °C in the presence of cADPR in NG108–15 cells. The arrow indicates an application of cADPR. J, Time courses of changes in $[Ca^{2+}]_i$ with (rhomb) or without 50 μM cADPR (triangle) at 40°C in NG108–15 cells. The arrow indicates the application of cADPR. K, Effects of oxytocin on $[Ca^{2+}]_i$ at 35 °C (triangle) or 40 °C (rhomb) in control (OTR-) or transformed NG108–15 cells to express human OT receptors (OTR+). The arrow indicates the addition of 200 nM OT (final concentration, 100 nM) from a calibrated micropipette into the recording medium. L, $[Ca^{2+}]_i$ elevation induced by extracellular application of 50 μM cADPR or 100 μM β-NAD$^+$ at 40 °C in the presence or absence of 50 μM 8-bromo-cADPR is expressed as percentage change over those before stimulation. See color version in plates section.

in responses to 100 μM β-NAD$^+$ and heat: 138 ± 5% ($n = 5$) and 118 ± 13% ($n = 4$) in the absence and presence of 8-bromo-cADPR ($P < 0.05$), respectively (Amina et al., 2010). These results suggest that cADPR contributes to Ca^{2+} mobilization from internal Ca^{2+} pools.

To examine whether cADPR acts as a second messenger downstream of OT receptors and whether TRPM2-like heat-sensitive channels are activated by OT receptor stimulation, we used NG108–15 cells transformed with a human OT receptor cDNA. Functional OT receptors were detected by GFP fluorescence in the transformed cells. As shown in Figure 3.2K, application of 100 nM OT elevated $[Ca^{2+}]_i$ to 134 ± 3% ($n = 7$) and 151 ± 4% ($n = 7$) at 35 °C and 40 °C, respectively. Therefore, this response was significantly enhanced (1.5-fold) when the cells were heated at 40 °C ($P < 0.001$). These results indicate that TRPM2 induced Ca^{2+} influx in response to cADPR (refer to Figure 3.4), though we still are missing data on cADPR- and heat-induced OT release in the hypothalamus.

3.3 Plasma oxytocin levels in wild-type and CD38 knockout mice

3.3.1 Plasma oxytocin level in adult mice

Supporting a link between CD38 and OT, *Cd38* knockout (*Cd38*$^{-/-}$) mice have reduced plasma OT levels in comparison with wild-type, but no changes in AVP levels (Figures 3.2A and B). In contrast, tissue levels in the hypothalamus and pituitary were higher in the knockout mice (Jin et al., 2007). These observations indicate that, although OT is produced and packaged into vesicles in the hypothalamic neurons and posterior pituitary nerve endings in *Cd38*$^{-/-}$ mice, it is not efficiently released into the brain and bloodstream. In both genotypes plasma OT levels were recovered by a single subcutaneous OT injection as shown in Figures 3.2C and D. Accordingly, although OT could not effectively function in *Cd38*$^{-/-}$ mice with impairment in social behavior, the behavioral phenotype of *Cd38*$^{-/-}$ mice could be normalized, as OT is transported into the brain, probably through the blood–brain barrier (Jin et al., 2007). We also used a genetic approach by infusion of a lentivirus carrying the human CD38 gene into the third ventricle of knockout mice. This procedure resulted in normalization of the plasma (Figure 3.2A) and CSF OT levels and thereby a normalization of social memory. These results indicate that the mechanisms underlying social behavior require CD38-dependent OT secretion and that AVP is released in a CD38-independent fashion.

3.3.2 Plasma oxytocin levels in infant mice

The degree of disruption of pup behavior was much milder in *Cd38*$^{-/-}$ than in OT or OT receptor knockout mice (Higashida et al., 2010). This result prompted us to measure plasma OT levels at various postnatal ages. Surprisingly, the plasma OT concentration in *Cd38*$^{-/-}$ pups at 1–3 weeks of age was not decreased, but was similar to that in *Cd38*$^{+/+}$ mice of the same age (Figure 3.2E). However, following weaning, at 2 months of age (young adult), plasma concentrations of OT were significantly lower in *Cd38*$^{-/-}$ than in *Cd38*$^{+/+}$ mice (Liu et al., 2008).

The decrease in plasma OT concentration seen after weaning in *Cd38*$^{-/-}$ mice suggests that there is an important critical period in which OT plasma levels switches from the juvenile stage to the adult stage. As breast milk is the only food source in lactating pups, we speculated that OT is taken in from the breast milk. The milk curd found in the stomachs of the offspring born to *Cd38*$^{-/-}$ females was quite different from that seen in pups born to OT and OT receptor knockout females (Nishimori et al., 1996; Takayanagi et al., 2005). OT was similarly abundant in the mammary gland tissue and breast milk of lactating *Cd38*$^{+/+}$ and *Cd38*$^{-/-}$ females (Figures 3.2G and H; Liu et al., 2008). During the three main stages of growth and development, plasma OT seems to be obtained from different sources. (1) Foetal stage: The foetus can obtain OT from the placenta, (Malek et al., 1996). (2) Infant stage (breastfeeding stage): the dams

provide OT to the pups from breast milk until weaning. (3) Adult stage (weaning stage): during this period, plasma OT is derived entirely from internal synthesis and secretion with no exogenous supply.

3.3.3 ADP-ribosyl cyclase activity

In the mouse central nervous system, ADP-ribosyl cyclase activity corresponding to CD38 was detected as early as embryonic day 15 (Ceni et al., 2006). The endogenous brain cADPR content is highest in the developing brain and declines in the late stage of adulthood.

Hypothalamic ADP-ribosyl cyclase activity in 1-week-old $Cd38^{+/+}$ mice was 6% of that seen in 2-month-old mice. This activity in $Cd38^{+/+}$ mice was significantly higher than that seen in age-matched knockout mice (Figure 3.2F). From the second week of life onwards, $Cd38^{+/+}$ mice showed significantly increased levels of ADP-ribosyl cyclase activity in the hypothalamus, with the difference in levels with respect to $Cd38^{-/-}$ mice also increased markedly (Liu et al., 2008).

ADP-ribosyl cyclase activity in the pituitary was lower than that in the hypothalamus in both the $Cd38^{+/+}$ and $Cd38^{-/-}$ mice (Liu et al., 2008). $Cd38^{-/-}$ mice showed little or no increase in ADP-ribosyl cyclase activity in the pituitary during development.

The role of CD38 in the regulation of OT secretion through cADPR-mediated intracellular calcium signalling has been demonstrated in adult mice. Lower and similar levels of ADP-ribosyl cyclase activity were found in the hypothalamus and pituitary of both 1-week-old $Cd38^{+/+}$ and $Cd38^{-/-}$ pups. The level of ADP-ribosyl cyclase activity is also relatively low at the foetal stage. This suggests that the maintenance of plasma OT relies mainly on the exogenous source from the placenta and breast milk. When the CD38 knockout pups suddenly lose their exogenous OT supply after weaning, the intrinsic activity of ADP-ribosyl cyclase remains at a low level. The relative shortage of endogenous OT release during the young adult stage results in diminished social recognition in adult animals.

3.4 Plasma oxytocin levels in control and autism human subjects

At the start of our examination of the relation between OT release and CD38 in humans, the expression of CD38 in the hypothalamus was immunohistochemically examined. As shown in Figure 3.3A, oxytocinergic neurons expressed CD38, suggesting that CD38 plays an important role in OT release in humans, as well as in mice (Jin et al., 2007). Armed with this information in humans, we performed a single-nucleotide polymorphism analysis of *CD38* DNA in control and in subjects with autism spectrum disorder (ASD). All 8 exons of human *CD38* were sequenced, and identified 2 different heterozygous mutations in 3 male ASD probands (Munesue et al., 2010). We found that a nucleotide substitution (4441C $>$ T) in exon 3 that leads to an arginine (Arg or R) to tryptophan (Trp or W) substitution at amino acid 140 (R140W) in one affected individual diagnosed with Asperger's syndrome and the two with the typical autism (associated with lower intelligence). In their families, the SNP of R140W was present in the proband's brothers and fathers. Judging from score of the autism-spectrum quotient (AQ) of the individuals with the mutation, they showed a tendency for ASD, even though adults' AQ scores were below the cut-off point for diagnosis. These results show that R140W-CD38 mutation is associated with ASD and familially accumulated.

Given these results, we obtained blood samples from brothers, parents, and the grandmother to further study the connection with the mutation and plasma OT and AVP levels. There was a significant difference in OT level between ASD patients without R140W mutation in CD38 (147.7 ± 15.0 ng/ml; $n = 26$) and with the mutation (79.2 ± 16.6 ng/ml; $n = 3$) (Figure 3.3B). As expected, there was no decrease in AVP level is the patients with R140W mutation (38.7 ± 2.9 ng/ml vs 28.2 ± 4.5 ng/ml), as shown in the distribution graph in Figure 3.3C. The relationship between plasma OT and AVP levels in individuals with and without the R140W mutation are plotted in Figure 3.3D. This graph reveals that there

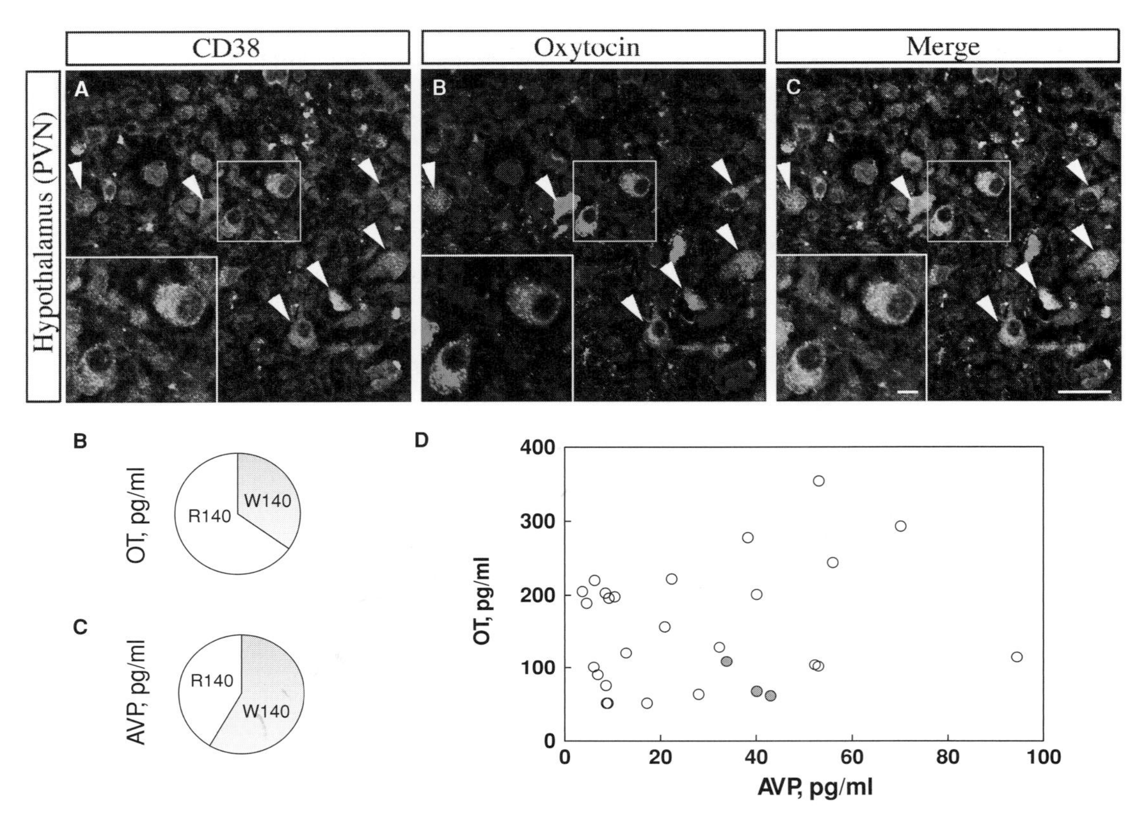

Figure 3.3 Studies in human subjects. A, Immunohistochemical analysis of CD38 (A) and oxytocin (B) in the human brain. Cell montages of panels were taken from the paraventricular nucleus (PVN) in the hypothalamus of autopsy subjects from the USA. Arrowheads indicate extensive colabeling. The insets in panels are enlarged images of neurons showing coexpression of CD38 and OT. B, C, Plasma oxytocin (B) and vasopressin (C) levels in ASD subjects with R140 or W140 allele. D, Scatchard plot of plasma concentrations of OT and AVP levels in 29 ASD patients with (filled circle) or without (open circle) the W140 allele. See color version in plates section.

is also a group of subjects with low plasma OT levels arising from factors other than the CD38 R140W mutation.

3.5 Conclusions

The OT-stimulated ADP-ribosyl cyclase activity and increases of $[Ca^{2+}]_i$ were susceptible to PKC in the hypothalamus and/or posterior pituitary. OT-induced elevations in $[Ca^{2+}]_i$ and subsequent OT release were both PKC-sensitive and cADPR-dependent (Figure 3.4). This supports the suggestion that OT-induced OT release, autoregulation of the positive feedback, is a PKC- and cADPR-dependent process in the hypothalamus and/or pituitary. These findings from male adult mice suggest that this PKC- and cADPR-dependent autoregulation of OT release is not related to the female's reproductive processes, but to social recognition or social behavior found in this mouse strain (Jin et al., 2007; Liu et al., 2008), as in causing uterus contraction during labor and triggering milk release

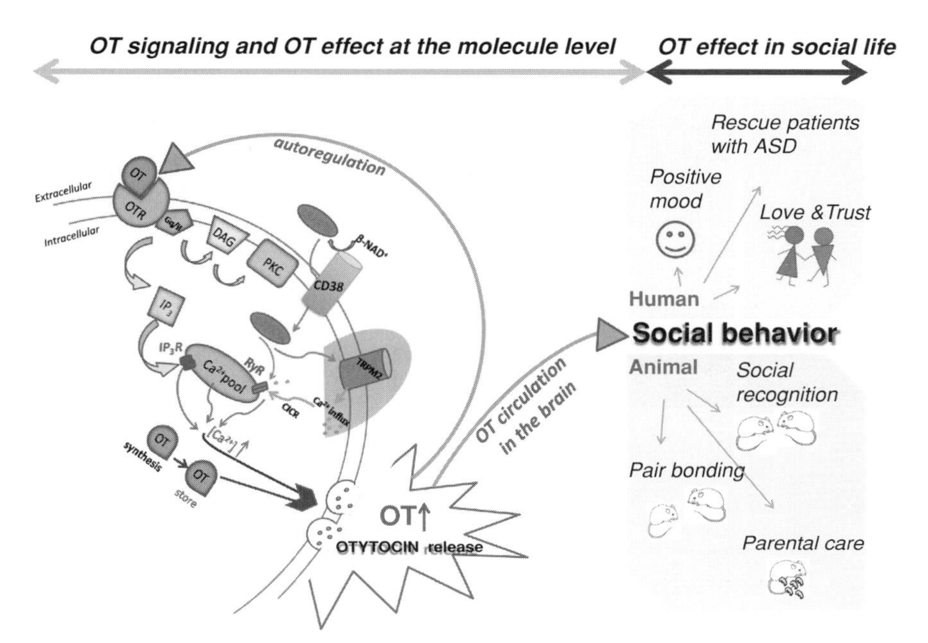

Figure 3.4 A schematic model of oxytocin signaling and oxytocin effect by cyclic ADP-ribose (cADPR) and heat that induces $[Ca^{2+}]_i$ increases. cADPR, extracellularly applied or converted by the ADP-ribosyl cyclase activity of CD38, is transported into the cell, and activates the ryanodine receptor (RyR). Consequently, binding of cADPR to melastatin-related transient receptor potential channel 2 (TRPM2) channels initiates Ca^{2+} influx. OT binding with OTR and through the cascade also initiates Ca^{2+} influx and OT release. These molecular events modulate social events and social life in rodent and human. The molecular event associated with the heat sensitivity is shaded in red. Targets of PKC are not shown. cADPR is also produced intracellulaly, but not illustrated. OT, oxytocin; OTR, oxytocin receptor; DAG, diacylglycerol; IP_3, inositol trisphosphate; IP_3R, inositol trisphosphate receptor; PKC, protein kinase C; CICR, Ca^{2+}-induced Ca^{2+} release. See color version in plates section.

from the breast tissue when infants are nursed (Moos et al., 1984; Neumann et al., 1994; 1996). The results of the recent series of studies showed that OT also has an important role in human social behavior as shown in Chapters 17–20 of this book. Taken together, these findings indicate that positive feedback of PKC- and cADPR-dependent OT release in the hypothalamus and pituitary is important for the proper display of social behavior (Figure 3.4).

We have demonstrated that the rise of $[Ca^{2+}]_i$ in the isolated nerve endings after OT stimulation was dually controlled by IP_3 and cADPR, respectively, resembling that seen in human myometrium cells (Barata et al., 2004). The initial part of $[Ca^{2+}]_i$ increases seems to be composed of IP_3-mediated $[Ca^{2+}]_i$ increases without Ca^{2+} influx, as the IP_3 antagonist delays the initial peak. In contrast, the cADPR is largely responsible for the sustained Ca^{2+} signal for up to 5 min, as shown by its blockade by the pretreatment of nerve endings with 8-bromo-cADPR. Thus, we can postulate that the OT-induced Ca^{2+} elevation is due to a release mediated by cADPR through ryanodine receptors, which is PKC dependent, followed by an initial Ca^{2+} mobilization by activation of the IP_3 receptors, which is not PKC sensitive. A similar susceptibility pattern to inhibitors was obtained for OT-mediated OT release in the isolated nerve endings, indicating that OT release is dependent upon the sustained phase of Ca^{2+} increases. The ADP-ribosyl cyclase activity

and/or CD38 is activated by PKC in downstream of OT receptor signaling and may play an important role in autoregulation of OT release both in the hypothalamus and neurohypophysis, and thereby influence social behavior (Figure 3.4). Our study also supports the idea that cADPR is a second messenger that modifies cellular Ca^{2+} signaling in OT reactions.

A series of recent studies suggested that OT may be related to autism (Insel, 2010). It has been reported that plasma OT levels in autistic children are lower than those in controls (Munesue et al., 2010) or age-matched normal controls (Modahl et al., 1998). Tanoue and Oda (1989) reported that children in their control group breast-fed significantly longer than autistic infants, and provided evidence that OT is critical for regulating social behavior during early experience in infants. Together with our data in mice, these data suggest that lack of adequate exogenous OT during the infant period would affect the normal development of the brain in genetically susceptible infants, thereby increasing the risk of autism.

OT treatment as a compensatory method has been started in several hospitals around the world. In one subject with low plasma OT level caused by the R140W SNP, we have observed improvements in social behavior, such as increased eye contact and positive communication by nasal infusion of OT (Munesue et al., 2010). Recently, Ebstein et al. (2009) suggested some interesting links between ASD and CD38 in humans. Genetic polymorphisms in the AVP-OT pathway, notably the AVP receptor 1a, the OT receptor, neurophysin I and II, and CD38 may contribute to deficits in socialization skills in ASD patients.

REFERENCES

Adan, R. A. H., Van Leeuwen, F. W., Sonnemans, M. A. F., et al. (1995). Rat oxytocin receptor in brain, pituitary, mammary gland, and uterus: partial sequence and immunocytochemical localization. *Endocrinol*, 136, 4022–4028.

Amina, S., Hashii, M., Ma, W. J., et al. (2010). Intracellular calcium elevation induced by extracellular application of cyclic-ADP-ribose or oxytocin is temperature-sensitive in rodent NG108–15 neuronal cells with or without exogenous expression of human oxytocin receptors. *J Neuroendocrinol*, 22, 460–466.

Barata, H., Thompson, M., Zielinska, W., et al. (2004). The role of cyclic-ADP-ribose-signaling pathway in oxytocin-induced Ca2+ transients in human myometrium cells. *Endocrinol*, 145, 881–889.

Beck. A., Kolisek, M., Bagley, L. A., Fleig, A., and Penner, R. (2006). Nicotinic acid adenine dinucleotide phosphate and cyclic ADP-ribose regulate TRPM2 channels in T lymphocytes. *FASEB J*, 20: 962–964.

Boittin, F. X., Dipp, M., Kinnear, N. P., Galione, A., and Evans, A. M. (2003). Vasodilation by the calcium-mobilizing messenger cyclic ADP-ribose. *J Biol Chem*, 278, 9602–9608.

Carter, C. S. (2003). Developmental consequences of oxytocin. *Physiol Behav*, 79: 383–397.

Ceni, C., Pochon, N., Villaz, M., et al. (2006). The CD38-independent ADP-ribosyl cyclase from mouse brain synaptosomes: a comparative study of neonate and adult brain. *Biochem J*, 395, 417–426.

De Flora, A., Zocchi, E., Guida, L., Franco, L., and Bruzzone, S. (2004). Autocrine and paracrine calcium signaling by the CD38/NAD+/cyclic ADP-ribose system. *Ann NY Acad Sci* 1028, 176–191.

Donaldson, Z. R. and Young, L. J. (2008). Oxytocin, vasopressin, and the neurogenetics of sociality. *Science*, 322, 900–904.

Ebstein, R. P., Israel, S., and Lerer, E. (2009). Arginine vasopressin and oxytocin modulate human social behavior. *Ann NY Acad Sci*, 1167: 87–102.

Freund-Mercier, M. J., Stoeckel, M. E., and Klein, M. J. (1994). Oxytocin receptors on oxytocin neurons: histoautoradiographic detection in the lactating rat. *J Physiol Lond*, 480, 155–161.

Gimpl, G. and Fahrenholz, F. (2001). The oxytocin receptor system: structure, function, and regulation. *Physiol Rev*, 81, 629–683.

Graeff, R. M., Franco, L., De Flora, A., and Lee, H. C. (1998). Cyclic GMP-dependent and -independent effects on the synthesis of the calcium messengers cyclic ADP-ribose and nicotinic acid denine dinucleotide phosphate. *J Biol Chem*, 273, 118–125.

Graeff, R. M., Walseth, T. F., Fryxell, K., Branton, W. D., and Lee, H. C. (1994). Enzymatic synthesis and characterizations of cyclic GDP-ribose. A procedure for distinguishing enzymes with ADP-ribosyl cyclase activity. *J Biol Chem*, 269, 30260–30267.

Hashii, M., Minabe, Y., and Higashida, H. (2000). cADP-ribose potentiates cytosolic Ca2+ elevation and Ca2+ entry via L-type voltage-activated Ca2+ channels in NG108–15 neuronal cells. *Biochem J*, 345, 207–215.

Higashida, H., Bowden, S. E., Yokoyama, S., et al. (2007). Overexpression of human CD38/ADP-ribosyl cyclase enhances acetylcholine-induced Ca2+ signalling in rodent NG108–15 neuroblastoma cells. *Neurosci Res*, 57, 339–346.

Higashida, H., Egorova, A., Higashida, C., et al. (1999). Sympathetic potentiation of cyclic ADP-ribose formation in rat cardiac myocytes. *J Biol Chem*, 274, 33348–33353.

Higashida, H., Lopatina, O., Yoshihara, T., et al. (2010). Oxytocin signal and social behaviour: comparison among adult and infant oxytocin, oxytocin receptor and CD38 gene knockout mice. *J Neuroendocrinol*, 22, 373–379

Higashida, H., Salmina, A. B., Olovyannikova, et al. (2007). Cyclic ADP-ribose as a universal calcium signal molecule in the nervous system. *Neurochem Int*, 51, 192–199.

Higashida, H., Yokoyama, S., Hashii, M., et al. (1997). Muscarinic receptor-mediated dual regulation of ADP-ribosyl cyclase in NG108–15 neuronal cell membranes. *J Biol Chem*, 227, 31272–31277.

Insel, T. R. (2010). The challenge of translation in social neuroscience: a review of oxytocin, vasopressin, and affiliative behavior. *Neuron*, 65, 768–779.

Jin, D., Liu, H. X., Hirai, H., et al. (2007). CD38 is critical for social behaviour by regulating oxytocin secretion. *Nature*, 446, 41–45.

Kim, S. Y., Cho, B. H., and Kim, U. H. (2010). CD38-mediated Ca2+ signaling contributes to angiotensin II-induced activation of hepatic stellate cells: attenuation of hepatic fibrosis by CD38 ablation. *J Biol Chem*, 285, 576–582.

Kosfeld, M., Heinrichs, M., Zak, P. J., Fischbacher, U., and Fehr, E. (2005). Oxytocin increases trust in humans. *Nature*, 435, 673–676.

Lambert, R. C., Dayanithi, G., Moos, F. C., and Richard, P. (1994). A rise in the intracellular Ca2+ concentration of isolated rat supraoptic cells in response to oxytocin. *J Physiol (Lond)*, 478, 275–287.

Lee, H. C. (2001). Physiological functions of cyclic ADP-ribose and NAADP as calcium messengers. *Annu Rev Pharmacol Toxicol*, 41, 317–345.

Liu, H. X., Lopatina O., Higashida C., et al. (2008). Locomotor activity, ultrasonic vocalization and oxytocin levels in infant CD38 knockout mice. *Neurosci Lett*, 448, 67–70.

Lopatina, O., Liu, H. X., Amina, S., Hashii, M., and Higashida, H. (2010). Oxytocin-induced elevation of ADP-ribosyl cyclase activity, cyclic ADP-ribose or Ca(2+) concentrations is involved in autoregulation of oxytocin secretion in the hypothalamus and posterior pituitary in male mice. *Neuropharmacology*, 58, 50–55.

Malek, A., Blann, E., and Mattison, D. (1996). Human placental transport of OXT. *J Matern Fetal Med*, 5, 245–255.

McGregor, I. S., Callaghan, P. D., and Hunt, G. E. (2008). From ultrasocial to antisocial: a role for oxytocin in the acute reinforcing effects and long-term adverse consequences of drug use? *Brit J Pharmacol*, 154, 358–368.

Modahl, C., Green, L., Fein, D., et al. (1998). Plasma OXT levels in autistic children. *Biol Psychiatry*, 43, 270–277.

Moos, F., Freund-Mercier, M. J., Guerné, Y., et al. (1984). Release of oxytocin and vasopressin by magnocellular nuclei in vitro: specific facilitatory effect of oxytocin on its own release. *J Endocrinol*, 102, 63–72.

Munesue, T., Yokoyama, S., Nakamura, K., et al. (2010). Two genetic variants of CD38 in subjects with autism spectrum disorder and controls. *Neurosci Res*, 67, 181–191.

Neumann, I., Douglas, A. J., Pittman, Q. J., Russell, J. A., and Landgraf, R. (1996). Oxytocin released within the supraoptic nucleus of the rat brain by positive feedback action is involved in parturition-related events. *J Neuroendocrinol*, 8, 227–233.

Neumann, I. D. (2008). Brain oxytocin: a key regulator of emotional and social behaviours in both females and males. *J Neuroendocrinol*, 20, 858–865.

Neumann, I., Koehler, E., Landgraf, R., and Summy-Long, J. (1994). An oxytocin receptor antagonist infused into the supraoptic nucleus attenuates intranuclear and peripheral release of oxytocin during suckling in conscious rats. *Endocrinol*, 134, 141–148.

Nirenberg, M., Wilson, S., Higashida, H., et al. (1983). Modulation of synapse formation by cyclic adenosine monophosphate. *Science*, 222, 794–799.

Nishimori, K., Young, L. J., Guo, Q., et al. (1996). Oxytocin is required for nursing but is not essential for parturition or reproductive behavior. *Proc Natl Acad Sci USA*, 93, 11699–704.

OuYang, W., Wang, G., and Hemmings, H. C. Jr. (2004). Distinct rat neurohypophysial nerve terminal populations identified by size, electrophysiological properties and neuropeptide content. *Brain Res*, 1024, 203–211.

Perraud, A. L., Fleig, A., and Dunn, C. A. (2001). ADP-ribose gating of the calcium-permeable LTRPC2 channel revealed by Nudix motif homology. *Nature*, 411, 595–599.

Rah, S. Y., Park, K. H., Han, M. K., Im, M. J., and Kim, U. H. (2005). Activation of CD38 by interleukin-8 signaling regulates intracellular Ca2+ level and motility of

lymphokine-activated killer cells. *J Biol Chem*, 280, 2888–2895.

Richard, P., Moos, F., and Freund-Mercier, M. J. (1991). Central effects of oxytocin. *Physiol Rev*, 71, 331–370.

Skuse, D. H. and Gallagher, L. (2009). Dopaminergic-neuropeptide interactions in the social brain. *Trends Cogn Sci*, 13, 27–35.

Sternfeld, L., Krause, E., Guse, A., and Schulz, I. (2003). Hormonal control of ADP-ribosyl cyclase activity in pancreatic acinar cells from rats. *J Biol Chem*, 278, 33629–33636.

Tanoue, Y., Oda, S. (1989). Weaning time of children with infantiole autism. *J Autism Dev Disord*, 19, 425–434.

Takayanagi. Y., Yoshida, M., and Bielsky, I. F. (2005). Pervasive social deficits, but normal parturition, in OXT receptor-deficient mice. *Proc Natl Acad Sci USA*, 102, 16096–16101.

Togashi, K., Hara, Y., Tominaga, T., et al. (2006). TRPM2 activation by cyclic ADP-ribose at body temperature is involved in insulin secretion. *EMBO J*, 25, 1804–1815.

Togashi, K., Inada, H., and Tominaga, M. (2008). Inhibition of the transient receptor potential cation channel TRPM2 by 2-aminoethoxydiphenyl borate (2-APB). *Br J Pharmacol*, 153, 1324–1330.

Tobin, V. A., Leng, G., Ludwig, M., and Douglas, A. J. (2010). Increased sensitivity of monoamine release in the supraoptic nucleus in late pregnancy: region- and stimulus-dependent responses. *J Neuroendocrinol*, 22, 430–437.

Young, L. J., Muns, S., Wang, Z., and Insel, T. R. (1997). Changes in oxytocin receptor mRNA in rat brain during pregnancy and the effects of estrogen and interleukin-6. *J Neuroendocrinol*, 9, 859–865.

Zak, P. J., Stanton, A. A., and Ahmadi, S. (2007). Oxytocin increases generosity in humans. *PLoS ONE*, 2, e1128.

4

The organizational effects of oxytocin and vasopressin

Behavioral implications

Bruce S. Cushing

4.1 Introduction

The critical role of the neuropeptides oxytocin (OT) and arginine vasopressin (AVP) in regulating social and socio-sexual behavior, from social recognition to mating to pair-bond formation to parental care, has been a major research focus for the past 30 years. The importance and wealth of information that has resulted makes for great reading, as evidenced by most of Part II, Chapters 5–16, being committed to covering this "modest" subject. Not to give the punch line away, but Part II reveals at least two critical things, first that the vast majority of research has focused on adults and secondly that while both OT and AVP may regulate many of the same behaviors in males and females the effects are often sexually dimorphic with OT playing a greater role in females and AVP in males (Cushing et al., 2001; Insel and Young, 2001). In Chapter 1 (Caldwell et al., this book) another critical aspect of the effects of OT and AVP in adults is discussed, which is that many of the behavioral effects, especially the regulation of social and socio-sexual behavior, are steroid dependent. Increasing steroid levels either enhance or are necessary for the behavioral effects of OT (estrogen) and AVP (estrogen and dihydrotestosterone) (see Caldwell et al., Chapter 1). This relationship is perhaps not surprising given that many of the behavioral and even physiological responses regulated by OT and AVP are associated either directly or indirectly with reproductive effort, that is, investigation of novel individuals, social recognition, social memory (for a review see Chapter 13), mating (Chapter 14), pair-bond formation (adult and maternal infant), parturition, maternal (Chapters 8–10) and paternal behavior (Chapter 8), in which sex steroids typically play a major role.

In contrast to this large body of research there is an emerging interest in the effects of these neuropeptides during development, both pre- and postnatal, on the ultimate expression of physiological and social behavior. While the hypothesis that AVP and OT may have an organizational effect within the brain producing long-term/permanent changes was proposed in the mid to late 1980s (AVP: Boer, 1985; OT: Noonan et al., 1989) the study of this phenomenon did not spark the same level of interest or funding as neuropeptide regulation in adults and languished compared to examination of postpubertal affects. However, the end of the twentieth and the beginning of the twenty-first century have seen a rebirth of interest in developmental effects producing increasing primary research and a flourish of review papers examining the role of neuropeptides during early critical periods and their potential relevance to social deficit disorders (Carter et al., 2009). Although I may be a bit biased, these studies and papers have revealed potentially critical findings, which have lead to

Oxytocin, Vasopressin, and Related Peptides in the Regulation of Behavior, ed. E. Choleris, D. W. Pfaff, and M. Kavaliers.
Published by Cambridge University Press.

rethinking the role of neuropeptides. Major findings include: (1) demonstration of an "organizational" effect of neuropeptides, (2) redefinition of the relationship between neuropeptides and steroids, and (3) demonstration of the critical importance of the early environment, including, the early social environment, in epigenetic/non-genomic regulation of not only the expression of adult behavior, but also in transgenerational transmission, and (4) stimulation of research on neuropeptides as a mechanism of, and possible treatment for, social deficit disorders, such as autism, schizophrenia, depression, and PTSD. Therefore, the goal of this chapter is to summarize the essential findings of this research discussing long-term behavioral and sexually dimorphic effects, the underlying mechanisms of effects/actions, the non-genomic effects, and the future direction of research of OT and AVP during development.

The understanding and implications of developmental effects vary between OT and AVP for several reasons: (1) OT has only one known, highly conserved, receptor. In contrast AVP has multiple receptors, V1 and V2, which have additional subtypes, that is, V1A and V1B. While the distribution of these receptors varies, V2 being found primarily in the periphery, all types are found in the brain and both of the V1 subtypes regulate behavior. This means that sorting out the effects and actions of vasopressin requires receptor-specific antagonists and agonists. (2) In knockout models there could be compensatory receptor expression for vasopressin receptor knockout mice, but not OT receptor knockout mice. (3) It is difficult to directly manipulate hormones within the CNS during development. In adults, OT, AVP, and their antagonists can be delivered directly via injection into ventricles or to site-specific regions in the brain. Although this has been accomplished in a few acute studies (Chen et al., 1988; Boer et al., 1994; Stribley and Carter, 1999) direct manipulation within the CNS is much more difficult, and therefore peripheral injections have been typically used for developmental studies. With a reduced blood–brain barrier in fetuses and neonates (Vorbrodt, 1993) this is an effective method for manipulating CNS affects. However, the peripheral administration of AVP has the potential to have a more profound effect than OT. Although both AVP and OT can affect a number of systems and responses peripheral manipulations of AVP has the potential to cause much greater adverse effects in both the adult and during development. Acting via the V2 receptor AVP plays a major role in water balance and therefore circulating levels can and do have a deleterious effect on an individual's physiology. In contrast, OT tends to affect smooth muscle contraction and many of the effects are related to adult behavioral responses.

4.1.1 Historical perspective

The vasopressin-deficient Brattleboro rat provided early indication of the potential role of vasopressin during development. Neonatal peripheral treatment with vasopressin permanently alters some systemic responses, including cardiac, and from this Boer (1985) hypothesized that AVP might also have an organizational effect within the central nervous system. The finding that during embryonic development treatment of pregnant females with vasopressin altered the development of the brain (Boer et al., 1988) supported the concept of an organizational effect. The development of the brain was associated with vasopressin acting via the V1 type receptor, as brain development in Wistar rats was only affected by a V1 antagonist, but not a V2 antagonist (Snijdewint and Boer, 1988). Neonatal treatment of pups resulted in increased emotionality in females and increased ambulatory behavior in males in an open field test (Boer et al., 1994). Results from this study provided some of the first empirical evidence of long-term behavioral changes associated with vasopressin during postnatal development, and support for an organizational effect. It also demonstrated that neonatal effects are sexually dimorphic. This study also tested the effect of repeated treatment with OT and found that as with a single injection of OT (Noonan et al., 1989, see below) neonatal treatment with OT did not

affect behavior in an open field. These results indicated that neonatal effects were sexually dimorphic and that vasopressin and OT regulate different behaviors.

Noonan et al. (1989) proposed that OT could have long-term effects on behavior through organizational effects within the brain and tested this by treating rats with a single dose of OT on postnatal day three. The results supported the hypothesis, with neonatal OT increasing novelty induced grooming in both males and females. In 1989, Shaprio and Insel described the ontogeny of OT receptors in the rat and demonstrated that during postnatal development the expression of OT was transitory in many regions of the brain. Based upon this transitory nature they hypothesized a limited time period for organizational effects that predicted a significant difference between responses to OT during the neonatal period and adulthood. The ontogeny and topography of vasopressinergic and oxytocinergic neurons using mRNA was described in the rat in 1990 (Bloch et al., 1990), followed shortly thereafter by a description of ontogeny of vasopressin receptor expression in the fetal and infant rat brain (Tribollett et al., 1991).

While these studies demonstrated a potential organizational effect of OT and AVP they did not provide evidence as to the system or systems being affected. Boer had suggested that vasopressin could be altering responses by organizing the central vasopressinergic system. This hypothesis was based in part upon the finding that neonatal treatment has long-term effects on peripheral receptor expression in the heart and kidney. Brattleboro rats, which displayed vasopressin deficiency, also displayed higher than normal OT levels throughout life and offered the possibility that one neuropeptide might affect the sister neuropeptide (Boer et al., 1988). But it would be almost a decade before it would be shown that OT and AVP had major effects on the long-term expression of social behavior, and we began to demonstrate the systems being impacted by these neuropeptides during development.

4.2 Behavioral effects

The goal of this to provide an understanding of the impact that OT or AVP may have on the "ultimate" expression of behavior not to elucidate nor detail all of the behavioral effects associated with these neuropeptides during development. The behavioral effects of OT and AVP can be divided into two categories, those that are relevant to the neonatal period versus long-term effects. While it may be in part due to the nature of the research questions that were asked, the majority of studies indicate that the behavioral effects associated with OT and AVP influence or regulate behaviors that are relevant to adult social interactions rather than those associated with the developmental period. To a large extent these findings provide the support for and are the basis of the hypothesis of an organizational effect of AVP and OT. Age appropriate effects for OT include altered ultrasound production in response to maternal isolation in rats (Insel and Winslow, 1991) and prairie voles (Kramer et al., 2003). In adults OT has been shown to be anxiolytic, reducing responses to stress, (McCarthy et al., 1997) and these findings suggest that it may play the same role during the neonatal period. In fetal rats intrathecal injections of AVP increased fetal activity, including mouthing, licking, wiping, and intracisternal injection of V1 antagonists reduced wiping and oral grasping of an artificial nipple (Valinskaya et al., 1994). In contrast, intrahemispheric injections of a V1 antagonist increased wiping and grasping of an artificial nipple. AVP deficient rat pups displayed a different group dynamic, reduced huddling and proximity to other pups, which may be involved in social interaction (Schank, 2009), but could ultimately be involved in thermoregulation and water balance. Central administration of vasopressin in rat pups decreased ultrasonic vocalizations and locomotor acitivty (Winslow and Insel, 1995).

The majority of studies have examined the subsequent effects of neonatal treatment in adults. Neonatal manipulations have subsequent long-term effects on many of the same behavioral and

physiological responses that these neuropeptides regulate in adults, indicating that during the neonatal period they function to establish adult behavior. Given the nature of the critical role of OT in regulating social behavior much of the research on neonatal effects of these peptides has been conducted using the highly social prairie vole. In prairie voles neonatal manipulation of OT affects the subsequent expression of a number of socio-sexual behaviors in adults. These included, but are not limited to, aggression (Bales and Carter, 2003a), alloparental behavior (Bales et al., 2004b), formation of partner preferences (Bales and Carter, 2003b), mating (Cushing et al., 2005), and reproductive success (Bales et al., 2004a; Cushing et al., 2005). In the Mandarin vole (*Microtus mandarinus*) neonatal treatment increased the probability that females, but not males, would form a partner preference, and increased reproductive activity in males (Jia et al., 2008). There are only a few studies examining the effects of neonatal AVP on the expression of adult behavior. This may be due in part to the constraints of treating with AVP during the neonatal period (see historical perspective). Neonatal manipulation of vasopressin, and selective inhibition of V1a receptor have been shown to have long-term effects on aggression (Stribley and Carter, 1999), mating (Meyerson et al., 1988), and open field behavior (Boer et al., 1993). While the effects of OT and AVP are typically sexually dimorphic, in adults, with AVP playing a greater role in males and OT in females (Cushing et al., 2001; Insel and Young, 2001), in neonatal treatment the degree of sexual dimorphism depends upon the behavior being studied. In some cases one may regulate the behavior in males and the other in females. For example in prairie voles neonatal treatment with AVP increased aggression in adult males but not females (Stribley and Carter, 1999), while neonatal OT increased aggression in females but not males (Bales and Carter, 2003a). In other cases they have similar effects, for example, in the prairie vole early treatment with OT enhanced the formation of pair-bonds and partner-preference in both male (Bales and Carter, 2003b) and female (Bales et al., 2007). However, the response may also be species specific as in Mandarin voles neonatal treatment, without mating, only enhanced partner preference in females (Jia et al., 2008). Finally for some behaviors the effect of OT may be limited to males. Male prairie voles typically display higher levels of spontaneous alloparental behavior than females (Roberts et al., 1998). Use of selective OT antagonist during neonatal development significantly reduced male, but not female, (Bales et al., 2004a), indicating that inhibition of endogenous OT has sexually dimorphic effects.

4.3 Early social environment

The early postnatal environment, especially in mammals, is a period of intense social interaction, maternal, sibling, and in some species paternal or with other relatives. These early experiences can and do influence the subsequent expression of social behavior. This critical period is a time of bidirectional affects, where the early environment may effect the development and sensitivity to hormones and well as the hormones affecting the neonatal social interactions. Early social experience affects the oxytocinergic and/or vasopressinergic system. In rats, the early social environment altered the subsequent expression of adult social behavior in both males and females (Francis et al., 1999). This was associated with site-specific increases in OTR in females and V1a in males (Francis et al., 2002). Maternal care may be directly affecting endogenous OT levels as contacts between mother and infants are known to increase OT production (Uvnäs-Moberg et al., 1998). Conversely, maternal separation reduced the subsequent expression of socio-sexual behavior. In rats maternal separation resulted in a decrease in OT neurons in the paraventricular nucleus of the hypothalamus (PVN) of females and AVP in males (Todeschin et al., 2009), which was associated with increased aggression and reduced social investigation by males. In another study long-term maternal separation produced a

decrease in OT immunoreactivity in the hypothalamus and amygdala of three-week-old male rat pups (Oreland et al., 2010). Maternal separation also affects AVP and OT receptors in males (Lukas et al., 2010).

The effect of the early social environment may not only be expressed in the adult, but also has been shown to be transgenerational in both rats (for review see Champagne, 2008) and prairie voles (Stone and Bales, 2010), with implications for cross-generational OT transmission in humans (Feldman et al., 2010). The transgenerational and long-term effects have been associated with epigenetic effects on estrogen receptor alpha (ERα) expression and the interplay of estrogen and OT and AVP (for review, see Champagne, 2008; Shepard et al., 2009). These reviews bring together a significant body of literature on early environmental, genetic, and epigenetic effects with an emphasis on estrogen and ERα. In turn, they argue very eloquently that, at least some of the behavioral and physiological changes may then occur via the oxytocinergic or vasopressinergic systems because of the steroid-dependent nature of OT and AVP. They discuss possible mechanisms including corticotrophin releasing hormone (CRH) and other regulatory hormones. I would also suggest the intriguing hypothesis that it is the effects on OT and AVP of the early social experiences that could be involved in the epigenetic regulation of ERα. This might be the case as OT has been shown to have an organizational effect on ERα (see Section 4.4.1, Yamamoto et al., 2004; Kramer et al., 2007; Perry et al., 2009). This also makes logical sense from the standpoint that contact and social interaction can have a fairly rapid effect on OT release/production, while gonadal steroid responses are generally much slower and very limited in females with inactive ovaries.

4.4 Organizational effects

The long-term behavioral effects of manipulation of endogenous or exogenous OT during the neonatal period provide strong support for the hypothesis that OT has an organizational effect on the CNS (Noonan et al., 1989; Shapiro and Insel, 1989; Yoshimura et al., 1996). The mechanism or mechanisms of OT actions or the systems that are being affected are not clear and are a work in progress. One approach to investigating long-term organizational effects is to ask what response is being affected and to then determine if the underlying mechanisms associated with these behaviors or physiological responses have been affected. While the long-term effects of neonatal OT cover a wide variety of responses they can be classified into two major categories : (1) Socio-sexual behavior, including aggression, pair-bond formation, mating, parental care, etc., and (2) Stress, such as social isolation, novel encounters, cardiac response, and corticosterone levels. Not surprisingly these are many of the same behaviors and responses that have been shown to be regulated/influenced in adults (see Chapters 5–16).

4.4.1 OT and estrogen

In adults, especially females, the behavioral effects of OT are estrogen dependent (Choleris et al., 2003). However, in mammals the ovaries are inactive during the fetal and neonatal period and therefore little or no estrogen is produced. This means that the dependence on steroids is highly unlikely during the neonatal period. In contrast, during the neonatal period many of OT's long-term effects are on estrogen-dependent behaviors, suggesting the hypothesis that during the neonatal period OT could be organizing subsequent response to estrogen. In rats, OT-treated females weighed significantly more after puberty and the increase was associated with fat depositions regulated by estradiol (Uvnäs-Moberg et al., 1998) and placental and fetal growth during pregnancy as adults (Sohlström et al., 2002). Neonatal OT influences the onset of first estrus and vaginal opening in female rats (Withuhn et al., 2003; Parent et al., 2008). In female prairie voles a single neonatal treatment on the day of birth affected sexual receptivity and the probability of successfully producing a litter (Cushing et al., 2005).

Female prairie voles undergo induced estrus with exposure to males, increasing estrogen levels and stimulating mating. Mating is also associated with increased female/female aggression and neonatal treatment with OT simulates the effects of estrogen by an increase in aggression following exposure to a male in both prairie voles (Bales and Carter, 2003a) and Mandarin voles (Jia et al., 2008b).

Studies in undifferentiated cancer cell lines and the developing ovary suggested that the developmental stage may play a role in the ability of OT to regulate the response/sensitivity to estrogen. In MCF7 breast cancer cells treatment with OT inhibited the ability of estadiol to stimulate mitosis (Cassoni et al., 1997). OT had a direct effect on several aspects of ERα expression including production of ERα mRNA, binding affinity, and transcriptional activity (Cassoni et al., 2002). Cancer cells are less differentiated than other cells, and "capable" of undergoing the equivalent of organizational effects. In prairie voles, OT treatment increased sensitivity to estradiol, lowering the threshold dose required to trigger estrus in sexually naïve females, but had no effect in sexually experienced females (Cushing and Carter, 1999). Unlike most female mammals, the completion of sexual development in prairie voles requires chemical and social cues from males (Carter et al., 1987). Regardless of chronological age, females are not sexually mature so that a sexually experience female and a sexually naïve female represent significantly different developmental stages. The differential response to OT suggests that the stage of development is associated with the response to OT manipulation and that puberty may alter the relationship between OT and estrogen, with the behavioral effects becoming steroid dependent, as observed in rats (Ivell and Walther, 2002). Outside of the central nervous system (CNS) there is evidence that during the neonatal period OT can affect development. In rats, neonatal OT affected apoptosis in the developing ovary (Marzona et al., 2003). Since the ovary is the primary site of estrogen production changes in the cellular composition of the ovary during development could alter the subsequent production of estrogen and/or ovarian sensitivity to LH, FSH, and estrogen. Finally, in rats maternal behavior altered central OT (Francis et al., 2002) and expression of ERα in females (Champagne et al., 2003) and vasopressin receptors in males (Francis et al., 2002).

The direct effect of endogenous and exogenous OT on the expression of estrogen receptors in the neonatal period was determined in two rodent species with distinctly different social systems, the highly social prairie vole and Sprague Dawley rats. In two studies, male and female prairie voles received a single injection of OT, a selective OT antagonist (OTA), or saline control on the day of birth (P1). The expression of ERα was then examined on P1, P8, P21, and in adults. While not discussed in detail here it should be noted that the expression of ERα is sexually dimorphic in adult prairie voles with females expressing significantly more ERα than males (Hnatczuk et al., 1994; Cushing et al., 2004) and this difference is apparent by P21 (Yamamoto et al., 2006). The results support the hypothesis that OT has an organizational effect of ERα expression and also indicate that the effects of OT are sexually dimorphic. In females, by P21 site- and treatment-specific effects were apparent with OT producing a significant increase in ERα in the ventromedial hypothalamus (VMH) (figures 2.1 and 2.2), while OTA treatment produced a significant decrease in ERα in the medial preoptic area (figures 2.1 and 2.2) (Yamamoto et al., 2006). In contrast, effects in males were not observed until P60, with OTA increasing the expression of ERα in the bed nucleus of the stria terminalis (BST) (Kramer et al., 2007). The prairie vole is a valuable model for studying human relevant social behavior in part because it is socially monogamous, monogamy is a rare trait found in only 3 to 5% of mammalian species (Kleiman, 1977). To determine if the organizational effects of OT on ERα is a general phenomenon or specific to highly social species the effects of neonatal OT manipulation were investigated in female rats. The effects of neonatal OT manipulation in female rats was the same as that observed in female prairie voles with OT increasing ERα expression in the VMH (figures 2.1 and 2.3)

and OTA decreasing ERα expression in the (MPOA) (figures 2.1 and 2.3)(Perry et al., 2009). These results clearly demonstrated that the organizational effects of OT are not species specific and while the effects on the ultimate expression of behavior may differ, the effects on the underlying mechanisms are the same. Further results indicate there may be a dose-dependent effect of OT during the neonatal period as female prairie voles treated with a lower dose of OT on P1 also displayed an increase in ERα in the lateral septum (LS) and central amygdala in addition to the VMH (Kramer et al., 2007). The fact that OT has an organizational effect on ERα could be highly significant as it suggests the possibility that non-genomic events, such as early social experiences that influence neonatal OT expression, can have a major impact on the ultimate expression of social behavior and influence social interactions.

Although the mechanism by which OT alters the expression/organizes ERα is unknown findings from studies of non-neural tissue provide potential options. In breast cancer cells, which are less differentiated than adult cells, OT treatment directly affected the expression of ERα. Effects of OT included altering ERα mRNA production, binding affinity, and transcriptional activity (Cassoni et al., 2002), while in the neonatal ovary OT affected the rate of apoptosis (Marzona et al., 2003). *In vivo* studies in prairie voles support the hypothesis that neonatal OT can directly affect the expression of ERα, as within 2 h of OT treatment on P1 the expression of ERα mRNA was altered. OT treatment significantly increased ERα mRNA in the hypothalamus and hippocampus, but not the cortex, while OTA decreased ERα mRNA in the hippocampus (figure 2.5, Pournajafi-Nazarloo et al., 2007a). These effects were mRNA specific as ERβ mRNA expression was unaffected. The direction of the effect is consistent with the effect of neonatal manipulation on ERα-IR (Yamamoto et al., 2006; Perry et al., 2009) with OT increasing and OTA decreasing ERα. The effect of OT on ERα mRNA was not limited to the brain as OT treatment also altered the expression of ERα mRNA in the heart of both prairie voles (Pournajafi-Nazarloo et al., 2007b) and rats (Pournajafi-Nazarloo et al., 2007c). The effect in the heart suggests two things. First, the organizational effect of OT on ERα could in part be responsible for changes in cardiovascular response/performance reported in response to prepubertal OT treatment (Uvnäs-Moberg et al., 1998; Holst et al., 2002; Grippo et al., 2007). Second, the organizational effects of OT are not limited to the CNS. If neurons responded to OT like breast cancer cells then OT could be altering affinity and transcription of ER, which could account in part for longer-term effects and could contribute to the ultimate expression of ERα and sensitivity to estradiol.

Our lab tested the hypothesis that neonatal OT may be altering the number of cells that express ERα through regulating apoptosis. Neonatal treatment of prairie voles did not support this hypothesis as there was no evidence that OT altered apoptosis or apoptosis in ERα expressing neurons (unpublished data). In females treated with OT or OTA on P1 the number of apoptotic cells, as indicated by staining for TUNEL, was low throughout the limbic system, on P8 and P14, and there was no colocalization with ERα-expressing neurons. This suggests that the effect of OT differs between ovarian tissue and CNS neurons, and further implies that in neurons OT is acting by directly affecting ERα expression.

Responses to OT and AVP are sexually dimorphic and have been linked to sexually dimorphic expression of several social deficit disorders, including depression, schizophrenia, and autism (Heinrichs et al., 2009). While I may admittedly be one of the few that find this perplexing, there is no indication that either during development or in adults that the oxytocinergic system is sexually dimorphic, so how then can OT be associated with sexually dimorphic responses? The organization of ERα could explain in large part the sexually dimorphic effect. During development the male gonad actively produces testosterone, which is converted intracellularly to estradiol by aromatase, while in the female there is little or no estrogen production until the ovaries become active at the onset of puberty. Therefore, changes in ERα expression would have differential effects in males and females. Additionally,

organization of ERα would have a significant impact in adult females when the oxytocinerigc system is estrogen dependent (see Chapters 1 and 9).

4.4.2 Other organizational effects

Organizational effects on ER may explain many of the long-term changes in behavior. However, there are a number of reasons why it is unlikely that changes in response to steroids are sufficient to explain all of the effects of OT. First, during pre- and postnatal development the ovaries are inactive. This means that neonatal females would show little or no response to changes in ERα because there is no steroid to bind to the receptors. As such, changes in ERα are unlikely to explain changes in behavior that occur during the neonatal period. Second, in adults as well as neonates, behavioral and physiological responses to stressors are frequently rapid, and these types of responses are not typically associated with changes in nuclear receptors. Activation of nuclear receptors, ERα, involves transcriptional and translational activity, which are not considered rapid responses. Thirdly, many of the affected behaviors are also influenced or regulated by other mechanisms, including OT and AVP (for a review, see Cushing and Kramer, 2005), which at the very least suggests there could be changes in these or other underlying mechanisms.

Although the evidence is still limited there are studies that indicate OT may affect/organize other systems. OT may affect both neonatal responses and have long-term effects through the oxytocinergic system. In both prairie voles (Yamamoto et al., 2004) (figure 4.4) and rats (Perry et al., 2009) neonatal OT and OTA treatment increased the number of cells expressing OT in the PVN. The long-term effect on OT production is species specific. In female rats the increase in the OT positive cells in the PVN continued into adulthood (Perry et al., 2009), while adult female prairie voles no longer displayed differential production of oxytocinergic neurons (Kramer et al., 2007). The variation could be the result of a differential species response or the fact that in the vole study females only received a single treatment on P1 while rats were treated from P1–P6. It seems most likely that the difference is species specific, as a single treatment on P1 in female rats produced long-term increased OT production in the posterior pituitary of adults (Young et al., 2005).

Only one study has examined the potential effect of neonatal OT on CNS neuropeptide receptor expression. In this study, P1 OT manipulation did not alter the expression of OTR but did affect V1a receptor binding in adult prairie voles (Bales et al., 2007). The effects were site specific and sexually dimorphic. In females, P1 treatment reduced V1a binding, with OT decreasing binding in the MPOA, BST, LS, medial dorsal thalamic nucleus, and the Cingulate Cortex, while OTA treatment produced a reduction in binding in the BST and Cingulate Cortex. In males, OT treatment increased V1a binding in the Cingulate Cortex, while OTA treatment had the opposite effect resulting in a decrease in V1a binding in the MPOA, BST, and LS. Most of these regions regulate social behavior and the medial dorsal thalamic nucleus is one of the regions that displayed increased neuronal activity in response to neonatal OT treatment (Cushing et al., 2003). This means that neonatal effects of OT could alter behavioral response through an organizational effect on the receptor of its "sister" nonapeptide. This could explain sexually dimorphic effects.

Interestingly, this same study did not find an organizational effect of OT on OTR expression in either males or females. This, however, does not rule out the possibility that OT does affect the expression of OTR. In the heart, OT manipulation on P1 resulted in a change in OTR, as measured by real-time PCR, in both P21 rats (Pournajafi-Nazarloo et al., 2007c) and P21 prairie voles (Pournajafi-Nazarloo et al., 2007b). However, these studies did not examine the effects in adult hearts, so it is unknown if these changes persisted. In the prairie voles, neonatal effects on OT production in the PVN were observed on P21 but not in adult females, indicating that, at least in prairie voles, while the effects are

significant they were not permanent. If the same thing happened in the brain then OTR expression would need to be looked for at an earlier age. In contrast, in rats the effects on PVN OT neurons was long term and still apparent in adult females. The difference between voles and rats suggests that it would be valuable to determine if there was also a long-term effect on OTR in female rats. If this were the case it would support the concept of OT playing a significant role in the expression of species-specific social behavior. The differential response also might be indicative of the change in the relationship between OT and estrogen that occurs in most adult females, but not prairie voles.

The study of the organizational effects of OT and AVP is a wide open field with tremendous potential and implications for understanding both basic regulatory mechanisms and possible prevention and treatment of social deficit disorders. The surface has barely been scratched in terms of the effects on ER, let alone OT and AVP. There are also indications that OT may have other major long-term organizational effects. In the heart, neonatal treatment affected endothelial nitric oxide synthase expression (Pournajafi-Nazarloo et al., 2007c) and we have recently demonstrated that neonatal OT has site-specific organizational effects on the serotonergic system (Eaton et al., 2011), which fits with the anxiolytic effects of OT and may explain, at least in part, the ability of OT treatment to prevent the onset of depression (Grippo et al., 2009; Heinrichs et al., 2009).

4.5 Ontological effects

Critical periods, when hormones, neurohormones, neurotransmitters, and other compounds have windows of time in which they can act to organize the brain, may occur throughout the life of an organism, but the developmental period, both embryonic and neonatal, typically represents a period when many changes are occurring. Therefore, the neonatal effects of OT and AVP are dependent upon the ontogeny of the oxytocinergic and vasopressinergic systems. While there are significant differences between the ontogeny of these systems, one thing that stands out is that there is little or no sexual dimorphism in these systems. In eutherian mammals the vasopressingeric system develops during the embryonic period (E), while the oxytocingeric system develops postnatally (P).

4.5.1 Oxytocinergic system

In mammals, central OT production either begins just prior to birth or shortly after. In mice, OT mRNA has been detected several days before birth on E18.5 (Jing et al., 1998), but peptide production is not observed until the early postnatal period, P1, in prairie voles (Yamamoto et al., 2004), and P4 or P7 in rats (Choy and Watkins, 1979; Altstein and Gainer, 1988). While OT was present in cell bodies of the PVN in prairie voles on P1 it was not until P8 that oxytocingeric fibers were observed emanating from the cell bodies (figure 2.4)(Yamamoto et al., 2004), suggesting that there may be a delay between production and release.

The ontogeny of OTR may explain much of the organizational effects of OT during the neonatal period and either the loss of effects or the dramatic changes in response observed in adults. Neonatal OTR expression is considered to be transitory with many regions expressing OTR only during early, preweaning, postnatal development. The trajectory of OTR expression also varies so that the period of maximum OTR expression differs by region across postnatal age (Choy and Watson, 1979; Shaprio and Insel, 1989; Jing et al., 1998; Chen et al., 2000). Taken together this means that first, there is definitely a neonatal critical period and second, the effects of OT should vary across the neonatal period, beginning on P1. Therefore, a single treatment on P1 would be predicted to have a different effect from one on P8, because different regions would be responsive. Paradoxically, a single treatment on P1 might not differ from repeated daily treatments starting on P1. A single treatment with OT during the neonatal period increased the

number of OT neurons in the PVN (Yamamoto et al., 2004) and the production of OT in the posterior pituitary (Young et al., 2005). The results of early treatment indicate an initial positive feedback system with increased OT increasing endogenous OT production, which could then interact with the changing pattern of OTR function in a manner similar to exogenous treatment. This may also explain, at least in part, long-term non-genomic effects of even single events, such as handling or maternal separation, which if they cause a change in endogenous levels of OT, stimulate a positive feedback on OT production.

The postnatal development of the oxytocinegric system may at least in part be due to the role of OT during pregnancy and labor in eutherian mammals. During pregnancy, circulating levels of OT are high and, depending upon the species, peak toward the end of pregnancy. At parturition there are additional pulses of OT, which facilitate uterine contractions. If OTR was present in the late-term fetus then circulating levels could have a significant impact on the organization of the fetal brain. Although limited in its scope, a comparative examination of ontogeny of the oxytocinergic system in the Brazilian opossum, a South American marsupial, supports this hypothesis. The ontogeny of the vasopressinergic and oxytocingeric systems are similar to eutherian mammals with AVP immunoreactivity occurring as early as E12 (Iqbal and Jacobson, 1995a), while OT-IR appears in the median eminence P1 with production occurring in the PVN and SON and posterior pituitary between P3 and P5 (Iqbal and Jacobson, 1995b). The development of the two systems in a primitive marsupial suggests that the developmental timing of the two systems is adaptive and may have evolved in the ancestral state. It also suggests birth is a critical aspect in the ontogeny of the oxytocinergic system. Compared to eutherian mammals birth in marsupials occurs at a significantly earlier stage of development. However, despite this chronological difference development of both systems occurs within a similar timeframe, suggesting that changes associated with parturition trigger development of the oxytocinergic system.

4.5.2 Vasopressinergic system

While Boer et al. (1994) speculated on an organizational effect of vasopressin there is little experimental evidence indicating pre- or postnatal organizational effects of AVP on the mechanisms regulating social behavior. At least two factors may have acted to limit study. First, as previously discussed, peripheral AVP plays a major role in water balance therefore manipulation of AVP can have significant impacts on physiological responses, which in turn may result in significant changes in behavior. Second, very early in the study of the role of AVP, in contrast to OT, it was discovered that not only are adults, but also the pre- and post-natal effects of AVP are testosterone dependent. Neonatal castration reduced the number of cells producing AVP in the brain and replacement restored it (DeVries et al., 1983). Testosterone has a direct effect on the expression of AVP mRNA in both the medial amygdala and the BST (Wang, 1994), two regions of the brain that are part of the social neural network and play a critical role in the expression of social behavior (Newman, 1999). Prenatal treatment with flutamide, an androgen receptor antagonist, eliminated the dependence of adult male rats on vasopressin for the formation of social memory and recognition (Axelson et al., 1999). Testosterone has the potential to organize the male vasopressinergic system from the onset as the vasopressinergic system begins embryonic development at the same time as testosterone increases in the fetal male. The steroidal regulation of AVP may also explain the sexually dimorphic role of AVP, with AVP playing a more significant role in male behavior (Cushing et al., 2001; Insel and Young, 2001). Combined, these studies suggest that many of the early developmental effects of AVP are under the regulation of testosterone, especially in males, and may have helped focus research on the role of testosterone on the vasopressinergic system rather than the organizational effects of AVP. Despite the lack of empirical evidence, given the close structural relationship of AVP to its sister nonapeptide OT, which has organizational effects, and the importance of

AVP in regulating social behavior, it is probable that AVP does have organizational effects. The question is not only whether AVP has organizational effects, but also if they are independent of testosterone? The organizational effects of AVP are a potentially significant area of research, but only time, and possibly funding, will tell.

4.6 Implications in social deficit disorders

Given the critical role OT and AVP play in the regulation of social behavior, it is not surprising that there is intense interest in, and rapidly expanding research programs to understand, the role of OT and AVP not only as underlying mechanisms of, but also for possible treatments of social deficit disorders. The potential importance/role of neuropeptides in social deficit disorders is discussed in detail Part III – Human studies; and there are a number of comprehensive reviews that include information about neuropeptides and social deficit disorders (e.g., Heinrich et al., 2009; Ryckmans, 2010). Some of which review specific conditions, such as autism (Chapters 19 and 20), schizophrenia, depression, and ADHD (see Carter, 2007; Marazziti and Dell'Osso, 2008; Neuhaus et al., 2010). The goal of this section is to highlight specifically some of the findings as associated with the early onset of social deficit disorders and the role of OT and AVP.

Literature reviews of the neurobiology of social deficit disorders and empirical human studies indicate that the same factors, the organizational and epigenetic effects of OT and AVP, as discussed in this chapter, may play a significant role in the expression of social deficit disorders (Carter, 2007; Gregory et al., 2010; Insel, 2010). It has also been hypothesized that the sexually dimorphic expression of social deficit disorders may be explained, at least in part, by the differential effects of AVP and OT. For example, it has been argued that the sexually dimorphic actions of AVP may increase the probability that a male will develop autism and that OT may potentiate or buffer the effects of AVP (Carter, 2007). In fact, individuals with ASD have significantly higher levels of AVP (Boso et al., 2007), while autistic children have reduced levels of OT (Modahl et al., 1998; Green et al., 2001). Differential receptor expression may also play a role, as polymorphisms in the OTR gene have been correlated with autism (Jacob et al., 2007) and attention deficit and hyperactivity disorder (Park et al., 2010).

In adults, low circulating levels of OT have been associated with autism and other social deficit disorders and the use of OT has been shown to temporarily reduce the symptomatic expression of social deficits (Hollander et al., 2007). It is tempting to speculate that if early onset conditions, such as autism, are subject to organizational influences of neuropeptides then early intervention may prevent permanent, or at least limit long-term, changes within the CNS, as opposed to temporarily reducing symptoms. One case in point is Prader–Willi Syndrome, which is associated with low circulating levels of OT (Hoybye, 2004) and fewer OT neurons (Swaab et al., 1995). In a potential mouse model for Prader–Willi Syndrome, neonatal feeding deficits were eliminated with a single injection of OT at birth (Schaller et al., 2010). The findings from studies of social deficit disorders underscore and emphasize the take home message of this chapter is the significance of understanding the roles of critical periods for the organizational effects of OT and AVP.

REFERENCES

Altstein, M. and Gainer, H. (1988). Differential biosynthesis and posttranslational processing of vasopressin and oxytocin in rat-brain during embryonic and postnatal-development. *Journal of Neuroscience*, 8, 3967–3977.

Axelson, J. F., Smith, M., and Duarte, M. (1999). Prenatal flutamide treatment eliminates the adult male rat's dependency upon vasopressin when forming social–olfactory memories. *Hormones and Behavior*, 36, 109–113.

Bales, K. L., Abdelnabi, M., Cushing, B. S., Ottinger, and M. A., Carter, C. S. (2004b). Effects of neonatal oxytocin manipulations on male reproductive potential in prairie voles. *Physiology and Behavior*, 81, 519–526.

Bales, K. L. and Carter, C. S. (2003a). Sex differences and developmental effects of oxytocin on aggression and

social behavior in prairie voles (*Microtus ochrogaster*). *Hormones and Behavior*, 44, 178–184.

Bales, K. L. and Carter, C. S. (2003b). Developmental exposure to oxytocin facilitates partner preferences in male prairie voles (*Microtus ochrogaster*). *Behavioral Neuroscience*, 117, 854–859.

Bales, K. L., Pfeifer, L. A., and Carter, C. S. (2004a). Sex differences and effects of manipulations of oxytocin on alloparenting and anxiety in prairie voles. *Developmental Psychobiology*, 44, 123–131.

Bales, K. L., van Westerhuyzen, J. A., Lewis-Reese, A. D., Grotte, N. D., Jalene, A., Lanter, J. A., and Carter, C. S. (2007). Oxytocin has dose-dependent developmental effects on pair-bonding and alloparental care in female prairie voles. *Hormones and Behavior*, b, 274–279.

Bloch, B., Guitteny, A. F., Chouham, S., Mougin, C., Roget, A., and Teoule R. (1990). Topography and ontogeny of the neurons expressing vasopressin, oxytocin, and somatostatin genes in the rat-brain – an analysis using radioactive and biotinylated oligonucleotides. *Cellular and Molecular Neurobiology*, 10, 99–112.

Boer, G. J. (1985). Vasopressin and brain-development – studies using the Brattleboro rat. *Peptides*, 6, 49–62.

Boer, G. J. (1993). Chronic oxytocin treatment during late gestation and lactation impairs development of rat offspring. *Neurotoxiology and Teratology*, 15, 383–389.

Boer, G. J., Quak, J., DeVries, M. C., and Heinsbroek, R. P. W. (1994). Mild sustained effects of neonatal vasopressin and oxytocin treatment on brain growth and behavior of the rat. *Peptides*, 15, 229–236.

Boer, G. J., VanHeerikhuize, J., and Vanderwoude, T. P. (1988). Elevated serum oxytocin of the vasopressin-deficient brattleboro rat is present throughout life and is not sensitive to treatment with vasopressin. *ACTA Endocrinology*, 117, 442–450.

Boso, M., Emanuele, E., Politi, P., Pace A., Arra, M., di Nemi, S. U., and Barale, F. (2007). Reduced plasma apelin levels in patients with autistic spectrum disorder. *Archives of Medical Research*, 38, 70–74.

Carter, C. S. (2007). Sex differences in oxytocin and vasopressin: Implications for autism spectrum disorders? *Behavioral Brain Research*, 176, 170–186.

Carter, C. S., Boone, E. M., Pournajafi-Nazarloo, H., and Bales, K. L. (2009). Consequences of early experiences and exposure to oxytocin and vasopressin are sexually dimorphic. *Developmental Neuroscience*, 31, 332–341.

Carter, C. S., Witt, D. M., Schnieder, J., Harris, Z. L., and Volkening, D. (1987). Male stimuli are necessary for female sexual-behavior and uterine growth in prairie voles (*Microtus-ochrogaster*). *Hormones and Behavior*, 21, 74–82.

Cassoni, P., Catalano, M. G., Sapino, A., Marrocco, T., Fazzari, A., Bussolati, G., and Fortunati, N. (2002). Oxytocin modulates estrogen receptor alpha expression and function in MCF7 human breast cancer cells. *International Journal of Oncology*, 21, 375–378.

Cassoni, P., Sapino, A., Fortunati, N., Munaron, L., Chini, B., and Bussolati, G. (1997). Oxytocin inhibits the proliferation of MDA-MB231 human breast-cancer cells via cyclic adenosine monophosphate and protein kinase A. *International Journal of Cancer*, 72, 340–344.

Champagne, F. A. (2008). Epigenetic mechanisms and the transgenerational effects of maternal care. *Frontiers in Neuroendocrinology*, 29, 386–397.

Champagne, F. A., Weaver, I. C. G., Diorio, J., Sharma, S., and Meaney, M. J. (2003). Natural variations in maternal care are associated with estrogen receptor a expression and estrogen sensitivity in the medial preoptic area. *Endocrinology*, 144, 4720–4724.

Chen, X. F., Chen, Z. F., Liu, R. Y., and Du, Y. C. (1988). Neonatal administration of a vasopressin analog (DDVAP) and hypertonic saline enhance learning-behavior in rats. *Peptides*, 4, 717–721.

Chen, Q., Schreiber, S. S., and Brinton, R. D. (2000). Vasopressin and oxytocin receptor mRNA expression during rat telencephalon development. *Neuropeptides*, 34, 173–180.

Choleris, E., Gustafsson, J. A., Korach, K. S., Muglia, L. J., Pfaff, D. W., and Ogawa, S. (2003). An estrogen-dependent four-gene micronet regulating social recognition: A study with oxytocin and estrogen receptor-alpha and -beta knockout mice. *Proceedings of the National Academy of Sciences USA*, 100, 6192–6197.

Choy, V. J. and Watkins, W. B. (1979). Maturation of the hypothalamo-neurohypophyseal system .1. localization of neurophysin, oxytocin and vasopressin in the hypothalamus and neural lobe of the developing rat-brain. *Cell and Tissue Research*, 197, 325–336.

Cushing, B. S. and Carter, C. S. (1999). Prior exposure to oxytocin mimics the effects of social contact and facilitates sexual behavior in females. *Journal of Neuroendocrinology*, 11, 765–769.

Cushing, B. S. and Kramer, K. M. (2005). Mechanisms underlying epigenetic effects of early social experience: the role of neuropeptides and steroids. *Neuroscience and Biobehavioral Reviews*, 29, 1089–1115.

Cushing, B. S., Levine, K., and Cushing, N. L. (2005). Neonatal manipulations of oxytocin affect reproductive behavior and reproductive success of adult female prairie voles (*Microtus ochrogaster*). *Hormones and Behavior*, 47, 22–28.

Cushing, B. S., Martin, J. O., Young, L. J., and Carter, C. S. (2001). The effects of neuropeptides on partner preference formation are predicted by habitat. *Hormones and Behavior*, 39, 48–58.

Cushing, B. S., Razzoli, M., Murphy, A. Z., Epperson, P. D., and Hoffman, G. E. (2004). Intraspecific variation in estrogen receptor alpha and the expression of male behavior in two populations of prairie voles. *Brain Research*, 1016, 247–254.

Cushing, B. S., Yamamoto, Y., Carter, C. S., and Hoffman, G. E. (2003). Central c-Fos expression in neonatal male and female prairie voles in response to treatment with oxytocin. *Developmental Brain Research*, 143, 129–136.

DeVries, G. J., Best, W., and Sluiter, A. A. (1983). The influence of androgens on the development of a sex difference in the vasopressinergic innervation of the rat lateral septum. *Developmental Brain Research*, 8, 377–380.

Eaton, J. L., Roche, L., Nguyen, K. N., Troyer, E., Papademetriou, E., Cushing, B. S., and Raghanti, M. A. (2011). *Developmental Psychobiology*, online May 18.

Feldman, R., Gordon, I., and Zagoory-Sharon, O. (2010). The cross-generation transmission of oxytocin in humans. *Hormones and Behavior*, 58, 669–676.

Francis, D., Diorio, J., Liu, D., and Meaney, M. J. (1999). Nongenomic transmission across generations of maternal behavior and stress responses in the rat. *Science*, 286, 1155–1158.

Francis, D. D., Young, L. J., Meaney, M. J., and Insel, T. R. (2002). Naturally occurring differences in maternal care are associated with the expression of oxytocin and vasopressin (V1a) receptors: gender differences. *Journal of Neuroendocrinology*, 14, 349–353.

Green, L., Fein, D., Modahl, C., Feinstein, C., Waterhouse, L., and Morris, M. (2001). Oxytocin and autistic disorder: Alterations in peptide forms. *Biological Psychology*, 50, 609–613.

Gregory, S. G., Connelly, J. J., Towers, A. J., Johnson, J., Biscocho, D., Markunas, C. A., Lintas, C., Abramson, R. K., Wright, H. H., Ellis, P., Langford, C. F., Worley, G., Delong, G. R., Murphy, S. K., Cuccaro, M. L., Persico, A., and Pericak-Vance, M. A. (2009). Genomic and epigenetic evidence for oxytocin receptor deficiency in autism. *BMC Medicine*, 7, 62.

Grippo, A. J., Cushing, B. S., and Carter, C. S. (2007). Depression-like behavior and stressor-induced neuroendocrine activation in female prairie voles exposed to chronic social isolation. *Psychological Medicine*, 69, 149–157.

Grippo, A. J., Trahanas, D. M., Zimmerman, R. R., Porges, S. W., and Carter, C. S. (2009). Oxytocin protects against negative behavioral and autonomic consequences of long-term social isolation. *Psychoneuroendocrinology*, 34, 1542–1553.

Heinrichs, M., von Dawans, B., and Domes, G. (2009). Oxytocin, vasopressin, and human social behavior. *Frontiers in Neuroendocrinology*, 30, 547–558.

Hnatczuk, O. C., Lisciotto, C. A., DonCarlos, L. L., Carter, C. S., and Morrell, J. I. (1994). Estrogen and progesterone receptor immunoreactivity (ER-IR and PR-IR) in specific brain areas of the prairie vole *(Microtus ochrogaster)* is altered by sexual receptivity and genetic sex. *Journal of Neuroendocrinology*, 6, 89–100.

Holst, S., Uvnas-Moberg, K., and Petersson, M. (2002). Postnatal oxytocin treatment and postnatal stroking of rats reduce blood pressure in adulthood. *Autonomic Neuroscience: Basic and Clinical*, 99, 85–90.

Hollander, E., Bartz, J., Chaplin, W., Phillips, A., Sumner, J., Soorya, L., Anagnostou, E., and Wasserman, S. (2007). Oxytocin increases retention of social cognition in autism. *Biological Psychology*, 61, 498–503.

Hoybye, C. (2004). Endocrine and metabolic aspects of adult Prader–Willi syndrome with special emphasis on the effect of growth hormone treatment. *Growth Hormone & IGF Research*, 14, 1–15.

Insel, T. R. (2010). The challenge of translation in social neuroscience: a review of oxytocin, vasopressin, and affiliative behavior. *Neuron*, 65, 768–779.

Insel, T. R. and Winslow, J. T. (1991). Central administration of oxytocin modulates the infant rats response to social-isolation. *European Journal of Pharmacology*, 203, 149–152.

Insel, T. R. and Young, L. J. (2001). The neurobiology of attachment. *Nature Review Neuroscience*, 2, 129–136.

Iqbal, J. and Jacobson, C. D. (1995a). Ontogeny of arginine vasopressin-like immunoreactivity in the Brazilian opossum brain. *Developmental Brain Research*, 89, 11–32.

Iqbal, J. and Jacobson, C. D. (1995b). Ontogeny of oxytocin-like immunoreactivity in the Brazilian opossum brain. *Developmental Brain Research*, 90, 1–16.

Ivell, R. and Walther, N. (2002). The role of sex steroids in the oxytocin hormone system. *Molecular and Cellular Endocrinology*, 151, 95–101.

Jacob, S., Brune, C. W., Carter, C. S., Leventhal, B. L., Lord, C., and Cook, E. H. (2007). Association of the oxytocin receptor gene (OXTR) in Caucasian children and adolescents with autism. *Neuroscience Letters*, 417, 6–9.

Jia, R., Tai, F., An, S., Broders, H., and Sun, S. (2008). Neonatal manipulation of oxytocin influences the partner preference in Mandarin voles (*Microtus mandarinus*). *Neuropeptides*, 42, 525–533.

Jing, X., Ratty, A. K., and Murphy, D. (1998). Ontogeny of the vasopressin and oxytocin RNAs in the mouse hypothalamus. *Neuroscience Research*, 30, 343–349.

Kleiman, D. G. (1977). Monogamy in mammals. *Quarterly Review of Biology*, 52, 39–69.

Kramer, K. M., Cushing, B. S., and Carter, C. S. (2003). Developmental effects of oxytocin on stress response: acute versus repeated exposure. *Physiology and Behavior*, 79, 775–782.

Kramer, K. M., Yoshida, S., Papademetriou, E., and Cushing, B. S. (2007). The organizational effects of oxytocin on the central expression of estrogen receptor alpha and oxytocin in adulthood. *BMC Neuroscience*, 8, 71.

Lukas, M., Bredewold, R., Neumann, I. D., and Veenema, A. H. (2010). Maternal separation interferes with developmental changes in brain vasopressin and oxytocin receptor binding in male rats. *Neuropharmacology*, 58, 78–87.

Marzona, L., Arletti, R., Benelli, A., Sena, P., and DePol, A. (2003). Effects of estrogens and oxytocin on the development of neonatal mammalian ovary. *In vivo*, 15, 271–279.

Marazziti, D. and Dell'Osso, M. C. (2008). The role of oxytocin in neuropsychiatric disorders. *Current Medicinal Chemistry*, 7, 698–704.

McCarthy, M. M., McDonald, C. H., Brooks, P. J., and Goldman, D. (1997). An anxiolytic action of oxytocin is enhanced by estrogen in the mouse. *Physiology and Behavior*, 60, 1209–1215.

Meyerson, B. J., Hoglund, U., Johansson, C., Blomqvist, A., and Ericson, H. (1988). Neonatal vasopressin antagonist treatment facilitates adult copulatory-behavior in female rats and increases hypothalamic vasopressin content. *Brain Research*, 437, 344–351.

Modahl, C., Green, L., Fein, D., Morris, M., Waterhouse, L., Feinstein, C., and Levin, H. (1998). Plasma oxytocin levels in autistic children. *Biological Psychology*, 43, 270–277.

Neuhaus, E., Beauchaine, T. P., and Bernier, R. (2010). Neurobiological correlates of social functioning in autism. *Clinical Psychology Review*, 30, 733–748.

Newman, S. W. (1999). The medial extended amygdala in male reproductive behavior – A node in the mammalian social behavior network. *Annals of the New York Academy of Science*, 877, 242–257.

Noonan, L. R., Continella, G., and Pedersen, C. A. (1989). Neonatal administration of oxytocin increases novelty-induced grooming in the adult-rat. *Pharmacology Biochemistry and Behavior*, 33, 555–558.

Oreland, S., Gustafsson-Ericson, L., and Nylander, I. (2010). Short- and long-term consequences of different early environmental conditions on central immunoreactive oxytocin and arginine vasopressin levels in male rats. *Neuropeptides*, 44, 391–398.

Parent, A. S., Rasier, G., Matagne, V., Lomniczi, A., Lebrethon, M. C., Gerard. A., Ojeda, S. R., and Bourguignon, J. P. (2008). Oxytocin facilitates female sexual maturation through a glia-to-neuron signaling pathway. *Endocrinology* 149, 1358–1365.

Park, J., Willmott, M., Vetuz, G., Toye, C., Kirley, A., Hawi, Z., Brookes, K. J., Gill, M., and Kent, L. (2010). Evidence that genetic variation in the oxytocin receptor (OXTR) gene influences social cognition in ADHD. *Progress in Neuro-Psychopharmacology and Biological Psychiatry*, 34, 697–702.

Perry, A. N., Paramadilok, A., and Cushing, B. S. (2009). Neonatal oxytocin alters subsequent estrogen receptor alpha protein expression and estrogen sensitivity in the female rat. *Behavioral Brain Research*, 205, 154–161.

Pournajafi-Nazarloo, H., Carr, M. S., Papademetriou, E., Schmidt, J. V., and Cushing, B. S. (2007a). Oxytocin selectively increases ERα mRNA in the neonatal hypothalamus and hippocampus of female prairie voles. *Neuropeptides*, 41, 39–44.

Pournajafi-Nazarloo, H., Papademeteriou, E., Saadat, H., Partoo, L., and Cushing, B. S. (2007b). Modulation of cardiac oxytocin receptor and estrogen receptor alpha mRNAs expression following neonatal oxytocin treatment. *Endocrine*, 31, 154–160.

Pournajafi-Nazarloo, H., Perry, A., Partoo, L., Papademeteriou, E., Azizi, F., Carter, C. S., and Cushing, B. S. (2007c). Neonatal oxytocin treatment modulates oxytocin receptor, atrial natriuretic peptide, nitric oxide synthase and estrogen receptor mRNAs expression in rat heart. *Peptide*, 28, 1170–1177.

Roberts, R. L., Williams, J. R., Wang, A. K., and Carter, C. S. (1998). Cooperative breeding and monogamy in prairie voles: Influence of the sire and geographical variation. *Animal Behaviour*, 55, 1131–1140.

Ryckmans, T. (2010). Modulation of the vasopressin system for the treatment of CNS diseases. *Current Opinions in Drug Discovery and Development*, 13, 538–547.

Schaller, F., Watrin, F., Sturny, R., Massacrier, A., Szepetowski, P., and Muscatelli, F. (2010). A single postnatal injection of oxytocin rescues the lethal feeding behaviour in mouse newborns deficient for the imprinted Magel2 gene. *Human Molecular Genetics*, 19, 4895–4905.

Schank, J. C. (2009). Early locomotor and social effects in vasopressin deficient neonatal rats. *Behavioral Brain Research*, 197, 166–177.

Shapiro, L. E. and Insel, T. R. (1989). Ontogeny of oxytocin receptors in rat forebrain – A quantitative study. *Synapse*, 3, 259–266.

Shepard, K. N., Vasiliki Michopoulos, V., Toufexis, D. J., and Wilson, M. E. (2009). Genetic, epigenetic and environmental impact on sex differences in social behavior. *Physiology and Behavior*, 97, 157–170.

Snijdewint, F. G. M. and Boer G. J. (1988). Neonatal treatment with vasopressin antagonist dp[tyr(me)2]avp, but not with vasopressin antagonist d(ch2)5[tyr(me)2]avp, inhibits body and brain-development and induces polyuria in the rat. *Neurotoxicological Teratology*, 10, 321–325.

Sohlström, A., Olausson, H., Brismar, K., and Uvnäs-Moberg, K. (2002). Oxytocin treatment during early life influences reproductive performance in *ad libitum* fed and food-restricted female rats. *Biology of the Neonate*, 81, 132–138.

Stone, A. L. and Bales, K. L. (2010). Intergenerational transmission of the behavioral consequences of early experience in prairie voles. *Behavioral Process*, 84, 732–738.

Stribley, J. M. and Carter, C. S. (1999). Developmental exposure to vasopressin increases aggression in adult prairie voles. *Proceedings of the National Academy of Sciences USA*, 96, 12601–12604.

Swaab, D. F., Purba, J. S., and Hofman, M. A. (1995). Alterations in the hypothalamic paraventricular nucleus and its oxytocin neurons (putative satiety cells) in Prader–Willi syndrome: a study of five cases. *Journal of Clinical Endocrinology and Metabolism*, 80, 573–579.

Tribollet, E., Goumaz, M., Raggenbass, M., Dubois-Dauphin, M., and Dreifuss, J. J. (1991). Early appearance and transient expression of vasopressin receptors in the brain of rat fetus and infant. An autoradiographical and electrophysiological study. *Developmental Brain Research*, 58, 13–24.

Todeschin, A. S., Winkelmann-Duarte, E. C., Jacob, M. H. V., Aranda, B. C. C., Jacobs, S., Fernandes, M. C., Ribeiro, M. F. M., Sanvitto, G. L., and Lucion, A. B. (2009). Effects of neonatal handling on social memory, social interaction, and number of oxytocin and vasopressin neurons in rats. *Hormones and Behavior*, 56, 93–100.

Uvnäs-Moberg, K., Alster, P., Petersson, M., Sohlström, A., and Björkstrand, E. (1998). Postnatal oxytocin injections cause sustained weight gain and increased nociceptive thresholds in male and female rats. *Pediatric Research*, 43, 1–5.

Varlinskaya, E. I., Petrov, E. S. Robinson, S. R., and Smotherman, W. P. (1994). Behavioral-effects of centrally administered arginine-vasopressin in the rat fetus. *Behavioral Neuroscience*, 108, 395–409.

Vorbrodt, A. W. (1993). Morphological evidence of the functional polarization of brain microvascular epithelium. In *The Blood–Brain Barrier*, Raven Press, Ltd., New York, NY: W.M. Pardridge, pp. 137–164.

Wang, Z. X. (1994). Testosterone effects on development of vasopressin messenger-RNA expression in the bed nucleus of the stria terminalis and medial amygdaloid nucleus in male-rats. *Developmental Brain Research*, 79, 147–150.

Winslow, J. T. and Insel, T. R. (1993). Effects of central vasopressin administration to infant rats. *European Journal of Pharmacology*, 233, 101–107.

Withuhn, T., Kramer, K. M., and Cushing, B. S. (2003). Early exposure to oxytocin affects the age of vaginal opening and first estrus in female rats. *Physiology and Behavior*, 8, 135–138.

Yamamoto Y, Carter, C. S., and Cushing, B. S. (2006). Neonatal manipulation of oxytocin effects expression of estrogen receptor alpha. *Neuroscience*, 137, 157–164.

Yamamoto, Y., Cushing, B. S., Kramer, K. M., Epperson, P., Hoffman, G. E., and Carter, C. S. (2004). Neonatal manipulations of oxytocin alter expression of oxytocin and vasopressin immunoreactive cells in the paraventricular nucleus of the hypothalamus in a gender specific manner. *Neuroscience*, 125, 947–955.

Yoshimura, R., Kimura, T., Watanabe, D., and Kiyama, H. (1996). Differential expression of oxytocin receptor mRNA in the developing rat brain. *Neuroscience Research*, 24, 291–304.

Young, E., Carter, C. S., Cushing, B. S., and Caldwell, J. D. (2005). Neonatal manipulation of oxytocin alters oxytocin levels in the pituitary of adult rats. *Hormone and Metabolic Research*, 37, 397–401.

PART II

Behavioral studies – Comparative approach

5

Social regulatory functions of vasotocin and isotocin in fish

Richmond R. Thompson and James C. Walton

Peptide behavioral regulatory mechanisms have been the subject of intense research in the last 20 years. Most of this research has focused on the ability of arginine-vasopressin- (AVP) and oxytocin- (OT) related peptides made of 9 amino acids (nonapeptides), to affect complex social behaviors in mammals. However, work on related molecules in species across major vertebrate groups is helping us better understand the fundamental mechanisms through which these peptides influence social behaviors, as well as how changes in these neurochemical systems have been associated with, and perhaps allowed for, the evolution of complex patterns of social organization. This includes work in fish, primarily teleosts (ray-finned, bony fish), a group with unparalleled diversity among vertebrates in social organization: there are species in which individuals spend most of their lives in solitude and species in which individuals form massive schools with hundreds of others, species in which no parental care is provided and species with sex-role reversal in which males provide all of the care, polygynous species in which males court multiple females and monogamous species in which long-term pair bonds form. There are even species in which some males masquerade as females, and some in which adult animals actually change sex in particular social conditions! Thus, it is an excellent group in which to study how animals have adapted in response to complex social pressures, and increasing evidence that will be reviewed in this chapter indicates that arginine vasotocin (AVT), the ancestral molecule that gave rise to all peptides within the AVP/OT family, is at the heart of that evolutionary plasticity.

At the same time, the AVT system in many modern fish has retained some of the primitive characteristics of the AVT system present in ancient vertebrates, as inferred by comparisons with jawless fish (Hoheisel et al., 1978), that were the foundation upon which all subsequent complexity in nonapeptide systems evolved. These include the confinement of AVT-producing cell bodies to the preoptic area / hypothalamus and a preponderance of descending projections to the pituitary and to the tegmentum and hindbrain. Thus, fish also represent an excellent group in which we can identify ancestral, potentially conserved mechanisms through which this family of molecules affects social behavior.

In the present chapter, we will first discuss recent molecular studies conducted in fish that have expanded our understanding of how various peptide forms and receptor subtypes have evolved in vertebrates. However, we will primarily focus on anatomical and pharmacological work implicating AVT and isotocin (IT), the teleost homologue of OT, in social regulation in this group of animals. We will highlight cases where functional similarities exist between fish and species from other vertebrate

Oxytocin, Vasopressin, and Related Peptides in the Regulation of Behavior, ed. E. Choleris, D. W. Pfaff, and M. Kavaliers.
Published by Cambridge University Press.

groups, as such similarities make it possible to identify highly conserved, fundamental mechanisms associated with social regulation. We will close by outlining what we believe are critical future directions for work on nonapeptide regulatory mechanisms in fish to take.

5.1 Molecular characteristics of fish nonapeptides and their receptors

Studies in fish, including jawless, cartilaginous, and ray- and lobe-finned bony fish, have made it possible to begin tracing the evolutionary history of the nonapeptide genes. Jawless fish, like cartilaginous and bony fish, amphibians, reptiles, and birds all contain a gene coding for a basic peptide, AVT, which indicates this is the ancestral nonapeptide gene in vertebrates (Gwee et al., 2008; Gwee et al., 2009, see Figure 5.1). Tandem duplication occurred in the common ancestor of jawed vertebrates, giving rise to two genes, oriented tail to head on the same DNA strand, the state still found in cartilaginous fish, lobe-finned fish, amphibians, birds, and marsupials. One of the copies continued coding for AVT in most vertebrates, though subsequent mutations eventually gave rise to AVP or lysine vasopressin in mammals, whereas mutations in the other copy gave rise to a family of neutral peptides across different vertebrate groups, including isotocin (IT; ray-finned fishes and some cartilaginous fishes), mesotocin (MT; lobe-finned fishes, amphibians, reptiles, birds, and marsupials), and OT (most mammals and some cartilaginous fish). Interestingly, there are at least 8 OT-related peptide genes in different cartilaginous fish, with some species exhibiting multiple forms, which suggests that extra gene duplications and mutations have occurred within that ancient group.

In tetrapods, there are multiple genes that code for receptors with specificity for AVT/AVP, including the AVPR1a-, AVPR1b-, and AVPR2-related receptors, and a single receptor with specificity for neutral peptides in the OT family, though both ligands can bind the heterologous receptors, albeit with lower affinities. AVPR1a- and IT/OT-related receptor genes have been sequenced from several teleost species (Heierhorst et al., 1990; Heierhorst et al., 1989; Hyodo et al., 1997; Suzuki et al., 1992; Venkatesh and Brenner, 1995), as well as in lungfish, a lobe-finned fish representative of the group from which tetrapods evolved, as was a AVPR2-related receptor gene (Konno et al., 2009). Likewise, a AVPR2-related gene was recently sequenced for the first time from several teleosts, including the birchir and medaka (Konno et al., 2010) and the river pupfish (Lema, 2010). In pupfish, not only was an IT/OT-related receptor gene also sequenced, but so were two AVPR1a-related receptor genes. Phylogenetic analysis of deduced amino acid sequences from pupfish AVPR1a genes showed that both are more closely related to AVPR1a receptor genes from other taxa than to AVPR1b receptor genes. The AVPR2-related receptor gene was expressed primarily in gills, heart, and kidney and was upregulated in hyperosmotic conditions, suggesting that it, like the mammalian AVPR2 receptor, codes for a receptor that mediates AVT effects on hydromineral balance. The AVPR1a1 and IT receptor genes were primarily expressed in brain, pituitary, and testes, and the AVPR1a2 primarily in brain, heart and muscle. Thus, one or both of the AVPR1a-related receptors, which were suggested to be paralogs resulting from genome duplication (tetraploidy), as well as the IT receptor, likely play key roles in mediating nonapeptide behavioral effects in this species and perhaps in other teleosts (see further discussion below). A PCR fragment generated using AVPR1b-selective primers was also recently reported in zebrafish (Filby et al., 2010), though it is unclear if the full AVPR1b sequence was amplified and subjected to nucleotide or deduced amino acid BLAST sequencing in that study. Thus, it is not yet possible to conclude that the PCR product was truly AVPR1b related. Together, these studies suggest that AVPR1a-, AVPR2-, and IT/OT-related receptor genes were present in early, bony fish, if not earlier (there are not yet any reported receptor sequences from any cartilaginous or jawless fish).

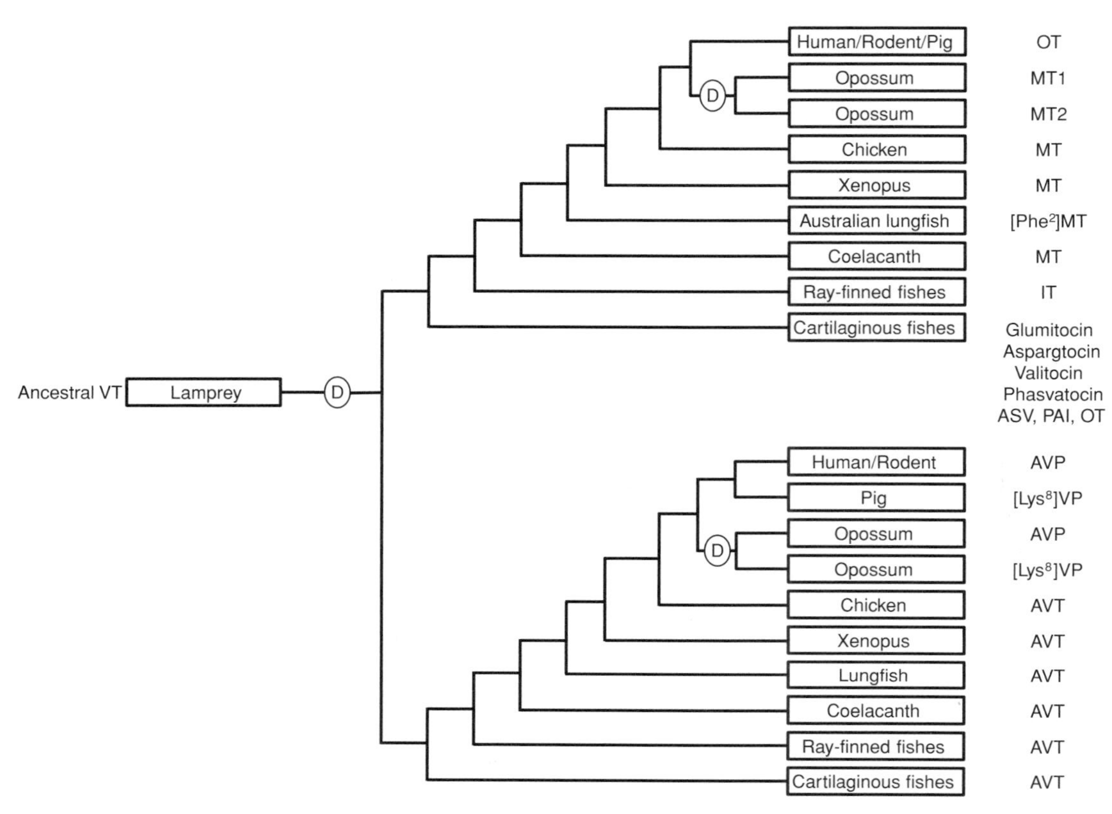

Figure 5.1 Vasopressin/Oxytocin family of nonapeptides. Vasotocin (AVT) is the ancestral nonapeptide gene in jawed vertebrates. A gene duplication event, followed by subsequent mutations in those duplicates, gave rise to vasopressin (AVP) in mammals from one duplicate; whereas mutations in the other duplicate gave rise to oxytocin (OT) in mammals, isotocin (IT) in teleosts, and mesotocin (MT) in birds, amphibians, marsupials and reptiles. Gene duplication event (D), asvatocin (ASV), phasitocin (PAI). Reproduced with permission from Gwee et al., 2008.

5.2 AVT/IT neuroanatomy: basic distributions in fish

AVT and IT neurons in all fish studied to date, including jawless, cartilaginous, and bony fish, are found almost exclusively within the preoptic area of the hypothalamus (POA), although a AVT cell population has also been identified more caudally in the anterior tuberal hypothalamus in several species, including plainfin midshipmen (*Goodson, Evans and Bass,* 2003), an African cichlid, *Astatotilapia burtoni* (Greenwood et al., 2008), and in two species of butterfly fish, the multiband and milletseed butterfly fish (Dewan et al., 2008). Developmentally, it has been shown in zebrafish that the IT system develops first and is followed by the AVT system, which originates in two separate regions, the ventral hypothalamus and dorsal preoptic area, under different transcriptional regulation (Eaton et al., 2008). The ventral hypothalamic cell group becomes difficult to discern as it migrates rostrally during ontogeny, but these cells may be the origin of the sparse AVT cells reported in the tuberal hypothalamus. The AVT and IT cells in the POA form an inverted L-shaped paired nucleus adjacent to the third ventricle. The POA is delineated by cell size;

anterior ventrolateral aspects are composed of the small parvocellular neurons (PPa). The dorsomedial and caudal portion of the POA make up the magnocellular preoptic nucleus (PM), which is further subdivided into three regions by cell size; parvocellular (PMp), magnocellular (PMm), and gigantocellular (PMg). IT and AVT cells are intermingled within the POA but have not been observed to be coexpressed within neurons of any species studied to date.

While some of the specific POA cell populations have been implicated in the regulation of various behaviors and phenotypes in some species, studies tracing the projections of these cells in teleosts are very limited. Both IT and AVT neurons from the PM have been shown to project simultaneously to the pituitary and other widespread extrahypothalamic sites throughout the brain in rainbow trout (Saito et al., 2004), and general fiber distributions for AVT and IT cells have been described in numerous species, though it has been difficult to determine the precise pathways associated with PPa cells or of specific PM subgroups. We will summarize some of the general fiber distributions here, and species-specific differences in relation to behavior will be discussed in more detail later. Generally, IT and AVT share conserved projection pathways from the preoptic areas to distributed brain sites, with IT generally having greater fiber densities than AVT in most regions (plainfin midshipmen, Goodson et al., 2003; rainbow trout, Saito et al., 2004; Van den Dungen et al., 1982), although exceptions do exist (green molly, Batten et al., 1990). Rostrally, AVT and IT fibers densely innervate the ventral telencephalon, including regions homologous to the septum (ventral and lateral regions), subpallial amygdala/bed nucleus of the stria terminalis (supracommissural region), and nucleus accumbens (dorsal region), all areas implicated in peptide social regulation in other vertebrates, and sparsely innervate the dorsal telencephalon, particularly the medial region thought to be homologous to the pallial amygdala (see O'Connell and Hofmann, 2011, for a review of telencephalic homologies in fish). AVT fibers have also been identified in the lateral region of the dorsal telencephalon that is thought to be homologous to the hippocampus, at least in goldfish (Thompson and Walton, 2009). Only sparse, if any, fibers project rostrally out into the olfactory bulbs in most species, although the AVPR1a2 gene and protein are expressed there in at least one species, the rock hind (Kline et al., 2011). In the diencephalon, dense fibers project throughout the POA, with the densest fibers coursing laterally and merging into the preopticohypophysial tract projecting to the pituitary. Fibers also project dorsally and ventrally, coursing along the midline to innervate the thalamus and habenula. Conserved descending pathways course both medially and laterally through the midbrain tegmentum. The lateral descending pathway runs through the midbrain tegmentum, innervating paralemniscal vocal acoustic regualtory regions in some species, and then turns dorsally to innervate the torus semicircularis, with sparse projections continuing up into the anterior tectum. The medial pathway descends through the tegmentum and into the hindbrain, where both IT and AVT sparsely innervate the cerebellum. Heavier projections innervate the reticular formation, octavolateralis nucleus, and multiple vagal nuclei, with the densest hindbrain projections going to the dorsal motor nucleus and area postrema. Both AVT and IT fibers continue on, descending into the spinal cord, where their final projection sites remain unidentified.

To date, there are only two papers describing AVT binding sites, receptor protein distributions, and/or receptor expression patterns in the brains of fish. In sea bass, H^3 AVT binding sites have been observed in the ventral telencephalon, the dorsal telencephalon, pars centralis, the hypothalamus, including the preoptic area and the anterior tuberal hypothalamus, and in the optic tectum (Moons et al., 1989). More recently, the distribution of the AVPR1a2 protein and AVPR1a2 mRNA have been described in the rock hind (Kline et al., 2011). The distributions of both are widespread in the brain and include numerous areas implicated in social regulation, including the olfactory bulbs, telencephalic areas homologous to those in the social brain network, the preoptic area and hypothalamuus, midbrain sensory regions, and

hindbrain regions implicated in AVT effects on social approach responses in teleosts (see further discussion below).

5.3 Reproductive strategies and AVT neuronal phenotypes

Males and females exhibit different behavioral strategies when mating that are typically controlled by sexually dimorphic circuits within the brain, including nonapeptide circuits. As in many other vertebrates (reviewed in Goodson and Bass, 2001), males in several teleost species have larger and/or more numerous AVT cells or higher levels of AVT gene expression in those cells than females, though in some cases only at certain periods during the breeding season (bluehead wrasse, Godwin et al., 2000; chum salmon, Hiraoka et al., 1997; Amargosa populations of river pupfish, Lema and Nevitt, 2004b; damselfish, Maruska, 2009; half-spotted goby, Maruska et al., 2007; medaka, Ohya et al., 2006). Although less often measured, males in at least one species, the sergeant damselfish, also have more dense AVT terminals in at least one target area, the vagal motor nucleus (VMN) in the hindbrain (Maruska, 2009). Such findings suggest, as has been observed for AVT/AVP systems in other vertebrates, that these peptides play important roles in the regulation male-typical reproductive functions, including courtship and aggressive behaviors, as will be discussed further below. However, there are exceptions; in some species there are no sexual dimorphisms in AVT cell density, number and/or size (miltiband and milliseed butterfly fish, Dewan et al., 2008; Big Spring populations of river pupfish, Lema and Nevitt, 2004b; goldfish, Parhar et al., 2001). In still others the reverse is true; female half-spotted gobies have more and larger AVT cells than males during the reproductive season (Maruska et al., 2007), and even though they have fewer AVT preoptic cells than males, female peacock blennies have higher levels of AVT expression per cell than territorial males (Grober et al., 2002), as do female grass puffers (Motohashi et al., 2008). Additionally, female damselfish have more AVT immunoreactive fibers in several midbrain sensory regions, which has prompted the hypothesis that AVT may be involved in the processing of male reproductive signals in this species (Maruska, 2009). Together, these studies suggest that AVT may not only participate in the regulation of male-typical reproductive behaviors in some teleosts, but also in the regulation of social behaviors that are common to both sexes and, in some species, in the regulation of female reproductive behaviors. In fact, examples of all three types of effect will be highlighted in this chapter.

In some teleosts, there are also alternative male reproductive strategies that correlate with differences in AVT phenotypes. In masu salmon, some males mature early and spawn in the first autumn of their life, while others take at least 3 years to mature. Levels of AVT gene expression peak in the autumn spawning period of the first year in the precocial males, whereas they do not in the "normal" males that do not mature and spawn until 3 years of age (Ota et al., 1999). This suggests that AVT may play some role in the regulation of precocial reproduction in this species, potentially modulating behavioral events associated with early spawning. In an African cichlid, *A. burtoni*, some males defend territories and form pair bonds with females, while others adopt a subordinate, non-territorial phenotype until the opportunity to take over a territory arises. AVT mRNA expression is higher in the parvocellular cells in subordinates than in territorial males, but higher in gigantocellular neurons in territorial males than in subordinates (Greenwood et al., 2008). Furthermore, while levels of AVT mRNA in gigantocellular cells were positively correlated with aggression and courtship, AVT expression in parvocellular cells was positively correlated with fleeing behavior (see Figure 5.2). Similarly, subordinate male zebrafish have more AVT-immunoreactive parvocellular cells than dominants, which actually have none, and dominants have more AVT-immunoreactive magnocellular cells than subordinates, which have none (Larson et al., 2006; no distinction was made between magnocellular and

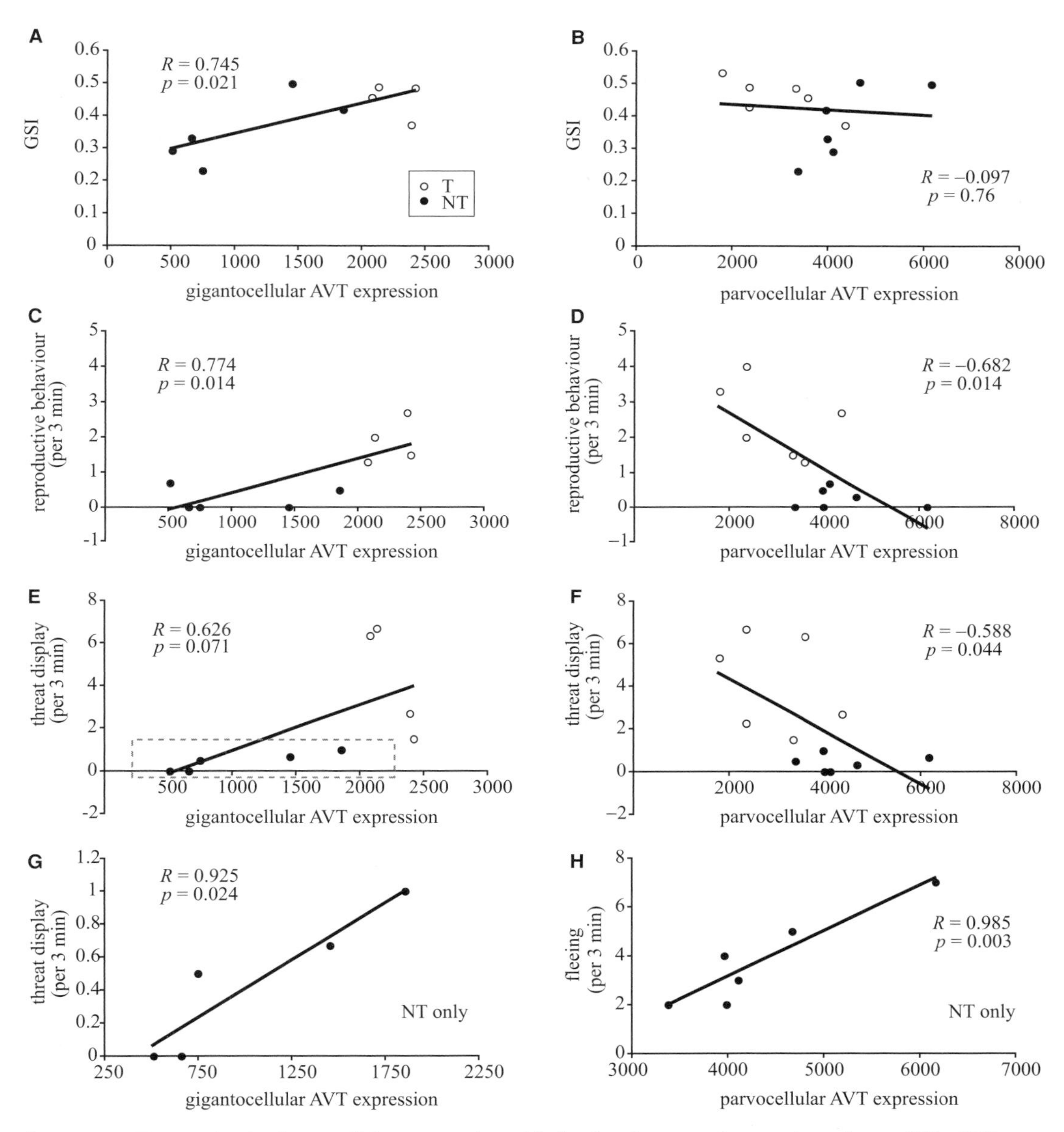

Figure 5.2 AVT expression in gigantocellular neurons is positively related to sex and aggression, whereas AVT mRNA expression in parvocellular cells with fleeing and the inhibition of sex and aggression. The figure shows the relationship between AVT mRNA expression in gigantocellular and parvocellular cells and gonadal size (GSI; A,B), reproductive behavior (C,D), threat displays in territorial (filled circles) and subordinate males (open circles; E,F), threat displays in territorial animals only (G), and fleeing (H). AVT expression in gigantocellular cells was positively correlated with gonadal size (A, $R = 0.75$, $p = 0.02$), reproductive behavior (C, $R = 0.77$, $p = 0.01$), and aggression, particularly in territorial males (G, $R = 0.93$, $p = 0.02$). In contrast, AVT expression in parvocellular neurons was negatively correlated with reproductive behavior (D, $R = -0.68$, $p = 0.01$) and threats (F, $R = -0.56$, $p = 0.04$), but positively correlated with fleeing (H, $R = 0.99$, $p = 0.003$). Reproduced with permission from Greenwood et al., 2008.

gigantocellular). Together, these studies suggest that caudal, magnocellular, and/or gigantocellular cells may be involved in the regulation of courtship and aggressive behavior in male teleosts, whereas parvocellular cells may be involved in the regulation of social withdrawal and/or subordinance (see further discussion below). In contrast, a study in false clown anemonefish did find that subordinate animals had more magnocellular AVT cells, and not parvocellular cells, after correcting for size differences (Iwata et al., 2010). It should be noted, though, that this study was done during a period of socially regulated sexual differentiation. Soon after the time when fish were sacrificed, dominant animals would have become females, second-ranked animals would have become males, and subordinates would have remained in a non-reproductive state. In fact, there was a strong trend for dominant and second-ranked individuals to have begun sexual differentiation at the time of sacrifice. Thus, the different numbers of magnocellular cells found in animals that differed in dominance status could have been related to their state of sexual maturation, and not dominance status, per se. Nonetheless, this study does indicate that it is not yet possible to make the generalization that AVT magno- and/or gigantocellular cells are associated with the production of male-typical aggression or courtship and parvocellular AVT neurons with the regulation of subordinance/social withdrawal, at least not across all developmental time points or in all species.

Even more radical differences in reproductive strategies between males in some teleosts – whether to adopt a male-typical phenotype or mimic females – are associated with variation in AVT neuronal phenotypes. In plainfin midshipmen, some males defend territories, court females, and care for the young, whereas others look and act like females, at least prior to spawning. This allows these female mimics to remain near the territorial males while they court real females and, when the pair enters the nest to spawn, to follow and release their own sperm, thus gaining fertilizations through sperm competition mechanisms. These "sneaker" males, which become sexually mature sooner than territorial males in this species, have more AVT immunoreactive cells, after correcting for body size, than territorial males or females (Foran and Bass, 1998). Likewise, male rock-pool blennies that adopt a sneaker phenotype become sexually mature earlier than territorial males and have more and larger AVT neurons, again after correcting for body size, than older, territorial males (Miranda et al., 2003). The functional significance of these phenotypic differences between territorial and sneaker males in these species remains unclear, though the increased numbers of AVT cells have been hypothesized, as previously proposed for increased AVT production in precocious male masu salmon, to play a role in the earlier initiation of reproductive behavior in the alternative, sneaker phenotypes. The adoption of the sneaker phenotype, at least in plainfin midshipmen, is also associated with decreased AVT sensitivity within the circuit that controls social vocalizations; central AVT infusions in territorial males inhibit neural responses that drive the contractions of the swim bladder that produce social vocalizations, but do not produce the same effect in the sneaker males, nor in females (Goodson and Bass, 2000a). In contrast, infusions of an AVPR1a antagonist stimulate those neural responses, but again only in territorial males. However, the contexts in which AVT normally modulates such responses have not yet been established, as these effects were observed in response to electrical stimulation of the brain in immobilized fish.

AVT also appears to be associated with the ability of some males in at least one species to adopt female-typical behaviors. Even though older, territorial male peacock blennies and younger sneaker males have similar numbers of AVT neurons (and both have more than females), the sneaker males, like females, have higher levels of AVT gene expression within those neurons than the territorial males (Grober et al., 2002). These males, like females, court territorial males. These findings suggest, in contrast to the more common idea that AVT/AVP are typically associated with the regulation of male-typical reproductive behaviors, that increased AVT

synthesis and, presumably, release, may actually stimulate female courtship behavior in this species. In fact, the administration of AVT does increase the courting of territorial males by sneaker males and tends to have the same effect in females, but it does not affect the courtship of females by territorial males (Carneiro et al., 2004). Interestingly, AVT can affect steroid synthesis in the ovaries in catfish (Singh and Joy, 2009b), suggesting the peptide could have indirect effects on female patterns of reproductive behavior in some species by changing levels of circulating steroids that ultimately reach the brain. If so, this could represent a novel mechanism through which peripheral peptides could influence social behavior in vertebrates.

Perhaps the most extreme cases in which variation in reproductive tactics have been linked to AVT neuronal phenotypes occur in species that exhibit a dramatic type of behavioral flexibility, adult sex change. Marine gobies can switch back and forth between male and female phenotypes as adults, depending on social context, and AVT cells are larger in fish that have adopted female phenotypes than in fish adopting male phenotypes (Grober and Sunobe, 1996). While these findings suggest that changes in the AVT system may contribute to changes in sexual phenotype, they are difficult to interpret because the functional significance of differences in cell size remains uncertain. A clearer picture has emerged from studies in bluehead wrasse, in which combined *in situ* hybridization and pharmacological approaches have more clearly established the relationship between changes in the AVT system and sex-role reversal. Within minutes to hours of the disappearance or death of the dominant male on a territory, the largest female begins to exhibit increased aggression towards males and to court other females. Eventually, the female develops testes and becomes a reproductively functional male, but levels of AVT gene expression in magnocellular neurons of the preoptic area increase in females during the early stages of behavioral sex change, before testes develop (Godwin et al., 2000), and the administration of an AVP1a antagonist prevents females from taking over a male's territory if he is removed (Semsar and Godwin, 2003). Thus, the increased AVT activity observed during sex change appears necessary for the emergence of male-typical behavior patterns. However, those changes in AVT expression do not appear sufficient to drive the behavioral sex role reversal, as AVT injections do not induce the adoption of territorial courtship and/or aggressive behaviors in females, even when territorial males are removed from the territory.

While studies demonstrating sex differences in AVT phenotypes, variation in AVT phenotypes that are related to alternative male phenotypes, and changes in AVT phenotypes that are associated with adult sex change highlight AVT functions related to the regulation of sex- and/or phenotype-specific reproductive behaviors, there are species in which the peptide may play similar roles in the regulation of social behaviors in both sexes. For example, paired male and female *Neolamprologus pulcher*, a cooperatively breeding cichlid in which both sexes participate in territorial defense, both have higher levels of AVT gene expression than subordinates (Aubin-Horth et al., 2007). Likewise, dominant female zebrafish, like dominant males, express more AVT mRNA in the preoptic area than subordinates, and AVT administration similarly affects aggression in both sexes (Filby et al., 2010, see further discussion below). Finally, AVT similarly inhibits, and an AVPR1a antagonist stimulates, the approach of conspecifics during the breeding season in male and female goldfish (Walton et al., 2010). Although AVT/AVP are often portrayed as mediators of male behavior and OT-related molecules as mediators of female behaviors, these studies are consistent with those beginning to emerge in species from other vertebrate groups, particularly mammals, showing that such classifications may not be as rigid as once thought. For example, AVP stimulates maternal behaviors in female rats (Bosch and Neumann, 2008) and may, in some contexts, stimulate affiliative responses in human females (Thompson et al., 2006), and OT has numerous effects on social functions in males (reviewed in Goodson and Thompson, 2010). Together, then, studies in teleosts indicate that AVT has functions related to

reproductive regulation across sexes and phenotypes, and they highlight how plasticity within the AVT system may underlie the evolution of flexible reproductive tactics.

5.4 Seasonal/cyclical changes in AVT neuronal phenotypes

5.4.1 In males

Seasonal changes in AVT systems in males in several teleost species, like the male-biased sex differences discussed above, suggest an involvement in the regulation of male reproductive functions, including behavior, though such interpretations must be made cautiously because seasonal changes in reproductive functions are often confounded with changes in environmental conditions that may influence AVT systems. However, in precocious male masu salmon that spawn at 1 year of age instead of 3, maximal levels of AVT expression occur in autumn, when the fish would normally spawn, despite being maintained in water conditions that minimize changes in salinity (Ota et al., 1999). Thus, a developmental trigger, likely in the form of large increases in steroid production the previous spring, appears to alter the development of the AVT system such that it peaks at the height of the early breeding season for those fish. Likewise, total brain AVT content is highest early in the breeding season and then declines across the season and reaches nadir levels outside of the breeding season in 3-spined sticklebacks (Gozdowska et al., 2006) and air-breathing catfish (Singh and Joy, 2008). In the catfish, the numbers of AVT neurons and the density of fibers in the neurohypophyseal tract also increase during the breeding season, and plasma levels of AVT actually peak during spawning, suggesting AVT may affect peripheral processes related to reproduction in this species. Similarly, male damselfish have more AVT-immunoreactive gigantocellular neurons during peak reproductive season than at any other time of year, suggesting these specific neurons may be involved in courtship and/or nest defense in this otherwise gregarious, shoaling species (Maruska, 2009). Additionally, fiber densities in the vagal motor nucleus (VMN) of the hindbrain increase during the breeding season in damselfish, and that increase is correlated with the increased numbers of gigantocellular neurons, suggesting those neurons may be the source of projections to that target. Although the behavioral significance of AVT release in the VMN is unknown in this species, AVT likewise innervates the VMN in goldfish (Thompson and Walton, 2009), and like AVT innervation of the VMN in damselfish, the expression of AVPR1a-like receptors in the goldfish hindbrain, including the VMN, is highest during the breeding season (Walton et al., 2010). In goldfish, hindbrain AVT appears to be involved in the regulation of social spacing in reproductive contexts (Thompson et al., 2008, see further discussion below), which could likewise occur in male sergeant damselfish.

Changes in AVT levels have also been observed in association with paternal functions in at least two teleost species, suggesting that it, like AVP in mammals (Wang et al., 1994), may also participate in the regulation of male parental behavior. In northern and dusky pipefish (in the same family as seahorses), males take exclusive care of the young in brood pouches, and in both species AVT levels increase in males during brooding (Ripley and Foran, 2010). Although it is unclear if AVT is related to physiological or behavioral demands associated with parenting in males in these species, this study is, to our knowledge, the first outside of mammals suggesting a role for AVT in paternal functions.

5.4.2 In females

There are also changes in AVT systems across reproductive seasons and/or cycles in females in some species. In female chum salmon, levels of AVT gene expression decrease during the final stages of spawning (Hiraoka et al., 1997). Although these changes could result from changing osmoregulatory demands on the fish as they move from salt to fresh water, this is unlikely because similar decreases in AVT gene expression are not observed in males

during the same migration. However, decreases are observed in females that have not yet ovulated as well as in those that have, so these changes do not appear tightly linked to the reproductive cycle. On the other hand, a tight link between nonapeptide expression levels and spawning does occur in female medaka, in which there was a decrease in the number of AVT/IT immunoreactive parvocellular neurons (the AVT antibody cross-reacted with IT in this study) immediately after spawning in controlled experiments in which females were kept in constant osmotic conditions (Ohya and Hayashi, 2006). Similarly, AVT gene expression is elevated in prespawning and spawning female grass pufferfish relative to females captured outside of the breeding season (Motahashi et al., 2008), and levels of brain AVT are elevated in prespawning periods of the breeding season in female air-breathing catfish, as are the numbers of preoptic AVT cells and the density of fibers in the neurohypophoseal tract (Singh and Joy, 2008). As in males of this species, plasma levels of AVT also peak at spawning. The similar seasonal changes in AVT profiles in males and females of this species suggest the peptide plays a role in the regulation of reproductive functions that are common to both sexes. AVT gene expression also increases during the breeding season in female 3-spined sticklebacks (Gozdowska et al., 2006), and the number of AVT immunoreactive gigantocellular cells increases in female sergeant damselfish during the breeding season, as do fiber densities in the torus semicircularis, an auditory region in the midbrain where AVT could modulate the processing of male reproductive vocalizations (Maruska, 2009). Together, these studies suggest that AVT plays a role in the regulation of female spawning processes, potentially including behavior, in numerous species. However, AVT has thus far only been shown to affect female-typical patterns of reproductive behavior in one species, the rock-pool blenny (Carneiro, et al., 2004), and in that study AVT significantly affected such behaviors in sneaker males that imitate females, but not in actual females. Clearly, additional work testing AVT effects on female behavior in this and other species is needed.

More complex patterns of seasonal change in AVT cell numbers and size have been observed in several species that suggest functions for this peptide may extend beyond the regulation of reproductive processes. In half-spotted gobies, the number and size of AVT-immunoreactive neurons increase during the peak breeding season and during the nonreproductive months relative to the immediate pre- and postspawn periods in some cell populations, primarily in females (Maruska et al., 2007). Similarly, the numbers of AVT-immunoreactive parvocellular cells increase outside of the breeding season in male damselfish (Maruska, 2009), as do the numbers of AVT-immunoreactive gigantocellular cells in female milletseed butterflyfish (Dewan et al., 2008). Finally, male and female plainfin midshipmen captured in the winter have AVT immunoreactive cells in the anterior tuberal hypothalamus, whereas fish captured during the spring/summer breeding season do not (Goodson et al., 2003). Together, these studies suggest that some AVT cell populations could influence non-reproductive behaviors outside of the breeding season, although it is also possible that some or all of these changes reflect responses to changing environmental conditions that are important for physiological regulatory processes, or that reproductive-related functions during the breeding season lead to increased peptide-release patterns that actually make it more difficult to visualize the peptide at those times of year with immunohistochemical methods. Very few studies in teleosts have examined AVT effects on behaviors outside of reproductive seasons/contexts, though it could be an exciting area to explore. To our knowledge, the only study that has done this was done in goldfish, in which AVT appears to lose its effects on social approach behavior outside of the breeding season (Walton et al., 2010; see further discussion below).

5.5 Steroid sensitivity of AVT neuronal phenotypes

Although numerous studies have noted the relationships between changing steroid levels across

reproductive seasons/cycles and changes in AVT/IT phenotypes, only a few have carefully examined the precise timing of those relationships or directly manipulated steroid levels to causally determine their effects on these peptide systems in fish. AVT gene expression in magnocellular and gigantocellular preoptic cells increases at the time when testosterone and estradiol reach peak levels in normally developing male masu salmon (Ota et al., 1999). In contrast, AVT expression levels are dissociated in time from increasing steroids in precocious maturing males, and rather occur at the time of spawning, months after a dramatic peak in steroid secretion that likely triggers the developmental changes associated with early maturation. This dissociation is important because it suggests that the AVT system may be programmed by prior patterns of steroid exposure. Such mechanisms could, in part, explain the different developmental trajectories observed within some sexes in some teleosts. Increased levels of testosterone are also closely associated in time with increased AVT gene expression in dominant males and females in the cooperatively breeding African cichlid, *N. pulcher* (Aubin-Horth et al., 2007), which suggests that similar gene sets, including AVT, may be upregulated by testosterone to increase aggression in males and females in some teleosts.

Surprisingly, sex steroids have been directly shown to affect the AVT system in only one species, the air-breathing catfish. In this species, ovariectomy decreases brain and isolated ovarian AVT in female catfish, an effect that can be reversed by low doses of estradiol (Singh and Joy, 2009a). Interestingly, the gonads can produce AVT in fish (Bobe et al., 2006; Singh and Joy, 2008), and its local production in ovaries can be increased by the *in vitro* application of estradiol, as well as progesterone. In contrast, AVT systems appear unresponsive to steroids in several of the other species in which direct steroid manipulations have been performed. For example, androgen manipulations that masculinize reproductive behavior in female goldfish do not affect the density of AVT cells in this species (Parhar et al., 2001). Of course, this is not surprising because the AVT system is not sexually dimorphic in this species, and AVT appears to similarly affect behaviors in males and females (Walton et al., 2010). There are differences in the number and size of AVT cells between male and female peacock blennies, yet 11-ketotestosterone, a non-aromatizable androgen that stimulates male-typical reproductive behaviors in females and inhibits female-typical behaviors in territorial males, does not affect AVT cell number or size (Oliveira et al., 2001). Similarly, although there are sexual dimorphisms in AVT gene expression in bluehead wrasse, these differences are not a function of circulating steroids, as they, like behavioral dimorphisms in this species, persist even after gonadectomy (Semsar and Godwin, 2003). Rather, social changes drive the sex differences in AVT gene expression in this species, although there do appear to be effects of testicular secretions on the size of gigantocellular AVT neurons, which is dependent on the presence of the testes. Manipulations of 11-ketotestosterone have no effect on the size of those neurons, suggesting their size may be influenced by the aromatization of testosterone into estradiol or by other gonadal factors. However, the behavioral significance of hormonally mediated changes in the size of these neurons is unclear, because male reproductive behaviors are not dependent on the gonads in bluehead wrasse. Nonetheless, the differential effects of gonadal secretions and social cues indicate that some parameters of the AVT neuronal system are differentially regulated in this and perhaps other species.

5.6 Comparison of the behavioral effects of AVT across fish species

As discussed above, anatomical and gene expression studies highlight the potential role that AVT plays in the regulation of behaviors related to reproduction, and in some cases pharmacological manipulations have provided causal evidence that AVT does play such a role. However, as has been observed in other vertebrates, AVT has different effects on

reproductive behaviors across, and sometimes even within, some species. We will first summarize those diverse findings, and then relate them to a model that has been proposed to account for context-dependent nonapeptide behavioral effects in vertebrates, a model that has, in part, been shaped by these very studies.

The first effects of AVT on behaviors related to reproduction were actually observed in a teleost, the killifish, in which AVT infusions were shown to stimulate stereotypical swimming patterns associated with a spawning reflex (Pickford and Strecker, 1977; Wilhemi, 1956). AVT has since been shown to stimulate male courtship behaviors in several species, including white perch (Salek et al., 2002) and bluehead wrasse (Semsar et al., 2001). Although exogenous AVT does not affect courtship in male damselfish, an AVPR1a antagonist does inhibit such behaviors, indicating that endogenous AVT does play a role in the display of male courtship in this species (Santangelo and Bass, 2010). AVT administration also selectively stimulates electrical signals used for courtship communication in two weakly electric South American fish from the family Gymnotiformes, *Apteronotus leptorhynchus* (Bastian et al., 2001) and *Brachyhypopomus gauderio* (Perrone et al., 2010). In the latter study in *B. gauderio*, an AVPR1a antagonist partially blocked the effects of AVT and, more impressively, completely blocked the increase in electrical discharges typically produced by males in the presence of a female, thereby demonstrating that endogenous AVT normally stimulates courtship communication. That AVT stimulates various forms of courtship behavior across numerous, unrelated teleosts, as well as in other vertebrates (reviewed in Goodson and Bass, 2001), suggests that such functions were present in the common ancestors of modern jawed vertebrates, if not earlier, as AVT-related peptides also play a role in reproductive regulation in invertebrates; (Fujino et al., 1999; Wagenaar et al., 2010). Additionally, a non-selective nonapeptide antagonist was recently shown to inhibit affiliative behaviors related to pair bonding in male convict cichlids, *Amatitlania nigrofasciata*, suggesting AVT may, like AVP in male prairie voles, play a role in pair-bond formation in this teleost (Oldfield and Hofmann, 2011). If so, then similar nonapeptide mechanisms independently evolved to facilitate pair bonding in these distantly related organisms. However, it should be noted that the antagonist delayed, but did not prevent, pair-bond formation in the convict cichlids, and the antagonist was not selective for AVT receptors, so further work needs to be done to more clearly elucidate its and/or IT's role in pair bonding in this species.

It should also be noted that AVT does not uniformly stimulate male courtship responses. In several species, AVT administration has no effect on male courtship behaviors, including pupfish (Lema and Nevitt, 2004a) and peacock blennies (Carneiro et al., 2004), nor does it affect electric organ discharges in reproductive (or other) contexts in another weakly electric Gymnotiform, *Gymnotus omarorum* (Perrone et al., 2010). Although all of these studies used intraperitoneal injections, which may be less effective than central infusions, all of them either found that the same manipulations did affect aggressive behavior (Lema and Nevitt, 2004a, see further discussion below), female courtship (Carneiro et al., 2004) or, in the case of the study by Perrone et al. (2010) in *G. omarorum*, that the same manipulation stimulated courtship vocalizations in a related species *B. gauderio*. The finding that AVT differentially affects courtship vocalizations in two related Gymnotiformes is particularly remarkable because it suggests that AVT regulatory mechanisms have changed in association with the evolution of different reproductive strategies; *G. omarorum* is territorial and highly aggressive, whereas *B. gauderio* is typically gregarious, though males do act aggressively towards other males in reproductive contexts. AVT can even have different effects on reproductive communication in animals with different behavioral phenotypes within a sex in a single species, as the example discussed earlier in plainfin midshipmen, in which central AVT affects social vocalizations in territorial males, but not in sneaker males or females,

illustrates (Goodson and Bass, 2000a). Thus, sensitivity of the AVT system appears to have diverged between sexes and even phenotypes within a sex in this species.

AVT's different effects in species with different life histories and in animals that have adopted different reproductive phenotypes is generally consistent with avian studies demonstrating different patterns of AVT neuronal activation in response to social stimuli and different levels of AVPR1a receptor binding in the brains in species with different patterns of social organization (Goodson et al., 2006; Goodson and Wang, 2006), as well as studies directly showing that AVT has different effects on aggression across such species, stimulating it in gregarious, colonial zebra finches but inhibiting it in territorial field sparrows (Goodson, 1998; Goodson and Adkins-Regan, 1999). However, more recent work has shown that AVT can produce different effects on aggression within the same species in different social contexts, and that it can similarly affect aggression in species with different social organizations, as long as the social contexts associated with the tests are similar (Goodson et al., 2009; Kabelik et al., 2009). These recent findings suggest that some caution should be used when interpreting differential AVT effects across species or phenotypes in studies that do not take the social context in which tests are conducted into account. For example, AVT may not have affected electric organ discharges in *G. omarorum* because the fish were removed from their territories before testing, and AVT may modulate social communication in this species only in territorial contexts. Together, these studies in teleosts, like those in birds, indicate that AVT has a range of effects on courtship-related behaviors, with the variability across species and/or phenotypes likely to reflect differing social contexts in which tests are done and/or differences in patterns of AVT sensitivity in brain circuits that mediate courtship-related behavior.

Several of the studies just discussed are also remarkable for beginning to elucidate where within the brain this molecule exerts its effects on reproductive behaviors in teleosts. In killifish, AVT's ability to stimulate courtship behaviors may be mediated within the preoptic area, as its effect disappears in animals in which the that area is ablated (Macey et al., 1974). However, it is also possible that a functional preoptic area is simply necessary for the behavior to occur, even if AVT's ability to stimulate the behavior occurs elsewhere. The specific central sites where AVT affects social communication processes have been more clearly established. The electrically induced firing of motor neurons that stimulate fictive social vocalizations in territorial plainfin midshipmen is inhibited by direct infusions of AVT into the anterior hypothalamus (Goodson and Bass, 2000a) and, further back in the brain, the paralemniscal midbrain, a vocal-acoustic processing area (Goodson and Bass, 2000b). Effects have also been observed even more caudally in *B. gauderio*, in which the direct application of AVT onto hindbrain pacemaker neurons that drive the electric signals in an *in vitro* slice preparation increased their firing rates (Perrone et al., 2010). These studies in plainfin midshipmen and *B. gauderio*, both species with highly derived social communication systems, in one case contractions of the swim bladder that produce a hum and in the other the production of an electric field around the animal's body, indicate that AVT can directly influence descending motor output pathways associated with the generation of stereotypical patterns of social behavior. Similarly, AVT can affect hindbrain sensorimotor processes associated with the regulation of a stereotypical courtship behavior in at least one other vertebrate, the roughskin newt (Rose and Moore, 2002), suggesting that the elaboration of descending AVT projections in responses to social pressures associated with reproduction is not unique to fish. The sites where AVT influences other courtship behaviors, including affiliative behavioral responses in monogamous cichlids, have not yet been identified. Recently, the ventral and lateral portions of the ventral telencephalon of the teleost forebrain have been homologized to the septum (O'Connell and Hofmann, 2011), a site where AVP produces some affiliative effects in male prairie voles (Liu et al., 2001), and AVT immunoreactive

fibers are more dense in this area in monogamous than polygynous butterfly fish (Dewan et al., 2011). Hopefully, future studies will be able to specifically manipulate AVT within this forebrain area to see if it too is where AVT affects affiliative behaviors in teleosts. AVP also promotes affiliative responses related to pair bonding via actions in the ventral pallidum in male prairie voles (Lim and Young, 2004), but a putative homolog of the ventral pallidum has not yet been identified in fish.

AVT also affects aggression in reproductive contexts, as it and AVP do in numerous other vertebrates. Thus, AVT stimulates aggression in male damselfish, whereas a AVP1a antagonist inhibits such behavior (Santangelo and Bass, 2006), as does a non-selective nonapeptide antagonist in male convict cichlids during pair-bond formation (Oldfield and Hofmann, 2011). IP injections of AVT also stimulate male-typical patterns of aggression in non-territorial male bluehead wrasse that otherwise display low levels of aggression, but an AVPR1a antagonist does not inhibit aggression in territorial males that typically display high levels of aggression (Semsar et al., 2001). In fact, AVT tends to decrease aggression in such males, suggesting that AVT has different effects on aggression in the two phenotypes. IP injections of AVT also inhibit the production of electric signals typically emitted during agonistic interactions in the gymnotiforme *A. leptorhynchus* (Bastian et al., 2001), as well as overt aggression in male river pupfish (Lema and Nevitt, 2004a), a polygynous, territorial species, and in male and female zebrafish (Filby et al., 2010), a less-aggressive, group-living species in which males and females only show aggression during the formation of dominance hierarchies. However, as in territorial male bluehead wrasse, an AVPR1a antagonist does not affect aggression in male pupfish, nor does it block exogenous AVT's inhibition of aggression in zebrafish. Expression of a putative AVPR1b-related receptor was higher in dominant male and female zebrafish in the study by Filby et al. (2010), leading them to suggest that AVT's ability to inhibit aggression in this species, and perhaps other teleosts, could be mediated by an AVPR1b-like receptor. AVPR1b receptors do appear to be involved in the regulation of aggression in mice, though in that species the effects of its activation are stimulatory (Wersinger et al., 2002).

Thus, no clear pattern has yet emerged that would explain why AVT sometimes stimulates and sometimes inhibits aggressive behavior in different species or even different phenotypes within the same species, nor has the role of endogenous AVT in aggressive modulation been conclusively determined in some species, even some in which exogenous AVT does affect aggressive responses. It is possible that the levels of endogenous release within the system at the time of testing are critical, and that those have differed in the individuals tested in different experiments, perhaps as a function of the contexts in which the tests have been conducted. Thus, exogenous AVT may stimulate aggressive responses if endogenous release is low, but if endogenous release is high the "extra" AVT may lead to levels of the peptide that inhibit aggression relative to control animals already releasing AVT at rates that maximally stimulate such responses. In fact, such an inverted U dose–response relationship for AVT's effects on aggression has been demonstrated in at least one teleost species (Santangelo and Bass, 2006). Of course, antagonist manipulations would fail to affect aggression in tests done in fish with low endogenous AVT release. Furthermore, the role that endogenous AVT plays in aggressive modulation may have been overlooked in some species because most studies have used a mammalian AVPR1a selective antagonist (Manning compound) and, as proposed by Filby et al. (2010), receptors other than AVPR1a may mediate AVT's effects on aggression in fish. Those receptors might not only include AVPR1b-like receptors, as they proposed but that have not yet been fully sequenced in any teleost, but also specific patterns of activation of the AVPR1a1 or AVPR1a2 subtypes that have been identified in several teleosts and which Manning compound may not equally antagonize. Finally, it is possible that AVT influences on aggression are, in some cases, not the result of direct effects within circuits that control stereotypical aggressive responses, but rather

indirect effects resulting from actions in other circuits that mediate approach/avoidance responses. Consistent with this possibility, the central administration of AVT in male rainbow trout does not directly affect aggression, but does inhibit the establishment of dominance (Backstrom and Winberg, 2009). Although an AVPR1a antagonist does not increase the probability that an animal will become dominant, it does increase the time it takes for a losing animal to retreat from the interaction. These data suggest that endogenous AVT does not directly influence stereotypical patterns of aggression in this species, but rather that it may promote social withdrawal in aggressive contexts. A similar mechanism could, in part, explain AVT's inhibition of aggression in some species, depending on how the tests are conducted. For example, such a mechanism might not affect measures of overt aggression in laboratory pair tests in which withdrawal is not an easy option, but would dramatically reduce aggression in field tests in which more escape options are available. This also highlights the possibility that AVT could differentially affect offensive and defensive forms of aggression, distinctions that are commonly made in the mammalian literature but that are rarely considered in tests of aggression in fish.

In fact, there are likely multiple AVT circuits that influence different social responses in fish, and the behavior of an animal in a given social interaction would therefore depend upon the relative activity within those circuits. Support for this idea comes from the previously discussed studies in cichlids, in which AVT mRNA expression in gigantocellular AVT cells is positively correlated with levels of courtship and aggression in males, but levels of AVT mRNA expression within parvocellular AVT cells are positively correlated with fleeing, a subordinate behavioral phenotype, and negatively with courtship and aggression (Greenwood et al., 2008). Most of the behavioral pharmacology studies done in fish have used intraperitoneal or, in some cases, intraventricular methods of delivery, both of which likely target multiple circuits simultaneously and thus are unlikely to reflect the normal balance of activity within them. The differing behavioral effects of exogenous AVT administration across fish species, sexes, and/or phenotypes may therefore also depend on which circuit is most sensitive within a given species, sex or phenotype, and context-dependent effects may arise if different levels of activation occur within those circuits in different social contexts. A similar explanation has been proposed to account for the variable effects of AVT in birds (Goodson and Kabelik, 2009).

Of course, we have discussed AVT effects on several courtship and aggressive responses in numerous teleosts that are similar to effects that have been observed in species from several other vertebrate groups. Thus, the idea that AVT/AVP circuits directly modulate stereotypical patterns of courtship and aggression is widely accepted. There are many fewer studies showing that these peptides can also affect social withdrawal responses potentially related to fleeing and/or subordinance. However, our lab has demonstrated that central infusions of AVT do promote social withdrawal in male and female goldfish, whereas an AVPR1a antagonist stimulates social approach (Thompson and Walton, 2004; Walton et al., 2010). Furthermore, the AVT pathway to the hindbrain, which is one of the most dense central VT projections in this species (Thompson and Walton, 2009), appears to mediate this effect, as fourth ventricle infusions, close to the terminal zone of that pathway, are more potent than third ventricle infusions (Thompson et al., 2008). Surprisingly, these behavioral effects appear to depend on changes in body state likely induced by interactions with peripherally projecting Substance P cells in the dorsal motor vagus, which AVT terminals encapsulate (see Figure 5.3), as the peripheral blockade of tachykinin receptors completely blocks central AVT's behavioral effects, whereas the central infusion of a tachykinin antagonist does not. The circuit is most sensitive in fish in full reproductive condition (Walton et al., 2010), indicating that while AVT affects a simple social response common to both sexes, it does so in reproductive contexts, perhaps regulating the spacing of animals within a group during the breeding season. Given the conserved nature of hindbrain projections, it is

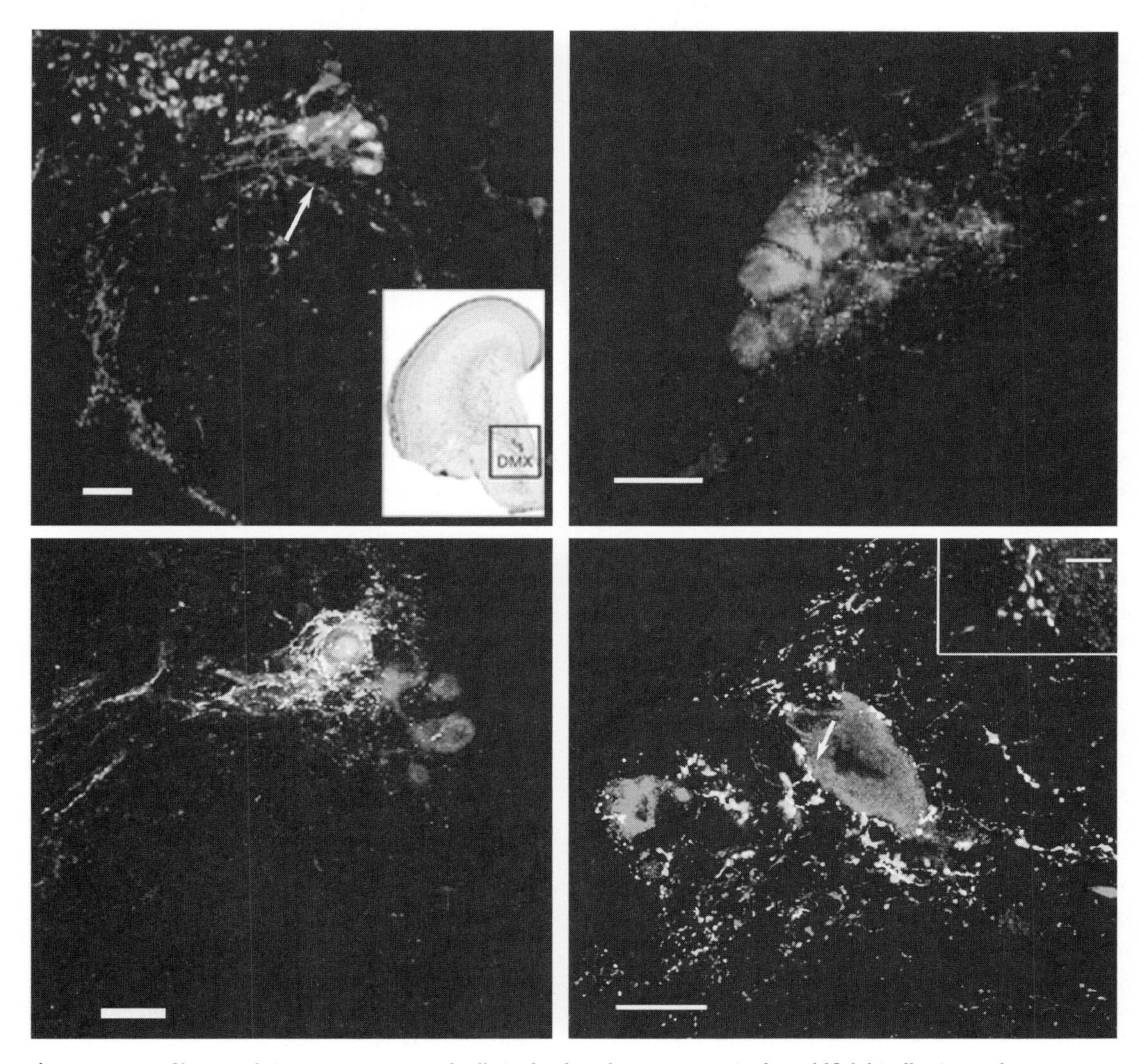

Figure 5.3 AVT fibers (red) innervate a group of cells in the dorsal motor vagus in the goldfish hindbrain (A) that are backfilled by intraperitoneal injections of a retrograde tracer (B; backfilled cells are blue/gold), indicating they project out the vagus nerve into the periphery. Those cells are immunoreactive for Substance P (C; Substance P cells are shown in red, AVT fibers in green). AVT fibers appear to directly contact the cell bodies of those neurons (D). From Thompson et al., 2008. Blocking peripheral Substance P receptors prevents central AVT from inhibiting social approach behavor, suggesting this connection is part of a central/peripheral feedback loop involved in social regulation in this species (see text). See color version in plates section.

possible that similar mechanisms involving central-peripheral feedback loops initiated by peptide influences in the hindbrain that affect social withdrawal may be widespread not only in other fish, but possibly across vertebrate taxa. Consistent with this possibility, paraventricular AVP neurons in mammals, which contribute projections to hindbrain autonomic regulatory regions (Nilaver et al., 1980; Sawchenko and Swanson, 1982), are activated by social stressors, and their activation is

correlated with subordinate behaviors (Goodson et al., 2005).

5.7 IT neuronal phenotypes and behavioral functions

Very few studies have determined if there are differences in IT neuronal phenotypes related to sex and/or alternative reproductive strategies in fish, if there are changes in IT parameters across reproductive seasons, if sex steroids influence IT systems, or if IT manipulations affect social behavior. We have therefore combined our discussion of such studies into a single section. Female *Lythrypnus dalli*, a sex-changing species, have more IT cells in the preoptic area than males (Black et al., 2004), and the numbers of IT cells appear to decrease as females change into males, a process that takes about a week. In female and "sneaker" male plainfin midshipmen, but not territorial males, IT administration inhibits neuronal responses related to the production of male-typical vocalizations, whereas an OT-receptor antagonist increases such responses (Goodson and Bass, 2000a). In 3-spined sticklebacks and goldfish, IT gene expression increases at the height of the breeding season in females, but not in males (Gozdowska et al., 2006; Zhang et al., 2009). Together, these studies suggest IT may stimulate female reproductive behaviors and/or inhibit male behaviors in some teleost species. Interestingly, peripheral IT injections in female goldfish can upregulate estradiol production (Mennigen et al., 2008), indicating that IT could affect such processes indirectly in some species by affecting levels of circulating estrogens. On the other hand, there are species like the grass pufferfish in which no sexual dimorphisms or seasonal changes in IT gene expression have been observed (Motohashi et al., 2008).

Seasonal changes in IT content in the brains of salmon suggest this peptide may also play a role in the regulation of reproductive processes in males. In precocious male masu salmon, the amount of IT in preoptic cells, but not levels of IT gene expression, appear to peak at the time of year (autumn) when they would normally spawn (Ota et al., 1999). The dissociation between peptide content and gene expression may reflect lower release rates and thus peptide accumulation in these cells at the time of year when spawning would normally occur. If so, this would either mean that IT is not involved in the regulation of spawning, or that its release is typically driven by social/environmental stimuli associated with spawning, which fish in this study were not exposed to because they were housed in single-sex groups until sacrifice. Social control of IT release in salmon would be consistent with studies showing that the release of OT can be socially regulated in mammals (Feldman et al., 2010).

No studies have directly tested the effects of steroid manipulations on IT gene expression or peptide production, but IT gene expression, like AVT gene expression, is highest in normally developing male masu salmon at times of year when testosterone and estradiol are highest (Ota et al., 1999). The same is not true in precocious males or females of the same species, suggesting there could be differences in the steroidal regulation of IT systems between alternative male phenotypes and sexes.

In sum, there is a limited amount of molecular, physiological and anatomical data suggesting IT may play a role in the regulation of reproductive processes, particularly in females, but there is not yet any causal evidence that IT affects reproductive behaviors in natural contexts in any species. On the other hand, intraventricular infusions of IT do appear to facilitate approach responses towards other males in male goldfish (Thompson and Walton, 2004), suggesting it could, like OT in humans (reviewed in Goodson and Thompson, 2010), have general affiliative effects, at least in this species. However, an OT/IT receptor antagonist was not used in that study, so the effects of endogenous IT remain unknown. Clearly, a great deal of work is needed to elucidate if and how IT affects social behavior in teleosts.

5.8 Conclusions and future directions

In this chapter we have summarized work done in fish that has begun to clarify the early stages of

nonapeptide gene evolution, outlined primitive nonapeptide circuits in fish brains that were likely the foundations upon which subsequent complexity in behavioral regulatory mechanisms evolved in vertebrates, and summarized some of the major effects of nonapeptides, particularly of AVT, on reproductive behaviors in fish. We have suggested that in fish, as in other vertebrates, there are likely distinct neural circuits with unique effects on social behaviors, and that these circuits are likely differentially developed in different species and/or phenotypes and differentially activated in different social contexts, thus giving rise to a complex variety of behavioral actions of AVT.

We would now like to highlight areas where we believe further research is necessary in fish. Although the nonapeptide cell populations and axonal projections have been generally described in numerous species, there are no tracing studies that conclusively demonstrate the precise projection pathways from individual cell populations. Additionally, there is only a single study showing AVT binding sites in the brain of one species (Moons et al., 1989), and a single study demonstrating the distribution of AVT-related receptor proteins and their gene expression in a single species (Kline et al., 2011) Clearly, the development of selective and sensitive techniques to visualize AVT binding and/or receptor expression in the brains of more species is needed. Furthermore, we know little about the social/environmental triggers that drive AVT and IT systems. Immediate early gene techniques have been developed in fish, but thus far none have been used to determine what stimuli activate AVT and IT neuronal populations and thus provide the contexts for endogenous peptide behavioral modulation. Additionally, although numerous pharmacological studies have now demonstrated an array of AVT behavioral effects, very few have actually determined where within the brain those effects are mediated. Site-specific infusions are difficult in many fish species due to their small size and aquatic habitat, but efforts to develop such approaches, or site-specific genetic manipulations, are needed in order to causally determine exactly where within the brain these peptides induce their complex behavioral effects. Finally, a great deal of work is needed to determine if and how IT affects social behavior in teleosts. We believe the coming years will be exciting ones as work progresses in these directions in fish, and that such work will ultimately help us better understand the fundamental mechanisms through which these molecules affect behavior in vertebrates, as well as how adaptations in these systems helped shape the evolution of complex patterns of social organization.

REFERENCES

Aubin-Horth, N., Desjardins, J. K., Martei, Y. M., Balshine, S., and Hofmann, H. A. (2007). Masculinized dominant females in a cooperatively breeding species. *Mol Ecol*, 16, 1349–1358.

Backstrom, T. and Winberg, S. (2009). Arginine-vasotocin influence on aggressive behavior and dominance in rainbow trout. *Physiol Behav*, 96, 470–475.

Bastian, J., Schniederjan, S., and Nguyenkim, J. (2001). Arginine vasotocin modulates a sexually dimorphic communication behavior in the weakly electric fish Apteronotus leptorhynchus. *J Exp Biol*, 204, 1909–1923.

Batten, T. F., Cambre, M. L., Moons, L., and Vandesande, F. (1990). Comparative distribution of neuropeptide-immunoreactive systems in the brain of the green molly, Poecilia latipinna. *J Comp Neurol*, 302, 893–919.

Black, M. P., Reavis, R. H., and Grober, M. S. (2004). Socially induced sex change regulates forebrain isotocin in Lythrypnus dalli. *Neuroreport*, 15, 185–189.

Bobe, J., Montfort, J., Nguyen, T., and Fostier, A. (2006). Identification of new participants in the rainbow trout (Oncorhynchus mykiss) oocyte maturation and ovulation processes using cDNA microarrays. *Reprod Biol Endocrinol*, 4, 39.

Bosch, O. J. and Neumann, I. D. (2008). Brain vasopressin is an important regulator of maternal behavior independent of dams' trait anxiety. *Proc Natl Acad Sci USA*, 105, 17139–17144.

Carneiro, L., Oliveira, R. F., Canario, A. V. M., and Grober, M. S. (2004). The effect of arginine vasotocin on courtship behavior in a blenniid fish with alternative reproductive tactics. *Fish Physiol Biochem*, 28, 241–243.

Dewan, A. K., Maruska, K. P., and Tricas, T. C. (2008). Arginine vasotocin neuronal phenotypes among congeneric territorial and shoaling reef butterflyfishes: species, sex and reproductive season comparisons. *J Neuroendocrinol*, 20, 1382–1394.

Dewan, A. K., Ramey, M. L., and Tricas, T. C. (2011). Arginine vasotocin neuronal phenotypes, telencephalic fiber varicosities, and social behavior in butterflyfishes (Chaetodontidae): Potential similarities to birds and mammals. *Horm Behav*, 59, 56–66.

Eaton, J. L., Holmqvist, B., and Glasgow, E. (2008). Ontogeny of vasotocin-expressing cells in zebrafish: selective requirement for the transcriptional regulators orthopedia and single-minded 1 in the preoptic area. *Dev Dyn*, 237, 995–1005.

Feldman, R., Gordon, I., Schneiderman, I., Weisman, O., and Zagoory-Sharon, O. (2010). Natural variations in maternal and paternal care are associated with systematic changes in oxytocin following parent-infant contact. *Psychoneuroendocrinology*, 35, 1133–1141.

Filby, A. L., Paull, G. C., Hickmore, T. F., and Tyler, C. R. (2010). Unravelling the neurophysiological basis of aggression in a fish model. *BMC Genomics*, 11, 498.

Foran, C. M. and Bass, A. H. (1998). Preoptic AVT immunoreactive neurons of a teleost fish with alternative reproductive tactics. *Gen Comp Endocrinol*, 111, 271–282.

Fujino, Y., Nagahama, T., Oumi, T., Ukena, K., Morishita, F., Furukawa, Y., Matsushima, O., Ando, M., Takahama, H., Satake, H., Minakata, H., and Nomoto, K. (1999). Possible functions of oxytocin/vasopressin-superfamily peptides in annelids with special reference to reproduction and osmoregulation. *J Exp Zool*, 284, 401–406.

Godwin, J., Sawby, R., Warner, R. R., Crews, D., and Grober, M. S. (2000). Hypothalamic Arginine Vasotocin mRNA Abundance Variation Across Sexes and with Sex Change in a Coral Reef Fish. *Brain Behav Evol*, 55, 77–84.

Goodson, J. L. (1998). Territorial aggression and dawn song are modulated by septal vasotocin and vasoactive intestinal polypeptide in male field sparrows (Spizella pusilla). *Horm Behav*, 34, 67–77.

Goodson, J. L. and Adkins-Regan, E., (1999). Effect of intraseptal vasotocin and vasoactive intestinal polypeptide infusions on courtship song and aggression in the male zebra finch (Taeniopygia guttata). *J Neuroendocrinol*, 11, 19–25.

Goodson, J. L. and Bass, A. H. (2000a). Forebrain peptides modulate sexually polymorphic vocal circuitry. *Nature*, 403, 769–772.

Goodson, J. L. and Bass, A. H. (2000b). Vasotocin innervation and modulation of vocal-acoustic circuitry in the teleost porichthys notatus [In Process Citation]. *J Comp Neurol*, 422, 363–379.

Goodson, J. L. and Bass, A. H. (2001). Social behavior functions and related anatomical characteristics of vasotocin/vasopressin systems in vertebrates. *Brain Res Brain Res Rev*, 35, 246–265.

Goodson, J. L., Evans, A. K., and Bass, A. H. (2003). Putative isotocin distributions in sonic fish: relation to vasotocin and vocal-acoustic circuitry. *J Comp Neurol*, 462, 1–14.

Goodson, J. L., Evans, A. K., Lindberg, L., and Allen, C. D. (2005). Neuro-evolutionary patterning of sociality. *Proc Biol Sci*, 272, 227–235.

Goodson, J. L., Evans, A. K., and Wang, Y. (2006). Neuropeptide binding reflects convergent and divergent evolution in species-typical group sizes. *Horm Behav*, 50, 223–236.

Goodson, J. L. and Kabelik, D. (2009). Dynamic limbic networks and social diversity in vertebrates: from neural context to neuromodulatory patterning. *Front Neuroendocrinol*, 30, 429–441.

Goodson, J. L., Kabelik, D., and Schrock, S. E. (2009). Dynamic neuromodulation of aggression by vasotocin: influence of social context and social phenotype in territorial songbirds. *Biol Lett*, 5, 554–556.

Goodson, J. L. and Thompson, R. R. (2010). Nonapeptide mechanisms of social cognition, behavior and species-specific social systems. *Curr Opin Neurobiol*, 20, 784–794.

Goodson, J. L. and Wang, Y. (2006). Valence-sensitive neurons exhibit divergent functional profiles in gregarious and asocial species. *Proc Natl Acad Sci USA*, 103, 17013–17017.

Gozdowska, M., Kleszczynska, A., Sokolowska, E., and Kulczykowska, E. (2006). Arginine vasotocin (AVT) and isotocin (IT) in fish brain: diurnal and seasonal variations. *Comp Biochem Physiol B Biochem Mol Biol*, 143, 330–334.

Greenwood, A. K., Wark, A. R., Fernald, R. D., and Hofmann, H. A. (2008). Expression of arginine vasotocin in distinct preoptic regions is associated with dominant and subordinate behaviour in an African cichlid fish. *Proc Biol Sci*, 275, 2393–2402.

Grober, M. S., George, A. A., Watkins, K. K., Carneiro, L. A., and Oliveira, R. F. (2002). Forebrain AVT and courtship in a fish with male alternative reproductive tactics. *Brain Res Bull*, 57, 423–425.

Grober, M. S. and Sunobe, T. (1996). Serial adult sex change involves rapid and reversible changes in forebrain neurochemistry. *Neuroreport*, 7, 2945–2949.

Gwee, P. C., Amemiya, C. T., Brenner, S., and Venkatesh, B. (2008). Sequence and organization of coelacanth neurohypophysial hormone genes: evolutionary history of the vertebrate neurohypophysial hormone gene locus. *BMC Evol Biol*, 8, 93.

Gwee, P. C., Tay, B. H., Brenner, S., and Venkatesh, B. (2009). Characterization of the neurohypophysial hormone gene loci in elephant shark and the Japanese lamprey: origin of the vertebrate neurohypophysial hormone genes. *BMC Evol Biol*, 9, 47.

Heierhorst, J., Mahlmann, S., Morley, S. D., Coe, I. R., Sherwood, N. M., and Richter, D. (1990). Molecular cloning of two distinct vasotocin precursor cDNAs from chum salmon (Oncorhynchus keta) suggests an ancient gene duplication. *FEBS Lett*, 260, 301–304.

Heierhorst, J., Morley, S. D., Figueroa, J., Krentler, C., Lederis, K., and Richter, D. (1989). Vasotocin and isotocin precursors from the white sucker, Catostomus commersoni: cloning and sequence analysis of the cDNAs. *Proc Natl Acad Sci USA*, 86, 5242–5246.

Hiraoka, S., Ando, H., Ban, M., Ueda, H., and Urano, A. (1997). Changes in expression of neurohypophysial hormone genes during spawning migration in chum salmon, Oncorhynchus keta. *J Mol Endocrinol*, 18, 49–55.

Hoheisel, G., Ruhle, H. J., and Sterba, G. (1978). The reticular formation of lampreys (Petromyzonidae) – a target area for exohypothalamic vasotocinergic fibres. *Cell Tissue Res*, 189, 331–345.

Hyodo, S., Ishii, S., and Joss, J. M. (1997). Australian lungfish neurohypophysial hormone genes encode vasotocin and [Phe2]mesotocin precursors homologous to tetrapod-type precursors. *Proc Natl Acad Sci USA*, 94, 13339–13344.

Iwata, E., Nagai, Y., and Sasaki, H. (2010). Social rank modulates brain arginine vasotocin immunoreactivity in false clown anemonefish (Amphiprion ocellaris). *Fish Physiol Biochem*, 36, 337–345.

Kabelik, D., Klatt, J. D., Kingsbury, M. A., and Goodson, J. L. (2009). Endogenous vasotocin exerts context-dependent behavioral effects in a semi-naturalistic colony environment. *Horm Behav*, 56, 101–107.

Kline, R. J., O'Connell, L. A., Hofmann, H. A., Holt, G. J., and Khan, I. A. (2011). The distribution of an AVT V1a receptor in the brain of a sex changing fish, Epinephelus adscensionis. *J Chem Neuroanat*, 42, 72–88.

Konno, N., Hyodo, S., Yamaguchi, Y., Kaiya, H., Miyazato, M., Matsuda, K., and Uchiyama, M. (2009). African lungfish, Protopterus annectens, possess an arginine vasotocin receptor homologous to the tetrapod V2-type receptor. *J Exp Biol*, 212, 2183–2193.

Konno, N., Kurosawa, M., Kaiya, H., Miyazato, M., Matsuda, K., and Uchiyama, M. (2010). Molecular cloning and characterization of V2-type receptor in two ray-finned fish, gray bichir, Polypterus senegalus and medaka, Oryzias latipes. *Peptides*, 31, 1273–1279.

Larson, E. T., O'Malley, D. M., and Melloni, R. H., Jr. (2006). Aggression and vasotocin are associated with dominant-subordinate relationships in zebrafish. *Behav Brain Res*, 167, 94–102.

Lema, S. C. (2010). Identification of multiple vasotocin receptor cDNAs in teleost fish: sequences, phylogenetic analysis, sites of expression, and regulation in the hypothalamus and gill in response to hyperosmotic challenge. *Mol Cell Endocrinol*, 321, 215–230.

Lema, S. C. and Nevitt, G. A. (2004a). Exogenous vasotocin alters aggression during agonistic exchanges in male Amargosa River pupfish (Cyprinodon nevadensis amargosae). *Horm Behav*, 46, 628–637.

Lema, S. C. and Nevitt, G. A. (2004b). Variation in vasotocin immunoreactivity in the brain of recently isolated populations of a death valley pupfish, Cyprinodon nevadensis. *Gen Comp Endocrinol*, 135, 300–309.

Lim, M. M. and Young, L. J. (2004). Vasopressin-dependent neural circuits underlying pair bond formation in the monogamous prairie vole. *Neuroscience*, 125, 35–45.

Liu, Y., Curtis, J. T., and Wang, Z. (2001). Vasopressin in the lateral septum regulates pair bond formation in male prairie voles (Microtus ochrogaster). *Behav Neurosci*, 115, 910–919.

Macey, M. J., Pickford, G. E., and Peter, R. E. (1974). Forebrain localization of the spawning reflex response to exogenous neurohypophysial hormones in the killfish, Fundulus heteroclitus. *J Exp Zool*, 190, 269–280.

Maruska, K. P. (2009). Sex and temporal variations of the vasotocin neuronal system in the damselfish brain. *Gen Comp Endocrinol*, 160, 194–204.

Maruska, K. P., Mizobe, M. H., and Tricas, T. C. (2007). Sex and seasonal co-variation of arginine vasotocin (AVT) and gonadotropin-releasing hormone (GnRH) neurons in the brain of the halfspotted goby. *Comp Biochem Physiol A Mol Integr Physiol*, 147, 129–144.

Mennigen, J. A., Martyniuk, C. J., Crump, K., Xiong, H., Zhao, E., Popesku, J., Anisman, H., Cossins, A. R.,

Xia, X., and Trudeau, V. L. (2008). Effects of fluoxetine on the reproductive axis of female goldfish (Carassius auratus). *Physiol Genomics*, 35, 273–282.

Miranda, J. A., Oliveira, R. F., Carneiro, L. A., Santos, R. S., and Grober, M. S. (2003). Neurochemical correlates of male polymorphism and alternative reproductive tactics in the Azorean rock-pool blenny, Parablennius parvicornis. *Gen Comp Endocrinol*, 132, 183–189.

Moons, L., Cambre, M., Batten, T. F., and Vandesande, F. (1989). Autoradiographic localization of binding sites for vasotocin in the brain and pituitary of the sea bass (Dicentrarchus labrax). *Neurosci Lett*, 100, 11–16.

Motohashi, E., Hamabata, T., and Ando, H. (2008). Structure of neurohypophysial hormone genes and changes in the levels of expression during spawning season in grass puffer (Takifugu niphobles). *Gen Comp Endocrinol*, 155, 456–463.

Nilaver, G., Zimmerman, E. A., Wilkins, J., Michaels, J., Hoffman, D., and Silverman, A. J. (1980). Magnocellular hypothalamic projections to the lower brain stem and spinal cord of the rat. Immunocytochemical evidence for predominance of the oxytocin-neurophysin system compared to the vasopressin-neurophysin system. *Neuroendocrinology*, 30, 150–158.

O'Connell, L. A. and Hofmann, H. A. (2011). The vertebrate mesolimbic reward system and social behavior network: A comparative synthesis. *J Comp Neurol*, 519(18), 3599–3639.

Ohya, T. and Hayashi, S. (2006). Vasotocin/isotocin-immunoreactive neurons in the medaka fish brain are sexually dimorphic and their numbers decrease after spawning in the female. *Zoolog Sci*, 23, 23–29.

Oldfield, R. G. and Hofmann, H. A. (2011). Neuropeptide regulation of social behavior in a monogamous cichlid fish. *Physiol Behav*, 102, 296–303.

Oliveira, R. F., Carneiro, L. A., Goncalves, D. M., Canario, A. V., and Grober, M. S. (2001). 11-Ketotestosterone inhibits the alternative mating tactic in sneaker males of the peacock blenny, Salaria pavo. *Brain Behav Evol*, 58, 28–37.

Ota, Y., Ando, H., Ueda, H., and Urano, A. (1999). Differences in seasonal expression of neurohypophysial hormone genes in ordinary and precocious male masu salmon. *Gen Comp Endocrinol*, 116, 40–48.

Parhar, I. S., Tosaki, H., Sakuma, Y., and Kobayashi, M. (2001). Sex differences in the brain of goldfish: gonadotropin-releasing hormone and vasotocinergic neurons. *Neuroscience*, 104, 1099–1110.

Perrone, R., Batista, G., Lorenzo, D., Macadar, O., and Silva, A. (2010). Vasotocin actions on electric behavior: interspecific, seasonal, and social context-dependent differences. *Front Behav Neurosci*, 4. DOI: 10.3389/fnbeh.2010.0052.

Pickford, G. E. and Strecker, E. L. (1977). The spawning reflex response of the killifish, Fundulus heteroclitus: isotocin is relatively inactive in comparison with arginine vasotocin. *Gen Comp Endocrinol*, 32, 132–137.

Ripley, J. L. and Foran, C. M. (2010). Quantification of whole brain arginine vasotocin for two Syngnathus pipefishes: elevated concentrations correlated with paternal brooding. *Fish Physiol Biochem*, 36, 867–874.

Rose, J. D. and Moore, F. L. (2002). Behavioral neuroendocrinology of vasotocin and vasopressin and the sensorimotor processing hypothesis. *Front Neuroendocrinol*, 23, 317–341.

Saito, D., Komatsuda, M., and Urano, A. (2004). Functional organization of preoptic vasotocin and isotocin neurons in the brain of rainbow trout: central and neurohypophysial projections of single neurons. *Neuroscience*, 124, 973–984.

Salek, S. J., Sullivan, C. V., and Godwin, J. (2002). Arginine vasotocin effects on courtship behavior in male white perch (Morone americana). *Behav Brain Res*, 133, 177–183.

Santangelo, N. and Bass, A. H. (2006). New insights into neuropeptide modulation of aggression: field studies of arginine vasotocin in a territorial tropical damselfish. *Proc Biol Sci*, 273, 3085–3092.

Santangelo, N. and Bass, A. H. (2010). Individual behavioral and neuronal phenotypes for arginine vasotocin mediated courtship and aggression in a territorial teleost. *Brain Behav Evol*, 75, 282–291.

Sawchenko, P. E. and Swanson, L. W. (1982). Immunohistochemical identification of neurons in the paraventricular nucleus of the hypothalamus that project to the medulla or to the spinal cord in the rat. *J Comp Neurol*, 205, 260–272.

Semsar, K. and Godwin, J. (2003). Social influences on the arginine vasotocin system are independent of gonads in a sex-changing fish. *J Neurosci*, 23, 4386–4393.

Semsar, K., Kandel, F. L., and Godwin, J. (2001). Manipulations of the AVT system shift social status and related courtship and aggressive behavior in the bluehead wrasse. *Horm Behav*, 40, 21–31.

Singh, V. and Joy, K. P. (2008). Immunocytochemical localization, HPLC characterization, and seasonal

dynamics of vasotocin in the brain, blood plasma and gonads of the catfish Heteropneustes fossilis. *Gen Comp Endocrinol*, 159, 214–225.

Singh, V. and Joy, K. P. (2009a). Effects of hCG and ovarian steroid hormones on vasotocin levels in the female catfish Heteropneustes fossilis. *Gen Comp Endocrinol*, 162, 172–178.

Singh, V. and Joy, K. P. (2009b). Relative in vitro seasonal effects of vasotocin and isotocin on ovarian steroid hormone levels in the catfish Heteropneustes fossilis. *Gen Comp Endocrinol*, 162, 257–264.

Suzuki, M., Hyodo, S., and Urano, A. (1992). Cloning and sequence analyses of vasotocin and isotocin precursor cDNAs in the masu salmon, Oncorhynchus masou: evolution of neurohypophysial hormone precursors. *Zoolog Sci*, 9, 157–167.

Thompson, R. R., George, K., Walton, J. C., Orr, S. P., and Benson, J. (2006). Sex-specific influences of vasopressin on human social communication. *Proc Natl Acad Sci USA*, 103, 7889–7894.

Thompson, R. R. and Walton, J. C. (2004). Peptide effects on social behavior: effects of vasotocin and isotocin on social approach behavior in male goldfish (Carassius auratus). *Behav Neurosci*, 118, 620–626.

Thompson, R. R. and Walton, J. C. (2009). Vasotocin immunoreactivity in goldfish brains: characterizing primitive circuits associated with social regulation. *Brain Behav Evol*, 73, 153–164.

Thompson, R. R., Walton, J. C., Bhalla, R., George, K. C., and Beth, E. H. (2008). A primitive social circuit: vasotocin-substance P interactions modulate social behavior through a peripheral feedback mechanism in goldfish. *Eur J Neurosci*, 27, 2285–2293.

Van den Dungen, H. M., Buijs, R. M., Pool, C. W., and Terlou, M. (1982). The distribution of vasotocin and isotocin in the brain of the rainbow trout. *J Comp Neurol*, 212, 146–157.

Venkatesh, B. and Brenner, S. (1995). Structure and organization of the isotocin and vasotocin genes from teleosts. *Adv Exp Med Biol*, 395, 629–638.

Wagenaar, D. A., Hamilton, M. S., Huang, T., Kristan, W. B., and French, K. A. (2010). A hormone-activated central pattern generator for courtship. *Curr Biol*, 20, 487–495.

Walton, J. C., Waxman, B., Hoffbuhr, K., Kennedy, M., Beth, E., Scangos, J., and Thompson, R. R. (2010). Behavioral effects of hindbrain vasotocin in goldfish are seasonally variable but not sexually dimorphic. *Neuropharmacology*, 58, 126–134.

Wang, Z., Ferris, C. F., and De Vries, G. J. (1994). Role of septal vasopressin innervation in paternal behavior in prairie voles (Microtus ochrogaster). *Proc Natl Acad Sci USA*, 91, 400–404.

Wersinger, S. R., Ginns, E. I., O'Carroll, A. M., Lolait, S. J., and Young, W. S., III. (2002). Vasopressin V1b receptor knockout reduces aggressive behavior in male mice. *Mol Psychiatry*, 7, 975–984.

Wilhemi, A. E., Pickford, G. E., and Sawyer, W. H. (1956). Initiation of the spawning reflex response in Fundulus by the administration of fish and mammalian neurohypophyseal preparations and synthetic oxytocin. *Endocrinology*, 57, 243–252.

Zhang, D., Xiong, H., Mennigen, J. A., Popesku, J. T., Marlatt, V. L., Martyniuk, C. J., Crump, K., Cossins, A. R., Xia, X., and Trudeau, V. L. (2009). Defining global neuroendocrine gene expression patterns associated with reproductive seasonality in fish. *PLoS One*, 4, e5816.

6

Vasotocin modulation of social behaviors in amphibians

Sunny K. Boyd

6.1 Introduction

Arginine vasotocin (AVT) belongs to a family of closely related peptides that are released from the neurohypophysis and also found broadly distributed within the central nervous system (CNS). AVT itself is generally considered the ancestral peptide of the family and it has been identified in representatives from all vertebrate classes (Hoyle, 1999; Urano et al., 1992). In amphibians, reptiles and birds, AVT and mesotocin (MT) are the two neurohypophysial peptides. Isotocin replaces MT in most fish. Homologous peptides in mammals are arginine vasopressin (AVP) and oxytocin (OT). Major elements of the neurohypophysial peptide system have all been largely conserved across vertebrates, including peptide gene structure, genes and structures of receptors, and the distribution of peptides and receptors (Acharjee et al., 2004; Goodson and Bass, 2001). Neuropeptides in this family have been consistently implicated in the control of remarkably similar social behaviors in diverse vertebrates (Goodson and Bass, 2001).

In amphibians, AVT is a potent modulator of behavior (Moore et al., 2005; Wilczynski et al., 2005). Research has focused on social behaviors closely associated with reproduction, including courtship, consummatory sexual behaviors and related aggressive behaviors. Although amphibians show other interesting social behaviors, including pair bonding and parental care (Brown et al., 2010), the effects of neuropeptides on these behaviors are unstudied. One current focus is on amplectic clasping behavior that is common across the two major groups of amphibians – anurans (frogs and toads) and urodeles (salamanders and newts). Interestingly, although AVT has profound effects on clasping in some urodeles, it has not been reported to influence clasping in any anuran (Moore et al., 2005; Propper and Dixon, 1997). The second focus of behavioral research in amphibians has been vocal communication, where anurans typically excel (vocal communication in urodeles is rare). AVT strongly influences the display of vocal signals and the behavioral responses of conspecifics to those signals in anuran amphibians. AVT modulates very similar behaviors in birds (this volume), but whether AVT influences any reptile social behavior is unknown.

6.2 AVT control of social behaviors in anuran amphibians

6.2.1 Anuran vocal behaviors

Anuran amphibians typically rely on a small set of stereotyped vocalizations for conspecific communication (Gerhardt, 1994). The advertisement call of males is the most prominent call and serves

Oxytocin, Vasopressin, and Related Peptides in the Regulation of Behavior, ed. E. Choleris, D. W. Pfaff, and M. Kavaliers.

for mate attraction and intermale spacing in most species. Both sexes give a weaker release call when inappropriately clasped. Evidence for modulation of anuran vocalizations by AVT is compelling. Neurohypophysial peptides also modulate vocal behavior in representatives from other vertebrate classes, including mammals, birds and fish (Goodson and Bass, 2000a; Maney et al., 1997; Winslow et al., 2000).

Advertisement calling is stimulated by exogenous AVT in seven frog species so far investigated (Boyd, 1994a; Burmeister et al., 2001; Chu et al., 1998; Kime et al., 2007; Klomberg and Marler, 2000; Marler et al., 1995; Penna et al., 1992; Propper and Dixon, 1997; Semsar et al., 1998; Ten Eyck, 2005; Tito et al., 1999; Trainor et al., 2003). One common theme across these studies is that AVT appears to increase motivation to call, by increasing the likelihood of calling at all, decreasing latency to first call, and/or increasing call rate or duration.

For release calling, the pattern is not so clear. Effects of AVT vary with the sex and species of the animal. In female frogs of two species, AVT decreases release calling (Boyd, 1992; Diakow, 1978). In males, AVT administration may increase, decrease, or not alter release call rates (Boyd, 1992; Raimondi and Diakow, 1981; Tito et al., 1999), depending on the frog species. Interaction of AVT with multiple neuroendocrine factors, including steroids, prostaglandins and prolactin, seems likely for the control of release calling (Boyd, 1992; Diakow, 1978; Weintraub et al., 1985). Such interactions may account for the species and sex differences observed.

Aggressive behaviors are modulated by neurohypophysial peptides across a variety of vertebrates. Aggressive calling and related territorial behaviors are enhanced by AVT in the gray treefrog (*Hyla versicolor*) and coqui frog (*Eleutherodactylus coqui*) (Klomberg and Marler, 2000; Semsar et al., 1998; Ten Eyck, 2005; Tito et al., 1999). Most commonly, the species-typical advertisement call is stimulated and then used in male–male interactions, rather than for signaling to females. However, the distinctive aggressive call of the gray treefrog, which is used only during competitive interactions between two males, is also increased by AVT (Tito et al., 1999). In contrast, in the cricket frog (*Acris crepitans*), AVT treatment produces calls typical of less-aggressive males (Marler et al., 1995). Thus, effects of AVT on aggression in male frogs may vary with the species and/or the social context. AVT similarly alters aggression in other non-mammalian vertebrates, including fish (Bastian et al., 2001; Lema and Nevitt, 2004; Oldfield and Hofmann, 2011; Santangelo and Bass, 2006; Semsar et al., 2001), and birds (Goodson et al., 2009; Maney et al., 1997). Importantly, whether AVT alters aggression in urodele amphibians is unknown, although urodeles show many interesting aggressive behaviors (Markman et al., 2009; Tornick, 2010).

6.2.2 Sites and mechanisms of AVT action on vocalizations

Anuran vocalizations are ultimately produced by contraction of laryngeal muscles (Emerson and Boyd, 1999; Zornik and Kelley, 2007). These muscles are controlled by motor neurons located in the brain stem in the laryngeal motor nucleus and axons from these cells travel to the larynx in cranial nerve X. Cells in the motor nucleus receive input from three primary sources: the pretrigeminal nucleus, which serves as the vocal pattern generator (PTN), reticular nuclei (trigeminal and hypoglossal) which coordinate breathing with vocalization, and the contralateral motor nucleus. The PTN, in turn, receives input primarily from the striatum, amygdala, thalamus, and preoptic area (POA). The neural pathway controlling anuran vocalization behavior is thus relatively well known.

The involvement of AVT in anuran amphibian vocalization is well supported by the distribution of AVT and AVT receptors in the CNS (Figure 6.1). Both the peptide and its receptors are found in every brain area implicated in control of frog vocalization. AVT cells and fibers are widespread in anuran brain, consisting of multiple distinct cell populations with extensive hypothalamic and extrahypothalamic fiber projections (Boyd, 1994b; Boyd

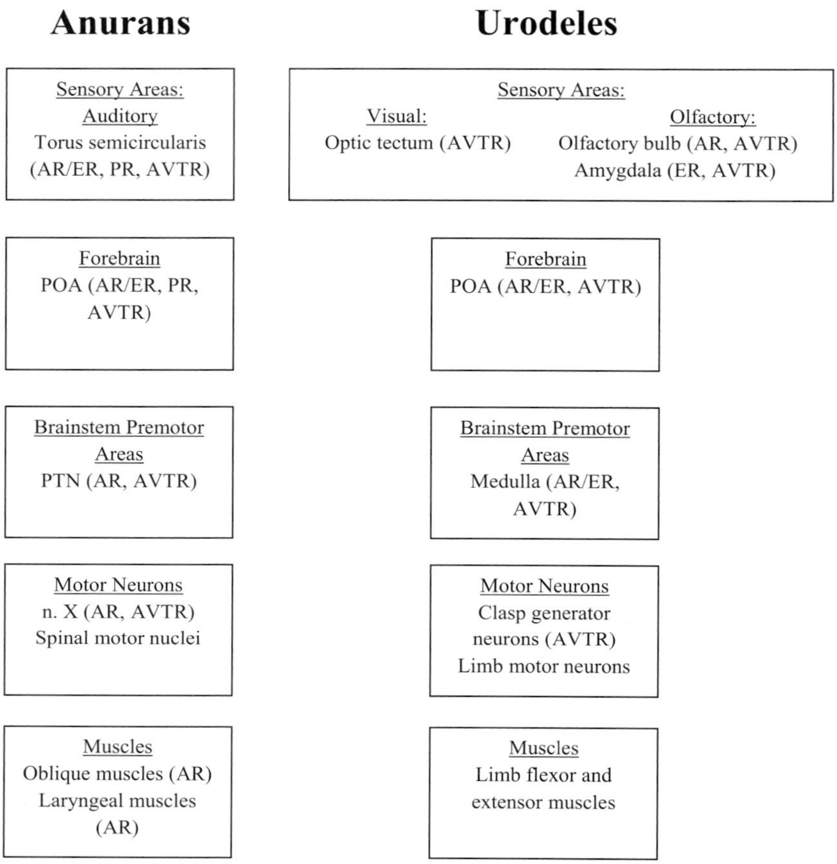

Anurans	Urodeles
Sensory Areas: Auditory Torus semicircularis (AR/ER, PR, AVTR)	Sensory Areas: Visual: Optic tectum (AVTR) Olfactory: Olfactory bulb (AR, AVTR) Amygdala (ER, AVTR)
Forebrain POA (AR/ER, PR, AVTR)	Forebrain POA (AR/ER, AVTR)
Brainstem Premotor Areas PTN (AR, AVTR)	Brainstem Premotor Areas Medulla (AR/ER, AVTR)
Motor Neurons n. X (AR, AVTR) Spinal motor nuclei	Motor Neurons Clasp generator neurons (AVTR) Limb motor neurons
Muscles Oblique muscles (AR) Laryngeal muscles (AR)	Muscles Limb flexor and extensor muscles

Figure 6.1 Schematic diagram comparing the key neural areas involved in displays of social behaviors in the two amphibian groups; the distribution of receptors for modulators known to alter these behaviors is shown. Based on (Emerson and Boyd, 1999) and (Wilczynski et al., 2005), with additional details from (Acharjee et al., 2004; Chakraborty and Burmeister, 2010; Guerriero et al., 2005; Hasunuma et al., 2010; Lewis et al., 2005; Lewis et al., 2004; O'Connell et al., 2011). Abbreviations: AR, androgen receptor; ER, estrogen receptor; PR, progesterone receptor; AVT-R, vasotocin receptor; POA, preoptic area; PTN, pretrigeminal nucleus; n. X, motor nucleus of cranial nerve X.

et al., 1992; Gonzalez and Smeets, 1992a; Gonzalez and Smeets, 1992b; Moore and Lowry, 1998). AVT-producing cells occur in three specific locations in the anuran vocal motor pathway: the amygdala, preoptic area and pretrigeminal nucleus (Boyd, 1997). The broad distribution of vasotocinergic pathways supports at least two primary but non-exclusive theories for AVT mechanism of action in the anuran vocal system.

First, some evidence suggests that AVT modulates frog vocalization via a mechanism at the motor output level. In a recent behavioral study of tungara frogs (*Physalaemus pustulosus*), AVT stimulation of advertisement calls appeared to alter airflow, based on the temporal and spectral changes in the calls (Kime et al., 2010). Similar conclusions can be reached for effects of AVT on green treefrog (*Hyla cinerea*) calls (Penna et al., 1992). The

Table 6.1 Intracranial injection of AVT into the laryngeal motor nucleus (LMN) stimulates advertisement calling in male bullfrogs.

Parameter	Vasotocin 0.1 ng	Vasotocin 1 ng	Vasotocin 10 ng	Systemic vasotocin 500 μg
Number calling/total male frogs	1/8	9/10	9/9	7/7
Call latency (min)	10	6.4 ± 2.0*	1.3 ± 0.6	1.2 ± 0.3
Calls/min	1	5.2 ± 1.1*	18.9 ± 5.2	14.4 ± 1.5
Call duration (s)	0.25 ± 0.30	0.30 ± 0.1	0.28 ± 0.1	1.2 ± 0.60
Intercall Interval (in bout; s)	0.25	0.51 ± 0.82	0.55 ± 0.1	0.6 ± 0.3
Calls/bout	2	5.2 ± 1.0*	14.0 ± 2.3	9.0 ± 3.3

Male bullfrogs with bilateral cannulae chronically implanted into the LMN were injected with artificial CSF alone (no frogs vocalized so data not shown) or AVT in the doses shown. Advertisement calling evoked by chorus playbacks was recorded for the first 15 min after injection. Published (Boyd, 1994a) and unpublished data from 30 min following a systemic intraperitoneal injection are shown in the last column for comparison. *Asterisks indicate significant differences between 1 ng and 10 ng treatment groups (paired t-test; $p < 0.05$). Artificial CSF and 0.1 ng doses were not statistically compared due to the large number of cells with zeros.

strongest support for an effect of AVT on motor output regions of the frog brain comes from bullfrogs (*Rana catesbeiana*, now known as *Lithobates catesbeianus*) that received intracranial injections of AVT directly into the laryngeal motor nuclei (Table 6.1). Bilateral injections of AVT into the LMN stimulated calling and significantly altered some call parameters. Doses of 10 ng and 1 ng per frog significantly increased the number of frogs that vocalized at all, compared to saline or a dose of 0.1 ng. For frogs that called following motor nucleus injection, the dominant frequency, bout structure, intercall interval and calls/min fell within the normal range of natural bullfrog advertisement calls. Vocalizations of intracranially injected frogs were, however, unusually short, staccato calls. At every dose of AVT injected into the LMN, calls were shorter than for normal calling. These results showed that exogenous injection into this one part of the vocal pathway is sufficient to stimulate calling in male frogs, but some elements of call structure were not species typical.

Anatomical support for modulation at the motor output level also comes specifically from bullfrogs. AVT fibers, terminal fields and AVT receptors are present in the laryngeal motor nuclei of bullfrogs (Figure 6.2; (Boyd, 1997; Boyd and Moore, 1992)). In addition, the vocal central pattern generator in the PTN is also a likely site of AVT action. This nucleus contains a seasonally variable, steroid-sensitive population of AVT cells (Boyd, 1994c; Boyd and Moore, 1992; Boyd et al., 1992). AVT receptors in the bullfrog PTN are also sexually dimorphic in concentration and are steroid sensitive (Boyd, 1997). These findings support the hypothesis that AVT may act in the laryngeal motor nucleus and/or PTN to modulate frog calling behavior. There is support for this hypothesis in other vertebrates as well. For example, AVT modulates vocalization in a fish in homologous areas (Goodson and Bass, 2000b). AVP modulates motor neuron activity in similar regions of the rat brainstem, although AVP action on vocalization via these regions has not been shown (e.g., Wrobel et al., 2010).

The second hypothesis suggests that AVT may modulate anuran vocal behavior via effects on sensory processing of social stimuli. Effects of AVT vary, for some species, depending upon the social context (Chu et al., 1998; Trainor et al., 2003). Although this hypothesis has strong experimental support in urodele amphibians, this is not the case for anurans. However, there are intriguing hints that this

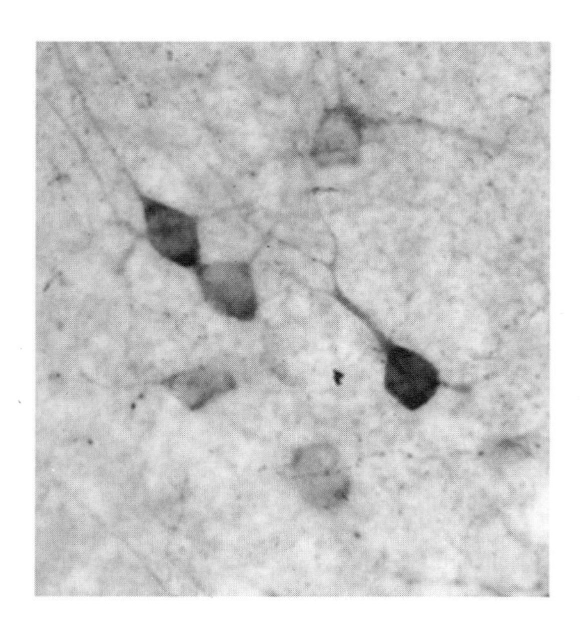

Figure 6.2 Cells and fibers in the male bullfrog laryngeal motor nucleus are immunopositive for the AVPR1a receptor subtype. An antibody against a peptide corresponding to the first 19 amino acids of the rat AVPR1a was used to label putative AVT receptors in bullfrog brainstem. An avidin-biotin based detection system was used with DAB-Ni for visualization of immunoreactivity.

mechanism may also be important in frogs and toads. For example, AVT changes auditory processing in the torus semicircularis region in the brain of the green treefrog (Penna et al., 1992). In addition, AVT levels in the nucleus accumbens of cricket frogs are negatively correlated with calling behavior (Marler et al., 1999). Although the function of the amphibian nucleus accumbens is not clear, involvement in olfactory processing is likely (Marin et al., 1997). Lastly, AVT infusion into the POA of a fish modulates vocal output (Goodson and Bass, 2000b). Investigation of effects of AVT in the POA of anurans would be a fruitful area of study. The POA is not only an acoustically sensitive region but it has been recently shown to be modulated by socially relevant stimuli in a treefrog (Almli and Wilczynski, 2009).

6.2.3 Interaction of AVT with gonadal and adrenal steroids

The vasotocinergic pathways in the amphibian brain are sexually dimorphic and sensitive to changes in gonadal steroid hormones, as are neurohypophysial peptide pathways in other vertebrates (De Vries and Panzica, 2006; O'Bryant and Wilczynski, 2010). In the amygdala pars lateralis of bullfrogs (likely homolog to the bed nucleus of the stria terminalis of mammals), males have significantly more AVT cells and fibers, compared to females (Boyd et al., 1992). AVT concentrations, when measured by radioimmunoassay, are significantly higher in males in six brain areas: amygdala pars lateralis, septal nucleus, habenula, optic tectum, PTN, and tegmentum (Boyd, 1994c). Only in the auditory dorsolateral nucleus was AVT higher in females. It is noteworthy that the amygdala and PTN are specifically part of the vocal pathway and the dorsolateral nucleus is necessary for phonotaxis, given the modulation of vocal and phonotaxis behaviors by AVT.

Gonadectomy and steroid replacement studies show that the gonads maintain AVT levels in multiple brain areas of both sexes in bullfrogs (Boyd, 1994c). Following gonad removal, the effect of replacement with the non-aromatizable androgen dihydrotestosterone or estradiol is site and sex specific. AVT receptors (homologous to the mammalian AVPR1a; (Acharjee et al., 2004; Searcy et al., 2011)) are also sexually dimorphic in their distribution, specifically in the amygdala pars lateralis, hypothalamus, pretrigeminal nucleus and dorsolateral nucleus (Boyd, 1997). Estradiol modulates receptor levels in the amygdala of both sexes in bullfrogs and both estradiol and dihydrotestosterone modulate receptor levels in the PTN of males only. These studies suggest that daily, seasonal, sex-specific, and socially mediated differences in gonadal steroids may profoundly change the modulatory effect of AVT on particular social behaviors.

These findings support the hypothesis that androgens are required for the behavioral effects of AVT to be observed but direct evidence in anurans is scant. In one study, castration of males reduced

the ability of AVT to stimulate advertisement calling in the green treefrog (Penna et al., 1992). In addition, effects of AVT on calling in bullfrogs are sexually dimorphic and seasonally variable, as are plasma steroid concentrations (Boyd, 1992). Certainly, androgens are required for the display of advertisement calling in frogs and toads, but the complex relationship between androgens and AVT is not yet clear (Moore et al., 2005; Wilczynski et al., 2005). Further study on interactions between estradiol and progesterone, in the control of social behaviors by AVT, would also be valuable.

In urodele amphibians, there is strong evidence for a corticosterone–AVT interaction in the control of social behaviors. There is little such evidence for anurans; however, there are suggestions that such interactions may exist. Certainly, corticosterone alone often decreases anuran amphibian vocalization, but not always (Leary, 2009). One prevalent hypothesis is that corticosterone causes a decrease in androgens required for maintenance of the vasotocinergic system (Emerson and Hess, 2001). This proves not to be the case for some species, such as the Woodhouse's and Great Plains toads, however (*Bufo woodhousii* and *cognatus; Leary et al.*, 2004). Thus, there is support for the alternative hypothesis that glucocorticoids influence AVT synthesis or release directly, rather than via effects on androgens (Leary, 2009; Leary et al., 2004). Glucocorticoid receptors occur in multiple areas implicated in control of vocalization, including the nucleus accumbens, amygdala, bed nucleus of the stria terminalis, POA, and torus semicircularis, so direct effects are certainly possible (Yao et al., 2008). On the other hand, AVT may promote calling by overcoming a glucocorticoid-mediated stress response (Marler et al., 1995), although this is not supported in the green treefrog (Burmeister et al., 2001). The details of the interactions among AVT, glucocorticoids, and androgens thus remain an open question.

6.2.4 Phonotaxis behaviors

AVT and other neurohypophysial peptides modulate appetitive sexual behaviors and, more generally, affiliative behaviors across many vertebrate species (e.g., Bielsky and Young, 2004; Boyd, 1994a; Castagna et al., 1998; Insel and Hulihan, 1995; Lim et al., 2004a; Lim et al., 2004b; Wang and Aragona, 2004; Williams et al., 1994). The most prevalent of such affiliative behaviors in anurans is positive phonotaxis. Females move toward the advertisement calls of conspecific males and clasping usually follows. Of possible endogenous factors that may control female anuran phonotaxis, AVT is a strong contender. AVT stimulates advertisement call phonotaxis in American toads (*Bufo americanus*; now *Anaxyrus americanus*) and bullfrogs by increasing the speed and decreasing the latency of females to approach a call (Boyd, 1994a; Schmidt, 1984; Schmidt, 1985).

AVT cells, fibers, and receptors are widespread in brain areas involved in auditory-evoked behaviors, such as phonotaxis. Lesion studies emphasize the importance of the mesencephalic torus semicircularis, a structure homologous to the mammalian inferior colliculus (Endepols et al., 2003). Even lesions that disturb less than 10% of the torus abolish phonotaxis behavior in gray treefrogs. AVT fibers and terminal fields are found in the torus, along with AVT receptors (Boyd, 1994c; Boyd, 1997; Boyd et al., 1992). The anuran torus semicircularis is thus a possible site of action of exogenous AVT in control of phonotaxis. In this location, AVT likely modulates primarily auditory processing related to display of the behavior, as shown for the green treefrog (Penna et al., 1992).

Diencephalic brain regions have been specifically linked to audio-motor integration in anurans. Analysis of immediate-early gene *egr-1* expression patterns in frogs shows multiple sites to be important for such integration (Hoke et al., 2005; Hoke et al., 2007). In treefrogs, phonotaxis behavior is negatively affected or abolished by lesions in some diencephalic nuclei, including the thalamus, POA, and suprachiasmatic nucleus (Endepols et al., 2003; Endepols et al., 2004; Walkowiak et al., 1999). The anuran diencephalon contains three populations of AVT-producing cells (magnocellular preoptic area, suprachiasmatic nucleus, dorsal and ventral hypothalamus), dense fiber projections,

and some of the highest concentrations of AVT receptors found in the brain (Boyd, 1994c; Boyd, 1997; Boyd et al., 1992; Gonzalez and Smeets, 1992a; Gonzalez and Smeets, 1992b). Current evidence thus supports the hypothesis that AVT may mediate anuran female phonotaxis via the diencephalon. In contrast, lesions of the dorsomedial pallium do not alter phonotaxis (Endepols et al., 2003; Walkowiak et al., 1999). This area is thus unlikely to be the site of AVT action, despite the high concentration of receptors detected in the pallium of anurans (Boyd, 1997).

6.3 AVT control of social behaviors in urodele amphibians

6.3.1 Courtship behaviors in the rough-skinned newt

Courtship behaviors in urodele amphibians show great diversity but the AVT modulation of such behaviors has been investigated only in two species from a single family. In the rough-skinned newt (*Taricha granulosa*), AVT involvement in modulation of the dorsal amplectic clasp of females by males has strong support (Moore et al., 2005). Exogenous AVT treatment increases clasping behaviors in male *Taricha*, while intracranial treatment with an antagonist or anti-AVT serum decreases the same behaviors. Determining the site of action of endogenous AVT has been complicated because AVT and its receptors are broadly distributed across the CNS of urodele amphibians (Boyd and Moore, 1991; Hasunuma et al., 2010; Hollis et al., 2005; Lowry et al., 1997; Moore and Lowry, 1998; Smeets and Gonzalez, 2001). However, as in anuran amphibians, there is evidence for two non-exclusive mechanisms for AVT in control of amplectic clasping – one related to processing of sensory stimuli and the other mechanism related to generation of motor output.

A proposed motor output mechanism is based on findings of AVT distribution, receptor distribution, and specific effects of the peptide on behavior, when locally applied, in motor regions of the rough-skinned newt CNS. First, motor aspects of the clasp depend critically on medullary and spinal regions in the CNS of amphibians (Rose and Moore, 2002). Exogenous AVT, applied directly to the medulla, increases the number of responsive neurons and the magnitude of response to clasp-triggering cloacal pressure in these newts (Rose et al., 1995). This mechanism of action is supported by the finding of AVT fibers and receptors in the medulla of rough-skinned newts (Boyd and Moore, 1991; Hollis et al., 2005; Lowry et al., 1997). In addition, a labeled AVT conjugate is internalized in about 70% of medulla reticulospinal neurons, suggesting a large population of brainstem neurons can respond to this peptide (Lewis et al., 2005).

A mechanism of action related to processing of sensory stimuli is supported by several studies. In the rough-skinned newt, high AVT concentrations in the optic tectum are correlated with the breeding season and sexually active males show higher concentrations in the dorsal preoptic area, optic tectum, ventral infundibulum and cerebrospinal fluid, compared to sexually inactive males (Zoeller and Moore, 1986; Zoeller and Moore, 1988). Given the prominence of visual cues in the courtship behavior of this newt, the link to the optic tectum is especially intriguing. Later behavioral and electrophysiological studies show that AVT increases responses of male newts to visual stimuli and also olfactory and tactile sexual stimuli (Rose et al., 1995; Thompson and Moore, 2000). An AVT antagonist does not influence medulla neuronal responsiveness to cloacal pressure, unless combined with pheromone exposure (Thompson et al., 2008). Thus, it is proposed for rough-skinned newts that AVT couples olfactory and tactile systems together to elicit behaviors uniquely suited for the social context. This supports the hypothesis that AVT modulates rough-skinned newt clasping behavior by influencing combined sensorimotor processing (Rose and Moore, 2002).

As in anuran amphibians, there is ample evidence for interaction of AVT systems and steroid hormones in the modulation of urodele behaviors. Androgens are required for exogenous AVT to stimulate clasping behavior in rough-skinned newts (Zoeller and Moore, 1982). AVT immunoreactivity

is sexually dimorphic, which is consistent with the hypothesis that androgens may modulate AVT (Moore et al., 2000). This hypothesis has great support in anurans and other vertebrates. In addition, AVT receptor density is decreased by castration in the amygdala of male rough-skinned newts (Boyd and Moore, 1991). Thus, it is proposed that androgens maintain multiple elements of vasotocinergic pathways and may promote clasping behavior via this mechanism.

AVT also interacts with the amphibian glucocorticoid, corticosterone, in modulating clasp behaviors (Rose and Moore, 2002). Corticosterone alone rapidly suppresses rough-skinned newt clasping behavior, but not when preceded by AVT injection (Coddington and Moore, 2003). The same result is seen for neuronal activity in the male newt medulla, where pretreatment with AVT also prevents the inhibitory effects of corticosterone. In reciprocal fashion, pretreatment with corticosterone modifies the responsiveness of medulla cells then treated with AVT: effects of AVT on medullary neurons are enhanced when AVT is administered 10 min after corticosterone but the opposite occurs when administered 30 min later (Rose et al., 1995). This represents a novel example of a very rapid effect of a glucocorticoid and suggests a mechanism for rapid and adaptive behavioral plasticity via corticosterone-AVT interactions (Orchinik, et al., 1991).

In this field, the research emphasis has been on the behavior of males and there are few reports on neurohypophysial modulation of behaviors in females. Interestingly, female newts show egg-laying behaviors with motor patterns very similar to clasping. AVT can modulate egg-laying behaviors in female rough-skinned newts, when combined with estradiol treatment but male-typical clasping is induced when females are treated with androgen and AVT (Moore et al., 1992). Ovariectomized female newts do not show egg-laying behavior when injected with AVT, suggesting that estradiol is required for modulation of this behavior. Ovariectomized female newts also show a 20% decrease in the concentration of putative AVT receptors in the amygdala, but there is no information on the role the amygdala might play in egg-laying behavior (Boyd and Moore, 1991). The combination of androgen and AVT treatment of female newts also induces male-typical behavioral responses toward female olfactory stimuli, culminating in females that will spend more time with female-scented newt models and clasping those models (Thompson and Moore, 2003). In both behavioral studies, androgen treatment alone was not sufficient to elicit male-typical behavior but the combination of androgen and AVT together was necessary.

6.3.2 Courtship behaviors in the Japanese red-bellied newt

Only one other urodele species shows AVT-sensitive behaviors, as so far reported. The courtship behavior of the Japanese red-bellied newt (*Cynops pyrrhogaster*) includes male displays with tail vibrations and the use of pheromones, but not amplectic clasping. AVT has been shown to modulate several aspects of red-bellied newt courtship – some aspects via central mechanisms and some via peripheral mechanisms. AVT treatment increases the incidence and frequency of tail vibration behavior by male red-bellied newts and spontaneous courtship behaviors are inhibited by an AVPR1 antagonist (Toyoda et al., 2003). Because intracranial injections are significantly more potent than intraperitoneal injections, a CNS site of action is suggested. Prolactin is also a modulator of red-bellied newt tail vibrations and evidence supports the hypothesis that prolactin acts to cause the release of AVT (Kikuyama et al., 2009).

Male red-bellied newts release a potent female-attracting pheromone from the abdominal gland of the cloaca (Kikuyama et al., 2005). AVT causes the release of the pheromone sodefrin from this gland, thus promoting reproduction via a peripheral, non-neural mechanism (Toyoda et al., 2003). A peripheral effect is also likely for AVT-induced spermatophore deposition (Toyoda et al., 2003). Thus, peripheral and central AVT mechanisms work

in concert in this amphibian to facilitate complex behavioral displays.

Acknowledgments

The support of the National Science Foundation is gratefully acknowledged, especially most recently #0725187 and #0235903.

REFERENCES

Acharjee, S., Do-Rego, J. L., Oh, D. Y., Moon, J. S., Ahn, R. S., Lee, K., Bai, D. G., Vaudry, H., Kwon, H. B., and Seong, J. Y. (2004). Molecular cloning, pharmacological characterization, and histochemical distribution of frog vasotocin and mesotocin receptors. *Journal of Molecular Endocrinology*, 33, 293–313.

Almli, L. M. and Wilczynski, W. (2009). Sex-specific modulation of cell proliferation by socially relevant stimuli in the adult green treefrog brain (*Hyla cinerea*). *Brain Behavior and Evolution*, 74, 143–154.

Bastian, J., Schniederjan, S., and Nguyenkim, J. (2001). Arginine vasotocin modulates a sexually dimorphic communication behavior in the weakly electric fish *Apteronotus leptorhynchus*. *Journal of Experimental Biology*, 204, 1909–1924.

Bielsky, I. F. and Young, L. J. (2004). Oxytocin, vasopressin, and social recognition in mammals. *Peptides*, 25, 1565–1574.

Boyd, S. K. (1992). Sexual differences in hormonal control of release calls in bullfrogs. *Hormones and Behavior*, 26, 522–535.

Boyd, S. K. (1994a). Arginine vasotocin facilitation of advertisement calling and call phonotaxis in bullfrogs. *Hormones and Behavior*, 28, 232–240.

Boyd, S. K. (1994b). Development of vasotocin pathways in the bullfrog brain. *Cell and Tissue Research*, 276, 593–602.

Boyd, S. K. (1994c). Gonadal steroid modulation of vasotocin concentrations in the bullfrog brain. *Neuroendocrinology*, 60, 150–156.

Boyd, S. K. (1997). Brain vasotocin pathways and the control of sexual behaviors in the bullfrog. *Brain Research Bulletin*, 44, 345–350.

Boyd, S. K. and Moore, F. L. (1991). Gonadectomy reduces the concentrations of putative receptors for arginine vasotocin in the brain of an amphibian. *Brain Research*, 541, 193–197.

Boyd, S. K. and Moore, F. L. (1992). Sexually dimorphic concentrations of arginine vasotocin in sensory regions of the amphibian brain. *Brain Research*, 588, 304–306.

Boyd, S. K., Tyler, C. J., and De Vries, G. J. (1992). Sexual dimorphism in the vasotocin system of the bullfrog (*Rana catesbeiana*). *Journal of Comparative Neurology*, 325, 313–325.

Brown, J. L., Morales, V., and Summers, K. (2010). A key ecological trait drove the evolution of biparental care and monogamy in an amphibian. *American Naturalist*, 175, 436–446.

Burmeister, S., Somes, C., and Wilczynski, W. (2001). Behavioral and hormonal effects of exogenous vasotocin and corticosterone in the green treefrog. *General and Comparative Endocrinology*, 122, 189–197.

Castagna, C., Absil, P., Foidart, A., and Balthazart, J. (1998). Systemic and intracerebroventricular injections of vasotocin inhibit appetitive and consummatory components of male sexual behavior in Japanese quail. *Behavioral Neuroscience*, 112, 233–250.

Chakraborty, M. and Burmeister, S. S. (2010). Sexually dimorphic androgen and estrogen receptor mRNA expression in the brain of tungara frogs. *Hormones and Behavior*, 58, 619–627.

Chu, J., Marler, C. A., and Wilczynski, W. (1998). The effects of arginine vasotocin on the calling behavior of male cricket frogs in changing social contexts. *Hormones and Behavior*, 34, 248–261.

Coddington, E. and Moore, F. L. (2003). Neuroendocrinology of context-dependent stress responses: Vasotocin alters the effect of corticosterone on amphibian behaviors. *Hormones and Behavior*, 43, 222–228.

De Vries, G. J. and Panzica, G. C. (2006). Sexual differentiation of central vasopressin and vasotocin systems in vertebrates: Different mechanisms, similar endpoints. *Neuroscience*, 138, 947–955.

Diakow, C. (1978). Hormonal basis for breeding behavior in female frogs – vasotocin inhibits release call of *Rana pipiens*. *Science*, 199, 1456–1457.

Emerson, S. B. and Boyd, S. K. (1999). Mating vocalizations of female frogs: Control and evolutionary mechanisms. *Brain Behavior and Evolution*, 53, 187–197.

Emerson, S. B. and Hess, D. L. (2001). Glucocorticoids, androgens, testis mass, and the energetics of vocalization in breeding male frogs. *Hormones and Behavior*, 39, 59–69.

Endepols, H., Feng, A. S., Gerhardt, H. C., Schul, J., and Walkowiak, W. (2003). Roles of the auditory midbrain and thalamus in selective phonotaxis in female gray treefrogs (*Hyla versicolor*). *Behavioural Brain Research*, 145, 63–77.

Endepols, H., Schul, J., Gerhardt, H. C., and Walkowiak, W. (2004). 6-hydroxydopamine lesions in anuran amphibians: A new model system for Parkinson's disease? *Journal of Neurobiology*, 60, 395–410.

Gerhardt, H. C. (1994). The evolution of vocalization in frogs and toads. *Annual Review of Ecology and Systematics*, 25, 293–324.

Gonzalez, A. and Smeets, W. (1992a). Comparative analysis of the vasotocinergic and mesotocinergic cells and fibers in the brain of two amphibians, the anuran *Rana ridibunda* and the urodele *Pleurodeles waltlii*. *Journal of Comparative Neurology*, 315, 53–73.

Gonzalez, A. and Smeets, W. (1992b). Distribution of vasotocin-like and mesotocin-like immunoreactivities in the brain of the South African clawed frog *Xenopus laevis*. *Journal of Chemical Neuroanatomy*, 5, 465–479.

Goodson, J. L. and Bass, A. H. (2000a). Forebrain peptides modulate sexually polymorphic vocal circuitry. *Nature*, 403, 769–772.

Goodson, J. L. and Bass, A. H. (2000b). Vasotocin innervation and modulation of vocal-acoustic circuitry in the teleost *Porichthys notatus*. *Journal of Comparative Neurology*, 422, 363–379.

Goodson, J. L. and Bass, A. H. (2001). Social behavior functions and related anatomical characteristics of vasotocin/vasopressin systems in vertebrates. *Brain Research Reviews*, 35, 246–265.

Goodson, J. L., Kabelik, D., and Schrock, S. E. (2009). Dynamic neuromodulation of aggression by vasotocin: Influence of social context and social phenotype in territorial songbirds. *Biology Letters*, 5, 554–556.

Guerriero, G., Prins, G. S., Birch, L., and Ciarcia, G. (2005). Neurodistribution of androgen receptor immunoreactivity in the male frog, *Rana esculenta*. *Annals of the New York Academy of Sciences*, 1040, 332–336.

Hasunuma, I., Toyoda, F., Kadono, Y., Yamamoto, K., Namiki, H., and Kikuyama, S. (2010). Localization of three types of arginine vasotocin receptors in the brain and pituitary of the newt *Cynops pyrrhogaster*. *Cell and Tissue Research*, 342, 437–457.

Hoke, K. L., Ryan, M. J., and Wilczynski, W. (2005). Social cues shift functional connectivity in the hypothalamus. *Proceedings of the National Academy of Sciences of the United States of America*, 102, 10712–10717.

Hoke, K. L., Ryan, M. J., and Wilczynski, W. (2007). Integration of sensory and motor processing underlying social behaviour in tungara frogs. *Proceedings of the Royal Society B-Biological Sciences*, 274, 641–649.

Hollis, D. M., Chu, J., Walthers, E. A., Heppner, B. L., Searcy, B. T., and Moore, F. L. (2005). Neuroanatomical distribution of vasotocin and mesotocin in two urodele amphibians (*Plethodon shermani* and *Taricha granulosa*) based on *in situ* hybridization histochemistry. *Brain Research*, 1035, 1–12.

Hoyle, C. H. V. (1999). Neuropeptide families and their receptors: Evolutionary perspectives. *Brain Research*, 848, 1–25.

Insel, T. R. and Hulihan, T. J. (1995). A gender-specific mechanism for pair bonding – oxytocin and partner preference formation in monogamous voles. *Behavioral Neuroscience*, 109, 782–789.

Kikuyama, S., Hasunuma, I., Toyoda, F., Haraguchi, S., and Tsutsui, K. (2009). Hormone-mediated reproductive behavior in the red-bellied newt. *Annals of the New York Academy of Sciences*, 1163, 179–186.

Kikuyama, S., Nakada, T., Toyoda, F., Iwata, T., Yamamoto, K., and Conlon, J. M. (2005). Amphibian pheromones and endocrine control of their secretion. *Annals of the New York Academy of Sciences*, 1040, 123–130.

Kime, N. M., Whitney, T. K., Davis, E. S., and Marler, C. A. (2007). Arginine vasotocin promotes calling behavior and call changes in male tungara frogs. *Brain Behavior and Evolution*, 69, 254–265.

Kime, N. M., Whitney, T. K., Ryan, M. J., Rand, A. S., and Marler, C. A. (2010). Treatment with arginine vasotocin alters mating calls and decreases call attractiveness in male tungara frogs. *General and Comparative Endocrinology*, 165, 221–228.

Klomberg, K. F. and Marler, C. A. (2000). The neuropeptide arginine vasotocin alters male call characteristics involved in social interactions in the grey treefrog, *Hyla versicolor*. *Animal Behaviour*, 59, 807–812.

Leary, C. J. (2009). Hormones and acoustic communication in anuran amphibians. *Integrative and Comparative Biology*, 49, 452–470.

Leary, C. J., Jessop, T. S., Garcia, A. M., and Knapp, R. (2004). Steroid hormone profiles and relative body condition of calling and satellite toads: Implications for proximate regulation of behavior in anurans. *Behavioral Ecology*, 15, 313–320.

Lema, S. C. and Nevitt, G. A. (2004). Exogenous vasotocin alters aggression during agonistic exchanges in male

amargosa river pupfish (*Cyprinodon nevadensis amargosae*). *Hormones and Behavior*, 46, 628–637.

Lewis, C. M., Dolence, E. K., Hubbard, C. S., and Rose, J. D. (2005). Identification of roughskin newt medullary vasotocin target neurons with a fluorescent vasotocin conjugate. *Journal of Comparative Neurology*, 491, 381–389.

Lewis, C. M., Dolence, E. K., Zhang, Z. J., and Rose, J. D. (2004). Fluorescent vasotocin conjugate for identification of the target cells for brain actions of vasotocin. *Bioconjugate Chemistry*, 15, 909–914.

Lim, M. M., Hammock, E. A. D., and Young, L. J. (2004a). The role of vasopressin in the genetic and neural regulation of monogamy. *Journal of Neuroendocrinology*, 16, 325–332.

Lim, M. M., Wang, Z. X., Olazabal, D. E., Ren, X. H., Terwilliger, E. F., and Young, L. J. (2004b). Enhanced partner preference in a promiscuous species by manipulating the expression of a single gene. *Nature*, 429, 754–757.

Lowry, C. A., Richardson, C. F., Zoeller, T. R., Miller, L. J., Muske, L. E., and Moore, F. L. (1997). Neuroanatomical distribution of vasotocin in a urodele amphibian (*Taricha granulosa*) revealed by immunohistochemical and *in situ* hybridization techniques. *Journal of Comparative Neurology*, 385, 43–70.

Maney, D. L., Goode, C. T., and Wingfield, J. C. (1997). Intraventricular infusion of arginine vasotocin induces singing in a female songbird. *Journal of Neuroendocrinology*, 9, 487–491.

Marin, O., Gonzalez, A., and Smeets, W. (1997). Basal ganglia organization in amphibians: Afferent connections to the striatum and the nucleus accumbens. *Journal of Comparative Neurology*, 378, 16–49.

Markman, S., Hill, N., Todrank, J., Heth, G., and Blaustein, L. (2009). Differential aggressiveness between fire salamander (*Salamandra infraimmaculata*) larvae covaries with their genetic similarity. *Behavioral Ecology and Sociobiology*, 63, 1149–1155.

Marler, C. A., Boyd, S. K., and Wilczynski, W. (1999). Forebrain arginine vasotocin correlates of alternative mating strategies in cricket frogs. *Hormones and Behavior*, 36, 53–61.

Marler, C. A., Chu, J., and Wilczynski, W. (1995). Arginine vasotocin injection increases probability of calling in cricket frogs, but causes call changes characteristic of less aggressive males. *Hormones and Behavior*, 29, 554–570.

Moore, F. L., Boyd, S. K., and Kelley, D. B. (2005). Historical perspective: Hormonal regulation of behaviors in amphibians. *Hormones and Behavior*, 48, 373–383.

Moore, F. L. and Lowry, C. A. (1998). Comparative neuroanatomy of vasotocin and vasopressin in amphibians and other vertebrates. *Comparative Biochemistry and Physiology C-Toxicology and Pharmacology*, 119, 251–260.

Moore, F. L., Richardson, C., and Lowry, C. A. (2000). Sexual dimorphism in numbers of vasotocin-immunoreactive neurons in brain areas associated with reproductive behaviors in the roughskin newt. *General and Comparative Endocrinology*, 117, 281–298.

Moore, F. L., Wood, R. E., and Boyd, S. K. (1992). Sex steroids and vasotocin interact in a female amphibian (*Taricha granulosa*) to elicit female-like egg-laying behavior or male-like courtship. *Hormones and Behavior*, 26, 156–166.

O'Bryant, E. L. and Wilczynski, W. (2010). Changes in plasma testosterone levels and brain AVT cell number during the breeding season in the green treefrog. *Brain Behavior and Evolution*, 75, 271–281.

O'Connell, L. A., Ding, J. H., Ryan, M. J., and Hofmann, H. A. (2011). Neural distribution of the nuclear progesterone receptor in the tungara frog, *Physalaemus pustulosus*. *Journal of Chemical Neuroanatomy*, 41, 137–147.

Oldfield, R. G. and Hofmann, H. A. (2011). Neuropeptide regulation of social behavior in a monogamous cichlid fish. *Physiology and Behavior*, 102, 296–303.

Orchinik, M., Murray, T. F., and Moore, F. L. (1991). A corticosteroid receptor in neural membranes. *Science*, 252, 1848–1851.

Penna, M., Capranica, R. R., and Somers, J. (1992). Hormone-induced vocal behavior and midbrain auditory sensitivity in the green treefrog, *Hyla cinerea*. *Journal of Comparative Physiology a-Sensory Neural and Behavioral Physiology*, 170, 73–82.

Propper, C. R. and Dixon, T. B. (1997). Differential effects of arginine vasotocin and gonadotropin releasing hormone on sexual behaviors in an anuran amphibian. *Hormones and Behavior*, 32, 99–104.

Raimondi, D. and Diakow, C. (1981). Sex dimorphism in responsiveness to hormonal induction of female behavior in frogs. *Physiology and Behavior*, 27, 167–170.

Rose, J. D., Kinnaird, J. R., and Moore, F. L. (1995). Neurophysiological effects of vasotocin and corticosterone on medullary neurons – implications for hormonal control of amphibian courtship behavior. *Neuroendocrinology*, 62, 406–417.

Rose, J. D. and Moore, F. L. (2002). Behavioral neuroendocrinology of vasotocin and vasopressin and the

sensorimotor processing hypothesis. *Frontiers in Neuroendocrinology*, 23, 317–341.

Santangelo, N. and Bass, A. H. (2006). New insights into neuropeptide modulation of aggression: Field studies of arginine vasotocin in a territorial tropical damselfish. *Proceedings of the Royal Society B-Biological Sciences*, 273, 3085–3092.

Schmidt, R. S. (1984). Mating call phonotaxis in the female american toad: Induction by hormones. *General and Comparative Endocrinology*, 55, 150–156.

Schmidt, R. S. (1985). Prostaglandin-induced mating call phonotaxis in female american toad: Facilitation by progesterone and arginine vasotocin. *Journal of Comparative Physiology A-Neuroethology Sensory Neural and Behavioral Physiology*, 156, 823–829.

Searcy, B. T., Bradford, C. S., Thompson, R. R., Filtz, T. M., and Moore, F. L. (2011). Identification and characterization of mesotocin and V1a-like vasotocin receptors in a urodele amphibian, *Taricha granulosa*. *General and Comparative Endocrinology*, 170, 131–143.

Semsar, K., Kandel, F. L. M., and Godwin, J. (2001). Manipulations of the AVT system shift social status and related courtship and aggressive behavior in the bluehead wrasse. *Hormones and Behavior*, 40, 21–31.

Semsar, K., Klomberg, K. F., and Marler, C. (1998). Arginine vasotocin increases calling site acquisition by nonresident male grey treefrogs. *Animal Behaviour*, 56, 983–987.

Smeets, W. and Gonzalez, A. (2001). Vasotocin and mesotocin in the brains of amphibians: State of the art. *Microscopy Research and Technique*, 54, 125–136.

Ten Eyck, G. R. (2005). Arginine vasotocin activates advertisement calling and movement in the territorial Puerto Rican frog, *Eleutherodactylus coqui*. *Hormones and Behavior*, 47, 223–229.

Thompson, R. R., Dickinson, P. S., Rose, J. D., Dakin, K. A., Civiello, G. M., Segerdahl, A., and Bartlett, R. (2008). Pheromones enhance somatosensory processing in newt brains through a vasotocin-dependent mechanism. *Proceedings of the Royal Society B-Biological Sciences*, 275, 1685–1693.

Thompson, R. R. and Moore, F. L. (2000). Vasotocin stimulates appetitive responses to the visual and pheromonal stimuli used by male roughskin newts during courtship. *Hormones and Behavior*, 38, 75–85.

Thompson, R. R. and Moore, F. L. (2003). The effects of sex steroids and vasotocin on behavioral responses to visual and olfactory sexual stimuli in ovariectomized female roughskin newts. *Hormones and Behavior*, 44, 311–318.

Tito, M. B., Hoover, M. A., Mingo, A. M., and Boyd, S. K. (1999). Vasotocin maintains multiple call types in the gray treefrog, *Hyla versicolor*. *Hormones and Behavior*, 36, 166–175.

Tornick, J. K. (2010). Factors affecting aggression during nest guarding in the eastern red-backed salamander (*Plethodon cinereus*). *Herpetologica*, 66, 385–392.

Toyoda, F., Yamamoto, K., Ito, Y., Tanaka, S., Yamashita, M., and Kikuyama, S. (2003). Involvement of arginine vasotocin in reproductive events in the male newt *Cynops pyrrhogaster*. *Hormones and Behavior*, 44, 346–353.

Trainor, B. C., Rouse, K. L., and Marler, C. A. (2003). Arginine vasotocin interacts with the social environment to regulate advertisement calling in the gray treefrog (*Hyla versicolor*). *Brain Behavior and Evolution*, 61, 165–171.

Urano, A., Hyodo, S., and Suzuki, M. (1992). Molecular evolution of neurohypophyseal hormone precursors. *Progress in Brain Research*, 92, 39–46.

Walkowiak, W., Berlinger, M., Schul, J., and Gerhardt, H. C. (1999). Significance of forebrain structures in acoustically guided behavior in anurans. *European Journal of Morphology*, 37, 177–181.

Wang, Z. X. and Aragona, B. J. (2004). Neurochemical regulation of pair bonding in male prairie voles. *Physiology and Behavior*, 83, 319–328.

Weintraub, A. S., Kelley, D. B., and Bockman, R. S. (1985). Prostaglandin E2 induces receptive behaviors in female *Xenopus laevis*. *Hormones and Behavior*, 19, 386–399.

Wilczynski, W., Lynch, K. S., and O'Bryant, E. L. (2005). Current research in amphibians: Studies integrating endocrinology, behavior, and neurobiology. *Hormones and Behavior*, 48, 440–450.

Williams, J. R., Insel, T. R., Harbaugh, C. R., and Carter, C. S. (1994). Oxytocin administered centrally facilitates formation of a partner preference in female prairie voles (*Microtus ochrogaster*). *Journal of Neuroendocrinology*, 6, 247–250.

Winslow, J. T., Hearn, E. F., Ferguson, J., Young, L. J., Matzuk, M. M., and Insel, T. R. (2000). Infant vocalization, adult aggression, and fear behavior of an oxytocin null mutant mouse. *Hormones and Behavior*, 37, 145–155.

Wrobel, L. J., Reymond-Marron, I., Dupre, A., and Raggenbass, M. (2010). Oxytocin and vasopressin enhance synaptic transmission in the hypoglossal motor nucleus

of young rats by acting on distinct receptor types *Neuroscience*, 165, 723–735.

Yao, M., Hu, F., and Denver, R. J. (2008). Distribution and corticosteroid regulation of glucocorticoid receptor in the brain of *Xenopus laevis*. *Journal of Comparative Neurology*, 508, 967–982.

Zoeller, R. T. and Moore, F. L. (1982). Duration of androgen treatment modifies behavioral response to arginine vasotocin in *Taricha granulosa*. *Hormones and Behavior*, 16, 23–30.

Zoeller, R. T. and Moore, F. L. (1986). Correlation between immunoreactive vasotocin in optic tectum and seasonal changes in reproductive behaviors of male rough-skinned newts. *Hormones and Behavior*, 20, 148–154.

Zoeller, R. T. and Moore, F. L. (1988). Brain arginine vasotocin concentrations related to sexual behaviors and hydromineral balance in an amphibian. *Hormones and Behavior*, 22, 66–75.

Zornik, E. and Kelley, D. B. (2007). Breathing and calling: Neuronal networks in the *Xenopus laevis* hindbrain. *Journal of Comparative Neurology*, 501, 303–315.

7

Nonapeptide mechanisms of avian social behavior and phenotypic diversity

James L. Goodson

7.1 Why are studies of nonapeptide systems in birds useful and important?

Nonapeptide circuits arising from parvocellular and magnocellular neurons of the preoptic area (POA) and hypothalamus are found across all vertebrate animals, and strong conservation is likewise observed for the evolutionarily "newer" cell groups that first appeared in stem tetrapods, including accessory magnocellular populations in the POA and anterior hypothalamus (AH), which produce arginine vasotocin (AVT) and/or mesotocin (MT) in non-mammalian tetrapods, and their respective homologs arginine vasopressin (AVP) and oxytocin (OT) in most mammals, plus parvocellular AVT/AVP populations in the suprachiasmatic nucleus and medial bed nucleus of the stria terminalis (BSTm). These newer populations are nearly ubiquitous across mammals, birds, amphibians and reptiles (Moore and Lowry, 1998; Goodson and Bass, 2001; De Vries and Panzica, 2006; Goodson, 2008). Projections from several nonapeptide cell groups extensively innervate a core social behavior network of the limbic forebrain and midbrain that is common to all vertebrates and exceptionally similar across amniotes (i.e. birds, reptiles and mammals) (Goodson, 2005; Goodson and Kabelik, 2009).

Nonapeptides are therefore a conserved and common feature of the vertebrate social brain, and hence studies in almost any species may have the potential to yield broadly relevant insights into the behavioral functions of nonapeptide systems. However, certain topics are more tractable in some taxa than others, and in several respects, birds offer opportunities that cannot be matched in other vertebrate classes. As described in the following sections, birds offer truly exceptional opportunities to study nonapeptide mechanisms of social communication, grouping, and aggression, and given that more than 80% of avian species are socially monogamous and biparental (Cockburn, 2006), birds offer extensive opportunities to study mechanisms of mate choice, pair bonding and parental care (most notably, *paternal* care), and also cooperative care by extended families, which is rare in mammals but common in birds (Reichard and Christophe, 2003; Hatchwell, 2009). Much of this potential has yet to be realized, although substantial progress has been made with respect to nonapeptide mechanisms that titrate group size preferences and produce species differences in flocking and territoriality. This line of inquiry has capitalized on the social diversity of the family Estrildidae, which contains many finch species that are very similar except in their grouping behavior (Goodwin, 1982; Goodson et al., 2006; Goodson and Wang, 2006; Goodson et al., 2009c).

The estrildid family also includes the zebra finch (*Taeniopygia guttata*), a species that exhibits an extraordinary range of robust and quantifiable behaviors. Virtually the full range of species-typical

Oxytocin, Vasopressin, and Related Peptides in the Regulation of Behavior, ed. E. Choleris, D. W. Pfaff, and M. Kavaliers.
Published by Cambridge University Press.

zebra finch behavior is easily quantified in the lab, given that zebra finches are diurnal and employ primary modes of communication that are directly observable by humans. We regularly quantify over 20 behaviors while zebra finches are interacting in a colony environment, and are able to track the natural formation of monogamous pair bonds as birds interact with each other in groups (Goodson et al., 1999; Kabelik et al., 2009). Furthermore, unselected domestic zebra finches are behaviorally indistinguishable from wild-caught zebra finches (Morris, 1958).

From a biomedical standpoint, zebra finches are also excellent models for humans, given (1) their primary reliance upon visual and acoustic communication, (2) their human-like social organization, which is based upon long-term monogamy and biparental nuclear families that are embedded within larger social groups, and (3) the absence of cyclical and obligate reproductive quiescence (i.e., like humans, zebra finches can reproduce continuously under permissive conditions) (Zann, 1996).

Robustness of the vasotocin system. AVT/AVP neurons are present in the BSTm of all land vertebrates that have been examined thus far, with the exception of Syrian hamsters (*Mesocricetus auratus*) (Moore and Lowry, 1998; Goodson and Bass, 2001; De Vries and Panzica, 2006; Bolborea et al., 2010). However, AVT/AVP peptide is often difficult to detect in the BSTm, and at least in prairie voles (*Microtus ochrogaster*), increased AVP mRNA is associated with decreased immunodetection of AVP, presumably because of peptide release (Bamshad et al., 1993; Wang et al., 1994). In addition, AVT/AVP neurons in the BSTm are steroid dependent and/or seasonally variable in many species (Goodson and Bass, 2001), potentially making AVT/AVP peptide even more difficult to detect in animals that are in non-reproductive condition. Fortuitously, zebra finches exhibit a variety of characteristics that make them ideal organisms for the study of these neurons. First, the number of AVT-ir neurons in the BSTm of zebra finches is not influenced by hormonal manipulations (Kabelik et al., 2010; but see Kimura et al., 1999), and thus we reliably observe large numbers of these neurons, regardless of whether the birds have been housed under reproductively stimulatory or non-stimulatory conditions. This is consistent with the fact that zebra finches breed opportunistically under a wide range of ecological conditions (and if housed under permissive conditions, captive zebra finches will breed almost continuously) (Zann, 1996). In addition, immunodetection of the BSTm neurons in zebra finches is not influenced by social interaction and does not correlate with constitutive or socially elicited Fos expression, indicating that we can explore the functional properties of these cells without the confound of conditional changes in their immunodetectability (Goodson and Wang, 2006; Goodson et al., 2009b; Kabelik et al., 2010). We know of no other species that offers such a readily studied AVT/AVP system in the BSTm.

7.2 Functional organization of avian nonapeptide systems

As in other amniotes, the major populations of nonapeptide neurons in birds lie within the supraoptic and paraventricular nuclei of the hypothalamus (SON and PVN, respectively) (Mikami et al., 1978; Kiss et al., 1987; Panzica et al., 1999; D'Hondt et al., 2000). The SON contains only magnocellular neurons, whereas the PVN expresses both magnocellular and parvocellular cell types. Additional magnocellular neurons are located in accessory cell groups that tend to be associated with the hypophyseal tract; these are more numerous for AVT than MT (Barth et al., 1997). The magnocellular neurons release nonapeptides via the posterior pituitary and thereby regulate hydromineral balance, cardiovascular tone, oviposition, and likely androgen production (Saito et al., 1987; Robinzon et al., 1988; Grossmann et al., 1995; Sharp et al., 1995; Sasaki et al., 1998; Harding and Rowe, 2003; Nephew et al., 2005a; Nephew et al., 2005b; Madison et al., 2008). These peripheral physiological effects likely feed back to the brain to influence behavior or may be otherwise coordinated with the central modulation of behavior. For instance, AVT exerts

inhibitory control over courtship in male zebra finches, likely via the inhibition of androgen production (Harding and Rowe, 2003); nonapeptides induce bearing down (laying) behavior in addition to effects on oviposition in chickens (*Gallus gallus*) (Takahashi and Kawashima, 2003); and in European starlings (*Sturnis vulgaris*), subcutaneous injections of AVT attenuate aggressive responses to crowding concomitantly with reductions in heart rate (Nephew et al., 2005a).

Nonapeptides may also influence peripheral body states, and thereby behavior, via hypothalamic projections to autonomic regions of the brainstem (Kiss et al., 1987; Panzica et al., 1999) and to the median eminence, where AVT induces release of adrenocorticotropin from the anterior pituitary (Castro et al., 1986; Romero et al., 1998). These projections almost certainly arise from parvocellular neurons of the PVN, as in mammals (De Vries and Buijs, 1983), although the behavioral relevance of this circuitry has not yet been investigated in birds. Likewise, the majority of nonapeptide functions that we will discuss in the sections below are likely attributable, at least primarily, to the parvocellular neurons of the PVN (which produce AVT and MT) and BSTm (AVT only). However, volumetric peptide release from the magnocellular populations of the PVN and SON may also occur, as shown for mammals (Landgraf and Neumann, 2004; Ludwig and Leng, 2006).

In most species that have been investigated, the AVT cell group of the BSTm is sexually dimorphic and sex steroid dependent in adulthood (Jurkevich et al., 1996; Goodson and Bass, 2001; De Vries and Panzica, 2006), and females may have virtually no AVT-immunoreactive (-ir) cells at all, as in chickens and Japanese quail (*Coturnix japonica*) (Jurkevich et al., 1997; Jurkevich et al., 1999; Panzica et al., 2001; Klein et al., 2006). Similar observations are made for AVT-ir fiber densities within putative projection targets of the BSTm AVT neurons, including the lateral septum (LS), medial preoptic nucleus, habenula, and ventral pallidum (Jurkevich et al., 1996; Goodson and Bass, 2001; De Vries and Panzica, 2006). Interestingly, organizational effects of estradiol on this circuitry are feminizing in quail, which is the opposite of organizational effects in rodents, whereas activational effects of estradiol are masculinizing in both adult quail and rodents. Thus, the endpoints are the same in quail and rodents, but the developmental mechanisms are somewhat different (De Vries and Panzica, 2006).

Although songbirds such as canaries (*Serinus canaria*) and dark-eyed juncos (*Junco hyemalis*) also exhibit hormonal and/or seasonal regulation of the BSTm neurons and their projections (Voorhuis et al., 1988; Voorhuis et al., 1991b; Plumari et al., 2004), not all songbirds follow this pattern. For instance, highly opportunistic estrildid finch species do not show reproductive condition-dependent variation in AVT-ir cell numbers (Kabelik et al., 2010), and consistent with this observation, combined aromatase inhibition and androgen receptor blockade does not influence the number or optical density of BSTm AVT-ir neurons in zebra finches. However, this manipulation nonetheless reduces constitutive colocalization of Fos with AVT in the BSTm, suggesting that transcriptional activity of the AVT-ir neurons is still under hormonal control. Given that aromatase inhibition alone does not produce this effect, hormonal effects on transcription are most likely androgenic. Finally, in sharp contrast to other birds, estrildid finches also exhibit very modest or no sexual dimorphism in the BSTm cell group (Kabelik et al., 2010).

Nonapeptide receptor types in birds are both similar and dissimilar to mammals. The cloned receptors in chickens include a single V_{1b}-like receptor (VT2) that is expressed solely in the pituitary, a single OT-like receptor (VT3), and two receptors that have properties similar to the mammalian V_{1a} receptor (V1aR), the VT1 and VT4 (Baeyens and Cornett, 2006; Cornett et al., 2007). Relative to the VT4, the VT1 receptor exhibits relatively low sequence homology with the mammalian V1aR and greater sequence identity with V_2-like receptors (Leung et al., in press). However, the VT1 nonetheless exhibits key amino acid residues that may confer V1aR-like binding properties (Acharjee et al., 2004), and unlike the mammalian V_2 receptor, the VT1 is expressed in the

brain (Leung et al., in press). Avian nonapeptide receptors appear to be promiscuous (Leung et al., 2009), but even so, they must exhibit differential sensitivities to AVT and MT, given that the behavioral and physiological effects of MT and AVT are not identical (Romero and Wingfield, 1998; Goodson et al., 2004; Takahashi and Kawashima, 2008; Goodson et al., 2009c). Furthermore, VT1 receptors expressed in COS7 cells exhibit a greater affinity for AVT and AVP than for MT and OT (Tan et al., 2000).

7.3 Peptide modulation of song

Somewhat surprisingly, only a handful of studies have examined the effects of nonapeptides on song. These studies uniformly support a role for AVT in vocal communication, but have employed several different manipulations and several different species. The earliest demonstration that nonapeptides influence singing came from a study in which testosterone-treated canaries were given subcutaneous injections of an AVT analog at several time points during the year. The analog increased and decreased song duration in a seasonal manner, leading the authors to conclude that AVT-mediated seasonal transitions in singing (Voorhuis et al., 1991a). As already noted, peripheral AVT delivered via minipumps for 48 h decreases courtship singing in male zebra finches, but this effect was eliminated by testosterone treatment, suggesting that AVT decreased androgen production (Harding and Rowe, 2003). On a more short-term basis, intramuscular injections of AVT in male Japanese quail (which are fowl, not songbirds) rapidly and potently reduce crowing and intramuscular injections of a V_1 antagonist reverse this effect (Castagna et al., 1998). Crowing is a non-song vocalization that appears to function solely as a mate attractant (Goodson and Adkins-Regan, 1997).

In contrast, experiments using central administrations of peptides and antagonists have provided no evidence that courtship singing is influenced by direct actions within the brain. In male zebra finches, directed song is unaffected by acute injections of AVT or a V1aR antagonist into the lateral ventricles or LS; by acute injections of MT or an OT antagonist into the lateral ventricles; by peripheral administrations of a novel V1aR antagonist that crosses the blood–brain barrier; or by chronic intraventricular administrations of a V1aR antagonist during the first three days of colony establishment, a time when courtship activity is very high (Goodson and Adkins-Regan, 1999; Goodson et al., 2004; Goodson et al., 2009a; Kabelik et al., 2009). However, these studies reliably demonstrated effects of AVT and the V1aR antagonist on aggression (these are context dependent; see Section 7.5).

In male field sparrows that were singly housed in large aviaries then placed in their natural habitat (one sparrow per field), predawn infusions of AVT into the LS significantly increased the use of the agonistic, "complex" song type during the dawn singing period, but had no effect on the multipurpose, "simple" song type that is used for mate attraction (Goodson, 1998a). Notably, this was the first study to demonstrate that social behavior circuitry of the basal forebrain controls the choice of *what* songs are produced by the telencephalic song system. Intraventricular infusions of AVT also induced agonistic singing in female white-crowned sparrows (*Zonotrichia leucophrys*), whereas no singing was exhibited following infusions of vehicle (Maney et al., 1997).

Male Lincoln's sparrows (*Melospiza lincolnii*) that were exposed to playback of high-quality song versus low-quality song displayed more singing and showed a significant decrease in AVT immunoreactivity within the BSTm, LS, and POA (Sewall et al., 2010). Whether these decreases reflect greater release or lower production remains to be determined, but either way, these findings underscore the importance of extrahypothalamic AVT circuitry for agonistic communication. AVT immunoreactivity in the BSTm and ventrolateral LS is also modestly greater in male white-throated sparrows (*Zonotrichia albicollis*) of the white-striped morph relative to the tan-striped morph (Maney et al., 2005). The white-striped morph is more aggressive, sings more and exhibits less paternal

care than the tan-striped morph (Knapton and Falls, 1983; Kopachena and Falls, 1993). These findings are at least partially consistent with findings based on pharmacological manipulations of septal AVT in field sparrows, given that agonistic song is promoted by AVT infusions (Goodson, 1998a). However, the same manipulations decrease overt aggression in field sparrows, suggesting that the contributions of septal AVT to agonistic communication do not reflect a general increase in agonistic motivation. An alternative hypothesis is that septal AVT is anxiogenic in field sparrows, as typically found for rodents (Landgraf et al., 1995), which could account for the pattern of observed effects (Goodson, 2008).

7.4 Nonapeptides and the evolution of social group sizes in finches

7.4.1 A comparative approach to species-typical group size

Our early studies showed that septal AVT modulates aggression differentially in male zebra finches and field sparrows: Septal AVT infusions increase aggression in the highly gregarious zebra finch (which was tested in a mate competition context in order to promote fighting) (Goodson and Adkins-Regan, 1999), but decrease resident–intruder aggression in the territorial field sparrow (Goodson, 1998a). We further obtained field sparrow-like effects in a territorial estrildid species that is otherwise very similar to zebra finches, the violet-eared waxbill (*Uraeginthus granatina*) (Goodson, 1998b), but as described in Section 7.5, we now know that the different AVT effects on aggression are related to the context of testing and do not reflect species-typical social structure. Nonetheless, through these experiments we began to ponder the basis for species differences in grouping, a phenotypic dimension that is of profound and obvious importance, but an aspect of behavior that had never been subjected to neurobiological analysis.

The basic plan that we generated was deceptively straightforward: We would simply collect several related species that exhibit a wide range of species-typical group sizes (preferably with cases of convergent evolution), but those that do not differ in mating system, patterns of parental care, or other major aspects of behavior and ecology. To our surprise, we found that this was not possible to achieve in rodents, due to confounds from mating system and other variables (King, 1968; Wolff, 1985), and after an exhaustive analysis of North American songbirds, we ended up back where we started – with the estrildid finch family. We have collected five estrildid finch species that are all socially monogamous and biparental, exhibit long-term (likely life-long) pair bonds, live in arid or semi-arid grassland scrub, and breed opportunistically or semi-opportunistically in relation to rainfall (Skead, 1975; Goodwin, 1982; Zann, 1996; Goodson et al., 2006; Goodson and Kingsbury, 2011). These include two territorial African species that live as male–female pairs year-round (violet-eared waxbill and melba finch, *Pytilia melba*); two highly gregarious, colonially breeding species that exhibit modal group sizes of approximately 100 (zebra finch and spice finch, *Lonchura punctulata*); and a moderately gregarious species, the Angolan blue waxbill (*Uraeginthus angolensis*), which exhibits a modal group size of approximately 20. The two territorial species have evolved their territorial behavior independently and the two colonial species have also evolved their extreme sociality independently (Goodson and Kingsbury, 2011). Our laboratory population of the Melba finch and the two waxbill species was established through the breeding of birds that were caught from a single location in the Kalahari thornscrub of South Africa in 2001. Zebra finches and wild-caught spice finches (likely of an Indian subspecies) have been obtained from commercial suppliers.

7.4.2 Vasotocin circuitry of the BSTm and LS differentiates territorial and flocking finch species

In order to identify brain areas that may be important for differentiating territorial and flocking finch species, we designed a simple behavioral paradigm

to employ in conjunction with analyses of Fos and egr-1 (ZENK) induction. Subjects were acclimated to a quiet room and then remained in that room overnight in a small cage. The next morning, a wire barrier was inserted into the middle of the cage, and a same-sex conspecific was introduced to the side opposite from the subject (both male and female subjects were used). Control subjects were exposed only to the wire barrier. This manipulation did not reliably elicit overt behavioral response, thus allowing us to examine species differences in immediate early gene (IEG) response that should primarily reflect species differences in sensory or motivational processes, not species differences in behavior (note that this first study was conducted prior to collections of the Melba finch) (Goodson et al., 2005). IEG responses in the territorial violet-eared waxbill were significantly greater than all three flocking species in several brain areas that also appear to collectively promote pup aversion behavior in female rats (Sheehan et al., 2000; Sheehan et al., 2001) – medial extended amygdala (including the BSTm), ventrolateral LS, AH and the lateral portion of the ventromedial hypothalamus (Goodson et al., 2005).

These findings showed that species differences in territoriality and grouping are experimentally tractable, but at this level of analysis, we observed no response patterns that appeared to be positively associated with gregariousness. However, a scattering of findings has shown that intercalated neuron phenotypes in the amygdala can exhibit very different, and even opposite, responses to stimuli (Nishijo et al., 1988; Choi et al., 2005; Paton et al., 2006), and thus we began a series of studies to neurochemically identify specific neuronal phenotypes that may be important for grouping. We were particularly interested in the AVT circuitry of the BSTm and LS, given that several studies in rodents, particularly microtine voles, had indirectly linked AVP circuitry of the BSTm and LS to species-specific social behaviors, particularly pair bonding, social recognition and affiliation (Bamshad et al., 1993; Bamshad et al., 1994; Wang et al., 1994; Liu et al., 2001; Landgraf et al., 2003). However, despite more than 20 years of research on the AVT/AVP cells in the BSTm (mostly focused on their sexual differentiation and regulation by hormones), the IEG responses of the BSTm AVT/AVP neurons had not yet been examined and the cells themselves had not been directly manipulated. A single study had shown that AVP mRNA increases significantly in the BSTm of male prairie voles following overnight cohabitation with a female (Wang et al., 1994), but this represented the sum of knowledge regarding the short-term responses of the BSTm AVT/AVP neurons to social stimuli.

Using the testing paradigm described above, we first quantified immunocytochemical double-labeling for AVT and Fos following exposure to same-sex conspecifics (Goodson and Wang, 2006). Relative to handled controls, AVT-ir BSTm neurons in the highly gregarious species *increased* their Fos activity in response to same-sex stimuli (e.g., Figure 7.1A), whereas the territorial estrildid species tended to *decrease* their Fos expression over a period of 90 min following exposure to a same-sex conspecific (e.g., Figure 7.1B; note that 90 min represents two half-lives of the Fos protein, at least in rodents). The modestly gregarious Angolan blue waxbill showed only a slight and non-significant increase in AVT-Fos colocalization, but a sufficient increase to yield a significant Species*Condition interaction when compared with the closely related, territorial violet-eared waxbill (Goodson and Wang, 2006). No sex differences were observed in any of these results and similar effects were not observed for AVT-Fos colocalization in the PVN (J. L. Goodson and Y. Wang, unpublished observations).

This pattern of results suggested the hypothesis that the AVT neurons of the BSTm are sensitive to the valence of social stimuli, such that the neurons increase their Fos activity in response to social stimuli that are of a positive, affiliation-related nature (e.g., a same-sex conspecific for a gregarious bird) and decrease their Fos activity in response to negative stimuli that normally elicit aversion or aggression (e.g., a same-sex conspecific for a territorial bird). In support of this hypothesis, we found that whereas territorial violet-eared waxbills decrease their AVT-Fos colocalization in response to a same-sex conspecific, as just described, they exhibit a

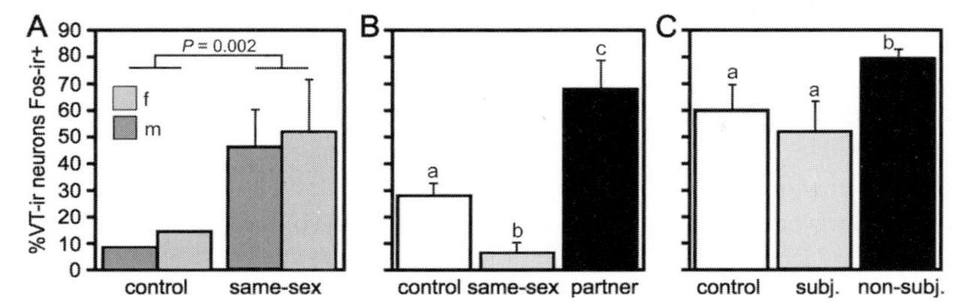

Figure 7.1 Valence sensitivity of the BSTm AVT neurons as demonstrated by context-dependent, socially induced changes in the immunocolocalization of AVT and Fos (a proxy marker of neural activity). (A) Following isolation in a quiet room, exposure to a same-sex conspecific through a wire barrier produces a robust increase in AVT neuronal activity (% of AVT neurons colocalizing Fos) in both male and female zebra finches. Zebra finches are highly gregarious. Total $n = 10$. (B) This same manipulation produces a significant decrease in AVT-Fos colocalization in the territorial violet-eared waxbill, a species that does not naturally exhibit same-sex affiliation, but exposure to the subject's pair-bond partner (a presumably positive stimulus), produces a robust increase in neuronal activity. Sexes are shown pooled. Total $n = 16$. (C) AVT-Fos colocalization increases in zebra finches following competition with a same-sex conspecific for courtship access to an opposite-sex bird, but not if the subject is paired with a highly aggressive partner and intensely subjugated. Given that subjugated animals were aggressively displaced or attacked 71–210 times during a 10-min interaction, social arousal alone does not increase AVT-Fos colocalization in the BSTm. Sexes are shown pooled. Total $n = 15$. Figures are modified from Goodson and Wang (2006).

very robust increase in colocalization after being reunified with their pair-bond partner (Figure 7.1B). Similarly, in zebra finches, AVT-Fos colocalization in the BSTm increases following courtship interactions in which a same-sex competitor is present, but not if the subject is intensely subjugated by their competitor and not allowed to court (Figure 7.1C) (Goodson and Wang, 2006). This latter observation is important, because it demonstrates that Fos responses of the BSTm AVT neurons do not reflect simple social arousal.

Finally, an additional experiment suggests that BSTm neurons do not respond to positive stimuli that are non-social (Goodson et al., 2009b). For this experiment, male zebra finches were deprived of bathing water for a month and then allowed to either take a bath or court a female. Zebra finches will bathe frequently if an open water source is available, and eight of nine subjects in the bath group bathed during the 15 min test (the other subject was discarded). Despite this robust appetitive response to the bath, no change in AVT-Fos colocalization was observed, whereas colocalization significantly increased in males that were allowed to court. The bath and female stimuli elicited comparable Fos responses in the medial preoptic area, indicating that the bath stimulus was sufficient to drive significant Fos responses. Although additional tests using other non-social stimuli would be informative, these results strongly suggest that Fos activity within the BSTm AVT neurons is influenced primarily by social stimuli. Finally, as an adjunct to this experiment, we examined AVT-Fos colocalization in male finches that reliably failed to court during prescreenings one month prior to sacrifice, and they showed no change in AVT-Fos colocalization in response to a female. Furthermore, these males exhibit only about a third as many AVT-ir neurons in the BSTm as the normal males (Goodson et al., 2009b). As already noted, nonapeptides do not influence directed singing per se (Goodson et al., 2004), and thus the abnormal anatomy of the non-courters may reflect more general social deficits.

AVT-ir cell numbers also vary across the five finch species, with the highly gregarious species exhibiting approximately 10 times more AVT-ir cells in the BSTm than the less-social species (Goodson and Wang, 2006), and similarly, the density of V_{1a}-like

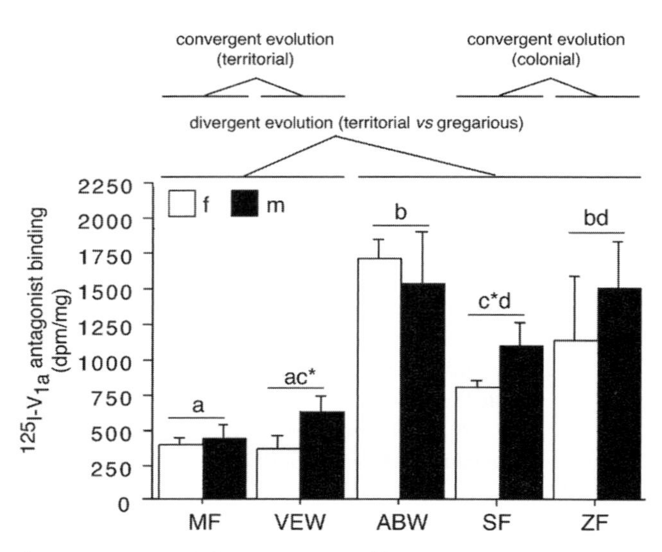

Figure 7.2 Species differences in linear ^{125}I-V1aR antagonist binding in the LS reflect evolutionary convergence and divergence in flocking and territoriality. Linear ^{125}I- V1aR antagonist binding in the dorsal (pallial) LS of the territorial Melba finch (MF), territorial violet-eared waxbill (VEW), moderately gregarious Angolan blue waxbill (ABW), colonial spice finch (SF) and colonial zebra finch (ZF), shown as decompositions per minute/mg (dpm/mg; means ± SEM). Different letters above the error bars denote significant species differences (Fisher's PLSD following significant ANOVA; $P < 0.0001$). Asterisks denote near-significant species differences ($P = 0.06$). Total $n = 40$. Modified from Goodson et al. (2006).

binding sites in the LS is significantly greater in all three flocking species than in the two territorial species (Figure 7.2) (Goodson et al., 2006). These anatomical features may effectively enhance positive responses to conspecific stimuli in the gregarious species, a hypothesis that is addressed below.

7.4.3 Vasotocin circuitry of the BSTm and LS promotes gregariousness

The experiments just described suggest the hypotheses that gregariousness is promoted by (1) activation of V_{1a}-like receptors in the LS, and (2) activity of AVT neurons in the BSTm. In order to test these hypotheses, we recently conducted behavioral assays of gregariousness (group-size preference) and social contact in male zebra finches following either intraseptal infusions of a V1aR antagonist (with vehicle controls) or infusion of AVT antisense oligonucleotides into the BSTm (with scrambled oligonucleotides controls). We additionally conducted assays of anxiety-like behavior using exploration and novelty-suppressed feeding tests (Kelly et al., 2011).

For behavioral quantification of social contact and gregariousness, we place subjects singly into a 1-meter-wide cage that adjoins two smaller cages, containing two and ten other males (Figure 7.3A). Seven perches are distributed along the width of the cage and the two perches closest to the sides are approximately 6 cm from the stimulus cages. The amount of time that subjects spend on these two side perches yields a measure of contact behavior, and the percent of that contact time that subjects spend next to the larger group yields a measure of gregariousness.

Using this paradigm, we found that blockade of septal V_{1a}-like receptors decreased preferences for the larger of two social groups by approximately 80% (median effect), while producing no effects on the choice to spend time in social contact. Antisense oligonucleotides likewise decreased gregariousness

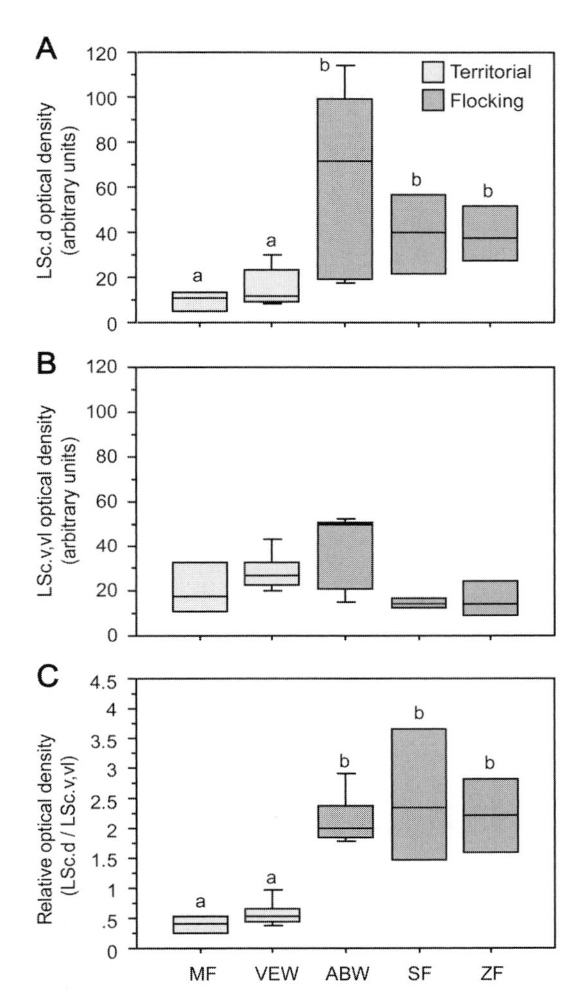

Figure 7.3 Species-specific distributions of OT-like binding sites reflect evolutionary convergence and divergence in flocking and territoriality. (A) Densities of binding sites in the dorsal (pallial) LS of the territorial Melba finch (MF), territorial violet-eared waxbill (VEW), moderately gregarious Angolan blue waxbill (ABW), and two highly gregarious, colonial species – the spice finch (SF) and colonial zebra finch (ZF). No sex differences are observed and sexes were pooled. Different letters above the boxes denote significant species differences (Mann–Whitney $P < 0.05$) following significant Kruskal–Wallis. (B) Binding densities tend to reverse in the subpallial LS ($P = 0.06$), suggesting that species differences in sociality are most closely associated with the relative (*cont.*) densities of binding sites along a dorso-ventral gradient, as confirmed in the bottom panel (C) using a dorsal:ventral ratio. Comparable results are obtained using dorsal-ventral difference measures. Total $n = 23$. Modified from Goodson et al. (2009c).

by a median of ~80%, but also produced a median *increase* in social contact of approximately 25%. These combined results strongly confirm our hypotheses, and further suggest that nonapeptides titrate gregariousness via processes other than those that promote social contact (at least in this particular species, which is extremely gregarious) (Kelly et al., 2011).

Both of our manipulations also produced strong increases in anxiety-like behavior, particularly in the novelty-suppressed feeding test, demonstrating that peptide effects on social group sizes are mechanistically linked to anxiety-like processes (Kelly et al., 2011). These effects are especially intriguing because they are in a direction opposite to the majority of comparable data in rodents, in which septal AVP is generally found to be anxiogenic (Landgraf et al., 1995). Whether this reflects broad phyletic differences between birds and mammals or, more interestingly, active selection in relation to the extreme gregariousness of zebra finches, remains to be determined.

7.4.4 OT-like receptors mediate peptide effects on grouping and exhibit species-specific distributions in territorial and flocking finch species

In mammals, OT is almost always found to promote affiliation (Carter et al., 2008; Heinrichs et al., 2009; Ross and Young, 2009; Goodson and Thompson, 2010), but until recently, the only relevant non-mammalian data on OT-like functions derived from teleost fish. IT inhibits agonistic vocal-motor activity in female and sneak-spawning male plainfin midshipman (*Porichthys notatus*) (Goodson and Bass, 2000) and, of greatest relevance here,

promotes social approach behavior in goldfish (*Carassius auratus*) (Thompson and Walton, 2004).

Using the choice apparatus shown in Figure 7.4A, we recently showed that peripheral injections of the selective OT receptor antagonist desGly–NH2,d(CH2)5[Tyr(Me)2, Thr4] ornithine vasotocin (OTA) decrease gregariousness in zebra finches in a female-specific manner (Figures 7.4B and D) (Goodson et al., 2009c). In contrast, intracerebroventricular infusions of OTA decrease gregariousness in both sexes (Figures 7.4C and E), but infusions of MT increased gregariousness only in females. However, unlike goldfish, these manipulations did not influence the amount of time spent in close proximity to conspecifics. Rather, reductions in the amount of time spent with the larger group were accompanied by approximately equal increases in the amount of time spent with the small group. Similar female-biased effects were observed for novel-familiar preferences, such that peripheral and central administrations of OTA reduced the amount of time that females spent in close proximity to a group of five familiar same-sex birds, and tended to increase the amount of time spent in close proximity to five novel birds (Goodson et al., 2009c).

An intriguing result from these experiments is that ventricular infusions of AVT produced no effect on gregariousness in either males or females. This result is surprising in light of the strong effects that we have obtained using AVT antisense oligonucleotides and intraseptal infusions of a V1aR antagonist (Kelly et al., 2011). However, the number of AVT/AVP populations is relatively large as compared to MT/OT cell groups (Barth et al., 1997) and the various hypothalamic and extrahypothalamic AVT/AVP cell groups exhibit very different Fos responses to social stimuli (see next section). Thus, if the numerous AVT cell groups produce distinct effects on behavior via differential patterns of neuromodulation, then it is perhaps not surprising that clear effects are not obtained using ventricular AVT infusions. Regardless, we observed significantly different effects for MT and AVT in both the novel-familiar and grouping tests (Goodson et al., 2009c), providing good evidence that avian nonapeptide receptors exhibit differential sensitivities to AVT and MT, despite being somewhat promiscuous.

The behavioral effects of MT and the OT receptor antagonist suggest that OT-like receptor distributions may be important contributors to species differences in grouping. In order to test this hypothesis, we examined binding of an iodinated OTA in the five estrildid species using receptor autoradiography. Although species differences were observed in multiple brain areas, those differences distinguished territorial from flocking species only in the LS. As shown in Figure 7.4A, all three flocking species exhibit much higher densities of OT-like binding sites in the dorsal (pallial) LS, whereas this pattern tends to reverse ventrally ($P = 0.06$; Figure 7.4B). If this dorsoventral pattern is an important contributor to gregariousness, we predicted that the *relative* density in the pallial and subpallial LS should more clearly distinguish territorial and flocking species. Indeed, whether expressed as a ratio or a difference measure, the relative density of OT-like binding sites in the pallial and subpallial LS differentiates territorial and flocking species in an exceptionally clear manner (Figure 7.4C). We therefore infused OTA directly into the LS of female zebra finches and found that gregariousness was significantly reduced relative to vehicle control tests. Comparable effects were not obtained with OTA infusions into striatal tissue laterally adjacent to the ventricle (Goodson et al., 2009c).

7.5. Vasotocin modulation of aggression varies with context and phenotype

Although AVT/AVP influence aggression across a wide range of vertebrates (Goodson and Bass, 2001), the relationship between AVT/AVP and aggression is highly complex and not yet fully understood (Veenema et al., 2010). For instance, intraseptal AVT infusions in male field sparrows selectively increase the use of a strictly territorial song type during the "dawn song" period, but inhibit overt, resident–intruder aggression (Goodson, 1998a), and in rats, central AVP release during aggression both

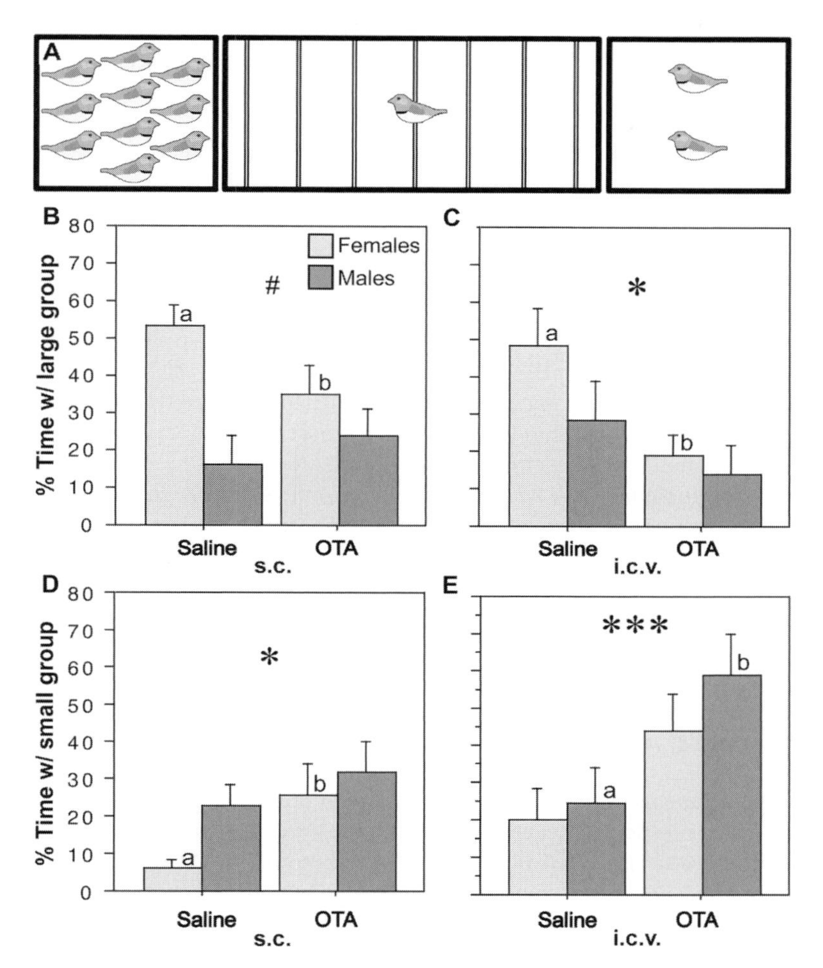

Figure 7.4 Endogenous peptide binding at oxytocin-like receptors promotes preferences for larger groups. (A) Choice apparatus design. A 1-m wide testing cage was subdivided into zones by seven perches (thin lines). Subjects were considered to be within close proximity when they were within 6 cm of a stimulus cage (i.e. on the perches closest to the sides of the testing cage). The stimulus cages contained either two or ten same-sex conspecifics. (B–E) Relative to vehicle treatments, subcutaneous (s.c.) or intracerebroventricular (i.c.v.) administrations of the oxytocin antagonist desGly–NH_2,$d(CH_2)_5$[Tyr(Me)^2, Thr^4]OVT (OTA), reduce the amount of time that zebra finches spend in close proximity to the large group (B-C) and increase the time in close proximity to the small group (D and E). $^*P < 0.05$, $^{***}P < 0.001$, main effect of treatment; $\#P < 0.5$ sex*treatment; $n = 12$ m, 12 f. Letters above the error bars denote significant within-sex effects. Modified from Goodson et al., 2009c.

increases and decreases, depending upon the brain area (Veenema et al., 2010). This complexity may reflect the involvement of many different AVT/AVP cell groups. For instance, at least eight distinct cell groups (mostly hypothalamic, and not including the BSTm population) alter their Fos activity following an aggressive encounter in male mice, but in virtually all cases, AVP-Fos colocalization is more

pronounced in subordinate animals and/or negatively correlated with aggression (Ho et al., 2010). Similarly, in territorial song sparrows, AVT-Fos colocalization in the PVN correlates negatively with aggression, and dominant animals may actually suppress AVT/AVP neuronal activity during aggressive encounters (Goodson and Kabelik, 2009). Consistent with these data, we find that administrations of a V1aR antagonist increase territorial aggression in less-aggressive violet-eared waxbill males that are typically subordinate (i.e., the males that we would expect to show increased activation of the PVN AVT neurons), whereas no effect is observed in more aggressive, dominant animals (Goodson et al., 2009a).

BSTm AVT/AVP neurons do not exhibit Fos responses to resident–intruder encounters in mice (Ho et al., 2010) and they do not respond to simulated territorial intrusions in song sparrows (Goodson and Kabelik, 2009). However, endogenous AVT actually *promotes* male aggression in other social contexts that should activate the BSTm cell group. For instance, endogenous AVT facilitates aggression in the context of mate competition in both zebra finches and violet-eared waxbills (Goodson et al., 2004; Goodson and Kabelik, 2009; Kabelik et al., 2009).

A recent experiment in male zebra finches provides an excellent parallel to the violet-eared waxbill experiments just described, and further demonstrates the functional complexity of endogenous AVT (Kabelik et al., 2009). Male zebra finches were introduced to large nesting cages in groups of four, and five females were provided as potential partners. Subjects were administered twice-daily lateral ventricular infusions of saline vehicle or a AVP antagonist cocktail containing a selective V1aR antagonist and a less-selective V_1 antagonist, and behavioral observations were conducted twice per day for three days. High levels of aggression were exhibited at the time of introduction, mostly focused on competition for mates. As in other mate-competition contexts (see above), aggression was significantly lower on the first day in males that received central V1aR antagonist infusions. However, this effect was completely reversed over subsequent days in the colony environment, such that the AVP antagonist instead increased aggression as most males paired and began to nest (Kabelik et al., 2009). Notably, aggression in paired zebra finches is largely focused on nest defense (Zann, 1996), a behavior that is similar in many ways to territorial aggression.

7.6 Concluding remarks

Of all the vertebrate classes, birds exhibit the highest percentage of social structures that are based on complex affiliative interactions such as long-term monogamy, biparental care, and cooperative breeding within philopatric, extended family groups (Reichard and Christophe, 2003; Cockburn, 2006; Alcock, 2009; Hatchwell, 2009; Goodenough et al., 2010). Birds are vocal, visual, typically conspicuous, and readily observed in the wild, and they exhibit seemingly endless social and ecological diversity. The rapidly expanding literature on avian social cognition and tool use has also documented some exceptional abilities that rival those of apes (Emery et al., 2007; Emery and Clayton, 2009b, Emery and Clayton, 2009a), such as the use of transitive social inference in jays (Paz et al., 2004). Birds also exhibit extreme diversity in social group sizes, and in species that are suitable for laboratory experimentation. We might therefore expect that birds would be heavily represented in the social neuroscience literature, as found for animal behavior literature more generally, but this is far from being the case, perhaps because of historical misunderstandings about the organization of the avian brain (Nieuwenhuys et al., 1998; Reiner et al., 2004). However, it is increasingly clear that social behavior circuits of the basal forebrain and midbrain are similar across all vertebrate groups, and remarkably conserved across amniotes (Goodson, 2005; Martinez-Garcia et al., 2008). This includes strong conservation in the basic features of nonapeptide systems, despite the various forms of species-specific diversity (Goodson and Thompson, 2010).

In the present review of avian nonapeptides, we have focused primarily on a series of studies that have elucidated nonapeptide mechanisms that evolve in relation to species-typical group size and that titrate group-size preferences (gregariousness) in zebra finches. An important revelation from these studies is that the AVT neurons of the BSTm are sensitive to the valence of social stimuli (Goodson and Wang, 2006). We have subsequently found that in male C57BL/6J mice, BSTm AVP neurons exhibit robust Fos responses to copulation (clearly a positive, affiliation-related stimulus) and very modest responses to nonaggressive same-sex chemoinvestigation, but show no greater Fos response to aggressive interactions than simple chemoinvestigation (Ho et al., 2010). This pattern of results demonstrates an outstanding predictive validity for the findings in birds, and suggests that avian studies may illuminate functional features of nonapeptide systems that are common across all amniote classes.

Notably, although the sensitivity of amygdala neurons to valence has been extensively characterized through neurophysiological and neuroimaging studies in mammals (Nishijo et al., 1988; Paton et al., 2006; Goodson and Thompson, 2010), specific cell types that process valence have not previously been identified. Neurophysiological studies in monkeys demonstrate that neurons exhibiting stable preferences for positive or negative stimuli are intercalated within the amygdala (Paton et al., 2006), and therefore any systems-level analysis of valence processing will first require the identification of markers for valence-sensitive cell types. Hence, the discoveries described here for BSTm AVT neurons may represent an important step toward understanding how social value is encoded and used by the brain to regulate behavior, including the titration of gregariousness.

Acknowledgments

The author acknowledges essential support for this work from the National Institutes of Mental Health, grant R01 MH062656; the University of California, San Diego; and Indiana University.

REFERENCES

Acharjee, S., Do-Rego, J. L., Oh da, Y., Ahn, R. S., Choe, H., Vaudry, H., Kim, K., Seong, J. Y., and Kwon, H. B. (2004). Identification of amino acid residues that direct differential ligand selectivity of mammalian and nonmammalian V1a type receptors for arginine vasopressin and vasotocin. Insights into molecular coevolution of V1a type receptors and their ligands. *Journal of Biological Chemistry*, 279, 54445–54453.

Alcock, J. (2009). *Animal Behavior: An Evolutionary Approach*. Sunderland, MA: Sinauer Associates.

Baeyens, D. A. and Cornett, L. E. (2006). The cloned avian neurohypophysial hormone receptors. *Comparative Biochemistry and Physiology B – Biochemistry and Molecular Biology*, 143, 12–19.

Bamshad, M., Novak, M. A., and De Vries, G. J. (1993). Sex and species differences in the vasopressin innervation of sexually naive and parental prairie voles *Microtus ochrogaster* and meadow voles *Microtus pennsylvanicus*. *Journal of Neuroendocrinology*, 5, 247–255.

Bamshad, M., Novak, M. A., and De Vries, G. J. (1994). Cohabitation alters vasopressin innervation and paternal behavior in prairie voles (*Microtus ochrogaster*). *Physiology and Behavior*, 56, 751–758.

Barth, S. W., Bathgate, R. A., Mess, A., Parry, L. J., Ivell, R., and Grossmann, R. (1997). Mesotocin gene expression in the diencephalon of domestic fowl: cloning and sequencing of the MT cDNA and distribution of MT gene expressing neurons in the chicken hypothalamus. *Journal of Neuroendocrinology*, 9, 777–787.

Bolborea, M., Ansel, L., Weinert, D., Steinlechner, S., Pèvet, P., and Klosen, P. (2010). The bed nucleus of the stria terminalis in the Syrian hamster (*Mesocricetus auratus*): absence of vasopressin expression in standard and wild-derived hamsters and galanin regulation by seasonal changes in circulating sex steroids. *Neuroscience*, 165, 819–830.

Carter, C. S., Grippo, A. J., Pournajafi-Nazarloo, H., Ruscio, M. G., and Porges, S. W. (2008). Oxytocin, vasopressin, and sociality. *Progress in Brain Research*, 170, 331–336.

Castagna, C., Absil, P., Foidart, A., and Balthazart, J. (1998). Systemic and intracerebroventricular injections of

vasotocin inhibit appetitive and consummatory components of male sexual behavior in Japanese quail. *Behavioral Neuroscience*, 112, 233–250.

Castro, M. G., Estivariz, F. E., and Iturriza, M. J. (1986). The regulation of the corticomelanotropic cell activity in Aves: II. Effect of various peptides on the release of ACTH from dispersed, perfused duck pituitary cells. *Comparative Biochemistry and Physiology, A*, 83, 71–75.

Choi, G. B., Dong, H. W., Murphy, A. J., Valenzuela, D. M., Yancopoulos, G. D., Swanson, L. W., and Anderson, D. J. (2005). Lhx6 delineates a pathway mediating innate reproductive behaviors from the amygdala to the hypothalamus. *Neuron*, 46, 647–660.

Cockburn, A. (2006). Prevalence of different modes of parental care in birds. *Proceedings of the Royal Society B-Biological Sciences*, 273, 1375–1383.

Cornett, L. E., Jacobi, S. E., and Mikhailova, M. V. (2007). Molecular cloning of an avian vasotocin receptor with homology to themammalian V1a-vasopressin receptor. Direct submission to Genbank 21 Aug 2007, accession number ABV24997.

D'Hondt, E., Eelen, M., Berghman, L., and Vandesande, F. (2000). Colocalization of arginine-vasotocin and chicken luteinizing hormone-releasing hormone-I (cLHRH-I) in the preoptic-hypothalamic region of the chicken. *Brain Research*, 856, 55–67.

De Vries, G. J. and Buijs, R. M. (1983). The origin of the vasopressinergic and oxytocinergic innervation of the rat brain with special reference to the lateral septum. *Brain Research*, 273, 307–317.

De Vries, G. J. and Panzica, G. C. (2006). Sexual differentiation of central vasopressin and vasotocin systems in vertebrates: Different mechanisms, similar endpoints. *Neuroscience*, 138, 947–955.

Emery, N. J. and Clayton, N. S. (2009). Comparative social cognition. *Annual Review of Psychology*, 60, 87–113.

Emery, N. J. and Clayton, N. S. (2009). Tool use and physical cognition in birds and mammals. *Current Opinion in Neurobiology*, 19, 27–33.

Emery, N. J., Seed, A. M., von Bayern, A. M. P., and Clayton, N. S. (2007). Cognitive adaptations of social bonding in birds. *Philosophical Transactions of the Royal Society B: Biological Sciences*, 362, 489–505.

Goodenough, J., McGuire, B., and Jakob, E. (2010). *Perspectives on Animal Behavior*. Hoboken, NJ, John Wiley and Sons, Inc.

Goodson, J. L. (1998a). Territorial aggression and dawn song are modulated by septal vasotocin and vasoactive intestinal polypeptide in male field sparrows (*Spizella pusilla*). *Hormones and Behavior*, 34, 67–77.

Goodson, J. L. (1998b). Vasotocin and vasoactive intestinal polypeptide modulate aggression in a territorial songbird, the violet-eared waxbill (Estrildidae: *Uraeginthus granatina*). *General and Comparative Endocrinology*, 111, 233–244.

Goodson, J. L. (2005). The vertebrate social behavior network: Evolutionary themes and variations. *Hormones and Behavior*, 48, 11–22.

Goodson, J. L. (2008). Nonapeptides and the evolutionary patterning of sociality. *Progress in Brain Research*, 170, 3–15.

Goodson, J. L. and Adkins-Regan, E. (1997). Playback of crows of male Japanese quail elicits female phonotaxis. *Condor*, 99, 990–993.

Goodson, J. L. and Adkins-Regan, E. (1999). Effect of intraseptal vasotocin and vasoactive intestinal polypeptide infusions on courtship song and aggression in the male zebra finch (*Taeniopygia guttata*). *Journal of Neuroendocrinology*, 11, 19–25.

Goodson, J. L. and Bass, A. H. (2000). Forebrain peptides modulate sexually polymorphic vocal circuitry. *Nature*, 403, 769–772.

Goodson, J. L. and Bass, A. H. (2001). Social behavior functions and related anatomical characteristics of vasotocin/vasopressin systems in vertebrates. *Brain Research Reviews*, 35, 246–265.

Goodson, J. L., Eibach, R., Sakata, J., and Adkins-Regan, E. (1999). Effect of septal lesions on male song and aggression in the colonial zebra finch (*Taeniopygia guttata*) and the territorial field sparrow (*Spizella pusilla*). *Behavioral Brain Research*, 98, 167–180.

Goodson, J. L., Evans, A. K., Lindberg, L., and Allen, C. D. (2005). Neuro-evolutionary patterning of sociality. *Proceedings of the Royal Society of London series B*, 272, 227–235.

Goodson, J. L., Evans, A. K., and Wang, Y. (2006). Neuropeptide binding reflects convergent and divergent evolution in species-typical group sizes. *Hormones and Behavior*, 50, 223–236.

Goodson, J. L. and Kabelik, D. (2009). Dynamic limbic networks and social diversity in vertebrates: From neural context to neuromodulatory patterning. *Frontiers in Neuroendocrinology*, 30, 429–441.

Goodson, J. L., Kabelik, D., and Schrock, S. E. (2009a). Dynamic neuromodulation of aggression by vasotocin: influence of social context and social phenotype in territorial songbirds. *Biology Letters*, 5, 554–556.

Goodson, J. L. and Kingsbury, M. A. (2011). Evolution of songbird sociality and the avian social brain. *Frontiers in Neuroanatomy*, in press.

Goodson, J. L., Lindberg, L., and Johnson, P. (2004). Effects of central vasotocin and mesotocin manipulations on social behavior in male and female zebra finches. *Hormones and Behavior*, 45, 136–143.

Goodson, J. L., Rinaldi, J., and Kelly, A. M. (2009b). Vasotocin neurons in the bed nucleus of the stria terminalis preferentially process social information and exhibit properties that dichotomize courting and non-courting phenotypes. *Hormones and Behavior*, 55, 197–202.

Goodson, J. L., Schrock, S. E., Klatt, J. D., Kabelik, D., and Kingsbury, M. A. (2009c). Mesotocin and nonapeptide receptors promote estrildid flocking behavior. *Science*, 325, 862–866.

Goodson, J. L. and Thompson, R. R. (2010). Nonapeptide mechanisms of social cognition, behavior and species-specific social systems. *Current Opinion in Neurobiology*, 20, 784–794.

Goodson, J. L. and Wang, Y. (2006). Valence-sensitive neurons exhibit divergent functional profiles in gregarious and asocial species. *Proceedings of the National Academy of Sciences*, 103, 17013–17017.

Goodwin, D. (1982). *Estrildid Finches of the World.* Ithaca, NY: Cornell University Press.

Grossmann, R., Kisliuk, S., Xu, B., and Muhlbauer, E. (1995). The hypothalamo-neurohypophyseal system in birds. *Advances in Experimental Medicine and Biology*, 395, 657–666.

Harding, C. F. and Rowe, S. A. (2003). Vasotocin treatment inhibits courtship in male zebra finches; concomitant androgen treatment inhibits this effect. *Hormones and Behavior*, 44, 413–418.

Hatchwell, B. J. (2009). The evolution of cooperative breeding in birds: kinship, dispersal and life history. *Proceedings of the Royal Society B-Biological Sciences*, 364, 3217–3227.

Heinrichs, M., von Dawans, B., and Domes, G. (2009). Oxytocin, vasopressin, and human social behavior. *Frontiers in Neuroendocrinology*, 30, 548–557.

Ho, J. M., Murray, J. H., Demas, G. E., and Goodson, J. L. (2010). Vasopressin cell groups exhibit strongly divergent responses to copulation and male-male interactions in mice. *Hormones and Behavior*, 58, 368–377.

Jurkevich, A., Barth, S. W., Aste, N., Panzica, G. C., and Grossmann, R. (1996). Intracerebral sex differences in the vasotocin system in birds: Possible implication in behavioral and autonomic functions. *Hormones and Behavior*, 30, 673–681.

Jurkevich, A., Barth, S. W., and Grossmann, R. (1997). Sexual dimorphism of arg-vasotocin gene expressing neurons in the telencephalon and dorsal diencephalon of the domestic fowl. An immunocytochemical and in situ hybridization study. *Cell and Tissue Research*, 287, 69–77.

Jurkevich, A., Barth, S. W., Kuenzel, V. J., Kohler, A., and Grossmann, R. (1999). Development of sexually dimorphic vasotocinergic system in the bed nucleus of stria terminalis in chickens. *Journal of Comparative Neurology*, 408, 46–60.

Kabelik, D., Klatt, J. D., Kingsbury, M. A., and Goodson, J. L. (2009). Endogenous vasotocin exerts context-dependent behavioral effects in a semi-naturalistic colony environment. *Hormones and Behavior*, 56, 101–107.

Kabelik, D., Morrison, J. A., and Goodson, J. L. (2010). Cryptic regulation of vasotocin neuronal activity but not anatomy by sex steroids and social stimuli in opportunistic desert finches. *Brain, Behavior, and Evolution*, 75, 71–84.

Kelly, A. M., Kingsbury, M. A., Hoffbuhr, K., Schrock, S. E., Waxman, B., Kabelik, D., Thompson, R. R., and Goodson, J. L. (2011). Vasotocin neurons and septal V1a-like receptors potently modulate finch flocking and responses to novelty. *Hormones and Behavior*, 60, 12–21.

Kimura, T., Okanoya, K., and Wada, M. (1999). Effect of testosterone on the distribution of vasotocin immunoreactivity in the brain of the zebra finch, *Taeniopygia guttata castanotis. Life Sciences*, 65, 1663–1670.

King, J. A., Ed. (1968). *Biology of Peromyscus (Rodentia).* Stillwater, MN: American Society of Mammologists.

Kiss, J. Z., Voorhuis, T. A. M., Van Eekelen, J. A. M., De Kloet, E. R., and De Wied, D. (1987). Organization of vasotocin-immunoreactive cells and fibers in the canary brain. *Journal of Comparative Neurology*, 263, 347–364.

Klein, S., Jurkevich, A., and Grossmann, R. (2006). Sexually dimorphic immunoreactivity of galanin and colocalization with arginine vasotocin in the chicken brain (*Gallus gallus domesticus*). *Journal of Comparative Neurology*, 499, 828–839.

Knapton, R. W. and Falls, J. B. (1983). Differences in parental contribution among pair types in the polymorphic white-throated sparrow. *Canadian Journal of Zoology*, 61, 1288–1292.

Kopachena, J. G. and Falls, J. B. (1993). Re-evaluation of morph-specific variations in parental behavior of the white-throated sparrow. *Wilson Bulletin*, 105, 48–59.

Landgraf, R., Frank, E., Aldag, J. M., Neumann, I. D., Sharer, C. A., Ren, X., Terwilliger, E. F., Niwa, M., Wigger, A., and Young, L. J. (2003). Viral vector-mediated gene transfer of the vole V1a vasopressin receptor in the rat septum: improved social discrimination and active social behaviour. *European Journal of Neuroscience*, 18, 403–411.

Landgraf, R., Gerstberger, R., Montkowski, A., Probst, J. C., Wotjak, C. T., Holsboer, F., and Engelmann, M. (1995). V1 vasopressin receptor antisense oligodeoxynucleotide into septum reduces vasopressin binding, social discrimination abilities, and anxiety-related behavior in rats. *Journal of Neuroscience*, 15, 4250–4258.

Landgraf, R. and Neumann, I. D. (2004). Vasopressin and oxytocin release within the brain: a dynamic concept of multiple and variable modes of neuropeptide communication. *Frontiers in Neuroendocrinology*, 25, 150–176.

Leung, C. H., Abebe, D., Earp, S. E., Goode, C. T., Grozhik, A. V., Mididoddi, P., and Maney, D. L. Neural distribution of vasotocin receptor mRNA in two species of songbird. *Endocrinology*, in press.

Leung, C. H., Goode, C. T., Young, L. J., and Maney, D. L. (2009). Neural distribution of nonapeptide binding sites in two species of songbird. *Journal of Comparative Neurology*, 513, 197–208.

Liu, Y., Curtis, J. T., and Wang, Z. (2001). Vasopressin in the lateral septum regulates pair bond formation in male prairie voles (*Microtus ochrogaster*). *Behavioral Neuroscience*, 115, 910–919.

Ludwig, M. and Leng, G. (2006). Dendritic peptide release and peptide-dependent behaviours. *Nature Reviews Neuroscience*, 7, 126–136.

Madison, F. N., Jurkevich, A., and Kuenzel, W. J. (2008). Sex differences in plasma corticosterone release in undisturbed chickens (*Gallus gallus*) in response to arginine vasotocin and corticotropin releasing hormone. *General and Comparative Endocrinology*, 155, 566–573.

Maney, D. L., Erwin, K. L., and Goode, C. T. (2005). Neuroendocrine correlates of behavioral polymorphism in white-throated sparrows. *Hormones and Behavior*, 48, 196–206.

Maney, D. L., Goode, C. T., and Wingfield, J. C. (1997). Intraventricular infusion of arginine vasotocin induces singing in a female songbird. *Journal of Neuroendocrinology*, 9, 487–491.

Martinez-Garcia, F., Novejarque, A., and Lanuza, E. (2008). Two interconnected functional systems in the amygdala of amniote vertebrates. *Brain Research Bulletin*, 75, 206–213.

Mikami, S., Tokado, H., and Farner, D. S. (1978). The hypothalamic neurosecretory systems of the Japanese quail as revealed by retrograde transport of horseradish peroxidase. *Cell and Tissue Research*, 194, 1–15.

Moore, F. L. and Lowry, C. A. (1998). Comparative neuroanatomy of vasotocin and vasopressin in amphibians and other vertebrates. *Comparative Biochemistry and Physiology. Part C, Pharmacology, Toxicology and Endocrinology*, 119, 251–260.

Morris, D. (1958). The comparative ethology of Grassfinches (Erythrurae) and Mannikins (Amadinae). *Proceedings of the Zoological Society of London*, 131, 389–439.

Nephew, B. C., Aaron, R. S., and Romero, L. M. (2005). Effects of arginine vasotocin (AVT) on the behavioral, cardiovascular, and corticosterone responses of starlings (*Sturnus vulgaris*) to crowding. *Hormones and Behavior*, 47, 280–289.

Nephew, B. C., Reed, L. M., and Romero, L. M. (2005). A potential cardiovascular mechanism for the behavioral effects of central and peripheral arginine vasotocin. *General and Comparative Endocrinology*, 144, 156–166.

Nieuwenhuys, R., ten Donkelaar, H. J., and Nicholson, C., Eds. (1998). *The Central Nervous System of Vertebrates.* New York: Springer-Verlag.

Nishijo, H., Ono, T., and Nishino, H. (1988). Single neuron responses in amygdala of alert monkey during complex sensory stimulation with affective significance. *Journal of Neuroscience*, 8, 3570–3583.

Panzica, G. C., Aste, N., Castagna, C., Viglietti-Panzica, C., and Balthazart, J. (2001). Steroid-induced plasticity in the sexually dimorphic vasotocinergic innervation of the avian brain: Behavioral implications. *Brain Research Reviews*, 37, 178–200.

Panzica, G. C., Plumari, L., Garcia-Ojeda, E., and Deviche, P. (1999). Central vasotocin-immunoreactive system in a male passerine bird (*Junco hyemalis*). *Journal of Comparative Neurology*, 409, 105–117.

Paton, J. J., Belova, M. A., Morrison, S. E., and Salzman, C. D. (2006). The primate amygdala represents the positive and negative value of visual stimuli during learning. *Nature*, 439, 865–870.

Paz, Y.M.C.G., Bond, A. B., Kamil, A. C., and Balda, R. P. (2004). Pinyon jays use transitive inference to predict social dominance. *Nature*, 430, 778–781.

Plumari, L., Plateroti, S., Deviche, P., and Panzica, G. C. (2004). Region-specific testosterone modulation of the vasotocin-immunoreactive system in male dark-eyed juncos, *Junco hyemalis*. *Brain Research*, 999, 1–8.

Reichard, U. H. and Christophe, B., Eds. (2003). *Monogamy: mating strategies and partnerships in birds, humans and other mammals.* Cambridge: Cambridge University Press.

Reiner, A., Perkel, D. J., Bruce, L. L., Butler, A. B., Csillag, A., Kuenzel, W., Medina, L., Paxinos, G., Shimizu, T., Striedter, G., Wild, M., Ball, G. F., Durand, S., Gunturkun, O., Lee, D. W., Mello, C. V., Powers, A., White, S. A., Hough, G., Kubikova, L., Smulders, T. V., Wada, K., Dugas-Ford, J., Husband, S., Yamamoto, K., Yu, J., Siang, C., Jarvis, E. D., and Guturkun, O. (2004). Revised nomenclature for avian telencephalon and some related brainstem nuclei. *Journal of Comparative Neurology*, 473, 377–414.

Robinzon, B., Koike, T. I., Neldon, H. L., Kinzler, S. L., Hendry, I. R., and el Halawani, M. E. (1988). Physiological effects of arginine vasotocin and mesotocin in cockerels. *British Poultry Science*, 29, 639–652.

Romero, L. M., Soma, K. K., and Wingfield, J. C. (1998). Hypothalamic-pituitary-adrenal axis changes allow seasonal modulation of corticosterone in a bird. *American Journal of Physiology*, 274, R1338–1344.

Romero, L. M. and Wingfield, J. C. (1998). Seasonal changes in adrenal sensitivity alter corticosterone levels in Gambel's white-crowned sparrows (*Zonotrichia leucophrys gambelii*). *Comparative Biochemistry and Physiology. Part C, Pharmacology, Toxicology and Endocrinology*, 119, 31–36.

Ross, H. E. and Young, L. J. (2009). Oxytocin and the neural mechanisms regulating social cognition and affiliative behavior. *Frontiers in Neuroendocrinology*, 30, 534–547.

Saito, N., Shimada, K., and Koike, T. I. (1987). Interrelationship between arginine vasotocin prostaglandin and uterine contractility in the control of oviposition in the hen *Gallus domesticus*. *General and Comparative Endocrinology*, 67, 342–347.

Sasaki, T., Shimada, K., and Saito, N. (1998). Changes of AVT levels in plasma, neurohypophysis and hypothalamus in relation to oviposition in the laying hen. *Comparative Biochemistry and Physiology. Part A, Molecular and Integrative Physiology*, 121, 149–153.

Sewall, K. B., Dankoski, E. C., and Sockman, K. W. (2010). Song environment affects singing effort and vasotocin immunoreactivity in the forebrain of male Lincoln's sparrows. *Hormones and Behavior*, 58, 544–553.

Sharp, P. J., Li, Q., Talbot, R. T., Barker, P., Huskisson, N., and Lea, R. W. (1995). Identification of hypothalamic nuclei involved in osmoregulation using fos immunocytochemistry in the domestic hen (*Gallus domesticus*), ring dove (*Streptopelia risoria*), Japanese quail (*Coturnix japonica*) and zebra finch (*Taeniopygia guttata*). *Cell and Tissue Research*, 282, 351–361.

Sheehan, T. P., Cirrito, J., Numan, M. J., and Numan, M. (2000). Using c-Fos immunocytochemistry to identify forebrain regions that may inhibit maternal behavior in rats. *Behavioral Neuroscience*, 114, 337–352.

Sheehan, T. P., Paul, M., Amaral, E., Numan, M. J., and Numan, M. (2001). Evidence that the medial amygdala projects to the anterior/ventromedial hypothalamic nuclei to inhibit maternal behavior in rats. *Neuroscience*, 106, 341–356.

Skead, D. M. (1975). Ecological studies of four Estrildines in the central Transvaal. *Ostrich*, 11 (Suppl.), 1–55.

Takahashi, T. and Kawashima, M. (2003). Arginine vasotocin induces bearing down for oviposition in the hen. *Poultry Science*, 82, 345–346.

Takahashi, T. and Kawashima, M. (2008). Mesotocin increases the sensitivity of the hen oviduct uterus to arginine vasotocin. *Poultry Science*, 87, 2107–2111.

Tan, F., Lolait, S. J., Brownstein, M. J., Saito, N., MacLeod, V., Baeyens, D. A., Mayeux, P. R., Jones, S. M., and Cornett, L. E. (2000). Molecular cloning and functional characterization of a vasotocin receptor subtype that is expressed in the shell gland and brain of the domestic chicken. *Biology of Reproduction*, 62, 8–15.

Thompson, R. R. and Walton, J. C. (2004). Peptide effects on social behavior: Effects of vasotocin and isotocin on social approach behavior in male goldfish (*Carassius auratus*). *Behavioral Neuroscience*, 118, 620–626.

Veenema, A. H., Beiderbeck, D. I., Lukas, M., and Neumann, I. D. (2010). Distinct correlations of vasopressin release within the lateral septum and the bed nucleus of the stria terminalis with the display of intermale aggression. *Hormones and Behavior*, 58, 273–281.

Voorhuis, T. A. M., De Kloet, E. R., and De Wied, D. (1991a). Effect of a vasotocin analog on singing behavior in the canary. *Hormones and Behavior*, 25, 549–559.

Voorhuis, T. A. M., De Kloet, E. R., and De Wied, D. (1991b). Ontogenetic and seasonal changes in immunoreactive

vasotocin in the canary brain. *Developmental Brain Research*, 61, 23–32.

Voorhuis, T. A. M., Kiss, J. Z., De Kloet, E. R., and De Wied, D. (1988). Testosterone-sensitive vasotocin-immunoreactive cells and fibers in the canary brain. *Brain Research*, 442, 139–146.

Wang, Z., Smith, W., Major, D. E., and De Vries, G. J. (1994). Sex and species differences in the effects of cohabitation on vasopressin messenger RNA expression in the bed nucleus of the stria terminalis in prairie voles (*Microtus ochrogaster*) and meadow voles (*Microtus pennsylvanicus*). *Brain Research*, 650, 212–218.

Wolff, J. O. (1985). Behavior. In *Biology of New World Microtus.* Stillwater, OK: American Society of Mammologists, 340–372.

Zann, R.A. (1996). *The Zebra Finch: A Synthesis of Field And Laboratory Studies.* Oxford: Oxford University Press.

8

Oxytocin, vasopressin, and the evolution of mating systems in mammals

Sara M. Freeman and Larry J. Young

8.1 Introduction

"Safety in numbers." "Pack mentality." "All for one and one for all." Our language is full of phrases like these, expressing the almost instinctual human understanding of the benefit of being social. We would argue that the majority of the 6.7 billion people in our collective human society do not realize that the socially motivated nature of our behavior is not the default way of life for many other species. Our high degree of sociality seems so natural to us that we are unaware that it is a rarity in the animal world. We are surrounded by other individuals and depend on them for the goods and services we need for our survival. We cooperate with others and even exhibit altruistic behavior. The choices we make every day, from traffic decisions to those made in conversations or in business, are all made with a subconscious consideration for how that decision might affect other people in our lives, even if those people are complete strangers. To look at it another way, a life of solitude is an unimaginable horror for an individual of our gregarious species. However, for many animal species, a solitary life is the norm, and conspecifics only come into contact in aggressive interactions over territory or resources, or to mate.

"Survival of the fittest." "Every man for himself." These ideas have a selfish and negative connotation in our culture but are the behavioral rule for many mammalian species. This begs the question, if self-promoting rather than affiliative behaviors are so prevalent, then why, and how, would behaviors that facilitate social interactions beyond fighting and mating ever evolve? In theory, the benefit of living in a social group is obvious; simply reread the first few words of this chapter. Group living, a level of social organization that is found in many species, requires that individuals at the very least tolerate close contact between conspecifics in the group, and at the very most behave in a prosocial manner, that is to behave in a way that promotes social contact and affiliation between individuals. In many cases of group living among mammals, the individuals are related members of an extended family, so acting prosocially makes sense evolutionarily. But our interest in this chapter is to delve into a level of social organization farther beyond familial prosocial behavior on the spectrum of sociality: monogamy.

Monogamy can come in many different forms, ranging from strict genetic monogamy, where the male and female pair to mate and procreate exclusively with each other, to social monogamy, where the male and female share a territory, mate, and take care of the offspring, but also may engage in extra-pair copulations on occasion. In either case, unrelated adult conspecifics will come to prefer to spend time together, to develop an attachment, to be "bonded." Unlike the case in more than 90% of mammalian species, those exhibiting a "monogamous" mating strategy develop a

Oxytocin, Vasopressin, and Related Peptides in the Regulation of Behavior, ed. E. Choleris, D. W. Pfaff, and M. Kavaliers. Published by Cambridge University Press.

relationship between the mated pair that extends well beyond orgasm. This pair bonding of a male and a female after mating can endure for their lifetime or merely for a breeding season (Kleiman, 1977). However, to limit one's mating opportunities to the reproductive status of one individual for any length of time is counterintuitive in the context of evolutionary theory and fitness. So, what are the proximate and ultimate causes for monogamy in mammals? This chapter will delve into the theoretical and empirical basis for the evolution of monogamy in mammals, with a focus on the two neuropeptides that have been shown to be particularly important in the regulation of pair bonding: oxytocin and vasopressin.

Mammals are a unique class of vertebrates with a distinct set of features that sets them apart from other classes of animals. Remembering lessons from middle-school biology, there are five main characteristics of mammals that set them apart from other vertebrate organisms: fur/body hair, three middle-ear bones, a neocortex in the brain, mammary glands, and (in all taxa except monotremes) giving birth to live young. These last two – giving birth to live young and nursing them with milk from mammary glands – are of particular interest here for two reasons. First, in the context of evolution, these two processes involve the reproductive system, a potent locus for natural selection. And second, these two characteristics of mammals are also both modulated by oxytocin (OT), a neurohypophyseal peptide hormone intimately involved in female mammalian reproductive physiology. Circulating OT binds to receptors in the uterus, where it stimulates uterine contractions and facilitates labor, and in the mammary glands, where it stimulates the milk let-down reflex in response to tactile stimulation from nursing (Burbach et al., 2006). This uniquely mammalian system also acts in the brain to initiate maternal nurturing behavior and mother–infant bonding after parturition (see Chapters 9 and 10) (Pedersen et al., 1985; Kendrick et al., 1986; Kendrick et al., 1987; Kendrick et al., 1988a, b; Pedersen et al., 1994). More recently, the oxytocinergic system has been examined for its role in modulating pair-bond formation in the females of monogamous mammalian species (Carter et al., 1995; Young and Wang, 2004). This new direction in OT research suggests some interesting evolutionary relationships between the neural mechanisms of mother–infant bonding and pair bonding in monogamous species.

Only 3–5% of mammals are considered to be monogamous, whereas up to 90% of birds display some degree of monogamy (Orians, 1969; Kleiman, 1977; Clutton-Brock, 1989). Theories about the reasons for this disparity have been based on the fact that, because only the mother lactates in mammals, only she can feed the offspring after birth, but in birds, both the mother and the father can contribute equally to feeding the young (Orians, 1969; Kleiman, 1977). Due to this unshared parental investment in mammals where females are "physiologically capable of providing for her own offspring before and after birth," it is often a more adaptive strategy for the male to depart shortly after mating to find another female to impregnate (Kleiman, 1977). This strategy results in an increased prevalence of non-monogamous mating systems in mammals as compared to birds and other classes of vertebrates with parental care. However, monogamy and biparental care of the young do exist in the mammalian phylogeny, usually in species where there is, or has been, significantly increased offspring fitness when the father contributes as well, such as in harsh environmental conditions like low food availability and high predation rate, or in low population densities (Getz and Hofmann, 1986). Also, high levels of mate guarding exhibited by the male after mating, to ensure paternity, have been hypothesized to be a corollary to monogamous mating systems in mammals as well (Brotherton et al., 1997; Schuiling, 2003). Mate guarding is a form of male-specific territorial behavior in which the male either physically or behaviorally prevents the female with whom he has just mated from subsequently mating with any other male after him. Male aggression and territorial behavior are strongly influenced by arginine vasopressin (AVP), a peptide hormone that is phylogenetically and physiologically very similar to OT. While most of the roles of OT have

been elucidated in females, the behavioral functions of AVP have mostly been demonstrated in males. However, it is becoming increasingly clear that the notion that OT regulates female behaviors while AVP regulates male behavior is a severe oversimplification and is simply incorrect. Nevertheless, AVP activates territorial scent-marking behavior in male hamsters when injected directly into the brain (Ferris et al., 1984). Arginine vasotocin (AVT), a non-mammalian homolog of AVP, has also been shown to mediate male-specific territorial behavior in species other than mammals, such as vocalization in frogs and birds (see Chapters 6 and 7) (Boyd, 1994; Maney et al., 1997; Goodson, 1998). AVP can also induce aggression in hamsters and voles (Winslow et al., 1993; Ferris et al., 1997; Young et al., 1997). Based on this evidence for the involvement of AVP in territorial behavior, as well as recent work examining the role of AVP in mediating pair-bond formation in males in monogamous voles, the AVP system is emerging as an integral player in the evolution of monogamy in mammals.

In this chapter, we will review the research exploring the role of AVP and OT in regulating pair bonding in monogamous prairie voles and speculate about an explanation for how a complex social behavior such as monogamy could evolve *de novo*. We hypothesize that the evolutionarily ancient neurohypophyseal systems of AVP and OT, which carry out other essential functions in the body like regulating water balance and initiating parturition, respectively, have undergone modifications in the central nervous system of monogamous mammalian species to promote the formation of the pair bond after mating. We will explore the neuroanatomical specializations in the OT and AVP systems in monogamous species and discuss how interactions between these neuropeptide systems and the brain's reward systems can lead to a pair bond.

8.2 Vasopressin

8.2.1 Antidiuretic hormone

Arginine vasopressin is a neurohypophyseal nonapeptide hormone that is closely related to OT, differing in only two positions in the nine amino acid sequence. The genes for these two peptides are thought to have arisen early in the vertebrate lineage from a duplication event; the genes are located on the same chromosome, separated by a small intergenic region, and transcribed in opposite directions (Hara et al., 1990; Burbach et al., 2001; Donaldson et al., 2008). The evolutionarily ancient functions of AVP are in the periphery, where it acts at the level of the blood vessels via the vasopressin 1a receptor (AVPR1a) to control vasoconstriction, and at the level of the kidney via the V2 receptor to control water balance (Barberis et al., 1992; Jard, 1998; Hibert et al., 1999). Thus, AVP is also referred to by endocrinologists as antidiuretic hormone (ADH). AVP has also been found to act at AVPR1a in the brain to influence male-typical mammalian social behaviors such as aggression (Ferris et al., 1997; Young et al., 1997), paternal behavior (Wang et al., 1998), and territoriality (Ferris et al., 1984; Albers et al., 1986). At first, the peripheral and behavioral functions of AVP may seem entirely unrelated; how does maintaining proper blood osmolarity relate to male social behavior? Urine concentration is a byproduct of water balance and blood osmolarity regulation, and scent marking with urine is one of the most common ways that male mammals establish and maintain a territory, within which all resources, food, and mates are his. Thus, it is logical that this system could have been "hijacked" by natural selection to develop a new function in territorial behavior: chemical communication (personal conversation, Elliott Albers).

8.2.2 Territoriality and aggression in hamsters

Territorial behavior is a common male-typical communication behavior that is used to delimit and defend the boundaries of an individual's territory. While territorial behavior can come in many forms, one that is often seen in mammals is scent marking, in which "excretions of urine, feces, sweat, or glandular secretions are disseminated in the environment" (Ferris et al., 1985). In hamsters, scent

marking is performed in a highly stereotyped manner called *flank marking* where the male rubs secretions of the flank glands on objects in his territory. This behavior is usually triggered by odors of other male hamsters, but flank marking can be stimulated in the absence of olfactory cues by injecting AVP into the medial preoptic area/anterior hypothalamus (MPOA-AH) of males (see Chapter 12) (Ferris et al., 1984). Injection of AVP into the hypothalamus also increases aggression in hamsters (Ferris et al., 1997). The behavioral effects on both aggression and scent marking are mediated by the AVPR1a (Ferris et al., 1984; Ferris et al., 1985).

8.2.3 From territoriality to pair bonding: Vasopressin and the male prairie vole

Prairie voles are hamster-sized rodents that display high levels of sociality, including social monogamy and biparental care of the young (Gruder-Adams and Getz, 1985). These rodents provide an ideal system for the study of complex social behavior, especially the neural circuitry of bonding between adults in a monogamous pair (Aragona and Wang, 2004; Young et al., 2004; McGraw and Young, 2010). Research on this species has been strengthened by comparative work performed in two vole species closely related to the prairie vole, the meadow vole and the montane vole, which are both solitary, non-affiliative rodents that do not exhibit monogamy or paternal care (Gruder-Adams and Getz, 1985).

8.2.3.1 Vasopressin and social behavior in male voles

Comparative studies using these monogamous and non-monogamous vole species have demonstrated species-specific roles for AVP in social behavior. Infusions of AVP directly into the brain increase intermale aggression in male prairie voles, but not in male montane voles (Young et al., 1997). Furthermore, central infusions of AVP enhance affiliative behavior toward a female in male prairie voles, but not in male montane voles (Young et al., 1999). Male prairie voles, but not male montane or meadow voles, develop long-lasting partner preferences, or pair bonds, following cohabitation and mating with a female. Pair-bond formation is accompanied by the development of a selective aggression toward novel females they encounter, one component in a suite of social behaviors related to the monogamous mating strategy. Mating-induced, territorial-like selective aggression can be blocked by injecting an AVPR1a antagonist either intracerebroventricularly (ICV) (Winslow et al., 1993) or into the anterior hypothalamus (Gobrogge et al., 2009).

The development of a pair bond in prairie voles can be tested in a laboratory using the partner preference test, in which the male, after a period of cohabitation with a female, is given a choice to either spend time with his familiar female partner, with a novel female, or in the neutral empty middle between the two (Figure 8.1) (Williams et al., 1992; Williams et al., 1994). The partner preference behavior exhibited by male prairie voles is in contrast to the "Coolidge effect" in which after mating, male rodents generally display a preference for a novel female (Wilson et al., 1963). In male prairie voles, mating or extended cohabitation time leads to the development of a partner preference, but cohabitation without mating or for less than 24 h does not typically lead to partner preference formation (Insel et al., 1995). AVP mediates the formation of a partner preference in male prairie voles since infusion of an AVPR1a antagonist into the brain blocks the formation of a partner preference, and infusion of AVP induces a partner preference in the absence of mating (Winslow et al., 1993). Thus, it appears as though the neural systems that mediate territorial behavior, including scent marking and territorial aggression, have been co-opted to mediate the formation of a pair bond in the monogamous prairie vole.

8.2.3.2 Neuroanatomy and pharmacology of vasopressin and pair-bond formation in male prairie voles

The first clues about the brain regions involved in AVP-mediated pair bonding came from comparative studies examining the distribution of AVPR1a in

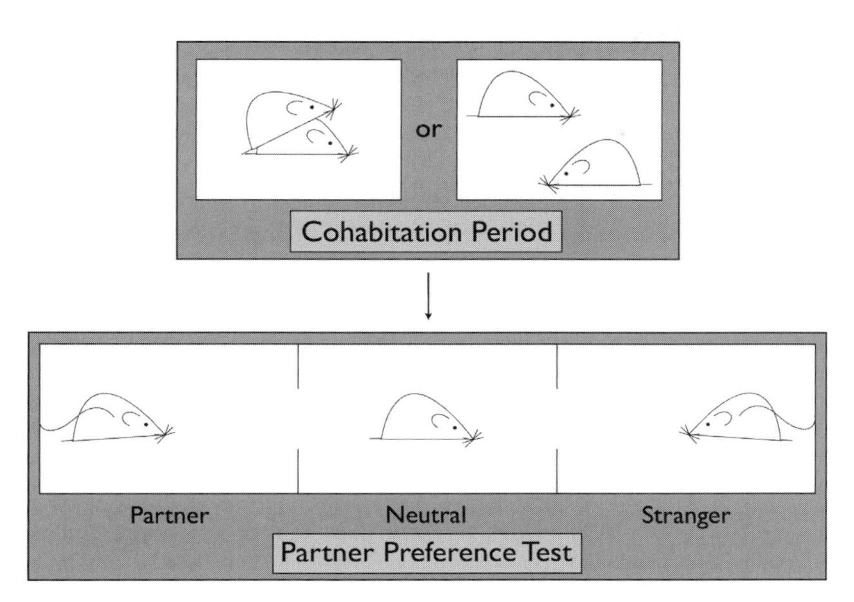

Figure 8.1 Partner preference paradigm. First, the experimental animal, which can be a male or a female, is allowed to freely interact with an opposite sex conspecific for an extended period of time, called the cohabitation period. During this period, if there is mating or extended cohabitation time (18–24 h), the animals form a social bond; if mating is prevented or cohabitation duration is shortened (6 h), no pair bond will form (Insel, Preston et al., 1995). Next, the formation of a social bond is assayed in the laboratory using the partner preference test. In this test phase, the familiar opposite sex conspecific from the cohabitation phase ("partner") is tethered to one end of a three-chambered arena, and a novel opposite sex conspecific of equal stimulus value ("stranger") is tethered to the opposite end of the arena. The experimental animal is placed in the center of the arena and allowed to freely wander for 3 hours. The amount of time that the experimental animal spends in social proximity, or huddling, with either the partner or the stranger is recorded.

the brains of prairie, montane and meadow voles. Receptor autoradiography for AVPR1a reveals striking species-specific differences in the pattern of binding. For example, prairie voles have higher densities of AVPR1a in several regions of the brain, including the ventral pallidum (VP), when compared to montane and meadow voles (Figures 8.2C and D) (Insel et al., 1994; Lim, Murphy, et al., 2004). Furthermore, blocking specific populations of AVPR1a, including the VP and lateral septum (LS), prior to cohabitation and mating with a female blocks the development of a partner preference (Figures 8.2E and F) (Liu et al., 2001; Lim and Young, 2004). Infusion of AVP directly into the LS leads to a partner preference following cohabitation without mating (Figure 8.2F) (Liu et al., 2001). These results indicate that the LS and VP are critical sites for vasopressin's action in mediating pair-bond formation in male prairie voles.

The observation of species differences in AVPR1a distribution, particularly in the VP, led to investigations of the AVPR1a gene (*avpr1a)* in an effort to detect potential genetic mechanisms that could account for the species differences in brain expression and behavior. It has been suggested that changes in the distribution of AVPR1a in the brain evolutionarily, for example, increases in expression of *avpr1a* in the VP, resulted in the capacity to form selective social bonds. This hypothesis was tested by recreating this redistribution of the receptor expression in the lab using viral vector mediated gene transfer. Quite astoundingly, increasing *avpr1a*

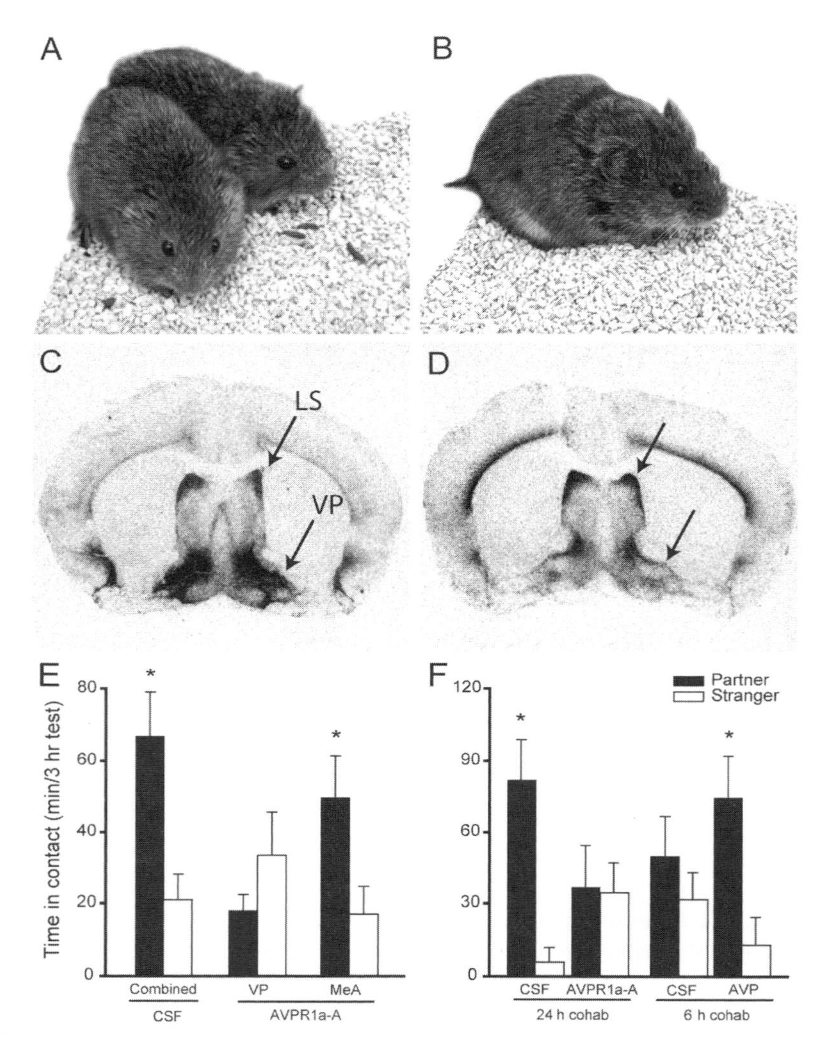

Figure 8.2 Vasopressin 1a receptor (AVPR1a) and social behavior in voles. (A) Socially monogamous prairie voles. (B) Solitary and promiscuous meadow vole. (C and D) Autoradiograms of AVPR1a binding in the ventral pallidum (VP) in coronal sections through the forebrain of a prairie (C) or meadow (D) vole. (E) Infusion of an AVPR1a antagonist (AVPR1a-A) into the VP but not the medial amygdala (MeA) blocks partner preference in male prairie voles, but infusion of cerebrospinal fluid (CSF) in these regions does not affect partner preference behavior (Combined). (F) Infusion of a AVPR1aA into the lateral septum (LS) of male prairie voles disrupts partner preference after a cohabitation phase lasting 24 h, and infusion of vasopressin (AVP) into the LS during a shortened 6-h cohabitation induces a partner preference, which is not seen if CSF is infused (Liu, Curtis et al., 2001). Figure adapted from Young and Wang, 2004 and Lim, Wang et al., 2004.

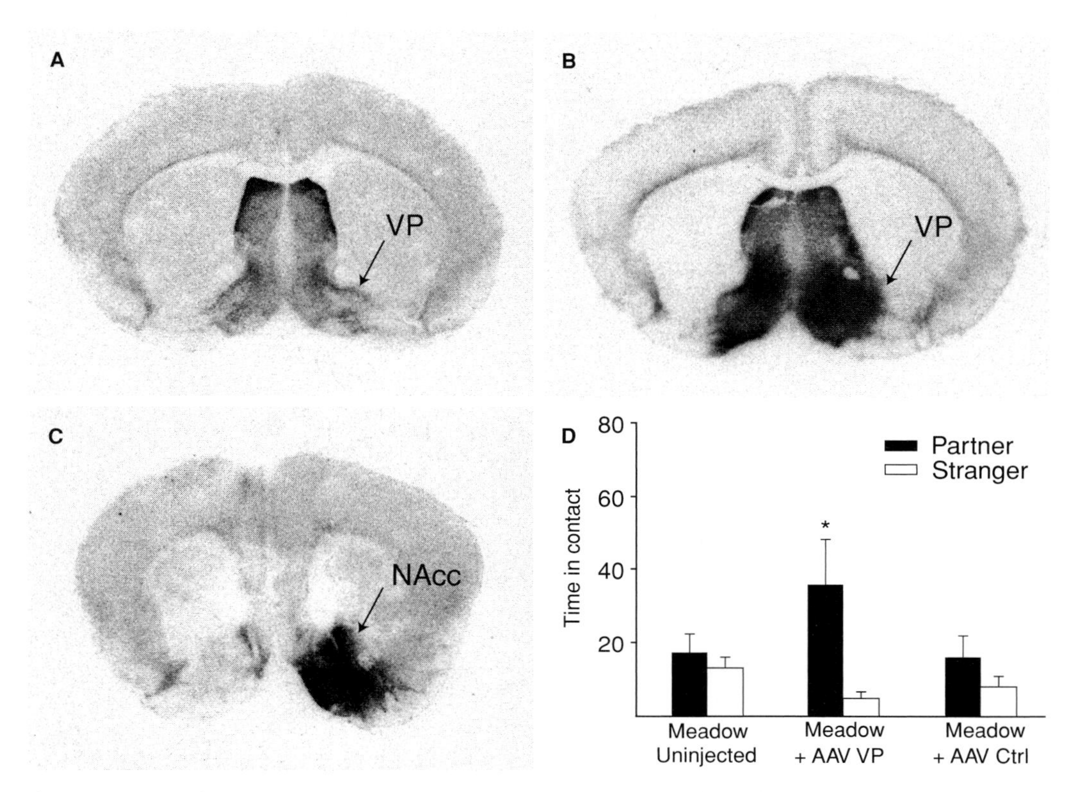

Figure 8.3 Manipulating vasopressin 1a receptor (AVPR1a) binding in meadow voles using viral vector mediated gene transfer. (A) Autoradiogram showing AVPR1a binding in a typical meadow vole. (B) Overexpression of the AVPR1a gene in the ventral pallidum (VP) by adeno-associated viral vector (AAV) mediated gene transfer (Meadow +AAV VP). (C) A stereotactic injection inadvertently placed too rostral to the VP, in this case located just ventral to the nucleus accumbens, which serves as an anatomical control (Meadow +AAV Ctrl). (D) When AVPR1a levels are artificially increased within the VP using AAV gene transfer (Meadow +AAV VP), male meadow voles display a partner preference, preferring social contact with their partner than with a stranger. This behavior is not seen in uninjected meadow voles (Meadow Uninjected) or meadow voles receiving a viral vector injection that missed the VP (Meadow +AAV Ctrl) (Lim, Wang et al., 2004). Time in contact is given in minutes per 3-h test. Figure adapted from Lim, Wang et al., 2004.

expression in the VP of non-monogamous meadow vole males results in the development of partner preference toward their mates (Figure 8.3) (Lim et al., 2004). This result demonstrates that variation in the expression of a single gene in a single brain region can drastically affect the social behavior of a species, which suggests that the *avpr1a* is a potential locus for the evolution of monogamy in prairie voles from a previously non-monogamous mating system.

8.2.3.3 Genetic variation and pair bonding in male voles

To better elucidate the potential molecular mechanism for the species-specific differences in the distribution of AVPR1a in the brains of voles, the coding and regulatory regions of the *avpr1a* gene were examined in monogamous and non-monogamous species. The analysis indicated that the coding region of the *avpr1a* gene in prairie voles and

montane voles are 99% homologous (Young, Nilsen, et al., 1999). However, the degree of homology is quite different in the 5′ flanking region of the gene, a region that likely determines expression patterns. In prairie voles, there is a ~400 base pair (bp) polymorphism in the length of a repetitive DNA element, referred to as a microsatellite, approximately 700 bp upstream of the transcription start site, and in the non-monogamous meadow and montane voles, this microsatellite is rudimentary (~50 bp) (Young, Nilsen, et al., 1999; Fink et al., 2006). Repetitive microsatellite elements are evolutionarily unstable and prone to mutation, which can in turn produce diversity in the expression of the gene (Kashi and King, 2006; Young and Hammock, 2007). Therefore, this microsatellite polymorphism was hypothesized to be a potential causative polymorphism for the divergence in brain AVPR1a expression patterns and social behavior seen between species.

Several studies were conducted to acquire empirical evidence in support of this idea. First, individual variation in the length of this microsatellite in male prairie voles correlated with individual variation in levels of AVPR1a binding in some areas of the brain and also predicted social behavior in laboratory tests. Males homozygous for the long microsatellite allele had increased AVPR1a density in the LS and olfactory bulb, and they displayed a reduced approach latency and increased duration and frequency of olfactory investigation of a social odor (but not a non-social odor) as well as a reduced latency to social investigation of a novel animal (Hammock and Young, 2005). In a partner preference test, the homozygous long males developed partner preferences more readily than the homozygous short males (Hammock and Young, 2005). These neuroanatomical and behavioral results suggested that variation in the length of the microsatellite region in the *avpr1a* gene could be altering both global AVPR1a expression patterns in the brain and the repertoire of social behavior displayed in these animals. However, the relationship between microsatellite length and *avpr1a* expression and behavior is not likely to be as simple as first hypothesized.

To further investigate the role that this microsatellite region might play in determining species-specific social behavior, prairie voles were taken into the field to examine the behavioral ecology of this system. However, the first study to evaluate the relationship between microsatellite length, AVPR1a binding patterns, and male prairie vole behavior in mixed-sex, semi-natural field enclosures found results that are inconsistent with those in the laboratory experiments discussed above. Ophir et al. also found that AVPR1a binding was associated with *avpr1a* microsatellite length, but the brain regions exhibiting the effect were different than those reported in Hammock et al., 2005. Furthermore, they found no association between *avpr1a* microsatellite length and social behavior in the field (Ophir, Campbell, et al., 2008). In another field study using semi-natural conditions, Solomon et al. evaluated the relationship between the length of the microsatellite and "indicators of social and genetic monogamy" in male prairie voles who were living in mixed-sex enclosures. These authors found that the length of the microsatellite did not correlate with social monogamy but did correlate with genetic monogamy (Solomon et al., 2009).

Thus, as is the case with many elegantly simple hypotheses, the data suggest that the mere expansion of the microsatellite is not responsible for the evolution of monogamy in prairie voles. A comparative analysis of *avpr1a* from 21 *Microtus* species revealed that the presence or absence of the microsatellite did not predict mating strategy within the genus (Fink et al., 2006). Moreover, when eight species of deer mouse within the genus *Peromyscus*, which includes both monogamous and promiscuous species, were examined for *avpr1a* microsatellite length, for AVPR1a distribution, and for mating system, there were no significant correlations found between any of these features (Turner et al., 2010). Thus, while early work suggested that instability in the microsatellite drove the diversity in receptor distribution, more careful analysis suggests that simple variation in the length of the microsatellite is not the primary cause of variation in expression

patterns. Rather, individual alleles of the microsatellite may be linked to functional polymorphisms that contribute to variation in expression.

8.2.3.4 Vasopressin receptors and fitness in the wandering male prairie vole

Individual differences in AVPR1a binding levels in various brain structures contribute to both territorial and social behaviors of male prairie voles. Male prairie voles can adopt one of two different mating strategies: either what is called a "wanderer," where the male does not reside with a single female but instead mates opportunistically in a larger home range that overlaps those of multiple females and males, or what is called a "resident," where the male shares and defends a territory with a female with whom he could be considered socially monogamous (Ophir, Wolff, et al., 2008). Ophir et al. hypothesized that in field enclosures occupied by males and females, the mating strategy and sexual behavior of the males would associate with individual variation in AVPR1a expression levels and patterns (Ophir, Wolff, et al., 2008). When measures of social and sexual fidelity in males were correlated with AVPR1a binding in various brain regions, the authors found that neither social nor sexual fidelity was associated with AVPR1a levels in regions that have been implicated in pair-bond formation or maintenance, namely the VP and LS (Ophir, Wolff, et al., 2008). However, they did find that AVPR1a binding in two brain regions involved in spatial memory, the posterior cingulate/retrosplenial cortex (PCing) and the laterodorsal thalamus, was inversely correlated with breeding success. Specifically, "wandering" males with low levels of binding in the PCing were more successful at siring offspring than those with high binding (Ophir, Wolff, et al., 2008). This result not only supports the notion that variation in receptor binding can predict some social behaviors, but it also links brain regions that are involved in spatial ability with a more flexible mating strategy and less-stringent territorial boundaries.

8.3 Oxytocin

8.3.1 A maternal hormone

As mentioned in the introduction, OT is a hormone that is involved in the reproductive physiology of female mammals. Like AVP, this nine amino acid neuropeptide is synthesized in the paraventricular nucleus (PVN) and the supraoptic nucleus (SON) of the hypothalamus. Cells from these regions send axon projections to the posterior pituitary, where they release OT in large amounts into the bloodstream at the neurohypophysis (Burbach et al., 2006). While OT acts in the periphery to induce uterine contractions during labor and milk ejection during nursing, the OT-synthesizing neurons in the PVN and SON also release OT in the brain, where it acts centrally at oxytocin receptors (OTR) to modulate social cognitive processes like social memory, maternal behavior, and affiliative behavior, to name a few (Landgraf and Neumann, 2004; Ludwig and Leng, 2006; Ross and Leng, 2009).

8.3.1.1 Oxytocin and maternal behavior

OT not only facilitates parturition and nursing, but it also mediates the onset of maternal nurturing by transforming the female's behavior after giving birth. Central infusion of OT in virgin rats results in the rapid onset of full maternal behavior (Pedersen and Prange, 1979; Pedersen et al., 1982), while blocking OTR delays the onset of maternal responsiveness and diminishes the expression of species-specific maternal behavior (see Chapter 9) (Pedersen et al., 1985; Pedersen et al., 1994).

In sheep, elevated levels of OT in the brain at parturition trigger not only the onset of maternal nurturing behavior but also the selective mother–infant bond (see Chapter 10) (Kendrick et al., 1986; Kendrick et al., 1987; Kendrick et al., 1988a). Sheep live in herds and many females give birth at the same time during the breeding season. Lambs are ambulatory soon after birth so ewes must quickly form a selective bond to their lamb in order to distinguish it from all other lambs.

After bonding to their lamb, ewes will show aggression toward foreign lambs but nurture their own. However, ICV infusions of OT in estrogen-primed ewes increases maternal behaviors and decreases aggressive behaviors toward novel lambs (Kendrick et al., 1987). Furthermore, both infusion of OT and vaginocervical stimulation (VCS), which stimulates release of OT in the brain, results in a ewe bonding to a novel lamb (Kendrick et al., 1988b).

To further support this evidence from sheep and rats, two mutant mouse strains have been generated to analyze the oxytocinergic system: an oxytocin peptide knockout mouse (OTKO), and an oxytocin receptor knockout mouse (OTRKO). Both of these mutant strains exhibit deficits in various social parameters, including aggression, social recognition, parental care, and reproductive behavior. Specifically, OTRKO postpartum mothers and virgins show disruptions in maternal behavior as assayed by latency to retrieve pups, latency to crouch over pups, and time spent crouching over pups (Takayanagi et al., 2005). This deficit is not as robust in OT peptide knockout animals, perhaps because the OTR is being activated by structurally and functionally related AVP (Nishimori et al., 1996; Young et al., 1996; Winslow and Insel, 2002; Pedersen et al., 2006).

Although the role of OTR expression in maternal behavior has yet to be explored in prairie voles, there is evidence that OTR plays a role in regulating alloparental behavior in this highly social species. Alloparental behavior is defined as parental-like behavior expressed toward infants that are not the offspring of that animal; in prairie voles, it is often used to describe the behavior of juveniles, which will often take care of their parents' next litter if they do not leave the nest after weaning (Carter and Getz, 1993). Individual variation in the density of OTR binding in the nucleus accumbens (NAcc) of both juvenile and adult female prairie voles is positively correlated with levels of alloparental behavior (Olazabal and Young, 2006a; Olazabal and Young, 2006b). Furthermore, infusion of an OTR antagonist into the NAcc of adult sexually naïve females reduces the percentage of animals that display alloparental behavior toward novel pups (Olazabal, 2006b). Thus in rats, mice, sheep and prairie voles, there is evidence that the OT system plays a role in the regulation of maternal nurturing behavior and, at least in sheep, in the development of the maternal bond.

8.3.2 Oxytocin and pair bonding in female voles

Based on this evidence that OT is required for mother–infant bonding in sheep and that OTR activation is necessary for appropriate maternal nurturing behavior in rodents, this social attachment system became a focus for research into the neural basis of pair bonding between the female and male prairie vole. The first study to examine the role of OT in prairie vole pair bonding found that in the absence of mating, central infusion of OT was sufficient to stimulate the formation of a partner preference in female prairie voles (Williams et al., 1994). Furthermore, infusion of an OTR antagonist ICV was able to block mating-induced partner preferences in the female prairie vole (Insel and Winslow, 1995). These studies suggest that common neuroendocrine mechanisms to some degree regulate maternal bonding and bonding between the female and the male partner.

In an effort to begin to understand the neural circuitry underlying OT-dependent partner preference formation, and to explore the nature of the variation in mating strategies among vole species, the neuroanatomical distribution of OTR binding was examined in prairie voles and non-monogamous montane voles. As previously discussed for the AVPR1a, there are also remarkable species differences in OTR distribution in the brain, which may account in part for the species differences in pair bonding between monogamous and non-monogamous voles. There is a high density of OTR expression in the NAcc in the monogamous prairie vole that is absent in the promiscuous montane vole (Figure 8.4) (Insel and Shapiro, 1992; Young and Wang, 2004). OTRs are present in both species in the prefrontal cortex (PFC), which, along with NAcc, is

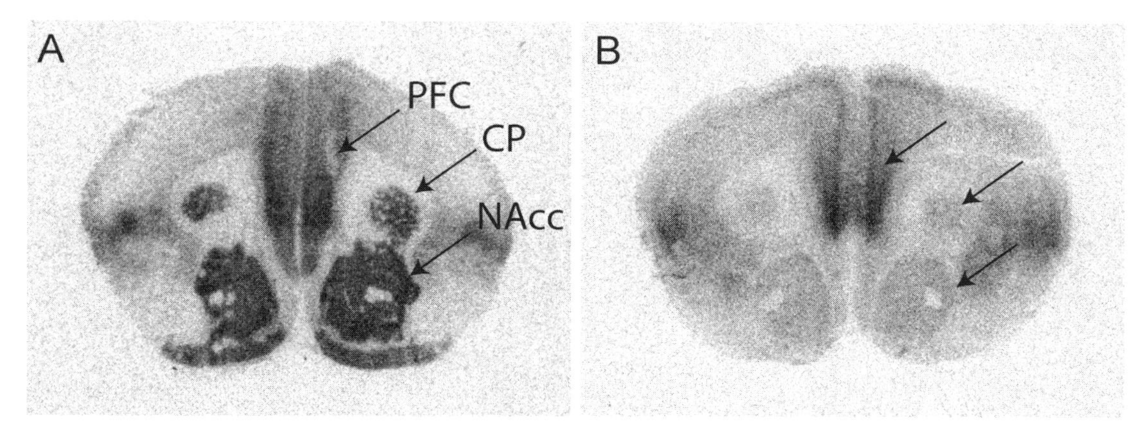

Figure 8.4 Oxytocin receptor distribution in voles. (A and B) Autoradiograms of oxytocin receptor binding densities in coronal sections through the forebrain of a prairie vole (A) and a meadow vole (B), showing the prefrontal cortex (PFC), caudate putamen (CP), and nucleus accumbens (NAcc). Figure adapted from Young and Wang, 2004.

part of the brain's reward and reinforcement pathway (Kelley, 2004; Wallace et al., 2008). Blocking OTR in the PFC or the NAcc by infusing an OTR antagonist site specifically into the brains of female prairie vole prevents partner preference formation, indicating that OTR activation in these regions is necessary for development of partner preferences in female prairie voles (Young et al., 2001). Furthermore, increasing OTR density in the NAcc by site-specific viral vector mediated gene transfer facilitates the formation of a partner preference in female prairie voles as well (Ross, Freeman, et al., 2009).

In vivo microdialysis to monitor OT release in the NAcc in female prairie voles during cohabitation with males showed that an increase in extracellular OT during mating (Ross, Cole, et al., 2009). Mating in prairie voles involves multiple intromissions, which likely results in significant VCS, a potent OT-releasing stimulus. Thus, it appears that while VCS during labor initiates OT release leading to the formation of a maternal bond in mother sheep, VCS during mating initiates OT release leading to the formation of a pair bond in the female prairie vole. Based on the data described above, the female's bond to a male is mediated by similar neural underpinnings as the bond of a mother to her offspring. Therefore, social monogamy as a mating strategy for females may result from an adaptation of the maternal circuitry in mammals.

8.4 Social recognition and reward: The pathway to bonding

We have discussed that both AVP and OT acting within specific brain regions stimulate the formation of pair bonds in male and female prairie voles. But how does activation of these neuropeptide receptors lead to a social preference for the partner, and hence, to a pair bond? Two components of social cognition that are critical to the formation of a pair bond are reinforcement of social interactions via the reward systems in the brain, and individual recognition based on social memory circuits. In rodents, social memory is tested in the laboratory by quantifying the duration of social investigation after repeated exposures to the same individual. Rodents will show decreased olfactory investigation time after each trial, and investigators interpret this habituation to indicate that

the rodent has formed a memory of that individual. Exposure to a novel animal after this habituation process results in a return to the original high levels of investigation seen during the first presentation.

Paradigms such as this one have been commonly used to assess social memory in various rodent species. Experiments using knockout mice have been instrumental in the investigation of the role of AVP and OT in social memory (see Chapter 13). Transgenic mice lacking the *avpr1a* gene, called AVPR1a knockout mice, exhibit social amnesia but have normal spatial memory and normal memory for non-social odors, indicating that both learning ability and olfactory function in these animals is still intact (Bielsky et al., 2004). Recovery of social memory in these animals can be attained by experimentally replacing AVPR1a using viral vector mediated gene transfer in the LS (Bielsky et al., 2005). Also, studies in rats have examined the role of AVP in social memory. In male rats, AVPR1a activation in the LS has been shown to be involved in retention of social memory; injecting AVP into the LS increases social memory, while injection of an AVPR1a antagonist impairs it (Dantzer et al., 1987; Dantzer et al., 1988). This result is important in the context of social memory and pair bonding, because LS is one of the brain regions that has been shown to be essential for the AVP-mediated formation of a partner preference in male prairie voles. Selective amnesia for social information is also seen in OTKO mice and can be induced in wild-type mice by treatment with an OT antagonist (Ferguson et al., 2000). This deficit in knockout mice can be rescued by infusions of OT ICV as well as into the MeA (Ferguson et al., 2000; Ferguson et al., 2001). Social memory deficits have also been documented in OTRKO male mice (Takayanagi et al., 2005). These experiments support the role of OT and AVP in individual recognition, a process that partner preference formation depends on.

The other essential component to the formation of a pair bond besides individual recognition is the reinforcement of social interactions by activation of the brain's reward circuits.

In female prairie voles, the formation of a partner preference relies on the expression of OTR in the NAcc, a region in the brain's mesolimbic reward pathway (Young et al., 2001). Another region in this pathway is the VP, an area critical for partner preference in male prairie voles. The VP has also been shown to be important for the reinforcing properties of drugs of abuse. When injected stereotaxically into the VP of a rat, psychostimulants like cocaine and amphetamine induce a conditioned place preference (CPP) for the area of the test arena where the rat received the dose, and this cocaine-induced CPP can be prevented if dopamine (DA) is depleted from the VP (Gong and Justice, 1996; Gong and Justice, 1997). In a CPP task, animals receive a reinforcing stimulus only when they cross into a certain area of a place preference arena; after a few conditioning trials, the animal will prefer to spend time in this part of the arena. This process is analogous to the partner preference exhibited by pair-bonded prairie voles: the partner is the conditioned stimulus reinforced by the rewarding aspects of mating.

This reward is believed to be mediated by DA. In female prairie voles, microinjection of a dopamine D2-like receptor antagonist into the NAcc, but not the PFC, blocked the development of a partner preference, and a dopamine D2-like receptor agonist injected in the NAcc caused females that had cohabitated with a male in the absence of mating to form a partner preference (Gingrich et al., 2000). This result implies that perhaps the rewarding aspect of mating, a behavior that increases dopamine (DA) levels in the NAcc, could be combining with activation of OTR in that region to cause a positive association between the social stimuli from that specific partner and the reward of mating (Young and Wang, 2004). Thus, it seems that in female prairie voles, the mating-induced increases in OT and DA converging on the NAcc may be causing natural reward learning and reinforcing the characteristics of the individual partner after mating to result in a partner preference.

Therefore, we suggest that with the convergence of reward systems, learning and memory circuits,

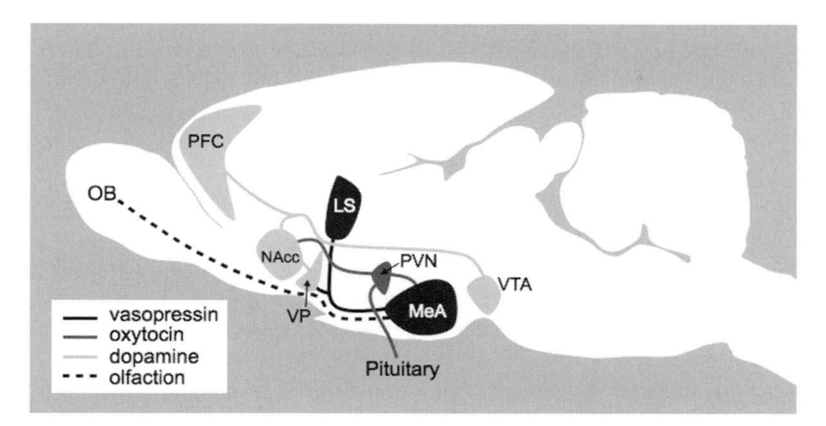

Figure 8.5 Sagittal view of a prairie vole brain illustrating a proposed neural circuit model for pair bonding. In this model, mating activates the ventral tegmental area (VTA), resulting in increased dopamine activity in the prefrontal cortex (PFC) and nucleus accumbens (NAcc). Concurrently, olfactory signals from the mate are transmitted via the olfactory bulb (OB) to the medial amygdala (MeA). Oxytocin (OT) acts in the MeA, and vasopressin (AVP) acts in the lateral septum (LS) to facilitate olfactory learning and memory. Mating also stimulates increased extracellular concentrations of OT in the PFC and NAcc of females, and of AVP in the ventral pallidum (VP) of males. AVP fibers in the LS and VP originate from cell bodies in the MeA. The source of OT projections to the NAcc, MeA, and PFC most likely originate from a population of cell bodies in the paraventricular nucleus of the hypothalamus (PVN). The concurrent activation of the dopaminergic system and the OT or AVP systems in the NAcc or VP, respectively, potentially results in the development of a conditioned partner preference. Figure adapted from Young and Wang, 2004.

and the activation of OT and AVP inputs in response to social stimuli could result in an increased desire to spend time with an opposite sex conspecific with which an individual has previously mated, or a "conditioned partner preference." This proposed neural circuitry of pair bonding is illustrated in Figure 8.5.

8.5 Beyond monogamy: Comparative work in other mammalian social systems

8.5.1 Oxytocin and African mole rats

The OT system has also been examined in African mole rats, a taxonomic group of rodents that exhibit a wide variety in their social organization (Kalamatianos et al., 2009). Despite being in the same phylogenetic family, the eusocial naked mole rat and the solitary caped mole rat represent opposite ends of the spectrum of sociality. The eusocial naked mole rat, *Heterocephalus glaber*, lives in colonies that can contain hundreds of individuals, and like other eusocial species, have a unique system of reproduction that includes one reproductively active queen and her one or few male mates. The other male and female individuals in the colony are reproductively suppressed, provide cooperative care of the young, and work together to burrow, forage, and protect the colony (Brett, 1991; Jarvis, 1991). On the other side of the spectrum, the caped mole rat, *Georychus capensis*, is a solitary species with interaction between conspecifics limited to mating and contact between the mother, pups, and littermates (Bennett and Jarvis, 1988). In caped mole rats, even the interaction between the mother and pups is limited; the mother becomes aggressive toward her offspring and drives them out of the burrow shortly after weaning (Bennett and Jarvis, 1988).

This difference in sociality was theorized by Kalamatianos et al. to be mediated in part by

neurochemical alterations in the OT system (Kalamatianos et al., 2009). When these authors performed receptor autoradiography for the OTR in both species, they found that the naked mole rats have higher levels of expression than the caped mole rats in multiple brain regions that also have high OTR binding in the prairie vole, most notably the NAcc (Kalamatianos et al., 2009). Drawing a parallel between monogamous prairie voles and eusocial naked mole rats is not as direct as comparing two solitary rodent species, like the caped mole rat and the meadow vole, but the eusocial organization of naked mole rats involves similar prosocial behaviors as seen in the monogamous prairie vole, such as extended periods of time in close side-by-side contact huddling together (Jarvis, 1991; Lacey and Sherman, 1991). Thus, authors hypothesize that the high levels of OTR seen in the NAcc of naked mole rats might mediate the high levels of prosocial contact and alloparental behavior required to maintain such an extended colonial social structure in this species.

8.5.2 Vasopressin and monogamy in deer mice

Another rodent species that has been examined in the context of social bonding and neurohypophyseal hormones is the monogamous California mouse, *Peromyscus californicus*. This species is an excellent model for the study of paternal behavior and male aggression, especially in comparison with the closely related promiscuous white-footed mouse, *Peromyscus leucopus*, which exhibits much less paternal care and lower levels of aggression than the California mouse (Bester-Meredith et al., 1999). It was hypothesized that there would be high levels of AVPR1a binding in the California mouse in the brain regions that have high AVPR1a binding in the prairie vole. While the authors found many unexpected differences in the AVPR1a pattern of binding between the monogamous California mouse and the prairie vole, there was dense AVPR1a binding in the LS and VP, two notable areas also high in AVPR1a in the prairie vole (Insel et al., 1991; Bester-Meredith et al., 1999). Both of these brain areas were also found to have much lower or even undetectable levels of AVPR1a binding in the promiscuous white-footed mouse as compared to the California mouse (Bester-Meredith et al., 1999). These results suggest that there are similarities in the underlying neurological systems for social behaviors such as monogamy, paternal care, and aggression, and the role that these systems play more generally in the mammalian brain may be conserved across phylogenetic groups.

8.5.3 Oxytocin, vasopressin, and primate neuroanatomy and social behavior

While the comparative study of OT and AVP in different mammalian social systems has expanded greatly since the first rodent studies were conducted, there is still a considerable lack of knowledge about the role that these systems play in the evolution and expression of primate social organization. This is largely due to limitations not only in access to primate brain tissue but, more importantly, in the lack of reliable detection and localization methods for OT and AVP receptors in the brains of primates. The pharmacological tools that have been used to study these systems in rodents do not have the same selectivity in primate tissue as they do in rodents; recent work from our own lab has shown that the antagonists and agonists used to manipulate these systems, as well as the radioligands used to detect the receptors in brain sections, have a mixed affinity for the human OTR and human AVPR1a (unpublished data). In fact, these results indicate that published AVPR1a distribution map in the rhesus macaque brain may in fact reflect AVPR1a that has been contaminated with non-specific OTR binding (Young, Toloczko, et al., 1999). More work is currently underway to establish the actions of these compounds in primate brain tissue, so that the necessary neuroanatomical background knowledge is available for future studies of the role of OT and AVP in primate social cognition.

8.5.3.1 Variation in oxytocin and social behavior in macaques

Despite these issues for neuroanatomical research, the role of OT in primate social behavior has recently begun to be investigated. In 2003, it was shown in rhesus macaques that OT levels in cerebrospinal fluid (CSF) correlate significantly with the level of expressed affiliative behavior (Winslow et al., 2003). Rosenblum et al. took advantage of the natural variation in social behavior within the macaque genus and compared CSF OT between the affiliative bonnet macaque and the aggressive pigtail macaque (Rosenblum et al., 2002). Bonnets spend much of their time in prosocial contact, such as grooming and huddling; they are more likely than pigtails to engage in social interaction with conspecific strangers; and they do not have a strong social hierarchy. Pigtails on the other hand have a strict hierarchy, maintain large individual distances, and are not likely to interact positively with strangers. As hypothesized, the affiliative bonnets were found to have higher CSF OT and lower CSF levels of corticotropin releasing factor (CRF; mediates response to stress) than the more aggressive pigtails (Rosenblum et al., 2002).

8.5.3.2 Monogamous primates: Marmosets, titi monkeys, and tamarins

Although monogamy is rare in primates, a few socially monogamous species of New World monkey, including the common marmoset, *Callithrix jacchus*, the black-pencilled marmoset, *Callitrhix pencillata*, the coppery titi monkey, *Callicebus cupreus*, and the cotton-top tamarin, *Saguinus Oedipus*, have become new animal models for the neuroendocrine basis of social bonding and parental care. In marmosets, there is preliminary evidence that OT may play a role in regulating some aspects of pair bonding. Daily intranasal (IN) delivery of OT in black-pencilled marmosets during a 3-week period in which males and females were paired with opposite sex conspecifics resulted in increased huddling with their partner, and daily delivery of an OTR antagonist caused decreased proximity to their partner as well as decreased food sharing, a highly social and cooperative behavior in these monkeys (Smith, 2010). Although there is no non-monogamous species for comparison in this model like there is in voles, the neuroanatomical map of OTR and AVPR1a in the socially monogamous common marmoset has been published (Schorscher-Petcu et al., 2009). Common marmosets have high levels of OTR binding in the NAcc and high levels of AVPR1a binding in the VP, which is consistent with the findings in voles. However, due to the pharmacological limitations described above, these results need to be corroborated by future studies in order to ensure their reliability.

Coppery titi monkeys, which have been used to study pair-bonding behavior in primates, have recently been used to investigate the role of AVP on pair bonding in males. In one experiment, males that have been paired with a female for at least one year were given either saline, a low dose of IN AVP, or a high dose of IN AVP before being presented in turn with either its female partner or an unfamiliar female stranger. When given saline, males contacted the unfamiliar female more frequently than their partner, which is expected because most mammalian males are attracted to novel females. But when they were given the high dose of AVP, males contacted their partner more frequently than the unfamiliar female (Jarcho et al., 2011). This result supports the idea that AVP underlies affiliative behavior in highly social primates.

The relationship of OT in pair-bonding behavior in cotton-top tamarins has also begun to be examined. Urinary levels of OT and various affiliative behaviors were measured in male–female pairs over the course of 3 weeks. OT levels within each pair was directly related to sexual behavior and the frequency of affiliative behaviors between male–female pairs, specifically initiation of huddling by males and initiation of sex by females (Snowdon et al., 2010).

8.6 Future and conclusions

In this chapter, we have reviewed the evidence that supports one possible mechanism by which the physiologically ancient functions of OT and AVP in mammals could have changed under evolutionary pressure to result in social bonding from a previously solitary system of social organization. Our proposed theory for the evolutionary basis for monogamy in voles is not necessarily generalizable to all monogamous mammalian species, but does provide some insight into the neural basis for attachment and bonding in these species. It seems that over evolutionary time the neural and genetic correlates of AVP-dependent male territorial behavior have been modified to give rise to the behavioral patterns associated with social monogamy in prairie voles. We hypothesize that in prairie voles, when this system interacts with those involved in social recognition, spatial memory, and reward, previously generalized territorial behaviors can become focused on the sensory signature of the female mate, resulting in the partner becoming a valued aspect of his territory. Thus, the male's pair bond to the female can be viewed in the context of an AVP-mediated territorial behavior, especially considering the selective aggression that the male displays after becoming pair bonded to a female. In females, the circuits that mediate the onset of maternal nurturing and infant attachment after parturition and during nursing have been exapted to give rise to the pair bond. These circuits converge via activation of OTRs with those involved in social recognition and reward, resulting in pair bonding to a male after mating.

While studies with marmosets, macaques, and tamarins have begun to move this field toward investigations in primate social behavior, it is still unknown whether the same neural circuits so eloquently mapped in rodents also apply to highly social primates, such as humans. The work described in this chapter and the rest of this book has paved the way for understanding how the human brain functions socially, but there is much left to be discovered here. What are the implications for the neural basis of social bonding in our species? While this chapter did not discuss whether humans should be considered monogamous or not, it is undeniable that men and women develop a relationship in our species that goes beyond orgasm.

There has been a growing number of studies examining both the AVP and OT systems in humans. A recently published study examined whether polymorphisms that exist in the human *AVPR1A* gene correlated with pair-bonding behavior among couples. Among men in long-term, live-in relationships lasting at least 5 years, those with one specific variant at a repeat polymorphism, called RS3, were more likely to be uncommitted in their relationship by remaining unmarried, and if married, were more likely to report a crisis in their marriage (Walum, 2008). Also, the partners of males with this variant were more likely to report dissatisfaction in the relationship (Walum et al., 2008). Studies delivering OT intranasally in humans have shown that OT can affect some aspects of interpersonal relationships, including face perception, emotion recognition, and trust (Kirsch et al., 2005; Kosfeld et al., 2005; Domes, Heinrichs, Glascher, et al., 2007; Domes Heinrichs, Michel, et al., 2007; Guastella et al., 2008).

Thus, it is apparent that understanding the roles of AVP and OT in regulating monogamy-related behaviors in animal models can provide important insights into the evolution of our own social structure.

REFERENCES

Albers, H. E., J. Pollock, W. H. Simmons, and C. F. Ferris (1986). A V1-like receptor mediates vasopressin-induced flank marking behavior in hamster hypothalamus. *J Neurosci*, 6(7): 2085–2089.

Aragona, B. J. and Z. Wang (2004). The prairie vole (Microtus ochrogaster): an animal model for behavioral neuroendocrine research on pair bonding. *ILAR J*, 45(1), 35–45.

Barberis, C., S. Audigier, T. Durroux, et al. (1992). Pharmacology of oxytocin and vasopressin receptors in the

central and peripheral nervous system. *Ann NY Acad Sci*, 652, 39–45.

Bennett, N. C. and J. U. M. Jarvis (1988). The reproductive biology of the Cape mole-rat, Georychus capensis (Rodentia, Bathyergidae). *Journal of Zoology*, 214(1), 95–106.

Bester-Meredith, J. K., L. J. Young, and C. A. Marler (1999). Species differences in paternal behavior and aggression in peromyscus and their associations with vasopressin immunoreactivity and receptors. *Horm Behav*, 36(1), 25–38.

Bielsky, I. F., S. B. Hu, X. Ren, E. F. Terwilliger, and L. J. Young (2005). The V1a vasopressin receptor is necessary and sufficient for normal social recognition: a gene replacement study. *Neuron*, 47(4), 503–513.

Bielsky, I. F., S. B. Hu, K. L. Szegda, H. Westphal, and L. J. Young (2004). Profound impairment in social recognition and reduction in anxiety-like behavior in vasopressin V1a receptor knockout mice. *Neuropsychopharmacology*, 29(3), 483–493.

Boyd, S. K. (1994). Arginine vasotocin facilitation of advertisement calling and call phonotaxis in bullfrogs. *Horm Behav*, 28(3), 232–240.

Brett, R. A. (1991). The population structure of naked mole-rat colonies. *The Biology of the Naked Mole-Rat*. P. W. Sherman, J. U. M. Jarvis and R. D. Alexander. New York, Princeton University Press, 97–136.

Brotherton, P. N., J. M. Pemberton, P. E. Komers, and G. Malarky (1997). Genetic and behavioural evidence of monogamy in a mammal, Kirk's dik-dik (Madoqua kirkii). *Proc Biol Sci*, 264(1382), 675–681.

Burbach, J. P., S. M. Luckman, D. Murphy, and H. Gainer (2001). Gene regulation in the magnocellular hypothalamo-neurohypophysial system. *Physiol Rev*, 81(3), 1197–1267.

Burbach, J. P. H., L. J. Young, and J. A. Russell (2006). Oxytocin: Synthesis, Secretion, and Reproductive Functions. *Knobil and Neill's Physiology of Reproduction*. J. D. Neill, Elsevier Academic Press, pp. 3055–3126.

Carter, C. S. A. C. DeVries, and L. L. Getz (1995). Physiological substrates of mammalian monogamy: the prairie vole model. *Neurosci Biobehav Rev*, 19(2), 303–314.

Carter, C. S. and L. L. Getz (1993). Monogamy and the prairie vole. *Sci Am*, 268(6), 100–106.

Clutton-Brock, T. H. (1989). Mammalian mating systems. *Proc R Soc Lond B Biol Sci*, 236(1285), 339–372.

Dantzer, R., R. M. Bluthe, G. F. Koob, and M. Le Moal (1987). Modulation of social memory in male rats by neurohypophyseal peptides. *Psychopharmacology (Berl)*, 91(3), 363–368.

Dantzer, R., G. F. Koob, R. M. Bluthe, and M. Le Moal (1988). Septal vasopressin modulates social memory in male rats. *Brain Res*, 457(1), 143–147.

Domes, G., M. Heinrichs, J. Glascher, et al. (2007). Oxytocin attenuates amygdala responses to emotional faces regardless of valence. *Biol Psychiatry*, 62(10), 1187–1190.

Domes, G., M. Heinrichs, A. Michel, C. Berger, and S. C. Herpertz (2007). Oxytocin improves "mind-reading" in humans. *Biol Psychiatry*, 61(6), 731–733.

Donaldson, Z. R., F. A. Kondrashov, A. Putnam, et al. (2008). Evolution of a behavior-linked microsatellite-containing element in the 5′ flanking region of the primate AVPR1A gene. *BMC Evol Biol*, 8, 180.

Ferguson, J. N., J. M. Aldag, T. R. Insel, and L. J. Young (2001). Oxytocin in the medial amygdala is essential for social recognition in the mouse. *J Neurosci*, 21(20), 8278–8285.

Ferguson, J. N., L. J. Young, E. F. Hearn, et al. (2000). Social amnesia in mice lacking the oxytocin gene. *Nat Genet*, 25(3), 284–288.

Ferris, C. F., H. E. Albers, S. M. Wesolowski, B. D. Goldman, and S. E. Luman (1984). Vasopressin injected into the hypothalamus triggers a stereotypic behavior in golden hamsters. *Science*, 224(4648), 521–523.

Ferris, C. F., R. H. Melloni, Jr., G. Koppel, et al. (1997). Vasopressin/serotonin interactions in the anterior hypothalamus control aggressive behavior in golden hamsters. *J Neurosci*, 17(11), 4331–4340.

Ferris, C. F., J. Pollock, H. E. Albers, and S. E. Leeman (1985). Inhibition of flank-marking behavior in golden hamsters by microinjection of a vasopressin antagonist into the hypothalamus. *Neurosci Lett*, 55(2), 239–243.

Fink, S., L. Excoffier, and G. Heckel (2006). Mammalian monogamy is not controlled by a single gene. *Proc Natl Acad Sci USA*, 103(29), 10956–10960.

Getz, L. L. and J. E. Hofmann (1986). Social organization in free-living prairie voles, Microtus ochrogaster. *Behavioral Ecology and Sociobiology*, 18(4), 275–282.

Gingrich, B., Y. Liu, C. Cascio, Z. Wang, and T. R. Insel (2000). Dopamine D2 receptors in the nucleus accumbens are important for social attachment in female prairie voles (Microtus ochrogaster). *Behav Neurosci*, 114(1), 173–183.

Gobrogge, K. L., Y. Liu, L. J. Young, and Z. Wang (2009). Anterior hypothalamic vasopressin regulates

pair-bonding and drug-induced aggression in a monogamous rodent. *Proc Natl Acad Sci USA*, 106(45), 19144–19149.

Gong, W., D. Neill, and J. B. Justice, Jr. (1996). Conditioned place preference and locomotor activation produced by injection of psychostimulants into ventral pallidum. *Brain Res*, 707(1), 64–74.

Gong, W., D. Neill, and J. B. Justice, Jr. (1997). 6-Hydroxydopamine lesion of ventral pallidum blocks acquisition of place preference conditioning to cocaine. *Brain Res*, 754(1–2), 103–112.

Goodson, J. L. (1998). Territorial aggression and dawn song are modulated by septal vasotocin and vasoactive intestinal polypeptide in male field sparrows (Spizella pusilla). *Horm Behav*, 34(1), 67–77.

Gruder-Adams, S. and L. L. Getz (1985). Comparison of the Mating System and Paternal Behavior in Microtus ochrogaster and M. pennsylvanicus. *Journal of Mammalogy*, 66(1), 165–167.

Guastella, A. J., P. B. Mitchell, and M. R. Dadds (2008). Oxytocin increases gaze to the eye region of human faces. *Biol Psychiatry*, 63(1), 3–5.

Hammock, E. A. and L. J. Young (2005). Microsatellite instability generates diversity in brain and sociobehavioral traits. *Science*, 308(5728), 1630–1634.

Hara, Y., J. Battey, and H. Gainer (1990). Structure of mouse vasopressin and oxytocin genes. *Brain Res Mol Brain Res*, 8(4), 319–324.

Hibert, M., J. Hoflack, S. Trumpp-Kallmeyer, et al. (1999). Functional architecture of vasopressin/oxytocin receptors. *J Recept Signal Transduct Res*, 19(1–4), 589–596.

Insel, T. R. R. Gelhard, and L. E. Shapiro (1991). The comparative distribution of forebrain receptors for neurohypophyseal peptides in monogamous and polygamous mice. *Neuroscience*, 43(2–3), 623–630.

Insel, T. R., S. Preston, and J. T. Winslow (1995). Mating in the monogamous male: behavioral consequences. *Physiol Behav*, 57(4), 615–627.

Insel, T. R. and L. E. Shapiro (1992). Oxytocin receptor distribution reflects social organization in monogamous and polygamous voles. *Proc Natl Acad Sci USA*, 89(13), 5981–5985.

Insel, T. R., Z. X. Wang, and C. F. Ferris (1994). Patterns of brain vasopressin receptor distribution associated with social organization in microtine rodents. *J Neurosci*, 14(9), 5381–5392.

Jarcho, M. R., S. P. Mendoza, W. A. Mason, X. Yang, and K. L. Bales (2011). Intranasal vasopressin affects pair bonding and peripheral gene expression in male Callicebus cupreus. *Genes Brain Behav*, 10(3), 375–383.

Jard, S. (1998). Vasopressin receptors. A historical survey. *Adv Exp Med Biol*, 449, 1–13.

Jarvis, J. U. M. (1991). Reproduction of naked mole-rats. *The Biology of the Naked Mole-Rat*. P. W. Sherman, J. U. M. Jarvis, and R. D. Alexander. New York, Princeton University Press, pp. 384–425.

Kalamatianos, T., C. G. Faulkes, M. K. Oosthuizen, et al. (2009). Telencephalic binding sites for oxytocin and social organization, a comparative study of eusocial naked mole-rats and solitary cape mole-rats. *J Comp Neurol*, 518(10), 1792–1813.

Kashi, Y. and D. G. King (2006). Simple sequence repeats as advantageous mutators in evolution. *Trends Genet*, 22(5), 253–259.

Kelley, A. E. (2004). Memory and addiction: shared neural circuitry and molecular mechanisms. *Neuron*, 44(1), 161–179.

Kendrick, K. M., E. B. Keverne, and B. A. Baldwin (1987). Intracerebroventricular oxytocin stimulates maternal behaviour in the sheep. *Neuroendocrinology*, 46(1), 56–61.

Kendrick, K. M., E. B. Keverne, B. A. Baldwin, and D. F. Sharman (1986). Cerebrospinal fluid levels of acetylcholinesterase, monoamines and oxytocin during labour, parturition, vaginocervical stimulation, lamb separation and suckling in sheep. *Neuroendocrinology*, 44(2), 149–156.

Kendrick, K. M., E. B. Keverne, C. Chapman, and B. A. Baldwin (1988). Intracranial dialysis measurement of oxytocin, monoamine and uric acid release from the olfactory bulb and substantia nigra of sheep during parturition, suckling, separation from lambs and eating. *Brain Res*, 439(1–2), 1–10.

Kendrick, K. M., E. B. Keverne, C. Chapman, and B. A. Baldwin (1988). Microdialysis measurement of oxytocin, aspartate, gamma-aminobutyric acid and glutamate release from the olfactory bulb of the sheep during vaginocervical stimulation. *Brain Res*, 442(1), 171–174.

Kirsch, P., C. Esslinger, Q. Chen, et al. (2005). Oxytocin modulates neural circuitry for social cognition and fear in humans. *J Neurosci*, 25(49), 11489–11493.

Kleiman, D. G. (1977). Monogamy in mammals. *Q Rev Biol*, 52(1), 39–69.

Kosfeld, M., M. Heinrichs, P. J. Zak, U. Fischbacher, and E. Fehr (2005). Oxytocin increases trust in humans. *Nature*, 435(7042), 673–676.

Lacey, E. A. and P. W. Sherman (1991). Social organization of naked mole-rat colonies: evidence for divisions of labor. *The Biology of the Naked Mole-Rat.* P. W. Sherman, J. U. M. Jarvis, and R. D. Alexander. New York, Princeton University Press, 275–336.

Landgraf, R. and I. D. Neumann (2004). Vasopressin and oxytocin release within the brain: a dynamic concept of multiple and variable modes of neuropeptide communication. *Front Neuroendocrinol,* 25(3–4), 150–176.

Lim, M. M., A. Z. Murphy, and L. J. Young (2004). Ventral striatopallidal oxytocin and vasopressin V1a receptors in the monogamous prairie vole (Microtus ochrogaster). *J Comp Neurol,* 468(4), 555–570.

Lim, M. M., Z. Wang, D. E. Olazabal, et al. (2004). Enhanced partner preference in a promiscuous species by manipulating the expression of a single gene. *Nature,* 429(6993), 754–757.

Lim, M. M. and L. J. Young (2004). Vasopressin-dependent neural circuits underlying pair bond formation in the monogamous prairie vole. *Neuroscience,* 125(1), 35–45.

Liu, Y., J. T. Curtis, and Z. Wang (2001). Vasopressin in the lateral septum regulates pair bond formation in male prairie voles (Microtus ochrogaster). *Behav Neurosci,* 115(4), 910–919.

Ludwig, M. and G. Leng (2006). Dendritic peptide release and peptide-dependent behaviours. *Nat Rev Neurosci,* 7(2), 126–136.

Maney, D. L., C. T. Goode, and J. C. Wingfield (1997). Intraventricular infusion of arginine vasotocin induces singing in a female songbird. *J Neuroendocrinol,* 9(7), 487–491.

McGraw, L. A. and L. J. Young (2010). The prairie vole: an emerging model organism for understanding the social brain. *Trends Neurosci,* 33(2), 103–109.

Nishimori, K., L. J. Young, Q. Guo, et al. (1996). Oxytocin is required for nursing but is not essential for parturition or reproductive behavior. *Proc Natl Acad Sci USA,* 93(21), 11699–11704.

Olazabal, D. E. and L. J. Young (2006). Oxytocin receptors in the nucleus accumbens facilitate "spontaneous" maternal behavior in adult female prairie voles. *Neuroscience,* 141(2), 559–568.

Olazabal, D. E. and L. J. Young (2006). Species and individual differences in juvenile female alloparental care are associated with oxytocin receptor density in the striatum and the lateral septum. *Horm Behav,* 49(5), 681–687.

Ophir, A. G., P. Campbell, K. Hanna, and S. M. Phelps (2008). Field tests of cis-regulatory variation at the prairie vole avpr1a locus: association with V1aR abundance but not sexual or social fidelity. *Horm Behav,* 54(5), 694–702.

Ophir, A. G., J. O. Wolff, and S. M. Phelps (2008). Variation in neural V1aR predicts sexual fidelity and space use among male prairie voles in semi-natural settings. *Proc Natl Acad Sci USA,* 105(4), 1249–1254.

Orians, G. H. (1969). On the Evolution of Mating Systems in Birds and Mammals. *American Naturalist,* 103(934), 15.

Pedersen, C. A., J. A. Ascher, Y. L. Monroe, and A. J. Prange, Jr. (1982). Oxytocin induces maternal behavior in virgin female rats. *Science,* 216(4546), 648–650.

Pedersen, C. A., J. D. Caldwell, M. F. Johnson, S. A. Fort, and A. J. Prange, Jr. (1985). Oxytocin antiserum delays onset of ovarian steroid-induced maternal behavior. *Neuropeptides,* 6(2), 175–182.

Pedersen, C. A., J. D. Caldwell, C. Walker, G. Ayers, and G. A. Mason (1994). Oxytocin activates the postpartum onset of rat maternal behavior in the ventral tegmental and medial preoptic areas. *Behav Neurosci,* 108(6), 1163–1171.

Pedersen, C. A. and A. J. Prange, Jr. (1979). Induction of maternal behavior in virgin rats after intracerebroventricular administration of oxytocin. *Proc Natl Acad Sci USA,* 76(12), 6661–6665.

Pedersen, C. A., S. V. Vadlamudi, M. L. Boccia, and J. A. Amico (2006). Maternal behavior deficits in nulliparous oxytocin knockout mice. *Genes Brain Behav,* 5(3), 274–281.

Rosenblum, L. A., E. L. Smith, M. Altemus, et al. (2002). Differing concentrations of corticotropin-releasing factor and oxytocin in the cerebrospinal fluid of bonnet and pigtail macaques. *Psychoneuroendocrinology,* 27(6), 651–660.

Ross, H. E., C. D. Cole, Y. Smith, et al. (2009). Characterization of the oxytocin system regulating affiliative behavior in female prairie voles. *Neuroscience,* 162(4), 892–903.

Ross, H. E., S. M. Freeman, L. L. Spiegel, et al. (2009). Variation in oxytocin receptor density in the nucleus accumbens has differential effects on affiliative behaviors in monogamous and polygamous voles. *J Neurosci,* 29(5), 1312–1318.

Ross, H. E. and L. J. Young (2009). Oxytocin and the neural mechanisms regulating social cognition and affiliative behavior. *Front Neuroendocrinol,* 30(4), 534–547.

Schorscher-Petcu, A., A. Dupre, and E. Tribollet (2009). Distribution of vasopressin and oxytocin binding sites in the brain and upper spinal cord of the common marmoset. *Neurosci Lett,* 461(3), 217–222.

Schuiling, G. A. (2003). The benefit and the doubt: why monogamy? *J Psychosom Obstet Gynaecol*, 24(1), 55–61.

Smith, A. S., A. Agmo, A. K. Birnie, and J. A. French (2010). Manipulation of the oxytocin system alters social behavior and attraction in pair-bonding primates, Callithrix penicillata. *Horm Behav*, 57(2), 255–262.

Snowdon, C. T., B. A. Pieper, C. Y. Boe, et al. (2010). Variation in oxytocin is related to variation in affiliative behavior in monogamous, pairbonded tamarins. *Horm Behav*, 58(4), 614–618.

Solomon, N. G., A. R. Richmond, P. A. Harding, et al. (2009). Polymorphism at the avpr1a locus in male prairie voles correlated with genetic but not social monogamy in field populations. *Mol Ecol*, 18(22), 4680–4695.

Takayanagi, Y., M. Yoshida, I. F. Bielsky, et al. (2005). Pervasive social deficits, but normal parturition, in oxytocin receptor-deficient mice. *Proc Natl Acad Sci USA*, 102(44), 16096–16101.

Turner, L. M., A. R. Young, H. Rompler, et al. (2010). Monogamy evolves through multiple mechanisms: evidence from V1aR in deer mice. *Mol Biol Evol*, 27(6), 1269–1278.

Wallace, D. L., V. Vialou, L. Rios, et al. (2008). The influence of DeltaFosB in the nucleus accumbens on natural reward-related behavior. *J Neurosci*, 28(41), 10272–10277.

Wang, Z., L. J. Young, G. J. De Vries, and T. R. Insel (1998). Voles and vasopressin: a review of molecular, cellular, and behavioral studies of pair bonding and paternal behaviors. *Prog Brain Res*, 119, 483–499.

Walum, H., L. Westberg, S. Henningsson, et al. (2008). Genetic variation in the vasopressin receptor 1a gene (AVPR1A) associates with pair-bonding behavior in humans. *Proc Natl Acad Sci USA*, 105(37), 14153–14156.

Williams, J. R., K. C. Catania, and C. S. Carter (1992). Development of partner preferences in female prairie voles (Microtus ochrogaster): the role of social and sexual experience. *Horm Behav*, 26(3), 339–349.

Williams, J. R., T. R. Insel, C. R. Harbaugh, and C. S. Carter (1994). Oxytocin administered centrally facilitates formation of a partner preference in female prairie voles (Microtus ochrogaster). *J Neuroendocrinol*, 6(3), 247–250.

Wilson, J. R., R. E. Kuehn, and F. A. Beach (1963). Modification in the sexual behavior of male rats produced by changing the stimulus female. *J Comp Physiol Psychol*, 56, 636–644.

Winslow, J. T., N. Hastings, C. S. Carter, C. R. Harbaugh, and T. R. Insel (1993). A role for central vasopressin in pair bonding in monogamous prairie voles. *Nature*, 365(6446), 545–548.

Winslow, J. T. and T. R. Insel (2002). The social deficits of the oxytocin knockout mouse. *Neuropeptides*, 36(2–3), 221–229.

Winslow, J. T., P. L. Noble, C. K. Lyons, S. M. Sterk, and T. R. Insel (2003). Rearing effects on cerebrospinal fluid oxytocin concentration and social buffering in rhesus monkeys. *Neuropsychopharmacology*, 28(5), 910–918.

Young, L. J. and E. A. Hammock (2007). On switches and knobs, microsatellites and monogamy. *Trends Genet*, 23(5), 209–212.

Young, L. J., M. M. Lim, B. Gingrich, and T. R. Insel (2001). Cellular mechanisms of social attachment. *Horm Behav*, 40(2), 133–138.

Young, L. J., R. Nilsen, K. G. Waymire, G. R. MacGregor, and T. R. Insel (1999). Increased affiliative response to vasopressin in mice expressing the V1a receptor from a monogamous vole. *Nature*, 400(6746), 766–768.

Young, L. J., D. Toloczko, and T. R. Insel (1999). Localization of vasopressin (V1a) receptor binding and mRNA in the rhesus monkey brain. *J Neuroendocrinol*, 11(4), 291–297.

Young, L. J. and Z. Wang (2004). The neurobiology of pair bonding. *Nat Neurosci*, 7(10), 1048–1054.

Young, L. J., J. T. Winslow, R. Nilsen, and T. R. Insel (1997). Species differences in V1a receptor gene expression in monogamous and nonmonogamous voles: behavioral consequences. *Behav Neurosci*, 111(3), 599–605.

Young, W. S., III, E. Shepard, J. Amico, et al. (1996). Deficiency in mouse oxytocin prevents milk ejection, but not fertility or parturition. *J Neuroendocrinol*, 8(11), 847–853.

9

Oxytocin regulation of maternal behavior

From rodents to humans

Cort A. Pedersen

9.1 Evolutionary background

Oxytocin is a member of a family of small proteins (nonapeptides) composed of a 6 amino acid ring and a 3 amino acid tail. Nonapeptides are found in invertebrate and vertebrate species and are among the most ancient peptide classes (Acher et al., 1995; Minakata, 2010). Oxytocin is synthesized almost exclusively in mammals and appears to have evolved from earlier nonapeptides found in lower vertebrates, none of which differs from oxytocin by more than 2 amino acids in the ring and tail structure. As is described in other chapters in this volume authored by Drs. Thompson (Chapter 5), Boyd (Chapter 6), and Goodson (Chapter 7), phylogenetically more ancient nonapeptides regulate a wide range of reproductive behaviors in fish, amphibians, reptiles and birds. Oxytocin facilitation of mammalian maternal behavior, parturition and milk ejection constitutes a relatively recent chapter in this evolutionary history.

Reproduction in placental mammals has unique characteristics that markedly enhance the survival of offspring. These include: pregnancy during which fetuses develop within a uterus located inside the mother's body where they draw nutrients directly from the maternal circulation through placentae attached to the uterine wall; expulsion of fetuses through the birth canal at maturation by contractions of the uterus (parturition); and feeding of newborns with milk, a rich and steady source of nutrition, secreted from the mother's mammary glands (lactation). Essential for the evolution of these aspects of placental mammalian reproduction was the coevolution of mechanisms in the placental mammalian brain that assure the rapid activation of avid and sustained maternal care of neonates as soon as they are delivered. Prompt initiation of maternal behavior at parturition was essential so newborns would be promptly cleaned of birth fluids and membranes that obstruct respiration and cause evaporative hypothermia as well as kept in close physical contact with the mother's body to facilitate nursing and to provide body heat. It is a marvel of evolution that oxytocin was selected to simultaneously initiate parturition and maternal care of newborns as well as the ejection of milk, the universal baby food of all mammalian species.

9.2 Brain oxytocin: The basics

Oxytocin has long been known to be released into the peripheral circulation during labor and nursing as well as to stimulate uterine contraction and milk ejection. Vaginocervical stimulation during parturition and nipple stimulation during suckling activate multisynaptic neural pathways that activate oxytocin neurons in the supraoptic and paraventricular nuclei (SON, PVN) in the anterior hypothalamus

Oxytocin, Vasopressin, and Related Peptides in the Regulation of Behavior, ed. E. Choleris, D. W. Pfaff, and M. Kavaliers.

that project to the posterior pituitary gland thereby causing the peripheral release of oxytocin.

Immunohistochemical methods have further clarified where oxytocin neurons are located and unequivocally demonstrated that oxytocin neurons project to numerous sites within the brain (Buijs et al., 1985; Sawchenko and Swanson, 1985; Sofroniew, 1980). Most oxytocin is synthesized in cell bodies in the SON and PVN. Oxytocin-synthesizing cell bodies are also located, depending upon the species, in smaller accessory nuclei in the anterior hypothalamus, medial preoptic area (mPOA), bed nucleus of the stria terminalis (BnST), the preoptic and anterior hypothalamic periventricular zone, and the lateral amygdala (Jirikowski et al., 1988; 1989; 1990; Young and Gainer, 2009). In the SON, all oxytocinergic cell bodies are large (magnocellular). In the PVN, oxytocin is synthesized in magnocellular neuron cell bodies as well as in smaller (parvocellular) cell bodies. Almost all magnocellular oxytocinergic neurons project to the posterior pituitary (the neurohypophyseal projection) where oxytocin is released into the peripheral circulation. A small number of parvocellular and perhaps magnocellular neurons project to the median eminence where oxytocin is secreted into the portal hypophyseal blood vessels that course through the anterior pituitary gland where oxytocin regulates the release of several anterior pituitary hormones including prolactin (Kotwica et al., 2006). Extensive oxytocin projections outside of the hypothalamus originate from parvocellular neurons within the PVN. Some extrahypothalamic projections may originate from the SON as well as accessory nuclei. The most prominent parvocellular projections traverse caudally through the brainstem and upper spinal cord with fibers terminating in many nuclei along the way. Other parvocellular oxytocinergic neurons project to numerous limbic and cortical areas. Oxytocin appears to be released from terminal projections and dendrites within the PVN and SON neurons in a manner similar to classical neurotransmitters. Estrogen treatment increased the number of oxytocin-immunostaining cell bodies in regions of the preoptic area-anterior hypothalamus region other than the PVN and SON and the intensity of staining of oxytocinergic fibers in this region and extrahypothalamic areas (Jirikowski et al., 1988). Estrogen receptor β mediates estradiol upregulation of oxytocin gene expression in magnocellular nuclei (Nomura et al., 2002; Shughrue et al., 2002).

Oxytocin appears to have only one receptor that belongs to the rhodopsin-type (class 1) G protein-coupled receptor family that is coupled to phospholipase C through $G_{\alpha q11}$ (Gimpl and Fahrenholz, 2001). Oxytocin receptors are expressed in many of the brain regions that receive oxytocin projections (although not all) as well as within the PVN and SON. Estradiol treatment effects differ considerably among the rat brain regions that exhibit the most oxytocin receptor binding. Binding is increased markedly in the ventromedial nucleus (VMN), more modestly in the BnST and possibly the central amygdala and not at all in the anterior olfactory nucleus and the ventral subiculum (Bale et al., 1995; de Kloet et al., 1986; Insel et al., 1997; Johnson et al., 1989). At parturition, oxytocin binding is significantly increased in the BnST and the VMN (Insel, 1990). Oxytocin receptor mRNA peaked during parturition in the SON (but not the PVN), mPOA, BnST, olfactory bulbs and the medial amygdala (Meddle et al., 2007). The number of cells double-labeled for oxytocin receptor and c-fos was also significantly increased in the SON, mPOA and BnST of parturient rats. The distribution of oxytocin receptors within the brain varies considerably among species and this variation has been linked to contrasting social behavior characteristics (Young et al., 2008).

9.3 Maternal behavior: The basics

Despite the undeniably critical importance of parenting in human development and motivation later in life, only a relatively few scientists have taken on the daunting task of investigating this very complex behavior. Despite their small number, researchers dedicated to this effort have made major advances

over the past several decades in understanding the neurobiological basis of maternal behavior. Most of this work has been conducted in rats and sheep. Excellent, detailed reviews of the neurobiology of maternal behavior have been published in recent years (Bridges, 2008; Numan and Insel, 2003). This section draws upon these sources to briefly summarize those aspects of this body of knowledge that are relevant in a chapter on oxytocin regulation of maternal behavior.

In rats, and probably most subprimate mammalian species, the brain contains a latent program for directing species typical maternal behavior toward newborns. In adult males and in females in all reproductive states other than parturition and lactation, this program is potently inhibited by olfactory (or other stimuli) from newborns that are perceived as aversive. Indeed, males and females in these other reproductive states will usually kill or, at minimum, actively avoid newborns. Aversion to pup odors does not emerge until after weaning in rats. This appears to be one manifestation of a broader increase in fearfulness that develops postweaning that probably has survival value after juveniles leave the protection of the mother's nest. Preweaning juveniles don't avoid newborns and start exhibiting components of maternal behavior after relatively brief periods of sustained contact with them.

Active components of maternal behavior in the rat (retrieving pups to a nest site, licking pups, building a nest) are activated by a motivational circuit at the core of which are projections from the mPOA to the ventral tegmental area (VTA) that stimulate the ascending mesolimbic pathway that releases dopamine in the nucleus accumbens thereby causing the expression of maternal behavior to be rewarding. Aversion to pups in adults other than parturient and lactating females is generated by processing of olfactory stimuli from pups in the corticomedial amygdala. Signals from this area to the MPOA prevent activation of the motivational pathway that permits expression of the hard-wired maternal behavior program. Hormone changes during late pregnancy (elevated estrogen and rapidly declining progesterone) reverse the inhibition of maternal behavior expression. This may involve elevated estrogen suppression of the aversive perception of pup stimuli, thus releasing the inhibition of the motivational circuit. Prolactin plays a vital role in mediating the maternal behavior-activating effects of elevated estrogen and rapidly declining progesterone at the end of pregnancy. Elevated estrogen stimulates the mPOA directly and possibly by blocking inhibitory input, thus initiating activation of the motivational circuit. Perioral stimulation when dams sniff and lick pups appears to reflexively activate retrieval behavior. Stimulation of the dam's ventrum by pups rooting for and latching onto nipples produces an increasingly quiescent state and eventually sleep in the mother during which she reflexively adopts an arched-back posture (kyphosis) over her litter for prolonged periods of time, which gives pups optimal access to her milk line. The kyphotic posture is accentuated during milk ejection, which can only occur when the mother is asleep. The periaqueductal gray area plays a key role in generating and maintaining the kyphotic nursing posture.

Once maternal behavior is initiated and has been expressed, even briefly, maintenance of the behavior rapidly becomes independent of the hormonal factors (sex hormones and prolactin) that were necessary for postpartum onset. Dopamine release in the nucleus accumbens from mesolimbic projections originating in the VTA, however, remains necessary to sustain maternal responses toward newborns. It is hypothesized that plastic changes in the MPOA occur after the onset of maternal behavior that permit pup stimuli to directly activate projections to the VTA resulting in sustained stimulation of the ascending dopaminergic mesolimbic system such that expressing maternal care remains highly rewarding.

The hormonal changes of late pregnancy and parturition that activate maternal behavior also have profound effects on the emotionality of dams. Their behavioral responses when tested in stressful situations indicate significantly reduced anxiety. A concomitant increase in aggression, most dramatically exhibited toward unfamiliar conspecifics placed in the maternal home cage, also occurs in the early postpartum period. Decreased anxiety and

increased aggressiveness in postpartum mothers are, like maternal behavior itself, sustained by physical contact with neonates. Both anxiety and aggressive responses rapidly return to baseline levels after separation from pups.

Postpartum maternal behavior experience produces what has been referred to as "maternal memory." Mothers allowed as little as 30 min of interaction with pups in the immediate postpartum period exhibit short latency resurgence of maternal behavior upon return of pups after a separation period of up to 10 days. In some, but not all, rat strains, females that have given birth to and reared pups show much shorter latencies of resurgent maternal behavior if reintroduced to pups, even after long periods of separation from pups.

Mothering experience also has other apparently permanent effects. Female rats that have nursed are "smarter" than nulliparous females, at least at spatial memory tasks Kinsley et al., 1999). This is thought to be an adaptation that provides a selective advantage in finding and remembering food sources over a larger territory, thus better sustaining the increased nutritional needs associated with lactation. Also, multiparous female rats exhibit significantly higher levels of maternal aggression than primiparous females, especially early in the postpartum period (Nephew et al., 2009).

The hormonal requirements for the activation of maternal behavior are quite different in laboratory mouse strains compared to rat strains. Nulliparous females of most laboratory strains of mice exhibit rapid onset (< 30 min) of all components of species-typical maternal behavior after introduction to young pups. This contrasts markedly with non-parturient wild mice that are usually infanticidal and cannibalistic towards newborns. Wild mice, however, rapidly become avidly maternal upon delivery of pups. Presumably, as in the rat, the hormone changes of late pregnancy that precipitate parturition contribute to the dramatic change in perception of pup stimuli and the immediate onset of maternal behavior in wild mice. However, this has not been studied. Apparently, in the course of breeding over hundreds of successive generations in captivity, non-parturient laboratory mice have lost the mechanisms that block activation of the motivational systems that mobilize the expression of maternal behaviors toward newborns. There have been no investigations into how such profound changes have come about.

The degree to which prepartum hormone changes as well as parturition and offspring-associated stimuli contribute to the onset of maternal behavior varies among other mammalian species. Approximately half of nulliparous females in the monogamous prairie vole are spontaneously maternal in their responses to newborns while the other half are infanticidal. Postpartum females, however, are all highly maternal. The role of prepartum hormonal events in the onset of vole maternal behavior remains unclear. Sheep form selective bonds to their own lambs that entails each ewe learning the distinctive odor of her lamb. Exposure to olfactory stimuli from their lambs during the first several hours after delivery is essential for ewes to form this social memory and maternal bond. Non-parturient ewes are hostile toward newborn lambs. In sheep, late gestational sex hormone changes and vaginocervical stimulation during parturition play essential roles in initiating postpartum maternal behavior and establishing the mother's selective bond to her lamb.

In primates and especially humans, the formation of maternal–infant bonds is much less dependent on the hormonal events of late pregnancy and parturition. Also, attachment to and recognition of individual offspring are mediated by integrated, multisensory cues rather than just olfactory stimuli (Broad et al., 2006).

9.4 Oxytocin regulation of maternal behavior in the rat

9.4.1 Oxytocin and the postpartum activation of maternal behavior

Klopfer (1971) was the first to hypothesize that oxytocin release during parturition may be important in activating maternal behavior. The first test of this theory was not supportive. Herrenkohl and

Rosenburg (1974) found that lesions of the neurohypophyseal projections from the PVN and SON to the posterior pituitary gland that eliminated oxytocin release into the peripheral circulation during parturition had no effect on the postpartum onset of maternal behavior. However, shortly after publication of reports of immunohistochemical visualization of oxytocin projections from the paraventricular nucleus to brain sites outside the hypothalamus (summarized in Buijs et al., 1985; Sawchenko and Swanson, 1985), Pedersen and Prange (1979) hypothesized that release of the neuropeptide within the brain may occur during parturition and trigger the onset of mothering. In the first experimental test of this theory, intact, nulliparous female rats received intracerebroventricular (ICV) injections of oxytocin (400 ng) or saline vehicle after which 3 neonates were introduced into each test cage. A significantly higher percentage of oxytocin recipients exhibited all components of rat maternal behavior within the ensuing test period, most within 20–40 min after exposure to newborns (Pedersen and Prange, 1979). Females in the oxytocin treatment group that exhibited rapid onset of maternal behavior were in phases of the estrous cycle in which estrogen levels were rising or had recently been elevated (diestrus, proestrus, estrus) suggesting that oxytocin efficacy may be estrogen dependent. This was confirmed in ovariectomized females in which ICV infusion of oxytocin stimulated the onset of maternal behavior in a high percentage of animals injected SC with high dose estradiol benzoate (100 μg/kg) 48 h prior to central treatment but not in controls injected with oil vehicle only (Pedersen and Prange, 1979). In subsequent experiments, Pedersen et al. (1982) examined dose–response relationships and the specificity of oxytocin induction of maternal behavior. The percentage of animals that began to exhibit maternal behavior after ICV administration was dose dependent between 100 and 400 ng. ICV doses (equimolar to 400 ng of oxytocin) of tocinoic acid, the 6 amino acid ring of the oxytocin molecule, and arginine vasopressin, which differs from the oxytocin molecule at only 2 amino acid positions, also significantly stimulated maternal behavior albeit at lower rates and, in the case of vasopressin, more slowly than oxytocin.

In retrospect, Pedersen and colleagues (Pedersen and Prange, 1979; Pedersen et al., 1982) were very fortunate to have conducted their experiments under conditions in which oxytocin was effective in stimulating maternal behavior. Subsequent studies conducted by other investigators (Rubin et al., 1983; Bolwerk and Swanson, 1984) failed to confirm that ICV administration of oxytocin activated maternal behavior in ovariectomized, estradiol-primed animals. Their test conditions and sources of animals differed from those of the original Pedersen and Prange (1979) and Pedersen et al. (1982) experiments. When Fahrbach et al. (1984; 1985a;b) precisely duplicated the test conditions of the original studies, including purchase of Sprague–Dawley females from the same breeder, Zivic–Miller Laboratories, they obtained results virtually identical to those of Pedersen and Prange (1979) and Pedersen et al. (1982) including confirmation that oxytocin induction of maternal behavior is estrogen dependent. Fahrbach et al. (1986) later reported that habituating animals to observation cages for less or more (0, 24 h) than the 2-h period used in the original studies by Pedersen and colleagues eliminated sensitivity to induction of maternal behavior by ICV injection of oxytocin. Wamboldt and Insel (1987) shed further light on the conditions required for administration of exogenous oxytocin to stimulate maternal behavior. They found that oxytocin had no facilitating effect when injected ICV in olfactory-intact, nulliparous Sprague–Dawley rats obtained from Taconic Farms that had been ovariectomized and estradiol-primed. However, after rendering females temporarily anosmic by intranasal administration of zinc sulfate, ICV injection of oxytocin was effective in stimulating the onset of maternal behavior in a significant percentage of females. The Wamboldt and Insel (1987) findings supported the conclusion that oxytocin was effective in the Pedersen and Prange (1979) and Pedersen et al. (1982) studies because the Zivic–Miller rats used in those experiments, which at the

Table 9.1 Oxytocin activation of maternal behavior: supportive evidence from rodent studies.

Study	Subjects	Hormone State	Experimental Manipulation	Outcome
Pedersen and Prange, 1979	Rat virgins Zivic–Miller	Intact, estrous cycling OVX, SC EB vs. oil	ICV OT vs. veh 2-h test cage habituation; ICV OT.	Rapid onset (20–40 min) of FMB, OT > veh, associated with higher E phases of estrous cycle; OT activation of FMB is E dependent.
Pedersen et al., 1982	Rat virgins Zivic–Miller	OVX, SC EB vs. oil	ICV OT doses vs. veh; ICV OT fragments, VP.	Rapid onset of FMB, OT dose related; Significant MB activation by tocinoic acid (OT ring); slower activation by VP.
Fahrbach et al., 1984	Rat virgins Zivic–Miller	OVX, SC EB vs. oil	ICV OT vs. veh.	Rapid onset of FMB, OT > veh; OT effect was E dependent.
Fahrbach et al., 1985b	Rat virgins Zivic–Miller	OVX, SC EB vs. oil	Brain site injections, OT vs.veh.	Rapid onset of MB, OT > veh, in VTA but not CNA or midbrain central gray.
Fahrbach et al., 1986	Rat virgins Zivic–Miller	OVX, SC EB vs. oil	ICV OT vs. veh; 0, 2, 24-h test-cage habituation.	OT activation of MB after 2, but not 0 or 24-h habituation.
Wamboldt and Insel, 1987	Rat virgins Taconic	OVX, SC EB vs. oil	ICV OT vs. veh; intranasal zinc sulfate vs. veh.	OT activation of MB in anosmic but not olfactory-intact animals.
Peterson et al.,1991	Rat juveniles	Intact	Intracisternal OT vs. veh.	OT increased holding/licking of pups in pre- but not postweanling juveniles.
Yu et al., 1996a	Rat virgins	OVX, SC EB	OT vs, veh into OB.	Rapid onset of FMB.
McCarthy et al.,1986	Wild mice, infanticidal	Intact virgins or late Preg	SC OT vs. veh.	OT inhibited infanticide in most animals, stimulated MB in some.
McCarthy, 1990	Wild mice, infanticidal	OVX, SC EB vs. oil or intact virgins	SC OT vs. veh. ICV OT vs. veh.	OT infanticide inhibition not E dependent; OT inhibits infanticide in most animals.

CNA = central amygdala; E = estradiol; EB = estradiol benzoate; FMB = full MB; MB = maternal behavior; OB = olfactory bulbs; OT = oxytocin; OVX = ovariectomy; Preg = pregnancy; veh = vehicle; VP = vasopressin; VTA = ventral tegmental area.

time bore a significant burden of upper respiratory pathogens, were essentially anosmic. Indeed, after Zivic–Miller took steps to rid their rat colonies of infections, ICV injections of oxytocin were no longer effective in stimulating maternal behavior in Sprague–Dawleys from that breeder, even when ovariectomized and estradiol-primed (unpublished observations).

The experiments examining effects of central administration of oxytocin on activation of maternal behavior in adult female rats (summarized in Table 9.1) have important implications. The consistent finding that oxytocin induction of mothering required estradiol pretreatment suggested that estrogen somehow increased central oxytocin sensitivity, possibly by increasing the numbers and/or

binding affinity of oxytocin receptors, and that endogenous oxytocin may be a factor mediating estrogen activation of maternal behavior in parturient rats. The ineffectiveness of ICV administration of oxytocin in olfactory-intact, estradiol-primed nulliparous female rats also indicated that either (1) the endogenous nonapeptide does not play a significant role in reversing perception of pup odors from aversive to attractive that is critical for the postpartum onset of rat maternal behavior, or (2) oxytocin administered by this route does not penetrate into sites involved in aversive perception of pup odors (e.g., the centromedial amygdala).

Oxytocin effects on maternal-like behavior in juvenile rats have also been reported (Peterson et al., 1991). Intracisternal injection of oxytocin (2 μg) compared to saline or no treatment significantly increased holding and licking of neonates in preweanling (18–22 days old) but not postweanling (26–34 days) juvenile rats during a 2-h observation period. Oxytocin did not increase other components of maternal behavior in juveniles. These results suggest that oxytocin may play a role in the short latency of onset of maternal behavior that is exhibited by juvenile rats prior to weaning.

In contrast to the variable results of central oxytocin administration summarized above, disruption of central oxytocin activity has consistently been found to inhibit the postpartum and ovarian steroid-induced onset of rat maternal behavior (Table 9.2). Pedersen et al. (1985) found that ICV administration of antiserum that selectively bound oxytocin significantly inhibited the onset of maternal responses to neonates in ovariectomized, nulliparous rats given a 21-day course of estradiol and progesterone previously shown to produce short latency onset maternal behavior (Bridges, 1984). Fahrbach et al. (1985a) demonstrated that ICV injection of either oxytocin antiserum or a selective oxytocin receptor antagonist delayed the usually rapid onset of mothering in rats that underwent hysterectomy and ovariectomy and high dose estradiol treatment on pregnancy day 16. Oxytocin antagonist administration on postpartum day 5 had no inhibitory effect on established maternal behavior. Van Leengoed et al. (1987) were the first to report that ICV injection of an oxytocin antagonist markedly inhibited the postpartum onset of maternal behavior. Insel and Harbaugh (1989) found that bilateral electrolytic lesions of the PVN, the source of most central oxytocin projections, on pregnancy day 15 profoundly disrupted postpartum initiation of maternal behavior. Lesioning of the PVN on postpartum day 4, however, had little effect on dams' behavior. The latter finding confirmed an earlier report from Numan and Corodimas (1985) that radiofrequency lesions of the PVN in lactating rats had no discernible effect on established mothering behavior. Significant, albeit waning, oxytocin control of some components of maternal behavior may persist for a few days postpartum. Kainic acid lesioning of the PVN on postpartum day 2 produced deficits in pup retrieval but not other components of maternal behavior during testing on postpartum day 4 (Olazabal and Ferreira 1997). Pedersen and colleagues found that ICV administration of oxytocin antagonist 24 h after the onset of parturition blocked the expression of all components of maternal behavior in approximately 50% of primiparturient dams during a 1-h reunion with pups after 2 h of separation (unpublished observations).

A limited number of studies have investigated where in the rat brain endogenous oxytocin plays a physiologically significant role in the activation of maternal behavior (Tables 9.1 and 9.2). Fahrbach et al. (1985b) found a significant increase in maternal behavior in ovariectomized, estradiol-primed Zivic–Miller rats after oxytocin injection into the VTA but not into the central amygdala or the midbrain central gray (although there was a trend toward an activating effect in the latter site). The effects of oxytocin antagonist injection into the latter two areas on the onset of maternal have not been tested. Bilateral injections of oxytocin antagonist into the mPOA or VTA inhibited the postpartum onset of maternal behavior (Pedersen et al., 1994). Furthermore, saturation binding assays in tissue dissected at intervals between pregnancy day 15 and postpartum day 7 found that oxytocin

Table 9.2 Central oxytocin disruption: effects on activation of maternal behavior.

Study	Subjects	Hormone State	Experimental Manipulation	Outcome
Pedersen et al., 1985	Rats	Virgin, OVX, SC EB+P × 21 days	ICV OT antiserum.	Delayed onset of MB.
Fahrbach et al., 1985a	Rats	H+OVX pregnancy day 16, SC EB	ICV OT antiserum or OTA.	Delayed onset of MB.
van Leengoed et al., 1987	Rats	Primiparturient	ICV OTA.	Delayed onset of MB.
Insel and Harbaugh, 1989	Rats	Primiparturient	PVN lesion Preg day 15.	Blocked onset of MB.
Pedersen et al., 1994	Rats	Primiparturient	OTA injection into mPOA or VTA.	Delayed onset of MB.
Yu et al., 1996a	Rats	Premiparturient	OTA injection into OB.	Delayed onset of MB.
Nishimori et al., 1996	Mice	Primiparturient	OTKO, standard cage.	Normal onset of MB.
Ragnauth et al., 2005	Mice	Virgin	OTKO and WT females in semi-natural environment.	All OTKOs killed test pups vs. 17% of WTs.
Takayanagi et al., 2005; MacBeth et al., 2010	Mice	Primiparturient Virgin	OTRKO.	Initial pup retrieval, crouched nursing deficits.
Olazabal and Young, 2006a	Prairie Voles	Nulliparous	OTA injection into NA.	Blocked usual 50% rate of spontaneous MB.
Boccia et al., 2007	Rhesus Monkey	Nulliparous	IV injection of L368, 899, CNS-penetrating human OTA.	Dose-related decrease in contact-seeking behavior and proximity to infant.

H+OVX = hysterectomy and ovariectomy; IV = intravenous; mPOA = medial preoptic area; NA = nucleus accumbens; OTA = oxytocin antagonist; OTKO = oxytocin gene knockout; OTRKO = oxytocin receptor gene knockout; PVN = hypothalamic paraventricular nucleus; WT = wild-type. See Table 9.1 for definitions of other acronyms and abbreviations.

binding (B_{max}) in the mPOA was significantly higher on pregnancy days 20, 22, 23 (before and during parturition) and postpartum day 1 (Pedersen, 1997). Increased oxytocin binding in this site coincided with the late pregnancy rise of estradiol concentrations. In the VTA, significant elevations in oxytocin binding occurred only on pregnancy day 23. Yu et al. (1996a) reported that OTA injection into the olfactory bulbs blocked the onset of maternal behavior in parturient rats. In addition, oxytocin injection into this site in ovariectomized, estradiol-primed virgin rats stimulated all components of maternal behavior in 50% of subjects as well as pup retrieval and crouching over pups in 70% and pup-licking in 90% of subjects within 2 h. These behavior effects may be related to oxytocin effects on the processing of pup olfactory stimuli. High-frequency stimulation of the PVN decreased mitral cell firing and increased granule cell (inhibitory interneuron) activity in the olfactory bulb. Oxytocin antagonist injection into the bulbs blocked these effects of PVN stimulation and oxytocin administration by microiontophoresis into the bulbs or ICV injection simulated the effects of PVN stimulation on mitral and granule cell activation (Yu et al., 1996b).

The significance of oxytocin in the activation of maternal behavior remains to be tested in numerous brain sites in rats. These include the BnST, ventral septum, magnocellular nuclei,

VMN and piriform cortex in which peripartum rats exhibit increased oxytocin binding, oxytocin receptor mRNA, numbers of activated (c-fos or fosB-immunostained) oxytocin receptor-expressing cells or oxytocin release (Insel, 1986;1990; Landgraf et al., 1991; Lin et al., 2003; Meddle et al., 2007) as well as the ventral subiculum that contains high concentrations of oxytocin receptors (Gimpl and Fahrenholz, 2001). Examination of the effects of oxytocin antagonist injection into the PVN is of particular importance given the evidence from sheep (see below) that increased oxytocin activity in this nucleus produces simultaneous release of oxytocin from projections to numerous brain sites involved in the activation of maternal behavior. Projections from the basolateral and basomedial amygdala stimulate active components of maternal behavior while projections from the medial amygdala are inhibitory (Numan et al., 2010). These regions within the amygdala all contain oxytocin receptors or are excited by local administration of oxytocin (Condes-Lara et al., 1994; Terenzi and Ingram, 2005; Veinante and Freund-Mercier, 1997). Oxytocin may promote the onset of maternal behavior by exerting opposite effects on the activity of discrete regions of the amygdala.

9.4.2 Oxytocin and established maternal behavior during lactation

Offspring-directed behaviors (Table 9.3)

The studies summarized above demonstrated that endogenous oxytocin plays a critical role in mediating sex hormone activation of maternal behavior in parturient rats but, like the reproductive hormones of pregnancy, is not necessary for sustaining established maternal behavior. Nonetheless, reduced yet significant enhancing effects of central oxytocin on pup-licking and quiescent, upright nursing (often referred to as arched-back nursing, ABN) have been demonstrated in nursing rats. Acute ICV injection of oxytocin antagonist at time points during the first week postpartum significantly reduced the frequencies of pup-licking (Champagne et al., 2001; Pedersen and Boccia, 2003) and quiescent upright nursing (Pedersen and Boccia, 2003). The goal of the Champagne et al. (2001) study was to examine the mechanisms underlying earlier demonstrations that the amount of maternal licking and ABN received during the early postnatal period determined the frequencies of these behaviors exhibited by lactating rats toward their own pups (Francis et al., 1999). Oxytocin antagonist administration lowered pup-licking frequencies only in mothers that exhibited high baseline frequencies, a consequence of having received high-frequency maternal licking as young pups. These findings were linked to greater oxytocin binding in the mPOA, BnST, lateral septum, PVN and central amygdala in mothers that had received high-frequency maternal licking in infancy. Injections of antagonist into these sites to determine where oxytocin stimulates pup-licking and quiescent, upright nursing in lactating rats remains to be done. However, Shahrokh et al. (2010) reported that oxytocin antagonist injection into the VTA abolished greater dopamine release in the nucleus accumbens in high compared to low-licking mothers during pup-licking bouts. The comprehensive behavior analysis conducted in the Pedersen and Boccia (2003) study permitted more detailed conclusions about oxytocin regulation of established maternal behavior in lactating rat mothers. Self-grooming episodes were clearly embedded in most pup-licking bouts. The effects of ICV injection of oxytocin antagonist indicated that central oxytocin increases pup licking by shifting the directional balance during oral grooming bouts away from self and toward pups. Because their methods included continuous observation of nursing behavior, Pedersen and Boccia (2003) were able to demonstrate that endogenous oxytocin raises the frequency of ABN by increasing the duration of bouts of quiescent upright nursing.

Febo and Ferris (2008) used fMRI to demonstrate that numerous brain areas are activated or deactivated during suckling in postpartum day 4–8 rat mothers. In addition to the expected effects on the PVN, they found that ICV injection of oxytocin increased and oxytocin antagonist decreased

Table 9.3 Oxytocin involvement in maintaining offspring-directed maternal behavior: supporting evidence.

Study	Experimental Treatment	Outcome
Pedersen et al., 1995	OTA vs. veh ICV or into VTA followed by introduction of pups after 2–6 days of proximal separation (PS) beginning on PPD 5 in rats.	OTA into VTA after 4–6 days of PS blocks resurgent MB, ICV OTA less effective; OTA into VTA less effective after 3 days of PS, ineffective after 2 days of PS.
Champagne et al., 2001	ICV OTA vs. veh in high vs. low pup-licking (PL) rat mothers on PPD 3.	OTA decreases PL frequencies of high PL-mothers to frequencies exhibited by low PL- mothers (no effect in latter group).
Pedersen and Boccia, 2003	ICV OTA vs. veh in rat mothers on PPDs 2/3 or 6/7.	OTA decreases PL, ABN and sustained ($\geq$ 2 min) upright, quiescent nursing bout frequencies.
Pedersen et al., 2006	MB directed toward stimulus neonates measured during 4-h periods on 3 successive days in nulliparous OTKO vs. WT mice.	OTKO exhibit lower frequencies of PL on test days 1 and 3; less nesting in corners of test cages and pup retrieval to nest sites over all test days.
Bosch and Neumann, 2008	Continuous ICV OT vs. veh infusion or daily ICV OTA vs. veh injection on PPDs 1–5 in rats.	OT increases and OTA decreases ABN but not PL frequencies.
Febo and Ferris, 2008	ICV OT or OTA vs. veh in PPDs 4–8 rat mothers undergoing fMRI while receiving suckling stimulation from pups.	OT increases and OTA decreases activity in the mPOA and NA, areas involved in sustaining established MB.

ABN = arched-back nursing; fMRI = functional magnetic resonance imaging; PPD = postpartum day. See Tables 9.1 and 9.2 for definitions of other acronyms and abbreviations.

local activity during suckling in areas involved directly or indirectly in processing olfactory stimuli from pups (olfactory tubercle, anterior olfactory nucleus, the insular cortex and the cortical amygdala) and areas critical to maintaining maternal behavioral motivation (mPOA, nucleus accumbens). Given this evidence that oxytocin continues to be released by suckling in these areas in established maternal behavior, Febo and Ferris (2008) speculated that oxytocin may be involved in sustaining the attractive perception of olfactory stimuli from pups and maintaining maternal motivation possibly by acting in the VTA to release dopamine in the nucleus accumbens and prefrontal cortex. They further speculated that prolonged blockade of oxytocin receptors may be necessary to demonstrate these putative roles of oxytocin. Bosch and Neumann (2008) reported that ICV injection of oxytocin antagonist in the morning of the first 5 days postpartum significantly reduced ABN but not pup licking in unselected Wistar rats, while chronic ICV infusion of oxytocin over the same postpartum time period increased ABN frequencies in rats selected for low anxiety that exhibit low baseline ABN. It should be noted that Bosch and Neumann (2008) found that vasopressin plays a more prominent role than oxytocin in maintaining maternal behavior during the early postpartum period.

Pedersen et al. (1995) conducted experiments that raised intriguing questions about the relationship between pup stimuli during the nursing period and oxytocin control over maternal behavior. In the first experiment, primiparous females were subjected to 6 days of total separation, no separation or proximal separation from pups beginning on postpartum

day 5. Proximal separation involved placing pups in a metal wash basket that had openings in the mesh that were sufficiently large for the mothers to contact pups with their paws, tongues and snouts but did not permit pup contact with mothers' ventral surfaces or nipples. Freshly nourished, 5-day-old pups replaced used pups in the wire cages every 8–16 h. Mothers persistently tried to retrieve pups from the wire cages throughout the proximal separation period. Brains were harvested from some mothers after 6 days of proximal separation, sectioned and immunostained for oxytocin. The numbers of discernible oxytocin immunoreactive cell bodies in the periventricular nuclei, mPOA, lateral POA and lateral hypothalamus were significantly or trended toward being significantly lower in the proximal compared to the total and/or no separation groups. After 4–6 days of proximal, no or total separation, other mothers were given full access to 8 freshly nourished 5-day-old pups. Their behavior was recorded during two subsequent observation periods. The number of pups retrieved and the total duration of crouched nursing in the proximal separation group were comparable to the no-separation group and significantly greater than the total separation group. In a third experiment, the effects on resurgent maternal behavior of oxytocin antagonist or saline vehicle administration ICV or into the VTA were compared between dams that had undergone proximal or no separation from pups for 4–6 days. In the proximal separation group, oxytocin antagonist but not saline injection into the VTA significantly inhibited resurgent maternal behavior. ICV-administered antagonist was less effective in suppressing maternal behavior in the proximal separation group. Oxytocin antagonist injection into the VTA had no inhibitory effect on maternal behavior in the no-separation group. A final experiment found that oxytocin antagonist infusion into the VTA was significantly more effective in inhibiting resurgent maternal behavior after 4 days of proximal separation compared to 2 or 3 days of proximal separation.

These experiments suggest that in the proximal separation condition in which maternal motivation to contact pups remains high but consummation of that motivated state through ventral contact with pups and nursing is not possible, oxytocin release from neurons in the medial preoptic area and anterior hypothalamus increases (as indicated by the marked decline in immunostaining of oxytocin cell bodies in these areas) and oxytocin activation of maternal behavior is reinstated (as indicated by the increasing effectiveness of oxytocin antagonist in blocking resurgent maternal behavior between 2 and 4 days of proximal separation). The overall implication is that pup stimulation of dams' ventral surface and nipples during nursing replaces (possibly even suppresses) oxytocin as the activator of maternal behavior during the maintenance phase of mothering. Under conditions in which other pup stimuli remain proximal (odor, taste, vocalizations) and maintain a high level of maternal motivation but ventral stimulation and nursing are disrupted, oxytocin may again emerge as a necessary activator of maternal behavior, possibly because suppression of oxytocin activity by pup stimulation of the dam's ventrum no longer occurs.

Maternal aggression (Table 9.4)

The most extensively investigated role of oxytocin in lactating rats is regulation of the heightened aggressiveness exhibited by rat mothers during the postpartum period. Neumann, whose group has published most extensively on the neuroendocrine basis of these behavioral changes, summarizes much of their work on oxytocin in this volume (Chapter 2). However, the main findings of this body of research bear repetition here. Neumann and colleagues have investigated the role of oxytocin in maternal aggression in Wistar mothers from an unselected line as well as mothers of lines selected over numerous generations for high and low anxiety (HAB, LAB) (Bosch et al., 2004; 2005; Bosch and Neumann, 2010). Maternal aggression was elicited and measured by placing a novel female in each mother's home cage for a 10-min period on postpartum day 3. Bosch et al. (2004) used microdialysis to study oxytocin

Table 9.4 Oxytocin regulation of maternal aggression (MA).

Study	Experimental Manipulation	Outcome
Evidence that oxytocin stimulates maternal aggression		
Ferris et al., 1992	OT or OTA vs. veh infusion into the CNA every other day × 3 starting on PPD day 5; MA test after each infusion.	OT increased biting of and contact time with male intruder during MA test compared to OTA or veh. No treatment effects on MB.
Consiglio and Lucion, 1996	Electrolytic lesions of the entire PVN on PPD 5.	Decreased frequency and duration of attacks on male intruder during MA testing.
Bosch et al., 2004	OT release measured in specific brain sites during MA testing on PPD 3.	OT release increased significantly in PVN but not in CNA or medio-lateral septum during MA test.
Bosch et al., 2005	Comparison of MA on PPD 3 between high and low anxiety rats (HAB, LAB), central OT release during MA tests, effects of site injections of OT or OTA on MA toward female intruder.	Higher MA in HAB mothers; OT release in PVN increased in HAB, decreased in LAB mothers during MA test; increase in OT release in CNA greater in HAB than LAB during MA test; OTA into PVN, CNA decreased MA in HAB; OT into PVN increased MA in LAB (latter effect not significant).
Evidence that oxytocin inhibits maternal aggression		
Giovenardi et al., 1998	Injections into the parvocellular PVN of ibotenic acid on PPD 2 or of OT antisense on PPD 5, 4 h before MA test.	Ibotenic acid treatment increases biting of intruder male and antisense treatment increases biting and frontal attacks.
Lubin et al., 2003	Injection of OTA vs. veh into the CNA or VTA 4 h before MA test on PPD 6.	OTA into the CNA but not VTA increased frequency of attacks on and duration of fighting with intruder males.
Consiglio et al., 2005	Injection of OT vs. veh into CNA or BnST on PPDs 5–7	OT injection decreased male intruder-directed biting and frontal attacks in CNA and biting in BnST.
Nephew et al., 2009	MA, OT and OTR mRNA in PVN compared between multipara and primipara on PPD 5 and 15.	Greater MA in multipara on PPD 5 associated with lower OT and OTR mRNA; significant decline in MA in multipara by PPD 15 associated with significant rise in OTR mRNA.

BnST = bed nucleus of the stria terminalis; h = hour; OTR = oxytocin receptor. See Tables 9.1–9.3 for definitions of other acronyms and abbreviations.

release in specific brain sites in unselected mothers during maternal aggression tests. Oxytocin release increased in the PVN but not the central amygdala or the medio-lateral septum. Subsequently, Bosch et al. (2005) reported that HAB mothers exhibited significantly greater maternal aggression than LAB mothers. During maternal aggression testing, oxytocin release in the PVN increased in HAB mothers but declined in LAB mothers. In both rat lines, oxytocin release increased in the central amygdala but significantly more in HAB compared to LAB mothers. Release in both sites during testing correlated directly with levels of offensive aggression exhibited by HAB and LAB mothers. Bilateral infusion of oxytocin antagonist into the PVN or central amygdala of HAB mothers significantly decreased their

offensive aggression. Infusion of oxytocin into the PVN increased some components of maternal aggression in LAB mothers but not significantly. There were no differences between HAB and LAB mothers in central oxytocin receptor mRNA expression or autoradiographic binding, indicating that differences in oxytocin release rather than receptor concentrations produced the contrasting levels of maternal aggression in these selected lines.

The results of two additional studies also indicate that oxytocin stimulates maternal aggression. In a small preliminary experiment (4 animals/treatment group), Ferris et al. (1992) found that oxytocin stimulated significantly more maternal aggression toward an adult male intruder than oxytocin antagonist or saline when repeatedly injected bilaterally into the central amygdala of lactating golden hamsters. Consiglio and Lucion (1996) reported that electrolytic lesioning of the PVN on postpartum day 5 in Wistar rats reduced the frequency and duration of attacks on an intruder male.

Evidence from other rat studies, however, suggests that central oxytocin inhibits maternal aggression. Giovenardi et al. (1998) found in Wistar rats that ibotenic acid injections selectively targeting neurons in the parvocellular PVN as well as suppression of oxytocin synthesis by local injection of oxytocin antisense resulted in increased rat maternal aggression on postpartum day 5. During the early postpartum period, injections of oxytocin into the central amygdala and the BnST (Consiglio et al., 2005) diminished maternal aggression while injections of oxytocin antagonist into the central amygdala had the opposite effect (Lubin et al., 2003). Nephew et al. (2009) demonstrated that maternal aggression in Sprague–Dawley rats was significantly higher in multiparous compared to primiparous rats on postpartum day 5. By postpartum day 15, however, aggression levels in multipara had declined to levels comparable to primipara. In multiparous mothers, greater maternal aggression on postpartum day 5 was associated with significantly less oxytocin and oxytocin receptor mRNA in the PVN than in primiparous mothers. Moreover, the decline in aggression in multipara by postpartum day 15 was associated with a significant increase in oxytocin receptor mRNA in the PVN compared to postpartum day 5.

The conflicting results summarized above suggest that the role of oxytocin in the regulation of maternal aggression varies substantially among rat lines even within the same strain as well as among species.

Postpartum reduction of anxiety (Table 9.5)

Stress-induced anxiety and activation of the hypothalamic–pituitary–adrenal (HPA) axis are reduced during pregnancy and lactation (Hard and Hansen, 1985; Neumann et al., 2000; 2001; Stern et al., 1973; Windle et al., 2010). Numerous studies have found that endogenous or centrally administered oxytocin is anxiolytic and diminishes stress-induced HPA activation in female and male rodents (Amico et al., 2004; Bale et al., 2001; Mantella et al., 2003; McCarthy et al., 1996; Ring et al., 2006; Windle et al., 1997). Based on the effects of ICV administration of an oxytocin antagonist, Neumann et al. (2000) concluded that endogenous oxytocin plays a significant role in the decline in anxiety as measured in the elevated plus maze test in pregnant (day 20 or 21) and postpartum (days 7–11) rats but does not contribute to the reduction in stress activation of the HPA axis in these reproductive states. In contrast, oxytocin antagonist significantly inhibited HPA activation, but not anxiety, in virgin females. Similar findings were obtained in later experiments (Bosch and Neumann, 2008): mothers that received daily ICV injections of oxytocin antagonist after giving birth exhibited higher anxiety on postpartum day 3 while postpartum mothers given continuous ICV infusion of oxytocin were less anxious compared to controls. However, in Wistar rat lines that had been selected over numerous generations for high or low anxiety (HAB, LAB), central oxytocin antagonist or oxytocin administration had no effect on anxiety. Figueira et al. (2008) demonstrated that the ventrocaudal periaqueductal gray is a site where oxytocin regulates anxiety in postpartum rats. Bilateral injection of oxytocin antagonist into this site increased anxiety-like behavior in the elevated plus maze on

Table 9.5 Oxytocin involvement in other behavioral changes in mothers.

Study	Experimental Manipulation	Outcome
Decreased anxiety and fear in postpartum rats		
Neumann et al., 2000	ICV OTA in PPD 7–11 rats.	Increased anxiety (decreased open arm time in the elevated plus maze [EPM] test).
Bosch and Neumann, 2008	Continuous ICV OT vs. veh infusion or daily ICV OTA vs. veh injection on each PPD.	OT decreased, OTA increased anxiety in the EPM test conducted on PPD 3.
Figueira et al., 2008	Infusion of OT or OTA into the ventrocaudal periaqueductal gray on PPD 7.	OT decreased, OTA increased anxiety in the EPM test.
Febo et al., 2009	ICV OT vs. veh prior to TMT (predator scent) exposure.	OT decreased the duration of freezing.
Maternal memory: Persistent short latency resurgence of MB after brief postpartum mothering experience in rats		
D'Cunha et al., 2011	OTA vs. veh infusion into the NA following 1 h of PP MB experience in primiparturient rats.	OTA significantly increased latencies of resurgent MB after 10 days of pup separation.
Increased spatial learning and memory ability in multiparous mice		
Tomizawa et al., 2003. Building on Kinsley et al. (1999) findings in rats of greater spatial learning and memory ability in multipara compared to nullipara.	Compared spatial learning/memory in nullipara and multipara; tested OT perfusion on L-LTP formation in hippocampal slices from nullipara; tested 20 days of ICV OTA on spatial memory acquisition and L-LTP formation in multipara.	Spatial learning/memory > in multipara; OT perfusion permitted single train stimulation induced L-LTP in nulliparous hippocampus; ICV OTA decreased spatial learning/memory in multipara and inhibited single train stimulation of L-LTP in multiparous hippocampus.

L-LTP = long-lasting long-term potentiation; PP = postpartum. See Tables 9.1–9.3 for definitions of other acronyms and abbreviations.

postpartum day 7, while injection of oxytocin had the opposite effect. These treatments had no effect on anxiety in nulliparous females tested in diestrus. ICV pretreatment with oxytocin in postpartum day 4–8 rat mothers decreased the duration of freezing when exposed to trimethylthiazoline (TMT), a predator scent extracted from fox feces (Febo et al., 2009). Central oxytocin administration prior to TMT exposure in rat dams undergoing fMRI resulted in significantly greater percent increases in neuronal activity in the anterior cingulate, BnST, and perirhinal area as well as significantly lower percentage increases in neuronal activity in a far greater number of brain areas. While these results do not permit specific conclusions about mechanisms, they do suggest that oxytocin decreases fear responses in rat mothers by modulating directly or indirectly the activity of a multitude of widely distributed brain areas involved in sensory processing, emotional arousal as well as motor and autonomic regulation.

Maternal memory (Table 9.5)

Postpartum primiparous rats allowed as little as 30 min of maternal experience before removal of pups exhibit short latency resurgence of maternal behavior upon reintroduction of pups for at least 10 days after parturition (Orpen and Fleming, 1987).

D'Cunha et al. (2011) reported that oxytocin antagonist infusion bilaterally into the nucleus accumbens after 1 h of postpartum pup interaction resulted in significantly longer latencies of onset of maternal behavior after 10 days of pup separation. The effect of oxytocin infusion into the nucleus accumbens after 15 min of initial maternal experience was uncertain because control groups as well as oxytocin recipients all showed short latencies of resurgent maternal behavior. These findings suggest that oxytocin receptor activation in the nucleus accumbens in the early postpartum period contributes to the consolidation of maternal memory.

9.5 Oxytocin regulation of maternal behavior in the mouse

As stated above, nulliparous mice of most laboratory strains, unlike laboratory rats, begin to exhibit the full range of species typical maternal behavior shortly after introduction of young pups into their home cages. This contrasts starkly with nulliparous and pregnant wild house mice most of which rapidly kill neonates. These responses to young remain unaltered until parturition when wild mouse mothers suddenly become avidly nurturing toward their newborns. The role of sex hormone changes during pregnancy in the initiation of maternal behavior in parturient wild mice remains unexamined. However, the frequency of infanticide in nulliparous wild mice was unaltered by ovariectomy and adrenalectomy (McCarthy et al., 1986).

Nonetheless, it has been demonstrated that subcutaneous and ICV administration of oxytocin significantly decreases the rate of infanticide in nulliparous and pregnant wild mice, an effect that did not require pretreatment with estrogen (McCarthy et al., 1986; McCarthy, 1990; table 1). A minority of oxytocin-treated mice began to exhibit maternal behavior. This limited body of research suggests oxytocin regulation of maternal behavior differs substantially in mice compared to rats. Unlike rats, oxytocin in mice appears to potently suppress the aversive perception of pup stimuli, thereby decreasing infanticide but is less effective in activating pup-directed nurturing behavior. It remains unclear whether the infanticide-inhibiting or maternal behavior-enhancing effects of oxytocin are dependent on endogenous estrogen levels. The two studies that were conducted in mice did not determine if the oxytocin-treated animals that began to actively mother pups were in high-estrogen stages of the estrous cycle and there was no investigation of whether oxytocin was more effective in estrogen-primed animals.

Studies of oxytocin and oxytocin receptor gene knockout (OTKO, OTRKO) mice have shed light on the significance of oxytocin in the expression of maternal behavior in laboratory mice (Table 9.2). In the first publications describing OTKO mice, the investigators were surprised that postpartum females exhibited no discernible deficits in maternal behavioral responses to their newborns although they were unable to eject milk (Nishimori et al., 1996). Using more quantitative behavior measures, Pedersen et al. (2006; Table 9.3) found that nulliparous female OTKO mice licked foster pups less frequently than wild-type (WT) females over 3 successive days of testing. OTKO females also exhibited unusual patterns of nest location and retrieval of pups to that location. While all WT mice retrieved all pups to a corner of their test cages, a significant percentage of OTKO females did not and proceeded to exhibit maternal behavior toward pups in the more exposed location in the middle of the cage where they were initially placed by the investigators. Those OTKO females that did establish a nest in a cage corner retrieved a significantly lower percentage of pups to that location. Ragnauth et al. (2005) found that maternal and other social behavior differences between OTKO and WT mice were much greater when females of both genotypes were cohoused in a semi-natural environment. Some OTKO female mice exhibited exceptionally high levels of aggression toward other females, OTKO as well as WT. All OTKO females living under these conditions immediately killed and cannibalized newborn pups placed near their nests in contrast to WT females that exhibited a 17% rate of infanticide

without cannibalization and a 50% rate of pup retrieval to their nests. The authors concluded that oxytocin may be necessary for initiation and maintenance of mouse maternal behavior in the wild, possibly by buffering the effects of social stress that interfere with directing nurturing behavior towards pups.

Studies of OTRKO mice suggest the oxytocin receptor plays a somewhat greater role than oxytocin itself in the activation of maternal behavior in this species. Observations during the early postpartum period revealed that a significantly higher percentage of OTRKO compared to WT mothers had some of their newborn pups scattered outside their nest sites (Takayanagi et al., 2005). In subsequent testing with three foster pups, postpartum OTRKO mice exhibited longer latencies to retrieve and crouch over all pups and shorter durations of crouching (probably the consequences of their slower retrieval). Virgin OTRKO females exhibited similar deficits in pup retrieval and crouching compared to virgin WT mice. MacBeth et al. (2010) measured responses of virgin females to foster pups during brief tests conducted over 3 successive days and found that pup-retrieval performance improved in OTRKO females and became indistinguishable from WT virgins by day 3. Conditional OTRKO mice with profound reductions in oxytocin receptor expression in some forebrain sites (lateral septum, hippocampus and ventral pallidum) but not others (e.g., medial amygdala, olfactory bulb and nucleus, neocortex) exhibited no deficits in postpartum maternal behavior (MacBeth et al., 2008).

The results of studies of OTKO and OTRKO mice clearly show that oxytocin and the oxytocin receptor are of little importance for the onset of maternal behavior in laboratory mice. The most prominent role of oxytocin in rat maternal behavior is mediating the activating effects of estrogen and progesterone changes during late pregnancy. Because the initiation of maternal behavior in laboratory mice is not ovarian steroid dependent, it is not surprising that oxytocin activity is of little significance. The small deficits in pup licking in OTKO mice reported by Pedersen et al. (2006) over successive days of testing suggest that, as in rats (Champagne et al., 2001; Pedersen and Boccia, 2003), oxytocin may exert a modest enhancing effect on this component of maternal behavior after the onset of maternal responses.

Tomizawa et al. (2003) discovered that multiparous mice, just like multiparous rats (Kinsley et al., 1999), have superior spatial learning and memory compared to nulliparous females. Furthermore, they demonstrated that oxytocin activity in the hippocampus contributes to the enhanced spatial learning and memory ability of experienced mothers (Table 9.5). Induction of long-lasting long-term potentiation (L-LTP) in the monosynaptic excitatory pathways in the hippocampus has been linked to long-term memory storage (Huang et al., 1996). Induction of L-LTP in most of these pathways requires multiple trains of electrical stimulation. Tomizawa et al. (2003) found that, in hippocampal slices from nulliparous mice perfused with oxytocin solution, one train stimulation of the Schaffer collateral fibers that synapse on CA1 pyramidal cells was sufficient to induce L-LTP in those cells. This effect of oxytocin was mediated by activation of the MAP kinase cascade and subsequent CREB phosphorylation. ICV injection of oxytocin antagonist for 20 days significantly decreased acquisition of spatial memory in multiparous females and prevented single train stimulation-induced L-LTP as well as reduced phospho-CREB expression in hippocampal slices harvested on postpartum day 10.

9.6 Oxytocin regulation of maternal behavior in sheep

Studies in this species (Table 9.6) are perhaps more relevant to understanding human parent–infant attachment than rodent studies because parturient ewes form a selective bond with their newborn lambs. Compared to the body of research in the rat, there has been a more comprehensive delineation of the brain sites in which oxytocin is

Table 9.6 Oxytocin activation of maternal behavior: supportive evidence from sheep and primates.

Study	Subjects	Hormone State	Experimental Manipulation	Outcome
Keverne et al., 1983 Poindron et al., 1989	Sheep, multipara	Postpartum, non-pregnant SC EB+P or in estrous	Vaginocervical stimulation (VCS) × 5 min.	Rapid onset of all components of MB in non-pregnant ewes; reversal of alien lamb rejection in postpartum ewes.
Kendrick et al., 1986	Sheep, multipara	Parturient or non-pregnant	VCS.	Increased OT concentrations in CSF during labor, parturition and VCS.
Kendrick et al., 1987	Sheep, multipara	Non-pregnant, SC EB vs. oil	ICV OT (5, 10, 20 μg) vs. veh.	All OT doses stimulated very rapid onset MB (< 30 s) which was E dependent.
Keverne and Kendrick, 1991	Sheep, multipara and nullipara	OVX, SC EB × 3 days	ICV OT (10, 50 μg) vs. veh.	OT decreased lamb rejection (multipara and nullipara), increased acceptance in multipara.
Lévy et al., 1992	Sheep	Primiparturient	ICV OT (10 μg × 2) vs. veh, +/− peridural anesthesia.	OT prevented inhibition of the onset of MB by peridural anesthesia.
Kendrick et al., 1988a;b; 1992	Sheep	Parturient; OVX, SC EB × 3 days	Microdialysis from brain sites; VCS after EB.	OT release rose in OB, mPOA, BnST, SN during parturition, in OB during VCS.
Lévy et al., 1995	Sheep	Non-pregnant	OT retrodialysis into the OB.	Increased release of NE, ACh and GABA (mediate selective olfactory bond to lamb).
Kendrick et al., 1992; 1997	Sheep, multipara	SC EB × 2 days or EB+P	OT infusion into mPOA or OB.	Decreased lamb rejection but did not increase maternal acceptance behaviors.
Da Costa et al., 1996	Sheep	Parturient, non-pregnant EB+P	Microdialysis from PVN; OT retrodialysis into PVN.	Increased OT release during parturition; Activated all components of MB.
Holman and Goy, 1995	Rhesus nullipara	Intact, menstrual phase unknown	ICV OT (0.5 IU) vs. veh.	OT increased some components of MB.

ACh = acetylcholine; CSF = cerebral spinal fluid; GABA = gamma-aminobutyric acid; NE = norepinephrine; SN = substantia nigra. See Tables 9.1–9.4 for definitions of other acronyms and abbreviations.

released and regulates maternal behavior in parturient females as well as investigation of concomitant neurochemical events in those sites. This impressive work is reviewed in more detail in this volume by Kendrick (Chapter 10). However, several of the key findings in this species fit nicely into this chapter.

The first evidence implicating oxytocin in the activation of maternal behavior in sheep was the demonstration that a brief period of vaginocervical stimulation (VCS), which increases central concentrations of oxytocin in this species (Kendrick et al., 1986), very rapidly changed behavioral responses

toward newborns from hostile and threatening to nurturing and accepting in non-pregnant, estrogen and progesterone-primed or estrous ewes that had prior postpartum maternal experience (Keverne et al., 1983; Poindron et al., 1989). Furthermore, VCS in recently parturient ewes, after selective bonding to their own lambs, reversed rejection, and promoted maternal adoption of lambs born to other mothers (Keverne et al., 1983). Subsequently, ICV administration of oxytocin was found to stimulate components of sheep maternal behavior in ovariectomized, estrogen-treated multiparous ewes (Kendrick et al., 1987). This activation was much more rapid in sheep (< 30 s) than rats (approximately 20–40 min) and, as in rats, was estrogen dependent. Nulliparous females were less sensitive to oxytocin. After ICV administration of oxytocin, lamb acceptance behaviors were increased and rejections behaviors were decreased in estrogen-primed multiparous ewes, while nulliparous ewes only exhibited a reduction in lamb rejection (Keverne and Kendrick, 1991). After longer periods of progesterone and estrogen treatment, however, ICV oxytocin administration reliably stimulated full maternal behavior in nullipara (Kendrick, 2000). Also, inhibition of the onset of maternal behavior by epidural anesthesia (Krehbiel et al., 1987) in primiparturient ewes was significantly reversed by ICV administration of oxytocin (Lévy et al., 1992).

Microdialysis has been used to demonstrate oxytocin release in the olfactory bulb, mPOA, BnST and substantia nigra in sheep during parturition or VCS. This methodology has also revealed increases of other neurotransmitters concomitant with oxytocin release in many of these areas (Kendrick et al., 1988a;b; 1992; Lévy et al., 1995). Oxytocin infusion by reverse dialysis into some extrahypothalamic sites has been shown to stimulate some but not all components of maternal behavior as well as to suppress other behaviors incompatible with attentive maternal care. For example, infusion of oxytocin into the mPOA and olfactory bulbs reduced rejection of lambs and acceptance of suckling attempts (Kendrick et al., 1992; 1997). In the ventromedial hypothalamus, oxytocin administration inhibited sexual behavior, an undesirable distraction from focused care of the newborn lamb (Kendrick et al., 1993). The PVN is the only site in which oxytocin infusion activated all components of sheep maternal behavior, presumably because it produces simultaneous release of oxytocin in all brain sites in which this nonapeptide activates specific aspects of sheep maternal behavior (Da Costa et al., 1996; Kendrick, 2000). Interestingly, infusion of tocinoic acid, the ring structure of the oxytocin molecule, was also effective in this site. Oxytocin infusion into the olfactory bulbs increased local norepinephrine, acetylcholine, and GABA release, which may facilitate the formation of the odor-based memory that underlies the selective bond of ewes to their newborn lambs (Lévy et al., 1993; 1995).

Oxytocin receptor gene expression has been demonstrated by *in situ* hybridization to increase significantly at parturition in the anterior olfactory nucleus, Islands of Calleja, mPOA, ventromedial hypothalamus, lateral septum, medial amygdala, BnST and PVN (Broad et al., 1993; 1999). Increases in the PVN and the Islands of Calleja only occurred in parturient multiparous ewes. In ovariectomized ewes, a course of progesterone and estradiol treatment that simulates ovarian steroid changes during pregnancy increased oxytocin receptor mRNA in all of these areas except the PVN, lateral septum, and medial amygdala. These findings indicate that ovarian steroid upregulation of oxytocin receptor gene expression occurs independent of maternal experience in most, but not all, brain areas.

9.7 Evidence of oxytocin regulation of maternal behavior in other non-human species

9.7.1 Prairie voles (Table 9.2)

The spontaneous expression of maternal behavior (licking pups and hovering over them in a nursing

posture) exhibited by approximately 50% of adult, nulliparous voles has been linked to higher oxytocin activity in the nucleus accumbens. Females exhibiting spontaneous maternal behavior had significantly higher oxytocin binding in the shell region of the nucleus accumbens but not in other brain sites (Olazabal and Young, 2006a). Bilateral injection of oxytocin antagonist into the nucleus accumbens, but not the caudate putamen, prevented expression of maternal behavior in all females, while animals injected with artificial CSF alone into the nucleus accumbens exhibited a 50% rate of spontaneous maternal behavior. Olazabal and Young (2006b) also compared regional oxytocin binding among juveniles of various rodent species including prairie voles, rats (which exhibit short latency [1–3 days] but not spontaneous maternal behavior) as well as meadow voles and mice that do not respond maternally to neonates. In juveniles, oxytocin binding in the nucleus accumbens and the caudate putamen was highest in prairie voles, intermediate in rats, and low in meadow voles and mice. The opposite binding pattern was found in the lateral septum of juveniles; high in meadow voles and mice, low in prairie voles and rats.

9.7.2 Rhesus monkeys

ICV injection of oxytocin in 2 nulliparous females increased expression of some components of maternal behavior toward newborns (maintaining close proximity, lip smacking, watching, and establishing physical contact). One of the females groomed and carried a stimulus infant while the other exhibited prosocial behaviors intended to encourage the infant to approach her (Holman and Goy, 1995; Table 9.6). Untreated nulliparous females (ages not given) in their monkey colony were noted to exhibit virtually no positive social behavior toward very young newborns. In another study (Boccia et al., 2007; Table 9.2), a 4-year-old nulliparous rhesus monkey was given two series of 3 IV injections at 2-day intervals. Both series of IV injections were administered during the luteal phase and approximately 1 month apart. The first series of IV injections contained, in order, saline alone, 1 mg/kg of a human oxytocin receptor-selective antagonist (L-368,899 [Pettibone and Freidinger, 1997] provided by Merck) that had previously been shown to penetrate the CNS, and 3 mg/kg of the antagonist. Each injection in the second series contained saline alone. Sixty minutes after each injection, the female was observed for 30 min in a large test cage separated by a wire screen from a small cage containing a 9-month old infant. Gaps in the wire screen were large enough for the adult to touch but not fully grasp the infant. Contact-soliciting behavior (lip smacking, touching) directed toward the infant as well as duration in close proximity to the infant were high after the first saline injection, lower after the low dose of oxytocin antagonist and very low after the high dose of antagonist. In contrast, these behaviors increased during the test periods after each of the 3 successive saline injections. Literature cited in the Boccia et al. (2007) report indicates that 4-year-old nulliparous rhesus monkeys usually exhibit high levels of interest in infants.

9.8 Summarizing animal studies of oxytocin regulation of maternal behavior

Studies in rodents and sheep strongly indicate that central oxytocin is highly significant and possibly necessary for the ovarian steroid activation of maternal behavior at parturition. During nursing in rats, when maternal behavior expression is no longer ovarian steroid dependent, oxytocin reduces anxiety and regulates maternal aggression. The latter role may be facilitating or inhibitory for reasons that have yet to be determined. During nursing, oxytocin continues to facilitate some components of maternal behavior (pup licking, arched-back nursing) but is not necessary for their expression.

Studies of the effects of oxytocin manipulations in specific brain sites as well as consideration of the role in maternal behavior of other brain sites where oxytocin is known to affect neural activity

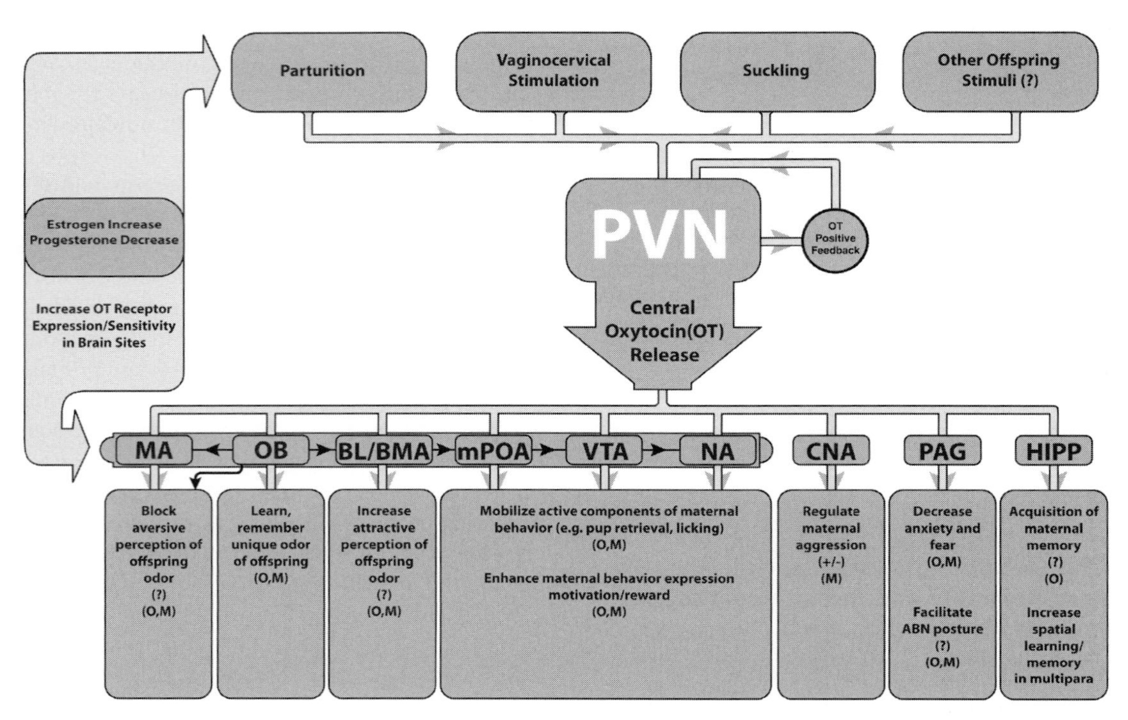

Figure 9.1 A model of oxytocin regulation of the onset (O) and maintenance (M) of maternal behavior.ABN = arched-back nursing; BL/BMA = basolateral and basomedial amygdala; CNA = central nucleus of the amygdala; HIPP = hippocampus; MA = medial amygdala; mPOA = medial preoptic area; NA = nucleus accumbens; OB = olfactory bulb; PAG = periaqueductal gray; PVN = hypothalamic paraventricular nucleus; VTA = ventral tegmental area. ? indicates no direct evidence that (1) other offspring stimuli activate PVN oxytocin projections, (2) oxytocin in the brain site is involved in the indicated aspect of maternal behavior regulation but the site receives oxytocin projections, expresses oxytocin receptors and/or has oxytocin responsive neurons. +/− indicates there is evidence of oxytocin facilitation and inhibition of maternal aggression in the central nucleus of the amygdala.

suggest a model of oxytocin regulation of maternal behavior summarized in Figure 9.1. The PVN is at the center of the proposed oxytocin neural circuitry because it is the hypothesized origin of oxytocin projections to extrahypothalamic brain sites involved in maternal behavior. Increasing estrogen and, in some species, decreasing progesterone concentrations during late pregnancy increase uterine contractions leading to parturition and vaginocervical stimulation that stimulates oxytocin release from PVN projections to brain sites where oxytocin contributes in site specific ways to the postpartum initiation of maternal behavior. Late-pregnancy ovarian steroid hormone changes also increase oxytocin receptor expression and perhaps sensitivity in brain sites where oxytocin facilitates maternal behavior activation. After the onset of maternal behavior, suckling, olfactory and perhaps other stimuli from offspring continue to stimulate oxytocin neurons in the PVN resulting in oxytocin release in brain sites involved in site specific regulation of behavior changes in nursing mothers (maintaining attraction to olfactory and perhaps other offspring stimuli, increasing aggression, decreasing anxiety as well as facilitating pup licking and arched-back nursing). Some components

of this model are speculative and will require validation by future investigations.

9.9 Methodological impediments to further elucidation of oxytocin mechanisms regulating maternal behavior

Detailed investigation of oxytocin mechanisms regulating offspring-directed maternal behavior in experimental animal species has stagnated in the past decade. This may be attributable to persistent, frustrating technical obstacles to studying oxytocin receptors in non-rodent species and delineating pathways activated by oxytocin in all species.

9.9.1 Selective antagonists and radioligands for oxytocin receptors in non-rodent species

Selective peptide antagonists for oxytocin receptors in rats (Manning et al., 1989; 1995) have enabled investigators to demonstrate in this species that activation of central oxytocin receptors is necessary for the initiation of maternal behavior and to identify specific brain sites where endogenous oxytocin exerts these effects. Among these, $d(CH_2)_5[Tyr(Me)^2, Thr^4, Tyr\text{-}NH_2^9]$ ornithine vasotocin has been employed most frequently to block central oxytocin receptors in behavior pharmacological studies, possibly because it was the first to be available commercially, although other antagonist analogs, such as des-Gly-$_2$, $d(CH_2)_5[\text{D-}Tyr^2, Thr^4]$ ornithine vasotocin, are quite effective (Pedersen and Boccia, 2002). Central administration of these antagonists have also contributed to our understanding of the role of central oxytocin in established pup-directed maternal behavior as well as other behaviors exhibited by lactating mothers such as heightened aggression. The development of the iodinated oxytocin antagonist, ^{125}I-$d(CH_2)_5$ $[Tyr(Me)^2,Thr^4,Tyr\text{-}NH_2^9]$ ornithine vasotocin, that exhibits high selectivity and affinity for the rat oxytocin receptor (Elands et al., 1988), has permitted the autoradiographic localization and quantification of oxytocin binding in the rat brain. As summarized above, these pharmacological tools have been used successfully to investigate the role of oxytocin in maternal behavior in other rodent species such as the prairie vole.

Unfortunately, the oxytocin antagonists and the Elands et al. (1988) radioligand that have been employed so successfully in rats and other rodents, have proven to be far less effective and selective in sheep and monkeys. This conclusion, which is supported by findings summarized in the sections above on oxytocin regulation of sheep and primate maternal behavior, suggests that these compounds will be of little or no use in investigating oxytocin mechanisms regulating maternal behavior in mammalian species other than rodents. This is not surprising because the amino acid sequence and, therefore, the conformation of the oxytocin receptor outside the agonist binding region vary considerably between rodents and other mammalian orders (Gimpl and Fahrenholz, 2001). Antagonists and radioligands that are efficacious and selective for central oxytocin receptors in non-rodent mammalian species have yet to be developed. Although other methods have convincingly demonstrated the importance of oxytocin in the activation of maternal behavior in sheep, no selective antagonists are available with which to definitively determine which aspects of maternal behavior are altered by blocking oxytocin receptors in each of the specific brain sites where oxytocin release has been demonstrated during parturition or VCS in this species. Also, in the absence of sheep oxytocin receptor-selective radioligands with which to conduct autoradiography, investigation of central receptors in this species has been limited to less-direct methods, i.e. visualization and measurement of mRNA using *in situ* hybridization and PCR. The unavailability of selective antagonists or radioligands is a major obstacle to testing the significance of central oxytocin in maternal behavior expression in non-rodent mammalian species as well as locating and quantifying oxytocin receptors in their brains. A notable exception to this dearth of pharmacological tools is the substantial number of selective human

oxytocin receptor antagonists that have been synthesized, including small molecule, non-peptide antagonists developed by pharmaceutical companies as potential treatments for premature labor (Manning et al., 2008; Pettibone and Freidinger, 1997). Other than the Boccia et al. (2007) study cited above, these compounds, some of which are commercially available, have yet to be employed to study oxytocin control of maternal behavior in primate species and none have been used to develop radioligands.

9.9.2 Oxytocin receptor immunohistochemistry

Adan et al. (1995) were the first to conduct oxytocin receptor immunohistochemistry in rats. In the brain, they found immunostaining in the BnST, ventromedial hypothalamus, ventral pallidum, PVN, and SON but not in other areas, such as the central nucleus of the amygdala and ventral hippocampus, in which dense oxytocin binding had been identified by autoradiography. Since this initial publication, only seven additional reports have appeared in the literature on oxytocin receptor immunostaining in rodents and none in other, non-primate species. Only two studies are relevant to oxytocin activation of maternal behavior (Lin et al., 2003; Meddle et al., 2007). Both combined oxytocin receptor and Fos or FosB immunohistochemistry to identify oxytocin receptor-expressing cells that were activated during parturition. Oxytocin receptor immunohistochemistry was employed in two additional studies of the brain (Liu et al., 2005; Murata et al., 2011) and three studies of the gastrointestinal tract (Feng et al., 2009; Qin et al., 2009; Welch et al., 2009). Among these studies; two (including Meddle et al., 2007) used the same antiserum as Adan et al. (1995), four used various lots of antiserum obtained from Santa Cruz Biotechnology, Sigma-Aldrich or MBL International and one (Lin et al., 2003) used a human oxytocin receptor-directed monoclonal antibody previously developed by Takemura et al. (1994). It is of note that there have been no follow-up studies from these groups using the same oxytocin receptor-immunostaining method (with the exception of Feng et al., 2009 and Qin et al., 2009, studies conducted by the same group presumably using the same lot of antiserum). Based on the experiences of numerous investigators who have attempted to conduct oxytocin receptor immunohistochemistry in rodent brain, the paucity of published studies in this area appears to be the result of frequent failure to obtain adequate immunostaining with the available oxytocin receptor-directed antisera. As an example of the experience of other investigators (personal communications), procedures for preparing sections of paraffin-embedded, fixed rat brain were initially worked out in the author's laboratory that provided excellent oxytocin receptor immunostaining using a particular lot of polyclonal antiserum obtained from Santa Cruz. Unfortunately, the next lot of oxytocin receptor antiserum generated by Santa Cruz produced no immunostaining using our previously successful tissue preparation methods. Multiple adjustments of those methods failed to re-establish immunostaining. Yoshida et al. (2009) have confirmed the unreliability of oxytocin immunohistochemistry using antiserum from several sources. They were unable to obtain acceptable immunostaining using the same procedures and antiserum as Adan et al. (1995) as well as three lots of antiserum from Santa Cruz and two lots of antiserum from MBL.

The lack of a reliable method of immunostaining oxytocin receptors is a major impediment to delineating the neural circuitry that mediates oxytocin activation and subsequent regulation of maternal behavior (and many other central effects of oxytocin). Combined oxytocin receptor and immediate early gene (e.g., Fos) immunohistochemistry would enable visualization of individual cells throughout the brain that are activated during the expression of maternal behavior. The addition of immunostaining for other substances would permit identification of the neurochemical phenotype of those activated cells and combining immunohistochemistry with anterograde or retrograde tract-tracing would determine their projections as well as the location of oxytocin neurons giving rise to

afferent connections (Deller et al., 2000). Application of these methods would vastly increase our understanding of the circuits regulated by oxytocin during the expression of maternal behavior. With oxytocin receptor immunohistochemistry, maternal behavior-regulating oxytocin receptor-expressing cells could be identified in brain areas, such as the VTA, where little or no binding is revealed by autoradiography. Sensitivity to oxytocin has been shown to be much greater in cells in the medial nucleus of the amygdala, where oxytocin binding is low, than in cells in the central nucleus in which oxytocin binding is very dense (Terenzi and Ingram, 2005). During lactation, the proportion of oxytocin-responsive cells increased in the medial nucleus but decreased in the central nucleus. These contrasting changes within amygdala nuclei in nursing mothers, which may be relevant to the onset and maintenance of maternal behavior, could be studied in greater detail and at the individual neuron level if reliable immunohistochemical methods were available.

The methodological roadblock resulting from the unavailability of reliable oxytocin receptor antiserum may be circumvented in mice using an exciting new technique developed by Yoshida et al. (2009). They developed an oxytocin receptor reporter knockin mouse in which part of the oxytocin receptor gene is replaced with Venus cDNA, which is fluorescent. Tryptophan hydroxylase immunohistochemistry conducted on brain sections from these animals demonstrated that approximately one-half of the immunostained neurons in the raphe nuclei are positive for Venus, thus indicating that those neurons express oxytocin receptors. Venus visualization combined with immunohistochemistry and/or tract tracing should enable identification in mice of oxytocin receptor-expressing neurons that are activated during the expression of maternal behavior as well as their neurochemical phenotypes and projections.

In contrast to the limited number of rat studies summarized above, many more human oxytocin receptor immunohistochemical studies have been published. The Kimura group (Takemura et al., 1994) was the first to report immunostaining of human oxytocin receptors using monoclonal antibodies that they had produced. Numerous reports by that group and others followed using monoclonal antibodies produced by Takemura et al. (1994) or polyclonal antiserum generated by Bussolati et al. (1996). These studies have examined oxytocin receptor expression in human or primate reproductive tissues or tumors (a partial summary includes Takemura et al., 1994; Bussolati et al., 1996; Cassoni et al., 2004; Cassoni et al., 1998; Einspanier et al., 1998; Frayne and Nicholson, 1998; Fuchs et al., 1998; Ito et al., 1996; Kimura et al., 1998; Lee et al., 1998; Viggnozi et al., 2005; Wakasa et al., 2009). The author's group (Boccia et al., 2001) successfully immunostained oxytocin receptors in cynomolgus macaque brain using 2F8, one of the monoclonal antibodies originally developed by Takemura et al. (1994). More recently, commercially obtained polyclonal antiserum has been employed to conduct oxytocin receptor immunohistochemistry in monkey and human ovaries (Saller et al., 2010). To date there are no studies that have utilized these monoclonal antibodies or polyclonal antisera to study activation of oxytocin receptor-expressing neurons in primate mothers.

9.10 Oxytocin correlates of human parental behavior

In recent years, studies of the relationships between peripheral oxytocin measures (plasma, saliva, urine) and human parent–infant interactions have been published at an accelerating rate (Table 9.7). Light et al. (2004) found that mothers who used cocaine during pregnancy tended to carry their babies less often when monitored at home, which is consistent with other evidence that these mothers exhibited poorer attachment to and care of their children than women without substance-abuse disorders (Burns et al., 1991; Rutherford et al., 2011; Tronick et al., 2005). The cocaine-abusing mothers studied by Light et al. (2004) had significantly

Table 9.7 Oxytocin correlates of human parental behavior.

Study	Subjects	Experimental Measures	Outcomes
Light et al., 2004	Mothers +/− cocaine use in pregnancy	Plasma OT; affect, MB and stress responses in lab × 2 days; MB at home.	Cocaine mothers had lower OT levels, greater hostility and depression, held babies less.
Levine et al., 2007	Mothers	Plasma OT early and late Preg; late Preg maternal–fetal bonding.	Rise in OT from early to late Preg related to greater maternal–fetal bonding.
Feldman et al., 2007	Mothers	Plasma OT during Preg and PP; behavior during MII, maternal interview.	OT levels during Preg and PP correlated positively with bonding behavior, thoughts.
Strathearn et al., 2009	Mothers	Plasma OT before, after MII at 7 months PP; Adult Attachment Interview.	OT levels rose after MII; increase greater in mothers with secure attachment.
Feldman et al., 2010a	Mothers and fathers	Plasma, saliva OT before and after PII; parental behavior during PII.	Postinteraction rise in OT in high affectionate touch mothers and high stimulatory touch fathers.
Feldman et al., 2010b	Mothers, fathers and infants	Plasma, saliva OT before and after PII; behavior and affect during PII.	OT levels correlated between parents and infants, and with PII affect synchrony and behavior engagement.
Gordon et al., 2010a	Mothers and fathers	Plasma OT before PII; behavior and affect during PII early PP and 6 months PP.	OT levels related between mothers and fathers, to mother affectionate and father stimulating behavior.
Gordon et al., 2010b	Fathers	Plasma OT pre-FII at 2 and 6 months PP; paternal behavior during FII at 6 months.	Fathers' OT levels positively associated with affect synchrony during interactions with infants.
Gordon et al., 2010c	Mothers, fathers and infants	Plasma OT pre-PII at 2 and 6 months PP; behaviors during PII at 6 months.	OT levels in parents related to coordinated proximity and touch among mothers, fathers, infants.
Bick and Dozier, 2010	Mothers	Urine OT post close physical interaction (computer game) with own or other child.	OT levels higher in mothers interacting with unfamiliar children.
Bakermans-Kranenburg and van IJzendoorn, 2008	Mothers	OTR gene variants; mothers' sensitive interactions with their toddlers during a series of problem solving tasks.	Mothers with presumed less efficient OTR allele dyads (AA/AG) exhibited lower levels of sensitive responsiveness to their children.

FII = father–infant interaction; MII = mother–infant interaction; PII = parent–infant interaction. See Tables 9.1–9.5 for definitions of other acronyms and abbreviations.

lower plasma oxytocin concentrations than non-drug-using mothers

Levine et al. (2007) reported greater maternal–fetal bonding in women whose plasma oxytocin concentrations increased from early to late pregnancy compared to women who exhibited different patterns of change in oxytocin concentrations over the course of pregnancy. Feldman et al. (2007) measured plasma oxytocin concentrations at approximately the same pre- and post-partum time points and reported that levels during early pregnancy and the first postpartum month were significantly related to several measures of maternal bonding including gaze, vocalizations, positive affect and affectionate touch as well as attachment-related thoughts and frequency of checking the infant.

Strathearn et al. (2009) reported significant increases in plasma oxytocin concentrations after intimate interaction of mothers with their infants at 7 months postpartum. Postinteraction elevations in oxytocin were higher in mothers with secure attachment based on their responses during the Adult Attachment Interview (van IJzendoorn, 1995) administered when they were pregnant.

Feldman et al. (2010a;b; 2011) examined, respectively, the relationship between oxytocin levels (plasma and salivary) and measures of maternal and paternal care exhibited during contact with their 4–6-month-old infants as well as cross-generational comparisons of oxytocin levels and their correlations with affect synchrony and social engagement between parents and their infants. Feldman et al. (2010a; 2011) found that oxytocin levels rose significantly following interactions with infants in mothers who exhibited high frequencies of affectionate touch but not in mothers who displayed lower amounts of affectionate contact. Among fathers, increases in oxytocin levels after infant interactions only occurred among those who exhibited high frequencies of stimulatory touch. The authors note that their findings parallel earlier observations that children prefer to play with their fathers and to be comforted by their mothers (Lamb, 2010). Feldman et al. (2010b; 2011) obtained baseline and postinfant interaction saliva from infants as well as saliva and plasma from mothers and fathers. Behaviors recorded during parent–infant interactions were coded to quantify affect synchrony and social engagement. Oxytocin levels rose significantly after parent–infant interaction in both parents and infants. Baseline and postcontact oxytocin concentrations as well as increases in oxytocin after interactions correlated significantly between parents and their infants. Greater oxytocin concentrations at each time point correlated in parents and infants with greater affect synchrony and greater social engagement. Feldman et al. (2010b) speculated that the strong relationship between parent and infant oxytocin concentrations may play a role in previously reported cross-generation transmission of attachment in humans (Bakermans-Kranenburg and van IJzendoorn, 2007).

Gordon et al. (2010a;b;c) investigated, respectively, the relationships between plasma oxytocin levels and the development of mothers' and fathers' parental behavior during the postpartum period, fathers' interactions with their infants, and the physical and affect synchrony among mothers, fathers and their infants. In these studies, parents' oxytocin concentrations were measured in the early puerperium (first few weeks, 2 months and 6 months postpartum). Unlike the Feldman et al. studies discussed above, oxytocin was not assayed after parent–infant interactions. In the first study (Gordon et al., 2010a), oxytocin was measured in 160 cohabitating mothers and fathers after delivery of their first infant. Records of interactions between each parent and her/his infant were recorded and coded in detail. Oxytocin concentrations in both parents increased overall across the postpartum period and mothers' and fathers' levels were similar and interrelated. Mothers' oxytocin levels correlated with the amount of affectionate parenting behaviors ("motherese" vocalizations, expression of positive effect, affectionate touch) while fathers' levels were related with amount of stimulatory parenting behaviors (proprioceptive contact, tactile stimulation and object presentation). The latter findings are similar to those of Feldman et al. (2010a). Gordon et al. (2010b) recorded and analyzed in detail

fathers' and infants' behaviors in 43 dyads during a social interaction session and then a toy exploration session, each 6 min in length. Averaged 2- and 6-month baseline oxytocin concentrations correlated significantly with the amount of time they engaged in behaviors scored as affect synchrony during the social interaction but did not correlate with the cumulative time when the dyads exhibited behaviors scored as father–infant coordinated exploratory play during the exploration session. Gordon et al. (2010c) scored and analyzed recordings made during free-play interactions among 35 mother–father–infant triads at 6 months. Average plasma oxytocin concentrations in mothers and fathers predicted total duration during the interaction period of triadic synchrony, defined as moments of coordination between physical proximity and affectionate touch between the parents as well as between parent and infant while both parent and child are synchronizing their social gaze.

Bick and Dozier (2010) assayed oxytocin concentrations in urine of mothers collected following a 25-min interaction with their own child and, on a separate occasion, with an unfamiliar child. Mothers and children played an interactive computer game during which the child sat in the mother's lap. Children were between 29 and 54 months of age. The order in which mothers were tested with own or another child was counterbalanced. In seeming contrast to the results of the other studies summarized above, Bick and Dozier (2010) found that urine oxytocin concentrations were higher in mothers that had interacted with unfamiliar children. Measures of maternal sensitivity, positivity, closeness and percentage of time engaged in physical contact did not differ between sessions when mothers interacted with their own versus an unfamiliar child. The authors speculate that the unanticipated results may be the result of the novelty of interacting with an unfamiliar child releasing oxytocin or the soothing effect of interacting with their own child during the novel test situation.

It is unknown if baseline levels or increases in peripheral oxytocin release in parents during interactions with infants or children are related to oxytocin activity or release within their brains. However, fMRI imaging studies in parents have found changes (mainly increases) in activity during exposure to own versus unfamiliar or distressed versus non-distressed infant stimuli in hypothalamic and limbic brain areas in which oxytocin receptors have been located in animals and in which oxytocin has been shown to enhance maternal behavior in those species (Swain et al., 2007). In these studies, altered activity has also been identified in reward-mediating (midbrain, ventral striatum) and areas of the human cortex that have been associated with empathy (anterior cingulate and insula).

In addition to the evidence cited above of significant relationships between peripheral oxytocin measures and human parental behavior, a rapidly increasing number of publications report that intranasal administration of oxytocin, which probably results in penetration of a significant amount of the neuropeptide into the brain (Born et al., 2002; Domes et al., 2007; 2010; Kirsch et al., 2005), has numerous prosocial effects and also reduces stress responses in human subjects (MacDonald and MacDonald 2010). Despite this mounting evidence that oxytocin exerts central effects in people, very little is known about oxytocin systems in the human brain. For example, studies of oxytocin CSF concentrations during pregnancy and parturition are few in number and contradictory (Altemus et al., 2004; Takeda et al., 1985; Takagi et al., 1985). Early autoradiographic studies of oxytocin binding in human brains used a radioligand, (^{125}I-d$(CH_2)_5$ [Tyr(Me)^2,Thr^4,Tyr-NH$_2$^{9}] ornithine vasotocin), that had proven to be highly selective for oxytocin receptors in rodent brain (Elands et al., 1988) as well as tritiated oxytocin identified binding in the basal nucleus of Meynert, the nucleus of the vertical limb of the diagonal band of Broca, the ventral lateral septal nucleus, throughout the preoptic-anterior hypothalamic area, the posterior hypothalamus, and in, in some brains, the globus pallidus and the ventral pallidum (Loup et al., 1991). In the brainstem, oxytocin binding was most intense in the substantia nigra pars compacta, the substantia gelantinosa of the caudal spinal trigeminal nucleus,

the dorsal horn of the upper spinal cord and the medial-dorso region of the nucleus of the spinal tract, and less prominent in the remainder of the solitary tract and other sites such as parts of the spinal trigeminal nucleus, the hypoglossal nucleus and the area postrema (Loup et al., 1989). Surprisingly, Loup et al. (1991) did not find radioligand binding in limbic sites of the human brain, such as the central amygdala, in which oxytocin binding has been shown to be prominent in rats and other rodents (Gimpl and Fahrenholz, 2001). Interestingly, Tolockzko et al. (1997) published evidence that the Elands et al. (1988) iodinated radioligand used by Loup et al. (1989; 1991) was more selective in rhesus monkey brain for vasopressin V1a than oxytocin receptors. The latter findings raise questions about the specificity of the Loup et al. (1989; 1991) finding.

More recently, Boccia et al. (2001) reported oxytocin receptor immunostaining of cell bodies and fibers in the preoptic area as well as immunostaining of fibers in the ventral septum in cynomolgus macaque brain sections using a monoclonal antibody selectively directed toward the human uterine OT receptor (Takemura et al., 1994). Using the same antibody, Boccia et al. (2003) found OT receptor-immunostaining in the human brain in cell bodies and/or fibers in the central and basolateral regions of the amygdala, medial preoptic area, anterior and ventromedial hypothalamus, olfactory nucleus, vertical limb of the diagonal band, ventrolateral septum, anterior cingulate gyrus and hypoglossal and solitary nuclei. The Loup et al. (1989; 1991) studies and especially the Boccia et al. (2003) findings suggest that oxytocin receptors may be located in many of the brain areas activated by infant stimuli in the fMRI studies discussed above including the anterior cingulate cortex, which has been implicated in empathy.

Studies of variance in the nucleotide sequences in a region of the third intron of the human oxytocin receptor gene (rs53576) have found that allele dyads associated with less efficient expression of the gene (AA, AG) compared to GG are associated with lower empathy (Rodrigues et al., 2009), insecure attachment (Costa et al., 2009) and less positive affect (Lucht et al., 2009). A higher frequency of AA expression was associated with autism in one study (Wu et al., 2005) but not another (Jacob et al., 2007). In addition, AA compared to GG has been linked to lower reward dependence subscores on the TPQ (Cloninger, 1987), a temperament characteristic associated with lower sociality (Gerretsen et al., 2010; Sizoo et al., 2009), as well as less activation of the amygdala by face emotional cues as measured by fMRI and lower gray matter volume in the hypothalamus in both sexes and greater volume in the amygdala of males (Tost et al., 2010). In the only study of its kind, Bakermans-Kranenburg and van IJzendoorn (2008) compared rs53576 variants with measures of maternal sensitivity based on behavior exhibited by mothers toward their 2-year-old toddlers during a series of problem solving tasks. Mothers with AA and AG variants showed less sensitivity toward their children during the tasks than mothers with the GG variant.

REFERENCES

Acher, R., Chauvet, J., and Chauvet, M. T. (1995). Man and the chimaera. Selective versus neutral oxytocin evolution. *Advances in Experimental and Medical Biology*, 395, 615–627.

Adan, R. A., van Leeuwen, F. W., Sonnemans, M. A., et al. (1995). Rat oxytocin receptor in brain, pituitary, mammary gland, and uterus: partial sequence and immunocytochemical localization. *Endocrinology*, 136, 4022–4028.

Altemus, M., Fong, J., Yang, R., et al. (2004). Changes in cerebrospinal fluid neurochemistry during pregnancy. *Biological Psychiatry*, 56, 386–392.

Amico, J. A., Mantella, R. C., Vollmer, R. R., and Li, X. (2004). Anxiety and stress responses in female oxytocin deficient mice. *Journal of Neuroendocrinology*, 16, 319–324.

Bakermans-Kranenburg, M. J. and Van Ijzendoorn, M. H. (2007). Research review: Genetic vulnerability or differential susceptibility in child development: the case of attachment. *Journal of Child Psychology and Psychiatry*, 48, 1160–1173.

Bakermans-Kranenburg, M. J. and Van Ijzendoorn, M. H. (2008). Oxytocin receptor (OXTR) and serotonin

transporter (5-HTT) genes associated with observed parenting. *Social Cognitive and Affective Neuroscience*, 3, 128–134.

Bale, T. L., Davis, A. M., Auger, A. P., Dorsa, D. M., and McCarthy, M. M. (2001). CNS region-specific oxytocin receptor expression: Importance in regulation of anxiety and sex behavior. *Journal of Neuroscience*, 21, 2546–2552.

Bale, T. L., Pedersen, C. A., and Dorsa, D. M. (1995). CNS oxytocin receptor mRNA expression and regulation by gonadal steroids. *Advances in Experimental and Medical Biology*, 395, 269–280.

Bick, J. and Dozier, M. (2010). Mothers' and children's concentrations of oxytocin following close, physical interactions with biological and non-biological children. *Developmental Psychobiology*, 52, 100–107.

Boccia, M. L., Goursand, A., Bachevalier, J., Anderson, K. D., and Pedersen, C. A. (2007). Peripherally administered non-peptide oxytocin antagonist, L368,899®, accumulates in limbic brain areas: A new pharmacological tool for the study of social motivation in non-human primates. *Hormones and Behavior*, 52, 344–351.

Boccia, M. L., Petrusz, P, Suzuki, K., Razzoli, M., and Pedersen, C. A. (2003). Immunostaining of oxytocin receptors in human female amygdala, preoptic area, hypothalamus, nucleus accumbens and cingulate gyrus. *Program No.* 943.11. 2003 Abstract Viewer/Itinerary Planner. Washington, DC: Society for Neuroscience.

Boccia, M. L., Panicker, A. K., Pedersen, C., and Petrusz, P. (2001). Oxytocin receptors in non-human primate brain visualized with monoclonal antibody. *Neuroreport*, 12, 1723–1726.

Bolwerk, E. L. M. and Swanson, H. H. (1984). Does oxytocin play a role in the onset of maternal behavior in the rat? *Journal of Endocrinology*, 101, 353–357.

Born, J., Lange, T., Kern, W., et al. (2002). Sniffing neuropeptides: a transnasal approach to the human brain. *Nature Neuroscience*, 5, 514–516.

Bosch, O. J., Kromer, S. A., Brunton, P. J., and Neumann, I. D. (2004). Release of oxytocin in the hypothalamic paraventricular nucleus, but not central amygdala or lateral septum in lactating residents and virgin intruders during maternal defense. *Neuroscience*, 124, 439–448.

Bosch, O. J., Meddle, S. L., Beiderbeck, D. I., Douglas, A. J., and Neumann, I. D. (2005). Brain oxytocin correlates with maternal aggression: Link to anxiety. *Journal of Neuroscience*, 25, 6807–6815.

Bosch, O. J. and Neumann, I. D. (2008). Brain vasopressin is an important regulator of maternal behavior independent of dams' trait anxiety. *Proceedings of the National Academy of Sciences of the United States of America*, 105, 17139–17144.

Bosch, O. J. and Neumann, I. D. (2010). Vasopressin released within the central amygdala promotes maternal aggression. *European Journal of Neuroscience*, 31, 883–891.

Bridges, R. S. (1984). A quantitative analysis of the roles of dosage, sequence, and duration of estradiol and progesterone exposure in the regulation of maternal behavior in the rat. *Endocrinology*, 114, 930–940.

Bridges, R. S. (2008). *Neurobiology of the Parental Brain*. Burlington, MA: Elsevier Academic Press.

Broad, K. D., Curley, J. P., and Keverne, E. B. (2006). Mother-infant bonding and the evolution of mammalian social relationships. *Philosophical Transactions of the Royal Society B*, 361, 2199–2214.

Broad, K. D., Kendrick, K. M., Sirinathsinghji, D. J., and Keverne, E. B. (1993). Changes in oxytocin immunoreactivity and mRNA expression in the sheep brain during pregnancy, parturition and lactation and in response to oestrogen and progesterone. *Journal of Neuroendocrinology*, 5, 711–719.

Broad, K. D., Levy, F., Evans, G., et al. (1999). Previous maternal experience potentiates the effect of parturition on oxytocin receptor mRNA expression in the paraventricular nucleus. *European Journal of Neuroscience*, 11, 3725–3737.

Buijs, R. M., De Vries, G. J., and Van Leeuwen, F. W. (1985). The distribution and synaptic release of oxytocin neurons. In Amico, J. A. and Robinson, A. G. (eds.) *Oxytocin: Clinical and Laboratory Studies*, Amsterdam: Elsevier Science Publishers B.V., pp. 77–86.

Burns, K., Chethik, L., Burns, W. J., and Clark, R. (1991). Dyadic disturbances in cocaine-abusing mothers and their infants. *Journal of Clinical Psychology*, 47, 316–319.

Bussolati, G., Cassoni, P., Ghisolfi, G., Negro, F., and Sapino, A. (1996). Immunolocalization and gene expression of oxytocin receptors in carcinomas and non-neoplastic tissues of the breast. *American Journal of Pathology*, 148, 1895–1903.

Cassoni, P., Marrocco, T., Sapino, A., Allia, E., and Bussolati, G. (2004). Evidence of oxytocin/oxytocin receptor interplay in human prostate gland and carcinomas. *International Journal of Oncology*, 25, 899–904.

Cassoni, P., Sapino, A., Stella, A., Fotunati, N., and Bussolati, G. (1998). Presence and significance of oxytocin

receptors in human neuroblastomas and glial tumors. *International Journal of Cancer*, 77, 695–700.

Champagne, F., Diorio, J., Sharma, S., and Meaney, M. J. (2001). Naturally occurring variations in maternal behavior in the rat are associated with differences in estrogen-inducible central oxytocin receptors. *Proceedings of the National Academy of Sciences of the United States of America*, 98, 12736–12741.

Cloninger, C. R. (1987). Neurogenetic adaptive mechanisms in alcoholism. *Science*, 236, 410–416.

Condes-Lara, M., Veinante, P., Rabai, M., and Freund-Mercier, M. J. (1994). Correlation between oxytocin neuronal sensitivity and oxytocin-binding sites in the amygdala of the rat: electrophysiological and histoautoradiographic study. *Brain Research*, 637, 277–286.

Consiglio, A. R., Borsoi, A., Pereira, G. A., and Lucion, A. B. (2005). Effects of oxytocin microinjected into the central amygdaloid nucleus and bed nucleus of stria terminalis on maternal aggressive behavior in rats. *Physiology and Behavior*, 85, 354–362.

Consiglio, A. R. and Lucion, A. B. (1996). Lesion of hypothalamic paraventricular nucleus and maternal aggressive behavior in female rats. *Physiology and Behavior*, 59, 591–596.

Costa, B., Pini, S., Gabelloni, P., et al. (2009). Oxytocin receptor polymorphisms and adult attachment style in patients with depression. *Psychoneuroendocrinology*, 34, 1506–1514.

Da Costa, A. P., Guevara-Guzman, R. G., Ohkura, S., Goode, J. A., and Kendrick, K. M. (1996). The role of oxytocin release in the paraventricular nucleus in the control of maternal behaviour in the sheep. *Journal of Neuroendocrinology*, 8, 163–177.

D'cunha, T. M., King, S. J., Fleming, A. S., and Levy, F. (2011). Oxytocin receptors in the nucleus accumbens shell are involved in the consolidation of maternal memory in postpartum rats. *Hormones and Behavior*, 59, 14–21.

Dekloet, E. R., Voorhuis, D. A., Boschma, Y., and Elands, J. (1985). Estradiol modulates density of putative "oxytocin receptors" in discrete rat brain regions. *Neuroendocrinology*, 44, 415–421.

Deller, T., Naumann, T., and Frotscher, M. (2000). Retrograde and anterograde tracing combined with transmitter identification and electron microscopy. *Journal of Neuroscience Methods*, 103, 117–126.

Domes, G., Heinrichs, M., Glascher, J., et al. (2007). Oxytocin attenuates amygdala responses to emotional faces regardless of valence. *Biological Psychiatry*, 62, 1187–1190.

Domes, G., Lischke, A., Berger, C., et al. (2010). Effects of intranasal oxytocin on emotional face processing in women. *Psychoneuroendocrinology*, 35, 83–93.

Einspanier, A., Bielefeld, A., and Kopp, J. H. (1998). Expression of the oxytocin receptor in relation to steroid receptors in the uterus of a primate model, the marmoset monkey. *Human Reproduction Update*, 4, 634–646.

Elands, J., Barberis, C., Jard, S., et al. (1988). 125I-labelled d(CH2)5[Tyr(Me)2,Thr4,Tyr-NH2(9)]OVT: a selective oxytocin receptor ligand. *European Journal of Pharmacology*, 147, 197–207.

Fahrbach, S. E., Morrell, J. I., and Pfaff, D. W. (1984). Oxytocin induction of short-latency maternal behavior in nulliparous, estrogen-primed female rats. *Hormones and Behavior*, 18, 267–286.

Fahrbach, S. E., Morrell, J. I., and Pfaff, D. W. (1985a). Possible role for endogenous oxytocin in estrogen-facilitated maternal behavior in rats. *Neuroendocrinology*, 40, 526–532.

Fahrbach, S. E., Morrell, J. I., and Pfaff, D. W. (1985b). Role of oxytocin in the onset of estrogen-facilitated maternal behavior. In J. A. Amico and A. G. Robinson (eds.) *Oxytocin: Clinical and Laboratory Studies*. Amsterdam: Elsevier Science Publishers B.V., pp. 372–388.

Fahrbach, S. E., Morrell, J. I., and Pfaff, D. W. (1986). Effect of varying the duration of pre-test cage habituation on oxytocin induction of short-latency maternal behavior. *Physiology and Behavior*, 37, 135–139.

Febo, M. and Ferris, C. F. (2008). Imaging the maternal rat brain. In Bridges, R. S. (ed.) *Neurobiology of the Parental Brain*. Burlington, MA: Elsevier Academic Press, pp. 61–74.

Febo, M., Shields, J., Ferris, C. F., and King, J. A. (2009). Oxytocin modulates unconditioned fear response in lactating dams: An fMRI study. *Brain Research*, 1302, 183–193.

Feldman, R., Gordon, I., Schneiderman, I., Weisman, O., and Zagoory-Sharon, O. (2010a). Natural variations in maternal and paternal care are associated with systematic changes in oxytocin following parent-infant contact. *Psychoneuroendocrinology*, 35, 1133–1141.

Feldman, R., Gordon, I., and Zagoory-Sharon, O. (2010b). The cross-generation transmission of oxytocin in humans. *Hormones and Behavior*, 58, 669–676.

Feldman, R., Gordon, I., and Zagoory-Sharon, O. (2011). Maternal and paternal plasma, salivary, and urinary oxytocin and parent-infant synchrony: considering stress

and affiliation components of human bonding. *Developmental Science*, 14, 752–761.

Feldman, R., Weller, A., Zagoory-Sharon, O., and Levine, A. (2007). Evidence for a neuroendocrinological foundation of human affiliation – Plasma oxytocin levels across pregnancy and the postpartum period predict mother-infant bonding. *Psychological Science*, 18, 965–970.

Feng, M., Qin, J., Wang C., et al. (2009). Estradiol upregulates the expression of oxytocin receptor in colon in rats. *American Journal of Physiology: Endocrinology and Metabolism*, 296, E1059–E1066.

Ferris, C. F., Foote, K. B., Meltser, M. G., Plenby, M. G., Smith, K. L., and Insel, T. R. (1992). Oxytocin in the amygdala facilitates maternal aggression. *Annals of the New York Academy of Sciences*, 652, 456–457.

Figueira, R. J., Peabody, M. F., and Lonstein, J. S. (2008). Oxytocin receptor activity in the ventrocaudal periaqueductal gray modulates anxiety-related behavior in postpartum rats. *Behavioral Neuroscience*, 122, 618–628.

Francis, D., Diorio, J., Liu, D., and Meaney, M. J. (1999). Nongenomic transmission across generations of maternal behavior and stress responses in the rat. *Science*, 286, 1155–1158.

Frayne, J. and Nicholson, H. D. (1998). Localization of oxytocin receptors in the human and macaque monkey male reproductive tracts: evidence for a physiological role of oxytocin in the male. *Molecular Human Reproduction*, 4, 527–532.

Fuchs, A. R., Behrens, O., Maschek, H., Kupsch, E., and Einspanier, A. (1998). Oxytocin and vasopressin receptors in human and uterine myomas during menstrual cycle and early pregnancy. *Human Reproduction Update*, 4, 594–604.

Gerretsen, P., Graff-Guerrero, A., Menon, M., et al. (2010). Is desire for social relationships mediated by the serotonergic system in the prefrontal cortex? An [18F]setoperone PET study. *Social Neuroscience*, 5, 375–383.

Gimpl, G. and Fahrenholz, F. (2001). The oxytocin receptor system: Structure, function, and regulation. *Physiological Reviews*, 81, 629–683.

Giovenardi, M., Padoin, M. J., Cadore, L. P., and Lucion, A. B. (1998). Hypothalamic paraventricular nucleus modulates maternal aggression in rats: effects of ibotenic acid lesion and oxytocin antisense. *Physiology and Behavior*, 63, 351–359.

Gordon, I., Zagoory-Sharon, O., Leckman, J. F., and Feldman, R. (2010a). Oxytocin and the development of parenting in humans. *Biological Psychiatry*, 68, 377–382.

Gordon, I., Zagoory-Sharon, O., Leckman, J. F., and Feldman, R. (2010b). Prolactin, oxytocin, and the development of paternal behavior across the first six months of fatherhood. *Hormones and Behavior*, 58, 513–518.

Gordon, I., Zagoory-Sharon, O., Leckman, J. F., and Feldman, R. (2010c). Oxytocin, cortisol, and triadic family interactions. *Physiology and Behavior*, 101, 679–684.

Hard, E. and Hansen, S. (1985). Reduced fearfulness in the lactating rat. *Physiology and Behavior*, 35, 641–643.

Herrenkohl, L. R. and Rosenberg, P. A. (1974). Effects of hypothalamic deafferentation late in gestation on lactation and nursing behavior in the rat. *Hormones and Behavior*, 5, 33–41.

Holman, S. D. and Goy, R. W. (1995). Experiential and hormonal correlates of care-giving in rhesus macaques. In C. R. Pryce, R. D. Martin, and D. Skuse (eds.) *Motherhood in Human and Nonhuman Primates*. Basel: Karger, pp. 87–92.

Huang, Y. Y., Nguyen, P. V., Abel, T., and Kandel, E. R. (1996). Long-lasting forms of synaptic potentiation in the mammalian hippocampus. *Learning and Memory*, 3, 74–85.

Insel, T. R. (1986). Postpartum increases in brain oxytocin binding. *Neuroendocrinology*, 44, 515–518.

Insel, T. R. (1990). Regional changes in brain oxytocin receptors postpartum: time-course and relationship to maternal behaviour. *Journal of Neuroendocrinology*, 2, 539–545.

Insel, T. R. and Harbaugh, C. R. (1989). Lesions of the hypothalamic paraventricular nucleus disrupt the initiation of maternal behavior. *Physiology and Behavior*, 45, 1033–1041.

Insel, T. W., Young, L., and Wang, Z. (1997). Central oxytocin and reproductive behaviours. *Reviews of Reproduction*, 2, 28–37.

Ito, Y., Kobayashi, T., Kimura, T., et al. (1996). Investigation of the oxytocin receptor expression in human breast cancer tissue using newly established monoclonal antibodies. *Endocrinology*, 137, 773–779.

Jacob, S., Brune, C. W., Carter, C. S., et al. (2007). Association of the oxytocin receptor gene (OXTR) in Caucasian children and adolescents with autism. *Neuroscience Letters*, 417, 6–9.

Jirikowski, G. F., Caldwell, J. D., Pedersen, C. A., and Stumpf, W. E. (1988). Estradiol influences oxytocin-immunoreactive brain systems. *Neuroscience*, 25, 237–248.

Jirikowski, G. F., Caldwell, J. D., Pilgrim, C., Stumpf, W. E., and Pedersen, C. A. (1989). Changes in immunostaining for oxytocin in the forebrain of the female rat during late

pregnancy, parturition and early lactation. *Cell and Tissue Research*, 256, 411–417.

Jirikowski, G. F., Caldwell, J. D., Stumpf, W. E., and Pedersen, C. A. (1990). Topography of oxytocinergic estradiol target neurons in the mouse hypothalamus. *Folia Histochemica et Cytobiologica*, 28, 3–9.

Johnson, A. E., Coirini, H., Ball, G. F., and Mcewen, B. S. (1989). Anatomical localization of the effects of 17β-estradiol on oxytocin receptor binding in the ventromedial hypothalamic nucleus. *Endocrinology*, 124, 207–211.

Kendrick, K. M. (2000). Oxytocin, motherhood and bonding. *Experimental Physiology*, 85, 111S-124S.

Kendrick, K. M., Da Costa, A. P. C., Broad, K. D., et al. (1997). Neural control of maternal behaviour and olfactory recognition of offspring. *Brain Research Bulletin*, 44, 383–395.

Kendrick, K. M., Fabrenys, C., Blache, D., Goode, J. A., and Broad, K. D. (1993). The role of oxytocin release in the mediobasal hypothalamus of the sheep in relation to female sexual receptivity. *Journal of Neuroendocrinology*, 5, 13–21.

Kendrick, K. M., Keverne, E. B., and Baldwin, B. A. (1987). Intracerebroventricular oxytocin stimulates maternal behavior in the sheep. *Neuroendocrinology*, 46, 56–61.

Kendrick, K. M., Keverne, E. B., Baldwin, B. A., and Sharman, D. F. (1986). Cerebrospinal fluid levels of acetylcholinesterase, monoamines and oxytocin during labour, parturition, vaginocerivical stimulation, lamb separation and suckling in sheep. *Neuroendocrinology*, 44, 149–156.

Kendrick, K. M., Keverne, E. B, Chapman, C., and Baldwin, B. A. (1988a). Intracranial dialysis measurement of oxytocin, monoamine and uric acid release from the olfactory bulb and substantia nigra of sheep during parturition, suckling, separation from lambs and eating. *Brain Research*, 439, 1–10.

Kendrick, K. M., Keverne, E. B., Chapman, C., and Baldwin, B. A. (1988b). Microdialysis measurement of oxytocin, aspartate, gamma-aminobutyric acid and glutamate release from the olfactory bulb of the sheep during vaginocervical stimulation. *Brain Research*, 442, 171–174.

Kendrick, K. M., Keverne, E. B., Hinton, M. R., and Goode, J. A. (1992). Oxytocin, amino acid and monoamine release in the region of the medial preoptic area and bed nucleus of the stria terminalis of the sheep during parturition and suckling. *Brain Research*, 569, 199–209.

Keverne, E. B. and Kendrick, K. M. (1991). Morphine and corticotrophin-releasing factor potentiate maternal acceptance in multiparous ewes after vaginocervical stimulation. *Brain Research*, 540, 55–62.

Keverne, E. B., Levy, F., Poindron, P., and Lindsay, D. T. (1983). Vaginal stimulation: An important determinant of maternal bonding in sheep. *Science*, 219, 81–83.

Kimura, T., Ito, Y., Einspanier, A., et al. (1998). Expression and immunolocalization of the oxytocin receptor in human lactating and non-lactating mammary glands. *Human Reproduction*, 13, 2645–2653.

Kinsley, C. H., Madonia, L., Gifford, G. W., et al. (1999). Motherhood improves learning and memory. *Nature*, 402, 137–138.

Kirsch, P., Esslinger, C., Chen, Q., et al. (2005). Oxytocin modulates neural circuitry for social cognition and fear in humans. *Journal of Neuroscience*, 25, 11489–11493.

Klopfer, P. H. (1971). Mother love: what turns it on? *American Scientist*, 59, 404–407.

Kotwica, G., Staszkiewicz, J., Skowronski, M. T., et al. (2006). Effects of oxytocin alone and in combination with selected hypothalamic hormones on ACTH, beta-endorphin, LH and PRL secretion by anterior pituitary cells of cyclic pigs. *Reproductive Biology*, 6, 115–131.

Krehbiel, D., Poindron, D., Lévy, F., and Prud'homme, M. J. (1987). Peridural anesthesia disturbs maternal behavior in primiparious and multiparous parturient ewes. *Physiology and Behavior*, 40, 463–472.

Lamb, M. E. (ed.) (1979). *The Role of the Father in Child Development*. 5th edn., New York: Wiley and Sons.

Landgraf, R., Neumann, I., and Pittman, Q. J. (1991). Septal and hippocampal release of vasopressin and oxytocin during late pregnancy and parturition in the rat. *Neuroendocrinology*, 54, 378–383.

Lee, K. H., Khan-Dawood, F. S., and Dawood, M. Y. (1998). Oxytocin receptor and its messenger ribonucleic acid in human leiomyoma and myometrium. *American Journal of Obstetrics and Gynecology*, 179, 620–627.

Levine, A., Zagoory-Sharon, O., Feldman, R., and Weller, A. (2007). Oxytocin during pregnancy and early postpartum: Individual patterns and maternal-fetal attachment. *Peptides*, 28, 1162–1169.

Lévy, F., Guevara-Guzman, R., Hinton, M. R., Kendrick, K. M., and Keverne, E. B. (1993). Effects of parturition and maternal experience on noradrenaline and acetylcholine release in the olfactory bulb in sheep. *Behavioral Neuroscience*, 107, 662–668.

Lévy, F., Kendrick, K. M., Goode, J. A., Guevara-Guzman, R., and Keverne, E. B. (1995). Oxytocin and vasopressin release in the olfactory bulb of parturient ewes: Changes with maternal experience and effects on acetylcholine, gamma-aminobutyric acid, glutamate and noradrenaline release. *Brain Research*, 669, 197–206.

Lévy, F., Kendrick, K. M., Keverne, E. B., Piketty, V., and Poindron, P. (1992). Intracerebral oxytocin is important for the onset of maternal behavior in inexperienced ewes delivered under peridural aenesthesia. *Behavioral Neuroscience*, 106, 427–432.

Light, K. C., Grewen, K. M., Amico, J. A., et al. (2004). Deficits in plasma oxytocin responses and increased negative affect, stress, and blood pressure in mothers with cocaine exposure during pregnancy. *Addictive Behavior*, 29, 1541–1564.

Lin, S. H., Kiyohara, T., and Sun, B. (2003). Maternal behavior: Activation of the central oxytocin receptor system in parturient rats? *NeuroReport*, 14, 1439–1444.

Liu, W. S., Pappas, G. D., and Carter, C. S. (2005). Oxytocin receptors in brain cortical regions are reduced in haploinsufficient (+/–) reeler mice. *Neurological Research*, 27, 339–345.

Loup, F., Tribollet, E., Duboisdauphin, M., and Dreifuss, J. J. (1991). Localization of high-affinity binding sites for oxytocin and vasopressin in the human brain. An autoradiographic study. *Brain Research*, 555, 220–232.

Loup, F., Tribollet, E., Dubois-Dauphin, M., Pizzolato, G., and Dreifuss, J. J. (1989). Localization of oxytocin binding sites in the human brainstem and upper spinal cord: an autoradiographic study. *Brain Research*, 500, 223–230.

Lubin, D. A., Elliott, J. C., Black, M. C., and Johns, J. M. (2003). An oxytocin antagonist infused into the central nucleus of the amygdala increases maternal aggressive behavior. *Behavioral Neuroscience*, 117, 195–201.

Lucht, M. J., Barnow, S., Sonnenfeld, C., et al. (2009). Associations between the oxytocin receptor gene (OXTR) and affect, loneliness and intelligence in normal subjects. *Progress in Neuropsychopharmacology and Biological Psychiatry*, 33, 860–866.

MacBeth, A. H., Stepp, J. E., Lee, H-J., Young, W. S., III, and Caldwell, H. K. (2010). Normal maternal behavior, but increased pup mortality, in conditional oxytocin receptor knockout females. *Behavioral Neuroscience*, 124, 677–685.

MacDonald, K. and MacDonald, T. M. (2010). The peptide that binds: a systematic review of oxytocin and its prosocial effects in humans. *Harvard Review of Psychiatry*, 18, 1–21.

Manning, M., Kruszynski, M., Bankowski, K., et al. (1989). Solid-phase synthesis of 16 potent (selective and nonselective) *in vivo* antagonists of oxytocin. *Journal of Medicinal Chemistry*, 32, 382–391.

Manning, M., Miteva, K., Pancheva, S., et al. (1995). Design and synthesis of highly selective *in vitro* and *in vivo* uterine receptor antagonists of oxytocin: comparisons with Atosiban. *International Journal of Peptide and Protein Research*, 46, 244–252.

Manning, M., Stoev, S., Chini, B., et al. (2008). Peptide and non-peptide agonists and antagonists for the vasopressin and oxytocin V1a, V1b, V2 and OT receptors: research tools and potential therapeutic agents. *Progress in Brain Research*, 170, 473–512.

Mantella, R. C., Vollmer, R. R., Li, X., and Amico, J. A. (2003). Female oxytocin-deficient mice display enhanced anxiety-related behavior. *Endocrinology*, 144, 2291–2296.

McCarthy, M. M. (1990). Oxytocin inhibits infanticide in female house mice (*Mus domesticus*). *Hormones and Behavior*, 24, 365–375.

McCarthy, M. M., Bare, J. E., and vom Saal, F. S. V. (1986). Infanticide and parental behavior in wild female house mice: Effects of ovariectomy, adrenalectomy and administration of oxytocin and prostaglandin F2α. *Physiology and Behavior*, 36, 17–23.

McCarthy, M. M., McDonald, C. H., Brooks, P. J., and Goldman, D. (1996). An anxiolytic action of oxytocin is enhanced by estrogen in the mouse. *Physiology and Behavior*, 60, 1209–1215.

Meddle, S. L., Bishop, V. R., Gkoumassi, E., Van Leeuwen, F. W., and Douglas, A. J. (2007). Dynamic changes in oxytocin receptor expression and activation at parturition in the rat brain. *Endocrinology*, 148, 5095–5104.

Minakata, H. (2010). Oxytocin/vasopressin and gonadotropin-releasing hormone from cephalopods to vertebrates. *Annals of the New York Academy of Science*, 1200, 33–42.

Murata, Y., Li, M. Z., and Masuko, S. (2011). Developmental expression of oxytocin receptors in the neonatal medulla oblongata and pons. *Neuroscience Letters*, 502, 157–161.

Nephew, B. C., Bridges, R. S., Lovelock, D. F., and Byrnes, E. M. (2009). Enhanced maternal aggression and associated changes in neuropeptide gene expression in multiparous rats. *Behavioral Neuroscience*, 123, 949–957.

Neumann, I. D., Russell, J. A., Douglas, A. J., Windle, R. J., and Ingram, C. D. (2001). Alterations in behavioral and

neuroendocrine stress coping strategies in pregnant, parturient and lactating rats. In J. A. Russell, A. J. Douglas, R. J. Windle, and C. D. Ingram (eds.). *The Maternal Brain: Neurobiological and Neuroendocrine Adaptation and Disorders in Pregnancy and Post Partum.* Amsterdam: Elsevier, pp. 143–152.

Neumann, I. D., Torner, L., and Wigger, A. (2000). Brain oxytocin: differential inhibition of neuroendocrine stress responses and anxiety-related behaviour in virgin, pregnant and lactating rats. *Neuroscience*, 95, 567–575.

Nishimori, K., Young, L. J., Guo, Q., et al. (1996). Oxytocin is required for nursing but is not essential for parturition or reproductive behavior. *Proceedings of the National Academy of Sciences of the United States of America*, 93, 11699–11704.

Nomura, M., Mckenna, E., Korach, K. S., Pfaff, D. W., and Ogawa, S. (2002). Estrogen receptor-β regulates transcript levels for oxytocin and arginine vasopressin in the hypothalamic paraventricular nucleus of male mice. *Molecular Brain Research*, 109, 84–94.

Numan, M., Bress, J. A., Ranker, L. R., et al. (2010). The importance of the basolateral/basomedial amygdala for goal-directed maternal responses in postpartum rats. *Behavioral Brain Research*, 214, 368–376.

Numan, M. and Corodimas, K. P. (1985). The effects of paraventricular hypothalamic lesions on maternal behavior in rats. *Physiology and Behavior*, 35, 417–425.

Numan, M. and Insel, T. (2003). *The Neurobiology of Parental Behavior*, New York: Springer-Verlag.

Olazabal, D. E. and Ferreira, A. (1997). Maternal behavior in rats with kainic acid-induced lesions of the hypothalamic paraventricular nucleus. *Physiology and Behavior*, 61, 779–784.

Olazabal, D. E. and Young, L. J. (2006a). Oxytocin receptors in the nucleus accumbens facilitate "spontaneous" maternal behavior in adult female prairie voles. *Neuroscience*, 141, 559–568.

Olazabal, D. E. and Young, L. J. (2006b). Species and individual differences in juvenile female alloparental care are associated with oxytocin receptor density in the striatum and the lateral septum. *Hormones and Behavior*, 49, 681–687.

Orpen, B. G. and Fleming, A. S. (1987). Experience with pups sustains maternal responding in postpartum rats. *Physiology and Behavior*, 40, 47–54.

Pedersen, C. A. (1997). Oxytocin control of maternal behavior. Regulation by sex steroids and offspring stimuli. *Annals of the New York Academy of Science*, 807, 126–145.

Pedersen, C. A., Ascher, J. A., Monroe, Y. L., and Prange, A. J., Jr. (1982). Oxytocin induces maternal behavior in virgin female rats. *Science*, 216, 648–650.

Pedersen, C. A. and Boccia, M. L. (2002). Oxytocin maintains as well as initiates female sexual behavior: Effects of a highly selective oxytocin antagonist. *Hormones and Behavior*, 41, 170–177.

Pedersen, C. A. and Boccia, M. L. (2003). Oxytocin antagonism alters rat dams' oral grooming and upright posturing over pups. *Physiology and Behavior*, 80, 233–241.

Pedersen, C. A., Caldwell, J. D., Johnson, M. F., Fort, S. A., and Prange, A. J. (1985). Oxytocin antiserum delays onset of ovarian steroid-induced maternal behavior. *Neuropeptides*, 6, 175–182.

Pedersen, C. A., Caldwell, J. D., Walker, C., Ayers, G., and Mason, G. A. (1994). Oxytocin activates the postpartum onset of rat maternal behavior in the ventral tegmental and medial preoptic areas. *Behavioral Neuroscience*, 108, 1163–1171.

Pedersen, C. A., Johns, J. M., Musiol, I., et al. (1995). Interfering with somatosensory stimulation from pups sensitizes experienced, postpartum rat mothers to oxytocin antagonist inhibition of maternal behavior. *Behavioral Neuroscience*, 109, 980–990.

Pedersen, C. A. and Prange, A. J. (1979). Induction of maternal behavior in virgin rats after intracerebroventricular administration of oxytocin. *Proceedings of the National Academy of Sciences of the United States of America*, 76, 6661–6665.

Pedersen, C. A., Vadlamudi, S. V., Boccia, M. L., and Amico, J. A. (2006). Maternal behavior deficits in nulliparous oxytocin knockout mice. *Genes, Brain and Behavior*, 5, 274–281.

Peterson, G., Mason, G. A., Barakat, A. S., and Pedersen, C. A. (1991). Oxytocin selectively increases holding and licking of neonates in preweanling but not postweanling juvenile rats. *Behavioral Neuroscience*, 105, 470–477.

Pettibone, D. J. and Freidinger, R. M. (1997). Discovery and development of non-peptide antagonists of peptide hormone receptors. *Biochemical Society Transactions*, 25, 1051–1057.

Poindron, P., Rempel, N., Troyer, A., and Krehbiel, D. (1989). Genital stimulation facilitates maternal behavior in estrous ewes. *Hormones and Behavior*, 23, 305–316.

Qin, J., Feng, M., Wang, C., et al. (2009). Oxytocin receptor expressed on the smooth muscle mediates the excitatory effect of oxytocin on gastric motility in rats. *Neurogastroenterology and Motility*, 21, 430–438.

Ragnauth, A. K., Devidze, N., Moy, V., et al. (2005). Female oxytocin gene-knockout mice, in a semi-natural environment, display exaggerated aggressive behavior. *Genes, Brain and Behavior*, 4, 229–239.

Ring, R. H., Malberg, J. E., Potestio, L., et al. (2006). Anxiolytic-like activity of oxytocin in male mice: behavioral and autonomic evidence, therapeutic implications. *Psychopharmacology (Berlin)*, 185, 218–225.

Rodrigues, S. M., Saslow, L. R., Garcia, N., John, O. P., and Keltner, D. (2009). Oxytocin receptor genetic variation relates to empathy and stress reactivity in humans. *Proceedings of the National Academy of Sciences of the United States of America*, 106, 21437–21441.

Rubin, B. S., Menniti, F. S., and Bridges, R. S. (1983). Intracerebroventricular administration of oxytocin and maternal behavior in rats after prolonged and acute steroid pretreatment. *Hormones and Behavior*, 17, 45–53.

Rutherford, H. J., Williams, S. K., Moy, S., Mayes, L. C., and Johns, J. M. (2011). Disruption of maternal parenting circuitry by addictive process: rewiring of reward and stress systems. *Frontiers in Psychiatry*, 2 (Article 37), 1–17.

Saller, S., Kunz, L., Dissen, G. A., et al. (2010). Oxytocin receptors in the primate ovary: molecular identity and link to apoptosis in human granulosa cells. *Human Reproduction*, 25, 969–976.

Sawchenko, P. E. and Swanson, L. W. (1985). Relationship of oxytocin pathways to the control of neuroendocrine and autonomic function. In Amico, J. A. and Robinson, A. G. (eds.) *Oxytocin: Clinical and Laboratory Studies*. Amsterdam: Elsevier Science Publishers B.V., pp. 87–103.

Shahrokh, D. K., Zhang, T. Y., Diorio, J., Gratton, A., and Meaney, M. J. (2010). Oxytocin-dopamine interactions mediate variations in maternal behavior in the rat. *Endocrinology*, 151, 2276–2286.

Shughrue, P. J., Dellovade, T. L., and Merchenthaler, I. (2002). Estrogen modulates oxytocin gene expression in regions of the rat supraoptic and paraventricular nuclei that contain estrogen receptor-β. In Poulain, D., Oliet, S., and Theodosis, D. (eds.) *Vasopressin and Oxytocin: From Genes to Clinical Applications*. Amsterdam: Elsevier Science Publishers BV, pp. 15–29.

Sizoo, B., van den Brink, W., Gorissen van Eenige, M., and van der Gaag, R. J. (2009). Personality characteristics of adults with autism spectrum disorders or attention deficit hyperactivity disorder with and without substance use disorders. *Journal of Nervous and Mental Disorders*, 197, 450–454.

Sofroniew, M. V. (1980). Projections from vasopressin, oxytocin, and neurophysin neurons to neural targets in the rat and human. *Journal of Histochemistry and Cytochemistry*, 28, 475–478.

Stern, J. M., Goldman, L., and Levine, S. (1973). Pituitary-adrenal responsiveness during lactation in rats. *Neuroendocrinology*, 12, 179–191.

Strathearn, L., Fonagy, P., Amico, J., and Montague, P. R. (2009). Adult attachment predicts maternal brain and oxytocin response to infant cues. *Neuropsychopharmacology*, 34, 2655–2666.

Swain, J. E., Lorberbaum, J. P., Kose, S., and Strathearn, L. (2007). Brain basis of early parent-infant interactions: psychology, physiology, and in vivo functional neuroimaging studies. *Journal of Child Psychology and Psychiatry*, 48, 262–287.

Takagi, T., Tanizawa, O., Otsuki, Y., et al. (1985). Oxytocin in the cerebrospinal-fluid and plasma of pregnant and nonpregnant subjects. *Hormone and Metabolic Research*, 17, 308–310.

Takayanagi, Y., Yoshida, M., Bielsky, I. F., et al. (2005). Pervasive social deficits, but normal parturition, in oxytocin receptor-deficient mice. *Proceedings of the National Academy of Sciences of the United States of America*, 102, 16096–16101.

Takeda, S., Kuwabara, Y., and Mizuno, M. (1985). Effects of pregnancy and labor on oxytocin levels in human plasma and cerebrospinal-fluid. *Endocrinologica Japonica*, 32, 875–880.

Takemura, M., Kimura, T., Nomura, S., et al. (1994). Expression and localization of human oxytocin receptor messenger-RNA and its protein in chorion and decidua during parturition. *Journal of Clinical Investigation*, 93, 2319–2323.

Terenzi, M. G. and Ingram, C. D. (2005). Oxytocin-induced excitation of neurones in the rat central and medial amygdaloid nuclei. *Neuroscience*, 134, 345–354.

Toloczko, D. M., Young, L., and Insel, T. R. (1997). Are there oxytocin receptors in the primate brain? *Annals of the New York Academy of Sciences*, 807, 506–509.

Tomizawa, K., Iga, N., Lu, Y. F., et al. (2003). Oxytocin improves long-lasting spatial memory during motherhood through MAP kinase cascade. *Nature Neuroscience*, 6, 384–390.

Tost, H., Kolachana, B., Hakimi, S., et al. (2010). A common allele in the oxytocin receptor gene (OXTR) impacts prosocial temperament and human hypothalamic-limbic structure and function. *Proceedings of the*

National Academy of Sciences of the United States of America, 107, 13936–13941.

Tronick, E. Z., Messinger, D. S., Weinberg, M. K., et al. (2005). Cocaine exposure is associated with subtle compromises of infants' and mothers' social-emotional behavior and dyadic features of their interaction in the face-to-face still-face paradigm. *Developmental Psychology*, 41, 711–722.

van IJzendoorn, M. H. (1995). Adult attachment representations, parental responsiveness, and infant attachment: a meta-analysis on the predictive validity of the Adult Attachment Interview. *Psychology Bulletin*, 117, 387–403.

van Leengoed, E., Kerker, E., and Swanson, H. H. (1987). Inhibition of postpartum maternal behavior in the rat by injecting an oxytocin antagonist into the cerebral ventricles. *Journal of Endocrinology*, 112, 275–282.

Veinante, P. and Freund-Mercier, M. J. (1997). Distribution of oxytocin- and vasopressin-binding sites in the rat extended amygdala: a histoautoradiographic study. *Journal of Comparative Neurology*, 383, 305–325.

Wakasa, T., Wakasa, K., Nakayama, M., et al. (2009). Change in morphology and oxytocin receptor expression in the uterine blood vessels during the involution process. *Gynecologic and Obstetric Investigation*, 67, 137–144.

Wamboldt, M. Z. and Insel, T. R. (1987). The ability of oxytocin to induce short latency maternal behavior is dependent on peripheral anosmia. *Behavioral Neuroscience*, 101, 439–441.

Welch, M. G., Tamir, H., Gross, K. J., et al. (2009). Expression and developmental regulation of oxytocin (OT) and oxytocin receptors (OTR) in the enteric nervous system (ENS) and intestinal epithelium. *Journal of Comparative Neurology*, 512, 256–270.

Windle, R. J., Shanks, N., Lightman, S. L., and Ingram, C. D. (1997). Central oxytocin administration reduces stress-induced corticosterone release and anxiety behavior in rats. *Endocrinology*, 138, 2829–2834.

Windle, R. J., Wood, S. A., Kershaw, Y. M., Lightman, S. L., and Ingram, C. D. (2010). Reduced stress responsiveness in pregnancy: relationship with pattern of forebrain c-fos mRNA expression. *Brain Research*, 1358, 102–109.

Wu, S., Jia, M., Ruan, Y., et al. (2005). Positive association of the oxytocin receptor gene (OXTR) with autism in the Chinese Han population. *Biological Psychiatry*, 58, 74–77.

Yoshida, M., Takayanagi, Y., Inoue, K., et al. (2009). Evidence that oxytocin exerts anxiolytic effects via oxytocin receptor expressed in serotonergic neurons in mice. *Journal of Neuroscience*, 29, 2259–2271.

Young, W. S. and Gainer, H. (2009). Vasopressin/oxytocin and receptors. In L. Squire (ed.) *Encyclopedia of Neuroscience*. Oxford: Academic Press, pp. 51–59.

Young, K. A., Liu, Y., and Wang, Z. (2008). The neurobiology of social attachment: A comparative approach to behavioral, neuroanatomical, and neurochemical studies. *Comparative Biochemistry and Physiology, C, Comparative Toxicology and Pharmacology*, 148, 401–410.

Yu, G. Z., Kaba, H., Okutani, F., Takahashi, S., and Higuchi, T. (1996a). The olfactory bulb: A critical site of action for oxytocin in the induction of maternal behaviour in the rat. *Neuroscience*, 72, 1083–1088.

Yu, G. Z., Kaba, H., Okutani, F., et al. (1996b). The action of oxytocin originating in the hypothalamic paraventricular nucleus on mitral and granule cells in the rat main olfactory bulb. *Neuroscience*, 72, 1073–1082.

Oxytocin regulation of sheep social and maternal behavior

Keith M. Kendrick

10.1 Introduction

A role for release of the neuropeptide oxytocin in the brain in facilitating the formation of affiliative bonds as well as maternal behavior was demonstrated for the first time in sheep 25 years ago (Kendrick et al., 1986; 1987) and as a model it has contributed much to our understanding of how this important prosocial peptide influences both brain and behavior. In this chapter, I will review the large number of behavioral, neuroanatomical and *in vivo* neurochemical experiments carried out in this species that have revealed some of the key aspects of how oxytocin acts within the brain to control both social and maternal behaviors.

10.1.1 Maternal behavior and mother offspring bonds in sheep

Sheep display a remarkable behavioral change immediately they give birth, exhibiting not only a range of maternal behaviors directed at caring for and communicating with their offspring, but also forming strong and exclusive attachment bonds with each of them. Although highly experienced multiparous ewes will sometimes show maternal behavior toward, and even try to steal lambs from other mothers, most pregnant ewes find lambs aversive prior to giving birth and will avoid them and even exhibit aggression toward them in the form of high-pitched protest bleats, front foot-stamping and head butts. However, immediately after they give birth they lick the lamb avidly and encourage it to get up and suckle using a low-pitched rumbling bleat, only used by maternal sheep communicating with their lambs. They also repeatedly nudge their lambs with their noses or prod it with their front feet to encourage them to stand up and suckle them. To help the lambs locate their teats they stand immobile and often push them in the right direction using their noses. During the first few hours of these intensive interactions between a maternal sheep and her lamb(s) she also learns to recognize each lamb's odor signature leading to the formation of an exclusive bond with them such that she will reject, often quite violently, any attempts by other lambs to suckle her. In this way the mother's own lambs have virtually exclusive access to her milk supply and care. The exclusivity of the bond between the mother and her lambs is also evidenced by the fact that she becomes far less concerned about seeking refuge in the flock, preferring to stay close to her lambs. She also becomes more aggressive toward intruders at this time and if she is physically separated from her lambs she will repeatedly vocalise using high-pitched protest bleats until reunited with them.

While there are a range of physiological changes occurring during pregnancy and the postpartum period which contribute to the spectrum of

Oxytocin, Vasopressin, and Related Peptides in the Regulation of Behavior, ed. E. Choleris, D. W. Pfaff, and M. Kavaliers.
Published by Cambridge University Press.

behavioral changes that occur in ewes after they give birth, it has become clear that release of oxytocin within the brain during birth and suckling contributes in an important way to most, if not all of them. The key physiological event for promoting both maternal behavior and bonding is feedback to the brain from vaginal and cervical stimulation occurring during the final stages of labor and particularly that during the expulsion of the lamb at birth. Maternal behavior toward and bonding with lambs can therefore be elicited in non-pregnant ewes treated with estrogen and progesterone simply by a few minutes of manual stimulation of the vagina and cervix (Kendrick and Keverne, 1991; Kendrick et al., 1991b; 1992a) and pregnant ewes delivered by caesarian section, or following epidural anesthesia, show a delayed onset of maternal behavior, particularly in primiparous animals (Krehbiel et al., 1987; Lévy et al., 1992b).

10.2 Cerebrospinal fluid oxytocin concentrations and effects of intracerebroventricular infusions on maternal behavior, offspring recognition, and bonding

Measurements of altered oxytocin concentrations in cerebrospinal fluid (CSF) during parturition or following artificial vaginal and cervical stimulation, together with subsequent investigations of the behavioral effects of intracerebroventricular ICV) infusions of the peptide on maternal behavior and bonding with lambs have provided important information about its functional roles (Kendrick, 2000).

10.2.1 Oxytocin concentrations in CSF and blood during parturition and suckling

Oxytocin concentrations increase markedly in both the blood and CSF of ewes during the latter stages of labor and reach their peak at birth. They are also increased to a much lesser extent during suckling (Kendrick et al., 1986; 1991a). Two important observations from these initial findings were that first oxytocin concentrations in the CSF around the time of birth are similar to or even higher than those in blood, and secondly that they remained elevated for a much longer duration (in CSF for up to 2 h compared with 15 min in plasma Kendrick et al., 1991a). The same pattern was seen in estrogen-primed ewes following a 5-min period of mechanical stimulation of the vagina and cervix, although in this case levels of evoked oxytocin release were considerably less than during labor and birth and only remained elevated in CSF for 20 min following stimulation, whereas plasma levels were only increased during actual stimulation. These findings both suggest that the half-life of oxytocin in CSF is considerably longer than in blood and also raise the possibility of a paracrine action for the peptide in the central nervous system via its circulation in the cerebroventricular system.

In general, the period of elevated oxytocin concentrations in the CSF of sheep following birth is similar to that where postpartum mothers will show maternal and bonding responses toward their lambs. In cases where ewes undergo difficult births and cannot interact with their lambs, or are separated from them for several hours after giving birth, then they are less likely to exhibit maternal responses towards or bonding with them.

Where pregnant ewes are delivered with a peridural anesthetic block in place then this has been shown to prevent significant increases in oxytocin concentrations as well as impairing maternal behavior induction (Lévy et al., 1992b).

10.2.2 Oxytocin and the blood–brain barrier

The sheep blood–brain barrier is relatively impermeable to oxytocin. Based on measurements of CSF concentrations of the peptide following a single intravenous injection we estimated that only around 0.01% crossed from the periphery into the brain ventricular compartment (Kendrick et al., 1986). This effectively ruled out any explanation of high concentrations of oxytocin in the CSF being derived from those in the peripheral circulation and made it unlikely that any centrally mediated

behavioral effects of the peptide could be promoted by peripheral injection.

10.2.3 Behavioral effects of ICV oxytocin treatment

Intracerebroventricular, but not intravenous, infusions of oxytocin were found to provoke a remarkably rapid (30 s – 1 min) induction of full maternal responses in non-pregnant multiparous sheep and also facilitate the subsequent olfactory recognition and bonding process with lambs (Kendrick et al., 1986; 1987; Kendrick, 2000; Keverne and Kendrick, 1992). Significant effects were seen with doses between 5 and 50 μg although clear dose–response relationships were not always found (Kendrick et al., 1987). Interestingly, ICV infusions of only the ring structure of oxytocin (tocinoic acid – the first six amino acids of oxytocin) were as effective as giving the whole peptide, suggesting that the ring structure is sufficient to activate brain oxytocin receptors in sheep (Keverne and Kendrick, 1992). The intensity of maternal responses following oxytocin or tocinoic acid infusions was equivalent, or even superior, to those induced by 5 min of vaginocervical stimulation. Furthermore, ICV oxytocin given in conjunction with vaginocervical stimulation did not facilitate maternal responses more than either treatment did alone (Keverne and Kendrick, 1990; 1992), suggesting that the effects of vaginocervical stimulation are likely to be primarily mediated by brain oxytocin release. At a dose of 1 μg only a third of animals showed evidence of maternal behavior induction and so this is probably around threshold.

Importantly, for the effects of ICV oxytocin on maternal behavior and bonding to occur the animals had to be primed first with sex steroids. In a first study using gonadally intact multiparous ewes it was found that estradiol-priming for 72 h was required (since the experiments used seasonal sheep outside of their breeding season their endogenous levels of estrogen would have been very low). In a subsequent study using ovariectomized ewes this requirement for estradiol priming was also found for the effects of both vaginocervical stimulation and ICV oxytocin (Kendrick and Keverne, 1991; Keverne and Kendrick, 1992).

The effects of ICV oxytocin on maternal behavior induction were found to be of relatively short duration, with treated ewes no longer showing maternal responses to lambs after an hour or so. This is approximately equivalent to the normal duration of elevated oxytocin concentrations in the CSF following birth and potentially explains why maternal responsiveness in postpartum ewes decreases markedly an hour or two after they give birth if they do not have a lamb to interact with. There is therefore a reasonable correspondence between the presence of high concentrations of oxytocin in the CSF and a ewe's potential to show maternal responses to any lambs it is exposed to.

Other behavioral effects of ICV oxytocin infusions have also been found in multiparous sheep. In the original study for example an increased amount of time spent eating was observed with higher doses (Kendrick et al., 1987). However, in a field-based preference test it was shown that following ICV oxytocin treatment ewes preferred to stay in close proximity with a lamb in a cage than to have access to food in another location. With control treatments ewes always preferred to access the food and avoided the lamb almost completely.

While no systematic study has been performed to assess the effects of ICV oxytocin on the behavior of male sheep, preliminary observations on four estrogen-treated castrated males failed to find any induction of maternal behavior or interest in lambs (Kendrick, unpublished observations).

10.2.4 Effects of maternal experience on responsiveness to ICV oxytocin treatment

In contrast to ovariectomized multiparous ewes, nulliparous ones receiving 3 days of estradiol priming only responded to either ICV oxytocin or vaginocervical stimulation by showing reduced aggression toward lambs and withdrawal from them (Keverne and Kendrick, 1991; Kendrick et al., 1991b). Since similar elevated levels of oxytocin

were found in the CSF of nulliparous and multiparous ewes after vaginocervical stimulation this suggested that either maternal experience was required for oxytocin to produce its full functional effects or that some aspect of the hormonal priming regime was suboptimal in nulliparous animals. Other studies suggest that it is the hormonal priming regime which is of more importance in this respect. Postpartum nulliparous (i.e., animals that have never given birth) ewes that are prevented from having a CSF increase in oxytocin or displaying maternal responses by a peridural anesthetic block can be induced to show full maternal responses to lambs by ICV oxytocin treatment. These animals would have had long-term priming with both increased progesterone and estrogen levels during pregnancy, resulting in full lactation. Long-term exogenous treatment of non-pregnant nulliparous ewes with both progesterone and estradiol to induce full lactation also results in full maternal behavior induction by either ICV oxytocin or vaginocervical stimulation (Kendrick et al., 1992a; Kendrick 2000). Thus, it would appear that one of the effects of maternal experience in sheep is to reduce the level of estrogen and progesterone priming required for oxytocin to evoke full maternal responses and subsequent bonding. These findings do, however, also suggest that there are different thresholds, and potentially mechanisms, for oxytocin's action in reducing aggression toward and withdrawal from lambs as opposed to facilitating positive maternal responses. In this respect it is interesting that the strength of some postpartum positive maternal behavior components are positively correlated with the ratio of estrogen to progesterone in sheep but not negative behaviors such as aggression (Dwyer, 2008).

10.2.5 Effects of ICV oxytocin agonists and antagonists, vasopressin, and other peptides

The oxytocin agonist Thr^4Gly^7 oxytocin has been found to stimulate maternal behavior and bonding with lambs with a similar potency to oxytocin itself following ICV administration. On the other hand, ICV infusions of the closely related peptide, arginine-vasopressin, were ineffective at doses of up to 50 μg (Kendrick, 2000). Unfortunately, as in monkeys, none of the oxytocin receptor antagonist ligands used to block behavioral and physiological actions of oxytocin in rodents have proved to be successful in doing so in sheep. They fail to demonstrate receptor binding in the sheep brain although they do bind peripherally in the kidney (Lévy et al., 1992a), raising the possibility that there may be some differences between central and peripheral oxytocin receptors. However, the cyclic hexapeptide receptor antagonist L-366,670 does partially block the effects of ICV oxytocin in inducing maternal behavior and bonding in estradiol-primed non-pregnant ewes (Kendrick, 2000).

There is evidence in sheep that endogenous opioids and corticotrophin releasing factor may have a modulatory role in mediating both central oxytocin release and maternal behavior and bonding. Both ICV morphine and corticotrophin releasing factor facilitate the release of oxytocin and induction of maternal behavior by vaginocervical stimulation (Keverne and Kendrick, 1990). Corresponding treatment with the mu opioid antagonist, naltrexone, reduces both oxytocin release and maternal behavior stimulation by vaginocervical stimulation (Kendrick and Keverne, 1989) and also interferes with the postpartum induction of maternal behavior and bonding (Caba et al., 1995). Expression of both preproenkephalin (Broad et al., 1993b) and corticotrophin factor (Broad et al., 1995) mRNA are also significantly increased in the sheep paraventricular nucleus at birth.

10.3 Localization of oxytocin and its receptors in the brain

Several studies have mapped the brain distribution of oxytocin and its receptor in the sheep brain and investigated effects of maternal experience or sex hormone priming (Broad et al., 1993a; 1999). There is an extensive overlap between these

regions and those that show increased immediate early gene expression (as a marker of neural activation) during birth, or following vaginocervical stimulation (da Costa et a., 1997; Kendrick et al., 1997a).

10.3.1 Distribution of oxytocin cells, fibers, and terminals in the sheep brain

Both imunocytochemistry and *in situ* hybridization histochemistry approaches have been used to map the distribution of cells and processes exhibiting oxytocin immunoreactivity or mRNA expression in the sheep brain (Broad et al., 1993a). As in other species, oxytocin cells are found primarily in the supraoptic and paraventricular nuclei of the hypothalamus although they are also present in the medial preoptic area, mediobasal hypothalamus, periventricular complex and bed nucleus of the stria terminalis. Oxytocin fibers and terminals can also be seen in these structures. Oxytocin fibers and/or terminals are also present in the olfactory bulb, piriform cortex, diagonal band, lateral septum, amygdala, nucleus accumbens, medial frontal, and anterior cingulate cortices.

10.3.2 Distribution of cells and processes in the sheep brain expressing an oxytocin receptor

The distribution of cells containing oxytocin receptors in the sheep brain has been demonstrated using immunohistochemical, digoxygenin riboprobe and *in situ* hybridization approaches (Broad et al., 1999). Cell bodies and processes expressing the oxytocin receptor are widely distributed in olfactory processing regions, limbic system, diencephalon, striatum, and brainstem and while original studies mapping the distribution of oxytocin itself were not as extensive it seems reasonable to conclude at this point that there are a number of brain regions that have oxytocin receptors but that do not appear to have cells, fibers, or terminals containing the peptide. This lends further support to a key paracrine role for oxytocin's effects on behavior via its circulation in the cerebroventricular system.

10.3.3 Effects of hormone priming, birth, and maternal experience on oxytocin and oxytocin receptor expression

Expression levels of mRNA for both oxytocin and its receptor are increased in the paraventricular nucleus during late pregnancy and reach peak levels at birth (Broad et al., 1993a; Broad et al., 1999; Kendrick and Keverne, 1992). While mRNA expression can be increased by estrogen and progesterone treatment, however, this does not reach levels seen during late pregnancy and birth and so it is likely that other factors such as glucocorticoids or thyroid hormone may play a role as well.

Primiparous ewes are known to have more problems that multiparous ones in developing maternal responses after giving birth and this may partly be due to the fact that upregulation of oxytocin receptor mRNA in the paraventricular nucleus is significantly less that that found in multiparous animals (Broad et al., 1999). Whether this is due to maternal experience inducing permanent changes in the morphology of the oxytocin neurons, as has been shown in rats (Hatton et al., 1992), is currently unknown.

10.3.4 Overlap between brain regions showing increased immediate early gene expression and expressing oxytocin receptors

Both birth and vaginocervical stimulation produce increased c-fos mRNA expression in most of the brain regions expressing oxytocin receptors (da Costa et al., 1997; Kendrick et al., 1997a; Kendrick 2000). Indeed, the only brain regions found with increased c-fos expression with no evidence for the presence of oxytocin receptors were the mediodorsal thalamus and the habenula. There is therefore a very strong concordance between brain regions activated by stimuli promoting maternal behavior and subsequent bonding, and those containing oxytocin receptors.

Reducing expression of the immediate early genes c-*fos* and c-*jun* in the paraventricular nucleus using infusions of antisense constructs has also been found to reduce the increases in oxytocin mRNA, and oxytocin release, normally observed in this region at birth and to slightly reduce one positive component of the maternal response, low-pitched bleats (da Costa et al., 1999). The oxytocin gene has so far not been reported to have an AP-1 binding site so it seems likely that c-fos and c-jun expression changes at birth are increasing those of oxytocin indirectly.

10.4 Release of oxytocin in specific regions of the brain during birth and suckling

A number of *in vivo* microdialysis sampling studies have been carried out to measure release of oxytocin in specific brain regions of sheep, during birth, vaginocervical stimulation and suckling (da Costa et al., 1999; Kendrick et al., 1988a; b; Kendrick and Keverne, 1992; Kendrick et al., 1992b, 1993; Lévy et al., 1995). Studies have also investigated neuromodulatory and behavioral effects of localized, bilateral retrodialysis infusions of the peptide.

10.4.1 Oxytocin release in specific brain regions

In vivo microdialysis sampling experiments have shown that oxytocin is released during birth in the hypothalamic paraventricular nucleus, medial preoptic area, bed nucleus of the stria terminalis and olfactory bulb (da Costa et al., 1996; 1999; Kendrick et al., 1988a; Kendrick et al., 1988b; Kendrick, 1991; Kendrick et al., 1992b; Lévy et al., 1995). In general birth-induced oxytocin release within these specific brain regions is of a similar duration (1–2 h) to that found in CSF, although concentrations immediately before and during birth are considerably higher than at other times (see da Costa et al., 1996).

Oxytocin release has also been demonstrated in the mediobasal hypothalamus of estrus ewes during mating and when they are shown face pictures of male sheep (Kendrick et al., 1993; Fabre-Nys et al., 1997).

10.4.2 Behavioral effects of oxytocin infusions in different brain regions

Bilateral retrodialysis infusions of oxytocin, or its ring structure, tocinoic acid, into the hypothalamic paraventricular nucleus of estrogen primed, ovariectomized multiparous ewes stimulates full maternal behavior responses and bonding (da Costa et al., 1996). The amount of oxytocin required to produce this effect is 150-fold lower than the minimum dose required to stimulate maternal behavior following ICV infusion so it is very unlikely that behavioral effects seen following paraventricular nucleus infusions are due to leakage into the cerebroventricular system. Retrodialysis infusions of high concentrations of vasopressin into the paraventricular nucleus did stimulate maternal behavior in 25% of the animals although given the high dose required it seems likely that this was due to its known weak agonist effects on the oxytocin receptor.

Similar retrodialysis infusions of oxytocin into other brain regions containing oxytocin terminals and receptors only influenced some specific components of maternal responses. Bilateral infusions in both the medial preoptic area and the olfactory bulbs reduced normal levels of high aggression by non-pregnant ewes toward lambs but did not stimulate any positive maternal behaviors (Kendrick et al., 1997a). This suggests that the olfactory projection from the olfactory bulb to the medial preoptic area via the corticomedial amygdala may be an important one for oxytocin's action in reducing the normal aversion ewes have toward lamb odors. It seems likely that this same pathway is also important in mediating formation of the olfactory recognition memory maternal ewes develop for their lambs since the olfactory bulb (Kendrick et al., 1992c; 1997a) and corticomedial amygdala (Keller et al., 2004) are important for this. While the effects of oxytocin in these regions on facilitating social recognition memory have not been directly

investigated in sheep they have been shown to be important in rodents (Dluzen et al., 2000; Larrazolo-López et al., 2008; Sanchez-Andrade and Kendrick 2009).

Retrodialysis infusions into the mediobasal hypothalamus have been shown to inhibit sexual receptivity in estrus sheep (Kendrick et al., 1993, which may also partly help to explain why, despite having similar hormone priming with estrogen and progesterone in late pregnancy, as in estrus, no sexual receptivity is exhibited toward males.

10.4.3 Neuromodulatory effects of oxytocin infusions in different brain regions

The microdialysis sampling and retrodialysis experiments in sheep were the first to reveal that oxytocin has *in vivo* modulatory actions on its own release as well as on a number of classical neurotransmitter systems, although it had been known to facilitate its own release via autoreceptors on magnocellular neurons *in vitro* for some time (Moos et al., 1984). In sheep retrodialysis experiments in the paraventricular nucleus showed that oxytocin infusions potently facilitate oxytocin release into the blood (da Costa et al., 1996), and presumably within the brain as well. Of the classical transmitters noradrenaline is the most widely modulated by oxytocin, with increased release being seen in the sheep olfactory bulb, medial preoptic area, mediobasal hypothalamus and paraventricular nucleus (Kendrick et al., 1997a; Kendrick et al., 1997b; Keverne et al., 1993). Acetylcholine release is also facilitated by oxytocin in the olfactory bulb, dopamine in the paraventricular nucleus and GABA in the medial preoptic area and olfactory bulb (Kendrick, 1991; Kendrick et al., 1992a; 1993; Lévy et al., 1995; da Costa et al., 1996). These neuromodulatory effects of oxytocin, or tocinoic acid, on classical transmitter release tended to occur within 15 min of the infusion and lasted for its duration (up to 1 h in some cases). While vasopressin tended to have a similar neuromodulatory profile its effects were of a shorter duration (generally less than 30 min (da Costa et al., 1996)). In the paraventricular nucleus oxytocin was also found to reduce aspartate and glutamate release 45–60 min after the start of retrodialysis infusions (da Costa et al., 1996). Since in most cases the neuromodulatory effects on amine release are in regions containing only terminals rather than cell bodies it seems likely that this primarily reflects modulation via presynaptic receptors. With GABA, aspartate, and glutamate, on the other hand, it could be either pre- or postsynaptic.

10.4.4 Are the behavioral effects of oxytocin dependent upon its neuromodulatory actions?

Unfortunately, no experiments have yet investigated directly in sheep the extent to which oxytocin's effects on maternal behavior and bonding are dependent upon it neuromodulatory actions. Work has shown that blockade of β-noradrenergic receptors in the olfactory bulb interferes with postpartum bonding behavior (Lévy et al., 1990), and since in rodents it is clear that oxytocin's promotion of olfactory-based social recognition memory is also dependent upon noradrenaline acting on noradrenergic receptors (Dluzen et al., 2000), it seems likely that this is also true for sheep. In sheep olfactory recognition is also impaired following treatment with the muscarinic antagonist scopolamine (Lévy et al., 1997) and so oxytocin's modulation of olfactory bulb acetylcholine may also be of functional importance.

Given the fact that in female voles oxytocin's effects in promoting partner bonds is dependent upon dopamine release acting on D2 receptors (Liu and Wang, 2003), once again it is probable that this is also the case in sheep, although it has yet to be proven.

At this stage the functional role for oxytocin's neuromodulatory actions on amino acid transmitters is unknown, although in terms of increased GABA release this might relate to inhibitory effects on locomotor behavior in particular as lambs suckle ewes. The inhibition of excitatory amino acid transmitters in the paraventricular nucleus occurs at the same time when oxytocin release following birth returns to baseline, and interestingly we have found

that oxytocin infusions into the PVN at this time have a much reduced modulatory effect on other classical transmitters. They may therefore reflect the effects of oxytocin receptor desensitization for example and be part of the necessary mechanism for breaking down the positive feedback cycle whereby oxytocin facilitates its own release.

10.5 How does oxytocin act within the brain to promote maternal, bonding, and recognition behaviors?

A potential model for the actions of oxytocin on socio-sexual behaviors in sheep has been described in several reviews (Kendrick et al., 1997; Kendrick 2000). This proposes that stimuli from the vagina and cervix during birth, or during suckling, promotes activation of brainstem norardrenergic and serotonergic pathways to facilitate activation of the paraventricular nucleus and both central and peripheral oxytocin release. Additional modulatory effects on paraventricular nucleus neurons are exerted via steroid hormones, corticotrophin releasing factor, and opioid systems and via dopaminergic inputs from the ventral tegmental area. Oxytocin acting to facilitate its own release via autoreceptors then ensures that high levels of release continue for an hour or so after birth. The activation of paraventricular nucleus oxytocin neurons results in a coordinated pattern of peptide release in a range of terminal regions via both direct projection pathways and via a paracrine action through transport in the cerebrospinal fluid. Oxytocin acts via its receptors in these different regions to modulate classical transmitter release, and it is the latter that drive the induction of maternal, social recognition, and bonding behaviors.

Although to date no experiments have investigated the importance of oxytocin for maintaining maternal and bonding behaviors in sheep it seems likely that, as in rodents (Insel and Harbaugh, 1989), the paraventricular nucleus is not essential for maintaining maternal behavior following its induction. At this stage the most likely scenario is that the function of the large and rapid coordinated release of oxytocin in the sheep brain at birth serves to kick-start maternal behavior and bonding via a facilitation of classical transmitter release in those regions controlling these behaviors. Thereafter the behaviors are maintained by the presence of offspring continuing to activate classical transmitter system although clearly oxytocin released through suckling can continue to facilitate this process but in a much more minor way.

REFERENCES

Broad, K. D., Kendrick, K. M., Sirinathsinghji, D. J. S., and Keverne E. B. (1993a). Changes in oxytocin immunoreactivity and mRNA expression in the sheep brain during pregnancy, parturition and lactation and in response to oestrogen and progesterone. *Journal of Neuroendocrinology*, 5, 435–444.

Broad, K. D., Kendrick, K. M., Sirinathsinghji, D. J. S., and Keverne E. B. (1993b). Changes in pro-opiomelanocortin and pre-proenkephalin mRNA levels in the ovine brain during pregnancy, paturition and lactation and in response to oestrogen and progesterone. *Journal of Neuroendocrinology*, 5, 711–719.

Broad, K. D., Keverne, E. B., and Kendrick, K. M. (1995). Corticotrophin releasing factor mRNA expression in the sheep brain during pregnancy, parturition and lactation and following exogenous progesterone and oestrogen treatment. *Molecular Brain Research*, 29, 310–316.

Broad, K. D., Lévy, F., Evans, G., Kimura, T., Keverne, E. B., and Kendrick, K. M. (1999). Previous maternal experience potentiates the effect of parturition on oxytocin receptor mRNA expression in the paraventricular nucleus. *European Journal of Neuroscience*, 11, 3725–3737.

Caba, M., Poindron, P., Krehbiel, D., Lévy, F., Romeyer, A., and Venier, G. (1995). Naltrexone delays the onset of maternal behavior in primiparous parturient ewes. *Pharmacology, Biochemistry and Behavior*, 52, 743–748.

Da Costa, A. P. C., Broad, K. D., and Kendrick, K. M. (1997). Olfactory memory and maternal behaviour-induced changes in c-*fos* and *zif*/268 mRNA expression in the sheep brain. *Molecular Brain Research*, 46, 63–76.

Da Costa, A. P. C., De La Riva, C., Guevara-Guzman, R., and Kendrick, K. M. (1999). C-Fos and c-jun in the paraventricular nucleus play a role in regulating peptide gene

expression, oxytocin release and maternal behaviour. *European Journal of Neuroscience*, 11, 2199–2210.

Da Costa, A. P. C., Guevara-Guzman, R., Ohkura, S., Goode, J. A., and Kendrick, K. M. (1996). The role of oxytocin release in the paraventricular nucleus in the control of maternal behaviour in the sheep. *Journal of Neuroendocrinology*, 8, 163–177.

Dluzen, D. E., Muraoka, S., Engelmann, M., Ebner, K., and Landgraf, R. (2000). Oxytocin induces preservation of social recognition in male rats by activating a-adrenoceptors of the olfactory bulb. *European Journal of Neuroscience*, 12, 760–766.

Dwyer, C. M. (2008). Individual variation in the expression of maternal behaviour: a review of the neuroendocrine mechanisms in the sheep. *Journal of Neuroendocrinology*, 20, 526–534.

Fabre-Nys, C., Ohkura, S., and Kendrick, K. M. (1997). Male faces and odors evoke differential patterns of neurochemical release in the mediobasal hypothalamus of female sheep during estrus. *European Journal of Neuroscience*, 9, 1666–1677.

Hatton, G. I., Modney, B. K., and Salm, A. K. (1992). Increases in dendritic bundling and dye coupling of supraoptic neurons after the induction of maternal behavior. *Annals of the New York Academy of Sciences*, 652, 142–155.

Insel, T. R. and Harbaugh, C. R. (1989). Lesions of the hypothalamic paraventricular nucleus disrupt the initiation of maternal behavior. *Behavioral Neuroscience*, 45, 1033–1041.

Keller, M., Perrin, G., Meurisse, M., Ferreira, G., and Lévy, F. (2004). Cortical and medial amygdale are both involved in the formation of olfactory offspring memory in sheep. *European Journal of Neuroscience*, 20, 3433–3441.

Kendrick, K. M. (1991). Interactions between peptides and conventional neurotransmitter systems in the brain. In H. Rollema, B. Westerink, and W. J. Drijfhout (eds.) *Monitoring Molecules in Neuroscience* University Centre for Pharmacy, Groningen, pp. 111–114.

Kendrick, K. M. (2000). Oxytocin, motherhood and bonding. *Experimental Physiology*, 85S, 111S–124S.

Kendrick, K. M., Da Costa, A. P. C., Broad, K. D., Ohkura, S., Guevara, R, Lévy, F., and Keverne, E. B. (1997). Neural control of maternal behaviour and olfactory recognition of offspring. *Brain Research Bulletin*, 44, 383–395.

Kendrick, K. M., da Costa, A. P., Hinton, M. R., and Keverne, E. B. (1992a). A simple method for fostering lambs using anoestrus ewes with artificially induced lactation and maternal behaviour. *Applied Animal Behaviour Science*, 34, 345–357.

Kendrick, K. M., Fabre-Nys, C., Blache, D., Goode, J. A., and Broad, K. D. (1993). The role of oxytocin release in the mediobasal hypothalamus of the sheep in relation to female sexual receptivity. *Journal of Neuroendocrinology*, 5, 13–21.

Kendrick, K. M., Guevara-Guzman, R., Zorilla, J., Hinton, M. R., Broad, K. D., Mimmack, M., and Ohkura, S. (1997b). Formation of olfactory memories mediated by nitric oxide. *Nature*, 388, 670–674.

Kendrick K. M. and Keverne E. B. (1989). Effects of intracerebroventricular infusions of naltrexone and phentolamine on central and peripheral oxytocin release and on maternal behaviour induced by vaginocervical stimulation in the ewe. *Brain Research*, 505, 329–332.

Kendrick, K. M. and Keverne, E. B. (1991). Importance of progesterone and estrogen priming for the induction of maternal behavior by vaginocervical stimulation in sheep: effects of maternal experience. *Physiology and Behavior*, 49, 745–750.

Kendrick, K. M. and Keverne, E. B. (1992b). Control of synthesis and release of oxytocin in the sheep brain. *Annals of the New York Academy of Sciences*, 652, 102–121.

Kendrick, K. M., Keverne, E. B., and Baldwin, B. A. (1987). Intracerebroventricular oxytocin stimulates maternal behaviour in the sheep. *Neuroendocrinology*, 46, 56–61.

Kendrick, K. M., Keverne, E. B., Baldwin, B. A., and Sharman, D. F. (1986). Cerebrospinal fluid levels of acetylcholinesterase, monoamines and oxytocin during labour, parturition, vaginocervical stimulation, lamb separation and suckling in sheep. *Neuroendocrinology*, 44 149–156.

Kendrick, K. M., Keverne, E. B., Chapman, C., and Baldwin, B. A. (1988a). Intra-cranial dialysis measurement of oxytocin, monoamine and uric acid release from the olfactory bulb and substantia nigra of sheep during parturition, suckling, separation from lambs and eating. *Brain Research*, 439 1–10.

Kendrick, K. M., Keverne, E. B., Chapman, C., and Baldwin, B. A. (1988b). Microdialysis measurement of oxytocin, aspartate, γ-aminobutyric acid and glutamate release from the olfactory bulb of the sheep during vaginocervical stimulation. *Brain Research*, 442, 171–174.

Kendrick, K. M., Keverne, E. B., Hinton, M. R., and Goode, J. A. (1991) Cerebrospinal fluid and plasma concentrations of oxytocin and vasopressin during parturition and

vaginocervical stimulation in the sheep. *Brain Research Bulletin*, 26, 803–807.

Kendrick, K. M., Keverne, E. B., Hinton, M. R., and Goode, J. A. (1992c). Oxytocin, amino acid and monoamine release in the medial preoptic area and bed nucleus of the stria terminalis of the sheep during parturition and suckling. *Brain Research*, 569, 199–209.

Kendrick, K. M., Lévy, F., and Keverne, E. B. (1991b). Importance of vaginocervical stimulation for the formation of maternal bonding in primiparous and multiparous parturient ewes. *Physiology and Behavior*, 50, 595–600.

Kendrick, K. M., Lévy, F., and Keverne, E. B. (1992b). Changes in the sensory processing of olfactory signals induced by birth in sheep. *Science*, 256, 833–836.

Keverne, E. B. and Kendrick, K. M. (1992). Oxytocin facilitation of maternal behaviour in sheep. *Annals of the New York Academy of Sciences*, 652, 83–101.

Keverne, E. B., and Kendrick, K. M. (1990). Morphine and corticotrophin releasing factor potentiate maternal acceptance in multiparous ewes after vaginocervical stimulation. *Brain Research*, 540, 55–62.

Keverne, E. B., Lévy, F., Guevara-Guzman, R., and Kendrick, K. M. (1993). Influence of birth and maternal experience on olfactory bulb neurotransmitter release. *Neuroscience*, 56, 557–565.

Krehbiel, D., Poindron, P., Lévy, F., and Prud'homme, M. J. (1987). Effects of peridural anaesthesia on maternal behaviour in primiparous and multiparous ewes. *Physiology and Behavior*, 40, 463–472.

Larrazolo-López, A., Kendrick, K. M., Aburto, M., Morimoto, S., and Guevara-Guzmán, R. (2008). Vaginocervical stimulation enhances social recognition memory in the rat via oxytocin release in the olfactory bulb. *Neuroscience*, 152, 585–593.

Lévy, F., Gervais, R., Kindermann, U., Orgeur, P., and Piketty, V. (1990). Importance of beta-noradrenergic receptors in the olfactory bulb of sheep for recognition of lambs. *Behavioral Neuroscience*, 104, 464–469.

Lévy, F., Kendrick, K. M., Goode, J. A., Guevara-Guzman, R., and Keverne, E. B. (1995). Oxytocin and vasopressin release in the olfactory bulb of parturient ewes: changes with maternal experience and effects on acetylcholine, γ-aminobutyric acid, glutamate and noradrenaline release. *Brain Research*, 669, 197–206.

Lévy, F., Dreifuss, J. J., Dubois-Dauphin, M., Bert, M., Barberis, C., and Tribollet, E. (1992a). Autoradiographic detection of vasopressin binding sites, but not oxytocin binding sites, in the olfactory bulb of sheep. *Brain Research*, 595, 154–159.

Lévy, F., Kendrick, K. M., Keverne, E. B., Piketty, V., and Poindron, P. (1992b). Intracerebral oxytocin is important for the onset of maternal Behavior in inexperienced ewes delivered under peridural anaesthesia. *Behavioural Neuroscience*, 106, 427–432.

Lévy, F., Richard, P., Meurisse, M., and Ravel, N. (1997). Scopolamine impairs the ability of parturient ewes to learn to recognise their lambs. *Psychopharmacology*, 129, 85–90.

Liu, Y. and Wang, Z. X. (2003). Nucleus accumbens oxytocin and dopamine interact to regulate pair bond formation in female prairie voles. *Neuroscience*, 121, 537–544.

Moos, F., Freund-Mercier M. J., Guerné, Y., Guerné, J. M., Stoeckel, M. E., and Richard, P. H. (1984). Release of oxytocin and vasopressin by magnocellular nuclei *in vitro*: specific facilitatory effect of oxytocin on its own release. *Journal of Endocrinology*, 102, 63–72.

Sanchez-Andrade, G., and Kendrick, K. M. (2009). The main olfactory system and social learning in mammals. *Behavioral Brain Research*, 200, 323–335.

11

The roles of vasopressin and oxytocin in aggression

Jerome H. Pagani, Scott R. Wersinger, and W. Scott Young, III

11.1 Introduction

Virtually every species exhibits behavior that is labeled aggression. The widespread nature of aggression suggests it has been strongly favored by natural and sexual selection. Aggression is highly adaptive for an individual because it may allow the animal to acquire limited resources, secure a mating opportunity, or protect its offspring. However, when these behaviors are displayed in other circumstances they may be maladaptive. For example, an adult male mouse that attacks a sexually receptive female has not only risked injury but also lost one of his few chances to reproduce. Expression of aggressive behavior under specific circumstances should be exquisitely regulated. The type of situation that will elicit aggressive behavior, as well as the behavior display, depends on the species and sex of the animal studied (Miczek et al., 2007).

In a large proportion of the literature, especially the human literature, violence and aggression are equated. Violence is qualitatively different from aggression. Aggression is highly adaptive. Violence is not, as it may result when aggression is not properly regulated. Recent work in humans using fMRI imaging suggests that the neurobiology underlying aggression and violence is different (reviewed in Dolan, 2010). It is critically important to define and use terms consistently.

In this chapter, we will review the role of the neuropeptides oxytocin (Oxt) and vasopressin (Avp) in the regulation of aggression. A consensus definition of aggression has proven elusive. One may regard aggression as simply the intention, whether conscious or not, to inflict or threaten harm as a response to the circumstances in which the animal finds itself (Moyer, 1968). Others regard the term **aggression** as having limited scientific value (Scott, 1966) and use the term agonistic behavior in its place. Agonistic behavior is defined as "a behavioral system composed of behavior patterns having the common function of adaptation to situations involving physical conflict between members of the same species" (Scott, 1966).

11.2 Commonly used testing paradigms to measure agonistic behavior in rodent models

The behavior of a subject is affected by many factors including the testing paradigm, the nature of the stimulus animal, and whether it is housed singly, with males, or with females. In our experience, the single biggest factor affecting behavior is the testing paradigm. The testing paradigms most commonly reported in the literature are variants of one of three general paradigms: the resident–intruder paradigm, the neutral cage paradigm, and the maternal paradigm.

Oxytocin, Vasopressin, and Related Peptides in the Regulation of Behavior, ed. E. Choleris, D. W. Pfaff, and M. Kavaliers.

11.2.1 The resident–intruder paradigm

In experiments that test hypotheses focusing on offensive attack behavior, the subject is the resident. Defensive aggression can also be studied using this same paradigm by reversing roles (Wersinger, Caldwell, Christiansen, and Young, 2007). In the resident–intruder paradigm a stimulus male is introduced into the home cage of a male subject. It is important that the subject have sufficient time to establish the testing arena as its home cage. Typically, the latency for the resident to attack the intruder is used as the measure of aggression. Clearly the definition of an attack is critical. As described in the previous section, even within a species, attack behavior can take several forms. Most often attack is defined as a lunge toward and/or biting of the stimulus animal (see (Crawley, 2000) and references therein).

The resident–intruder paradigm is most often used when the investigator seeks to measure attack behavior. The outcome of an interaction using this paradigm is relatively predictable. The intruder typically begins by displaying threat behavior. Although the intruder may display submissive behavior in response to these threats, the paradigm prevents the intruder from leaving. Consequently, the resident typically attacks the intruder. The intruder rarely attacks the resident.

11.2.2 The neutral cage paradigm

In this paradigm both the subject and the stimulus animal are placed into a clean testing arena. Neither animal has had the opportunity to deposit chemosensory cues or establish a territory. In this paradigm less fighting behavior is typically observed than in the resident–intruder paradigm (as shown in (Wersinger, Ginns, O'Carroll, Lolait, and Young, 2002)).

11.2.3 Thematernal paradigm

This paradigm allows for aggression to be assessed in females. Most rodent species do not attack conspecifics with the exception of hamsters, whose females are highly aggressive. Lactating female rodents quickly and vigorously attack intruders, however (Svare, Betteridge, Katz, and Samuels, 1981). In the maternal paradigm an intruder male is placed into the cage of a lactating female. This is most often done after the litter has been removed from the cage, although this is not always the case.

11.3 External factors affecting agonistic behavior

More external factors than can be reviewed here affect agonistic behavior. Some of these include housing condition, season, reproductive status, age, health (including degree of satiety), and previous experience. Given that the relationship between Oxt and Avp and various social behaviors, it is important to discuss the well-characterized influence of housing condition on aggression.

It has been long known that the housing condition affects agonistic behavior in many species, including mice, rats, and hamsters (e.g., (Blanchard and Blanchard, 1977)). For example, male rats and male mice housed alone attack an intruder more quickly than males housed with a mate (e.g., (Valzelli, 1969). Single-housing of rats and mice results in a constellation of neurochemical changes in the brain, including changes in serotonin and dopamine (Lapiz, Fulford, et al., 2001; Lapiz, Mateo, Durkin, Parker, and Marsden, 2001). The changes in agonistic behavior that result from single-housing is likely highly complex since many drugs including antidepressants, anticholinergics, antiserotonergics, and neuroleptics selectively antagonize the isolation-induced increase in aggression (Malick, 1979).

Although through a more indirect mechanism, housing condition affects agonistic behavior by allowing the formation of dominance hierarchies. When rodents are group housed, they form dominance hierarchies. These hierarchies have permanent effects on the behavior of the animal (e.g., (Blanchard, Blanchard, Takahashi, and Takahashi, 1977)). A dominant animal is more likely to display

attack behavior than a subordinate male irrespective of the testing paradigm used. Subordinate animals differ behaviorally and neurochemically from dominant animals in many other ways, including changes in the vasopressin system (Goodson and Bass, 2001). In addition, subordinate and dominant hamsters respond differently to exogenous administration of Avp (Ferris and Delville, 1994; Ferris et al., 2006; Ferris et al., 1997).

11.4 Vasopressin and oxytocin systems

11.4.1 Distributions of peptides in mammals

The majority of Avp within the CNS is expressed within the magnocellular neurons of the SON and PVN, from where it is transported to the posterior pituitary. The evidence for any extra-pituitary projections from the SON is scant (Alonso, Szafarczyk, and Assenmacher, 1986; Mason, Ho, and Hatton, 1984). In contrast, parvocellular neurons of the PVN provide robust projections, especially to the brainstem and spinal cord. Areas innervated by Avp fibers include the hippocampus and subiculum, diagonal band of Broca, locus coeruleus, solitary tract nucleus, dorsal motor nucleus of the vagus, medullary adrenergic groups and spinal cord (Buijs, Swaab, Dogterom, and van Leeuwen, 1978; de Vries and Buijs, 1983; Millan, Millan, Czlonkowski, and Herz, 1984; Sawchenko and Swanson, 1982).

Avp is also expressed within parvocellular neurons of the suprachiasmatic nucleus (SCN), bed nucleus of the stria terminalis (BNST), and medial amygdala (MeA) (Sofroniew, 1983). Within the BNST and MeA, Avp-expressing cells are sex steroid-dependent with the males having more Avp immunoreactive (Avp-ir) cells than females in some species (Buijs, 1987; Caffe and van Leeuwen, 1983; de Vries, Duetz, Buijs, van Heerikhuize, and Vreeburg, 1986; van Leeuwen, Caffe, and De Vries, 1985), but not all. For example, there appears to be no sex difference in Avp-ir within the BNST and MeA of Syrian hamsters; instead, galanin may have replaced Avp as the gender-dependent peptide (Miller, Ferris, and Kolb, 1999). Nevertheless, there are sexual dimorphisms in brain arginine vasotocin (AVT) in the bullfrogs and newts, suggesting that across phyla Avp and Avp-like compounds are sensitive to gonadal steroids (Boyd and Moore, 1992). The BNST and MeA send Avp fibers to the olfactory tubercle, nucleus of the diagonal band, ventral pallidum, lateral septum, ventral hippocampus, paraventricular thalamic nuclei, zona incerta, lateral habenula, ventral tegmental area, substantia nigra, periventricular gray, median and dorsal raphe nuclei and the locus coeruleus (de Vries, Buijs, Van Leeuwen, Caffe, and Swaab, 1985). Neurons immunoreactive for Avp have also been described within the medial and lateral septum, vertical limb of the nucleus of diagonal band of Broca and locus coeruleus (Sofroniew, 1985; Urban, Miller, Drake, and Dorsa, 1990), but only those in the diagonal band have been confirmed by hybridization histochemistry (Hallbeck, Hermanson, and Blomqvist, 1999; Planas, Kolb, Raskind, and Miller, 1995; Urban et al., 1990). The patterns of Avp immunostaining in four different vole species are similar to each other and to other rodents and also show similar gender dimorphisms regardless of their social behavior (see below and (Wang, Zhou, Hulihan, and Insel, 1996). A hybridization histochemical study in the rat has found Avp expression within several new areas, including pyramidal cells of the hippocampus, parabrachial nucleus and a portion of the mesencephalic reticular nucleus (Hallbeck et al., 1999).

Oxt synthesized in the magnocellular neurons of the PVN and SON project to the posterior pituitary. Parvocellular (or, at least smaller than the magnocellular) neurons in the PVN project to similar areas in the brainstem and spinal cord as the Avp neurons described previously. Parvocellular Oxt neurons outside the PVN have been described in mice (Castel and Morris, 1988; Jirikowski, Caldwell, Stumpf, and Pedersen, 1990) and various vole species (Wang et al., 1996). However, in the rat, it appears that the PVN is responsible for most, if not all, brain Oxt projections (de Vries and Buijs, 1983; Rinaman, 1998) although see (Jirikowski et al., 1990). Oxt is not expressed in the SCN and

there are very few or no Oxt neurons in the BNST and MeA.

11.4.2 Distributions of receptors in mammals

There are two principle classes of Avp receptors: Avpr1 and Avpr2 receptors, both of which are 7-transmembrane G-protein coupled receptors. There is no conclusive evidence for the expression of the Avpr2 within the CNS. Transcription of the ARHGAP4 gene within the CNS overlaps the Avpr2 and reverse transcriptase-PCR analysis must use specific conditions to avoid this "contaminant" (Foletta, Brown, and Young, 2002). There are two subtypes of the Avpr1: Avpr1a and Avpr1b. In the periphery, Avpr1a mediates the effects of Avp on vasoconstriction and can be found in liver, kidney, platelets, and smooth muscle (Ostrowski et al., 1992; Watters, Poulin, and Dorsa, 1998). The distribution of Avpr1a expression within the CNS has been primarily studied using receptor autoradiography and hybridization histochemistry. *In vitro* receptor autoradiography identifies the locations of binding to the receptor protein, whereas *in situ* hybridization histochemistry identifies the cells that transcribe the receptor gene. The former technique was greatly facilitated through the use of specific and potent ^{125}I-labeled Avpr1a antagonists (Barberis et al., 1995; Johnson et al., 1993; Kremarik, Freund-Mercier, and Stoeckel, 1993). Prominent Avpr1a binding is present in the rat LS, neocortical layer IV, hippocampal formation, amygdalostriatal area, BNST, various hypothalamic areas (including SCN), ventral tegmental area, substantia nigra, superior colliculus, dorsal raphe, nucleus of the solitary tract and inferior olive (Johnson et al., 1993). Avpr1a binding is moderate throughout the spinal cord, but with higher binding in the dorsolateral motoneurons in general and all motoneurons in the lumbar 5/6 levels where innervation to the perineal muscles originates (Tribollet, Barberis, and Arsenijevic, 1997).

Neurons containing Avpr1a transcripts are found extensively throughout the rat CNS, being especially prominent, for example, in the olfactory bulb, hippocampal formation, LS, SCN, PVN, anterior hypothalamic area, arcuate nucleus, lateral habenula, ventral tegmental area, substantia nigra (pars compacta), superior colliculus, raphe nuclei, locus coeruleus, inferior olive, area postrema, and nucleus of the solitary tract (Ostrowski, Lolait, and Young, 1994; Szot, Bale, and Dorsa, 1994). Transcripts are also detected in the choroid plexus and endothelial cells. The distributions of Avp (and Oxt) binding have been examined in a number of rodent species and they are remarkably similar. Differences in binding in selected areas may mediate important adaptations or behavioral traits, however.

The Avpr1b was originally described in the anterior pituitary where it facilitates the release of adrenocorticotropic hormone (ACTH) from the corticotropes (Antoni, Holmes, Makara, Karteszi, and Laszlo, 1984; Jard, Barberis, Audigier, and Tribollet, 1987). Avpr1b in the pituitary helps mediate the effects of Avp on the hypothalamic–pituitary–adrenal axis, which is the regulator of the stress response in mammals (Volpi, Rabadan-Diehl, and Aguilera, 2004). Avpr1b mRNA is also found in a variety of peripheral tissues including kidney, thymus, heart, lung, spleen, uterus, and breast (Lolait et al., 1995), although its role in these tissues remains unclear. Cell bodies were reported throughout the rat brain to contain Avpr1b-ir (Hernando, Schoots, Lolait, and Burbach, 2001; Stemmelin, Lukovic, Salome, and Griebel, 2005) or mRNA (Vaccari, Lolait, and Ostrowski, 1998). More recent evidence from *in situ* hybridization histochemistry with more specific probes suggests that the CA2 pyramidal neurons of the hippocampus are the most prominently site of expression in mice and rats (Young, Li, Wersinger, and Palkovits, 2006). This has been confirmed in a preliminary study using *in vitro* receptor autoradiography (Roper et al., 2009).

A single Oxt receptor (Oxtr) appears to transduce the actions of Oxt. This receptor was first isolated and identified by Kimura and colleagues (Kimura, Tanizawa, Mori, Brownstein, and Okayama, 1992). The Oxtr is also a member of the G-protein-coupled receptor family. Like other members of this family, the Oxtr contains seven transmembrane domains

and is similar in structure to the Avp receptors. Initially, the location of Oxtr expression has best been examined with *in vitro* receptor autoradiography by using a potent and specific ^{125}I-labeled antagonist (Elands et al., 1988; Kremarik et al., 1993; Veinante and Freund-Mercier, 1997). In the rat, Oxt binding is found in numerous regions, especially in the hippocampal formation (ventral subiculum particularly), LS, central amygdala (CeA), olfactory tubercle, accumbens nucleus shell, dorsal caudate-putamen, BNST, MeA, and ventromedial hypothalamus (VMH). Binding in the spinal cord is light and confined to the superficial dorsal horn (Tribollet et al., 1997).

Hybridization histochemistry reveals Oxtr transcripts in many areas of the rat CNS, including main and accessory olfactory bulbs, neocortical layers II and III, piriform cortical layer II, hippocampal formation, olfactory tubercle, BNST, medial habenula, VMH, PVN, and SON. Expression is lower in the midbrain, pons and medulla (Vaccari et al., 1998). Recently, an Oxtr-LacZ reporter mouse has shown additional Oxtr gene expression in the medial septum, parts of the amygdala and mammillary nuclei, and some brainstem nuclei (Gould and Zingg, 2003).

The distribution of the Oxtr is highly species specific, as is elegantly illustrated in receptor binding differences between two closely related species of voles, the polygamous montane vole and the monogamous prairie vole (Insel and Shapiro, 1992). Differences in the distributions of the Oxtr have also been shown among mice, rats, voles, hamsters, and guinea pigs (Insel, Young, Witt, and Crews, 1993). These differences in Oxtr distribution across species are thought to convey differing behavioral phenotypes, as discussed below.

A question remains with regard to the absence of Avp or Oxt in many regions in which their receptors are found. A recent finding suggests that perhaps dendrites or axon collaterals from magnocellular neurons of the hypothalamus (PVN and SON) might reach more distant targets (Ross and Young, 2009). Another possibility is that peptides released by the magnocellular dendrites into the third ventricular cerebrospinal fluid reach distant sites by bulk flow (Veening, de Jong, and Barendregt, 2010).

11.5 Vasopressin and aggression

Vasopressin (AVP) is implicated in agonistic behavior in taxonomically diverse groups including mammals (e.g., hamsters, voles) and fish (e.g., wrasse) (Bester-Meredith, Young, and Marler, 1999; Ferris and Delville, 1994; Goodson and Bass, 2001; Koolhaas, Everts, de Ruiter, de Boer, and Bohus, 1998; Lowry et al., 1997; Moore, Lowry, and Rose, 1994; Semsar, Kandel, and Godwin, 2001). In laboratory settings, rodents are often used to model aggressive interactions (Blanchard and Blanchard, 2003; Blanchard, Fukunaga, Blanchard, and Kelley, 1975; Malick, 1975; Miczek, Maxson, Fish, and Faccidomo, 2001). Even in humans, a life history of aggression against others correlates with increased CSF levels of VP and Oxt (Coccaro, Kavoussi, Hauger, Cooper, and Ferris, 1998).

The role of Avp in the regulation of aggression has been studied extensively in Syrian hamsters. Syrian hamsters, a solitary species, are highly aggressive towards conspecifics. They also exhibit a stereotypic form of scent-marking behavior, known as flank marking, to demonstrate dominance (Siegel, 1985). In 1984, it was discovered that Avp microinjected into the medial preoptic-anterior hypothalamic area (MPOA-AH) results in a dose-dependent increase in flank-marking behavior (Ferris, Albers, Wesolowski, Goldman, and Luman, 1984). Avpr1a antagonists administered orally or injected into the MPOA-AH block Avp-facilitated flank-marking behavior (Caldwell and Albers, 2003; Ferris, Delville, Grzonka, Luber-Narod, and Insel, 1993; Ferris et al., 2006; Ferris, Singer, Meenan, and Albers, 1988). However, the specific site where Avp stimulates aggression is different from the one that stimulates flank marking. Also, the doses of Avp used to induce aggression and flank marking differ (Albers, Liou, and Ferris, 1988; Caldwell and Albers, 2003, 2004; Ferris et al., 1997).

Aggression is facilitated when Avp is microinjected into the anterior hypothalamus (AH)

(Caldwell and Albers, 2004; Ferris et al., 1997) and inhibited when Avpr1a antagonists are microinjected there (Ferris and Potegal, 1988). Intracerebroventricular administration of an Avpr1a antagonist into juvenile golden hamsters reduces play fighting (Cheng and Delville, 2009). Superimposed on this system is the influence of gonadal steroids. Whereas Avp innervation of the MPOA-AH is not gonadal steroid-dependent (Albers, Rowland, and Ferris, 1991), Avpr1a biosynthesis is (Johnson, Barberis, and Albers, 1995; Young, Wang, Cooper, and Albers, 2000).

Other brain areas are involved in the regulation of aggression. Avpr1a binding within the ventrolateral hypothalamus is gonadal steroid-dependent, and a decrease there is correlated with reduced initiation of aggression (Delville, Mansour, and Ferris, 1996b). The Avpr1b also regulates aggression in Syrian hamsters. Hamsters administered an oral Avpr1b antagonist display less aggression than untreated controls (Blanchard et al., 2005), and further evidence for the role of the Avpr1b in the regulation of behavior is described below.

In prairie voles there is evidence that Avp may be important to increases in aggression seen following pair-bond formation. Specifically, compared to controls, pair-bonded males that show high levels of aggression toward unfamiliar conspecifics have increased c-fos activation within the AH; specifically, within cells that contain Avp and tyrosine hydroxylase (Gobrogge, Liu, Jia, and Wang, 2007).

Studies of mice that have been bred for either a long attack latency or short attack latency have provided valuable insights into critical differences between highly aggressive individuals and their less-aggressive counterparts. Short attack latency mice have less Avp-ir innervation in the lateral septum (LS) and fewer Avp-ir neurons in the bed nucleus of the stria terminalis (BNST) than long attack latency mice (Compaan, Buijs, Pool, De Ruiter, and Koolhaas, 1993). A negative correlation between Avp and aggression in the LS was reported subsequently (Everts, De Ruiter, and Koolhaas, 1997). Rats bred to have short attack latencies have decreased septal release of Avp during resident intruder tests but were not affected by intraseptal administration of Avp (or of Avpr1a antagonist in long-latency rats) (Veenema, Bredewold, and Neumann, 2007). However, in normal Wistar rats, those with higher aggression show increased LS and decreased BNST AVP release (Veenema, Beiderbeck, Lukas, and Neumann, 2010). Aggression was blocked by intra-LS administration of an Avp antagonist in these rats. This is similar to what is found in *Peromyscus californicus* (California mouse) that are more aggressive than *P. leucopus* (white-footed mouse): they show a positive correlation between aggression and Avp-ir in the BNST and LS (Bester-Meredith and Marler, 2001). In prairie voles and California mice, intracerebroventricular (i.c.v.) injections of Avpr1a antagonists inhibit aggression and in prairie voles injections of Avp increase aggression (Bester-Meredith, Martin, and Marler, 2005; Winslow, Hastings, Carter, Harbaugh, and Insel, 1993). However, the authors suggest that Avp antagonists do not block the expression of aggression per se, as breeder males with established aggression are unaffected; but, rather, affect the transition to aggression, that is, initiation of aggression. Nonetheless, intra-AH Avp increases and Avpr1a antagonist decreases aggression toward females, even if pair-bonded males (that have increased Avpr1a in the AH) in the latter case (Gobrogge, Liu, Young, and Wang, 2009).

The ability to display aggression is susceptible to manipulation depending on an individual animal's early life experience and/or social status. This plasticity in aggression is linked with changes in either Avp or Avp receptor distributions in a variety of mammalian species (Ferris et al., 1988; Frazier, Trainor, Cravens, Whitney, and Marler, 2006; Melloni, Connor, Hang, Harrison, and Ferris, 1997; Roche and Leshner, 1979; Stribley and Carter, 1999). While sexually naïve prairie voles generally do not show territorial aggression, early postnatal exposure to Avp can increase displays of aggression similar to levels seen in postmated animals (Stribley and Carter, 1999). Interestingly, aggression toward females in prairie voles correlates with the degree of paternal care (Ophir, Crino, Wilkerson,

Wolff, and Phelps, 2008), but the relationship to Avp is unknown. Newborn California mice that are retrieved more by their fathers show higher levels of aggression in adulthood than mice that are retrieved less. These animals also have increased Avp-ir fibers in the dorsal fiber tracts of the BNST (Frazier et al., 2006). Adult male rats that undergo maternal separation as pups have higher aggression, more Avp mRNA and Avp-ir in the PVN and SON, perhaps due to poor stress coping as these animals also show more depression-like activity during, and an increased ACTH response following, forced swim tests (Todeschin et al., 2009; Veenema, Blume, Niederle, Buwalda, and Neumann, 2006). Similarly, juvenile rats with increased maternal separation show increased offensive play fighting that is accompanied by increase Avp mRNA in the PVN and BNST (Veenema and Neumann, 2009). In mice that undergo maternal separation, males have longer latencies to attack intruders, while lactating mice have shorter latencies (Veenema et al., 2007). In these mice, the PVN was found to have increased Avp-ir in males, and decreased Oxt-ir in lactating females. Maternal separation in rats is accompanied by a number of changes in the developmental program of expression of Avpr1a and Oxtr in a number of brain regions (Lukas, Bredewold, Neumann, and Veenema, 2010). Mice raised in a communal setting demonstrate reduced maternal aggression compared to conventionally singly housed females (Curley, Davidson, Bateson, and Champagne, 2009), but the relationship to Avp is unknown.

Multiparous rats show increased maternal aggression despite reduced expression of Avp and Avpr1a (as well as Oxt and Oxtr) in the PVN and SON at postpartum day 5 (Nephew, Bridges, Lovelock, and Byrnes, 2009). Rats bred to have higher levels of anxiety exhibit higher levels of maternal aggression and of Avp release from the central nucleus of the amygdala and this aggression is inhibited by an Avpr1a antagonist (Bosch and Neumann, 2010). No changes in Avp binding were noticed there. The related BNST of the lactating Wistar rat has increased levels of Avpr1a binding and maternal aggression is inhibited by Avpr1 antagonist delivery there (Bosch, Pfortsch, Beiderbeck, Landgraf, and Neumann, 2010). Curiously, i.c.v. Avp decreases and Avpr1a antagonist increases aggression in lactating rat dams (Nephew and Bridges, 2008).

Exposure of hamsters to anabolic-androgenic steroids during adolescence increases adult aggression, but it also increases Avpr1a-ir within the AH, though the effects appear to be reversible following cessation of treatment (Grimes, Ricci, and Melloni, 2006). In anabolic-androgenic steroid-treated hamsters that have an Avpr1a antagonist microinjected into the AH, there is a reduction in the intensity, but not the initiation, of aggression (Harrison, Connor, Nowak, Nash, and Melloni, 2000; Melloni et al., 1997).

Initial studies suggested that Avp might have differential effects on aggression in dominant versus subordinate animals. Mice administered lysine-vasopressin just after a social defeat, increase their submissive behavior on subsequent tests compared to saline-treated animals (Roche and Leshner, 1979). Injections into the MPOA-AH results in transient reversals of dominate/subordinate relationships in Syrian hamsters, with subordinate animals displaying increased flank-marking behavior when treated with Avp and dominant animals less flank-marking behavior when treated with an Avpr1 antagonist (Ferris, Meenan, Axelson, and Albers, 1986). Consistent with these findings, subordinate hamsters have fewer Avp-ir neurons in the magnocellular nucleus circularis compared to dominant animals (Ferris, Axelson, Martin, and Roberge, 1989). Following repeated agonistic encounters, dominant hamsters have more Avpr1a binding in the ventromedial hypothalamus compared to their subordinate opponents (Cooper, Karom, Huhman, and Albers, 2005). In socially isolated hamsters, there is an increase in aggression that correlates with increased Avpr1a binding in the AH, PVN and lateral hypothalamus, whereas in socially experienced hamsters Avpr1a binding is significantly greater in the central amygdala (Albers, Dean, Karom, Smith, and Huhman, 2006). Even in Avp-facilitated aggression, several weeks of social isolation is required for Avp to be effective,

suggesting that social experience, or lack thereof, is critical for the effects of Avp on aggression (Caldwell and Albers, 2004; Ferris et al., 1997). Rats that are more successful in obtaining food have more Avpr1a binding in the LS compared to subordinates. These differences in receptor binding are not testosterone or corticosterone dependent, suggestive of experience-dependent changes in the Avp system (Askew, Gonzalez, Stahl, and Karom, 2006). Taken together, there is substantial evidence that the Avp system, and specifically Avpr1a biosynthesis, can be altered by social experience.

Inactivation of either the Avpr1a or Avpr1b genes in mice (knockouts, KO) has provided insights into their roles in the regulation of aggression. While it was expected that Avpr1a KO mice would show reduced aggression due to lack of Avp signaling, this was not the case (Wersinger, S. R., Temple, J. L., Caldwell, H. K., **and** Young, W. S., III, 2008), probably due to developmental compensation. Conversely, Avpr1b appears to be critical for proper expression of aggression, as Avpr1b KO mice show significant impairments in displays of aggression as compared to wild-type controls (Wersinger, Caldwell, Christiansen, and Young, 2007; Wersinger et al., 2002; Wersinger et al., 2004). In addition, an orally administered Avpr1b antagonist reduces aggression in mice and hamsters (Blanchard et al., 2005; Serradeil-Le Gal et al., 2005). Avpr1b KO mice have longer attack latencies and fewer attacks compared to wild-type controls, though the latencies and frequencies of attack can be increased with experience and competition for food (Wersinger, Caldwell, Christiansen, et al., 2007; Wersinger et al., 2002). These impairments are specific to social forms of aggression, as predatory aggression remains unaffected. Furthermore, only the attack component is affected, as the Avpr1b KO display defensive postures but do not initiate defensive or "retaliatory" attacks (Wersinger, Caldwell, Christiansen, et al., 2007). Inactivation of the Avrp1b in a more aggressive mouse strain (*Mus musculus castaneus*) also reduces aggression in those males (Caldwell and Young, 2009). Avpr1b KO mice show reduced maternal aggression as well (Wersinger, Caldwell, Christiansen, et al., 2007). In a setting where knockout mice are housed together, dominance hierarchies are established; but not through conventional aggression but perhaps through increased mounting (Caldwell, Dike, Stevenson, Storck, and Young, 2010). Avpr1b may not be involved in the transduction of the olfactory signals, but rather in coupling the social context with the appropriate behavioral response (Wersinger et al., 2004). Given the restricted distribution of Avpr1b mRNA within the brain, with highest expression within the CA2 field of the hippocampus (Young, Li, Wersinger, and Palkovits, 2006), and its role in memory for temporal order (DeVito et al., 2009) it may be that the Avpr1b is important for "social memory" for which an aggressive response would be appropriately triggered. Future studies that focus on the specific neuroanatomical areas involved in Avpr1b's mediation of aggression will be important for understanding the role of Avp in the regulation of aggression.

Recent use of magnetic resonance promises to expand our understanding of brain responses to potential aggressive encounters. Ferris and coworkers (Ferris, 2008; Ferris et al., 2008) showed that at least 16 regions, including the LH and various cortical areas, responded to presentation of an intruder to the restrained rat. Furthermore, the activation was greatly attenuated by administration of the orally active Avpr1a antagonist SRX251.

11.5.1 AVP and aggression in humans

In humans, heightened aggression is associated with increased impulsivity and is a characteristic of individuals with personality disorders. There is a positive correlation of CSF Avp concentration with a life history of non-directed general aggression as well as aggression toward individuals (Coccaro et al., 1998). However, Virkkunen and colleagues (1994) found no differences in CSF Avp between violent offenders and controls. These discrepancies, however, may be due to differences in the patient populations being studied. Intranasal Avp administration, at least in men, is reported to increase the emotional response to neutral stimuli resulting

in an increased perception of threat and facilitate evaluation of emotional facial images (Guastella, Kenyon, Alvares, Carson, and Hickie, 2010; Thompson, Gupta, Miller, Mills, and Orr, 2004; Thompson, George, Walton, Orr, and Benson, 2006). Recently, intranasal Avp was demonstrated to reduce connectivity between subgenual and supragenual cingulate cortices, part of an emotion-activated medial prefrontal-amygdala circuit (Zink, Stein, Kempf, Hakimi, and Meyer-Lindenberg, 2010). How this modulation of emotional responses relates precisely to aggression awaits further study.

11.6 Oxytocin and aggression

Generally, aggression in male rodents is believed to be heavily under the control of Avp (see Caldwell, Lee, Macbeth, and Young, 2008 for review). The role of Oxt in controlling aggressive behavior in males is ambiguous and likely depends upon the species used, the test animals' sexual status and the age at which Oxt levels are manipulated. In prairie voles, ventricular delivery of Oxt reduces sexual behavior but has no effect on aggression (Mahalati, Okanoya, Witt, and Carter, 1991). Oxt has been shown to affect aggressive behavior in prairie voles after mating, however, this effect was not seen in non-monogamous, montane voles (Winslow, Shapiro, Carter, and Insel, 1993).

Oxt significantly increases sexual and aggressive behaviors in dominant, but not subordinate pair-housed male squirrel monkeys, during interaction with a female (Winslow and Insel, 1991). The increase in aggression is blocked following concomitant administration of Oxt and the Oxt antagonist OVTA (Winslow and Insel, 1991).

Interestingly, male Wistar rats introduced as an intruder to the cage of a singly housed male rat have between two- and five-fold increases in Oxt levels in the SON and anterior ventrolateral portion of the hypothalamus (Engelmann, Ebner, Landgraf, Holsboer, and Wotjak, 1999), suggesting that Oxt's role in aggressive encounters may have more to do with the stress response to this type of social interaction than aggression per se. This mirrors the link between maternal aggression and an anxious phenotype reported in female rats (Bosch, Meddle, Beiderbeck, Douglas, and Neumann, 2005). It is worth noting that inducing subordination in males by long-term shared housing increases anxiety but does not result in a change in hypothalamic Oxt mRNA levels (Reber and Neumann, 2008).

In KO mice, one Oxt KO line is mildly less aggressive than WT or HET controls, and shows no difference in anxiety behavior in an open field (DeVries, Young, and Nelson, 1997). In a different line of Oxt KO mice, increased aggressive behavior is seen in the resident–intruder paradigm and decreased anxiety in the elevated plus maze (EPM) (Winslow et al., 2000). These effects were noted only in KO mice born to obligate mice (KO–KO matings); KO mice, and their WT controls, were cross-fostered to WT mothers. Non-obligates (KOs produced from HET–HET matings) show no reduction in anxiety and a small increase in aggression only on the third aggressive encounter (Winslow et al., 2000). This suggests that the effects on aggression and anxiety are due to the lack of Oxt in the prenatal environment, or an interaction of genotype and the stress of cross-fostering. These finding are consistent with the elevated levels of aggression reported in Oxtr knockouts generated from non-obligates, consistent with the idea that a lack of prenatal activation of the Oxt system results in increased adult aggression (Takayanagi et al., 2005).

Female mammals are most aggressive during the postpartum period. In rodents, the mother will attack an unfamiliar male introduced to the cage for several days after giving birth. This type of aggression, termed maternal aggression, is influenced by a variety of factors that have been extensively reviewed elsewhere (Lonstein and Gammie, 2002). Currently studies of maternal aggression in Oxt and Oxtr KO lines are lacking and this area remains ripe for exploration.

There are several brain areas that appear to be critical to the mediation of maternal aggression, including the amygdala, MPOA and the PVN. The expression of the immediate early gene c-Fos in the PVN is elevated in aggressive, but not

non-aggressive, lactating female rats following exposure to intruders, and the immediate early gene EGR-1 is elevated by aggressive experience above levels associated with lactation in lactating female rats (Gammie and Nelson, 2001; Hasen and Gammie, 2006). Aggressive encounters in female rats have been shown to elevate c-Fos and EGR-1 levels in several amygdalar nuclei (Gammie and Nelson, 2000; Hasen and Gammie, 2006).

The results from PVN lesions are less clear since not all findings report effects in the same direction. Most do suggest a role for the PVN in maternal aggression, however. Electrolytic lesions of the PVN have been shown to decrease maternal aggression in rats (Consiglio and Lucion, 1996). Ibotenic acid lesions directed at the parvocellular portion of the PVN increase maternal aggression, an effect also obtained by blockade of Oxt synthesis by injection of Oxt antisense mRNA into the same region (Giovenardi, Padoin, Cadore, and Lucion, 1998). Fiber sparing kainic acid lesions of the PVN fail to reduce aggressive behavior, however (Olazabal and Ferreira, 1997). Both studies used female Wistar rats, lesioned on the second day postpartum (selectively targeting the parvocellular region of the PVN) and tested for maternal aggression within five days of giving birth. Thus, it is difficult to reconcile these two findings. The preponderance of evidence does point to PVN involvement in maternal aggression.

In the amygdala, administration of bicuculline, a GABA antagonist, decreases aggression in lactating rats (Hansen and Ferreira, 1986). Infusion of a dual Oxtr/Avpr1a antagonist into the central nucleus of the amygdala (CeA) increases the number of attacks postpartum female rats made against intruders (Lubin, Elliott, Black, and Johns, 2003). Oxt infused into the CeA and BNST decreases the frequency of biting and frontal attacks (Consiglio, Borsoi, Pereira, and Lucion, 2005). Oxt has the opposite effect in golden hamsters, however. Administration of Oxt into the CeA increases aggression against a male intruder in postpartum females (Ferris et al., 1992). It is unclear whether this difference is species specific or somehow related to dosage.

Oxt may exert its influence on maternal aggression through its role in modulating anxiety. Oxtr are found in the PVN, BNST, and the CeA, areas that are part of the circuitry mediating anxiety responses. Administration of Oxt into some of these areas has been linked to increases in maternal aggression (Ferris et al., 1992; Harmon, Huhman, Moore, and Albers, 2002; Lubin et al., 2003); but see (Consiglio et al., 2005). Cocaine given to postpartum rats decreases Oxt binding in the BNST (Jarrett, McMurray, Walker, and Johns, 2006) as well as the VTA, MPOA, hippocampus (Johns, Lubin, Walker, Meter, and Mason, 1997) and amygdala (Elliott, Lubin, Walker, and Johns, 2001). A similar administration of cocaine increases maternal aggression in both lactating (Lubin, Meter, Walker, and Johns, 2001a) and virgin (Lubin, Meter, Walker, and Johns, 2001b) rats. Postpartum, non-lactating female rats given cocaine 10–30 days prior to testing have decreased levels of Oxt in the MPOA and increased levels of aggression (Johns et al., 2010). This suggests that cocaine may alter aggression by affecting Oxt levels in parts of the brain known to mediate anxiety. Moreover, increased levels of aggression and Oxt release are found in lactating rats bred for high levels of anxiety but not in a less-aggressive, low-anxiety strain (Bosch et al., 2005). This increase in aggressive behavior is blocked in the high-anxiety rats by administration of an Oxtr antagonist but has no effect on the low-anxiety group (Bosch et al., 2005). Aggression levels in the low-anxiety group are increased by delivery of Oxt to the PVN, however. Thus, Oxt may influence anxiety and aggression together in a manner dependent upon circulating levels of Oxt (Bosch, 2010).

11.6.1 Oxt and aggression in humans

The administration of oxytocin has been reported to increase prosocial feelings and behaviors. The intranasal administration of oxytocin prior to a money transfer game causes subjects to give more money to a fictitious second player than when given a placebo, but only if subjects believed they were giving money to a person and not a "project"

(Kosfeld, Heinrichs, Zak, Fischbacher, and Fehr, 2005). Intranasal oxytocin maintains trusting behavior, even when subjects learn that their trust has been breached (i.e., the other player fails to return money 50% of the time (Baumgartner, Heinrichs, Vonlanthen, Fischbacher, and Fehr, 2008) and increases generosity (Zak, Stanton, and Ahmadi, 2007).

Little is known about the role of Oxt in human aggression. Higher levels of autoantibodies reactive for Oxt are found in males with conduct disorder than in controls (Fetissov et al., 2006). Decreased levels of Oxt in the CSF of adult males and females have also been associated with higher levels of reported aggressive behavior (Lee, Macbeth, Pagani, and Young, 2009). Oxt administration has been shown to reduce amygdalar activity in response to fear-inducing visual stimuli (Kirsch et al., 2005) and may also increase feelings of trust and affiliation by reducing amygdala activation induced by social or novel situations (Baumgartner et al., 2008; Meyer-Lindenberg, 2008). Administration of oxytocin blunts the social stress of speaking in front of an audience, a situation that increases reported feelings of stress as well as cortisol levels; this effect is enhanced by concomitant social support from a friend prior to the stressor (Heinrichs, Baumgartner, Kirschbaum, and Ehlert, 2003). Oxt also has direct influences in the amygdala, where it attenuates amygdalar activation to fearful faces in people with generalized social anxiety disorder (Labuschagne et al., 2010). The oxytocin-induced reduction of social anxiety may have an impact on aggressive behavior given the link between anxiety levels and aggression in several animal models (Bosch et al., 2005; Bosch, Sartori, Singewald, and Neumann, 2007; Winslow et al., 2000). In humans, Oxt may also act to decrease anxiety by increasing recognition (Savaskan, Ehrhardt, Schulz, Walter, and Schachinger, 2008) and feelings of affiliation (Kosfeld et al., 2005).

Oxytocin is not involved in just positive social behaviors though: subjects display increased levels of envy following intranasal oxytocin (Shamay-Tsoory et al., 2009). It also promotes ethnocentrism by strengthening "in-group" connections and while at the same time weakening "out-group" connections (de Dreu et al., 2010, 2011). By affecting the perceived level of closeness to self-defined social categories, oxytocin may indirectly influence aggression (Otten and Stapel, 2007).

11.7 Internal factors affecting agonistic behavior

A large number of internal factors affect agonistic behavior. Two of these factors, testosterone and serotonin (5-HT), interact especially with the Avp system. The influences of gonadal steroids on Avp and Oxt expression are discussed in Chapter 1 by Dhakar et al. (this book). The proaggressive effects of testosterone are well known but the link between its effects on Avp expression and aggression is still a subject of research beyond the scope of this chapter. Pharmacological evidence has unequivocally shown that alteration of serotonergic tone influences agonistic behavior. Manipulations that enhance serotonergic activity typically inhibit offensive aggression in males while manipulations interfering with 5-HT typically facilitate offensive aggression in males (reviewed in (Wersinger, Caldwell, Christiansen, et al., 2007; Wersinger, Caldwell, Martinez, et al., 2007)). Within the ventrolateral hypothalamus, 5-HT blocks Avp-facilitated offensive aggression (Delville, Mansour, and Ferris, 1996a). In the AH, 1a receptor, but not a 1b, serotonin receptor agonist produces a dose-dependent inhibition of Avp facilitated offensive aggression (Ferris, Stolberg, and Delville, 1999). The 5-HT innervation to this area is traced to the dorsal, median, and caudal linear raphe nuclei (Ferris et al., 1999). Likewise, treatment with selective serotonin re-uptake inhibitors suppresses Avp-facilitated flank-marking behavior (Ferris, Rasmussen, Messenger, and Koppel, 2001). In rats that have undergone maternal separation, and who subsequently show higher levels of aggression, 5-HT-ir in the AH and SON negatively correlates with aggression (Veenema et al., 2006). With their opposing

effects on aggression, Avp and 5-HT are biologically interconnected, and studies focusing on their interactions are critical to increasing our understanding of the neural regulation of aggression.

11.8 Summary

Vasopressin and oxytocin clearly have important roles in regulating aggression. Further research is especially critical to understand their roles in human aggression for which much less is known. The continued and expanded applications of pharmacology genomics, genetic and viral manipulations of expression (including optogenetics), and various brain scanning techniques promise to expand our knowledge dramatically in both animals and humans.

Acknowledgments

This research was supported, on part, by the NIMH Intramural Research Program (Z01-MH-002498-21).

REFERENCES

Albers, H. E., Dean, A., Karom, M. C., Smith, D., and Huhman, K. L. (2006). Role of V1a vasopressin receptors in the control of aggression in Syrian hamsters. *Brain Research*, 1073–1074, 425–430.

Albers, H. E., Liou, S. Y., and Ferris, C. F. (1988). Testosterone alters the behavioral response of the medial preoptic-anterior hypothalamus to microinjection of arginine vasopressin in the hamster. *Brain Research*, 456(2), 382–386.

Albers, H. E., Rowland, C. M., and Ferris, C. F. (1991). Arginine-vasopressin immunoreactivity is not altered by photoperiod or gonadal hormones in the Syrian hamster (Mesocricetus auratus). *Brain Research*, 539(1), 137–142.

Alonso, G., Szafarczyk, A., and Assenmacher, I. (1986). Radioautographic evidence that axons from the area of supraoptic nuclei in the rat project to extrahypothalamic brain regions. *Neuroscience Letters*, 66(3), 251–256.

Antoni, F., Holmes, M., Makara, G., Karteszi, M., and Laszlo, F. (1984). Evidence that the effects of arginine-8-vasopressin (AVP) on pituitary corticotropin (ACTH) release are mediated by a novel type of receptor. *Peptides*, 5(3), 519–522.

Askew, A., Gonzalez, F. A., Stahl, J. M., and Karom, M. C. (2006). Food competition and social experience effects on V1a receptor binding in the forebrain of male Long-Evans hooded rats. *Hormones and Behavior*, 49(3), 328–336.

Barberis, C., Balestre, M., Jard, S., et al. (1995). Characterization of a novel, linear radioiodinated vasopressin antagonist: an excellent radioligand for vasopressin V1a receptors. *Neuroendocrinology*, 62(2), 135–146.

Baumgartner, T., Heinrichs, M., Vonlanthen, A., Fischbacher, U., and Fehr, E. (2008). Oxytocin shapes the neural circuitry of trust and trust adaptation in humans. *Neuron*, 58(4), 639–650.

Bester-Meredith, J. K. and Marler, C. A. (2001). Vasopressin and aggression in cross-fostered California mice (Peromyscus californicus) and white-footed mice (Peromyscus leucopus). *Hormones and Behavior*, 40(1), 51–64.

Bester-Meredith, J. K., Martin, P. A., and Marler, C. A. (2005). Manipulations of vasopressin alter aggression differently across testing conditions in monogamous and non-monogamous *Peromyscus* mice. *Aggressive Behavior*, 31, 189–199.

Bester-Meredith, J. K., Young, L. J., and Marler, C. A. (1999). Species differences in paternal behavior and aggression in peromyscus and their associations with vasopressin immunoreactivity and receptors. *Hormones and Behavior*, 36(1), 25–38.

Blanchard, D. C. and Blanchard, R. J. (2003). What can animal aggression research tell us about human aggression? *Hormones and Behavior*, 44(3), 171–177.

Blanchard, D. C., Blanchard, R. J., Takahashi, L. K., and Takahashi, T. (1977). Septal lesions and aggressive behavior. *Behavioral Biology*, 21(1), 157–161.

Blanchard, R. J. and Blanchard, D. C. (1977). Aggressive behavior in the rat. *Behavioral Biology*, 21(2), 197–224.

Blanchard, R. J., Fukunaga, K., Blanchard, D. C., and Kelley, M. J. (1975). Conspecific aggression in the laboratory rat. *Journal of Comparative and Physiological Psychology*, 89(10), 1204–1209.

Blanchard, R. J., Griebel, G., Farrokhi, C., et al. (2005). AVP V1b selective antagonist SSR149415 blocks aggressive behaviors in hamsters. *Pharmacology Biochemistry and Behavior*, 80(1), 189–194.

Bosch, O. J., Meddle, S. L., Beiderbeck, D. I., Douglas, A. J., and Neumann, I. D. (2005). Brain oxytocin correlates with maternal aggression: link to anxiety. *Journal of Neuroscience*, 25(29), 6807–6815.

Bosch, O. J. and Neumann, I. D. (2010). Vasopressin released within the central amygdala promotes maternal aggression. *European Journal of Neuroscience*, 31(5), 883–891.

Bosch, O. J., Pfortsch, J., Beiderbeck, D. I., Landgraf, R., and Neumann, I. D. (2010). Maternal behaviour is associated with vasopressin release in the medial preoptic area and bed nucleus of the stria terminalis in the rat. *Journal of Neuroendocrinology*, 22(5), 420–429.

Bosch, O. J., Sartori, S. B., Singewald, N., and Neumann, I. D. (2007). Extracellular amino acid levels in the paraventricular nucleus and the central amygdala in high- and low-anxiety dams rats during maternal aggression: regulation by oxytocin. *Stress*, 10(3), 261–270.

Boyd, S. K. and Moore, F. L. (1992). Sexually dimorphic concentrations of arginine vasotocin in sensory regions of the amphibian brain. *Brain Research*, 588(2), 304–306.

Buijs, R. M. (1987). Vasopressin localization and putative functions in the brain. In D. M. Gash and G. J. Boer (Eds.), *Vasopressin: Principles and Properties* (pp. 91–115). New York: Plenum Press.

Buijs, R. M., Swaab, D. F., Dogterom, J., and van Leeuwen, F. W. (1978). Intra- and extra-hypothalamic vasopressin and oxytocin pathways in the rat. *Cell and Tissue Research*, 186(3), 423–433.

Caffe, A. R. and van Leeuwen, F. W. (1983). Vasopressin-immunoreactive cells in the dorsomedial hypothalamic region, medial amygdaloid nucleus and locus coeruleus of the rat. *Cell and Tissue Research*, 233(1), 23–33.

Caldwell, H. K. and Albers, H. E. (2003). Short-photoperiod exposure reduces vasopressin (V1a) receptor binding but not arginine-vasopressin-induced flank marking in male Syrian hamsters. *Journal of Neuroendocrinology*, 15(10), 971–977.

Caldwell, H. K. and Albers, H. E. (2004). Effect of photoperiod on vasopressin-induced aggression in Syrian hamsters. *Hormones and Behavior*, 46(4), 444–449.

Caldwell, H. K., Dike, O. E., Stevenson, E. L., Storck, K., and Young, W. S., III. (2010). Social dominance in male vasopressin 1b receptor knockout mice. *Hormones and Behavior*, 58(2), 257–263.

Caldwell, H. K., Lee, H. J., Macbeth, A. H., and Young, W. S., III. (2008). Vasopressin: behavioral roles of an "original" neuropeptide. *Progress in Neurobiology*, 84(1), 1–24.

Caldwell, H. K. and Young, W. S., III. (2009). Persistence of reduced aggression in vasopressin 1b receptor knockout mice on a more "wild" background. *Physiology and Behavior*, 97(1), 131–134.

Castel, M. and Morris, J. F. (1988). The neurophysin-containing innervation of the forebrain of the mouse. *Neuroscience*, 24(3), 937–966.

Cheng, S. Y. and Delville, Y. (2009). Vasopressin facilitates play fighting in juvenile golden hamsters. *Physiology and Behavior*, 98(1–2), 242–246.

Coccaro, E. F., Kavoussi, R. J., Hauger, R. L., Cooper, T. B., and Ferris, C. F. (1998). Cerebrospinal fluid vasopressin levels: correlates with aggression and serotonin function in personality-disordered subjects. *Archives of General Psychiatry*, 55(8), 708–714.

Compaan, J. C., Buijs, R. M., Pool, C. W., De Ruiter, A. J., and Koolhaas, J. M. (1993). Differential lateral septal vasopressin innervation in aggressive and nonaggressive male mice. *Brain Research Bulletin*, 30(1–2), 1–6.

Consiglio, A. R., Borsoi, A., Pereira, G. A., and Lucion, A. B. (2005). Effects of oxytocin microinjected into the central amygdaloid nucleus and bed nucleus of stria terminalis on maternal aggressive behavior in rats. *Physiology and Behavior*, 85(3), 354–362.

Consiglio, A. R. and Lucion, A. B. (1996). Lesion of hypothalamic paraventricular nucleus and maternal aggressive behavior in female rats. *Physiology and Behavior*, 59(4–5), 591–596.

Cooper, M. A., Karom, M., Huhman, K. L., and Albers, H. E. (2005). Repeated agonistic encounters in hamsters modulate AVP V1a receptor binding. *Hormones and Behavior*, 48(5), 545–551.

Crawley, J. N. (2000). *What's Wrong with My Mouse*. New York: Wiley-Liss, pp. 171–174.

Curley, J. P., Davidson, S., Bateson, P., and Champagne, F. A. (2009). Social enrichment during postnatal development induces transgenerational effects on emotional and reproductive behavior in mice. *Frontiers of Behavioral Neuroscience*, 3, Article 25.

de Dreu, C. K. W., Greer, L. L., Handgraaf, M. J. J., et al. (2010). The neuropeptide oxytocin regulates parochial altruism in intergroup conflict among humans. *Science*, 32 (5984), 1408–1411.

de Dreu, C. K. W., Greer, L. L., Van Kleef, G. A., et al. (2011). Oxytocin promotes human ethnocentrism. *Proceedings of the National Academy of Sciences USA*, 108(4), 1262–1266.

de Vries, G. J. and Buijs, R. M. (1983). The origin of the vasopressinergic and oxytocinergic innervation of the rat brain with special reference to the lateral septum. *Brain Research*, 273(2), 307–317.

de Vries, G. J., Buijs, R. M., Van Leeuwen, F. W., Caffe, A. R., and Swaab, D. F. (1985). The vasopressinergic innervation of the brain in normal and castrated rats. *Journal of Comparative Neurology*, 233(2), 236–254.

de Vries, G. J., Duetz, W., Buijs, R. M., van Heerikhuize, J., and Vreeburg, J. T. (1986). Effects of androgens and estrogens on the vasopressin and oxytocin innervation of the adult rat brain. *Brain Research*, 399(2), 296–302.

Delville, Y., Mansour, K. M., and Ferris, C. F. (1996a). Serotonin blocks vasopressin-facilitated offensive aggression: interactions within the ventrolateral hypothalamus of golden hamsters. *Physiology and Behavior*, 59(4–5), 813–816.

Delville, Y., Mansour, K. M., and Ferris, C. F. (1996b). Testosterone facilitates aggression by modulating vasopressin receptors in the hypothalamus. *Physiology and Behavior*, 60(1), 25–29.

DeVito, L. M., Konigsberg, R., Lykken, C., et al. (2009). Vasopressin 1b receptor knock-out impairs memory for temporal order. *Journal of Neuroscience*, 29(9), 2676–2683.

DeVries, A. C., Young, W. S., III, and Nelson, R. J. (1997). Reduced aggressive behaviour in mice with targeted disruption of the oxytocin gene. *Journal of Neuroendocrinology*, 9(5), 363–368.

Dolan, M. C. (2010) What imaging tells us about violence in anti-social men. *Crim Behav Ment Health*, 20(3):199–214.

Elands, J., Barberis, C., Jard, S., et al. (1988). 125I-labelled d(CH2)5[Tyr(Me)2,Thr4,Tyr-NH2(9)]OVT: a selective oxytocin receptor ligand. *European Journal of Pharmacology*, 147(2), 197–207.

Elliott, J. C., Lubin, D. A., Walker, C. H., and Johns, J. M. (2001). Acute cocaine alters oxytocin levels in the medial preoptic area and amygdala in lactating rat dams: implications for cocaine-induced changes in maternal behavior and maternal aggression. *Neuropeptides*, 35(2), 127–134.

Engelmann, M., Ebner, K., Landgraf, R., Holsboer, F., and Wotjak, C. T. (1999). Emotional stress triggers intrahypothalamic but not peripheral release of oxytocin in male rats. *Journal of Neuroendocrinology*, 11(11), 867–872.

Everts, H. G., De Ruiter, A. J., and Koolhaas, J. M. (1997). Differential lateral septal vasopressin in wild-type rats: correlation with aggression. *Hormones and Behavior*, 31(2), 136–144.

Ferris, C. F. (2008). Functional magnetic resonance imaging and the neurobiology of vasopressin and oxytocin. *Progress in Brain Research*, 170, 305–320.

Ferris, C. F., Albers, H. E., Wesolowski, S. M., Goldman, B. D., and Luman, S. E. (1984). Vasopressin injected into the hypothalamus triggers a stereotypic behavior in golden hamsters. *Science*, 224(4648), 521–523.

Ferris, C. F., Axelson, J. F., Martin, A. M., and Roberge, L. F. (1989). Vasopressin immunoreactivity in the anterior hypothalamus is altered during the establishment of dominant/subordinate relationships between hamsters. *Neuroscience*, 29(3), 675–683.

Ferris, C. F. and Delville, Y. (1994). Vasopressin and serotonin interactions in the control of agonistic behavior. *Psychoneuroendocrinology*, 19(5–7), 593–601.

Ferris, C. F., Delville, Y., Grzonka, Z., Luber-Narod, J., and Insel, T. R. (1993). An iodinated vasopressin (V1) antagonist blocks flank marking and selectively labels neural binding sites in golden hamsters. *Physiology and Behavior*, 54(4), 737–747.

Ferris, C. F., Foote, K. B., Meltser, H. M., et al. (1992). Oxytocin in the amygdala facilitates maternal aggression. *Annals of the New York Academy of Science*, 652, 456–457.

Ferris, C. F., Lu, S. F., Messenger, T., et al. (2006). Orally active vasopressin V1a receptor antagonist, SRX251, selectively blocks aggressive behavior. *Pharmacology Biochemistry and Behavior*, 83(2), 169–174.

Ferris, C. F., Meenan, D. M., Axelson, J. F., and Albers, H. E. (1986). A vasopressin antagonist can reverse dominant/subordinate behavior in hamsters. *Physiology and Behavior*, 38(1), 135–138.

Ferris, C. F., Melloni, R. H., Jr., Koppel, G., et al. (1997). Vasopressin/serotonin interactions in the anterior hypothalamus control aggressive behavior in golden hamsters. *Journal of Neuroscience*, 17(11), 4331–4340.

Ferris, C. F. and Potegal, M. (1988). Vasopressin receptor blockade in the anterior hypothalamus suppresses aggression in hamsters. *Physiology and Behavior*, 44(2), 235–239.

Ferris, C. F., Rasmussen, M. F., Messenger, T., and Koppel, G. (2001). Vasopressin-dependent flank marking in golden hamsters is suppressed by drugs used in the treatment of obsessive-compulsive disorder. *BMC Neuroscience*, 2, 10.

Ferris, C. F., Singer, E. A., Meenan, D. M., and Albers, H. E. (1988). Inhibition of vasopressin-stimulated flank

marking behavior by V1-receptor antagonists. *European Journal of Pharmacology*, 154(2), 153–159.

Ferris, C. F., Stolberg, T., and Delville, Y. (1999). Serotonin regulation of aggressive behavior in male golden hamsters (Mesocricetus auratus). *Behavioral Neuroscience*, 113(4), 804–815.

Ferris, C. F., Stolberg, T., Kulkarni, P., et al. (2008). Imaging the neural circuitry and chemical control of aggressive motivation. *BMC Neuroscience*, 9, 111.

Fetissov, S. O., Hallman, J., Nilsson, I., et al. (2006). Aggressive behavior linked to corticotropin-reactive autoantibodies. *Biological Psychiatry*, 60(8), 799–802.

Foletta, V. C., Brown, F. D., and Young, W. S., III. (2002). Cloning of rat ARHGAP4/C1, a RhoGAP family member expressed in the nervous system that colocalizes with the Golgi complex and microtubules. *Molecular Brain Research*, 107(1), 65–79.

Frazier, C. R., Trainor, B. C., Cravens, C. J., Whitney, T. K., and Marler, C. A. (2006). Paternal behavior influences development of aggression and vasopressin expression in male California mouse offspring. *Hormones and Behavior*, 50(5), 699–707.

Gammie, S. C. and Nelson, R. J. (2000). Maternal and mating-induced aggression is associated with elevated citrulline immunoreactivity in the paraventricular nucleus in prairie voles. *Journal of Comparative Neurology*, 418(2), 182–192.

Gammie, S. C. and Nelson, R. J. (2001). cFOS and pCREB activation and maternal aggression in mice. *Brain Research*, 898(2), 232–241.

Giovenardi, M., Padoin, M.J., Cadore, L. P., and Lucion, A. B. (1998). Hypothalamic paraventricular nucleus modulates maternal aggression in rats: effects of ibotenic acid lesion and oxytocin antisense. *Physiology and Behavior*, 63(3), 351–359.

Gobrogge, K. L., Liu, Y., Jia, X., and Wang, Z. (2007). Anterior hypothalamic neural activation and neurochemical associations with aggression in pair-bonded male prairie voles. *Journal of Comparative Neurology*, 502(6), 1109–1122.

Gobrogge, K. L., Liu, Y., Young, L. J., and Wang, Z. (2009). Anterior hypothalamic vasopressin regulates pair-bonding and drug-induced aggression in a monogamous rodent. *Proceedings of the National Academy of Sciences USA*, 106(45), 19144–19149.

Goodson, J. L. and Bass, A. H. (2001). Social behavior functions and related anatomical characteristics of vasotocin/vasopressin systems in vertebrates. *Brain Research Reviews*, 35(3), 246–265.

Gould, B. R. and Zingg, H. H. (2003). Mapping oxytocin receptor gene expression in the mouse brain and mammary gland using an oxytocin receptor-LacZ reporter mouse. *Neuroscience*, 122(1), 155–167.

Grimes, J. M., Ricci, L. A., and Melloni, R. H., Jr. (2006). Plasticity in anterior hypothalamic vasopressin correlates with aggression during anabolic-androgenic steroid withdrawal in hamsters. *Behavioral Neuroscience*, 120(1), 115–124.

Guastella, A. J., Kenyon, A. R., Alvares, G. A., Carson, D. S., and Hickie, I. B. (2010). Intranasal arginine vasopressin enhances the encoding of happy and angry faces in humans. *Biological Psychiatry*, 67(12), 1220–1222.

Hallbeck, M., Hermanson, O., and Blomqvist, A. (1999). Distribution of preprovasopressin mRNA in the rat central nervous system. *Journal of Comparative Neurology*, 411(2), 181–200.

Hansen, S. and Ferreira, A. (1986). Food intake, aggression, and fear behavior in the mother rat: control by neural systems concerned with milk ejection and maternal behavior. *Behavioral Neuroscience*, 100(1), 64–70.

Harmon, A. C., Huhman, K. L., Moore, T. O., and Albers, H. E. (2002). Oxytocin inhibits aggression in female Syrian hamsters. *Journal of Neuroendocrinology*, 14(12), 963–969.

Harrison, R. J., Connor, D. F., Nowak, C., Nash, K., and Melloni, R. H., Jr. (2000). Chronic anabolic-androgenic steroid treatment during adolescence increases anterior hypothalamic vasopressin and aggression in intact hamsters. *Psychoneuroendocrinology*, 25(4), 317–338.

Hasen, N. S. and Gammie, S. C. (2006). Maternal aggression: new insights from Egr-1. *Brain Research*, 1108(1), 147–156.

Heinrichs, M., Baumgartner, T., Kirschbaum, C., and Ehlert, U. (2003). Social support and oxytocin interact to suppress cortisol and subjective responses to psychosocial stress. *Biological Psychiatry*, 54(12), 1389–1398.

Hernando, F., Schoots, O., Lolait, S. J., and Burbach, J. P. (2001). Immunohistochemical localization of the vasopressin V1b receptor in the rat brain and pituitary gland: anatomical support for its involvement in the central effects of vasopressin. *Endocrinology*, 142(4), 1659–1668.

Insel, T. R. and Shapiro, L. E. (1992). Oxytocin receptor distribution reflects social organization in monogamous and polygamous voles. *Proceedings of the National Academy of Sciences USA*, 89(13), 5981–5985.

Insel, T. R., Young, L., Witt, D. M., and Crews, D. (1993). Gonadal steroids have paradoxical effects on brain

oxytocin receptors. *Journal of Neuroendocrinology*, 5(6), 619–628.

Jard, S., Barberis, C., Audigier, S., and Tribollet, E. (1987). Neurohypophyseal hormone receptor systems in brain and periphery. *Progress in Brain Research*, 72, 173–187.

Jarrett, T. M., McMurray, M. S., Walker, C. H., and Johns, J. M. (2006). Cocaine treatment alters oxytocin receptor binding but not mRNA production in postpartum rat dams. *Neuropeptides*, 40(3), 161–167.

Jirikowski, G. F., Caldwell, J. D., Stumpf, W. E., and Pedersen, C. A. (1990). Topography of oxytocinergic estradiol target neurons in the mouse hypothalamus. *Folia et Histochemica Cytobiologica*, 28(1–2), 3–9.

Johns, J. M., Lubin, D. A., Walker, C. H., Meter, K. E., and Mason, G. A. (1997). Chronic gestational cocaine treatment decreases oxytocin levels in the medial preoptic area, ventral tegmental area and hippocampus in Sprague-Dawley rats. *Neuropeptides*, 31(5), 439–443.

Johns, J. M., McMurray, M. S., Joyner, P. W., et al. (2010). Effects of chronic and intermittent cocaine treatment on dominance, aggression, and oxytocin levels in post-lactational rats. *Psychopharmacology*, 211(2), 175–185.

Johnson, A. E., Audigier, S., Rossi, F., et al. (1993). Localization and characterization of vasopressin binding sites in the rat brain using an iodinated linear AVP antagonist. *Brain Research*, 622(1–2), 9–16.

Johnson, A. E., Barberis, C., and Albers, H. E. (1995). Castration reduces vasopressin receptor binding in the hamster hypothalamus. *Brain Research*, 674(1), 153–158.

Kimura, T., Tanizawa, O., Mori, K., Brownstein, M. J., and Okayama, H. (1992). Structure and expression of a human oxytocin receptor. *Nature*, 356(6369), 526–529.

Kirsch, P., Esslinger, C., Chen, Q., et al. (2005). Oxytocin modulates neural circuitry for social cognition and fear in humans. *Journal of Neuroscience*, 25(49), 11489–11493.

Koolhaas, J. M., Everts, H., de Ruiter, A. J., de Boer, S. F., and Bohus, B. (1998). Coping with stress in rats and mice: differential peptidergic modulation of the amygdala-lateral septum complex. *Progress in Brain Research*, 119, 437–448.

Kosfeld, M., Heinrichs, M., Zak, P. J., Fischbacher, U., and Fehr, E. (2005). Oxytocin increases trust in humans. *Nature*, 435(7042), 673–676.

Kremarik, P., Freund-Mercier, M. J., and Stoeckel, M. E. (1993). Histoautoradiographic detection of oxytocin- and vasopressin-binding sites in the telencephalon of the rat. *Journal of Comparative Neurology*, 333(3), 343–359.

Labuschagne, I., Phan, K. L., Wood, A., et al. (2010). Oxytocin attenuates amygdala reactivity to fear in generalized social anxiety disorder. *Neuropsychopharmacology*, 35(12), 2403–2413.

Lapiz, M. D., Fulford, A., Muchimapura, S., et al. (2001). Influence of postweaning social isolation in the rat on brain development, conditioned behaviour and neurotransmission. *Neuroscience and Behavioral Physiology*, 87(6), 730–751.

Lapiz, M. D., Mateo, Y., Durkin, S., Parker, T., and Marsden, C. A. (2001). Effects of central noradrenaline depletion by the selective neurotoxin DSP-4 on the behaviour of the isolated rat in the elevated plus maze and water maze. *Psychopharmacology*, 155(3), 251–259.

Lee, H. J., Macbeth, A. H., Pagani, J. H., and Young, W. S., III. (2009). Oxytocin: the great facilitator of life. *Progress in Neurobiology*, 88(2), 127–151.

Lolait, S. J., O'Carroll, A. M., Mahan, L. C., et al. (1995). Extrapituitary expression of the rat V1b vasopressin receptor gene. *Proceedings of the National Academy of Sciences USA*, 92(15), 6783–6787.

Lonstein, J. S. and Gammie, S. C. (2002). Sensory, hormonal, and neural control of maternal aggression in laboratory rodents. *Neuroscience and Biobehavioral Reviews*, 26(8), 869–888.

Lowry, C. A., Richardson, C. F., Zoeller, T. R., et al. (1997). Neuroanatomical distribution of vasotocin in a urodele amphibian (Taricha granulosa) revealed by immunohistochemical and in situ hybridization techniques. *Journal of Comparative Neurology*, 385(1), 43–70.

Lubin, D. A., Elliott, J. C., Black, M. C., and Johns, J. M. (2003). An oxytocin antagonist infused into the central nucleus of the amygdala increases maternal aggressive behavior. *Behavioral Neuroscience*, 117(2), 195–201.

Lubin, D. A., Meter, K. E., Walker, C. H., and Johns, J. M. (2001a). Dose-related effects of chronic gestational cocaine treatment on maternal aggression in rats on postpartum days 2, 3, and 5. *Progress Neuropsychopharmacology and Biological Psychiatry*, 25(7), 1403–1420.

Lubin, D. A., Meter, K. E., Walker, C. H., and Johns, J. M. (2001b). Effects of chronic cocaine administration on aggressive behavior in virgin rats. *Progress Neuropsychopharmacology and Biological Psychiatry*, 25(7), 1421–1433.

Lukas, M., Bredewold, R., Neumann, I. D., and Veenema, A. H. (2010). Maternal separation interferes with

developmental changes in brain vasopressin and oxytocin receptor binding in male rats. *Neuropharmacology*, 58(1), 78–87.

Mahalati, K., Okanoya, K., Witt, D. M., and Carter, C. S. (1991). Oxytocin inhibits male sexual behavior in prairie voles. *Pharmacology Biochemistry and Behavior*, 39(1), 219–222.

Malick, J. B. (1975). Effects of age and food deprivation on the development of muricidal behavior in rats. *Physiology and Behavior*, 14(2), 171–175.

Malick, J. B. (1979). The pharmacology of isolation-induced aggressive behavior in mice. *Curr Dev Psychopharmacology*, 5, 1–27.

Mason, W. T., Ho, Y. W., and Hatton, G. I. (1984). Axon collaterals of supraoptic neurones: anatomical and electrophysiological evidence for their existence in the lateral hypothalamus. *Neuroscience*, 11(1), 169–182.

Melloni, R. H., Jr., Connor, D. F., Hang, P. T., Harrison, R. J., and Ferris, C. F. (1997). Anabolic-androgenic steroid exposure during adolescence and aggressive behavior in golden hamsters. *Physiology and Behavior*, 61(3), 359–364.

Meyer-Lindenberg, A. (2008). Impact of prosocial neuropeptides on human brain function. *Progress in Brain Research*, 170, 463–470.

Miczek, K. A., de Almeida, R. M., Kravitz, E. A., et al. (2007). Neurobiology of escalated aggression and violence. *Journal of Neuroscience*, 27(44), 11803–11806.

Miczek, K. A., Maxson, S. C., Fish, E. W., and Faccidomo, S. (2001). Aggressive behavioral phenotypes in mice. *Behavioural Brain Research*, 125(1–2), 167–181.

Millan, M. J., Millan, M. H., Czlonkowski, A., and Herz, A. (1984). Vasopressin and oxytocin in the rat spinal cord: distribution and origins in comparison to [Met]enkephalin, dynorphin and related opioids and their irresponsiveness to stimuli modulating neurohypophyseal secretion. *Neuroscience*, 13(1), 179–187.

Miller, M. A., Ferris, C. F., and Kolb, P. E. (1999). Absence of vasopressin expression by galanin neurons in the golden hamster: implications for species differences in extrahypothalamic vasopressin pathways. *Molecular Brain Research*, 67(1), 28–35.

Moore, F. L., Lowry, C. A., and Rose, J. D. (1994). Steroid-neuropeptide interactions that control reproductive behaviors in an amphibian. *Psychoneuroendocrinology*, 19(5–7), 581–592.

Moyer, K. E. (1968). Kinds of aggression and their physiological basis. *Communications in Behavioral Biology*, 2A, 65–87.

Nephew, B. C. and Bridges, R. S. (2008). Central actions of arginine vasopressin and a V1a receptor antagonist on maternal aggression, maternal behavior, and grooming in lactating rats. *Pharmacology Biochemistry and Behavior*, 91(1), 77–83.

Nephew, B. C., Bridges, R. S., Lovelock, D. F., and Byrnes, E. M. (2009). Enhanced maternal aggression and associated changes in neuropeptide gene expression in multiparous rats. *Behavioral Neuroscience*, 123(5), 949–957.

Olazabal, D. E. and Ferreira, A. (1997). Maternal behavior in rats with kainic acid-induced lesions of the hypothalamic paraventricular nucleus. *Physiology and Behavior*, 61(5), 779–784.

Ophir, A. G., Crino, O. L., Wilkerson, Q. C., Wolff, J. O., and Phelps, S. M. (2008). Female-directed aggression predicts paternal behavior, but female prairie voles prefer affiliative males to paternal males. *Brain Behavior and Evolution*, 71(1), 32–40.

Ostrowski, N. L., Lolait, S. J., Bradley, D. J., et al. (1992). Distribution of V1a and V2 vasopressin receptor messenger ribonucleic acids in rat liver, kidney, pituitary and brain. *Endocrinology*, 131(1), 533–535.

Ostrowski, N. L., Lolait, S. J., and Young, W. S., III. (1994). Cellular localization of vasopressin V1a receptor messenger ribonucleic acid in adult male rat brain, pineal, and brain vasculature. *Endocrinology*, 135(4), 1511–1528.

Otten, S. and Stapel, D. A. (2007). Who is this Donald? How social categorization affects aggression-priming effects. *European Journal of Social Psychology*, 37, 1000–1015.

Planas, B., Kolb, P. E., Raskind, M. A., and Miller, M. A. (1995). Vasopressin and galanin mRNAs coexist in the nucleus of the horizontal diagonal band: a novel site of vasopressin gene expression. *Journal of Comparative Neurology*, 361(1), 48–56.

Reber, S. O. and Neumann, I. D. (2008). Defensive behavioral strategies and enhanced state anxiety during chronic subordinate colony housing are accompanied by reduced hypothalamic vasopressin, but not oxytocin, expression. *Annals of the New York Academy of Sciences*, 1148, 184–195.

Rinaman, L. (1998). Oxytocinergic inputs to the nucleus of the solitary tract and dorsal motor nucleus of the vagus in neonatal rats. *Journal of Comparative Neurology*, 399(1), 101–109.

Roche, K. E. and Leshner, A. I. (1979). ACTH and vasopressin treatments immediately after a defeat increase future submissiveness in male mice. *Science*, 204(4399), 1343–1344.

Roper, J. A., Grant, E., Craighead, M., et al. (2009). The role of the vasopressin Vl b receptor in the HPA axis response to stress: molecular and pharmacological studies, *World Conference on Neurohypophysial Hormones*. Kitakyushu, Japan.

Ross, H. E. and Young, L. J. (2009). Oxytocin and the neural mechanisms regulating social cognition and affiliative behavior. *Frontiers in Neuroendocrinology*, 30(4), 534–547.

Savaskan, E., Ehrhardt, R., Schulz, A., Walter, M., and Schachinger, H. (2008). Post-learning intranasal oxytocin modulates human memory for facial identity. *Psychoneuroendocrinology*, 33(3), 368–374.

Sawchenko, P. E. and Swanson, L. W. (1982). Immunohistochemical identification of neurons in the paraventricular nucleus of the hypothalamus that project to the medulla or to the spinal cord in the rat. *Journal of Comparative Neurology*, 205(3), 260–272.

Scott, J. P. (1966). Agonistic behavior of mice and rats: a review. *American Zoologist*, 6: 683–701.

Semsar, K., Kandel, F., and Godwin, J. (2001). Manipulations of the AVT system shift social status and related courtship and aggressive behavior in the bluehead wrasse. *Hormones and Behavior*, 40(1), 21–31.

Serradeil-Le Gal, C., Wagnon, J., Tonnerre, B., et al. (2005). An overview of SSR149415, a selective nonpeptide vasopressin V(1b) receptor antagonist for the treatment of stress-related disorders. *CNS Drug Reviews*, 11(1), 53–68.

Shamay-Tsoory, S. G., Fischer, M., Dvash, J., et al. (2009). Intranasal administration of oxytocin increases envy and schadenfreude (gloating). *Biological Psychiatry*, 66(9), 864–870.

Siegel, H. I. (1985). Aggressive Behavior. In H. I. Siegel (Ed.), *The Hamster: Reproduction and Behavior* (pp. 261–286). New York: Plenum Press.

Sofroniew, M. V. (1983). Morphology of vasopressin and oxytocin neurones and their central and vascular projections. *Progress in Brain Research*, 60, 101–114.

Sofroniew, M. V. (1985). Vasopressin- and neurophysin-immunoreactive neurons in the septal region, medial amygdala and locus coeruleus in colchicine-treated rats. *Neuroscience*, 15(2), 347–358.

Stemmelin, J., Lukovic, L., Salome, N., and Griebel, G. (2005). Evidence that the lateral septum is involved in the antidepressant-like effects of the vasopressin V1b receptor antagonist, SSR149415. *Neuropsychopharmacology*, 30(1), 35–42.

Stribley, J. M. and Carter, C. S. (1999). Developmental exposure to vasopressin increases aggression in adult prairie voles. *Proceedings of the National Academy of Sciences USA*, 96(22), 12601–12604.

Svare, B., Betteridge, C., Katz, D., and Samuels, O. (1981). Some situational and experiential determinants of maternal aggression in mice. *Physiology and Behavior*, 26(2), 253–258.

Szot, P., Bale, T. L., and Dorsa, D. M. (1994). Distribution of messenger RNA for the vasopressin V1a receptor in the CNS of male and female rats. *Molecular Brain Research*, 24(1–4), 1–10.

Takayanagi, Y., Yoshida, M., Bielsky, I. F., et al. (2005). Pervasive social deficits, but normal parturition, in oxytocin receptor-deficient mice. *Proceedings of the National Academy of Sciences USA*, 102(44), 16096–16101.

Thompson, R., Gupta, S., Miller, K., Mills, S., and Orr, S. (2004). The effects of vasopressin on human facial responses related to social communication. *Psychoneuroendocrinology*, 29(1), 35–48.

Thompson, R. R., George, K., Walton, J. C., Orr, S. P., and Benson, J. (2006). Sex-specific influences of vasopressin on human social communication. *Proceedings of the National Academy of Sciences USA*, 103(20), 7889–7894.

Todeschin, A. S., Winkelmann-Duarte, E. C., Jacob, M. H., et al. (2009). Effects of neonatal handling on social memory, social interaction, and number of oxytocin and vasopressin neurons in rats. *Hormones and Behavior*, 56(1), 93–100.

Tribollet, E., Barberis, C., and Arsenijevic, Y. (1997). Distribution of vasopressin and oxytocin receptors in the rat spinal cord: sex-related differences and effect of castration in pudendal motor nuclei. *Neuroscience*, 78(2), 499–509.

Urban, J. H., Miller, M. A., Drake, C. T., and Dorsa, D. M. (1990). Detection of vasopressin mRNA in cells of the medial amygdala but not the locus coeruleus by in situ hybridization. *Journal of Chemical Neuroanatomy*, 3(4), 277–283.

Vaccari, C., Lolait, S. J., and Ostrowski, N. L. (1998). Comparative distribution of vasopressin V1b and oxytocin receptor messenger ribonucleic acids in brain. *Endocrinology*, 139(12), 5015–5033.

Valzelli, L. (1969). Aggressive behavior induced by isolation. In S. Garattini and E. B. Sigg (Eds.), *Aggressive Behavior* (pp. 70–76). London: Excerta Medica Foundation.

van Leeuwen, F. W., Caffe, A. R., and De Vries, G. J. (1985). Vasopressin cells in the bed nucleus of the stria

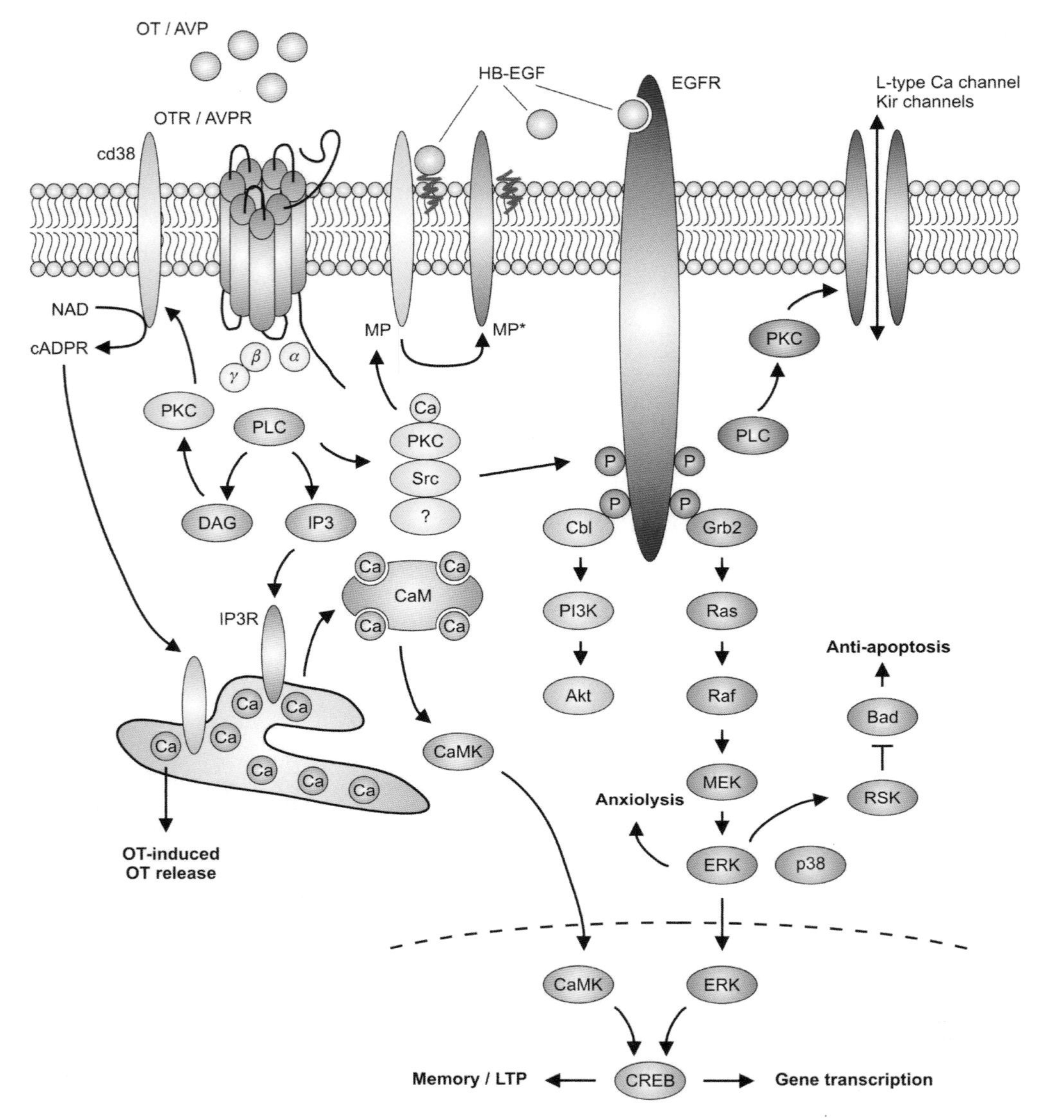

Figure 2.2 Pathways and associated processes controlled by OT and AVP. This scheme has been constructed on the basis of the publications cited in the text. Please note that these concern studies in different neurons in several brain regions, and that not all the processes depicted in the figure necessarily occur in every cell type, or are implicated in any type of behavior. *Blue module:* OT- / AVP-binding to their receptors (OTR / AVPR) activates phospholipase C (PLC), which mobilizes inositol trisphosphate (IP3) and diacylglycerol (DAG). IP3 recruits Ca^{2+} from intracellular stores by binding to the IP3 receptor (IP3R). The resulting increase of intracellular Ca^{2+} activates calmodulin (CaM) and Ca^{2+}/calmodulin-dependent kinases (CaMK); part of activated CaMK translocates to the nucleus. *Light green module:* Although DAG activates protein kinase C (PKC) following both OTR / AVPR receptor binding, PKC activates cd38 only in the case of OTR activation, leading to the formation of cADPR. cADPR stimulates the release of intracellular Ca^{2+} independently from IP3, resulting in autostimulation of OT release in the SON and PVN. *Purple module:* OTR and AVPR activation transactivate the epidermal growth factor receptor (EGFR) via two pathways. First, PLC may recruit Ca^{2+}, PKC, Src, or an unknown factor (cell-type dependent), which then activate a metalloproteinase (MP). Activated MP (MP*) liberates membrane-bound EGF (or heparin-binding EGF, HB-EGF), which binds to and activates the EGFR. Second, Src phosphorylates the EGFR directly. *Orange module:* Activated EGFR binds the adaptor molecule Cbl, which binds PI3K. PI3K activates Akt, which is a central mediator of many intracellular processes. *Brown module:* Activated EGFR binds the adaptor molecule complex Shc/Grb2/Sos (depicted as Grb2), which activates Ras. This activates the MAPK cascade Raf – MEK1/2 – ERK1/2, as well as p38. Activated MEK1/2 is necessary for the anxiolytic activity of OT. It also inhibits apoptosis when phosphorylated following AVPR activation via the phosphorylation of RSK, and subsequent phosphorylation deactivation of Bad. ERK1/2 and CaMK in the nucleus phosphorylate CREB to induce LTP and improvement of spatial memory (OT-mediated), and to control gene expression. *Dark green module:* OT and AVP exert fast effects on ion channels, including L-type Ca^{2+} and inwardly rectifying K^{+} channels (Kir channels), which might depend on a PLC – PKC sequence.

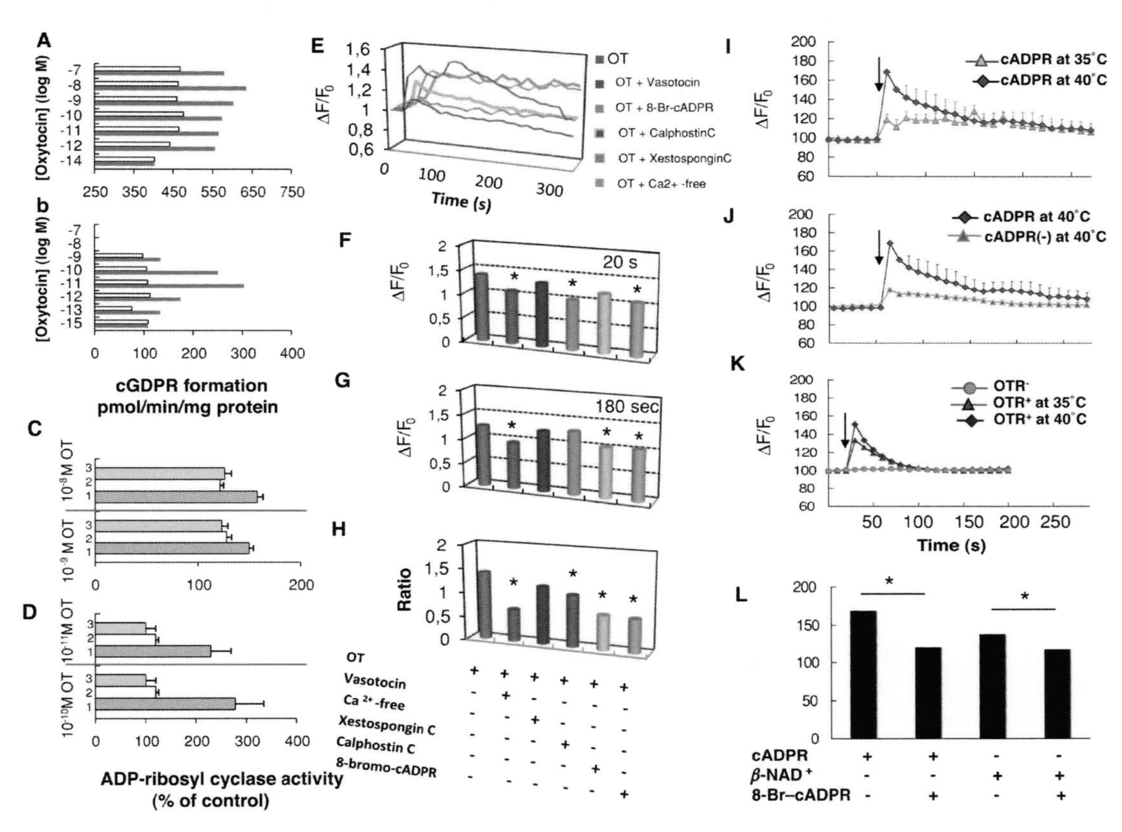

Figure 3.2 ADP-ribosyl cyclase activities and intracellular calcium concentrations under various conditions. A, B, ADP-ribosyl cyclase activities were measured as the rate of cyclic GDP-ribose formation by homogenates isolated from mouse hypothalamus (A) and posterior pituitary (B) for 5 min under various concentrations of OT with (open bars) or without 1 mM of the OT receptor antagonist, vasotocin (filled bars). C, D, ADP-ribosyl cyclase activity, presented as percentages of control cyclic GDP-ribose formation activity, were measured in the presence or absence of OT (4 different concentrations indicated), PKI-STSP (5 nM) or PKC-inhibitor calphostin C (100 nM) in the hypothalamus (C) and pituitary (D). 1 – OT present, 2 – OT and PKI-STSP, 3 – OT and calphostin C. E, Average time courses of changes in $[Ca^{2+}]_i$ elicited with 100 pM OT with or without 1 mM OT receptor antagonist vasotocin, 100 nM PKC-inhibitor calphostin C, 2 mM IP_3-inhibitor Xestospongin C, 100 μM 8-bromo-cADPR, and extracellular Ca^{2+}. F, G, Average increases in $[Ca^{2+}]_i$ measured at 20 s (F) and 180 s (G) after OT stimulation. Data are shown as changes in fluorescence divided by resting fluorescence, i.e. $\Delta F/F_0$. H, OT concentrations are presented as OT release ratio (arbitrary unit) in an isolated nerve endings under 100 pM OT stimulation (5 min) with or without 1 mM of the oxytocin receptor antagonist, vasotocin, 100 nM PKC-inhibitor calphostin C, 2 mM IP_3-Inhibitor Xestospongin C, 100 mM 8-bromo-cADPR, and extracellular Ca^{2+}, as indicated. I, Time courses of changes in $[Ca^{2+}]_i$ drawn from fluorescence imaging comparing stimulation at 35 and 40 °C in the presence of cADPR in NG108–15 cells. The arrow indicates an application of cADPR. J, Time courses of changes in $[Ca^{2+}]_i$ with (rhomb) or without 50 μM cADPR (triangle) at 40°C in NG108–15 cells. The arrow indicates the application of cADPR. K, Effects of oxytocin on $[Ca^{2+}]_i$ at 35 °C (triangle) or 40 °C (rhomb) in control (OTR-) or transformed NG108–15 cells to express human OT receptors (OTR+). The arrow indicates the addition of 200 nM OT (final concentration, 100 nM) from a calibrated micropipette into the recording medium. L, $[Ca^{2+}]_i$ elevation induced by extracellular application of 50 μM cADPR or 100 μM β-NAD^+ at 40 °C in the presence or absence of 50 μM 8-bromo-cADPR is expressed as percentage change over those before stimulation.

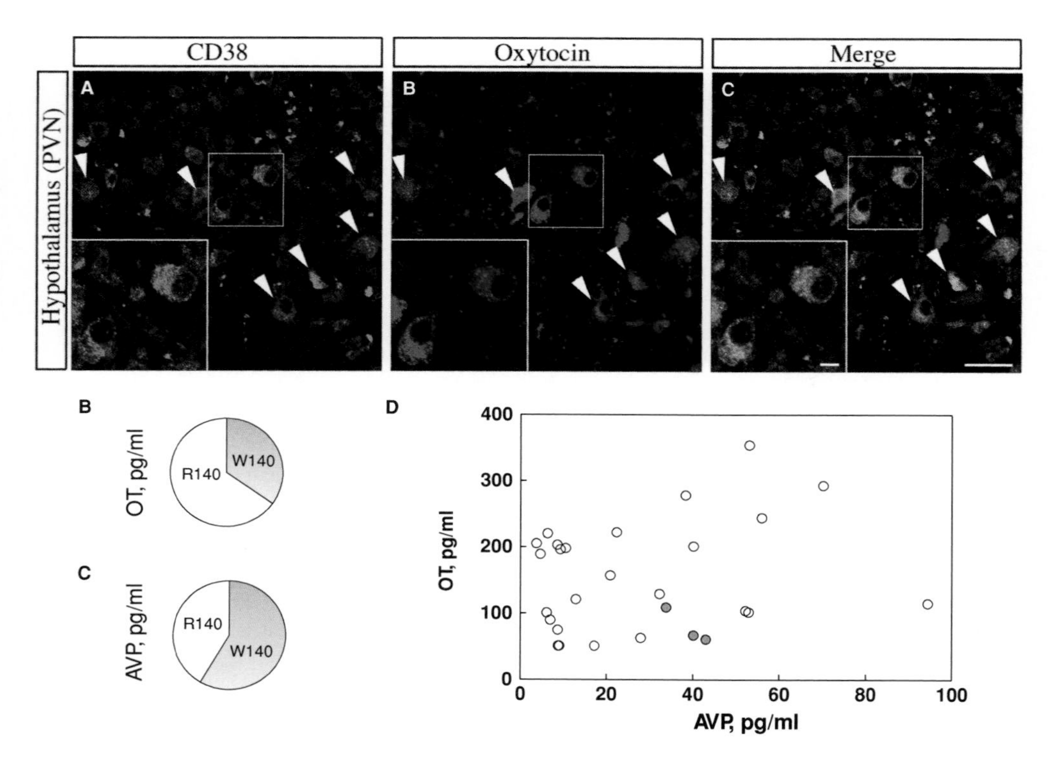

Figure 3.3 Studies in human subjects. A, Immunohistochemical analysis of CD38 (A) and oxytocin (B) in the human brain. Cell montages of panels were taken from the paraventricular nucleus (PVN) in the hypothalamus of autopsy subjects from the USA. Arrowheads indicate extensive colabeling. The insets in panels are enlarged images of neurons showing coexpression of CD38 and OT. B, C, Plasma oxytocin (B) and vasopressin (C) levels in ASD subjects with R140 or W140 allele. D, Scatchard plot of plasma concentrations of OT and AVP levels in 29 ASD patients with (filled circle) or without (open circle) the W140 allele.

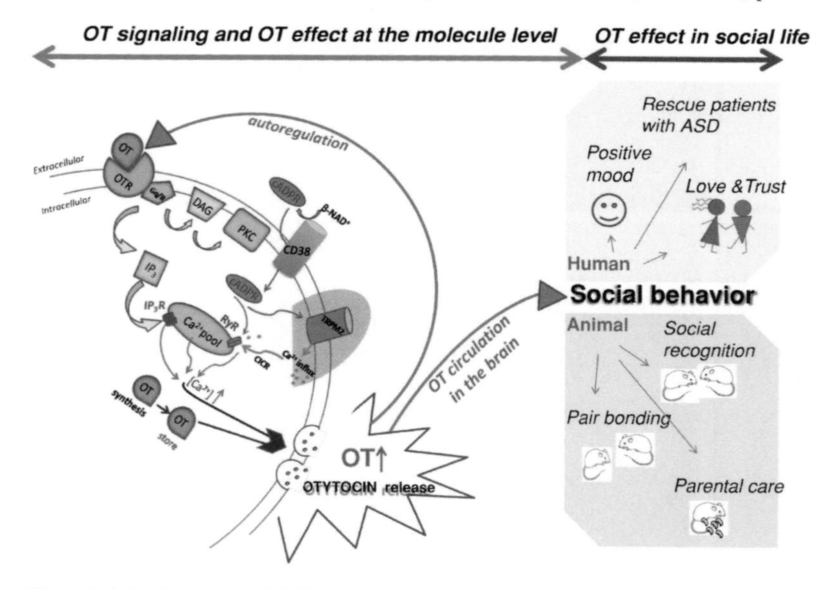

Figure 3.4 A schematic model of oxytocin signaling and oxytocin effect by cyclic ADP-ribose (cADPR) and heat that induces $[Ca^{2+}]_i$ increases. cADPR, extracellularly applied or converted by the ADP-ribosyl cyclase activity of CD38, is transported into the cell, and activates the ryanodine receptor (RyR). Consequently, binding of cADPR to melastatin-related transient receptor potential channel 2 (TRPM2) channels initiates Ca^{2+} influx. OT binding with OTR and through the cascade also initiates Ca^{2+} influx and OT release. These molecular events modulate social events and social life in rodent and human. The molecular event associated with the heat sensitivity is shaded in red. Targets of PKC are not shown. cADPR is also produced intracellulaly, but not illustrated. OT, oxytocin; OTR, oxytocin receptor; DAG, diacylglycerol; IP_3, inositol trisphosphate; IP_3R, inositol trisphosphate receptor; PKC, protein kinase C; CICR, Ca^{2+}-induced Ca^{2+} release.

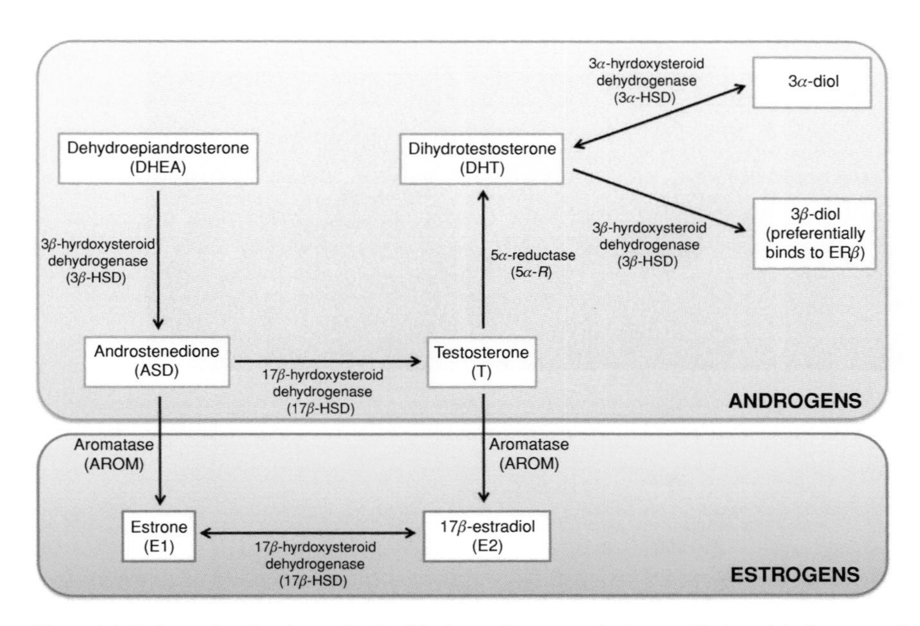

Figure 1.1 Pathway showing the synthesis of the key androgens and estrogens that regulate the expression of OXT and AVP peptides and receptors. Adapted and modified from Handa et al., 2009.

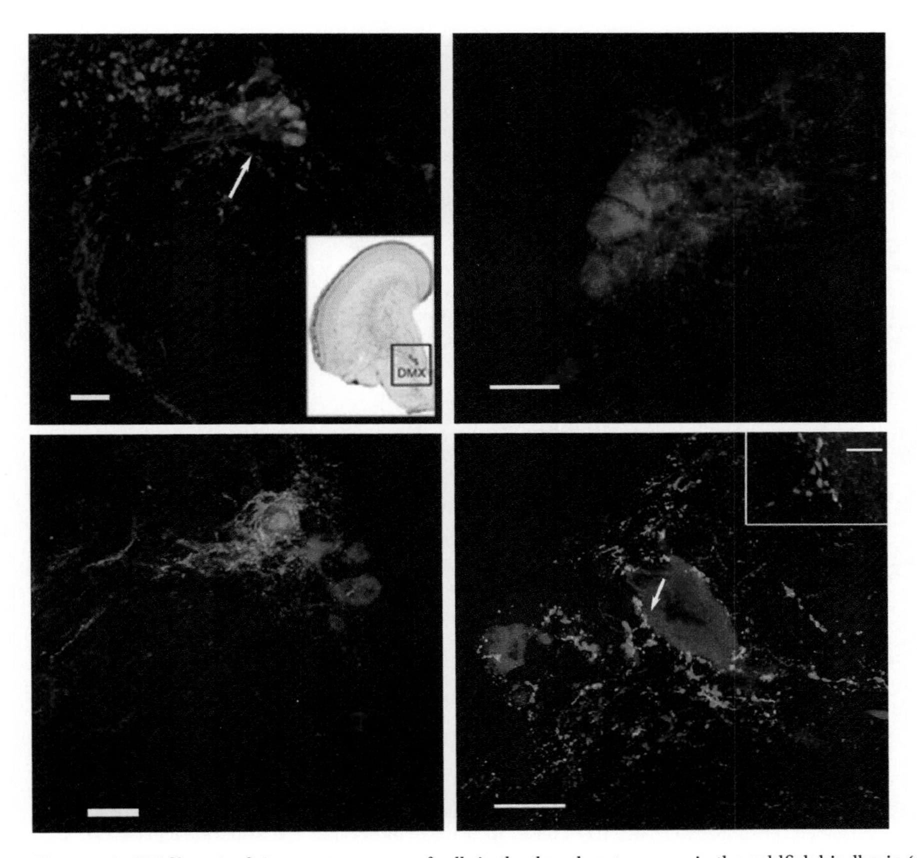

Figure 5.3 AVT fibers (red) innervate a group of cells in the dorsal motor vagus in the goldfish hindbrain (A) that are backfilled by intraperitoneal injections of a retrograde tracer (B; backfilled cells are blue/gold), indicating they project out the vagus nerve into the periphery. Those cells are immunoreactive for Substance P (C; Substance P cells are shown in red, AVT fibers in green). AVT fibers appear to directly contact the cell bodies of those neurons (D). From Thompson et al., 2008. Blocking peripheral Substance P receptors prevents central AVT from inhibiting social approach behavior, suggesting this connection is part of a central/peripheral feedback loop involved in social regulation in this species (see text).

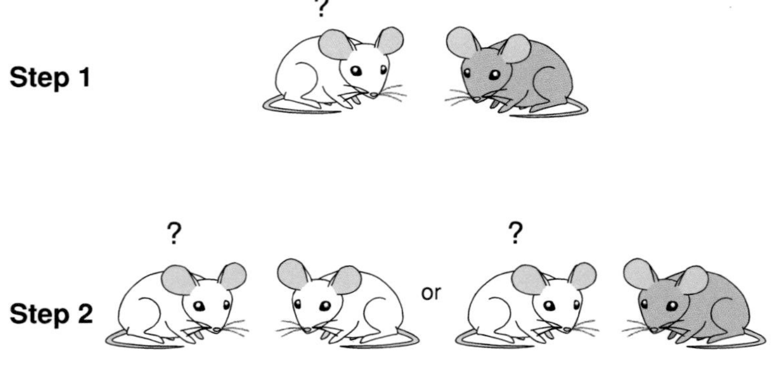

Figure 13.1 The habituation–dishabituation paradigm for the assessment of social recognition in laboratory rodents. In step 1, the experimental rodent (white) is presented with a conspecific (gray) with whom he/she can become familiar, either over a single or over multiple presentations. In step 2, the experimental rodent is either presented with the same (gray), now familiar, conspecific, or with a novel (white) unfamiliar conspecific. Typically, in step 2 the experimental rodent will show decreased social investigation of a familiar conspecific (habituation) and increased social investigation of a novel conspecific (dishabituation). This same paradigm can also be performed using only odor cues from conspecifics.

Figure 13.2 The discrimination paradigm for the assessment of social recognition in laboratory rodents. In step 1, the experimental rodent (white) is presented with either one (gray) or two (gray and white) conspecifics with whom he/she can become familiar, either over a single or over multiple presentations. In step 2, the experimental rodent is concurrently presented with a familiar (gray) and an unfamiliar (black) conspecific. Typically, in step 2 the experimental rodent will investigate the unfamiliar more than the familiar conspecific. This same paradigm can also be performed using only odor cues from conspecifics.

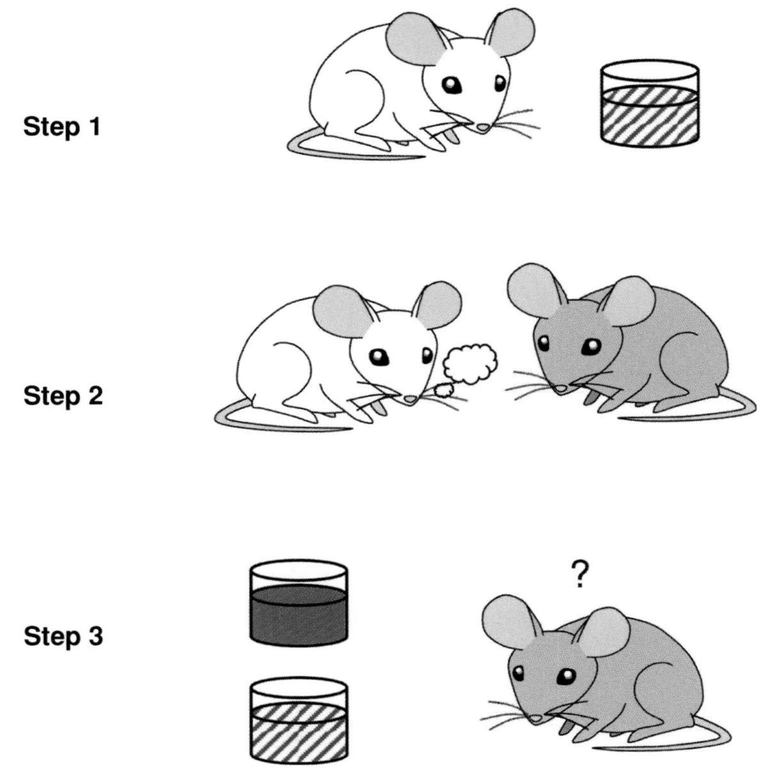

Figure 13.3 The social transmission of food preferences paradigm for the assessment of social learning in laboratory rodents. In step 1, the *demonstrator* rodent (white) consumes a novel flavored diet. In step 2, the demonstrator is allowed to interact with an *observer* rodent (gray) for a certain period of time. In step 3 the observer rodent is given a choice between two novel, differently flavored diets, one of which is the same that was present on their demonstrator's breath during the social interaction. Typically, in step 3 the observer rodent will prefer to consume the novel flavored diet that their demonstrator has previously eaten. This paradigm can be used to assess the mechanisms of acquisition, consolidation and retrieval of the socially acquired food related memory.

Figure 13.4 The mate-copying paradigm for the assessment of social learning in laboratory rodents. In step 1, the *observer* rodent (white female) witnesses two conspecifics mate. In step 2, the observer rodent is given a choice between two opposite sex conspecifics, one of which is the same they have witnessed mating. Typically, in step 2 an observer female rodent will prefer to mate with the male they have witnessed mating with another female. In male rodents, only lower-quality subordinate males were shown to copy the mating preferences of dominant males. This same paradigm can also be performed using only odor cues from conspecifics.

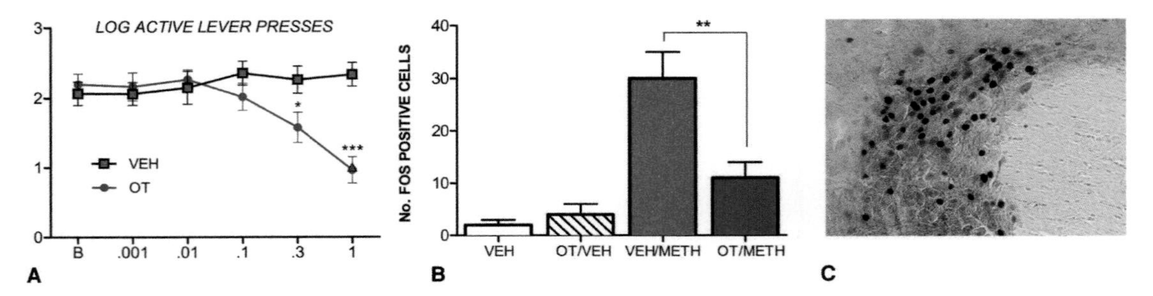

Figure 15.1 A. Peripheral oxytocin (doses on the *x*-axis) reduces intravenous METH self-administration in rats (Carson et al., 2010a). B. Peripheral oxytocin (2 mg/kg) reduces METH-induced activation of the Subthalamic Nucleus (Carson et al., 2010b). C. Peripheral oxytocin (2mg/kg) activates OT-positive neurons in the Supraoptic Nucleus (Carson et al., 2010b).

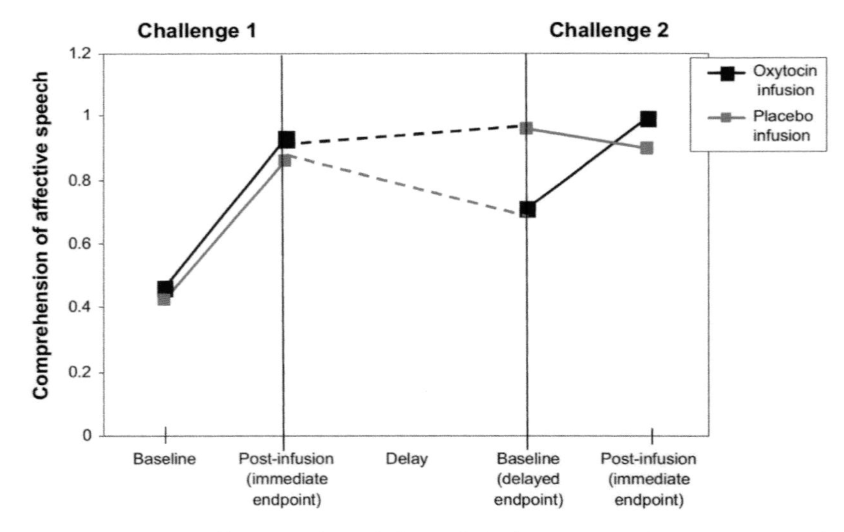

Figure 20.2 Estimated linear trends (on the basis of mixed regression analysis) across time in the dichotomized scores of the affective speech comprehension as a function of conditions (oxytocin vs. placebo) and order of administration (oxytocin first vs. placebo first). Dotted line represents interval between the first and second challenge procedures (days: mean = 16.07; SD = 14.26).

A

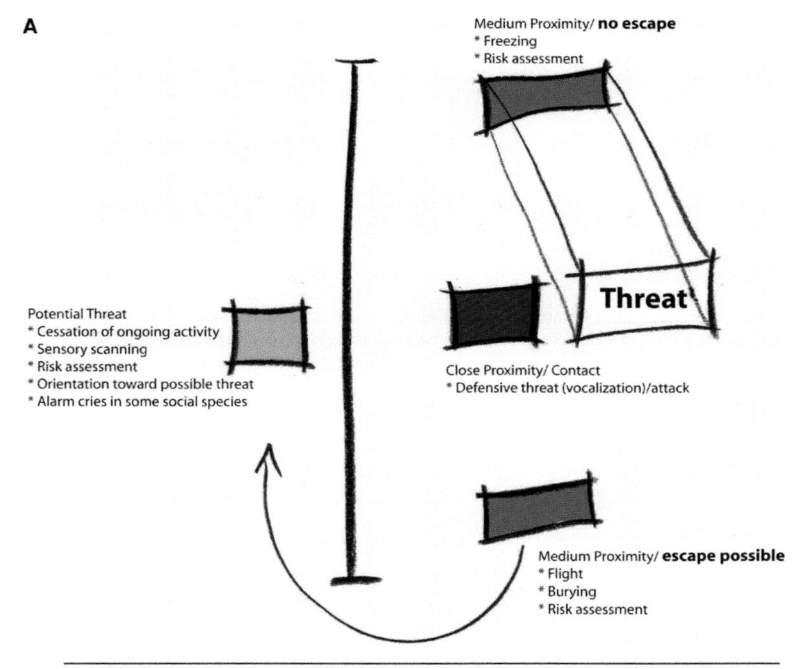

B

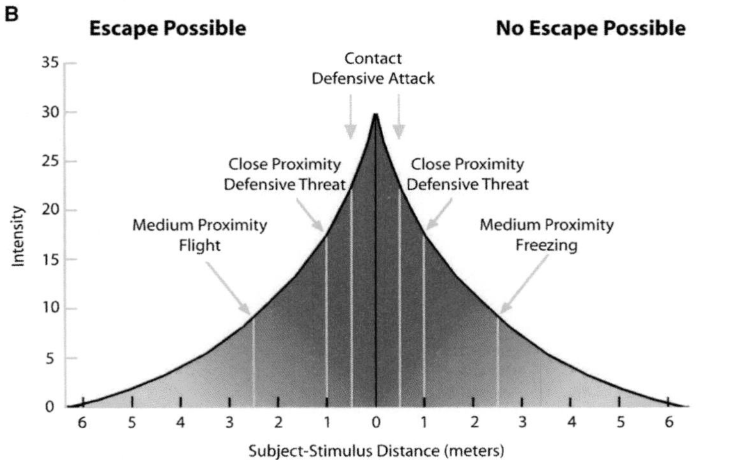

Figure 17.1 Fear- and anxiety-like defensive behaviors are modulated by subject-threat distance and context (e.g. availability of escape). (Graph B adapted from Blanchard et al., 2009).

terminalis of the rat: sex differences and the influence of androgens. *Brain Research*, 325(1–2), 391–394.

Veenema, A. H., Beiderbeck, D. I., Lukas, M., and Neumann, I. D. (2010). Distinct correlations of vasopressin release within the lateral septum and the bed nucleus of the stria terminalis with the display of intermale aggression. *Hormones and Behavior*, 58(2), 273–281.

Veenema, A. H., Blume, A., Niederle, D., Buwalda, B., and Neumann, I. D. (2006). Effects of early life stress on adult male aggression and hypothalamic vasopressin and serotonin. *European Journal of Neuroscience*, 24(6), 1711–1720.

Veenema, A. H., Bredewold, R., and Neumann, I. D. (2007). Opposite effects of maternal separation on intermale and maternal aggression in C57BL/6 mice: link to hypothalamic vasopressin and oxytocin immunoreactivity. *Psychoneuroendocrinology*, 32(5), 437–450.

Veenema, A. H. and Neumann, I. D. (2009). Maternal separation enhances offensive play-fighting, basal corticosterone and hypothalamic vasopressin mRNA expression in juvenile male rats. *Psychoneuroendocrinology*, 34(3), 463–467.

Veening, J. G., de Jong, T., and Barendregt, H. P. (2010). Oxytocin-messages via the cerebrospinal fluid: behavioral effects; a review. *Physiology and Behavior*, 101(2), 193–210.

Veinante, P. and Freund-Mercier, M. J. (1997). Distribution of oxytocin- and vasopressin-binding sites in the rat extended amygdala: a histoautoradiographic study. *Journal of Comparative Neurology*, 383(3), 305–325.

Virkkunen, M., Rawlings, R., Tokola, R., et al. (1994). CSF biochemistries, glucose metabolism, and diurnal activity rhythms in alcoholic, violent offenders, fire setters, and healthy volunteers. *Archives of General Psychiatry*, 51(1), 20–27.

Volpi, S., Rabadan-Diehl, C., and Aguilera, G. (2004). Regulation of vasopressin V1b receptors and stress adaptation. *Annals of the New York Academy of Sciences*, 1018, 293–301.

Wang, Z., Zhou, L., Hulihan, T. J., and Insel, T. R. (1996). Immunoreactivity of central vasopressin and oxytocin pathways in microtine rodents: a quantitative comparative study. *Journal of Comparative Neurology*, 366(4), 726–737.

Watters, J. J., Poulin, P., and Dorsa, D. M. (1998). Steroid hormone regulation of vasopressinergic neurotransmission in the central nervous system. *Progress in Brain Research*, 119, 247–261.

Wersinger, S. R., Caldwell, H. K., Christiansen, M., and Young, W. S., III. (2007). Disruption of the vasopressin 1b receptor gene impairs the attack component of aggressive behavior in mice. *Genes Brain and Behavior*, 6(7), 653–660.

Wersinger, S. R., Caldwell, H. K., Martinez, L., et al. (2007). Vasopressin 1a receptor knockout mice have a subtle olfactory deficit but normal aggression. *Genes Brain and Behavior*, 6(6), 540–551.

Wersinger, S. R., Ginns, E. I., O'Carroll, A. M., Lolait, S. J., and Young, W. S., III. (2002). Vasopressin V1b receptor knockout reduces aggressive behavior in male mice. *Molecular Psychiatry*, 7(9), 975–984.

Wersinger, S. R., Kelliher, K. R., Zufall, F., et al. (2004). Social motivation is reduced in vasopressin 1b receptor null mice despite normal performance in an olfactory discrimination task. *Hormones and Behavior*, 46(5), 638–645.

Wersinger, S. R., Temple, J. L., Caldwell, H. K., and Young, W. S., III. (2008). Inactivation of the oxytocin and the vasopressin (Avp) 1b receptor genes, but not the Avp 1a receptor gene, differentially impairs the Bruce effect in laboratory mice (Mus musculus). *Endocrinology*, 149(1), 116–121.

Winslow, J. T., Hastings, N., Carter, C. S., Harbaugh, C. R., and Insel, T. R. (1993). A role for central vasopressin in pair bonding in monogamous prairie voles. *Nature*, 365(6446), 545–548.

Winslow, J. T., Hearn, E. F., Ferguson, J., et al. (2000). Infant vocalization, adult aggression, and fear behavior of an oxytocin null mutant mouse. *Hormones and Behavior*, 37(2), 145–155.

Winslow, J. T. and Insel, T. R. (1991). Social status in pairs of male squirrel monkeys determines the behavioral response to central oxytocin administration. *Journal of Neuroscience*, 11(7), 2032–2038.

Winslow, J. T., Shapiro, L., Carter, C. S., and Insel, T. R. (1993). Oxytocin and complex social behavior: species comparisons. *Psychopharmacology Bulletin*, 29(3), 409–414.

Young, L. J., Wang, Z., Cooper, T. T., and Albers, H. E. (2000). Vasopressin (V1a) receptor binding, mRNA expression and transcriptional regulation by androgen in the Syrian hamster brain. *Journal of Neuroendocrinology*, 12(12), 1179–1185.

Young, W. S., Li, J., Wersinger, S. R., and Palkovits, M. (2006). The vasopressin 1b receptor is prominent in

the hippocampal area CA2 where it is unaffected by restraint stress or adrenalectomy. *Neuroscience*, 143(4), 1031–1039.

Zak, P. J., Stanton, A. A., and Ahmadi, S. (2007). Oxytocin increases generosity in humans. *PLoS One*, 2(11), e1128.

Zink, C. F., Stein, J. L., Kempf, L., Hakimi, S., and Meyer-Lindenberg, A. (2010). Vasopressin modulates medial prefrontal cortex-amygdala circuitry during emotion processing in humans. *Journal of Neuroscience*, 30(20), 7017–7022.

12

Role of vasopressin in flank marking and aggression

Craig F. Ferris, Richard H. Melloni, Jr., and H. Elliott Albers

12.1 Introduction

While investigating the effect of different neuropeptides on circadian timing in male golden hamsters (*Mesocricetus auratus*) we made the serendipitous discovery that arginine vasopressin (AVP) microinjected dorsal to the suprachiasmatic nucleus caused a robust stereotypic motor behavior – flank marking (Ferris et al., 1984b). Moments after microinjection, hamsters frantically groomed and wet their sebaceous glands on the dorso-lateral flanks, arched their back and vigorously rubbed the glands against the walls of their home cage. This behavior occurs naturally in hamsters and functions in social communication by disseminating pheromonal signals from the glands to objects in the environment (Johnston, 1975). This behavioral effect was unique to AVP and not other vasoactive peptides (Albers and Ferris, 1985). This was the first example of a neuropeptide having a discrete action in the brain on the central regulation of a complex communicative behavior. While this stereotyped flank-marking behavior is unique to hamsters, this chance discovery brought attention to AVP's potential as a chemical signal in the regulation of species-specific social behaviors.

12.2 Characterizing the effects of AVP on flank marking

Following these initial findings that injection of AVP into the hypothalamus could induce flank marking, we conducted a series of experiments to more fully characterize this phenomenon. AVP was found to stimulate high levels of marking in both male and female hamsters and these effects were anatomically discrete (Ferris et al., 1984b). The area sensitive to AVP extended from the caudal medial preoptic area (MPOA) to the caudal anterior hypothalamus (AH). Flank marking was not induced by other peptides including oxytocin (OT), although subsequent studies have demonstrated that high concentrations of OT can induced flank marking in male but not female hamsters (Harmon et al., 2002). We also quantified the effects of AVP and several other peptides on grooming of the flank gland region. AVP and OT were found to produce two to three times more grooming of the flank gland when injected into the MPOA-AH than vehicle or other peptides (Albers and Ferris, 1985). We also examined whether peripheral feedback from the flank gland (e.g., flank gland grooming) was necessary for AVP to induce flank marking (Albers and Ferris, 1986).

Oxytocin, Vasopressin, and Related Peptides in the Regulation of Behavior, ed. E. Choleris, D. W. Pfaff, and M. Kavaliers.
Published by Cambridge University Press.

Interestingly, surgical removal of the flank gland did not reduce flank marking in response to AVP or grooming of the flank gland region in response to AVP or OT. Taken together, these data suggested the AVP acted in a limited area of the hypothalamus to induce high levels of communicative behavior even in the absence of peripheral feedback.

12.3 Pharmacology

In a series of studies, we examined the pharmacology of AVP-induced flank marking. AVP-stimulated flank marking is dose dependent with an ED50 (50% of full behavioral response) of 0.9 μM when given in a dose of 0.1 ng in a microinjection volume of 100 nl (Ferris et al., 1988a). Flank marking occurs within 60–90 s of microinjection, lasts for 5–6 min and can be elicited the following day by an injection into the same brain area. With the aid of highly selective AVP receptor antagonists and agonists provided by Maurice Manning and Wilbur Sawyer, we reported that AVP-triggered flank marking was mediated by a AVP_{1a} receptor ($AVPR_{1A}$) (Albers et al., 1986b). Indeed, natural flank-marking behavior elicited by odors from conspecifics can be inhibited by the site-specific microinjection of a AVP_{1a} antagonist into the MPOA-AH of male hamsters (Ferris et al., 1985). It was subsequently shown that AVP_{1a} antagonists produced a dose-dependent reduction in AVP-stimulated flank marking that lasts for over 12 h and recovers by the next day (Ferris et al., 1988a) setting the stage for a series of studies using these tools to parse out the role of AVP in different social behaviors.

12.4 Behavior

Natural flank-marking behavior is easily elicited by placing a hamster into the recently vacated home cage of another hamster. In this context, the intruder investigates the novel cage, grooms their flank gland and proceeds to scent mark the sides and corners of the cage communicating their presence in this new environment. The frequency of scent marking in this context is usually 8–10 flank marks in a 10-min observation period. To elicit a higher frequency of flank marking, hamsters with no prior social history can be placed into a neutral arena and allowed to establish a dominant/subordinate relationship (Ferris et al., 1987). Interestingly, the initial confrontation presents with little flank marking but a high level of offensive aggression, that is, initiated attacks and bites, as combatants establish a dominant/subordinate relationship. On subsequent interactions over the following days the level of aggression goes down and the level of flank marking rises to between 30–40 flank marks for the dominant hamster and between 10–20 for its submissive conspecific. We interpreted this finding to mean that flank marking is used to communicate social status, eliminating the need for physical aggression. Indeed, removal of the flank glands from either one or both of the combatants results in a sustained level of aggression by the dominant hamster that does not abate with repeated interactions. Hamsters do not need flank glands to be dominant, but they do need them to communicate social status and minimize the risk of physical injury from aggressive interactions. The information conveyed by flank marking in established dominant/subordinate relationships persists for some time as familiar conspecifics brought together in a neutral arena weeks later show little or no aggression but a high level of flank marking particularly by the dominant hamster.

In an interesting study, we hypothesized that established dominant/subordinate relationships between hamsters could be reversed by blocking flank marking in the dominant hamster and stimulating flank marking in the submissive hamster during social encounters (Ferris et al., 1986b). Despite repeated treatments that caused robust differences in flank-marking behavior, upon cessation of treatment, hamsters returned to their original social status. From this study we concluded that once social status is determined, triggering or inhibiting stereotyped flank-marking behavior with AVP or AVP_{1a} antagonist; respectively, will not alter

communication and learned behavior between hamsters. Not to be discouraged, we hypothesized that dominance status between animals with no prior history could be predetermined by giving one animal AVP_{1a} antagonist prior to their first social encounter. In pilot studies, we were frustrated to see that when one or the other of a pair was treated with AVP_{1a} antagonist the level of aggression was extremely low between the dyad preventing any assignment of social status to either hamster.

From these unsuccessful attempts to predetermine social status by manipulating AVP neurotransmission, it occurred that AVP_{1a} antagonist might be blocking aggressive responding. To test this hypothesis, adult male hamsters were treated with V_{1a} antagonist and tested in a resident–intruder paradigm for offensive aggressive behavior (Ferris and Potegal, 1988). It should be noted that the site of microinjection for these aggression studies and future work on aggression was confined to the AH. Resident hamsters treated with antagonist showed a dose-dependent decrease in the number of biting attacks toward a smaller male intruder placed into their home cage. The latency to bite the intruder was also significantly increased with antagonist treatment but the amount of time the resident spent investigating the intruder was unaffected, showing that the reduction in aggression was not a generalized effect on social behavior. Sexual motivation was unaffected by AVP_{1a} antagonist treatment. In a subsequent study, the role of AVP in aggressive responding was tested in a neutral area paradigm (Potegal and Ferris, 1990). In this case, twenty-four adult male hamsters were paired and tested in three experimental conditions: (1) neither hamster was given antagonist, (2) one member of the pair was treated with antagonist, or (3) both members were treated with AVP_{1a} antagonist. When confronted in a neutral area, vehicle-treated hamsters showed a combined high number of attacks. The level of aggression was significantly reduced when one member of pair was treated with antagonist. When both members of the pair were treated with receptor antagonist there was little or no aggression. These two studies using the resident–intruder and neutral arena paradigms to study offensive aggression were the first reports in the literature of AVP having a significant role in intraspecific aggression. This finding in male hamsters has since been replicated in several different laboratories (Caldwell and Albers, 2004; Cheng and Delville, 2009; Jackson et al., 2005b) and extended to include rats (Elkabir et al., 1990; Ferris et al., 2008; Koolhaas et al., 1990), voles (Gobrogge et al., 2007; Stribley and Carter, 1999; Winslow et al., 1993), mice (Bester-Meredith et al., 1999; Compaan et al., 1992), pigs (D'Eath et al., 2005) and humans (Coccaro et al., 1998; Thompson et al., 2004). Indeed, there is a large body of literature reporting blockade of AVP V_{1a} receptors in a variety of animals suppresses offensive aggression (Ferris, 2005). Consequently, drugs that target and block the AVP AVP_{1a} are being developed as potential therapeutics for the treatment of impulsivity and violence. Recently, a new class of non-peptidic compounds targeted to the human AVP_{1a} was developed using a monocyclic beta lactam platform (Guillon et al., 2006). One of these potential drugs, SRX251, was tested for antiaggression activity in the hamster resident–intruder paradigm (Ferris et al., 2006). Oral administration of SRX251 caused a dose-dependent decrease in several measures of aggressive behavior without affecting motor activity, olfactory communication, and sexual motivation. Oral administration also blocks aggressive responding in rats and suppresses the distributed neural circuit responsible for aggressive motivation as identified through functional magnetic resonance imaging (Ferris et al., 2008).

Given the substantial body of evidence that V_{1a} antagonists inhibit aggression, one would expect that AVP would stimulate aggression. Indeed, injection of AVP into the AH of resident male hamsters with past experience of fighting significantly increases their aggression toward smaller intruders (Ferris et al., 1997b). AVP also stimulates aggression in male hamsters that are individually housed but not group housed (Caldwell and Albers, 2004). One interesting feature concerning the effects of AVP on aggression is that AVP must be injected in concentrations low enough that flank marking is not

induced (i.e., 0.09 μM). It has also been noted that hamsters exposed to environments that increase the potential for aggression are more sensitive to the aggression stimulating effects of AVP (Caldwell and Albers, 2004). One possibility is that certain environments can increase an individual's propensity for aggression by increasing the number of AVP_{1a} receptors in the AH. In support of this hypothesis is the finding that socially isolated male hamsters who display higher levels of aggression than those given social experience also have significantly higher levels of V_{1a} binding in the AH (Albers et al., 2006). While it is clear that AVP can regulate both aggression and flank marking by its action through V_{1a} receptors in the AH and MPOA it is not clear whether the identity of the postsynaptic neurons are the same. Are the neurons that organize and control flank marking the same as those that control aggression?

In female hamsters AVP appears to have distinctly different effects on aggression than in males. Injection of a V_{1a} antagonist into the AH significantly stimulates aggression in females, while injection of AVP inhibits aggression (Gutzler et al., 2010).

12.5 Neuroanatomy

To better understand the role of AVP in flank marking and aggression, studies were undertaken to identify the integrated neural network involved in these AVP-mediated behaviors. Site-specific microinjections of AVP given in a dose of 0.1 ng in the very small volume of 10 nl was used to map out the area of the hypothalamus involved in AVP-triggered flank marking (Ferris et al., 1986a). Sites that were responsive to AVP were subsequently lesioned with the neurotoxin kainate given in a comparably small volume of 20 nl. These site-specific lesions essentially eliminated odor-induced flank marking attesting to the significance of the injection site in the organization and expression of flank marking. This study identified a ventral medial area of the hypothalamus extending from the caudal MPOA to the caudal AH. This region of the hypothalamus has been shown in other mammals, for example, dogs, cats, gerbils to regulate scent-marking behavior (Hart, 1974; Hart and Voith, 1978; Yahr et al., 1982). Receptor binding studies showed V_{1a} receptors localized to this area of the hypothalamus supporting the pharmacological results of the AVP and V_{1a} antagonist microinjections (Ferris et al., 1993; Johnson et al., 1995).

As noted previously, site specific microinjections of AVP_{1a} antagonist into this area of the hypothalamus blocks flank-marking behavior and intraspecific aggression naturally elicited by the odors and interactions between conspecifics demonstrating that AVP is a fundamental chemical signal in the control of these behaviors. What is the source of this AVP innervation to the MPOA-AH? Mapping of the distribution, projection and number of AVP neurons in the golden hamster revealed a high density of AVP fibers coming from the magnocellular neurons localized to the paraventricular nucleus (PVN), nucleus circularis (NC) and medial supraoptic nucleus (mSON) (Mahoney et al., 1990). Interestingly, many of the AVP neurons localized to these areas do not project to the neurohypophysis as in other rodents (Ferris et al., 1992). This makes the organization of the AVP system in the hamster unique amongst other rodent species. Indeed, the small vasopressinergic neurons in the amygdala and bed nucleus of the stria terminalis (BNST), the source of extra-hypothalamic AVP in rats, are absent in the hamster, as is the highly organized number of magnocellular neurons of the PVN (Bolborea et al., 2010; Delville and Ferris, 1995). In rats and several other species the AVP projections from the medial amygdala and BNST are dependent on gonadal hormones and also display sex differences (de Vries, 2008). In contrast, in hamsters there is no evidence that gonadal hormones can alter the expression of AVP, although gonadal hormones do regulate the number of V_{1a} receptors in several brain regions (Albers, et al., 1991; Huhman and Albers, 1993; Johnson et al., 1995; Young, et al., 2000).

By microinjecting the suicide transport lectin, volkensin into the neurohypophysis it was possible to eliminate the AVP neurons in the hypothalamus projecting to the pituitary gland. The remaining AVP

neurons in the NC and mSON have axonal fibers projecting toward and around the MPOA-AH, suggesting they are the source of AVP regulating flank marking and aggression in hamsters. This hypothesis was supported by a study using kainate microinjections to lesion these different populations of AVP neurons while testing for odor-induced flank marking and correlating the loss of flank marking with loss of AVP immunostaining (Ferris et al., 1990b). To exclude any contribution by the small AVP neurons in the suprachiasmatic nucleus, we also lesioned this brain area (Delville et al., 1998b). Circadian running activity was disrupted but odor-induced flank marking was unaffected. Interestingly, many of the extra hypothalamic areas innervated by the small AVP neurons in the amygdala and BNST of the rat were innervated by the SCN of the hamster.

Indirect evidence to the significance of AVP neurotransmission associated with the NC was provided in a study comparing AVP perikarya, fibers, and neuropeptide concentrations between dominant animals and their submissive conspecifics (Ferris et al., 1989). Animals with a history of social subjugation and subordinate status have fewer AVP neurons in the NC and lower AVP levels in the AH than their dominant conspecific. Interestingly, dominant animals do not have a better or more robust AVP system than their subordinate partner as we had originally hypothesized. In fact, they have a comparable AVP system to control animals without any history of dominance status. Instead, it is the subordinate animal that presents with a change in their AVP system with social subjugation. The change is lower levels of AVP and suggests a reduction in AVP neurotransmission. This finding was corroborated in a study examining the neurobiological consequences of social subjugation during adolescence in golden hamsters (Delville et al., 1998a). Male hamsters were weaned on postnatal day 25 and over the subsequent two-week peripubertal period, exposed to daily subjugation by adult males. As adults, hamsters with a history of adolescent subjugation were highly aggressive toward smaller conspecifics but retreated from equal sized or bigger hamsters as compared to litter mates with no early subjugation. Again, as in the previous study on subordination and AVP in the AH, there was a significant decrease in AVP levels and immunostaining.

With the MPOA-AH as a focal point, studies were undertaken to identify the afferent and efferent connections and the potential involvement of AVP outside the hypothalamus. In a track-tracing study, horseradish peroxidase (HRP) and Phaseolus vulagaris-leucoagglutinin (PHAL) were injected into the lateral septum and retrograde-labeled neurons and anterograde-labeled nerve terminals mapped in the MPOA-AH (Ferris et al., 1990c). HRP positive neurons were localized along the periphery of the AH. None were AVP positive. PHAL positive fibers and terminals originating from the lateral septum displayed a pattern of innervation to the MPOA-AH that mirrored the distribution of AVP cell bodies and fiber density in the medial basal hypothalamus. The density of PHAL positive nerve terminals was particularly high in the NC and mSON. Unilateral ibotenate lesions in the lateral septum reduced odor-induced flank-marking behavior while microinjection of AVP and AVP_{1a} antagonist in the septum stimulates and inhibits flank marking, respectively (Irvin et al., 1990). This AVP-sensitive reciprocal relationship between the septum and MPOA-AH suggests a feedback mechanism involved in the regulation of flank marking. Indeed, AVP neurotransmission in the lateral septum is reported to regulate social memory in rodents (Dantzer et al., 1988). In a follow-up study, we tested which anatomical substrate, the AH or septum, was more critical in the regulation of flank marking (Ferris et al., 1994). Stimulating the septum with AVP in the presence of a unilateral lesion of the AH suppressed flank marking. Conversely, hamsters with AVP microinjection in the AH in the presence of a unilateral lesion of the septum showed robust flank marking behavior. Thus, the AH held a more significant place in the organization and expression of flank marking behavior. One of the more interesting things to come out of this study was the clear demonstration that a single unilateral injection of AVP in the AH was sufficient to elicit flank marking.

However, it was necessary to recruit the contralateral AH in the response since injection on one side while blocking the other suppressed flank marking. The precise anatomical connections between left and right sides of the MPOA-AH needed for the organization and expression of flank marking are unknown.

In addition to the lateral septum other areas having reciprocal connections with the MPOA-AH were examined for their involvement in the organization and expression of flank marking. Microinjection of AVP into the periaqueductal gray causes a dose-dependent increase in flank marking (Hennessey et al., 1992). The distributed neural network activated by AVP-stimulated flank marking was mapped using immunostaining for the immediate early gene *fos* (Bamshad and Albers, 1996). Robust increases in flank marking to AVP microinjection in the MPOA-AH were correlated with increased numbers of cells immunostained for Fos protein in the BNST, central nucleus of the amygdala, and periaqueductal gray (PAG). Lesioning the central nucleus of the amygdala suppresses odor-induced flank marking; however, this amygdaloid area is insensitive to the microinjection of AVP (Bamshad et al., 1997).

12.6 Vasopressin and the developmental onset of flank marking

Hamsters can be observed flank marking as early as postnatal day 20 but not before (Ferris et al., 1996b). This onset of flank marking is not associated with any change in V_{1a} receptors in the MPOA-AH as the distribution and density of V_{1a} binding at postnatal day 18 is comparable to adult hamsters (Delville et al., 1994b). Indeed, microinjection of AVP into the MPOA-AH on P-18 triggers flank-marking behavior (Ferris et al., 1996b). This would suggest that the postsynaptic transduction of AVP signaling contributing to the organization and expression of flank marking is intact prior to the onset of odor-induced flank marking. In other words, the receptor is in place but the endogenous ligand AVP is not available for release. This notion was confirmed when we found a robust expression of AVP neuropeptide around P-19 in the AH. What was unexpected was the constant level of AVP mRNA across P-18 to P-22, suggesting maturation in post-translational processing during this development period resulting in the elevated synthesis and release of AVP in the MPOA-AH facilitating the onset of flank-marking behavior.

Testosterone secreted during adolescence appears to play an organization role in the mechanisms controlling flank marking (Schulz et al., 2006). Adult testosterone treatment activates flank-marking behavior only in males that were exposed to testicular hormones during adolescence. In addition, males exposed to testicular hormones during adolescence exhibit significantly less AVP receptor binding within the lateral septum than males deprived of adolescent hormones. These data suggest that hormone-dependent remodeling of synapses normally occurs in the lateral septum during adolescence and demonstrate the importance of gonadal steroid hormone exposure during adolescence in the control of flank marking.

Recently, the role of AVP in the developmental onset of aggression in hamsters was examined by studying play fighting as a behavioral antecedent to adult offensive aggression (Cheng and Delville, 2009). Play fighting as compared to adult aggression is more repetitive in the attacks and focuses on the face and neck – something never seen in adult offensive aggression. Microinjection of AVP_{1a} antagonist into the AH causes a dose-dependent decrease in several aspects of play fighting.

12.7 Interactions between vasopressin and other chemical signals in the control of flank marking and aggression

While AVP is undoubtedly fundamental in the organization and expression of flank marking and aggression there are many other chemical signals that interact with AVP to affect behavior. In female hamsters norepinephrine (NE) modulates the response of the MPOA-AH to AVP.

Coadministration of NE with AVP inhibits AVP-induced flank marking in a dose-dependent manner. In contrast, coadministration of AVP with dopamine, serotonin (5-HT) or neuropeptide Y does not reduce AVP-induced marking. Interestingly, NE has no effect on AVP-induced marking in male hamsters (Whitman et al., 1992). The function of NE's inhibitory effect on AVP-induced flank marking in females is not clear, however, it may relate to the coordination of behavioral events during the estrous cycle. For example, it is possible that the ovarian hormones and social cues that induce sexual receptivity simultaneously inhibit other potentially competing behaviors such as flank marking and this inhibition is mediated by NE. In support of this hypothesis is the finding that hormonal and social stimuli that induce sexual receptivity in female hamsters can block the ability of AVP to stimulate flank marking (Albers and Rowland, 1989).

In male hamsters AVP-triggered flank marking is strongly affected by (5-HT). Systemic administration of the 5-HT reuptake inhibitor fluoxetine significantly inhibits flank marking stimulated by the odors of other hamsters (Ferris et al., 1997b). In addition, there is evidence that 5-HT receptors in the MPOA-AH can influence AVP-induced flank marking. Studies employing various 5-HT receptor agonists suggest that the effects of 5-HT on flank marking may be mediated by the several 5-HT receptor subtypes including 5-HT1A, 5-HT1B, 5-HT2A, 5-HT4 and 5-HT7 (Albers et al., 2002). Comparison of these data with the inability of 5-HT to inhibit AVP-induced flank marking in female hamsters suggests that there may be substantial sex differences in the effects of both NE and 5-HT on AVP activity in the MPOA-AH.

The neuropeptide galanin is coexpressed in the hamster magnocellular system (Miller et al., 1999). Indeed, galanin is colocalized and cosecreted with other chemical signals and usually has an opposite function to the other transmitter. Comicroinjection of AVP and galanin into the MPOA-AH suppresses AVP-induced flank marking (Ferris et al., 1999c). Glutamate, the ubiquitous excitatory amino acid also interacts with AVP at the MPOA-AH. Microinjection of AVP with AP-5, a NMDA antagonist, or GAMS, a non-NMDA antagonist, significantly reduces flank-marking behavior (Bamshad et al., 1996).

Various chemical signals are involved in the regulation of aggression. One in particular, 5-HT appears to have a seminal role in reducing aggression in all mammals studied including humans. Therefore, it is not surprising that 5-HT has some interaction with AVP in control of aggression in hamsters. Microinjection of 5-HT$_{1A}$ receptor agonist 8-OH-DPAT into the AH causes a dose-dependent reduction in offensive aggression (Ferris et al., 1999b). Peripheral treatment with fluoxetine inhibits AVP release in the AH, blocks offensive aggression of a resident toward an intruder and suppresses AVP facilitated attacks and bites (Delville et al., 1996; Ferris, 1996; Ferris et al., 1997b). These findings support the notion that peripheral fluoxetine can alter aggression by suppressing the release of AVP and/or by blocking the activity of AVP on afferent neurons. Neuroanatomical studies corroborate these neurochemical data and show an interaction between AVP and 5-HT. The AH has a high density of 5-HT $_{1A}$ receptors overlapping the V$_{1A}$ binding site in this area (Ferris et al., 1999b; Ferris et al., 1997b). There is a high density of 5-HT terminals and synaptic boutons throughout the AH. In the areas of the mSON and NC these synaptic boutons are closely associated with the primary dendrites of AVP neurons (Ferris et al., 1997b). Fluoro-Gold injections in AH label neurons in the dorsal, median, and caudal linear raphe nuclei, a portion of which also contains tryptophan hydroxylase immunoreactivity (Ferris et al., 1999b). Presumably, the release of 5-HT into the AH acts through 5-HT$_{1A}$ receptors to inhibit AVP-facilitated offensive aggression.

The notion that AVP facilitates aggression and 5-HT suppresses aggression, in part, by inhibiting the activity of the AVP system appears to translate to humans. Patients with personality disorder characterized by a history of fighting and assault show a negative correlation for prolactin release in response to d-fenfluramine challenge (Coccaro et al., 1998). The blunted prolactin response in these

patients is suggestive of a hyporesponsive 5-HT system. These patients show a significant positive correlation between cerebrospinal fluid levels of AVP and aggression. Hence, a hyporesponsive 5-HT system may contribute to enhanced central release of AVP facilitating impulsive and aggressive behavior in humans.

As noted earlier, adult hamsters with a history of adolescent stress have changes in the neurochemical systems regulating aggressive behavior. In addition to the decrease in AVP levels there is an increase in 5-HT innervation to the AH as compared to their sibling controls (Delville et al., 1998a). Since 5-HT decreases aggressive behavior, in part, by inhibiting the activity of the AVP system at the level of the AH, it is possible that the stress of threat and attack in adolescence alters the interaction between the AVP/5-HT systems affecting the regulation of aggression and possibly the context-dependent nature of the aggressive response (Ferris, 2000).

12.8 Gender and gonadal steroids affect vasopressin-mediated flank marking and aggression

The circulating levels of gonadal hormones have dramatic effects on expression of social communication in mammals. In hamsters, testosterone has a significant effects on the size of the flank gland and influences the amount of flank-marking behavior (Johnston, 1981). In males, castration significantly reduces odor-stimulated flank marking and testosterone replacement therapy restores precastration levels of the behavior. In females, odor-stimulated flank marking significantly varies over the estrous cycle with the lowest levels of marking occurring on the day of sexual receptivity (Albers and Rowland, 1989). Ovariectomy significantly reduces the levels of odor-stimulated flank marking while administration of estradiol increases odor-stimulated marking two-fold over vehicle controls. Injection of progesterone into estradiol-treated ovariectomized hamsters reduces marking.

As discussed above, testosterone alters AVP expression in several key limbic system structures in a variety of species. In hamsters, however, manipulation of testosterone has not been found to alter AVP expression in any brain areas (Albers et al., 1991). Nevertheless, gonadal hormones have a significant impact on the behavioral response to AVP. Castration significantly reduces the ability of AVP to stimulate flank marking in the MPOA-AH by 50% and replacement of testosterone to castrated males restores the precastration behavioral response to AVP (Ferris et al., 1988a). In females, comparison of the amount AVP expression in several key limbic structures over the estrous cycle reveals no significant changes suggesting that ovarian hormones do not modulate AVP expression (Huhman and Albers, 1993). However, like male hamsters gonadal hormones do appear to alter the behavioral response to AVP injected into the MPOA-AH. The amount of AVP-induced flank marking varies significantly over the estrous cycle with the lowest levels observed on estrus (Albers et al., 1996). Ovariectomized hamsters implanted with Silastic capsules containing estradiol mark at significantly higher levels in response to AVP injected into the MPOA-AH than ovariectomized hamsters or ovariectomized hamsters administered estradiol and progesterone (Albers et al., 1996; Huhman and Albers, 1993).

AVP not only stimulates flank marking when injected into the MPOA-AH but also when injected into the septum and the PAG. To determine whether the MPOA-AH was the only site where gonadal hormones influence the behavioral effects of AVP, the ability of AVP to stimulate flank marking was compared in intact and castrated male hamsters following injections into the septum or PAG (Johnson et al., 1995). While some subtle differences in the effects of AVP were observed between castrated males and males with circulating testosterone, these effects were only apparent at high concentration of peptide. Taken together, the existing data suggests that gonadal hormones act to modulate AVP-stimulated marking primarily in the MPOA-AH and not in the septum or PAG.

One way that gonadal hormones might alter the behavioral response to AVP is by altering the number of V_{1a} receptors. In males, castration significantly reduces V_{1a} binding in the medial preoptic nucleus (MPN) and the medial preoptic area (MPOA), but not in variety of other limbic structures ((Caldwell and Albers, 2003; Caldwell et al., 2008; Johnson et al., 1995; Young et al., 2000). The castration reduced V_{1a} binding in the MPN and MPOA is restored by administration of testosterone. Interestingly, the effects of testosterone on V_{1a} mRNA appear to be restricted to the MPN, despite the more extensive effects of testosterone on V_{1a} binding in the MPOA. In female hamsters significant variations in V_{1a} binding over the estrous cycle have not been detected. However, when females were exposed to short "winter-like" photoperiods where estrous cycles cease and there is a significant reduction of ovarian mass and circulating estradiol levels, there is a significant reduction in V_{1a} binding in several limbic structures including the MPN and MPOA (Caldwell and Albers, 2004).

There is a significant sex difference in flank marking and aggression, and unlike in most species, female hamsters mark more and are more aggressive than males. Females mark at significantly higher rates than do males when exposed to the odors of male hamsters (Hennessey et al., 1992) and sexually unreceptive females defeat male hamsters (Marques and Valenstein, 1977; Payne and Swanson, 1972). AVP injected into the MPOA-AH stimulates similar levels of flank marking in males and females across of range of concentrations (Hennessey et al., 1994). Although AVP_{1a} binding has not be compared in males and females in the MPOA-AH, sex differences have been identified in the ventro-lateral hypothalamus with males having significantly higher levels of binding than females (Delville and Ferris, 1995). The possibility of sex differences in AVP immunoreactivity has also been examined. In one study, no sex differences were identified (Hennessey et al., 1994) while in another males were found to have higher levels of AVP in the supraoptic nucleus (Delville et al., 1994a).

12.9 Developmental exposure to drugs of abuse affects vasopressin-mediated offensive aggression

Although the manipulation of physiologic levels of testosterone in the adult hamster appears not to alter AVP expression in brain (Albers et al., 1991), a considerable body of evidence indicates that pubertal exposure to supraphysiologic levels of testosterone (collectively termed anabolic-androgenic steroids in drug-abuse literature) produces significant alterations in AVP neural circuitry, expression and function in the AH that correlate with – and modulate – the generation of the highly aggressive phenotype in hamsters. For instance, exposure to moderate-to-high doses of testosterone and select synthetic testosterone derivatives during puberty (5 mg/kg/day SC × 28 days) show increased offensive aggression that is dependent upon the time and duration of exposure. In particular, hamsters treated with moderate-to-high doses of testosterone during early-to-mid puberty (i.e. 14 days – P27–P42) show greater aggression compared to sesame oil (vehicle)-treated littermates (Ferris et al., 1997a; Grimes et al., 2007; Melloni and Ferris, 1996; Melloni et al., 1997) however, testosterone treatment during the majority of the pubertal period (i.e. early-to-late adolescence – P27–P56) produces even greater aggression-stimulating effects; characterized by the large number of chases, lateral and rump attacks, upright offensive postures, and flank marks (Grimes and Melloni, 2005; Grimes et al., 2003; Harrison et al., 2000b). When examined for AVP, pubertal testosterone exposure was found to enhance the development of AVP afferent fibers to the AH (Harrison et al., 2000b; Melloni et al., 1997) and increase the available pool of AVP peptide in the AH (Harrison et al., 2000b) but not alter AH AVP mRNA expression (Harrison et al., 2000b) or AH AVP AVP_{1a} binding (DeLeon et al., 2002), suggesting that moderate increases in AVP neural development, production and activity within the AH brain region underlie pubertal testosterone-induced offensive aggression. This set of findings is in agreement with earlier work indicating that moderate increases in

AVP in the AH facilitate offensive aggression in hamsters (Ferris et al., 1997a). The highly aggressive phenotype observed in hamsters treated with supraphysiologic doses of testosterone during puberty could be blocked by the central application of AVP AVP_{1a} antagonists delivered directly to the AH by site-specific microinjection (Harrison et al., 2000b), further supporting a role for heightened AH AVP activity in the testosterone-induced aggressive response. Moreover, recent *in vivo* microdialysis studies reveal that aggressive, testosterone-treated pubertal hamsters release more AVP into the AH during an aggression test than non-aggressive controls (Melloni and Ricci, 2010), supporting the view that pubertal exposure to testosterone promotes AVP activity in this brain site during an agonistic encounter with conspecifics. Interestingly, the difference in release between the two treatment groups is characterized by a moderate, but statistically significant increase (~35%) in AH AVP release in testosterone-treated animals. These data indicate that AH AVP neurons in testosterone-treated animals release considerably more AVP than their vehicle-treated counterparts under *stimulated* conditions. Together, these data suggest that pubertal testosterone exposure increases offensive aggression by increasing development, production and release of AVP in the AH, supporting a direct evidence for a causal role of AH-AVP function in pubertal testosterone-induced offensive aggression.

The notion that increases in AH AVP production, neural development, and activity precipitate pubertal testosterone-induced offensive aggression is further supported by both developmental short-term testosterone exposure (Grimes et al., 2007) and long-term testosterone withdrawal (Grimes et al., 2006) studies. In the developmental short-term study, the temporal relationship between the onset of pubertal testosterone-induced offensive aggression and developmental alterations in the AH AVP neural system was examined. Animals treated with testosterone during the adolescent period displayed alterations in AH AVP afferent development and offensive aggression compared to controls, however; these differences emerged at different times during the treatment period. Specifically, animals examined for AH AVP after very short treatment periods (i.e., <1 week) showed no differences in AH AVP afferent innervation compared to controls. However, by 2 weeks of exposure pubertal animals treated with testosterone display significant increases in AVP afferent density within the AH. When aggression levels and extent of AH AVP afferent innervation were examined for each time point, a strong, positive correlation between AH AVP fiber density and levels of offensive aggression was observed, indicating that at times of increased AH AVP tone, animals were more aggressive than when levels of AH AVP are low. This correlation between aggression and AH AVP supports the notion that testosterone-induced increases in AVP production and afferent development within the AH underlie the generation of the highly aggressive phenotype. This notion was further supported by long-term withdrawal studies that showed the maintenance of the highly aggressive phenotype temporally correlates with AVP production and afferent development to the AH brain region (Grimes et al., 2006). Here, hamsters treated with testosterone during the adolescent period displayed increases in AH AVP afferent development and offensive aggression compared to controls, however, these increases return to control levels at different times during withdrawal from testosterone exposure. Specifically, significant increases in AH AVP afferent innervation remain in testosterone-treated animals compared to controls across short withdrawal periods (i.e., <2 weeks) show. However, increases in AH AVP afferent density in testosterone-treated hamsters were no longer observed at later times of withdrawal. When aggression levels and extent of AH AVP afferent innervation for each time point were collapsed across treatment, a strong, positive correlation between AH AVP fiber density and levels of offensive aggression was observed, again supporting the notion that at times of increased AH AVP tone, animals respond more aggressively than when levels of AH AVP are low. This correlation between aggression and AH AVP across withdrawal further strengthens the notion

that the interactions between pubertal testosterone exposure and the AH AVP neural system directly underlie the highly aggressive phenotype. Further, these data indicate that pubertal testosterone exposure has short-term, reversible effects on AH AVP in hamsters, and that changes in the AH AVP neural system are linked to the display of the aggressive or non-aggressive behavioral phenotype in hamsters during testosterone withdrawal.

As highlighted above, the AH AVP neural system modulating offensive aggression in hamsters is normally inhibited by 5HT (Ferris et al., 1999a; Ferris et al., 1997a) and the inhibitory nature of 5HT on aggression is due to action at 5HT1 receptors, most notably 5HT1A receptors expressed on neurons in the AH (Ferris et al., 1999a). Interestingly, in aggressive hamsters treated with testosterone during puberty, 5HT1A receptor expression was found to be reduced only on neurons located in the AH (Ricci et al., 2006), specifically on a select subset of magnocellular AVP neurons in the medial supraoptic nucleus (mSON) (Melloni and Ricci, 2010). The mSON is a principal source of the AVP afferent innervation to the AH brain region (Ferris et al., 1989; Ferris et al., 1990a; Mahoney, 1990), thus, this reduction in inhibitory 5HT1A receptor activity may serve to effectively disinhibit AH AVP neurons implicated in the aggressive response, facilitating the expression of the highly aggressive phenotype. In support of this notion, systemic treatment with the 5HT1A agonist R(+)-8-OH-DPAT dose dependently blocks the aggressive phenotype (Ricci et al., 2006), supporting a suppressive role for 5HT1A signaling through AVP neurons in the AH in pubertal testosterone-induced offensive aggression. Similarly, 5HT1B receptors have been localized to AVP neurons in the AH, specifically those in a small cluster centered directly within the AH proper in the nucleus circularis (NC), a second principal source of AVP afferent innervation to the AH (Ferris et al., 1989; Ferris et al., 1990a; Mahoney, 1990). In contrast to that seen with 5HT1A receptors, 5HT1B receptor expression was found to be increased on neurons located in the AH of aggressive hamsters treated with testosterone during puberty (Grimes and Melloni, 2005), specifically on magnocellular AVP neurons in the NC (Melloni and Ricci, 2010). Yet, systemic treatment with the 5HT1B agonist anpirtoline dose dependently blocks the aggressive phenotype (Grimes and Melloni, 2005), supporting an additional suppressive role for 5HT through signaling through 5HT1B receptors on AVP neurons in the AH in pubertal testosterone-induced offensive aggression. Together, these data suggested that pubertal testosterone exposure dramatically altered 5HT signaling through 5HT1A and 5HT1B receptors on AH AVP neurons, supporting the hypothesis that pubertal testosterone exposure alters the connectivity between the AH 5HT/AVP systems, perhaps disrupting by the capacity of the AH 5HT neural system to inhibit AH AVP and offensive aggression. The apparent mutually exclusive down- and up-regulation of 5HT1A and 5HT1B receptor activity on the two principal neuronal groups supplying AVP to the AH may be a direct response to pubertal testosterone, or a pharmacodynamic response to the significant reduction in 5HT afferent innervation to the AH observed in aggressive, pubertal testosterone-treated hamsters (Grimes and Melloni, 2002). However, pharmacologic challenge studies showing that increasing AH 5HT levels via fluoxetine administration selectively blocks pubertal testosterone-induced offensive aggression (Grimes and Melloni, 2002), indicate that although the development and activity of the 5HT afferent neural system is compromised following pubertal testosterone exposure, AH 5HT activity (via 5HT1A and 5HT1B receptors on AH AVP neurons) remains capable of modulating the testosterone-induced aggressive response.

The hypothalamic dopamine (DA) neural system has been shown to be sensitive to testosterone in a fashion consistent with the generation of the aggressive phenotype, so alterations in this neural system may participate in pubertal testosterone-induced offensive aggression. Recent studies showed that pubertal testosterone exposure increases tyrosine hydroxylase (i.e., TH, the rate-limiting enzyme in the biosynthesis of DA) and DA D2 subtype receptor expression in the AH, but no other brain region implicated in offensive aggression (Ricci et al.,

2009; Schwartzer et al., 2009). Pubertal testosterone-treated hamsters had more TH neurons and afferent fibers in the two principal AVP neuronal compartments, i.e. the NC and mSON, leading to the speculation that AVP and DA were coexpressed and released from the same population of AH neurons. When this possibility was examined, it was found that TH- and AVP-containing neurons did indeed localize to the NC and mSON, and in both areas TH localized to AVP and non-AVP neurons, while AVP also localized to a separate set of neurons in these regions (Melloni and Ricci, 2010). These data indicated that the source(s) of AVP and DA to the LAH may originate from AH neurons coexpressing these signals, as well as from separate and distinct sets of AVP and DA neurons, illustrating that there are heterogeneous populations of AVP/DA neurons in the AH. From a functional standpoint, these data support the notion that pubertal testosterone exposure facilitates offensive aggression by increasing the release/activity of both AVP and DA in the AH brain region. This concept is further supported by recent studies showing that central administration of the selective DA D2 receptor antagonist eticlopride to the AH via microinjection selectively and dose dependently blocks pubertal testosterone-induced aggression (Schwartzer and Melloni, 2010a) in a fashion nearly identical to that observed by using AVP V_{1A} antagonists (Harrison et al., 2000b). This response appears to be DA D2 receptor specific, as the selective DA D5 receptor antagonist SCH-23390 has no effect on the pubertal testosterone-induced aggressive response (Schwartzer and Melloni, 2010b).

Pubertal exposure to other drugs of abuse has also been shown to modulate AVP-mediated offensive aggression in hamsters. For instance, repeated exposure to exceptionally low doses of cocaine hydrochloride (0.5 mg/kg/day IP x 28 days) produces animals that display high levels of offensive aggression (Harrison et al., 2000a). The highly aggressive phenotype observed in hamsters treated with low doses of cocaine during puberty could be blocked by the central application of AVP AVP_{1a} antagonists delivered directly to the AH by site-specific microinjection (Jackson et al., 2005a), supporting a role for heightened AH AVP development, production and activity in the cocaine-induced aggressive response. When examined for AH AVP however, unlike pubertal testosterone-treated animals, pubertal cocaine exposure was not found to enhance the development of AVP afferent fibers to the AH or increase the levels of AVP peptide in the AH (Jackson et al., 2005a). However, *in vitro* superfusion studies using AH brain slices revealed significant differences in AH AVP release between aggressive, cocaine-treated hamsters compared to non-aggressive, saline-treated controls (Jackson et al., 2005a). In this study, no ostensible differences in basal AH AVP release were observed between brain slices from cocaine- and saline-treated hamsters. However, a significant difference in AH AVP release was observed between groups following the electrical stimulation of AH brain slices. Stimulated AH AVP release from brain slices of cocaine-treated animals was approximately five times that of saline-treated littermates. Interestingly, the difference in release between the two treatment groups is characterized by a moderate, but statistically significant increase (25–30%) in electrically evoked release from slices from cocaine-treated animals. These data indicate that AH AVP neurons in cocaine-treated animals release considerably more AVP than their saline-treated counterparts under *stimulated* conditions, which behaviorally may be best represented by a confrontation with a conspecific, supporting a direct evidence for a causal role of AH-AVP function in pubertal cocaine-induced offensive aggression. Under this assertion, cocaine animals would release more AH AVP when confronted with a stimulus animal in their home cage than saline-treated counterparts, facilitating the highly escalated aggressive phenotype observed in stimulant-treated animals. This scenario is consistent with that observed using *in vivo* microdialysis measuring AH AVP release in aggressive, pubertal testosterone-treated hamsters during an aggressive encounter (Melloni and Ricci, 2010). At a different level, these data suggest that moderate increases in AVP release can occur from

AH neurons in response to cocaine under stimulated conditions without requiring the *de novo* synthesis of AVP or enhancement of AH AVP development.

Together, data from these pubertal drug abuse models imply that the spectrum of AH AVP-mediated agonistic behaviors are tightly regulated, and that moderate disturbances in normal AVP signaling in the AH brain region can have significant behavioral effects in hamsters. The data suggest that exposure to testosterone or cocaine during pubertal development causes the "functional stimulation" of AH AVP neurons implicated in the control of offensive aggression. From a mechanistic standpoint, these studies indicate that moderate increases in AH AVP release (25–35% over basal) are necessary and sufficient for pubertal testosterone and cocaine-induced offensive aggression. These data are consistent with those examining the dynamics of AVP expression during development (Ferris et al., 1996a) where increases in offensive aggression were coincident with moderate elevations in the amount of AVP in the AH. Indeed, functional studies have shown that only moderate elevations of AVP in the AH (via direct microinjections) facilitate aggressive responding in hamsters (Ferris et al., 1997a). In fact, microinjections of low doses of AVP (0.09 μM) are necessary to stimulate aggressive responding in hamsters. Overstimulation of AVP signaling pathways in the AH by markedly increasing AVP content in this brain region (>10 times that required to stimulate aggressive responding) leads to an overwhelming tendency for the animal to flank mark (Albers et al., 1986a; Ferris et al., 1988b; Ferris et al., 1984a). Therefore, in both these pubertal drug-abuse models the aggressive behavioral phenotype observed may be explained by the tendency of these drugs to direct only moderate increases in AVP release into the AH brain region, predisposing the animal to heightened offensive aggression.

12.10 Seasonal variations

Syrian hamsters are seasonal breeders and display dramatic changes in social behavior over the annual cycle (Bartness and Goldman, 1989; Garrett and Campbell, 1980). During the summer breeding season when photoperiods are long, males and females have high levels of circulating gonadal hormones and display high levels of social communication and sexual behavior. During the summer male and female hamsters are also capable of exhibiting high levels of aggression toward other hamsters not in breeding condition. During the winter when no breeding occurs, the short photoperiods cause regression of the reproductive system, a dramatic reduction in circulating gonadal hormones and no sexual behavior. Interestingly, during the winter hamsters typically display higher levels of aggression than they do during the summer when gonadal hormone levels are high. In short, there is a negative correlation between the propensity for aggression and the circulating levels of gonadal hormones in hamsters.

During the breeding season the effects of testosterone on aggression are not entirely clear with some studies finding that testosterone promoting aggression, while others do not (Albers et al., 2002). Testosterone might influence aggression by acting on AVP receptors is in the ventro-lateral hypothalamus. Microinjecting AVP into the ventrolateral hypothalamus (VLH) of the hamster facilitates offensive aggression (Delville et al., 1996; Ferris and Delville, 1994). Interestingly, AVP receptor binding within the VLH is androgen dependent (Delville and Ferris, 1995). This raises the possibility that the diminished offensive aggression that can occur in castrated hamsters (Ferris et al., 1989; Payne and Swanson, 1972; Vandenbergh, 1971; Whitsett, 1975) might be caused by a loss of AVP responsiveness in the VLH.

During the non-breeding season when testosterone levels are low and the propensity for aggression is high there is a dramatic reduction in V_{1a} binding in both the MPN and MPOA that are quite similar to the reduction in binding seen in castrated hamsters during the breeding season (Caldwell and Albers, 2003; Caldwell et al., 2008). However, surprisingly, despite these low levels of V_{1a} binding in the short photoperiod exposed hamsters AVP injected

into the MPOA-AH produces flank-marking levels that are similar to those produced in intact males housed in "summer-like conditions" that have significantly higher levels of V_{1a} binding in this same region. In summary, although short photoperiod reduces AVP_{1a} binding in the MPOA-AH, it does not reduce the amount of flank marking produced by injection of AVP into this same region. Therefore, short photoperiod exposure overrides the reduction in V_{1a} receptors and maintains high levels of AVP-induced flank marking. The mechanisms responsible for these effects remain to be defined (Gutzler et al., 2010).

In male hamsters, short photoperiod increases the propensity for aggression despite producing significant reductions in circulating of testosterone (Bartness and Goldman, 1989; Garrett and Campbell, 1980). Interestingly, AVP does not appear to be involved in the increased aggression observed in short photoperiod-housed hamsters and, in fact, short photoperiod appears to eliminate the ability of AVP to modulate aggression by its action in the MPOA-AH (Caldwell and Albers, 2004). Neither injection of AVP nor a potent V_{1a} antagonist altered aggression when compared to vehicle-treated males. Thus, the effects of AVP within the MPOA-AH on aggression appear to be limited to the breeding season.

12.11 Conclusion

Since the original report by David de Vied that AVP had a significant effect on cognitive behavior in rats there has been a wealth of studies on the behavioral neurobiology of this neuropeptide in a range of animal species. In this chapter we have discussed the role of AVP in flank marking, a stereotyped communicative behavior used by hamsters for olfactory communication and in the role of AVP in aggression, a finding generalized to all mammals. AVP neurotransmission is fundamental to these behaviors. The microinjection of AVP and $AVPR_{1a}$ into specific areas of the medial basal hypothalamus causes dose-dependent changes in flank marking and aggression. The AVP neurons regulating behavior, while independent of the hypothalamic neurohypophysial system, are subpopulations of magnocellular neurons in the medial supraoptic nucleus, nucleus circularis and paraventricular nucleus. The expression of AVP in these neurons and the $AVPR_{1a}$ receptors throughout the brain, changes in response to development, gender, and environmental experiences, particularly with respect to social status and exposure to drugs of abuse. Perturbations in AVP neurotransmission can result in inappropriate aggressive responding in animals and humans and highlight its potential role in the pathophysiology of violence and mental illness.

Acknowledgments

This work was supported by grants from NIH and NSF including R01-DA029620 to RHM and NSF IOS 0923301 to HEA.

REFERENCES

Albers, H. E. and Ferris, C. F. (1985). Behavioral effects of vasopressin and oxytocin within the medial preoptic area of the golden hamster. *Regul Pept*, 12, 257–260.

Albers, H. E. and Ferris, C. F. (1986). Role of the flank gland in vasopressin induced scent marking behavior in the hamster. *Brain Res Bull*, 17, 387–389.

Albers, H. E. and Rowland, C. M. (1989). Ovarian hormones influence odor stimulated flank marking behavior in the hamster (Mesocricetus auratus). *Physiol Behav*, 45, 113–117.

Albers, H. E., Rowland, C. M., and Ferris, C. F. (1991). Arginine-vasopressin immunoreactivity is not altered by photoperiod or gonadal hormones in the Syrian hamster (Mesocricetus auratus). *Brain Res*, 539, 137–142.

Albers, H. E., Karom, M., and Whitman, D. C. (1996). Ovarian hormones alter the behavioral response of the medial preoptic anterior hypothalamus to arginine-vasopressin. *Peptides*, 17, 1359–1363.

Albers, H. E., Karom, M., and Smith, D. (2002). Serotonin and vasopressin interact in the hypothalamus to control communicative behavior. *Neuroreport*, 13, 931–933.

Albers, H. E., Pollock, J., Simmons, W. H., and Ferris, C. F. (1986a). A V1-like receptor mediates vasopressin-

induced flank marking behavior in hamster hypothalamus. *J Neurosci*, 6, 2085–2089.

Albers, H. E., Pollock, J., Simmons, W. H., and Ferris, C. F. (1986b). A V1-like receptor mediates vasopressin-induced flank marking behavior in hamster hypothalamus. *J Neurosci*, 6, 2085–2089.

Albers, H. E., Dean, A., Karom, M. C., Smith, D., and Huhman, K. L. (2006). Role of V1a vasopressin receptors in the control of aggression in Syrian hamsters. *Brain Res*, 1073–1074, 425–430.

Bamshad, M. and Albers, H. E. (1996). Neural circuitry controlling vasopressin-stimulated scent marking in Syrian hamsters (Mesocricetus auratus). *J Comp Neurol*, 369, 252–263.

Bamshad, M., Cooper, T. T., Karom, M., and Albers, H. E. (1996). Glutamate and vasopressin interact to control scent marking in Syrian hamsters (Mesocricetus auratus). *Brain Res*, 731, 213–216.

Bamshad, M., Karom, M., Pallier, P., and Albers, H. E. (1997). Role of the central amygdala in social communication in Syrian hamsters (Mesocricetus auratus). *Brain Res*, 744, 15–22.

Bartness, T. J. and Goldman, B. D. (1989). Mammalian pineal melatonin: a clock for all seasons. *Experientia*, 45, 939–945.

Bester-Meredith, J., Young, L., and Marker, C. (1999). Species differences in paternal behavior and aggression in peromyscus and their associations with vasopressin immunoreactivity and receptors. *Horm Behav*, 36, 25–38.

Bolborea, M., Ansel, L., Weinert, D., Steinlechner, S., Pevet, P., and Klosen, P. (2010). The bed nucleus of the stria terminalis in the Syrian hamster (Mesocricetus auratus): absence of vasopressin expression in standard and wild-derived hamsters and galanin regulation by seasonal changes in circulating sex steroids. *Neuroscience*, 165, 819–830.

Caldwell, H. K. and Albers, H. E. (2003). Short-photoperiod exposure reduces vasopressin (V1a) receptor binding but not arginine-vasopressin-induced flank marking in male Syrian hamsters. *J Neuroendocrinol*, 15, 971–977.

Caldwell, H. K. and Albers, H. E. (2004). Effect of photoperiod on vasopressin-induced aggression in Syrian hamsters. *Horm Behav*, 46, 444–449.

Caldwell, H. K., Smith, D. A., and Albers, H. E. (2008). Photoperiodic mechanisms controlling scent marking: interactions of vasopressin and gonadal steroids. *Eur J Neurosci*, 27, 1189–1196.

Cheng, S. Y. and Delville, Y. (2009). Vasopressin facilitates play fighting in juvenile golden hamsters. *Physiol Behav*, 98, 242–246.

Coccaro, E. F., Kavoussi, R. J., Hauger, R. L., Cooper, T. B., and Ferris, C. F. (1998). Cerebrospinal fluid vasopressin levels: correlates with aggression and serotonin function in personality-disordered subjects. *Arch Gen Psychiatry*, 55, 708–714.

Compaan, J. C., Koolhaas, J. M., Buijs, R. M., Pool, C. W., de Ruiter, A. J., and van Oortmerssen, G. A. (1992). Vasopressin and the individual differentiation in aggression in male house mice. *Ann NY Acad Sci*, 652, 458–459.

D'Eath, R. B., Ormandy, E., Lawrence, A. B., Sumner, B. E., and Meddle, S. L. (2005). Resident-intruder trait aggression is associated with differences in lysine vasopressin and serotonin receptor 1A (5-HT1A) mRNA expression in the brain of pre-pubertal female domestic pigs (Sus scrofa). *J Neuroendocrinol*, 17, 679–686.

Dantzer, R., Koob, G. F., Bluthe, R. M., and Le Moal, M. (1988). Septal vasopressin modulates social memory in male rats. *Brain Res*, 457, 143–147.

de Vries, G. J. (2008). Sex differences in vasopressin and oxytocin innervation of the brain. *Prog Brain Res*, 170, 17–27.

DeLeon, K. R., Grimes, J. M., and Melloni, R. H., Jr. (2002). Repeated anabolic-androgenic steroid treatment during adolescence increases vasopressin V(1A) receptor binding in Syrian hamsters: correlation with offensive aggression. *Horm Behav*, 42, 182–191.

Delville, Y. and Ferris, C. F. (1995). Sexual differences in vasopressin receptor binding within the ventrolateral hypothalamus in golden hamsters. *Brain Res*, 681, 91–96.

Delville, Y., Koh, E. T., and Ferris, C. F. (1994a). Sexual differences in the magnocellular vasopressinergic system in golden hamsters. *Brain Res Bull*, 33, 535–540.

Delville, Y., Mansour, K. M., and Ferris, C. F. (1996). Serotonin blocks vasopressin-facilitated offensive aggression: interactions within the ventrolateral hypothalamus of golden hamsters. *Physiol Behav*, 59, 813–816.

Delville, Y., Melloni, R. H., Jr., and Ferris, C. F. (1998a). Behavioral and neurobiological consequences of social subjugation during puberty in golden hamsters. *J Neurosci*, 18, 2667–2672.

Delville, Y., De Vries, G. J., Schwartz, W. J., and Ferris, C. F. (1998b). Flank-marking behavior and the neural distribution of vasopressin innervation in golden hamsters with suprachiasmatic lesions. *Behav Neurosci*, 112, 1486–1501.

Delville, Y., Mansour, K. M., Quan, E. W., Yules, B. M., and Ferris, C. F. (1994b). Postnatal development of the vasopressinergic system in golden hamsters. *Brain Res Dev Brain Res*, 81, 230–239.

Elkabir, D. R., Wyatt, M. E., Vellucci, S. V., and Herbert, J. (1990). The effects of separate or combined infusions of corticotrophin-releasing factor and vasopressin either intraventricularly or into the amygdala on aggressive and investigative behaviour in the rat. *Regul Pept*, 28, 199–214.

Ferris, C., Stolberg, T., Kulkarni, P., Murugavel, M., Blanchard, R., Blanchard, D., Febo, M., Brevard, M., and Simon, N. (2008). Imaging the neural circuitry and chemical control of aggressive motivation. *BMC Neuroscience*, 9, 111.

Ferris, C. F. (1996). Serotonin inhibits vasopressin facilitated aggression in the Syrian hamster. In: C. Ferris and T. Grisso (Eds.), *Understanding Aggressive Behavior in Children*, New York Academy of Sciences, New York, pp. 98–103.

Ferris, C. F. (2000). Adolescent stress and neural plasticity in hamsters: a vasopressin-serotonin model of inappropriate aggressive behaviour. *Exp Physiol*, 85 Spec No, 85S–90S.

Ferris, C. F. (2005). Vasopressin/oxytocin and aggression, in: G. Bock and J. Goode (Eds.), *Molecular Mechanisms Influencing Aggressive Behaviours*, Wiley, Chichester, UK, pp. 190–200.

Ferris, C. F. and Potegal, M. (1988). Vasopressin receptor blockade in the anterior hypothalamus suppresses aggression in hamsters. *Physiol Behav*, 44, 235–239.

Ferris, C. F. and Delville, Y. (1994). Vasopressin and serotonin interactions in the control of agonistic behavior. *Psychoneuroendocrinology*, 19, 593–601.

Ferris, C. F., Meenan, D. M., and Albers, H. E. (1986a). Microinjection of kainic acid into the hypothalamus of golden hamsters prevents vasopressin-dependent flank-marking behavior. *Neuroendocrinology*, 44, 112–116.

Ferris, C. F., Stolberg, T., and Delville, Y. (1999a). Serotonin regulation of aggressive behavior in male golden hamsters (Mesocricetus auratus). *Behav Neurosci*, 113, 804–815.

Ferris, C. F., Pollock, J., Albers, H. E., and Leeman, S. E. (1985). Inhibition of flank-marking behavior in golden hamsters by microinjection of a vasopressin antagonist into the hypothalamus. *Neurosci Lett*, 55, 239–243.

Ferris, C. F., Meenan, D. M., Axelson, J. F., and Albers, H. E. (1986b). A vasopressin antagonist can reverse dominant/subordinate behavior in hamsters. *Physiol Behav*, 38, 135–138.

Ferris, C. F., Axelson, J. F., Shinto, L. H., and Albers, H. E. (1987). Scent marking and the maintenance of dominant/subordinate status in male golden hamsters. *Physiol Behav*, 40, 661–664.

Ferris, C. F., Singer, E. A., Meenan, D. M., and Albers, H. E. (1988a). Inhibition of vasopressin-stimulated flank marking behavior by V1-receptor antagonists. *Eur J Pharmacol*, 154, 153–159.

Ferris, C. F., Singer, E. A., Meenan, D. M., and Albers, H. E. (1988b). Inhibition of vasopressin-stimulated flank marking behavior by V1- receptor antagonists. *Eur J Pharmacol*, 154, 153–159.

Ferris, C. F., Axelson, J. F., Martin, A. M., and Roberge, L. F. (1989). Vasopressin immunoreactivity in the anterior hypothalamus is altered during the establishment of dominant/subordinate relationships between hamsters. *Neuroscience*, 29, 675–683.

Ferris, C. F., Irvin, R. W., Potegal, M., and Axelson, J. F. (1990a). Kainic acid lesions of vasopressinergic neurons in the hypothalamus disrupts flank marking behavior in golden hamsters. *J Neuroendocrinol*, 2, 123–129.

Ferris, C. F., Irvin, R. W., Potegal, M., and Axelson, J. F. (1990b). Kainic Acid lesion of vasopressinergic neurons in the hypothalamus disrupts flank marking behavior in golden hamsters. *J Neuroendocrinol*, 2, 123–129.

Ferris, C. F., Gold, L., De Vries, G. J., and Potegal, M. (1990c). Evidence for a functional and anatomical relationship between the lateral septum and the hypothalamus in the control of flank marking behavior in Golden hamsters. *J Comp Neurol*, 293, 476–485.

Ferris, C. F., Delville, Y., Irvin, R. W., and Potegal, M. (1994). Septo-hypothalamic organization of a stereotyped behavior controlled by vasopressin in golden hamsters. *Physiol Behav*, 55, 755–759.

Ferris, C. F., Delville, Y., Bonigut, S., and Miller, M. A. (1999c). Galanin antagonizes vasopressin-stimulated flank marking in male golden hamsters. *Brain Res*, 832, 1–6.

Ferris, C. F., Albers, H. E., Wesolowski, S. M., Goldman, B. D., and Luman, S. E. (1984a). Vasopressin injected into the hypothalamus triggers a stereotypic behavior in golden hamsters. *Science*, 224, 521–523.

Ferris, C. F., Albers, H. E., Wesolowski, S. M., Goldman, B. D., and Luman, S. E. (1984b). Vasopressin injected into the hypothalamus triggers a stereotypic behavior in golden hamsters. *Science*, 224, 521–523.

Ferris, C. F., Pilapil, C. G., Hayden-Hixson, D. M., Wiley, R. G., and Koh, E. T. (1992). Functional and anatomically distinct populations of vasopressinergic magnocellular neurons in the female golden hamster. *Journal of Neuroendocrinology*, 4, 193–205.

Ferris, C. F., Delville, Y., Grzonka, Z., Luber-Narod, J., and Insel, T. R. (1993). An iodinated vasopressin (V1) antagonist blocks flank marking and selectively labels neural binding sites in golden hamsters. *Physiol Behav*, 54, 737–747.

Ferris, C. F., Delville, Y., Brewer, J. A., Mansour, K., Yules, B., and Melloni, R. H., Jr. (1996a). Vasopressin and developmental onset of flank marking behavior in golden hamsters. *J Neurobiol*, 30, 192–204.

Ferris, C. F., Delville, Y., Brewer, J. A., Mansour, K., Yules, B., and Melloni, R. H., Jr. (1996b). Vasopressin and developmental onset of flank marking behavior in golden hamsters. *J Neurobiol*, 30, 192–204.

Ferris, C. F., Melloni, R. H., Jr., Koppel, G., Perry, K. W., Fuller, R. W., and Delville, Y. (1997a). Vasopressin/serotonin interactions in the anterior hypothalamus control aggressive behavior in golden hamsters. *J Neurosci*, 17, 4331–4340.

Ferris, C. F., Melloni, R. H., Jr., Koppel, G., Perry, K. W., Fuller, R. W., and Delville, Y. (1997b). Vasopressin/serotonin interactions in the anterior hypothalamus control aggressive behavior in golden hamsters. *J Neurosci*, 17, 4331–4340.

Ferris, C. F., Lu, S.-F., Messenger, T., Guillon, C. D., Koppel, G. A., Miller, M. J., Heindel, N. D., and Simon, N. G. (2006). Orally active vasopressin V1a receptor antagonist, SRX251, selectively blocks aggressive behavior. *Pharm, Biochem Behav*, 83, 169–174.

Garrett, J. W. and Campbell, C. S. (1980). Changes in social behavior of the male golden hamster accompanying photoperiodic changes in reproduction. *Horm Behav*, 14, 303–318.

Gobrogge, K. L., Liu, Y., Jia, X., and Wang, Z. (2007). Anterior hypothalamic neural activation and neurochemical associations with aggression in pair-bonded male prairie voles. *Journal of Comparative Neurology*, 502, 1109–1122.

Grimes, J. M. and Melloni, R. H., Jr. (2002). Serotonin modulates offensive attack in adolescent anabolic steroid-treated hamsters. *Pharmacol Biochem Behav*, 73, 713–721.

Grimes, J. M. and Melloni, R. H. (2005). Serotonin 1B receptor activity and expression modulate the aggression-stimulating effects of adolescent anabolic steroid exposure in hamsters. *Behavioral Neuroscience*, 119, 1184–1194.

Grimes, J. M., Ricci, L. A., and Melloni, R. H., Jr. (2003). Glutamic acid decarboxylase (GAD65) immunoreactivity in brains of aggressive, adolescent anabolic steroid-treated hamsters. *Horm Behav*, 44, 271–280.

Grimes, J. M., Ricci, L. A., and Melloni, R. H. (2006). Plasticity in anterior hypothalamic vasopressin correlates with aggression during anabolic/androgenic steroid withdrawal. *Behav Neurosci*, 120, 115–124.

Grimes, J. M., Ricci, L. A., and Melloni, R. H. (2007). Alterations in anterior hypothalamic vasopressin, but not serotonin, correlate with the temporal onset of aggressive behavior during adolescent anabolic-steroid exposure in hamsters. *Behav Neurosci*, 121, 941–948.

Guillon, C. D., Koppel, G. A., Brownstein, M. J., Chaney, M. O., Ferris, C. F., Lu, S.-F., Fabio, K. M., Miller, M. J., Heindel, N. D., Hunden, D. C., Cooper, R. D. G., Kaldor, S. W., Skelton, J. J., Dressman, B. A., Clay, M. P., Steinberg, M. I., Bruns, R. F., and Simon, N. G. (2006). Azetidinones as vasopressin V1a antagonists. *Bioorganic Med Chem*, 15, 2054–2080.

Gutzler, S. J., Karom, M., Erwin, W. D., and Albers, H. E. (2010). Arginine-vasopressin and the regulation of aggression in female Syrian hamsters (Mesocricetus auratus). *Eur J Neurosci*, 31, 1655–1663.

Harmon, A. C., Moore, T. O., Huhman, K. L., and Albers, H. E. (2002). Social experience and social context alter the behavioral response to centrally administered oxytocin in female Syrian hamsters. *Neuroscience*, 109, 767–772.

Harrison, R. J., Connor, D. F., Nowak, C., and Melloni, R. H., Jr. (2000a). Chronic low-dose cocaine treatment during adolescence facilitates aggression in hamsters. *Physiol Behav*, 69, 555–562.

Harrison, R. J., Connor, D. F., Nowak, C., Nash, K., and Melloni, R. H., Jr. (2000b). Chronic anabolic-androgenic steroid treatment during adolescence increases anterior hypothalamic vasopressin and aggression in intact hamsters. *Psychoneuroendocrinology*, 25, 317–338.

Hart, B. L. (1974). Medial preoptic-anterior hypothalamic area and socio-sexual behavior of male dogs: a comparative neuropsychological analysis. *J Comp Physiol Psychol*, 86, 328–349.

Hart, B. L. and Voith, V. L. (1978). Changes in urine spraying, feeding and sleep behavior of cats following medial preoptic-anterior hypothalamic lesions. *Brain Res*, 145, 406–409.

Hennessey, A. C., Whitman, D. C., and Albers, H. E. (1992). Microinjection of arginine-vasopressin into the periaqueductal gray stimulates flank marking in Syrian hamsters (Mesocricetus auratus). *Brain Res*, 569, 136–140.

Hennessey, A. C., Huhman, K. L., and Albers, H. E. (1994). Vasopressin and sex differences in hamster flank marking. *Physiol Behav*, 55, 905–911.

Huhman, K. L. and Albers, H. E. (1993). Estradiol increases the behavioral response to arginine vasopressin (AVP) in the medial preoptic-anterior hypothalamus. *Peptides*, 14, 1049–1054.

Irvin, R. W., Szot, P., Dorsa, D. M., Potegal, M., and Ferris, C. F. (1990). Vasopressin in the septal area of the golden hamster controls scent marking and grooming. *Physiol Behav*, 48, 693–699.

Jackson, D., Burns, R., Trksak, G., Simeone, B., DeLeon, K. R., Connor, D. F., Harrison, R. J., and Melloni, R. H., Jr. (2005a). Anterior hypothalamic vasopressin modulates the aggression-stimulating effects of adolescent cocaine exposure in Syrian hamsters. *Neuroscience*, 133, 635–646.

Jackson, D., Burns, R., Trksak, G., Simeone, B., Deleon, K. R., Connor, D. F., Harrison, R. J., and Melloni, R. H., Jr. (2005b). Anterior hypothalamic vasopressin modulates the aggression-stimulating effects of adolescent cocaine exposure in Syrian hamsters. *Neuroscience*, 133, 635–646.

Johnson, A. E., Barberis, C., and Albers, H. E. (1995). Castration reduces vasopressin receptor binding in the hamster hypothalamus. *Brain Res*, 674, 153–158.

Johnston, R. E. (1975). Scent marking by male golden hamsters (Mesocricetus auratus) I. Effects of odors and social encounters. *Z Tierpsychol*, 37, 75–98.

Johnston, R. E. (1981). Testosterone dependence of scent marking by male hamsters (Mesocricetus auratus). *Behav Neural Biol*, 31, 96–99.

Koolhaas, J. M., Van den Brink, T. H. C., Roozendal, B., and Boorsma, F. (1990). Medial amygdala and aggressive behavior: Interaction between testosterone and vasopressin. *Agg Behav*, 16, 223–229.

Mahoney, P. D., Koh, E. T., Irvin, R. W., and Ferris, C. F. (1990). Computer-aided mapping of vasopressin neurons in the hypothalamus of the male golden hamster: Evidence of magnocellular neurons that do not project to the neurohypophysis. *Neuroendocrinology*, 2, 113–122.

Marques, D. M. and Valenstein, E. S. (1977). Individual differences in aggressiveness of female hamsters: response to intact and castrated males and to females. *Anim Behav*, 25, 131–139.

Melloni, R. H., Jr. and Ferris, C. F. (1996). Adolescent anabolic steroid use and aggressive behavior in golden hamsters. *Ann N Y Acad Sci*, 794, 372–375.

Melloni, R. H., Jr. and Ricci, L. A. (2010). Adolescent exposure to anabolic/androgenic steroids and the neurobiology of offensive aggression: A hypothalamic neural model based on findings in pubertal Syrian hamsters. *Horm Behav*, 58, 177–191.

Melloni, R. H., Jr., Connor, D. F., Hang, P. T., Harrison, R. J., and Ferris, C. F. (1997). Anabolic-androgenic steroid exposure during adolescence and aggressive behavior in golden hamsters. *Physiol Behav*, 61, 359–364.

Miller, M. A., Ferris, C. F., and Kolb, P. E. (1999). Absence of vasopressin expression by galanin neurons in the golden hamster: implications for species differences in extrahypothalamic vasopressin pathways. *Brain Res Mol Brain Res*, 67, 28–35.

Payne, A. P. and Swanson, H. H. (1972). The effect of sex hormones on the aggressive behaviour of the female golden hamster (Mesocricetus auratus Waterhouse). *Anim Behav*, 20, 782–787.

Potegal, M. and Ferris, C. F. (1990). Intraspecific aggression in male hamsters is inhibited by intrahypothalamic vasopressin-receptor antagonist. *Agg Behav*, 15, 311–320.

Ricci, L. A., Schwartzer, J. J., and Melloni, R. H., Jr. (2009). Alterations in the anterior hypothalamic dopamine system in aggressive adolescent AAS-treated hamsters. *Horm Behav*, 55, 348–355.

Ricci, L. A., Rasakham, S., Grimes, J. M., and Melloni, R. H. (2006). Serotonin 1A receptor activity and expression modulate adolescent anabolic/androgenic steroid induced aggression in hamsters. *Pharmacol Biochem Behav*, 85, 1–11.

Schulz, K. M., Menard, T. A., Smith, D. A., Albers, H. E., and Sisk, C. L. (2006). Testicular hormone exposure during adolescence organizes flank-marking behavior and vasopressin receptor binding in the lateral septum. *Horm Behav*, 50, 477–483.

Schwartzer, J. J. and Melloni, R. H., Jr. (2010a). Anterior hypothalamic dopamine D2 receptors modulate adolescent AAS-induced offensive aggression. *Behav Pharm*, in press.

Schwartzer, J. J. and Melloni, R. H., Jr. (2010b). Dopamine activity in the lateral anterior hypothalamus modulates AAS-induced aggression through D2 but not D5 receptors. *Behav Neurosci*, 124, 645–655.

Schwartzer, J. J., Ricci, L. A., and Melloni, R. H., Jr. (2009). Interactions between the dopaminergic and GABAergic

neural systems in the lateral anterior hypothalamus of aggressive AAS-treated hamsters. *Behav Brain Res*, 203, 15–22.

Stribley, J. and Carter, C. S. (1999). Neonatal vasopressin increases aggression in prairie voles. *Proc Natl Acad Sci USA*, 96, 12601–12604.

Thompson, R., Gupta, S., Miller, K., Mills, S., and Orr, S. (2004). The effects of vasopressin on human facial responses related to social communication. *Psychoneuroendocrinology*, 29, 35–48.

Vandenbergh, J. G. (1971). The effects of gonadal hormones on the aggressive behaviour of adult golden hamsters (Mesocricetus auratus). *Anim Behav*, 19, 589–594.

Whitman, D., Hennessey, A., and Albers, H. E. (1992). Norepinephrine inhibits vasopressin-stimulated flank marking in the Syrian hamster by acting within the medial preoptic-anterior hypothalamus. *Journal of Neuroendocrinology*, 4, 541–546.

Whitsett, J. M. (1975). The development of aggressive and marking behavior in intact and castrated male hamsters. *Horm Behav*, 6, 47–57.

Winslow, J. T., Hastings, N., Carter, C. S., Harbaugh, C. R., and Insel, T. R. (1993). A role for central vasopressin in pair bonding in monogamous prairie voles. *Nature*, 365, 545–548.

Yahr, P., Commins, D., Jackson, J. C., and Newman, A. (1982). Independent control of sexual and scent marking behaviors of male gerbils by cells in or near the medial preoptic area. *Horm Behav*, 16, 304–322.

Young, L. J., Wang, Z., Cooper, T. T., and Albers, H. E. (2000). Vasopressin (V1a) receptor binding, mRNA expression and transcriptional regulation by androgen in the Syrian hamster brain. *J Neuroendocrinol*, 12, 1179–1185.

13

The involvement of oxytocin and vasopressin in social recognition and social learning

Interplay with the sex hormones

Riccardo Dore, Anna Phan, Amy E. Clipperton-Allen, Martin Kavaliers, and Elena Choleris

13.1 Overview

Social behavior is widespread among vertebrate species, with various degrees of sociality being shown in different species. The commonly studied species of laboratory rodents, rats (*Rattus norvegicus*) and mice (*Mus musculus/Mus domesticus*), show high levels of social behaviors and as such are typically housed in groups in animal facilities (Latham and Mason, 2004). Social life comes with general and specialized cognitive skills. Among these, social recognition and social learning have been extensively investigated in the laboratory and their neurobiology is beginning to be elucidated. In particular, the nonapeptides oxytocin (OT) and arginine-vasopressin (AVP) have been implicated in both of these social cognitive skills. Intriguingly, a modulatory role for the sex hormones has been shown for both. We review here the literature dealing with OT, AVP, and sex hormone regulation of social recognition and social learning.

13.2 Social recognition

Social recognition and social memory enable humans and animals to distinguish conspecifics, an important skill for social life in most gregarious species. For example, social recognition is the basis for hierarchical organizations and social bonds (e.g., mother–offspring, pair bonds; (Carter and Keverne, 2002; Choleris, Kavaliers and Pfaff, 2004). Without this ability, an animal could not distinguish between familiar mate and intruder, friend and foe, determine an individual's social status, and utilize this information to display the appropriate behaviors (affiliative or aggressive) towards conspecifics. Thus, understanding the neurobiological basis of social recognition and the use of social information can allow full appreciation of not only the mechanisms of social behavior, but also of its significance from a functional and evolutionary point of view (Tang-Martinez, 2003).

The recognition of others occurs at various levels: social hierarchical status, sex, reproductive state (e.g., estrous phase), health (e.g., parasitic load), emotions (e.g., stress – empathy), genetic relatedness (kin recognition), familiarity (have I met you before?) and individual identity (yes, you are George). The latter, *true individual recognition*, is characterized by unique changes in an animal's behavior towards another individual following a specific social experience (Gheusi et al., 1994) and can be assessed by observing specific behavioral responses towards individuals with whom an animal has had different experiences (Lai et al., 2005; Johnston, 2003; Lai and Johnston, 2002). Individuals belonging to the same class can share characteristics (e.g., gender, relatedness or familiarity) that can be detected using cues that do not need to be

Oxytocin, Vasopressin, and Related Peptides in the Regulation of Behavior, ed. E. Choleris, D. W. Pfaff, and M. Kavaliers.

individual-specific, such as features related to hormonal levels (e.g., bright coloration or specific body odors in high-testosterone dominant individuals). On the basis of these class-shared cues an animal can then modulate its social behavior. For example, a mouse may show less social approach/interest and behave less aggressively toward all familiar conspecifics because they are not novel, independently of whether that mouse can actually distinguish between specific familiar individual mice. Often, animals and humans combine information from both individual-specific and class-shared cues when assessing others and adjusting their behavior towards them on the basis of social recognition processes. Hence, in order to distinguish "true individual recognition" from other forms of social recognition, all class-shared features (genetic relatedness, familiarity, health, hierarchical status, etc.) need to be kept equal in research studies. For example, true individual recognition has been demonstrated in hamsters. Male hamsters will selectively avoid specific individuals who have previously defeated them, but would not avoid other familiar males with similar class characteristics (i.e., being a dominant, confident and high-testosterone guy) who have not previously defeated them (Lai et al., 2005). Similarly, male rats show an individual specific memory for a dominant rat. In a food competition task they will readily give way to the individual that has previously defeated them in that competition, but not to another, equally dominant, individual (Cordero and Sandi, 2007; Timmer and Sandi, 2010; reviewed in van der Koij and Sandi, 2012).

True individual recognition and different types of class recognition processes are likely to involve at least partially different neurobiological mechanisms. Hence, the proper identification of the different types of social recognition can lead to a better understanding of the studies on the proximal mechanisms of social recognition. In most laboratory studies that have investigated the neurobiological bases of social recognition, the specific behavioral paradigms employed typically assess familiarity, rather than true individual recognition

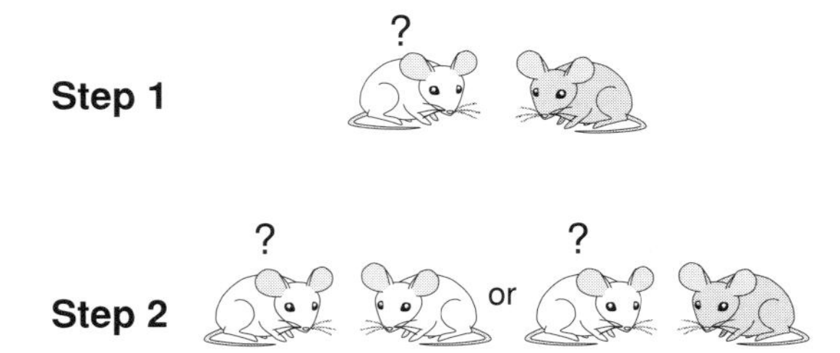

Figure 13.1 The habituation–dishabituation paradigm for the assessment of social recognition in laboratory rodents. In step 1, the experimental rodent (white) is presented with a conspecific (gray) with whom he/she can become familiar, either over a single or over multiple presentations. In step 2, the experimental rodent is either presented with the same (gray), now familiar, conspecific, or with a novel (white) unfamiliar conspecific. Typically, in step 2 the experimental rodent will show decreased social investigation of a familiar conspecific (habituation) and increased social investigation of a novel conspecific (dishabituation). This same paradigm can also be performed using only odor cues from conspecifics. See color version in plates section.

(reviewed in Choleris et al., 2009; Choleris et al., 2012; Gabor et al., 2011).

Familiarity recognition is most commonly measured by means of two different paradigms: the habituation–dishabituation paradigm and the social discrimination paradigm, both generally referred to as "social recognition." Both of these paradigms exploit the natural tendency of rodents to sniff, or investigate, an unfamiliar individual more than a familiar individual.

The *habituation–dishabituation paradigm* (Figure 13.1) is the most commonly utilized for studying recognition in rodents (Gheusi et al., 1994). In this paradigm, a subject animal is exposed to the same stimulus animal over either a single (e.g., Thor and Holloway, 1982; Thor, 1980) or repeated (e.g., Choleris et al., 2003; Choleris et al., 2007) trials (step 1). When the stimulus animal is removed and later reintroduced, subjects typically display a decreased duration of social investigation compared to the first encounter (*habituation*). Following habituation, a novel unfamiliar social

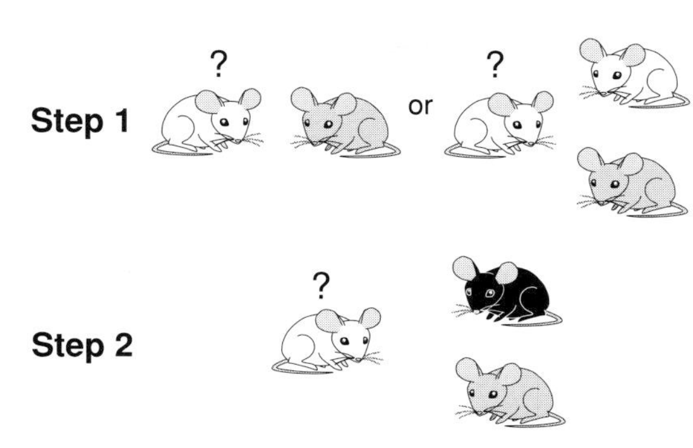

Figure 13.2 The discrimination paradigm for the assessment of social recognition in laboratory rodents. In step 1, the experimental rodent (white) is presented with either one (gray) or two (gray and white) conspecifics with whom he/she can become familiar, either over a single or over multiple presentations. In step 2, the experimental rodent is concurrently presented with a familiar (gray) and an unfamiliar (black) conspecific. Typically, in step 2 the experimental rodent will investigate the unfamiliar more than the familiar conspecific. This same paradigm can also be performed using only odor cues from conspecifics. See color version in plates section.

stimulus is presented, which typically results in increased social investigation (*dishabituation*) (step 2). One of the main disadvantages of this procedure is that repeated testing of the same animal can lead to non-specific behavioral changes (such as sensitization to the testing procedure) (Engelmann et al., 1995), which might interfere with monitoring of social memory.

In the *social discrimination paradigm* (Figure 13.2) an experimental animal is exposed to either one or two stimulus animals (step 1). After a predetermined period of time, it is re-exposed to the same stimulus animal and, at the same time, to a novel stimulus animal (step 2) (Engelmann et al., 2011). This procedure directly tests an animal's ability to discriminate between two social stimuli with a simultaneous binary choice between a familiar and an unfamiliar conspecific (Engelmann et al., 1995; Choleris et al., 2006). Thus, this is a more direct way of assessing social recognition. Although the social discrimination paradigm is generally less utilized than the habituation–dishabituation procedure, it has been shown to be more sensitive for assessing social recognition, as it can detect social recognition even when the habituation–dishabituation procedure fails to do so (Engelmann et al., 1995; Choleris et al., 2006).

While we have generally described two common means of testing social recognition, we need to emphasize that there are several variations of both tests that use different exposure times, intertrial intervals and stimulus animals. Since the results of habituation–dishabituation and social discrimination procedures can be driven by both familiarity recognition and true individual recognition, and both factors are typically manipulated in these paradigms, these paradigms usually do not allow for any conclusive demonstrations of true individual recognition.

13.2.1 Oxytocin and social recognition

OT has been implicated in the regulation of various social behaviors (reviewed in Donaldson and Young, 2008, and Chapters 7–14; 16, 17–20), including that of social recognition. Systemic studies with OT initially yielded mixed results. In a two-trial habituation–dishabituation paradigm, systemic injection of OT antagonists to adult male rats (Popik and Vetulani, 1991) facilitated the recognition of a previously presented juvenile rat, while male rats treated with supraphysiological doses of OT showed impairment in social memory (Popik and Vetulani, 1991; Dantzer et al., 1987). However, low doses of systemically administered OT after the first social exposure seemed to facilitate social memory in these male rats (Popik et al., 1992). These early results are difficult to interpret because the drugs used in these studies have a very limited degree of blood–brain barrier penetration (see discussion in Popik and van Ree, 1998). Subsequent investigations with intracranial administration confirmed an important role for OT in social recognition. Intracerebroventricular administration to male rats of low, but not high, doses of OT 5 min after the first encounter with a juvenile animal reduced the social investigation time of the same animal

during a second presentation 120 min later. This effect could be reversed by administration of an OT antagonist (Benelli et al., 1995). Similar results were also found in female rats; intracerebroventricular injection of an OT antagonist, but not of OT, impaired social recognition (Engelmann et al., 1998). Taken together, these studies support a facilitatory role for OT in social recognition in rats of both sexes at low doses, while supraphysiological OT doses do not seem to further enhance it, and may in fact, impair social recognition.

Studies with OT knockout (KO) mice (OTKO) have demonstrated and confirmed that this peptide is essential for the normal expression of social recognition in both sexes (Choleris et al., 2003; Choleris et al., 2006; Ferguson et al., 2000). When tested in the habituation–dishabituation paradigm, both male and female OTKO mice persisted in a high degree of investigation of a familiar mouse across presentations. These KO mice had the same social investigation duration as the wild-type (WT) mice initially, indicating unimpaired initial social responses, but did not habituate to the repeated presentation of the familiar mouse. Thus, their investigation times during the habituation sessions remained significantly higher than the WT mice, which habituated normally. Moreover, when a new mouse was introduced, OTKO mice failed to show enhanced investigation during the dishabituation phase (Choleris et al., 2003; Ferguson et al., 2000). The deficit in social recognition demonstrated by these KO mice seems to be due to a lack of activational, rather than developmental effects of OT. This is supported by the fact that OT infusion into the lateral ventricles and the medial amygdala before (but not after) the first social encounter can restore social recognition in adult OTKO mice (Ferguson et al., 2000; Ferguson et al., 2001). The social recognition impairment in OTKO female mice was confirmed using the social discrimination test variant, where a simultaneous choice between a familiar and an unfamiliar animal was presented. In this test, the OTKO mice spent an equal amount of time sniffing a novel mouse and a familiar one, suggesting that they were unable to discriminate between the two (Choleris et al., 2006). The social recognition impairment of the OTKO mice also affected their behavioral responses in the evolutionarily relevant context of recognition and avoidance of parasitized conspecifics; both male (Kavaliers et al., 2006) and female (Kavaliers et al., 2005; Kavaliers et al., 2003). OTKO mice failed to recognize and avoid the odors of infected conspecifics and to modulate their responses on the basis of prior familiarity (reviewed in Kavaliers and Choleris, 2011 and Chapter 14).

In support of OT involvement in social recognition, the males of an oxytocin receptor (OTR) knockout (OTRKO) line of mice as well as of a specific postweaning OTRKO in the forebrain ($OTRKO^{FB-}$) show impairments in social recognition (Lee et al., 2008; Takayanagi et al., 2005). The males of the OTRKO line were impaired in their ability to habituate to familiar individuals, with equally high levels of investigation of both novel and familiar conspecifics (Takayanagi et al., 2005), thus mirroring the results obtained with OTKO mice (Choleris et al., 2003; Choleris et al., 2006; Ferguson et al., 2000). Interestingly, the OTKO, OTRKO and $OTRKO^{FB-}$ male mice could still recognize females of different strains in a choice test (Macbeth et al., 2009), suggesting that OT/OTR mechanisms mediate the recognition of familiarity but not that of broader (i.e., strain) class features.

Importantly, the above-mentioned studies show that OT's effects on social recognition seem to be specific, since the OTKO mice and the $OTRKO^{FB-}$ mice are not impaired in general behaviors (i.e., activity levels, grooming, investigation of non-social stimuli), motor functions, spatial learning, or discrimination of social or non-social olfactory cues (Choleris et al., 2003; Ferguson et al., 2000; Kavaliers et al., 2003; Lee et al., 2008). Intriguingly, postlearning exogenous OT administration via the intranasal route specifically improved memory for faces in men and women (Rimmele et al., 2009; Savaskan et al., 2008), without affecting the recognition of non-social stimuli (Rimmele et al., 2009) or facial expressions (Savaskan et al., 2008), suggesting that even in humans OT may be specifically involved in social recognition.

Male OTRKO^{FB-} mice developed a reduction in OTR expression in the lateral septum (LS), hippocampus and ventral pallidum, while its expression remained normal in the medial amygdala and olfactory bulbs, olfactory nucleus and neocortex postweaning (Lee et al., 2008). These OTRKO^{FB-} mice were found to be impaired in social recognition when tested in a two-trial habituation paradigm. During the second social exposure, these OTRKO^{FB-} male mice showed a reduction in social investigation when presented with either a familiar or a novel social stimulus (Lee et al., 2008). This suggests that the social recognition deficit of these OTRKO^{FB-} mice may be different from that of the OTKO and OTRKO mice, who showed no reduction in social investigation when presented with either a familiar or novel mouse. Indeed, when OTRKO^{FB-} male mice were tested a week later in a habituation–dishabituation paradigm, consisting of four habituation sessions prior to dishabituation testing, they performed normally (Lee et al., 2008). Thus, the OTRKO^{FB-} mice appear to be better at recognizing familiar mice when repeatedly presented with a social stimulus rather than after a single presentation, suggesting that their social recognition deficit can be rescued with more extensive training. Furthermore, these results suggest forebrain OT may be able to facilitate performance on this paradigm, but it does not appear to be necessary for social recognition. In support of this, male rats receiving an OT infusion in the LS (Popik and van Ree, 1991; Popik et al., 1992; van Wimersma Greidanus and Maigret, 1996), ventral hippocampus (van Wimersma Greidanus and Maigret, 1996) and in the medial preoptic area of the hypothalamus (Popik and van Ree, 1991) demonstrated improved social recognition in a manner that was, however, not reversible by an OTR antagonist or anti-OT serum. This suggests that while these areas may be able to facilitate social recognition, OT effects on social recognition may be dependent on the receptors of other systems, such as those for arginine-vasopressin (AVPR) (see later in this chapter), to which OT can also bind (Gimpl and Fahrenholz, 2001).

The amygdala, instead, has been repeatedly implicated in OT/OTR mediation of social recognition. Male mice were found to have elevated activation of c-fos (an immediate-early gene) in the medial amygdala after a social encounter, while this c-fos activation was absent in OTKO mice (Ferguson et al., 2001). Perhaps most convincingly, it was found that infusion of OT into the medial amygdala was sufficient to restore social recognition in male OTKO mice but, only when OT was delivered before the social interaction (Ferguson et al., 2001). This suggests that OT may be especially important for the acquisition phase of social memory (but see Benelli et al., 1995). Moreover, inhibiting OT action in the medial amygdala seems sufficient to cause impairment in social recognition in male and female mice. Delivery of OT antagonists (Ferguson et al., 2001; Samuelsen and Meredith, 2011) or antisense DNA against the OTR gene (Choleris et al., 2007) into the medial amygdala of WT mice impaired their ability to habituate to a familiar conspecific, in a manner similar to the impairment of the OTKO and OTRKO mice (Choleris et al., 2003; Ferguson et al., 2000; Takayanagi et al., 2005). Similarly, social hierarchy-related memory for specific dominant males was reduced by infusion of an OTR antagonist in the medial amygdala immediately after memory acquisition (Timmer et al., 2011). Hence, OT/OTR mechanisms in the medial amygdala seem to regulate true individual recognition as well as familiarity recognition.

In rodents, the medial amygdala is the site of convergence of socially relevant (Beauchamp and Yamazaki, 2003; Dulac and Torello, 2003, Kang et al., 2009) inputs from the main and accessory olfactory systems. In a recent study, it was shown that male mice injected with an OT antagonist into the medial amygdala failed to show immediate early gene activation in this brain region in response to conspecific female urine odor (Samuelsen and Meredith, 2011). Interestingly, these mice also failed to show a response to heterospecific (predator and non-predator) odors, indicating that OT in the medial amygdala may play an important role in processing conspecific and heterospecific chemosensory signals (Samuelsen and Meredith, 2011). This suggests that the social recognition impairment seen in OTKO and OTRKO mice may be due to the loss

of OT-mediated selectivity in medial amygdala activation in response to conspecific chemical signals (Samuelsen and Meredith, 2009), or an inability to properly process biologically relevant chemical odors.

Consistent with anatomical connections between the medial amygdala and olfactory bulb, it has also been shown that OT administration in the olfactory bulb of male rats prolongs social recognition memory for a juvenile in the habituation–dishabituation paradigm in a manner that depends upon the enhancement of norepinephrine action at the α, but not β, adrenoceptors (Dluzen et al., 1998b; Dluzen et al., 2000). Similarly, olfactory bulb enhancement of OT, either via direct administration or vaginocervical stimulation, prolonged social recognition memory for a juvenile in the habituation–dishabituation paradigm in proestrus female rats via action of both α and β adrenoceptors (Larrazolo-Lopez et al., 2008). However, administration of OT antagonists into the olfactory bulbs impaired OT-facilitated social recognition in females (Larrazolo-Lopez et al., 2008) but not in male rats (Dluzen et al., 2000) suggesting a sex difference in olfactory bulb OT involvement in social recognition. In OTKO mice this was tested only in males, with administration of OT into the olfactory bulbs failing to restore social recognition (Ferguson et al., 2001). Overall, the results of these studies suggest that OT action in some areas of the brain, such as the olfactory bulbs (in males), medial preoptic area, and LS seem to modulate social recognition. OT/OTR action at the olfactory bulbs may only be required for social recognition in females (Larrazolo-Lopez et al., 2008), while medial amygdala OT/OTR appears to be necessary in both sexes (Choleris et al., 2007; Ferguson et al., 2001; Samuelsen and Meredith, 2011).

13.2.2 Arginine-vasopressin and social recognition

In the mid-1980s, results of different studies showed that peripheral administration of AVP or AVP-derived peptides that have no peripheral activity, following the initial exposure to a juvenile conspecific, prolonged social memory in a dose-dependent manner, with treated male rats showing a decline in the investigation of a previously encountered juvenile even 2 and 24 h after the initial exposure, when social memory is normally no longer present (Dantzer et al., 1987; Sekiguchi et al., 1991). In addition, these effects could be reversed/abolished by administration of an AVP antagonist (Dantzer et al., 1987). A role for endogenous AVP in social recognition is also supported by the deficit shown by the strain of naturally AVP-deficient Brattleboro male rats, whose investigation of a previously encountered and a novel conspecific was higher than that of normal Long Evans rats at both habituation and dishabituation (Engelmann and Landgraf, 1994; Feifel et al., 2009). Some differences between the involvement of OT and AVP in social recognition were found: while improved social recognition in a habituation paradigm was observed after systemic administration of lower doses of OT, AVP was not effective at the same low-dose range (Popik et al., 1992), suggesting a differential role for the two peptides in social memory. In addition, and differently from OT, a number of studies have pointed at a differential involvement of AVP in the social recognition of male and female rats. Generally, it seems that AVP is more important in social recognition in males than in females. While peripheral administration of AVP improved social recognition of a juvenile in the habituation paradigm in female and male rats (Bluthe et al., 1990; Bluthe and Dantzer, 1990), administration of AVP antagonist inhibited social recognition in males but had no effect in females (Engelmann et al., 1998; Bluthe et al., 1990; Bluthe and Dantzer, 1990), suggesting that even though it is not necessary, AVP can also enhance social recognition in females, likely via non-AVP receptor (AVPR) mechanisms. This sex difference in AVP involvement in social recognition is consistent with the greater AVP expression observed in the male than in the female brain in a number of species (reviewed in de Vries, 2008 and Chapters 1 and 2) and the fact that AVP seems to be generally involved in male-typical social behaviors such as aggression, territoriality and male reproduction (reviewed in

Donaldson and Young, 2008 and Chapters 7, 11, and 12).

Increasing brain AVP concentration in male rats through supraoptic nucleus osmotic stimulation enhanced acquisition of social recognition of a juvenile in a habituation paradigm (Engelmann, Ludwig and Landgraft, 1994). Also, intracerebroventricular AVP administration to adult male rats immediately after social exposure prolonged the retention of the memory for the familiar juvenile in a habituation paradigm (Bluthe and Dantzer, 1992; Le Moal et al., 1987). These effects on acquisition and retention of social memory could be countered/abolished by administration of an AVP antiserum (van Wimersma Greidanus and Maigret, 1996), supporting the idea of a brain mechanism of AVP mediation of social recognition.

Social recognition has been investigated in gene KO mice of AVPR1a and AVPR1b, the two AVPRs that are expressed in the brain (for details of their distribution, see Chapters 1 and 2). Male mice with KO of the gene coding for the vasopressin receptor AVPR1a (AVPR1aKO) were completely impaired in social recognition and did not show habituation after repeated exposures to an individual, with social investigation remaining high throughout the habituation–dishabituation test (Bielsky et al., 2004). Interestingly, this behavioral profile is the same as that seen in OTKO and OTRKO mice (Choleris et al., 2003; Ferguson et al., 2000; Takayanagi et al., 2005). These effects were not accompanied by any impairment in olfaction, locomotion, or spatial and olfactory learning tasks but they did also involve anxiety responses (Bielsky et al., 2004). Interestingly, differently derived male AVPR1aKO mice were found to have intact social recognition of ovariectomized females in the habituation–dishabituation test and disrupted recognition of social (urine) and non-social odors (Wersinger et al., 2007). Mice in which the gene coding for the AVPR1b was KO (AVPR1bKO) were only partially impaired in social recognition (Wersinger et al., 2002); in the habituation–dishabituation test they habituated more slowly to familiar females than did the WT males, and they did not retain the memory for a familiar female after a 30-min delay, whereas the WT mice did. Despite performing worse than the WT mice, the AVPR1bKO mice did habituate to a familiar individual and showed dishabituation with the introduction of a novel mouse (Wersinger et al., 2002). This suggests that although AVPR1bKO mice are impaired in social recognition compared to WT mice, they are still able to distinguish between familiar and unfamiliar conspecifics when multiple (rather than only one) habituation sessions are used and a short time elapses between habituation and testing, suggesting that additional training can rescue their social recognition deficit. The impairment of these mice appears to extend to other aspects of social behavior; when AVPR1bKO male mice were given a choice between investigating either male or female soiled bedding or clean bedding, they showed no preferences, while WT mice preferred to sniff soiled bedding and preferred the female's bedding to the male's (Wersinger et al., 2004). Similarly, the AVPR1bKO mice failed to show any preferences in both, a choice between a littermate and an empty compartment, and a choice between a littermate and a novel mouse (DeVito et al., 2009). However, the AVPR1bKO mice were able to discriminate between male and female urine and water, as well as distinguish between the male and female odor when an operant testing paradigm was used, and were not impaired in motor skills, sensory processing, spatial learning, anxiety, predatory behavior, sexual behavior, or defensive behavior within an aggressive encounter (Wersinger et al., 2004; DeVito et al., 2009). However, they were impaired in the temporal components of the object recognition and paired associative tests (DeVito et al., 2009). Therefore, these KO mice may have altered social motivation and episodic temporal memory rather than a deficit in social recognition *per se* (Wersinger et al., 2004; DeVito et al., 2009). Overall, the AVPR KO studies suggest a predominant role for the AVPR1a in the mediation of AVP regulation of social recognition (but see Wersinger et al., 2007).

The brain areas involved in AVP and OT mediation of social recognition only partially overlap,

with several interesting differences. While OT seems to mediate effects on social recognition through its action at the medial amygdala, AVP antagonists delivered to the medial amygdala of male mice failed to impair social recognition (Bielsky et al., 2005). Contrary to OT-mediated social recognition, the LS seems to be more important for AVP's effects on social recognition in male rats and mice, via AVPR1a mechanisms. AVP administration into the LS of male rats after the initial interaction with a juvenile improved the memory for that juvenile in both the habituation–dishabituation and the discrimination paradigms (Popik et al., 1992; Engelmann and Landgraf, 1994; Appenrodt et al., 2002; Dantzer et al., 1988; Everts and Koolhaas, 1997); endogenous release of AVP in the LS correlated with social discrimination in a microdialysis study, in a manner that could be partially blocked by infusion of an AVP antagonist (Engelmann et al., 1994). Similarly, early-life stress-induced impairment in social recognition in male rats was associated with a failure to show a rise in LS AVP levels during the acquisition phase of the social discrimination paradigm, and could be reversed by retrodialysis of AVP in the LS (Lukas et al., 2011). Consistently, administration of AVP antiserum (van Wimersma-Greidanus and Maigret, 1996), AVPR1 receptor antagonist (Appenrodt et al., 2002; Dantzer et al., 1988; Everts and Koolhaas, 1997; Everts and Koolhaas, 1999), AVPR1a specific antagonist (Bielsky et al., 2005), and antisense DNA targeting the AVPR1a gene (Landgraf et al., 1995) in the LS caused an impairment in social recognition and memory in habituation, habituation–dishabituation and discrimination paradigms, and prevented AVP-induced improvements in male rats (Popik et al., 1992; Landgraf et al., 1995). Furthermore, AVPR1 and AVPR2 antagonists in the LS reversed AVP-induced improvements in social memory (Popik et al., 1992; Appenrodt et al., 2002). These effects seem to be specific for social memory, since treatment with an AVP antagonist in the same area did not affect sensory processing (Bielsky et al., 2005), object recognition (Everts and Koolhaas, 1997), spatial memory or anxiety (Everts and Koolhaas, 1999, but see Landgraf et al., 1995) for effects on anxiety-like behaviors). Additional genetic studies further point to AVP/AVPR1a in the LS as a crucial mechanism of social recognition in male rodents. The social recognition impairment of the AVPR1aKO mice could be recued with re-expression of the gene in the LS; male AVPR1aKO mice that received a viral vector containing the AVPR1a gene in the LS showed normal habituation–dishabituation and in a subsequent discrimination test they still showed the social memory 2, 6, and 24 h after initial exposure, while the control mice showed it for only 30 min (Bielsky et al., 2005). Interestingly, AVPR1a overexpression in the LS of WT mice prolonged the social memory to 24 h in the social discrimination paradigm (Bielsky et al., 2005). Consistently, AVPR1a gene transfer in the LS of male rats enhanced social memory, as evidenced by the fact that the rats could discriminate between familiar and unfamiliar juveniles even 2 h after the initial exposure, while the control rats no longer showed social recognition. This effect was similar to that obtained after AVP infusion and could be reversed by a AVPR1a (but not an OTR) selective antagonist (Landgraf et al., 2003). These AVPR gene-specific effects and those of AVP receptor antagonists on AVP-induced social recognition allow one to rule out the possibility that AVP effects may be due to the activation of OTRs, which are also expressed in the LS (since AVP also binds to the OTRs, although with lower affinity than OT; Gimpl and Fahrenholz, 2001). Taken together, these studies support a specific role for AVP/AVP1aR in the LS in the mediation of social recognition.

Other brain areas that express AVP and its receptors (Ostrowski et al., 1994; Szot et al., 1994) and Chapters 1 and 2) have been investigated for their role in social recognition. The posthabituation administration of AVP antiserum in the dorsal and ventral hippocampus, but not in the nucleus olfactorius, impairs social recognition in male rats (van Wimersma Greidanus and Maigret, 1996). Infusion of AVP in the olfactory bulb, that expresses an intrinsic AVP system (Tobin et al., 2010), enhances social memory in male rats in a manner that is mediated

by an enhancement of norepinephrine action at the α, but not β, adrenoceptors (Dluzen et al., 1998b) and initially appeared to not be countered by either OT or AVP antagonists (Dluzen et al., 1998a; Dluzen et al., 1998b). More recent results with a different AVPR1 antagonist, siRNA targeting the AVPR1a, and selective AVP gene KO in the olfactory bulbs suggest an AVPR1a-dependent regulation of short-term (30 min) memory retention for juveniles in male mice (Tobin et al., 2010), suggesting an important role for olfactory bulb AVP/AVPR1a in social memory in males. Thus, it seems as though AVPR1a in the LS and in the olfactory bulbs is key for AVP mediation of social recognition, while other brain regions could play a role in modulatory mechanisms.

13.2.3 Sex hormones and social recognition

The performance of female mice (Sanchez-Andrade and Kendrick, 2009; Sanchez-Andrade and Kendrick, 2011), and possibly rats (Juraska et al., 2001; but see Engelmann et al., 1998), in the social recognition paradigm was shown to be better during the proestrous phase of the estrous cycle; thus, social recognition seems to be facilitated when levels of circulating estrogens and progesterone are high (Walmer et al., 1992). Further studies showed that ovariectomy induces impairment in memory for a familiar conspecific that could be reversed by treatment with estrogens (Hlinak, 1993) or estrogens and progesterone (Spiteri and Agmo, 2009) in female rats and mice (Tang et al., 2005). Pharmacological investigations revealed that treatment with low doses of an estrogen receptor (ER) alpha (ERα) selective agonist 48 h prior to testing improved social recognition in both gonadally intact and ovariectomized female mice, while higher doses seem to impair social recognition. Similarly, an ERβ selective agonist improved social recognition in gonadally intact and ovariectomized female mice, as well as in gonadally intact male mice (CL Cragg, E Fissore, DW Pfaff, E Choleris, unpublished data).

Using the habituation–dishabituation paradigm, early studies showed that ERαKO male (Imwalle et al., 2002) and both ERαKO and ERβKO female (Choleris et al., 2003) mice were impaired in social recognition. However, a later study with a binary choice discrimination paradigm confirmed that ERαKO and OTKO female mice were completely impaired in social discrimination, while the ERβKO mice were able to demonstrate social discrimination, although they did not perform as well as WT mice (Choleris et al., 2006). The acquisition and retention of social memory in ERαKO and ERβKO mice have been recently investigated using the habituation–dishabituation paradigm followed 24 h later with a social discrimination test; both paradigms used anesthetized same-sex conspecifics as the social stimuli (Sanchez-Andrade and Kendrick, 2011). These studies confirmed the impairment in social recognition of ERαKO female mice in both the social memory acquisition and retention tests. The male ERαKO mice, instead, showed normal social memory acquisition, but exhibited retention deficits. In the same study, male and female ERβKO mice showed no deficits compared to WT in either social recognition memory acquisition or 24 h retention (Sanchez-Andrade and Kendrick, 2011). A few differences between these (Sanchez-Andrade and Kendrick, 2011) results and those of others (Choleris et al., 2003; Choleris et al., 2006; Imwalle et al., 2002) could be the due to differences in the experimental protocols utilized, such as the type of social stimuli (anesthetized vs awake) accessibility of social stimuli (inside a container vs. freely accessible), testing environment (home vs novel) and number of trials (1, 4, 8). Together, the results with the KO mice suggest that ERα may be necessary for social memory formation in female and memory retention in male mice, while ERβ seems to play a smaller facilitatory role that may emerge only under more challenging testing conditions (see discussion in Sanchez-Andrade and Kendrick, 2011). Like the OTKO mice, the complete or partial social recognition impairment of the ERαKO and ERβKO mice affected their capability to recognize and avoid parasitized conspecifics (reviewed in Chapter 14); female ERαKO and ERβKO mice showed only reduced aversive responses to the odors of parasitized males

(Kavaliers et al., 2004), and male ERαKO and ERβKO mice could not modulate their behavior on the bases of previous familiarity with females (Kavaliers et al., 2008).

Recent pharmacological studies showed that, when administered 15 min before learning, estradiol (Phan et al., 2011) and a ERα agonist, but not a ERβ agonist, enhanced social recognition, as well as object and place recognition, in ovariectomized female mice in a very rapid time frame (40 min). This suggests that ERα, but not ERβ, affects social recognition through rapid, likely non-genomic, mechanisms (Phan et al., 2011). These estradiol (Phan et al., 2012) and ERα agonist behavioral effects were associated with parallel increases in hippocampal dendritic spines, suggesting that estrogens can rapidly affect learning processes in general, possibly via the mitogen-activated protein kinase-dependent intracellular signalling cascade, which is involved in synaptic plasticity and the formation of new memories (Thomas and Huganir, 2004; Zhao and Brinton, 2007).

The long-term effects that have emerged from pharmacological and gene KO studies (reviewed above), instead, suggest that estrogens may affect social recognition via their genomic action on the OT system (Chapter 1). Estrogens control OT production in the paraventricular hypothalamic nucleus (PVN) via ERβ (Dellovade et al., 1999) and, via ERα, they control OTR expression in the brain (Young et al., 1998). Levels of OT and OTR fluctuate with estrogen levels during the estrous cycle (Bale et al., 1995; Ho and Lee, 1992; Sarkar et al., 1992) and are reduced by ovariectomy. Administration of estrogens regulates OT production (Dellovade et al., 1999) and increases the electrical excitability of OT neurons in the PVN (Akaishi and Sakuma, 1985), where ERβ, but not ERα, is highly expressed (Mitra et al., 2003), and OT production is likely regulated by ERβ's activation (Patisaul et al., 2003). Consistently, ERβKO male (Nomura et al., 2002) and female (Patisaul et al., 2003) mice treated with estrogens failed to show increased OT expression in the PVN, even if they have normal baseline levels of OT and OT mRNA. In wild-type mice, instead, estrogen administration increases OT expression acting via ERβ. The fact that ERβKO mice have normal baseline levels of OT may explain why they are still able to recognize a familiar animal (Choleris et al., 2006; Sanchez-Andrade and Kendrick, 2011).

Estrogens regulate the density of OTRs and the transcription of the OTR gene in various areas of the brain (de Kloet et al., 1986; Quinones-Jenab et al., 1997), including the medial amygdala, where OT mediates social recognition (Ferguson et al., 2001) via its action at the OTR (Choleris et al., 2007). ERs are both highly expressed in the medial amygdala (Mitra et al., 2003) where ERα (Young et al., 1998) but not ERβ (Patisaul et al., 2003) is needed for the transcription of the OTR gene. These observations are in agreement with social recognition impairment of the ERαKO mice. In addition, medial posterodorsal amygdala suppression of ERα in female rats blocked social recognition, while a similar reduction of ERα in the ventromedial nucleus of hypothalamus (VMN) did not (Spiteri et al., 2010). Recently, high-social male mice were found to have more mRNA for OT and AVP in the PVN and more medial amygdala OTR, AVPR1a and ERα mRNA than low-social mice, with OT and AVP mRNAs positively correlated with ERβ and OTR, and AVPR1a mRNA with ERα (Murakami et al., 2011). These findings support the proposed 4-gene "micronet" model, where ERβ regulation of OT production in the PVN and ERα regulation of amygdala OTR (Choleris et al., 2003) are involved in medial amygdala processing of olfactory information relevant for social recognition (Johnston, 2003; Beauchamp and Yamazaki, 2003; Dulac and Torello, 2003; Kang et al., 2009), and suggest that estrogen-dependent AVP in the PVN and AVPR1a in the medial amygdala may be similarly involved in social behavior.

It has also been suggested that androgens play a role in social recognition. Male mice had improved social memory after castration; while gonadally intact males showed habituation to a familiar conspecific only when tested 30–60 min following the first exposure, castrated males still showed it after 2–3 h (Bluthe et al., 1993). Intact male performance could be brought up to that of castrated males only

with additional training (Bluthe et al., 1993). Similarly, in rats, castration seemed to improve social recognition, with only castrated males still demonstrating significant habituation to a familiar juvenile 2 h after the initial exposure (Sawyer et al., 1984). The performance of castrated male rats could be worsened to that of intact males by chronic infusion of AVP in the lateral ventricle, while infusion of an AVP antagonist improved performance (Bluthe and Dantzer, 1992). Hence, initial studies suggested that male gonadal hormones in combination with AVP interfere with social recognition. However, in other studies repeated prenatal treatment with an androgen receptor (AR) antagonist did not affect social recognition in rats, but it rendered it AVP independent (i.e., an AVP antagonist could no longer impair it), suggesting that organizational effects of androgens via the AR can affect AVP sensitivity, but not social recognition *per se* (Axelson et al., 1999). Additionally, shortly after castration rats showed a temporary disruption in social recognition that could be restored by additional training (Bluthe et al., 1990). These castrated rats also had lower BNST and medial amygdala AVP projections to the LS and became insensitive to the testosterone-reversible impairing effects on social recognition that an AVP antagonist had in gonadally intact rats (Bluthe et al., 1990). This is in agreement with the known hormone-dependent higher levels of AVP in males than females in various limbic brain areas including BNST, medial amygdala and LS, although whether these are mediated by androgenic or estrogenic metabolites of testosterone is still unclear (reviewed in de Vries, 2008 and Chapter 1).

Studies with KO mice seem to support the notion that testosterone affects social recognition via estrogenic mechanisms after being converted to estradiol by the aromatase enzyme. Male mice with KO of the gene that codes for aromatase (*cyp19)*, the ArKO mice, were impaired in social recognition in the habituation–dishabituation paradigm and showed a marked reduction in AVP in the LS, medial amygdala, BNST and the supraoptic nucleus (Pierman et al., 2008; Plumari et al., 2002). Treatment with a combination of two testosterone metabolites, the estrogen estradiol and the androgen dihydrotestosterone, restored social recognition in castrated ArKO mice, and restored AVP levels, particularly in the LS, suggesting that the observed impairment in the ArKO mice was due to activational and not organizational effects of the gene KO (Pierman et al., 2008). Because the ArKO mice have both almost undetectable estradiol and higher testosterone than WT (Fisher et al., 1998), their impairment in social recognition may be due to either the lack of estrogens or to AR-mediated effects of testosterone and metabolites. Further studies in which the effects of castration alone and those of treatment with estrogenic and androgenic hormones, alone or in combination, are assessed in ArKO mice would help clarify this still open question. In addition, more studies directly investigating the role of testosterone and the AR in social recognition in laboratory rats and mice are still needed.

Overall, it seems that both OT and AVP effects on social recognition are modulated by the sex hormones; OT effects by estradiol and AVP effects by testosterone, possibly via its conversion to estradiol. This may be important under natural conditions where social and sexual behavior as well as an animal's hormonal milieu are adaptively modulated by, and change in response to, varying environmental conditions.

13.3 Social learning

Social learning, where an "individual's behavior is influenced by observation of, or interaction with, another animal or its products" (e.g., odor cues, Kavaliers et al., 2006; Heyes, 1994) has been described across the animal kingdom, from honey bees (the famous dance, von Frisch, 1967) to humans (Whiten et al., 1996). Animals can learn a variety of adaptive information from others, both about the physical environment (e.g., feeding sites, nesting sites, dietary choices, fear and avoidance of predators; reviewed in (Shettelworth, 2010) and

about their social world (e.g., communication signals, hierarchy, aggressiveness, mating preferences and choices; see Choleris et al., 2009; Choleris et al., 2012; Gabor et al., 2011 for recent reviews). Social learning has been extensively investigated in laboratory rodent species where the social transmission of a number of behaviors/information has been shown. For example, mice, rats, and gerbils can acquire a food preference by interacting with a recently fed conspecific (e.g., Galef and Wigmore, 1983; Valsecchi and Galef, 1989; Choleris et al., 1998; Valsecchi et al., 1996); mice can learn from other mice specific food-gathering behavior with an artificial puzzle box (Valsecchi et al., 2002); rats can acquire specific food-processing behavior from other rats (Laland and Plotkin, 1990; Terkel, 1996); female mice can copy the mating preferences of other female mice (Kavaliers et al., 2006); and mice and deer mice can acquire fear and avoidance of micropredators, (i.e., biting flies) by observing others (Kavaliers et al., 2001; Kavaliers et al., 2005; Kavaliers et al., 2003). While the adaptiveness and evolution of social learning can vary depending upon the species and the type of socially acquired information, by "exploiting the expertise of others" (Russon, 1997) an animal can generally reduce or avoid the disadvantages and potential costs associated with individual learning.

The involvement of OT, AVP and estrogens in social learning has been investigated in the laboratory using two paradigms: the social transmission of food preferences (STFP) and female mate-choice copying. The STFP is typically run in 3 steps (Figure 13.3). In step 1, a *demonstrator* animal is allowed to consume a flavored food for a period of time. In step 2, the demonstrator interacts with a conspecific, the *observer*, who can acquire food-related information by smelling its odor on the demonstrator's breath. In the subsequent step 3, when given a choice, the observer preferentially consumes the same food the demonstrator ate in step 1 over another novel food (Galef and Wigmore, 1983; Valsecchi et al., 1996). Various control studies have demonstrated the social nature of the STFP; observer rats, mice and gerbils will not acquire a food preference by

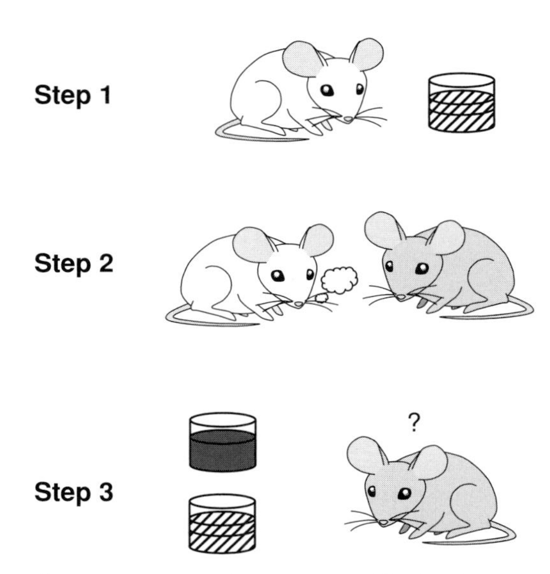

Figure 13.3 The social transmission of food preferences paradigm for the assessment of social learning in laboratory rodents. In step 1, the *demonstrator* rodent (white) consumes a novel flavored diet. In step 2, the demonstrator is allowed to interact with an *observer* rodent (gray) for a certain period of time. In step 3 the observer rodent is given a choice between two novel, differently flavored diets, one of which is the same that was present on their demonstrator's breath during the social interaction. Typically, in step 3 the observer rodent will prefer to consume the novel flavored diet that their demonstrator has previously eaten. This paradigm can be used to assess the mechanisms of acquisition, consolidation and retrieval of the socially acquired food related memory. See color version in plates section.

interacting with the odor of the food alone (Choleris et al., 1998; Valsecchi et al., 1996; Choleris et al., 2011), the odor of the food in the presence of a conspecific that has not consumed it (Choleris et al., 2011) or a rat- or mouse-shaped cotton ball covered in food (Valsecchi and Galef, 1989, Galef et al., 1988). The STFP has been used to study the acquisition, consolidation and retrieval phases of the socially acquired memory. For example, by manipulating an animal's internal state (e.g., Choleris et al., 2011; Clipperton et al., 2008b) or the quality of the social interaction at step 2 (e.g., Choleris et al., 1998)

Figure 13.4 The mate-copying paradigm for the assessment of social learning in laboratory rodents. In step 1, the *observer* rodent (white female) witnesses two conspecifics mate. In step 2, the observer rodent is given a choice between two opposite sex conspecifics, one of which is the same they have witnessed mating. Typically, in step 2 an observer female rodent will prefer to mate with the male they have witnessed mating with another female. In male rodents, only lower-quality subordinate males were shown to copy the mating preferences of dominant males. This same paradigm can also be performed using only odor cues from conspecifics. See color version in plates section.

one can investigate motivational and acquisition processes involved in social learning. Moreover, by introducing a delay between steps 2 and 3 (minutes, hours, days, weeks) one can specifically focus on consolidation and retrieval of the socially acquired memory (e.g., Clipperton et al., 2008a; Clipperton-Allen et al., 2010c; Choleris et al., 2009; Choleris and Kavaliers, 1999). Hence, the STFP is an excellent paradigm for the study of social learning and its neurobiological mechanisms.

The mate-choice copying paradigm, first developed in fish and birds (Dugatkin, 1996; Dugatkin and Godin. 1992; Galef and White, 1998), has been extended to mice, rats and even humans (Kavaliers et al., 2006; Galef et al., 2008; Jones et al., 2007; Little et al., 2011; Waynforth, 2007). Briefly, in step 1 (Figure 13.4) an observer female witnesses another female mating with a male. In step 2, when given a choice between mating with that same male or another unknown male, the observer female will typically show a preference for the male she has seen mating with the other female, thus copying her mating choices. In rodents, whose social behavior is largely driven by olfactory cues (Pruett-Jones, 1992; Brennan and Kendrick, 2006), this same paradigm can be run using the body odors of the animals, rather the whole animals; an observer female mouse will show a preference for the odor of a male that was previously paired with the odor of a sexually receptive female (Kavaliers, 2006). Unsurprisingly, as this type of social learning is related to reproductive behavior, mate-choice copying is observed only with reproduction-prone (i.e., sexually receptive) females (Kavaliers et al., 2006). In addition, sexually naïve inexperienced females were shown to rely on social learning more than experienced females (Kavaliers, unpublished results). Mate copying can even override genetics; the innate preferences of female guppies (Dugatkin, 1996; Dugatkin and Godin, 1992) and quail (Galef and White, 1998) could be reversed via mate copying. Similarly, in mice, when a male odor is aversive to begin with, because the male is infected with a parasite, mate copying can reduce or eliminate a female's natural avoidance and aversive responses (Kavaliers et al., 2006). Interestingly, the male quails show the opposite phenomenon: they will selectively *avoid* a previously preferred female if they have seen her being courted by another male (White and Galef, 1999). Hence, these males still use "public information" to guide their mating preferences, but they do not copy those of others. More recently, we found that male mice also can engage in mate-choice copying, though it appears to be restricted to lower quality subordinate males copying the preferences of dominant males (Kavaliers, unpublished results).

We review below the studies that have investigated the involvement of OT in female mate-choice copying in rodents and, and of OT, AVP, and the sex hormones in the social transmission of food preferences.

13.3.1 Oxytocin and mate-choice copying

OT involvement in mate copying by sexually naïve females has been investigated in mice using both pharmacological and genetic approaches. OTKO

females were impaired in the use of inadvertent social information (Kavaliers et al., 2006); unlike WT females, the OTKO females did not "copy" the mate choices of other females when in step 1 they were exposed to the combined odors of a male and that of an estrous female and subsequently presented, in step 2, with a choice between the odor of the female-paired male and that of another male (Figure 13.4). This impairment was also seen when the odor of a sexually receptive female was paired with that of a parasite-infected male. Through mate copying, WT females reversed their natural avoidance of the infected male odor, while OTKO females did not show such a reversal. Similarly, treatment of female mice with an OT antagonist prevented the development of a preference for the odor of a male that had been previously paired with the odor of a sexually receptive female, and prevented the social learning-mediated reversal of the natural avoidance of parasitized males (Kavaliers et al., 2008). Being able to modulate one's innate avoidance responses on the basis of others' experience can be adaptive, especially for sexually naïve females; preferring a parasite-free male may not always lead to the choice of a parasite-resistant male. Males might be parasite free because they have not encountered the parasite, and not because they have resisted it. Rather, high-quality males can be those who can cope with a certain level of infection better than other males (Getty, 2002). Hence, if other, experienced, females show a preference for a parasitized male, it may be adaptive for a sexually naïve female to copy such preferences. In this case, females may be considered to "trust" the mate preferences of other females, and OT involvement here would thus be consistent with the reported involvement of OT in human trust-based decision making (Baumgartner et al., 2008; Kosfeld et al., 2005 and Chapters 18–20).

13.3.2 Oxytocin, vasopressin, and social learning of food preferences

Using a "difficult" STFP paradigm, in which control observer male rats do not show a preference for the weak flavor of tea consumed by their demonstrator, systemic treatment with OT and AVP metabolite desglycinamide[Arg8]-vasopressin (DGAVP) improved performance when given immediately after the social interaction (step 2) or 2 h before the choice test at step 3 (Popik and van Ree, 1993). This suggests that both OT and AVP have a facilitatory effect on consolidation and retrieval of a memory that was acquired within a social context. However, treatment with another AVP metabolite (AVP_{4-9}) in female rats 6 or 8 days after the social interaction, when controls still show social learning, and 1 h before the choice test reduced the preference for the demonstrator food (Strupp et al., 1990; Bunsey and Strupp, 1990). AVP_{4-9} and vehicle treated female rats showed the same preference for demonstrator food when administration occurred 11 days after social interaction and 1 h before choice test (Bunsey and Strupp, 1990). Finally, AVP_{4-9} administration to female rats 1 h before receiving the choice test 14 or 16 days after the social interaction produced a preference for demonstrator food, at a time when control observers no longer showed it (Strupp et al., 1990; Bunsey and Strupp, 1990). Taken together, these results suggest that AVP affects the STFP differently depending on the memory demand of the paradigm (Strupp et al., 1990; Bunsey and Strupp, 1990); AVP improves performance on the difficult versions (14–16 day delays or with a weak flavor of tea), where control animals do not show preferences for the demonstrator food, and impairs performance on the easier task (6–8 day delay), which controls can perform well.

Additional evidence that AVP is involved in the STFP comes from a study using sodium-sensitive hypertensive Dahl S rats. These rats have a mutant Angiotensin II/vasopressin receptor (AngII/AVP), which renders them hyper-responsive to the effects of Angiotensin II on blood pressure and AVP in the presence of sodium (Ruiz-Opazo et al., 2002). When maintained on a low-sodium diet, male Dahl S rats did not show social learning in the STFP test. However, when fed a hypertensive higher-sodium diet, they showed a preference for the demonstrator food following a delay between steps 2 and 3 of 5 min or 3 h, but not 24 h (Ruiz-Opazo et al., 2004). The presence of sodium enhances the effect of endogenous AVP (Ruiz-Opazo et al., 2002), which

in turn may improve performance on a version of the STFP that is too challenging in the absence of sodium.

In summary, both OT and AVP improve performance on the STFP when control subjects are unable to show a preference for the demonstrator's food (Popik and van Ree, 1993; Strupp et al., 1990; Bunsey and Strupp, 1990; Ruiz-Opazo et al., 2004). When the task is easier, such that controls will show a socially acquired food preference, AVP impairs learning (Strupp et al., 1990; Bunsey and Strupp, 1990). It has been suggested that these results are due to the interaction of the exogenous AVP metabolite with endogenous changes in the chemical storage of memory related to its accessibility (Strupp et al., 1990; Bunsey and Strupp, 1990). Further investigations with specific brain manipulations of OT and AVP are needed for a full understanding of the involvement of these two peptides in the brain mechanisms underlying the acquisition, consolidation and retrieval of socially mediated memories. For example, to the best of our knowledge, whether OT and AVP affect the acquisition of STFP is presently unknown.

13.3.3 Gonadal hormones and social learning of food preferences

In mice, performance in the STFP has been linked to their endogenous hormonal state. Ovariectomized female mice show a socially acquired food preference only in the first 2 h of testing, both when testing occurred immediately after learning or at various intervals (from 90 min to 1 week) postacquisition (e.g., Clipperton et al., 2008b; Clipperton-Allen et al., unpublished results), suggesting a sex hormone modulation of the STFP. Data with gonadally intact females also point at a role for the sex hormones; gerbils who are pregnant at the time of a social interaction with their recently fed male partner show greater social learning than female gerbils who interact with their male partner while not being pregnant (Choleris et al., 2012). Mice in the proestrous phase of the estrous cycle, when estrogens and progesterone are higher (Walmer et al., 1992), showed the longest preference for the demonstrator food when there was no delay between the social interaction and choice test (Figure 13.3, steps 2 and 3); females in diestrus maintained the preference for a shorter time, and estrus females only showed a reduced preference, that did not reach statistical significance at any time point (Choleris et al., 2011). Similarly, with a delay of 24 h between steps 2 and 3, only mice trained in proestrus showed a preference for the demonstrator food (Sanchez-Andrade, James and Kendrick, 2005). Similar results have been obtained with rats; virgin rats in low-estrogen phases of the estrous cycle showed a preference for the demonstrator flavor in a 2 h test following an 8 h delay between social interaction and choice test (Fleming et al., 1994). In the same study, a higher proportion of postpartum than virgin rats showed a socially acquired food preference, which was also significantly stronger (Fleming et al., 1994). These results with rats and mice could indicate an adaptive improvement in this olfactory memory when females are reproductively active (Sanchez-Andrade and Kendrick, 2009).

Recent investigations on the involvement of the two ERs in the STFP show that activation of ERα with a specific agonist prior to training (with no delay between the two steps; (Clipperton et al., 2008b) and after training (either before or after memory consolidation; Clipperton-Allen et al., 2009; Clipperton-Allen et al., 2011b; Clipperton et al., 2008a; Clipperton et al., 2009), blocked social learning in ovariectomized mice without affecting feeding behavior *per se*, suggesting that ERα-mediated mechanisms that may inhibit memory encoding, consolidation, or retrieval processes in the STFP.

The effects of selective activation of ERβ with specific agonists depend upon the timing of drug administration; when given prior to the social interaction during which learning occurs, ERβ activation prolonged the preference for the demonstrator food to a level similar to that of gonadally intact females in proestrus (Choleris et al., 2011; Clipperton et al., 2008b). When activation of ERβ occurred immediately after learning, the STFP was blocked, while an ERβ agonist administered after

memory consolidation, 24 h postacquisition, either prolonged or blocked the socially learned preference, in a dose-dependent fashion (Clipperton et al., 2009; Clipperton-Allen et al., 2011b). The effects of ERβ activation were specific for the STFP, as they did not extend to food consumption in general. When administered prior to learning, the ERβ agonist also affected the social interaction, making the observer more subordinate to its demonstrator (Clipperton et al., 2008b; Clipperton-Allen et al., 2010b), and this may explain the prolonged STFP, since subordinate observers were shown to learn better from dominant demonstrators even in other social learning paradigms (Kavaliers, Colwell and Choleris, 2005). The postlearning effects of ERβ activation instead point at an interference with social learning mechanisms similar to that seen with ERα agonists (Clipperton et al., 2008b; Clipperton-Allen et al., 2008a; Clipperton-Allen et al., 2008b; Clipperton-Allen et al., 2011a).

Estrogenic effects on STFP and other cognitive functions may be mediated by its action on other systems. ER activation induces the transcription of several genes and can activate neuromodulatory systems, such as OT (Dellovade et al., 1999; de Kloet et al., 1986; Quinones-Jenab et al., 1997), dopamine (Lee and Mouradian, 1999; Liu and Xie, 2004; McDermott et al., 1994), acetylcholine (Miller et al., 1999; McEwen and Alves, 1999) and neurotrophic factors (Numakawa et al., 2010) such as brain-derived neurotrophic factor (BDNF) that promotes neuronal survival and differentiation, synaptic plasticity and long-term potentiation, a mechanism for learning and memory. Recently, it was shown that, like social bonds in voles (Chapter 8), the STFP is mediated by dopamine (Choleris et al., 2011). Whether, like for social recognition, the effects of estrogens in the STFP are mediated also by action upon OT and AVP has not been investigated.

Recently, even progesterone (P) and its metabolite allopregnanolone (AP) were shown to affect the STFP; when administered to ovariectomized mice preacquisition (before step 2) P reduced the duration of the social interactions and of the socially transmitted food preference, suggesting a generalized reduction in social behavior (Dore et al., 2011). In the same study, postacquisition P reduced the duration of the socially acquired food preference, suggesting impairing effects on memory. AP, both pre- and postsocial learning (step 2) either prolonged or reduced the preference for the demonstrator food in a dose-dependent manner. AP is a known allosteric modulator of the γ-aminobutyric acid type-A receptor with anxiolytic action (Schule et al., 2011), and its dose-dependent effects on the STFP are reminiscent of those of a benzodiazepine anxiolytic on male gerbils (Choleris et al., 1998). The different effects of P and AP suggest that P affects the STFP through mechanisms other than its conversion to AP.

To the best of our knowledge, whether testosterone is also involved in social learning is currently unknown. Only a few studies have tested both males and females in social learning in rodents and in most of them sex differences were not investigated (reviewed in Choleris and Kavaliers, 1999). In one study where both male and female observers (and female demonstrators) were used, there appeared to be a stronger preference for the demonstrator food in females than in the males, suggesting a stronger role for female, than male, sex hormones in the STFP. It must be noted, however, that the factor of sex was not directly examined (Boylan et al., 2007).

13.4 Conclusions

We showed here that OT and AVP, two nonapeptides that are highly involved in animal and human social behavior, are both implicated in two types of social cognition: social recognition and social learning. This is interesting, because even if these cognitive processes are both social, they have conceptually different functions. Social recognition may be considered a "prerequisite" for complex social interactions; for example, to form special bonds with specific individuals or know one's own and others' place within a hierarchy. Social learning instead appears to be a consequence of sociality; once others are

around, they can then become a source of adaptive information about the environment. In this scenario, social learning would have evolved secondarily to other social cognitive processes and it would not be surprising then, that social recognition and social learning share at least part of their underlying neurobiological mechanisms.

Another interesting observation is the dependence of social recognition and social learning upon the sex hormones. While this seems to make evolutionary sense for social recognition – knowing who is who becomes especially important for mate choices, reproduction and parental cares – the functional significance of sex hormone regulation of social learning is less clear. It may be that it reflects hormonal effects on learning in general; other nonsocial types of learning are also often found to be regulated/mediated by an animal's hormonal status, and this is believed to have evolutionary implications; for example, estrogens, which are high during the behavioral estrous (proestrus), sexually receptive, phase of the cycle, mediate an enhancement in spatial skills at a time when females become more active in their search for a mate. Alternatively, it may be that gonadal hormone involvement in social learning is secondary to their involvement in social recognition and other social cognitive processes in general. In this scenario, a hormone-mediated improvement in social "skill(s)" at a time when this is adaptive would result in enhanced processing of social information in general, both when this information is *about* others and when it originates *from* others. Therefore, in this view it is not surprising that hormone regulation of the neurohypophiseal peptides is found across all animal groups covered in this book; it may be an early trait in the evolution of vertebrates and their widespread sociality.

REFERENCES

Akaishi, T. and Sakuma, Y. (1985). Estrogen excites oxytocinergic, but not vasopressinergic cells in the paraventricular nucleus of female rat hypothalamus. *Brain Research*, 335, 302–305.

Appenrodt, E., Juszczak, M., and Schwarzberg, H. (2002). Septal vasopressin induced preservation of social recognition in rats was abolished by pinealectomy. *Behavioural Brain Research*, 134, 67–73.

Axelson, J. F., Smith, M., and Duarte, M. (1999). Prenatal flutamide treatment eliminates the adult male rat's dependency upon vasopressin when forming social-olfactory memories. *Hormones and Behavior*, 36, 109–118.

Bale, T. L., Dorsa, D. M., and Johnston, C. A. (1995). Oxytocin receptor mRNA expression in the ventromedial hypothalamus during the estrous cycle. *Journal of Neuroscience*, 15, 5058–5064.

Baumgartner, T., Heinrichs, M., Vonlanthen, A., Fischbacher, U., and Fehr, E. (2008). Oxytocin shapes the neural circuitry of trust and trust adaptation in humans. *Neuron*, 58, 639–650.

Beauchamp, G. K. and Yamazaki, K. (2003). Chemical signalling in mice. *Biochemical Society Transactions*, 31, 147–151.

Benelli, A., Bertolini, A., Poggioli, R., Menozzi, B., Basaglia, R., and Arletti, R. (1995). Polymodal dose-response curve for oxytocin in the social recognition test. *Neuropeptides*, 28, 251–255.

Bielsky, I. F., Hu, S. B., Ren, X., Terwilliger, E. F., and Young, L. J. (2005). The V1a vasopressin receptor is necessary and sufficient for normal social recognition: a gene replacement study. *Neuron*, 47, 503–513.

Bielsky, I. F., Hu, S. B., Szegda, K. L., Westphal, H., and Young, L. J. (2004). Profound impairment in social recognition and reduction in anxiety-like behavior in vasopressin V1a receptor knockout mice. *Neuropsychopharmacology*, 29, 483–493.

Bluthe, R. M. and Dantzer, R. (1992). Chronic intracerebral infusions of vasopressin and vasopressin antagonist modulate social recognition in rat. *Brain Research*, 572, 261–264.

Bluthe, R. M. and Dantzer, R. (1990). Social recognition does not involve vasopressinergic neurotransmission in female rats. *Brain Research*, 535, 301–304.

Bluthe, R. M., Gheusi, G., and Dantzer, R. (1993). Gonadal steroids influence the involvement of arginine vasopressin in social recognition in mice. *Psychoneuroendocrinology*, 18, 323–335.

Bluthe, R. M., Schoenen, J., and Dantzer, R. (1990). Androgen-dependent vasopressinergic neurons are involved in social recognition in rats. *Brain Research*, 519, 150–157.

Boylan, C. B., Blue, M. E., and Hohmann, C. F. (2007). Modeling early cortical serotonergic deficits in autism. *Behavioural Brain Research*, 176, 94–108.

Brennan, P. A. and Kendrick, K. M. (2006). Mammalian social odours: attraction and individual recognition. *Philosophical Transactions of the Royal Society of London. Series B, Biological Sciences*, 361, 2061–2078.

Bunsey, M. and Strupp, B. J. (1990). A vasopressin metabolite produces qualitatively different effects on memory retrieval depending on the accessibility of the memory. *Behavioral and Neural Biology*, 53, 346–355.

Carter, C. S. and Keverne, E. B. (2002). The neurobiology of social affiliation and pair bonding. In *Hormones, Brain and Behavior Vol. 1* (eds. D. W. Pfaff, A. P. Arnold, A. M. Etgen, S. E. Fahrbach, and R. T. Rubin), pp. 299–337. Academic Press, New York.

Choleris, E., Clipperton-Allen, A. E., Gray, D. G., Diaz-Gonzalez, S., and Welsman, R. G. (2011). Differential effects of dopamine receptor d1-type and d2-type antagonists and phase of the estrous cycle on social learning of food preferences, feeding, and social interactions in mice. *Neuropsychopharmacology*, 36, 1689–1702.

Choleris, E., Clipperton-Allen, A. E., Phan, A., and Kavaliers, M. (2009). Neuroendocrinology of social information processing in rats and mice. *Frontiers in Neuroendocrinology*, 30, 442–459.

Choleris, E., Gustafsson, J. A., Korach, K. S., Muglia, L. J., Pfaff, D. W., and Ogawa, S. (2003). An estrogen-dependent four-gene micronet regulating social recognition: a study with oxytocin and estrogen receptor-alpha and -beta knockout mice. *Proceedings of the National Academy of Sciences of the United States of America*, 100, 6192–6197.

Choleris, E. and Kavaliers, M. (1999). Social learning in animals: sex differences and neurobiological analysis. *Pharmacology, Biochemistry, and Behavior*, 64, 767–776.

Choleris, E., Kavaliers, M., and Pfaff, D. W. (2004). Functional genomics of social recognition. *Journal of Neuroendocrinology*, 16, 383–389.

Choleris, E., Little, S. R., Mong, J. A., Puram, S. V., Langer, R., and Pfaff, D. W. (2007). Microparticle-based delivery of oxytocin receptor antisense DNA in the medial amygdala blocks social recognition in female mice. *Proceedings of the National Academy of Sciences of the United States of America*, 104, 4670–4675.

Choleris, E., Ogawa, S., Kavaliers, M., Gustafsson, J. A., Korach, K. S., Muglia, L. J., and Pfaff, D. W. (2006). Involvement of estrogen receptor alpha, beta and oxytocin in social discrimination: A detailed behavioral analysis with knockout female mice. *Genes, Brain, and Behavior*, 5, 528–539.

Choleris, E., Phan, A., Clipperton-Allen, A. E., Valsecchi, P., and Kavaliers, M. (2012). Estrogenic involvement in social learning, social recognition and pathogen avoidance. *Frontiers in Neuroendocrinology*, 33, 140–159.

Choleris, E., Valsecchi, P., Wang, Y., Ferrari, P., Kavaliers, M., and Mainardi, M. (1998). Social learning of a food preference in male and female Mongolian gerbils is facilitated by the anxiolytic, chlordiazepoxide. *Pharmacology, Biochemistry, and Behavior*, 60, 575–584.

Clipperton-Allen, A. E., Almey, A., Melichercik, A., Allen, C. P., and Choleris, E. (2011a). Effects of an estrogen receptor alpha agonist on agonistic behaviour in intact and gonadectomized male and female mice. *Psychoneuroendocrinology*, 36, 981–995.

Clipperton-Allen, A. E., Cragg, C. L., Wood, A. J., Pfaff, D. W., and Choleris, E. (2010b). Agonistic behavior in males and females: effects of an estrogen receptor beta agonist in gonadectomized and gonadally intact mice. *Psychoneuroendocrinology*, 35, 1008–1022.

Clipperton-Allen, A. E., Flaxey, I., Webster, H., and Choleris, E. (2009). Pre-acquisition administration of estradiol benzoate prolongs the preference for a socially transmitted food preference in ovariectomized CD1 mice. *Society for Neuroscience Abstract*, 681.14.

Clipperton-Allen, A. E., Flaxey, I., Webster, H. K., Nediu-Mihalache, C., and Choleris, E. (2010c). Effects of immediate and delayed post-acquisition estradiol benzoate on a socially transmitted food preference in ovariectomized female CD1 mice. *Society for Neuroscience Abstract*, 296.18.

Clipperton-Allen, A. E., Roussel, V. R., Ying, H. L., Mikloska, K. V., and Choleris, E. (2011b). Effects of chronic ERα and ERβ agonists on a socially transmitted food preference in ovariectomized CD1 mice. *Society for Neuroscience Abstract*, 728.04

Clipperton, A. E., Brown, A., Hussey, B., Tam, C., and Choleris, E. (2008a). Estrogen receptor alpha agonist impairs memory but not acquisition of a socially learned food preference. *Society for Neuroscience Abstract*, 794. 21.

Clipperton, A. E., Flaxey, I. J. A., Foucault, J. N., Rush, S. T., and Choleris, E. (2009). Post-acquisition administration of an estrogen receptor beta agonist blocks the transmission of food preferences in ovariectomized CD1 mice. *Society for Behavioral Neuroendocrinology Abstract*, 13, P3.76.

Clipperton, A. E., Spinato, J. M., Chernets, C., Pfaff, D. W., and Choleris, E. (2008b). Differential effects of estrogen receptor alpha and beta specific agonists on social learning of food preferences in female mice. *Neuropsychopharmacology*, 33, 2362–2375.

Cordero, M. I. and Sandi, C. (2007). Stress amplifies memory for social hierarchy. *Frontiers in Neuroscience*, 1, 175–184.

Dantzer, R., Bluthe, R. M., Koob, G. F., and Le Moal, M. (1987). Modulation of social memory in male rats by neurohypophyseal peptides. *Psychopharmacology*, 91, 363–368.

Dantzer, R., Koob, G. F., Bluthe, R. M., and Le Moal, M. (1988). Septal vasopressin modulates social memory in male rats. *Brain Research*, 457, 143–147.

de Kloet, E. R., Voorhuis, D. A., Boschma, Y., and Elands, J. (1986). Estradiol modulates density of putative 'oxytocin receptors' in discrete rat brain regions. *Neuroendocrinology*, 44, 415–421.

de Vries, G. J. (2008). Sex differences in vasopressin and oxytocin innervation of the brain. *Progress in Brain Research*, 170, 17–27.

Dellovade, T. L., Zhu, Y. S., and Pfaff, D. W. (1999). Thyroid hormones and estrogen affect oxytocin gene expression in hypothalamic neurons. *Journal of Neuroendocrinology*, 11, 1–10.

DeVito, L. M., Konigsberg, R., Lykken, C., Sauvage, M., Young, W. S., III, and Eichenbaum, H. (2009). Vasopressin 1b receptor knock-out impairs memory for temporal order. *Journal of Neuroscience*, 29, 2676–2683.

Dluzen, D. E., Muraoka, S., Engelmann, M., Ebner, K., and Landgraf, R. (2000). Oxytocin induces preservation of social recognition in male rats by activating alpha-adrenoceptors of the olfactory bulb. *The European Journal of Neuroscience*, 12, 760–766.

Dluzen, D. E., Muraoka, S., Engelmann, M., and Landgraf, R. (1998a). The effects of infusion of arginine vasopressin, oxytocin, or their antagonists into the olfactory bulb upon social recognition responses in male rats. *Peptides*, 19, 999–1005.

Dluzen, D. E., Muraoka, S., and Landgraf, R. (1998b). Olfactory bulb norepinephrine depletion abolishes vasopressin and oxytocin preservation of social recognition responses in rats. *Neuroscience Letters*, 254, 161–164.

Donaldson, Z. R. and Young, L. J. (2008). Oxytocin, vasopressin, and the neurogenetics of sociality. *Science*, 322, 900–904.

Dore, R., Ervin, K., Perkins, C., Gallagher, N., Clipperton Allen, A. E., and Choleris, E. (2011). The effect of progesterone and the neuroactive steroid allopregnanolone on social learning in female mice. *Society for Behavioral Neuroendocrinology Abstracts*, 15, 3.02.

Dugatkin, L. A. (1996). Interface between culturally based preferences and genetic preferences: female mate choice in Poecilia reticulata. *Proceedings of the National Academy of Sciences of the United States of America*, 93, 2770–2773.

Dugatkin, L. A. and Godin, J. G. (1992). Reversal of female mate choice by copying in the guppy (Poecilia reticulata). *Proceedings. Biological Sciences / The Royal Society*, 249, 179–184.

Dulac, C. and Torello, A. T. (2003). Molecular detection of pheromone signals in mammals: from genes to behaviour. *Nature Reviews. Neuroscience*, 4, 551–562.

Engelmann, M., Ebner, K., Wotjak, C. T., and Landgraf, R. (1998). Endogenous oxytocin is involved in short-term olfactory memory in female rats. *Behavioural Brain Research*, 90, 89–94.

Engelmann, M., Hädicke, J., and Noack, J. (2011). Testing declarative memory in laboratory rats and mice using the nonconditioned social discrimination procedure. *Nature Protocols*, 6, 1152–1162.

Engelmann, M. and Landgraf, R. (1994). Microdialysis administration of vasopressin into the septum improves social recognition in Brattleboro rats. *Physiology and Behavior*, 55, 145–149.

Engelmann, M., Ludwig, M., and Landgraft, R. (1994). Simultaneous monitoring of intracerebral release and behavior: endogenous vasopressin improves social recognition. *Journal of Neuroendocrinology*, 6, 391–395.

Engelmann, M., Wotjak, C. T., and Landgraf, R. (1995). Social discrimination procedure: an alternative method to investigate juvenile recognition abilities in rats. *Physiology and Behavior*, 58, 315–321.

Everts, H. G. and Koolhaas, J. M. (1999). Differential modulation of lateral septal vasopressin receptor blockade in spatial learning, social recognition, and anxiety-related behaviors in rats. *Behavioural Brain Research*, 99, 7–16.

Everts, H. G. and Koolhaas, J. M. (1997). Lateral septal vasopressin in rats: role in social and object recognition? *Brain Research*, 760, 1–7.

Feifel, D., Mexal, S., Melendez, G., Liu, P. Y., Goldenberg, J. R., and Shilling, P. D. (2009). The brattleboro rat displays a natural deficit in social discrimination that is restored by clozapine and a neurotensin analog. *Neuropsychopharmacology*, 34, 2011–2018.

Ferguson, J. N., Aldag, J. M., Insel, T. R., and Young, L. J. (2001). Oxytocin in the medial amygdala is essential for

social recognition in the mouse. *Journal of Neuroscience*, 21, 8278–8285.

Ferguson, J. N., Young, L. J., Hearn, E. F., Matzuk, M. M., Insel, T. R., and Winslow, J. T. (2000). Social amnesia in mice lacking the oxytocin gene. *Nature Genetics*, 25, 284–288.

Fisher, C. R., Graves, K. H., Parlow, A. F., and Simpson, E. R. (1998). Characterization of mice deficient in aromatase (ArKO) because of targeted disruption of the cyp19 gene. *Proceedings of the National Academy of Sciences of the United States of America*, 95, 6965–6970.

Fleming, A. S., Kuchera, C., Lee, A., and Winocur, G. (1994). Olfactory-based social learning varies as a function of parity in female rats. *Psychobiology*, 22, 37–43.

Gabor, C. S., Phan, A., Clipperton-Allen, A. E., Kavaliers, M., and Choleris, E. (2012). Interplay of oxytocin, vasopressin, and sex hormones in the regulation of social recognition. *Behavioral Neuroscience*, 126, 97–109.

Galef, B. G., Jr. and White, D. J. (1998). Mate-choice copying in Japanese quail, Coturnix coturnix japonica. *Animal Behaviour*, 55, 545–552.

Galef, B. G., Jr, Lim, T. C. W., and Gilbert, G. S. (2008). Evidence of mate choice copying in Norway rats, *Rattus norvegicus*. *Animal Behaviour*, 75, 1117–1123.

Galef, B. G., Jr, Mason, J. R., Preti, G., and Bean, N. J. (1988). Carbon disulfide: a semiochemical mediating socially induced diet choice in rats. *Physiology and Behavior*, 42, 119–124.

Galef, B. G., Jr. and Wigmore, S. W. (1983). Transfer of information concerning distant foods: a laboratory investigation of the 'information-centre' hypothesis. *Animal Behaviour*, 31, 748–758.

Getty, T. (2002). Signaling health versus parasites. *The American Naturalist*, 199, 363–371.

Gheusi, G., Blunthé, R., Goodall, G., and Dantzer, R. (1994). Social and individual recognition in rodents: methodological aspects and neurobiological bases. *Behavioural Processes*, 33, 59–88.

Gimpl, G. and Fahrenholz, F. (2001). The oxytocin receptor system: structure, function, and regulation. *Physiological Reviews*, 81, 629–683.

Heyes, C. M. (1994). Social learning in animals: categories and mechanisms. *Biological Reviews of the Cambridge Philosophical Society*, 69, 207–231.

Hlinak, Z. (1993). Social recognition in ovariectomized and estradiol-treated female rats. *Hormones and Behavior*, 27, 159–166.

Ho, M. L. and Lee, J. N. (1992). Ovarian and circulating levels of oxytocin and arginine vasopressin during the estrous cycle in the rat. *Acta Endocrinologica*, 126, 530–534.

Imwalle, D. B., Scordalakes, E. M., and Rissman, E. F. (2002). Estrogen receptor alpha influences socially motivated behaviors. *Hormones and Behavior*, 42, 484–491.

Johnston, R. E. (2003). Chemical communication in rodents: from pheromones to individual recognition. *Journal of Mammalogy*, 84, 1141–1162.

Jones, B. C., DeBruine, L. M., Little, A. C., Burriss, R. P., and Feinberg, D. R. (2007). Social transmission of face preferences among humans. *Proceedings. Biological sciences / The Royal Society*, 274, 899–903.

Juraska, J. M., Markham, J. A., and Loweth, J. A. (2001). The social recognition memory of female rats is more efficient during proestrus than during estrus. *Society for Neuroscience Abstracts*, 23, 534.16.

Kang, N., Baum, M. J., and Cherry, J. A. (2009). A direct main olfactory bulb projection to the 'vomeronasal' amygdala in female mice selectively responds to volatile pheromones from males. *European Journal of Neuroscience*, 29, 624–634.

Kavaliers, M. and Choleris, E. (2011). Sociality, Pathogen Avoidance and the Neuropetides, Oxytocin and Vasopressin. *Psychological Science*, in press.

Kavaliers, M., Choleris, E., Agmo, A., Braun, W. J., Colwell, D. D., Muglia, L. J., Ogawa, S., and Pfaff, D. W. (2006). Inadvertent social information and the avoidance of parasitized male mice: a role for oxytocin. *Proceedings of the National Academy of Sciences of the United States of America*, 103, 4293–4298.

Kavaliers, M., Choleris, E., Ågmo, A., Muglia, L. J., Ogawa, S., and Pfaff, D. W. (2005). Involvement of the oxytocin gene in the recognition and avoidance of parasitized males by female mice. *Animal Behaviour*, 70, 693–702.

Kavaliers, M., Choleris, E., Agmo, A., and Pfaff, D. W. (2004). Olfactory-mediated parasite recognition and avoidance: linking genes to behavior. *Hormones and Behavior*, 46, 272–283.

Kavaliers, M., Choleris, E., and Colwell, D. D. (2001). Learning from others to cope with biting flies: social learning of fear-induced conditioned analgesia and active avoidance. *Behavioral Neuroscience*, 115, 661–674.

Kavaliers, M., Choleris, E., Ogawa, S., and Pfaff, D. W. (2008). "Trust" in Mice: Oxytocin Mediated Mate Copying by Female Mice. *Society for Neuroscience Abstracts*, 38, 279.11.

Kavaliers, M., Colwell, D. D., and Choleris, E. (2005). Kinship, familiarity and social status modulate social

learning about 'micropredators' (biting flies) in deer mice. *Behavioural Ecology and Sociobiology*, 58, 60–71.

Kavaliers, M., Colwell, D. D., and Choleris, E. (2003). Learning to fear and cope with a natural stressor: individually and socially acquired corticosterone and avoidance responses to biting flies. *Hormones and Behavior*, 43, 99–107.

Kavaliers, M., Colwell, D. D., Choleris, E., Agmo, A., Muglia, L. J., Ogawa, S., and Pfaff, D. W. (2003). Impaired discrimination of and aversion to parasitized male odors by female oxytocin knockout mice. *Genes, Brain, and Behavior*, 2, 220–230.

Kavaliers, M., Devidze, N., Choleris, E., Fudge, M., Gustafsson, J. A., Korach, K. S., Pfaff, D. W., and Ogawa, S. (2008). Estrogen receptors alpha and beta mediate different aspects of the facilitatory effects of female cues on male risk taking. *Psychoneuroendocrinology*, 33, 634–642.

Kosfeld, M., Heinrichs, M., Zak, P. J., Fischbacher, U., and Fehr, E. (2005). Oxytocin increases trust in humans. *Nature*, 435, 673–676.

Lai, W. S. and Johnston, R. E. (2002). Individual recognition after fighting by golden hamsters: a new method. *Physiology and Behavior*, 76, 225–239.

Lai, W. S., Ramiro, L. L., Yu, H. A., and Johnston, R. E. (2005). Recognition of familiar individuals in golden hamsters: a new method and functional neuroanatomy. *Journal of Neuroscience*, 25, 11239–11247.

Laland, K. N. and Plotkin, H. C. (1990). Social learning and social transmission of foraging information in Norway rats. *Animal Learning and Behavior*, 18, 246–251.

Landgraf, R., Frank, E., Aldag, J. M., Neumann, I. D., Sharer, C. A., Ren, X., Terwilliger, E. F., Niwa, M., Wigger, A., and Young, L. J. (2003). Viral vector-mediated gene transfer of the vole V1a vasopressin receptor in the rat septum: improved social discrimination and active social behaviour. *European Journal of Neuroscience*, 18, 403–411.

Landgraf, R., Gerstberger, R., Montkowski, A., Probst, J. C., Wotjak, C. T., Holsboer, F., and Engelmann, M. (1995). V1 vasopressin receptor antisense oligodeoxynucleotide into septum reduces vasopressin binding, social discrimination abilities, and anxiety-related behavior in rats. *Journal of Neuroscience*, 15, 4250–4258.

Larrazolo-Lopez, A., Kendrick, K. M., Aburto-Arciniega, M., Arriaga-Avila, V., Morimoto, S., Frias, M., and Guevara-Guzman, R. (2008). Vaginocervical stimulation enhances social recognition memory in rats via oxytocin release in the olfactory bulb. *Neuroscience*, 152, 585–593.

Latham, N. and Mason, G. (2004). From house mouse to mouse house: The behavioural biology of free-living Mus musculus and its implications in the laboratory. *Applied Animal Behaviour Science*, 86, 261–289.

Le Moal, M., Dantzer, R., Michaud, B., and Koob, G. F. (1987). Centrally injected arginine vasopressin (AVP) facilitates social memory in rats. *Neuroscience Letters*, 77, 353–359.

Lee, H. J., Caldwell, H. K., Macbeth, A. H., Tolu, S. G., and Young, W. S., III. (2008). A conditional knockout mouse line of the oxytocin receptor. *Endocrinology*, 149, 3256–3263.

Lee, S. H. and Mouradian, M. M. (1999). Up-regulation of D1A dopamine receptor gene transcription by estrogen. *Molecular and Cellular Endocrinology*, 156, 151–157.

Little, A. C., Jones, B. C., Debruine, L. M., and Caldwell, C. A. (2011). Social learning and human mate preferences: a potential mechanism for generating and maintaining between-population diversity in attraction. *Philosophical Transactions of the Royal Society of London. Series B, Biological Sciences*, 366, 366–375.

Liu, B. and Xie, J. (2004). Increased dopamine release in vivo by estradiol benzoate from the central amygdaloid nucleus of Parkinson's disease model rats. *Journal of Neurochemistry*, 90, 654–658.

Lukas, M., Bredewold, R., Landgraf, R., Neumann, I. D., and Veenema, A. H. (2011). Early life stress impairs social recognition due to a blunted response of vasopressin release within the septum of adult male rats. *Psychoneuroendocrinology*, 36, 843–853.

Macbeth, A. H., Lee, H. J., Edds, J., and Young, W. S., III. (2009). Oxytocin and the oxytocin receptor underlie intrastrain, but not interstrain, social recognition. *Genes, Brain, and Behavior*, 8, 558–567.

McDermott, J. L., Liu, B., and Dluzen, D. E. (1994). Sex differences and effects of estrogen on dopamine and DOPAC release from the striatum of male and female CD-1 mice. *Experimental Neurology*, 125, 306–311.

McEwen, B. S. and Alves, S. E. (1999). Estrogen actions in the central nervous system. *Endocrine Reviews*, 20, 279–307.

Miller, M. M., Hyder, S. M., Assayag, R., Panarella, S. R., Tousignant, P., and Franklin, K. B. (1999). Estrogen modulates spontaneous alternation and the cholinergic phenotype in the basal forebrain. *Neuroscience*, 91, 1143–1153.

Mitra, S. W., Hoskin, E., Yudkovitz, J., Pear, L., Wilkinson, H. A., Hayashi, S., Pfaff, D. W., Ogawa, S., Rohrer, S. P.,

Schaeffer, J. M., McEwen, B. S., and Alves, S. E. (2003). Immunolocalization of estrogen receptor beta in the mouse brain: comparison with estrogen receptor alpha. *Endocrinology*, 144, 2055–2067.

Murakami, G., Hunter, R. G., Fontaine, C., Ribeiro, A., and Pfaff, D. (2011). Relationships among estrogen receptor, oxytocin and vasopressin gene expression and social interaction in male mice. *European Journal of Neuroscience*, 34, 469–477.

Nomura, M., McKenna, E., Korach, K. S., Pfaff, D. W., and Ogawa, S. (2002). Estrogen receptor-beta regulates transcript levels for oxytocin and arginine vasopressin in the hypothalamic paraventricular nucleus of male mice. *Brain Research. Molecular Brain Research*, 109, 84–94.

Numakawa, T., Yokomaku, D., Richards, M., Hori, H., Adachi, N., and Kunugi, H. (2010). Functional interactions between steroid hormones and neurotrophin BDNF. *World Journal of Biological Chemistry*, 1, 133–143.

Ostrowski, N. L., Lolait, S. J., and Young, W. S., III. (1994). Cellular localization of vasopressin V1a receptor messenger ribonucleic acid in adult male rat brain, pineal, and brain vasculature. *Endocrinology*, 135, 1511–1528.

Patisaul, H. B., Scordalakes, E. M., Young, L. J., and Rissman, E. F. (2003). Oxytocin, but not oxytocin receptor, is regulated by oestrogen receptor beta in the female mouse hypothalamus. *Journal of Neuroendocrinology*, 15, 787–793.

Phan, A., Gabor, C. S., Favaro, K. J., Kaschak, S. L., Armstrong, J. N., MacLusky, N. J., and Choleris, E. (2012). Low doses of 17β-estradiol rapidly improve learning and increase hippocampal dendritic spines. *Neuropsychopharmacology*, 37, 2299–2309.

Phan, A., Lancaster, K. E., Armstrong, J. N., MacLusky, N. J., and Choleris, E. (2011). Rapid effects of estrogen receptor alpha and beta selective agonists on learning and dendritic spines in female mice. *Endocrinology*, 152, 1492–1502.

Pierman, S., Sica, M., Allieri, F., Viglietti-Panzica, C., Panzica, G. C., and Bakker, J. (2008). Activational effects of estradiol and dihydrotestosterone on social recognition and the arginine-vasopressin immunoreactive system in male mice lacking a functional aromatase gene. *Hormones and Behavior*, 54, 98–106.

Plumari, L., Viglietti-Panzica, C., Allieri, F., Honda, S., Harada, N., Absil, P., Balthazart, J., and Panzica, G. C. (2002). Changes in the arginine-vasopressin immunoreactive systems in male mice lacking a functional aromatase gene. *Journal of Neuroendocrinology*, 14, 971–978.

Popik, P. and van Ree, J. M. (1993). Social transmission of flavoured tea preferences: facilitation by a vasopressin analog and oxytocin, *Behavioral and Neural Biology*, 63–68.

Popik, P. and van Ree, J. M. (1998). Neurohypophyseal peptides and social recognition in rats. *Progress in Brain Research*, 119, 415–436.

Popik, P. and van Ree, J. M. (1991). Oxytocin but not vasopressin facilitates social recognition following injection into the medial preoptic area of the rat brain. *European Neuropsychopharmacology*, 1, 555–560.

Popik, P. and Vetulani, J. (1991). Opposite action of oxytocin and its peptide antagonists on social memory in rats. *Neuropeptides*, 18, 23–27.

Popik, P., Vetulani, J., and van Ree, J. M. (1992). Low doses of oxytocin facilitate social recognition in rats. *Psychopharmacology*, 106, 71–74.

Popik, P., Vos, P. E., and Van Ree, J. M. (1992). Neurohypophyseal hormone receptors in the septum are implicated in social recognition in the rat. *Behavioural Pharmacology*, 3, 351–358.

Pruett-Jones, S. (1992). Independent Versus Nonindependent Mate Choice: Do Females Copy Each Other? *The American Naturalist*, 140, 1000–1009.

Quinones-Jenab, V., Jenab, S., Ogawa, S., Adan, R. A., Burbach, J. P., and Pfaff, D. W. (1997). Effects of estrogen on oxytocin receptor messenger ribonucleic acid expression in the uterus, pituitary, and forebrain of the female rat. *Neuroendocrinology*, 65, 9–17.

Rimmele, U., Hediger, K., Heinrichs, M., and Klaver, P. (2009). Oxytocin makes a face in memory familiar. *Journal of Neuroscience*, 29, 38–42.

Ruiz-Opazo, N., Lopez, L. V., and Herrera, V. L. (2002). The dual AngII/AVP receptor gene N119S/C163R variant exhibits sodium-induced dysfunction and cosegregates with salt-sensitive hypertension in the Dahl salt-sensitive hypertensive rat model. *Molecular Medicine (Cambridge, Mass.)*, 8, 24–32.

Ruiz-Opazo, N., Lopez, L. V., and Tonkiss, J. (2004). Modulation of learning and memory in Dahl rats by dietary salt restriction. *Hypertension*, 43, 797–802.

Russon, A. E. (1997). Exploiting the expertise of others. In *Machiavellian Intelligence II Extension and Evaluations* (eds. A. Whiten and R. W. Byrne), pp. 174–231. Cambridge University Press, New York.

Samuelsen, C. L. and Meredith, M. (2011). Oxytocin antagonist disrupts male mouse medial amygdala response to chemical-communication signals. *Neuroscience*, 180, 96–104.

Samuelsen, C. L. and Meredith, M. (2009). Categorization of biologically relevant chemical signals in the medial amygdala. *Brain Research*, 1263, 33–42.

Sanchez-Andrade, G., James, B. M., and Kendrick, K. M. (2005). Neural encoding of olfactory recognition memory. *Journal of Reproduction and Development*, 51, 547–558.

Sanchez-Andrade, G. and Kendrick, K. M. (2011). Roles of alpha- and beta-estrogen receptors in mouse social recognition memory: effects of gender and the estrous cycle. *Hormones and Behavior*, 59, 114–122.

Sanchez-Andrade, G. and Kendrick, K. M. (2009). The main olfactory system and social learning in mammals. *Behavioural Brain Research*, 200, 323–335.

Sarkar, D. K., Frautschy, S. A., and Mitsugi, N. (1992). Pituitary portal plasma levels of oxytocin during the estrous cycle, lactation, and hyperprolactinemia. *Annals of the New York Academy of Sciences*, 652, 397–410.

Savaskan, E., Ehrhardt, R., Schulz, A., Walter, M., and Schachinger, H. (2008). Post-learning intranasal oxytocin modulates human memory for facial identity. *Psychoneuroendocrinology*, 33, 368–374.

Sawyer, T. F., Hengehold, A. K., and Perez, W. A. (1984). Chemosensory and hormonal mediation of social memory in male rats. *Behavioral Neuroscience*, 98, 908–913.

Schule, C., Eser, D., Baghai, T. C., Nothdurfter, C., Kessler, J. S., and Rupprecht, R. (2011). Neuroactive steroids in affective disorders: target for novel antidepressant or anxiolytic drugs? *Neuroscience*, 191, 55–77.

Sekiguchi, R., Wolterink, G., and van Ree, J. M. (1991). Analysis of the influence of vasopressin neuropeptides on social recognition of rats. *European neuropsychopharmacology*, 1, 123–126.

Shettelworth, S. J. (2010). Social learning. In *Cognition, Evolution, and Behavior*, pp. 466–506. Oxford University Press, New York.

Spiteri, T. and Agmo, A. (2009). Ovarian hormones modulate social recognition in female rats. *Physiology and Behavior*, 98, 247–250.

Spiteri, T., Musatov, S., Ogawa, S., Ribeiro, A., Pfaff, D. W., and Agmo, A. (2010). The role of the estrogen receptor alpha in the medial amygdala and ventromedial nucleus of the hypothalamus in social recognition, anxiety and aggression. *Behavioural Brain Research*, 210, 211–220.

Strupp, B. J., Bunsey, M., Bertsche, B., Levitsky, D. A., and Kesler, M. (1990). Enhancement and impairment of memory retrieval by a vasopressin metabolite: an interaction with the accessibility of the memory. *Behavioral Neuroscience*, 104, 268–276.

Szot, P., Bale, T. L., and Dorsa, D. M. (1994). Distribution of messenger RNA for the vasopressin V1a receptor in the CNS of male and female rats. *Brain Research. Molecular Brain Research*, 24, 1–10.

Takayanagi, Y., Yoshida, M., Bielsky, I. F., Ross, H. E., Kawamata, M., Onaka, T., Yanagisawa, T., Kimura, T., Matzuk, M. M., Young, L. J., and Nishimori, K. (2005). Pervasive social deficits, but normal parturition, in oxytocin receptor-deficient mice. *Proceedings of the National Academy of Sciences of the United States of America*, 102, 16096–16101.

Tang, A. C., Nakazawa, M., Romeo, R. D., Reeb, B. C., Sisti, H., and McEwen, B. S. (2005). Effects of long-term estrogen replacement on social investigation and social memory in ovariectomized C57BL/6 mice. *Hormones and Behavior*, 47, 350–357.

Tang-Martinez, Z. (2003). Emerging themes and future challenges: forgotten rodent, neglected questions. *Journal of Mammalogy*, 1212–1227.

Terkel, J. (1996). Cultural transmission of feeding behavior in the black rat (*Rattus rattus*). In *The Root of Culture* (eds. C. M. Heyes and B. G. Galef, Jr.), pp. 17–47. Academic Press, New York.

Thomas, G. M. and Huganir, R. L. (2004). MAPK cascade signalling and synaptic plasticity. *Nature Reviews. Neuroscience*, 5, 173–183.

Thor, D. H. and Holloway, W. R. (1982). Social memory of the male laboratory rat. *Journal of Comparative and Physiological Psychology*, 96, 1000–1006.

Thor, D. H. (1980). Testosterone and persistence of social investigation in laboratory rats. *Journal of Comparative and Physiological Psychology*, 94, 970–976.

Timmer, M., Cordero, M. I., Sevelinges, Y., and Sandi, C. (2011). Evidence for a role of oxytocin receptors in the long-term establishment of dominance hierarchies. *Neuropsychopharmacology*, 36, 2349–2356.

Timmer, M. and Sandi, C. (2010). A role for glucocorticoids in the long-term establishment of a social hierarchy. *Psychoneuroendocrinology*, 35, 1543–1552.

Tobin, V. A., Hashimoto, H., Wacker, D. W., Takayanagi, Y., Langnaese, K., Caquineau, C., Noack, J., Landgraf, R., Onaka, T., Leng, G., Meddle, S. L., Engelmann, M., and Ludwig, M. (2010). An intrinsic vasopressin system in the olfactory bulb is involved in social recognition. *Nature*, 464, 413–417.

Valsecchi, P., Bosellini, I., Sabatini, F., Mainardi, M., and Fiorito, G. (2002). Behavioral analysis of social effects on the problem-solving ability in the house mouse. *Ethology*, 108, 1115–1134.

Valsecchi, P., Choleris, E., Moles, A., Guo, C., and Mainardi, M. (1996). Kinship and familiarity as factors affecting social transfer of food preferences in adult Mongolian gerbils (Meriones unguiculatus). *Journal of Comparative Psychology (Washington, D.C.: 1983)*, 110, 243–251.

Valsecchi, P. and Galef, B. G., Jr. (1989). Social influences on the food preferences of house mice (*Mus musculus*). *International Journal of Comparative Psychology*, 2, 245–256.

van der Kooij, M. A. and Sandi, C. (2012). Social memories in rodents: Methods, mechanisms and modulation by stress. *Neuroscience Biobehavioral Reviews*, 36, 1763–1772.

van Wimersma Greidanus, T. B. and Maigret, C. (1996). The role of limbic vasopressin and oxytocin in social recognition. *Brain Research*, 713, 153–159.

von Frisch, K. (1967). *The Dance Language and Orientation of Bees.* Harvard University Press, Cambridge, Massachusetts.

Walmer, D. K., Wrona, M. A., Hughes, C. L., and Nelson, K. G. (1992). Lactoferrin expression in the mouse reproductive tract during the natural estrous cycle: correlation with circulating estradiol and progesterone. *Endocrinology*, 131, 1458–1466.

Waynforth, D. (2007). Mate choice copying in humans. *Human Naturalist*, 18, 264–271.

Wersinger, S. R., Caldwell, H. K., Martinez, L., Gold, P., Hu, S. B., and Young, W. S., III. (2007). Vasopressin 1a receptor knockout mice have a subtle olfactory deficit but normal aggression. *Genes, Brain, and Behavior*, 6, 540–551.

Wersinger, S. R., Ginns, E. I., O'Carroll, A. M., Lolait, S. J., and Young, W. S., III. (2002). Vasopressin V1b receptor knockout reduces aggressive behavior in male mice. *Molecular Psychiatry*, 7, 975–984.

Wersinger, S. R., Kelliher, K. R., Zufall, F., Lolait, S. J., O'Carroll, A. M., and Young, W. S., III. (2004). Social motivation is reduced in vasopressin 1b receptor null mice despite normal performance in an olfactory discrimination task. *Hormones and Behavior*, 46, 638–645.

White, D. J. and Galef B. G., Jr. (1999). Social effects on mate choices of male Japanese quail, Coturnix japonica. *Animal Behaviour*, 57, 1005–1012.

Whiten, A., Custance, D. M., Gomez, J. C., Teixidor, P., and Bard, K. A. (1996). Imitative learning of artificial fruit processing in children (Homo sapiens) and chimpanzees (Pan troglodytes). *Journal of Comparative Psychology*, 110, 3–14.

Young, L. J., Wang, Z., Donaldson, R., and Rissman, E. F. (1998). Estrogen receptor alpha is essential for induction of oxytocin receptor by estrogen. *Neuroreport*, 9, 933–936.

Zhao, L. and Brinton, R. D. (2007). Estrogen receptor alpha and beta differentially regulate intracellular Ca(2+) dynamics leading to ERK phosphorylation and estrogen neuroprotection in hippocampal neurons. *Brain Research*, 1172, 48–59.

14

Oxytocin, vasopressin, sociality, and pathogen avoidance

Martin Kavaliers and Elena Choleris

14.1 Introduction – pathogen detection and avoidance

Pathogen avoidance and recognition are vital components of animal evolutionary ecology and social cognition. Parasites, broadly defined to include microparasites (e.g., viruses and bacteria) and macroparasites (e.g., protozoa, helminths, nematodes, arthropods and even vertebrates) live in durable relationships with their hosts. The effects of parasitism are instrumental in influencing fundamental aspects of host biology. Parasites draw energy, shelter, transport and reproductive opportunities from their hosts. They exploit the proximate mechanisms that modulate host social behaviors to increase the likelihood of their transmission and in turn influence the recognition mechanisms and avoidance responses of their hosts (Kavaliers et al., 2004a; 2005). Animals of all taxa have evolved a range of elegant and interacting physiological, morphological, neurobiological, immunological, and behavioral responses to reduce their risk of contact with and infection by pathogens (Hart, 1990; Kavaliers and Choleris, 2011; Moore, 2002). In particular, the need to defend against infection has given rise to the complexity of the adaptive immune system. Different arms of the vertebrate immune system are activated in response to different types of infection (e.g., viral, bacterial and helminthic), with cytokine signaling molecules mediating the regulation of the different immune arms (Milinski, 2006).

Likewise, a rich repertoire of adaptive behavioral responses to pathogen threat are evident. Behaviors that are most commonly involved in infection avoidance include; selective defecation and fecal avoidance to avoid contamination, escape movements and changes in habitat location, alterations in the locations and timing of foraging, grooming, preening, and specific movements (e.g., tail swishing) to avoid biting arthropods, nest fumigation with natural insecticides and self-medication with natural herbicides, wound licking to prevent infection, and importantly changes in socially related behaviors and interactions (see examples in Hart, 1990; Kavaliers et al., 2004a; Moore, 2002). At least some of these strategies and behaviors are present in all of the vertebrate species that have been studied.

Humans as well are continually exposed to the threat of infection from a variety of pathogens. These can fall into three distinct categories based on their modes of transmission – zoonotic, multihost, and human specific. Zoonotic parasites develop and reproduce entirely in non-human hosts and can infect humans, but are not directly transmitted from human to human. The others, termed non-zoonotic, can be transmitted from human-to-human. Multihost parasites may be transmitted from human-to-human as well as through interspecies interactions, while human specific

Oxytocin, Vasopressin, and Related Peptides in the Regulation of Behavior, ed. E. Choleris, D. W. Pfaff, and M. Kavaliers.

parasites are transmitted only from human to human. Although the magnitude of infection threat is likely to have increased with the advent of large group settlements and animal domestication around 11,000 years ago, many pathogens are of considerable antiquity and are likely to have imposed selection pressures on humans for tens of thousands of years (Pearce-Duvet, 2006). A "parasite-stress" model of human sociality proposes that parasitic (infectious) diseases (primarily non-zoonotic) are responsible for shaping features of human psychology and behavior (Thornhill et al., 2010) As a result, humans, like many other species of vertebrates, have a rich repertoire of physiological, hormonal, immunological and behavioral responses that function in the recognition and avoidance of pathogen threat. For example, just as livestock show selective defecation and fecal avoidance to avoid contamination, humans display disgust and behavioral rejection responses to food and other items potentially contaminated with pathogens (e.g., Oaten et al., 2009). Similarly aversive behavioral responses, including that of disgust, as well as changes in social and sexual responses, are shown by men and women to individuals that can pose a threat of contamination or infection (e.g., Neuberg et at., 2011). As Giovanni Boccacio vividly records in the Decameron, in medieval times, the lethality of the bubonic plague caused people to "shun and flee from the sick and all that pertained to them and thus doing, each thought to secure immunity for himself" (Boccacio, 1352).

Alexander (1974) elegantly hypothesized that parasitism by directly (i.e., one individual to another) and indirectly (i.e., involving an intermediary such as a biting insect) transmitted parasites is a cost of sociality. There is now a growing interest in the important relationships between social behavior and the risk of exposure to parasites and infection in both humans and non-humans (Altizer et al., 2003; Kavaliers and Choleris, 2011; Nunn and Altizer 2006). Social interactions provide key opportunities for parasite and pathogen transmission, affecting the likelihood of the spread of infection and transmission of disease. Because social behaviors facilitate interactions between conspecifics, they increase the probability of parasite exposure and transmission from infected to infected individuals. In pioneering studies, Freeland (1976) suggested that aspects of primate social behaviors and interactions have evolved to reduce the spread of new and existing parasites and disease. As such, social behavior has been shaped by pathogen pressure, with animals having evolved a variety of mechanism and adaptive behavioral responses to recognize and minimize their exposure to individuals that can pose a threat of contagion. Examples of these include; recognition and avoidance of strangers; social distancing and territorial behavior to exclude conspecifics that may carry novel pathogens; modifications in sexual behavior to avoid mating with infected individuals; and specific behaviors to protect offspring from infection (examples in Kavaliers et al., 2004; Neuberg et al., 2011). In humans this has led to the postulation of a so-called behavioral immune system that presents a set of cognitive and affective responses to actual and potential infection threat (Schaller and Park, 2011).

Critical for the avoidance of pathogens is the ability to recognize, and minimize social interactions with, infected or potentially infected individuals. The evolution of sociality has simultaneously brought forth the need for effective and adaptive processing of social stimuli and coping with the increased threat of infection and parasite transmission. Individuals may modify their social interactions in response to the threat of infection, and this behavior may be adaptive in reducing the risk of infection.

The nonapeptides, oxytocin (OT) and arginine-vasopressin (AVP) have important roles in the mediation of various aspects of social behavior, including responses to social threats, that encompass pathogens (Choleris et al., 2009, 2012; Kavaliers and Choleris, 2011). In particular, from an evolutionary point of view OT in one form or another (isotocin, mesotocin or oxytocin) is present in an extensive range of species form fish, reptiles to primates, where it plays a key role in the control of social and reproductive behaviors

(Chapters 5–13; Goodson and Thompson, 2010). As briefly reviewed here there is now accumulating evidence that OT and AVP are also part of the modulatory systems involved in recognition and avoidance of the actual and potential pathogen threat posed by conspecifics.

14.2 Social information, odors, and infection avoidance

14.2.1 Odor detection and processing

The ability to use social information for the recognition and avoidance of infected individuals is crucial. The social behavior of many mammals is finely tuned to information provided by chemical signals. Recent fossil evidence has suggested that olfaction was the driving force for early mammalian development (Rowe et al., 2011) and likely social behavior (Ladeveze et al., 2011). Olfactory information not only modulates the social behavior of conspecifics, but also enables animals to adjust their own responses in behavioral interactions. Odors provides genetically encoded information on species, sex, individual, and class identity and kinship of the owner, as well as information on the individual's current reproductive, social and health, infection, and immune status (Brennan and Kendrick, 2006; Hurst and Benyon, 2004). Indeed, although most prominent in rodents, odor cues are also used by humans in the context of sexual attraction, mate choice, assessment of condition and the expression of disgust responses associated with pathogen presence (e.g., Havelick and Roberts, 2009; Lundstrom et al., 2008; Mille and Maner 2010; Moshkin et al., 2012).

Odor-based social recognition in rodents involves the use of volatile and involatile sources of chemical information. Mice and rats can distinguish volatile urinary odors associated with differences in the alleles borne at the major histocompatibility complex (MHC). The MHC is a large cluster of polymorphic genes coding for the molecules involved in the adaptive (as opposed to the innate) immune response (Milinski, 2006). Infection activates the release of specific MHC antigens and affects odor phenotypes. MHC products are expressed as glycoproteins and function to bind and present antigens that trigger the appropriate immune responses from T lymphocytes. There are two main classes of MHC that are responsive to different types of infection – MHC class I that react to intracellular parasites (e.g., viruses) and class II that react to extracellular parasites (e.g., nematodes, bacteria) The MHC – associated odors are adaptively used for mediating individual and kin recognition and guiding social and mating preferences (Brennan and Kendrick, 2006). MHC-dependent mate choice occurs across a variety of socio-sexual systems and has been shown to favor mates with either dissimilar, diverse, or specific MHC genotypes. The former is considered to be olfactory based while the latter two can involve other features and sensory cues.

Mice also use involatile odor cues provided by the highly polymorphic gene complex of the major urinary proteins (MUPs), for social, sexual, and individual recognition. Although primarily present in the urine, MUPs are also expressed in the salivary glands, lachrymal glands, and nasal tissue (Hurst and Benyon, 2004). Recently, darcin has been identified as a specific MUP that is consistently present in the urine of male, but not female mice, and functions as a sexual attractant to which response to other odor components can become associated (Roberts et al., 2010). The MUPs bind and release components that can be used for social recognition, with either the proteins themselves and/or protein–ligand complexes providing cues about identity and condition (Hurst and Benyon, 2004).

The two primary components of the rodent chemosensory system are the main and accessory olfactory systems and pathways (Baum and Kelliher, 2009). The main olfactory epithelium detects volatile odor constituents (largely, though not exclusively, volatile MHC associated compounds), likely at some distance from their source, while the accessory olfactory system detects volatile and involatile molecules that are pumped into the vomeronasal organ during close proximity with the odor source

(Hurst and Benyon, 2004). The accessory olfactory pathway includes sensory neurons in the vomeronasal organ (VNO) and their projections to the accessory olfactory bulb (AOB), along with secondary connections of AOB mitral cells to targets in the medial amygdala (MeA) and posterior medial amygdala. The MeA in turn projects to brain regions controlling many aspects of social behavior, including the medial prepoptic area, the anterior hypothalamus, the bed nucleus of the stria terminalis and the ventromedial nucleus of the hypothalamus. In contrast, olfactory sensory neurons of the main olfactory system project via the main olfactory bulb to the cortical regions of the amygdala. In addition, there is reported to be a direct projection from the main olfactory bulb to the MeA. Chemical signal information processing via the MeA is considered to be an important contribution to the hypothalamic circuits involved in rodent defensive and reproductive behaviors (Baum and Kelliher, 2009).

14.2.2 Odors, parasites, and sexual and social responses

Social interactions associated with mate selection and mating present an especially high risk for infection, with females of various species preferentially selecting parasite (disease)-free or -resistant males on the basis of condition-dependent cues. By choosing healthy mates, females gain direct benefits by decreasing the risk of acquiring an infection during mating. Choosing healthy males could also provide females with a fitness advantage because these males are likely to supply "good genes" that may enhance offspring disease resistance. In either case this requires that a female utilizes social information for the recognition and avoidance of either infected or potentially infected males

Hamilton and Zuk (1982) were the first to directly consider the evolutionary interactions between pathogens and their hosts as they relate to mate choice and sexual selection. The Hamilton–Zuk hypothesis suggests that secondary sexual characters ("ornaments"), such as colorful plumage and red skin, signal male health status, with females choosing the healthiest males according to these signals. As a result, the female is likely to obtain "good genes" for her offspring that ensure resistance against prevailing parasites. They also specifically suggested that animals should examine urine and faecal cues in an attempt to detect mates that are free from disease and parasites. Results of a variety of studies have now demonstrated that female rodents are able to use odor cues to recognize and avoid individuals infected with a variety of macroparasites, microparasites, viruses and bacteria components (see Arakawa et al., 2011; Kavaliers, et al., 2004a, 2005; Kavaliers and Choleris, 2011).

In simultaneous choice tests (infected vs. uninfected odors) sexually receptive female mice displayed an overall preference for, and initial choice of, the odors of uninfected males as well as aversive responses to, the odors of asymptomatic males subclinically infected (i.e., non-sick with no indications of poor condition or pathology) with a range of parasites (e.g., directly transmitted mouse specific nematode, protozoan, louse and influenza) (e.g., Ehman and Scott, 2001; Kavaliers and Colwell 1995a;b; Kavaliers et al., 1997 Kavaliers et al., 2003a;b;c; 2005; Penn and Potts 1998; Penn et al., 1998) Also, sexually receptive females avoided actual infected males when given a choice between infected and uninfected males. Likewise, in sequential choice test females recognized and avoided the odors of infected males. In addition, sexually receptive females avoided the odors of unfamiliar, but not familiar, uninfected males that were previously associated with the odors of a parasitized male and, thus, only potentially infective (Kavaliers et al., 1998a). Sexually receptive infected females also recognized and avoided males with infections that were different from their own (Kavaliers et al., 2005).

Odor-choice tests do not examine mate choice directly but, instead, examine an apparent choice and expression of social and sexual interest and the appetitive components of mate choice. The preference tests used here do not explicitly distinguish between social (being near a partner but not necessarily mating) versus sexual (actual mating) and it has been suggested that they should be more

appropriately termed partner preferences (Henley et al., 2011). There is, however, also evidence indicating that the results of odor tests are consistent with actual mate choices and mating and reproductive outcomes (reviewed in Kavaliers et al., 2004a), although the actual consummatory components of mate responses might be affected differently by the presence of an infected individual.

An additional consequence of exposure to the odors of the infected males was a decreased nociceptive or pain sensitivity and the induction of analgesia or antinociception. Analgesia is an important component of the suite of defensive responses to stimuli associated with real or potential threat, and is advantageous in situations in which a direct response to noxious stimuli might otherwise disrupt other adaptive behavioral responses (Kavaliers et al., 2005). In addition, the neurochemical correlates of analgesia facilitate the expression of associated behavioral responses.

Female mice also displayed analgesic responses to the odors of infected males, with the magnitude of responses depending on prior experience with the male. Sexually receptive females exposed to familiar infected males displayed attenuated analgesic (and likely reduced associated stress responses), whereas the odors of novel males elicited heightened analgesia (Kavaliers and Colwell, 1995a; Kavaliers et al., 1998a; 2003b;c). In all cases, however, the females avoided and selected against infected males in choice tests. This differential analgesic response necessitates the social recognition of, and memory for, the odors of specific infected males. The females may have learnt through a single exposure, that a particular infected male was to be directly avoided, thus minimizing the display of heightened analgesia and stress responses.

Female mice and rats also detected and avoided the odors of "sick" individuals as well as "sick" individuals themselves. Sick animals display a suite of physiological and behavioral responses (e.g., anorexia, adipsia, anhedonia, lethargy and reduced activity and likely diminished arousal, reduced social behavior) that help cope with the infection (Hart, 1988). Sickness behavior incorporates a motivational reorganization such that animals prioritize their behaviors to deal with infection, but are still able to deal with other threats (e.g., predators) and demands (Aubert, 1999). Classical sickness behaviors may not be adaptive in response to chronic parasitic infections. Many individuals face some parasite burden and as such a generalized behavioral depression associated with sickness response would compete with other critical behaviors (Kavaliers et al., 2005).

Acute infection by viruses or bacteria induces an activation of the immune system and the release of proinflammatory cytokines, such as interleukin-1β and tumor necrosis factor, and the display of sickness behaviors (Dantzer et al., 2008). Systemic endotoxin, lipopolysaccharide (LPS), a component of gram-negative bacterial cell walls, is commonly used to model bacterial infection because it mimics the effects of the bacteria without the risk of infection. Female mice can discriminate and avoid LPS treated individual, with these effects becoming more pronounced after prior priming with an environmental cue (e.g., stress odors of other individuals) suggestive of the presence of social and possibly pathogen threat (Renault et al., 2008). Likewise, odors from LPS-treated male rats elicited robust avoidance responses in female rats, as evidenced by decreased sniffing and active avoidance behaviors (Arkawa et al., 2012).

Odors also influence the competitive, social and sexual interactions and the behavioral responses of males. Accordingly, male rodents use odor cues to recognize actual and potential pathogen threat and to modulate their social and sexual preferences and responses. Male mice discriminated between the odors of parasitized and non-parasitized females displaying reduced interest in, and attraction to, the odors of subclinically infected females along with the expression of analgesic responses (Kavaliers et al., 1998b; 2004a). Male rats also showed a reduced interest in the odors of LPS-treated females (Arkawa et al., 2010) and exhibited a clear preference for a healthy female over a sick one and they performed significantly fewer mounts and intromissions and spent less time with an LPS-treated partner than a saline-injected female

(Arkawa et al., 2012; Avistur and Yirmyia, 1999). Male mice displayed an increased wariness to, and avoidance of, the odors of infected and potentially infected males (Kavaliers et al., 2004). In male rats, acute sickness induced by LPS suppressed their social and odor investigation by juvenile conspecific males (Arakawa et al., 2010). Central infusion of the proinflammatory cytokine, IL-β, produced an avoidance response to males similar to that elicited by LPS, while central administration of the anti-inflammatory, IL-1-, suppressed these avoidance responses (Arkawa et al., 2010). Hence, both males and females are capable of recognizing and avoiding both sick and non-sick infected individuals.

14.3 Oxytocin, vasopressin, and the recognition and avoidance of pathogen threat

14.3.1 Oxytocin, vasopressin, and social behavior

The nonapeptides, oxytocin and arginine vasopressin, have key influences on social functions in vertebrates, including humans (Choleris et al., 2009; Donaldson and Young, 2008; Insel, 2010; Meyer-Lindenberg et al., 2011) and have now also been shown to be directly associated with the recognition and avoidance of the social threat presented by an infected conspecific (reviewed in Kavaliers and Choleris, 2011). OT affects social affiliation, pair bonding, maternal behaviors, and social recognition consistent with an apparent overall "prosocial" effect (Donaldson and Young, 2008). Results of studies, with OT gene "knockout" mice (OTKO) and mice lacking the OT receptor (OTR) have shown that OT is necessary for social recognition and essential for familiarity recognition. Results of investigations with OT antisense and OT infusion have pointed at the MeA as the site of action. AVP also has a role in social recognition, though apparently more so in males than in females, and is known for promoting aggression and other male-typical behaviors (reviewed in Choleris et al., 2009; Chapter 13).

Prosocial effects of OT, and responses associated with social recognition, are also evident in humans (reviewed in Bartz et al., 2011; Meyer-Lindenberg et al., 2011). Intranasal OT administration increases perceptions of attractiveness and the trustworthiness of familiar faces in men and women (Theodoridu et al., 2009; 2011) and induces subsequent feelings of familiarity for affiliative (smiling) faces (Guastella et al., 2008). Intranasal OT also facilitates the recognition of words related to sex and relationships (Unkellbach et al., 2008). Intranasal OT also enhances cooperation and trustworthiness in financial games with a familiar social partner (Kosfeld et al., 2005). As in mice, these OT effects have been linked to actions at the MeA (Gamer et al., 2010; Chapters 19 and 20).

Although fewer studies have examined the roles of AVP in human social regulation, a variety of genetic investigations have suggested a link between AVP and social behavior (Insel, 2010; Meyer-Lindenberg et al., 2011). In addition, intranasal AVP promotes the encoding of both positive and negative facial stimuli in men (Gustella et al., 2010) and as in male rodents, AVP heightens aggressive responses to threatening stimuli in men (Chapter 19).

14.3.2 Oxytocin, vasopressin, and pathogen threat

Although OT is generally considered as "prosocial" promoting affiliation and bonding, OT does promote territoriality and maternal aggression towards unfamiliar intruders (Bosch et al., 2005; Donaldson and Young, 2008). Evidence from humans similarly indicates that OT is not just involved in the mediation of prosocial responses. Men primed for envy displayed after intranasal OT responses that were interpreted as being consistent with increased levels of envy and "schadenfreude" (Shamay-Tsoory et al., 2009). In addition, OT promoted giving in a financial game, but only if the subjects had prior contact and familiarity with their partners. OT had no effect if the incentive to cooperate was low and actually decreased giving if the partner was unfamiliar (Declerck et al., 2010). In studies using monetary transfer and an implicit association test an

"in-group" bias was seen in subjects receiving intranasal OT (De Dreu, 2012; De Dreu et al., 2010, 2011), with the authors also speculating on the possibility of "out-group" exclusion.

The results of these studies suggest that OT enhances the general saliency of social stimuli, with the nature of the behavioral responses being dependent on the social context and nature of the social stimulus (e.g., Baumgartner et al., 2008; Declerck et al., 2010; De Dreu, 2012; Shamay-Tsoory et al., 2009). This is consistent with OT both enabling social recognition and motivating in-group favoritism. An alternative social-approach/withdrawal model, again based on the results of studies with humans, has also been proposed (Kemp and Guastella, 2011). According to this model OT is not associated with avoidance or aversive responses, but rather OT promotes approach related behavior to positive salient stimuli and reduces withdrawal-related behaviors.

The social saliency hypothesis of OTs actions is consistent with the proposal that OT is involved in the mediation of the recognition of the social threat posed by actual and pathogen risk. This is directly supported by the results of investigations showing that OT is intimately associated with the detection and avoidance of infected mice and rats (Arkawa et al., 2010; Kavaliers et al., 2004a; Kavaliers and Choleris, 2011). In a mate-choice test, sexually receptive female OTKO mice were specifically impaired in their recognition and avoidance of the odors of males subclinically infected with either a nematode or louse, but not to other social odors and non-social (i.e., cat odor) threats (Kavaliers et al., 2004a). These females also displayed attenuated analgesic responses to the infected males and failed to distinguish between the odors of novel and familiar infected males. Interestingly, olfactory cues from OT injected male can also induce antinociception in their cagemates (Agren et al., 1997).

OT involvement in the recognition by female mice of the pathogen threat presented by parasitized male mice is further supported by the findings that peripheral treatment with the oxytocin antagonist, L368,899, resulted in impaired recognition and avoidance of infected males (Kavaliers et al., 2004a). Sexually receptive female mice that were treated with the OT antagonist also displayed reduced analgesic responses to the odors of infected males and displayed minimal discrimination between the odors of novel and familiar males. These findings indicate that OT is part of the central neuroendocrine mechanism mediating the recognition of pathogen threat in a partner preference and mate-choice scenario. The relative roles of involatile and volatile odors and MHC- and MUP-related cues that convey male health (i.e., immune status) and identity in determining the recognition and avoidance by females remains to be determined,

Evidence from men and women similarly suggests that OT affects responses to salient facial sexual cues, consistent with a role for OT in determining sexual preferences and mate choice (e.g., Theodoridou et al., 2011; Unkellbach et al., 2008). As in rodents, human female mate choice is also sensitive to, and modified by, pathogen threat (e.g., Little et al., 2011; Moshkin et al., 2012), thus raising the possibility of a role for OT also here.

The mate-choice decisions of females are also influenced by the mating decisions of other females, with such non-independent mate choice being termed "mate-choice copying." Sexually naïve females may use the interests of other females to obtain information about potential mate quality. It was shown that the presence of the odors of another female in behavioral estrous with that of a subclinically infected male attenuated the avoidance, corticosterone, and analgesic responses, and resulted in the subsequent choice of the odors of that specific infected individual and not other infected males (Kavaliers et al., 2006). While this may appear counterintuitive, "uninfected" does not always necessarily imply a more parasite-resistant and better quality male. A higher-quality male may be better able to resist or tolerate infection than a lower-quality uninfected male that has never encountered the infection. Therefore, copying the choices or preferences of a sexually experienced female may be adaptive for a naïve female.

OTKO females, and female mice treated with an OT antagonist were impaired in the use of this indirect social information and did not copy the mate choices of other females (reviewed in Chapter 13; Kavaliers et al., 2006). This is reminiscent of the role that OT has in enhancing trust and trust-related decision making and enhancing the positive social salience of male cues in humans (Baumgartner et al., 2008, Hurlemann et al., 2010). These findings indicate that OT is not only associated with the use and recognition of direct social information, but also with the use of indirect social information, both of which are likely to play a significant role in assessing pathogen and other social threats as they relate to mate choice and likely social preferences.

As indicated, males also face a significant threat of infection during social and sexual interactions, with the proportion of individuals infected and the severity of infection often higher in males than in females. OTKO males were also specifically impaired in their recognition of, and display of aversive responses to, the odors of infected and potentially infected males in a manner equivalent to that seen in OTKO females. OTKO males showed a reduced aversive response to the odors of both subclinically infected males, and uninfected males that been previously associated with infected males and, thus only potentially infected (in preparation) The OTKO males also displayed reduced analgesic responses to the odors of infected males and failed to distinguish between the odors of novel and familiar males (Kavaliers et al., 2004). This shows that male mice, like females, are able to both recognize infected individuals and modulate their responses on the basis of prior familiarity and that this involves OT.

There is also evidence from rodents linking OT to the recognition and avoidance of "sick" individuals. Adult male and female rats with reduced OT level arising from postweaning social isolation (Tanaka et al., 2010), had a reduced ability to discriminate the odors of males treated with LPS. Also, female prairie voles, with relatively high OT levels, displayed a significantly greater avoidance of LPS treated males than female meadow voles with reduced OT levels (Klein and Nelson, 1999).

Preliminary evidence also suggests that OTKO female mice have a reduced ability to discriminate the odors of LPS-treated males from that of untreated males (Kavaliers et al., 2004a). Elegant direct evidence for OT involvement in the mediation of responses to sickness-related odors in male rats has been presented (Arkawa et al., 2010). They showed that the expression of OT receptor mRNA in the MeA was increased when juvenile male rats were exposed to the volatile bedding-odor cues of other LPS-treated adult males. They further showed that bilateral infusion of an OT receptor antagonist into the MeA reduced approach behavior to healthy odor to the same levels as seen with "sickness" odors, consistent with a reduction of social recognition.

The studies of Arkawa et al. (2010) also showed the involvement of AVP in the mediation of avoidance responses to sickness odors. They showed that exposure to the odors of LPS-treated males increased AVP mRNA expression in the MeA and that bilateral infusion of a V1A receptor antagonist into the MeA-inhibited avoidance responses to the odors of LPS-treated individuals. Whether or not AVP has a role in mediating responses to other pathogens and in females remains to be determined.

OT and AVP interact with three receptors – OT, V1A, and V1B all belonging to the G protein-coupled receptor superfamily. There is evidence of interaction between OT and AVP as well data suggesting that the central nucleus of the amygdala OT and AVP may be differentially modulating excitatory inputs associated with various threats (Huber et al., 2005). Thus, it is possible that manipulations of either AVP or OT systems at the level of amygdala (e.g., Knobloch et al., 2012) will elicit shifts in responses to various threat, including that of social threats associated with the presence of pathogens.

14.3.3 Steroidal–neuropeptide interactions in the recognition of pathogen threat

The MeA where the main and accessory olfactory pathways converge, is critical for social recognition in mice and a target for the altered responses to

the odors of infected individuals. OT and AVP in the MeA have been associated with social recognition in mice (Choleris et al., 2009), along with evidence for a role for AVP and OT in the main olfactory bulb (Wacker et al., 2011). The regulation of OT at the level of the MeA involves estrogen receptors (ERα and ERβ) (Choleris et al., 2009; 2012). Under the influence of ERβ, OT synthesis is increased by estradiol in the paraventricular nucleus of the hypothalamus, while estradiol bound to ERα increases transcription of the OTR in the amygdala. Disruption at the level of either OT, OTR, ERα, or ERβ genes and their products could lead to impaired processing and/or integration of odor information, modifying socio-sexual motivation (reviewed in Chapter 13).

Estrogenic regulation of social behavior and the recognition of infected conspecifics has been investigated in male and female mice in which the genes encoding for ERα or ERβ had been disrupted. Female and male ERαKO, ERβKO, like OTKO males and females, were impaired in their social recognition and display of avoidance and aversive responses to the odors of parasitized males (Choleris et al., 2012; Kavaliers et al., 2004a). Female WTs displayed aversive responses to either the volatile + involatile, or just the volatile odor components of parasitized males, with the ERKOs displaying attenuated responses (in preparation). These effects of the ERs may be either through their roles in the regulation of OT and/or additional roles in the modulation of social recognition, with ERα likely being essential, and ERβ having a modulatory role (Choleris et al., 2009; 2012) This suggests that OT, OTR, ERα and ERβ genes are part of the neuroendocrine mechanisms by which males and females can distinguish and avoid parasitized individuals on the basis of odor cues.

Investigations with AVP have to date been restricted to the avoidance responses to male rats to sickness odors (Arkawa et al., 2010), Relatively less is known about the regulation of the actions of AVP and its potential roles in the mediation of responses to other infections. There is some evidence of a role for estradiol as well as testosterone (and its estrogenic metabolites) which have been linked to the aggression promoting and social recognition actions of AVP (Chiao and Blizinsky, 2010; Donaldson and Young, 2008; Jorgensen et al., 2003). Also, the roles of various neurotransmitters (e.g., serotonin, dopamine, glutamate, GABA) that influence OT/AVP activity (Donaldson and Young, 2008) and are also involved in the regulation social and sexual responses, and in the modulation of pathogen avoidance (Kavaliers et al., 2004a) remains to be determined.

14.3.4 OT, AVP, infection, and immune responses

OT has been associated with immune function. CD8, a transmembrane glycoprotein, that mediates oxytocin secretion and social behavior (Jin et al., 2007 and see Chapter 3), is also an important immune cell marker. CD8 is expressed in the majority of natural killer cells and macrophages and has a major role in triggering proliferation and immune responses in lymphocytes (Malavasi et al., 2008) suggesting a direct link between the regulation/expression of OT and immune function. Interesting, low CD38 expression in lymphoblastoid cells and impaired social processing were both associated with autism in a family-based study (Lerer et al., 2010) OT and AVP have also been associated with the modulation of steroid-dependent immune function and the facilitation of wound healing (Karelina and Devries, 2011). Also, oxytocin has been shown to alleviate the neuroendocrine and cytokine responses to LPS in healthy men (Clodi et al., 2008).

There is suggestive evidence linking social behavior, social recognition, immune function and OT. Social regulation of immune function in both rodents and non-humans has been shown. For example, exposure of male mice to female odors has been shown to enhance the resistance of male mice to influenza (Litvinova et al., 2010). This could entail OT mediation of both social recognition and immune modulation. There is also evidence for male and female mice that OT affects odor cues,

with both males and females being able to distinguish between the odors of OTWT and OTKO individuals on the basis of odor (Ågmo et al., 2007; Kavaliers et al., 2004b, 2012). These shifts in odor cues may well encompass changes in immune function.

In humans there is also evidence of socially regulated gene expression in leuckocytes (Cole et al., 2011). In lonely, socially isolated individuals there is a shift in immune activity away from antiviral activity towards adaptive immune responses. Viral infection is facilitated by social contact with conspecifics, including that associated with mate choice, thus necessitating enhanced antiviral activity. This again could entail OT mediated social regulation and immune functioning. In this regard, there is an intriguing recent report that in women the mere visual perception of diseased-looking individuals, that could potentially involve OT, leads to a more aggressive immune response (Schaller et al., 2010). Along with this, there is some evidence for an overall reduction in immune functions in isolated individuals, with social isolation being linked to decreased OT levels (Norman et al., 2010; Tanaka et al., 2010). The exact relations between social recognition, infection avoidance, immune activity and OT and AVP in both humans and non-humans remain to be unraveled.

14.4 Implications and future directions

The threat posed by infection and pathogens has important consequences for various features of human societies and social behavior (e.g., Hatemi et al., 2011; Fincher et al., 2008; Schaller and Murray, 2008). Variations in interindividual interactions, political attitudes, conformity, and the expression of personality traits such as introversion and extroversion, have all been linked to the risk of infection (Navarrete and Fessler, 2006; Schaller and Murray, 2008). Ethnocentrism, stigmatization and predijuical attitudes towards outsiders ("out-groups") have also been related to pathogen avoidance (see examples in Fincher et al., 2008; Schaller and Murray, 2008).

Evidence outlined here points at a role for OT and AVP in the modulation of the aforementioned human behaviors. OT has selective influences on the evaluative processing of threatening socially salient stimuli (Norman et al., 2011) and has been associated with decreased cooperation with, and avoidance of, unfamiliar individuals (Declerck et al., 2010). In-group favoritism, and negative social emotions such as envy (Shamay-Tsoory et al., 2009) were enhanced by OT (De Dreu 2012; De Dreu et al., 2011) Also, intranasal OT hindered trust and prosocial responses (reviewed in Bartz et al., 2011a). In addition, AVP heightens the responses to socially threatening stimuli in men as well as enhancing responses to sexual cues in men (Shalev et al., 2011; Zink et al., 2010). Together, this suggests that teasing apart the relations between OT and AVP, pathogen threat and sociality will lead to a fuller understanding of the regulation and expression of social behavior.

Acknowledgments

Supported by research grants from The Natural Sciences and Engineering Research Council of Canada.

REFERENCES

Ågmo, A., Choleris, E., Kavaliers, M., Pfaff, D. W., and Ogawa, S. (2007). Social and sexual incentive properties of estrogen receptor α, estrogen receptor β, or oxytocin knockout mice. *Genes Brain and Behavior*, 7, 70–77.

Agren, G., Unvas-Moberg, K., and Lundeberg, T. (1997). Olfactory cues from an oxytocin-injected male rat can induce anti-nociception in its cagemates. *NeuroReport*, 8, 3073–3076.

Alexander, R. D. (1974). The evolution of social behavior. *Annual review of Ecology and Systematics*, 5, 325–383.

Altizer, S., Nunn, C. L., Thrall, P. H., Gittleman, J. L., Antonovics, J., Cunningham, A. A., Dobson, A. P., Ezenwa, V., Jones, K. E., Pedersen, A. B., Poss, M., and Pulliam, J. R. C. (2003). Social organization and parasite risk in mammals: integrating theory and empirical studies. *Annual Review of Ecology, Evolution and Systematics*, 34, 517–547.

Arakawa, H., Arakawa, K., and Deak, T. (2010). Oxytocin and vasopressin in the medial amygdala differentially modulate approach and avoidance behavior towards illness-related social odor. *Neuroscience*, 171, 1141–1151.

Arakawa, H., Cruz, S., and Deak, T. (2011). From models to mechanisms: odorant communication as a key determinant of social behavior in rodents during illness-associated states. *Neuroscience and Biobehavioral Review*, 35, 1916–1928.

Arkawa, H., Cruz, S., and Deak, T. (2012). Attractiveness of illness-associated cues in females is modulated by ovarian hormones, but not associated with pro-inflammatory cytokine levels. *Brain Behavior and Immunity*, 26, 40–49.

Aubert, A. (1999). Sickness and behavior in animals: a motivational perspective. *Neuroscience and Biobehavioral Reviews*, 2, 1029–1036.

Avitsur, R. and Yirmiya, R. (1999). The immunobiology of sexual behavior: gender differences in the suppression of sexual activity during illness. *Pharmacology, Biochemistry and Behavior*, 64, 787–796.

Bartz, J. A., Zaki, J., Bolger, N., and Ochner, K. N. (2011). Social effects of oxytocin in humans: context and person matter. *Trends in Cognitive Sciences*, 15, 301–309.

Baum, M. J. and Kelliher, K. R. (2009). Complementary roles of the main and accessory olfactory systems in mammalian mate recognition. *Annual Review of Physiology*, 71, 141–160.

Baumgartner, T, Heinrichs, M., Vonlanthen, A., Fischbacher, U., and Fehr, E. (2008). Oxytocin shapes the neural circuitry of trust and trust adaptation in humans. *Neuron*, 58, 639–650.

Bocaccio, G. (1352). First Day, Intoduction. In *Decameroe* In: The Decameron: a new translation: 21 novelle. Contemporary reactions, modern criticisms; selected translated and edited by M. Musa, and P. E. Bondella, New York, Norton Press, 1977.

Brennan, P. A. and Kendrick, K. M. (2006). Mammalian social odours: attraction and individual recognition. *Philosophical Transactions of the Royal Society. B*, 361, 2061–2078.

Bosch, O. J., Medddle, S. L., Beiderbeck, D. I., Douglas, A. J., and Neumann, I. D. (2005). Brain oxytocin correlates with maternal aggression: Link to anxiety. *Journal of Neuroscience*, 25, 6807–6815.

Chiao, J. Y. and Blizinsky, K. D. (2010). Culture-gene coevolution of individualism-collectivism and the serotonin transporter gene. *Proceedings of the Royal Society B*, 277, 529–537.

Choleris, E., Clipperton-Allen, A. E., Phan, A., and Kavaliers, M. (2009). Neuroendocrinology of social information processing in rats and mice. *Frontiers in Neuroendocrinology*, 30, 442–458.

Choleris, E., Clipperton-Allen, A. E., Phan, A., Valsecchi, A., and Kavaliers, M. (2012). Estrogenic involvement in social learning, social recognition and pathogen avoidance. *Frontiers in Neuroendocrinology*, 33, 140–159.

Clodi, M., Vila, G., Geyeregger, R., Riedl, M., Stulning, T. M., Struck, J., and Luger, T. A., and Luger, A. (2008). Oxytocin alleviates the neuroendocrine and cytokine response to bacterial endotoxin in healthy men, *American Journal of Physiology Endocrinology and Metabolism*, 295, E686–E991.

Cole, S. W., Hawkley, L. C., Arevalo, J. M. G., and Cacioppo, J. T. (2011). transcript origin analysis identifies antigen-presenting cells as primary targets of socially regulated gene expression in leukocytes. *Proceedings of the National Academy of Sciences USA*, 108, 3080–3085.

Dantzer, R., O'Connor, J. C., Freund, G. G., Johnson, R. W., and Kelly, K. W. (2008). From inflammation to sickness and depression: when the immune system subjugates the brain. *Nature Reviews Neuroscience*, 9, 46–56.

Declerck, C. H., Boone, C., and Kiyonari, T. (2010). Oxytocin and conflict under conditions of uncertainty: The modulating role of incentives and uncertainty. *Hormones and Behavior*, 57, 368–374.

De Dreu, C. K. W. (2012). Oxytocin modulates cooperation within and competition between groups: an integrative review and research agenda. *Hormones and Behavior*, 61, 419–428.

De Dreu, C. K., Greer, L. L., Handgraaf, M. J., Shalvi, S., Van Kleef, G. A., Baas, M., Ten Velden, F. S., Van Dijik, E., and Feith, S. W. (2010). The neuropetide oxytocin regulates parochial altruism in intergroup conflict among humans. *Science*, 328, 1408–1411.

De Dreu, C. K. W., Greer, L. L., Van Kleef, G. A., Shalvi, S., and Handgraaf, M. J. (2011). Oxytocin promotes human ethocentrism, *Proceedings of the National Academy of Science USA*, 108, 1262–1266.

Donaldson, Z. R. and Young, L. J. (2008). Oxytocin, vasopressin, and the neurogenetics of sociality. *Science*, 322, 900–904.

Ehman, K. D. and Scott, M. E. (2001). Urinary odour preferences of MHC congeric female mice, *Mus domesticus*, implications for kin recognition and detection of parasitized males. *Animal Behaviour*, 62, 781–789.

Fincher, C. L., Thornhill, R., Murray, D. R., and Schaller, M. (2008). Pathogen prevalence predicts human cross-cultural variability in individualism/collectivism. *Proceedings of the Royal Society B*, 275, 1279–1284.

Freeland, W. J. (1976). Pathogens and the evolution of primate sociality. *Biotropica*, 8, 12–24.

Gamer, M., Zurowski, B., and Buchel, C. (2010). Different amygdala subregions mediate valence-related and attentional effects of oxytocin in humans. *Proceedings of the National Academy of Science USA*, 107, 9400–9405.

Goodson, J. L. and Thompson, R. D. (2010). Nonapeptide mechanisms of social cognition, behavior and species specific social systems. *Current Opinion in Neurobiology*, 20, 784–794.

Guastella, A. J., Mitchell, P. B., and Forgas, J. P. (2008). Oxytocin enhances the encoding of positive social memories in humans. *Biological Psychiatry*, 64, 256–258.

Guastella, A. J., Kenyon, A. R., Alvares, G. A., Carson, D. S., and Hickie, I. B. (2010). Intranasal arginine vasopressin enhances the encoding of happy and angry faces in humans. *Biological Psychiatry*, 67, 1220–1222.

Hamilton, W. D. and Zuk, M. (1982). Heritable true fitness and bright birds: a role for parasites. *Science*, 282, 384–386.

Hart, B. L. (1988). Biological bases of the behavior of sick animals. *Neuroscience and Biobehavioral Reviews*, 12, 123–135.

Hart, B. L. (1990). Behavioral adaptations to pathogens and parasites: five strategies. *Neuroscience and Biobehavioral Reviews*, 14, 273–291.

Hatemi, P. K., Gillespie, N. A., Eaves, L. J., Maher, B. S., Webb, B. T., Heath, A. C., Medland, S. E., Smyth, D. C., Beeby, H. N., Gordon, S. D., Montgomery, G. W., Zhu, G., Byrne, E. M., and Martin, N. G. (2011). A genome-wide analysis of liberal and conservative political attitudes. *Journal of Politics*, 73, 271–285.

Havlicek, J. and Roberts, S. (2009). MHC-correlated mate choice in humans: a review. *Psychoneuroendocrinology*, 34, 497–512.

Henley, C. L., Nunez, A. A., and Clemens, L. G. (2011). Hormones of choice: The neuroendocrinology of partner preference in animals. *Frontiers in Neuroendocrinology*, 32, 146–154.

Huber, D., Veinante, P., and Stoop, R. (2005). Vasopressin and oxytocin excite distinct neuronal populations in the central amygdala. *Science*, 308, 245–248.

Hurlemann, R., Patin, A., Onur, O. A., Cohen, M. X., Baumgartner T., Metzler, S., Dibeck, I., Galliant, J., Wagner, M., Maier, M., and Kendrick, K. M. (2010). Oxytocin enhances amygdala-dependent, socially reinforced learning and emotional empathy in humans. *Journal of Neuroscience*, 30, 4999–5007.

Hurst, J. L. and Beynon, R. J. (2004). Scent wars: the chemobiology of competitive signaling in mice. *Bioessays*, 26, 1288–1289.

Insel, T. R. (2010). The challenge of translation in social neuroscience: a review of oxytocin, vasopressin, and affiliative behavior. *Neuron*, 66, 768–779.

Jin, D., Liu, H. X., Hirai, H., et al. (2007). CD38 is critical for social behaviour by regulating oxytocin secretion. *Nature*, 446, 41–45.

Jorgensen, H., Riis, M., Knigge, U., Kajer, A., and Warberg, J. (2003). Serotonin receptors involved in vasopressin and oxytocin secretion. *Journal of Neuroendocrinology*, 15, 242–249.

Karelina, K. and De Vries, C. (2011). Modeling social influences in human health. *Psychosomatic Medicine*, 73, 67–74.

Kavaliers, M. and Colwell, D. D. (1995a). Odours of parasitized males induce aversive responses in female mice. *Animal Behaviour*, 50, 1164–1169.

Kavaliers, M. and Colwell, D. D. (1995b). Discrimination by female mice between the odours of parasitized and non-parasitized males. *Proceedings of the Royal Society B*, 261, 31–35.

Kavaliers, M., Colwell, D, D., Ossenkopp, K.-P., and Perrot-Sinal, T. S. (1997). Altered responses to female odors in parasitized male mice: neuromodulatory mechanisms and relations to female choice. *Behavioral Ecology and Sociobiology*, 40, 373–384.

Kavaliers, M., Colwell, D. D., and Choleris, E. (1998a). parasitized female mice display reduced aversive responses to the odours of infected males. *Proceedings of the Royal Society B*, 265, 111–118.

Kavaliers, M., Colwell, D. D., and Choleris, E. (1998b). Analgesic responses of male mice exposed to the odors of parasitized females: effects of male sexual experience and status. *Behavioral Neuroscience*, 112, 1001–1011.

Kavaliers, M., Fudge, M. A., Colwell, D. D., and Choleris, E. (2003a). Aversive and avoidance responses of female mice to the odors of males infected with an ectoparasite and the effects of prior familiarity. *Behavioral Ecology and Sociobiology*, 54, 423–430.

Kavaliers, M., Colwell, D. D., Choleris, E., Agmo, A., Muglia, L. J., Ogawa, S., and Pfaff, D. W. (2003b). Impaired discrimination of and aversion to parasitized male odors by female oxytocin knockout mice. *Genes Brain and Behavior*, 2, 220–230.

Kavaliers, M., Choleris, E., Ågmo, A., and Pfaff, D. W. (2004a). Olfactory-mediated parasite recognition and avoidance: linking genes to behavior. *Hormones and Behavior*, 46, 272–283.

Kavaliers, M., Agmo, Å., Choleris, E., Gustafsson, J-Å., Korach, K. S., Muglia, L. J., Pfaff, D. W., and Ogawa, S. (2004b). Oxytocin and estrogen receptor α and β knock-out mice provide discriminably different odor cues in behavioral assays. *Genes Brain and Behavior*, 3, 189–195.

Kavaliers, M., Choleris, E., and Pfaff, D. W. (2005). Genes, odours and the recognition of parasitized individuals by rodents. *Trends in Parasitology*, 21, 423–429.

Kavaliers, M., Choleris, E., Ågmo, A., Braun, W. J., Colwell, D. D., Muglia, L. J., Ogawa, S., and Pfaff. D. W. (2006). Inadvertent social information and the avoidance of parasitized male mice: a role for oxytocin. *Proceedings of the National Academy of Science USA*, 103, 4293–4298.

Kavaliers, M. and Choleris, E. (2011). Sociality, pathogen avoidance and the neuropeptides oxytocin and vasopressin. *Psychological Science*, 22, 1367–1374.

Kavaliers, M., Clipperton-Allen, A., Cragg, C. L., Gustaffson, J-Å., Korach, K. S., Muglia, L., and Choleris, E. (2012). Male risk taking, female odors, and the role of estrogen receptors. *Physiology and Behavior*, 107, 751–761.

Kemp. A. H. and Guastella, A. J. (2011). The roles of oxytocin in human affect: a novel hypothesis. *Current Directions in Psychological Science*, 20, 222–231.

Klein, S. L. and Nelson, R. J. (1999). Activation of the immune-endocrine system with lipopolysaacharide reduces affiliative behaviors in voles. *Behavioral Neuroscience*, 113, 1042–1045.

Knobloch, H. S., Charlet, A., Hoffmann, L. C., Eliava, M., Khrulev, S., Cetin, A. H., Osten, P., Schwarz, M. K., Seeburg, P. H., Stoop, R., and Grinevich, V. (2012). Evoked axonal oxytocin release in the central amygdale attenuates fear response. *Neuron*, 73, 553–566.

Kosfeld, M., Heinrichs, M., and Zak, P. J., Fischbacher, U., Fehr, E. (2005). Oxytocin increases trust in humans. *Nature*, 435, 673–676.

Ladeveze, S., de Muizonm C., Beck, R. M. D., Germain, D., and Cespedes-Paz, R. (2011). Earliest evidence of mammalian social behaviour in the basal tertiary of Bolivia. *Nature*, 474, 83–86.

Lerer, E., Levi, S., Israel, S., Yaari, M., Nemanov, L., Mankuta, D., Yirmiya, N., and Ebstein, R. P. (2010). Low CD8 expression in lymphoblastoid cells and haplotypes are both associated with autism in a family-based study. *Autism Research*, 3, 293–302.

Little, A. C., DeBruine, L. M., and Jones, B. C. (2011). Exposure to visual cues of pathogen contagion changes preferences for masculinity and symmetry in opposite-sex faces. *Proceedings of the Royal Society B*, 278, 202–209.

Litvinova, E. A., Goncharova, E. P., Zaydman, A. M., Zenkva, M. A., and Moshkin, M. P. (2010). Female scent signals enhance the resistance of male mice to influenza. *PLoS One*, 5, e9473–e9479.

Lundstrom, J. N., Boyle, J. A., Zatorre, R. J., and Jones-Gotman, M. (2008). Functional neuronal processing of body odors differs from that of similar common odors. *Cerebral Cortex*, 18, 1466–1474.

Malavasi, F., Deaglio, S., Funaro, A., Ferrero, E., and Horenstein, A. L., et al. (2008). Evolution and function of the ADP ribosyl cyclase/CD38 gene family in physiology and pathology. *Physiological Reviews*, 88, 841–846.

Meyer-Lindenberg, A., Domes, G., Kirsch, P., and Heinrichs, M. (2011). Oxytocin and vasopressin in the human brain: social neuropeptides for translational medicine. *Nature Reviews Neuroscience*, 12, 524–538.

Milinski, M. (2006). The major Histocompatibility complex, sexual selection and mate choice. *Annual Review of Ecology, Evolution and Systematics*, 7, 159–186.

Miller, S. L. and Maner, J. K. (2010). Scent of a woman: Men's testosterone responses to olfactory ovulation cues. *Psychological Science*, 21, 276–283.

Moore, J. (2002). *Parasites and the Behavior of Animals*, Oxford University Press, Oxford.

Moshkin, M., Litvinova, N., Litvinova, E. A., Bedareva, A., Lutsyuk, A., and Gerlinskaya, L. (2012). Scent recognition of infected status in humans. *Journal of Sexual Medicine*, 9, 3211–3218.

Navarette, C. D. and Fessler, D. M. T. (2006). Disease avoidance and ethocentrism: the effect of disease vulnerability and disgust sensitivity on intergroup attitude. *Evolution and Human Behavior*, 27, 270–282.

Neuberg, S. L., Kenrick, D. T., and Schaller, M. (2011). Human threat management systems: self protection and disease avoidance. *Neuroscience and Biobehavioral Reviews*, 35, 1042–1051.

Norman, G. J., Karelina, K., Morris, J. S., Zhana, N., Cochran, M., and DeVries, A. C. (2010). Social influences on neuropathic pain and depressive like behavior; a role for oxytocin. *Psychosomatic Medicine*, 72, 519–526.

Norman, G. J., Cacioppo, J. T., Morris, J. S., Karelina, K., Malarkey, W. B., Devries, A. C., and Berntson, G. G. (2011). Selective influences of oxytocin on the evaluative processing of social stimuli. *Journal of Psychopharmacology*, 25, 1313–1319.

Nunn, C. L. and Altizier, S. M. (2006). *Infectious Diseases in primates: behavior, Ecology and Evolution*, Oxford University Press, Oxford, 400pp.

Oaten, M., Stevenson, R. J., and Case, T. I. (2009). Disgust as a disease-avoidance mechanism. *Psychological Bulletin*, 135, 303–321.

Pearce-Duvet, J. M. C. (2006). The origin of human pathogens: evaluating the role of agriculture and domestic animals in the evolution of human disease. *Biological Reviews*, 81, 369–382.

Penn, D. and Potts, W. K. (1998). Chemical signals and parasite-mediated sexual selection. *Trends in Ecology and Evolution*, 13, 391–396.

Penn, D., Schneider, G., White, K., Slev, P., and Potts, W. K. (1998). Influenza infection neutralizes the attractiveness of male odours to female mice (*Mus musculus*). *Ethology*, 104, 685–694.

Renault, J., Gheusi, G., and Aubert, A. (2008). Changes in social exploration of a lipopolysaccharides-treated conspecific in mice: role of environmental cues. *Brain Behavior and Immunity*, 22, 1201–1207.

Roberts, S. A., Simpson, D. M., Armstrong, S. D., Davidson, A. J., Robertson, D. H., McLean, L., Beynon, R. J., and Hurst, J. L. (2010). Darcin: a male pheromone that stimulates female memory and attraction to an individual male's odour. *BMC Biology*, 8, 75–96.

Rowe, T. B., Macrini, T. E., and Luo, Z.-X., (2011). Fossil evidence on the origin of the mammalian brain. *Science*, 332, 955–957.

Schaller, M. and Murray, D. R. (2008). Pathogens, personality, and culture: disease prevalence predicts worldwide variability in socio-sexuality, extraversion, and openness to experience. *Journal of Personality and Social Psychology*, 95, 212–221.

Schaller, M., Miller, G. E., Gervais, W. M., Yager, S., and Chen, E. (2010). Mere visual perception of other people's disease symptoms facilitates a more aggressive immune response. *Psychological Science*, 21, 649–652.

Schaller, M. and Park, J. H. (2011). The behavioral immune system (and why it matters). *Current Directions in Psychological Science*, 20, 99–103.

Shalev, I., Israel, S., Uzefovsky, F., Gritsenko, I., Kaitz, M., and Ebstein, R. P. (2011). Vasopressin needs an audience: Neuropeptide elicited stress responses are contingent upon perceived social evaluative threats. *Hormones and Behavior*, 60, 121–127.

Shamay-Tsoory, S. G., Fischer, M., Dvash, J., Harai, H., Perach-Bloom, M., and Levkovitz, Y. (2009). Intranasal administration of oxytocin increases envy and schadenfreude (gloating). *Biological Psychiatry*, 66, 864–870.

Tanaka, K., Osako, Y., and Yuri, K. (2010). Juvenile social experience regulates central neuropeptides relevant to emotional and social behaviors. *Neuroscience*, 166, 1036–1042.

Theodoridou, A., Rowe, A. C., Penton-Voak, I. S., and Rogers, P. J. (2009). Oxytocin increases perceived facial trustworthiness and attractiveness. *Hormones and Behavior*, 56, 128–132.

Theodoridou, A., Rowe, A. C., Rogers, P. J., and Penton-Voak, I. S. (2011). Oxytocin administration leads to a preference for masculinized faces. *Psychoneuroendocrinology*, 36, 1257–1260.

Thornhill, R., Murray, D. R., and Schaller, M. (2010). Zoonotic and non-zoonotic diseases in relation to human personality and societal values: support for the parasite stress model. *Evolutionary Psychology*, 8, 151–169.

Unkelbach, C., Guastella, A. J., and Forgas, J. P. (2008). Oxytocin selectively facilitates recognition of positive sex and relationship words. *Psychological Science*, 19, 1092–1094.

Wacker, D. W., Engelmann, M., Tobin, V. A., Meddle, S. L., and Ludwig, M. (2011). Vasopressin and social odor processing in the olfactory bulb and anterior olfactory nucleus. *Annals of the New York Academy of Science*, 1220, 106–116.

Zink C. F., Stein, J. L., Kemf, L., Hakimi, S., and Meyer-Lindenberg, A. (2010). Vasopressin modulates medial prefrontal cortex-amygdala circuitry during emotion processing in humans. *Journal of Neuroscience*, 30, 7017–7022.

15

Oxytocin and addiction

Recent preclinical advances and future clinical potential

Iain S. McGregor and Michael T. Bowen

15.1 Introduction

The idea that oxytocin (OT) may be involved in addiction-relevant behaviors has had a long gestation (e.g., Kovacs et al., 1998; Sarnyai and Kovacs, 1994), although it is only in the last few years that the idea has been expanded with a series of interesting preclinical results and fledgling interest from clinicians dealing with addicted clients. At first glance, the idea that the "social neuropeptide" might be involved in addiction is appealing. In our last review of this topic we noted that the acute effects of recreational drugs are often intensely prosocial while their long term effects may be conversely antisocial (McGregor et al., 2008). Might modulation of OT and related neuropeptide systems underpin both sets of effects?

For many years classical psychoanalytic theory has forwarded the idea of oral fixation whereby certain forms of drug consumption resemble "suckling" behavior in infants. Addicts are seen as somehow being marooned in this infantile stage. The thirsty drinker sucks at the beer bottle with relish to obtain relief from his or her dysphoria, while the smoker sucks strongly at the cigarette in the hope of banishing their cravings and obtaining satisfaction. Does OT, as the regulator of infantile attachment and nourishment (e.g., Schaller et al., 2010), have a role in reinforcing such primitive behaviors in addicts, in which drugs replace milk?

As any drug and alcohol clinician will attest, there are high levels of antisocial personality disorder, borderline personality disorder, impulsivity and criminality amongst heavy users of drugs and alcohol (Chen et al., 2003; Darke et al., 2004; Dawe et al., 2009). While it is likely that such traits are sometimes constitutive, it is also likely that such personalities sometimes evolve from the corrosive effects of chronic drug abuse on the brain, and in particular through drug-induced modifications to the neural substrates of affiliative behavior. Addicts generally make poor decisions, and this is especially the case in the social sphere. Drug users become "bonded" to their drugs and to the drug-related cues that envelop their lives, often to the exclusion of other forms of reward, including other human beings (Edwards and Self, 2006; Volkow et al., 2011).

Effective treatments for addictions, of which there are relatively few, often have a magic "X" factor that comprises some form of social rehabilitation or social reintegration. Might the social support provided by Alcoholics Anonymous meetings, for example, achieve success by resetting dysfunctional OT pathways (Koerner, 2010)? Is it possible that cuddles can cure addictions? Can we beat drugs with hugs?

In the current chapter we will address some of these issues and provide an update on the possible roles of OT in addiction-relevant behaviors. We will also address exciting emerging evidence that OT

Oxytocin, Vasopressin, and Related Peptides in the Regulation of Behavior, ed. E. Choleris, D. W. Pfaff, and M. Kavaliers.

itself might provide a novel therapeutic for treating addictions.

15.2 Oxytocin and the acute prosocial effects of party drugs

An emerging body of evidence suggests that the desirable prosocial effects of a number of different recreational drugs may involve OT. Results are for the most part suggestive, rather than definitive, and outcomes have varied according to the drug under investigation. As with all OT research, there are often major difficulties in measuring the ebb and flow of the peptide in the brains of awake animals engaging in behavior. This seriously limits our knowledge of how various drugs impinge acutely on brain OT release. An additional problem is that there are currently no radioligands that allow us to probe central oxytocin receptor (OTR) dynamics in human clinical populations. Existing pharmacological probes for OTRs that are used in animal models often lack specificity, and there is also continuing uncertainty about the extent to which peripherally injected or intranasally administered OT penetrates the brain.

Nonetheless, there is convincing evidence implicating OT in the acute prosocial effects of the party drug 3,4-methylenedioxymethamphetamine (MDMA), or "Ecstasy." This work has transitioned nicely from basic research in laboratory animals to laboratory studies with human participants. MDMA is a fascinating drug for psychologists and neuroscientists given its capacity to engender strong feelings of love and closeness toward other people, increased trust, and greater openness to the views and feelings of others. A bumper sticker in California during the 1980s, when MDMA first arrived on the scene, proclaimed "don't get married for 6 months after taking Ecstasy," underscoring the capacity of the drug to produce prolonged "artificial chemical love" in *Homo Sapiens*. Psychopharmacologists now increasingly see MDMA as a tool with which to better understand human affiliative behavior (Bedi et al., 2009) and there is interest in the therapeutic potential of MDMA in treating post-traumatic stress disorder (PTSD) and other psychiatric problems (Mithoefer et al., 2011).

The powerful prosocial action of MDMA seen in humans is replicated in various animal models. For example, in unfamiliar pairs of rats meeting for the first time, MDMA markedly reduces aggression and increase a behavior known as adjacent lying (Morley and McGregor, 2000; Thompson et al., 2007; Thompson et al., 2008; Kirilly et al., 2006). This rat "cuddling" is reminiscent of the increased social contacts in animals given OT or OT-like agonists (Witt et al., 1992; Goodson et al., 2009; Hicks et al., 2012), with OTR antagonists provoking a reciprocal loss of preference for social stimuli (Lukas et al., 2011). It was therefore natural to ask whether MDMA-induced facilitation in rats might involve OT. Accordingly, MDMA was shown to powerfully induce Fos expression in the OT-releasing neurons in the supraoptic nucleus (SON) and paraventricular nucleus (PVN) of the hypothalamus, leading to dose-dependent increases in plasma OT, and activation of forebrain neural circuits known to regulate affiliative behavior (Thompson et al., 2007; Thompson et al., 2009). Importantly, the prosocial effects of MDMA were significantly attenuated by central administration of the OTR antagonist tocinoic acid (TOC, 20 μg ICV) (Thompson et al., 2007).

The important role of OTR activation in the effects of MDMA is further highlighted by recent drug discrimination experiments in which the OT antagonist atosiban selectively interfered with MDMA, but not amphetamine, appropriate responding in rats (Broadbear et al., 2011). The OT analog carbetocin partially generalized to the MDMA training cue, suggesting that the prosocial effects of MDMA are not only mediated by the OTR but that enhanced sociability is also one of the primary features that distinguishes the subjective effects of MDMA from other drugs such as amphetamine.

High ambient temperatures, and dehydration, increase OT release (Uvnas-Moberg et al., 1993) which could further amplify the oxytocinergic effects of MDMA and its subsequent social and neural effects. Accordingly, we reported a facilitation of MDMA's prosocial and rewarding effects

in rats under high ambient temperatures (Cornish et al., 2003; Hargreaves et al., 2007). This interaction might help to explain the tendency for human users to take MDMA under hot and sweaty conditions (Parrott, 2004). Recent work illuminates the capacity of hyponatremia (low plasma sodium), a condition frequently caused in humans by MDMA (Budisavljevic et al., 2003), to boost peripheral OT levels and social affiliation in rodents. Animals must lose their fear of other animals, perhaps, to visit the communal watering hole, and to facilitate this, osmolality and sociability merge in the actions of OT in the SON (Krause et al., 2011).

The evidence linking OT and MDMA is not restricted to rodents. Laboratory studies with MDMA have confirmed the powerful prosocial effects in humans, with humans given MDMA showing increased empathy and sociability, a diminished response to threatening social stimuli, and an augmented response to social reward (Bedi et al., 2009; Bedi et al., 2010). Raised peripheral OT levels were found in dance party patrons who had taken MDMA (Wolff et al., 2006; Bedi et al., 2010) as well as in a double-blind, randomized, placebo-controlled study of healthy volunteers given MDMA (Dumont et al., 2009). Strikingly, the subjective prosocial feelings induced by MDMA in humans were more highly correlated with plasma OT levels than plasma MDMA levels (Dumont et al., 2009).

MDMA is a potent releaser of 5-hydroxytryptamine (5-HT), and the OT-dependent prosocial effects of MDMA in both humans and rats appear linked to a cascade in which 5-HT release impacts on hypothalamic 5-HT_{1A} receptors, leading to OT release. The 5-HT containing terminals and the OT-containing cell bodies in the SON and PVN share close proximity (Emiliano et al., 2006) and 5-HT_{1A} receptors are localized on the perikarya of magnocellular OT containing neurons (de Souza Villa et al., 2008; Marvin et al., 2010). Accordingly, MDMA-like adjacent lying in rats is produced by the 5-HT_{1A} receptor agonist 8-OH-DPAT, with these prosocial effects attenuated by the OT antagonist TOC (Thompson et al., 2007). Further, in rats, the 5-HT_{1A} antagonist WAY 100,635 blocked both the prosocial effects and heightened OT release caused by MDMA (Thompson et al., 2007; Thompson et al., 2008) as well as associated MDMA-induced SON activation (Hunt et al., 2011). This fits nicely with observations that the anxiolytic drug buspirone, a partial 5-HT_{1A} receptor agonist that has been used clinically in treating anxiety disorders, also stimulates OT release (Uvnas-Moberg et al., 1996).

MDMA is not the only drug that may exert prosocial effects through stimulation of OT. Another popular dance party drug, gamma-hydroxybutyrate (GHB), is similarly prosocial to MDMA, with potent antiaggressive, anxiolytic as well as other prosocial properties (Pedraza et al., 2007; Schmidt-Mutter et al., 1998; Sumnall et al., 2008). GHB, like MDMA, causes strong activation of oxytocinergic circuitry in the SON and PVN (Van Nieuwenhuijzen et al., 2009). Indeed, GHB was once used to promote uterine contractions in childbirth, suggestive of a powerful OT-like tocogenic effect (Geldenhuys et al., 1968). Recently, we have shown that mephedrone (4-Methylmethcathinone, also known as "Meow") also causes substantial SON and PVN activation, and this might conceivably underlie the intense prosocial and entactogenic effects inherent in user reports with this emerging, increasingly popular dance party drug (Motbey et al., 2011).

There is clear support then for oxytocin as a common denominator through which certain party drugs may exert their prosocial effects. However, the story becomes more complex when a wider range of recreational drugs is considered. Alcohol is perhaps the most commonly used "party drug," but available evidence suggests that it inhibits, rather than stimulates, OT release in addition to its well-documented inhibitory effects on vasopressin (AVP) (Gibbens and Chard, 1976; Subramanian, 1999). It is important to be reminded here that peripheral OT levels can be a rather poor proxy for the activity of the central OT system, and it is therefore conceivable that central OT systems are stimulated by alcohol, perhaps by releasing central OT circuits from the reciprocal inhibition caused by AVP (Huber et al., 2005; Neumann et al., 2006). This important

point could be usefully addressed in future studies examining central OT levels in alcohol consuming rats by using microdialysis. Similarly, opiates inhibit peripheral OT release although central effects could be more closely scrutinized (Russell et al., 1989). Also confusing the issue are the recent laboratory studies indicating that methamphetamine has subtle prosocial effects in human participants, yet it is a drug that appears to be devoid of acute stimulatory effects on OT (Carson et al., 2010b; Bedi et al., 2010).

On the other hand, not all drugs that strongly stimulate OT release are necessarily prosocial. Lithium is a potent stimulator of hypothalamic oxytocinergic circuitry (Cui et al., 2001), and while a useful treatment for a number of psychiatric conditions, is not necessarily prosocial in its effects. The relationship between OT and antidepressant action could also use further clarification: similar to MDMA the action of SSRI, SNRI and tricyclic antidepressants on SERT and 5-HT_{1A} receptors will most likely stimulate OT and such actions could conceivably underlie the utility of these drugs in conditions such as social anxiety disorder and depression (Rotzinger et al., 2010; Uvnas-Moberg et al., 1999). Indeed, the low plasma levels of OT seen in depressed cohorts (Scantamburlo et al., 2007) might conceivably be rebalanced by SSRI and other antidepressant actions. It is notable that OT and OTR agonists have antidepressant-like and anxiolytic effects in animal models (Chaviaras et al., 2010; Grippo et al., 2009; Rotzinger et al., 2010; Ring et al., 2010).

15.3 Does oxytocin mediate the acute rewarding effects of drugs or only their social effects?

While the above discussion provides increasingly clear evidence that OT plays a role in the acute prosocial effects of certain drugs, it is unclear whether this is distinct from the rewarding, or euphorogenic, effects of these drugs. In other words, it is not clear whether OT release *per se* is sufficient to produce reward. The proliferation of studies involving administration of intranasal OT in humans provides strong evidence that OT administration is not euphorogenic, and is typically not even able to be discriminated from placebo by the recipient, despite sometimes causing profound modulation of social preference and social cognition (Macdonald et al., 2011). One study in rats reported a conditioned place preference with systemically administered OT, suggesting reward, yet the dose involved was extremely high (8 mg/kg) and would presumably have had major non-specific effects (Liberzon et al., 1997). Current indications are that OT is in little danger of becoming an abused drug in its own right, although the poor brain penetration and short half-life of the compound may contribute to this. It remains conceivable that emerging non-peptide OTR agonists (e.g., Ring et al., 2010) might have abuse potential, perhaps analogous to MDMA or benzodiazepines.

It is more likely, however, that OT modulates the reinforcing properties of social engagement and social situations, and primes social reward rather than providing the reward itself (Sarnyai, 2011). For example, rodent studies suggest that MDMA-induced activation of key forebrain reward sites such as the nucleus accumbens (NAcc) is greater when the drug is given under social conditions, suggesting that the drug amplifies the impact of social interaction on the neural substrates of reward (Thompson et al., 2009). This might involve interactions between OT and the classic "drug reward" pathways centered on the mesolimbic dopamine system. OTRs are localized in many sites that are relevant to drug-seeking behavior, including the NAcc, ventral tegmental area, bed nucleus of the stria terminalis, central amygdala, medial amygdala, hippocampus and ventral pallidum (Vaccari et al., 1998). An impressive body of work conducted with voles indicates that OTRs in key mesolimbic sites, such as the NAcc regulate social processes, such as monogamy, via an interaction with dopamine (Young et al., 2011). Interestingly, recent work has suggested not only that drug reward and social reward utilize some of the same pathways but has also indicated that animals with more

evolved social reward systems (such as monogamous prairie voles) may be more sensitive to drug reward (Anacker and Ryabinin, 2010). Additionally, oxytocinergic projections from the PVN innervate the dopaminergic neurons in the ventral tegmental area to cause penile erections, increased sexual motivation, and increased mesolimbic dopamine activity (Melis and Argiolas, 2011).

However, the exact interaction between dopamine and OT appears complex and may vary according to behavioral contexts (Baskerville and Douglas, 2010). For example, we recently reported that methamphetamine-induced Fos expression in the NAcc core is significantly reduced by systemic OT pretreatment, consolidating previous research findings that OT reduces methamphetamine-induced dopamine efflux in the NAcc (Qi et al., 2008). Interestingly, lithium – a drug that increases OT release in rodents (Cui et al., 2001) – also prevents methamphetamine-induced Fos expression in the prelimbic cortex, NAcc core, caudate-putamen, and central amygdala (Lee et al., 1999), possibly through an OT-dependent mechanism.

The environment of most animals contains both social and non-social reinforcers. At any given moment an animal can decide to seek social interaction with conspecifics or to pursue the rewards inherent in "objects," be they toys, computers, foodstuffs or drugs. Might there be parallel brain circuits for social and non-social rewards, with OT increasing the incentive motivational properties of social relative to non-social stimuli? While OT stimulates appetite in breastfeeding infants (with breastfeeding itself a form of social engagement for both infant and mother), it generally decreases appetite for both food and water in the adult animal (Olson et al., 1991; Verty et al., 2004) while priming social and sexual motivational systems. Extreme dopaminergic stimulation, in the absence of OT, produces intense object or self-oriented behaviors, as shown in stimulant-induced stereotypy in laboratory animals. The phenomenon of "punding" seen in heavy methamphetamine users, as well as Parkinsonian patients treated with dopaminergic agonists, refers to compulsive performance of repetitive, mechanical tasks, such as assembling and disassembling objects or, collecting, sorting or cleaning household objects (Fasano and Petrovic, 2010; Gescheidt and Bares, 2011). Intense stereotypy and fascination with mechanical objects is also characteristic of autism, and some results suggest that OT treatment can help break such stereotypy (Hollander et al., 2003; Sala et al., 2011). It is also striking that OTR knockout mice display impaired behavioral flexibility (Sala et al., 2011). We hypothesize here that dopamine acting alone in basal ganglia circuits might bias behavior toward object-oriented pursuits, but with the additional stimulation of OT, as is obtained with drugs such as MDMA and GHB, the bias is perhaps tilted towards intense focus on social stimuli, which may be coupled by an abrupt termination of object-oriented, goal-directed activity. This might constitute an important and sudden switching of behavioral modes from frenetic execution of object-oriented behavioral loops into a mode of social engagement.

The Polyvagal theory of Porges (2003) speaks to the neural substrates of a social engagement system in which brain circuitry used for defensive freezing is co-opted to facilitate the passive, non-defensive immobility, and parasympathetic dominance that is characteristic of social engagement. The ventrolateral periaqueductal gray, which is rich in OTRs, is seen as a major node in this network. High doses of OT are known to cause sedation and immobility in rodents, which is coupled to anxiolytic and prosocial effects (Uvnas-Moberg, 1998a). In the infant, such a system facilitates the passivity required to feed and fosters a sense of relaxation and contentment.

15.4 "Breaking the loop": Acute effects of oxytocin on drug self-administration

The suggestion that OT might provide an exit strategy from the narrow "behavioral loops" that characterize compulsive drug taking is supported by a number of observations in preclinical research.

The seminal research by the Kovacs group provided some important early signs of OT's efficacy in this regard. Exogenously administered OT in rodents decreased the hyperactivity, locomotor sensitization and stereotyped behaviors caused by cocaine (Sarnyai, 1998) and inhibited the self-administration of opiates in rats (Kovacs and Van Ree, 1985; Ibragimov et al., 1987).

In more recent studies, OT given ICV inhibited the formation of a conditioned place preference (CPP) to methamphetamine, facilitated the extinction of methamphetamine-induced CPP and prevented its stress-induced reinstatement in mice (Qi et al., 2009). Mechanistically, this was linked to OT inhibiting the enhanced dopamine utilization in the striatum caused by methamphetamine, recapitulating earlier findings that OT antagonizes cocaine-induced increases in dopamine utilization in the NAcc (Kovacs et al., 1990). Additionally, OT inhibited prefrontal glutamate release during stress-induced reinstatement of methamphetamine CPP (Qi et al., 2008) and it has been subsequently argued that OT might attenuate addiction-related changes in NMDAR1 expression in the prefrontal cortex (Yang et al., 2010).

Our own research extended these important findings by showing that OT (0.3–1 mg/kg IP) powerfully inhibits intravenous methamphetamine self-administration in rats (Figure 15.1A) (Carson et al., 2010a) and diminishes the capacity of non-contingent methamphetamine "primes" to reinstate methamphetamine-seeking behavior in abstinent rats. Fos immunohistochemistry revealed that systemic OT significantly reduced methamphetamine-induced neuronal activation in the NAcc core, and the subthalamic nucleus (Figure 15.1B). The subthalamic nucleus is seen as a key component in the circuitry underlying compulsive behaviors, and has recently been targeted with deep-brain stimulation as a means of alleviating severe obsessive compulsive disorder and modifying the aforementioned "punding" behavior caused by dopamine agonists in Parkinson's disease (Klavir et al., 2009; Mallet et al., 2008; Pallanti et al., 2010). We have also recently shown that systemic OT, or microinjections of OT directly into the NAcc core, or subthalamic nucleus, can reduce the CPP produced by methamphetamine in rats (Baracz et al., 2012).

Systemically administered OT (2 mg/kg), as well as reducing methamphetamine-induced Fos expression in key brain regions, also activates OT-positive cells in the SON (Figure 15.1C), consistent with earlier suggestions of a feedforward effect of OT on its own dendritic release, akin to "self-stimulation" (Ludwig and Leng, 2006; Rossoni et al., 2008). Systemic OT also activated a number of other central sites including the central amygdala, lateral parabrachial nucleus and the locus coeruleus as had been similarly reported in earlier Fos studies involving ICV OT administration (Kita et al., 2006; Popeski and Woodside, 2001). We believe that these findings help to settle the issue of whether peripherally administered OT penetrates the brain, at least with respect to rodents.

Other recent work from our laboratory has examined the interaction between OT and alcohol self-administration in rats. This follows on from earlier observations that OT inhibits the development of physiological tolerance to alcohol with respect to outcome measures such as hypothermia and ataxia (Kovacs et al., 1998; Jodogne et al., 1991; Szabo et al., 1987; Tirelli et al., 1992). Initially we examined how systemic OT affected consumption of a sweet alcohol-containing beverage that is very popular with young Australians (*Raspberry Vodka Cruiser*, 5% ethanol). Alcohol-preferring "P" rats were given a choice between this *Alcopops* beverage and a non-alcoholic sweet solution (3% sucrose) in daily sessions in a "lickometer" apparatus (Hargreaves and McGregor, 2007). Strikingly, a single administration of OT (1 mg/kg) resulted in a long lasting-decline (at least 6 weeks) in the intake of *Alcopops* relative to sucrose (Figure 15.1A). Overall intake of fluid was not affected by oxytocin treatment except a moderate decline in both *Alcopops* and sucrose intake during acute treatment, and a decrease in *Alcopops* consumption and increase in sucrose consumption over the 6 weeks after the administration of the single dose of oxytocin.

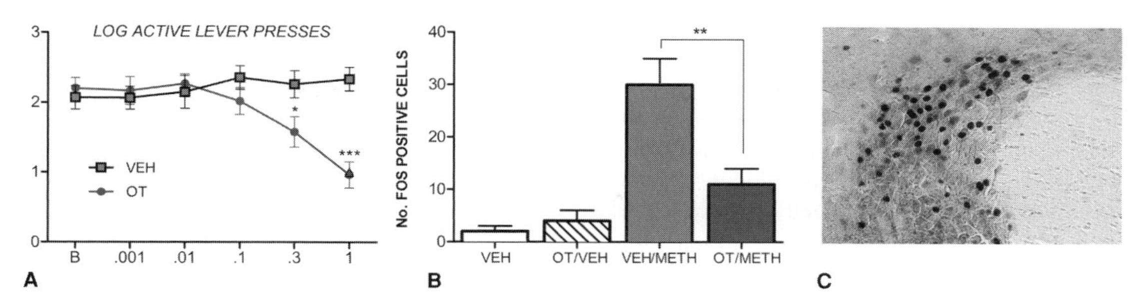

Figure 15.1 A. Peripheral oxytocin (doses on the x-axis) reduces intravenous METH self-administration in rats (Carson et al., 2010a). B. Peripheral oxytocin (2 mg/kg) reduces METH-induced activation of the Subthalamic Nucleus (Carson et al., 2010b). C. Peripheral oxytocin (2mg/kg) activates OT-positive neurons in the Supraoptic Nucleus (Carson et al., 2010b). See color version in plates section.

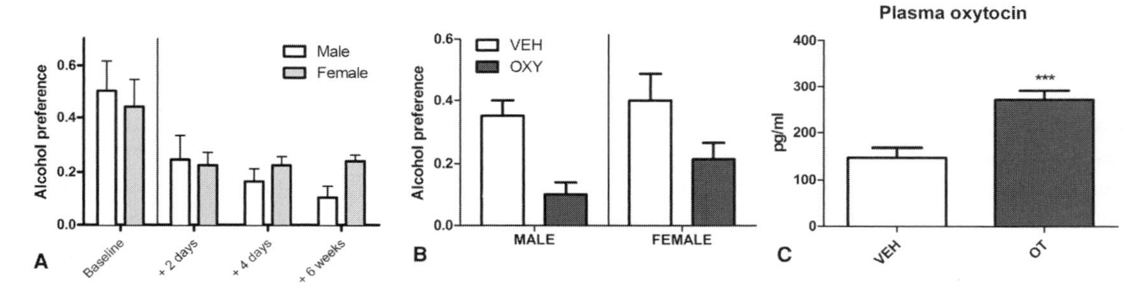

Figure 15.2 A. Lasting reduction in "Alcopops" preference (relative to 3% sucrose) in "P" rats after a single oxytocin dose (1 mg/kg). B. Repeated daily oxytocin pretreatment (1 mg/kg x 10 days) decreases "Alcopops" preference in "P" rats tested 1 month later. C. Repeated intermittent oxytocin treatment increases plasma oxytocin levels measured 2 weeks later.

To rule out an explanation of this effect in terms of conditioned taste aversion, we uncoupled OT administration from alcohol presentation by giving rats repeated doses of OT (1 mg/kg, IP) for 10 days prior to initiating *Alcopops* access 2 weeks after the final OT dose was administered. Here, a remarkably low *Alcopops* preference was evident in OT pretreated rats, particularly in male rats (Figure 15.2B).

These effects with alcohol, while striking, require further validation. For example, OT knockout mice consume greater amounts of sweet solutions than wild-types (Amico et al., 2005) and it would clearly be of interest to examine whether they also display excessive levels of alcohol consumption. It is also striking that Vasopressin V_{1A} (AVPR1a) knockout mice show increased alcohol preference, which has been linked to elevated glutamate in striatal circuits that are implicated in compulsive drinking (Sanbe et al., 2008).

These results also invite a possible reinterpretation of the therapeutic efficacy of baclofen in treating alcohol use disorders. As outlined in the excellent book by Olivier Ameisen, "The End of My Addiction," baclofen (a $GABA_B$ agonist that is often used to treat spasticity) has a striking ability to inhibit alcohol craving. This has lead to major worldwide interest and clinical trials of this substance for alcohol-use disorders, albeit with some mixed results (Ameisen, 2011; Addolorato et al., 2011; Garbutt et al., 2010; Liu and Wang, 2011). We have documented that high doses of baclofen, and GHB, both potently stimulate the SON in rats, most likely leading to increased levels of OT (Van

Nieuwenhuijzen et al., 2009). It is therefore possible that the ability of baclofen to inhibit alcohol and other drug craving rests on an indirect action on OT. Oxytocinergic effects might also be relevant to the anticraving effects of GHB, which is licensed in Italy for the treatment of alcohol withdrawal and craving (Caputo, 2011; Leone et al., 2010). These suggestions are worthy of further investigation and might further consolidate the potential efficacy of OT in treating alcohol use disorders.

15.5 Neuroadaptations in oxytocin during the development of addiction and tolerance to drugs

Addiction to drugs rarely happens overnight and preclinical and clinical research is often oriented around elucidation of the lasting neuroadaptations that occur as a result of chronic drug exposure that might drive compulsive drug seeking and neurocognitive and personality abnormalities in habitual drug users. Brain OT systems display profound neuroplasticity in response to a variety of environmental and pharmacological stimuli and in relation to developmental milestones such as puberty, pregnancy and parturition (Gimpl and Fahrenholz, 2001; Kramer et al., 2006; Neumann, 2007; Slattery and Neumann, 2008; Lukas et al., 2010). Several studies suggest that repeated exposure to drugs of abuse cause major neuroadaptations in various markers of the functionality of the endogenous OT system (summarized in Table 15.1). For example, repeated low-dose delta-9-tetrahydrocannabinol (THC) downregulated OTR expression and diminished OT innervation in the NAcc of rats (Butovsky et al., 2006). It is notable here that chronic exposure to cannabinoids causes lasting social deficits in rats (O'Shea et al., 2004; O'Shea et al., 2006; Quinn et al., 2008). Chronic morphine exposure decreased brain OT synthesis (You et al., 2000), while chronic ethanol exposure caused degeneration of OT-containing magnocellular neurons in the hypothalamus of both humans and rats (Silva et al., 2002; Sivukhina et al., 2006). Repeated cocaine administration decreased hippocampal, hypothalamic, and systemic OT levels (Sarnyai, 1998).

Recently, we examined changes in OT and OTR mRNA in rats showing residual social interaction deficits and increased anxiety after administration of MDMA or GHB, or their combination (van Nieuwenhuijzen et al., 2010). Although MDMA is not a drug that lends itself to repetitive daily use, users frequently escalate their preferred dose of MDMA over time to regain the same prosocial effects obtained when the drug was first experienced. Accordingly, rats that have been pre-exposed to MDMA show a blunted response to subsequent doses of MDMA and require a higher dose to stimulate the same level of social interaction induced by the initial lower dose (Thompson et al., 2008). When tested drug-free, several weeks after MDMA or GHB exposure, rats show reduced social interaction as well as residual increases in anxiety on the elevated plus maze and emergence tests, and also have poorer memory than controls (van Nieuwenhuijzen et al., 2010). MDMA pretreated rats are also impaired in their coping responses to acute stress (McGregor et al., 2003; Morley et al., 2001; Thompson et al., 2004).

Eight weeks after treatment, we found increased OT mRNA in the hypothalamus of MDMA-treated animals, increased OTR mRNA in the hypothalamus of GHB-treated animals and an intermediate effect in MDMA/GHB-treated animals. These changes might conceivably reflect upregulation of the endogenous OT system in response to MDMA or GHB-induced OTR stimulation. This is consistent with the plasticity of the OT system and its ability to establish a positive feedback loop whereby heightened OT release stimulates the OT system and leads to greater activity of the endogenous OT system. However, the social deficits suggest that there may be diminished functionality in the endogenous OT system, not augmented. As such, it is possible that exposure to MDMA and GHB may exhaust the endogenous OT system and that OT mRNA in the case of MDMA and OTR mRNA in the case of GHB are upregulated to compensate for this depletion.

Table 15.1 Oxytocin-related neuroadaptations following chronic exposure to drugs.

Species	Treatment	Finding	Reference
Human	Alcoholics (p.m.)	Decreased supraoptic OT-containing neurons	Sivukhina et al. (2006)
Rat	Chronic alcohol	Decreased hypothalamic OT-containing neurons	Silva et al. (2002)
Rat	Gestational alcohol/ nicotine	Decreased VTA and MPO OT in dams; Increased VTA OT in adolescent offspring; decreased VTA OT in adult offspring	McMurray et al. (2008)
Rat	Gestational alcohol / nicotine	Increased OTR binding in NA of adult offspring males but not females	Williams et al. (2009)
Rat	Chronic cocaine	Decreased hypothalamic and hippocampal OT	Sarnyai (1998)
Rat	Chronic GHB	Increased hypothalamic OTR mRNA	van Nieuwenhuijzen et al. (2010)
Rat	Chronic MDMA	Increased hypothalamic OT mRNA	van Nieuwenhuijzen et al. (2010)
Rat	Chronic morphine	Decreased hippocampal OT	Kovacs et al. (1987)
Rat	Chronic morphine	Decreased OT synthesis	You et al. (2000)
Rat	Chronic THC	Diminished OT immunoreactivity in accumbens	Butovsky et al. (2006)

Ongoing experiments are aimed at trying to further disentangle the exact mechanisms involved.

Other drugs causing lasting social deficits in rats, and presumably humans too, include methamphetamine, cannabinoids and NMDA antagonists (e.g., ketamine and phyncyclidine (PCP)) (summarized in McGregor et al., 2008). If neuroadaptations caused by repeated exposure to drugs of abuse involve endogenous OT pathways, then administration of exogenous oxytocin may have potent effects in preventing some key drug-induced neuroadaptations. Indeed, the social withdrawal caused by subchronic PCP administration is reversed by intra-amygdala administration of OT (Lee et al., 2005) or by AVPR1a agonists (Matsuoka et al., 2005).

In summary, lasting social dysfunction as a result of drug exposure is evident in studies with laboratory animals (McGregor et al., 2008) and the neuroadaptations that mediate such effects require better definition. Simple assessment of plasma OT in drug-abusing populations might be a useful first step. In this regard, a recent study from our group (Carson et al., 2012) failed to find any difference in plasma OT or AVP levels in current methamphetamine users meeting DSM IV criteria for dependence, but uncovered blunted cortisol levels. These participants, however, were not tested during a period of abstinence, and future studies will hopefully obtain data for dependent populations during withdrawal and abstinence.

15.6 Oxytocin might inoculate against future vulnerability to drug abuse

In the therapeutic setting, maintenance of long-term abstinence is currently seen by many as the biggest therapeutic hurdle in treating addictions (Heilig and Koob, 2007a; Barr et al., 2009). Given that OT appears to have acute inhibitory effects on the intake of alcohol, opiates and stimulants, it is interesting to speculate whether OT treatment might cause lasting changes in the disposition to consume drugs and alcohol. In some published studies the well-characterized anxiolytic and prosocial effects of OT are more apparent with repeated, rather than acute, administration. For example, chronic, but not acute, OT attenuated the pathological high anxiety of female rats selectively bred for high anxiety-related behavior (Slattery and Neumann, 2010) and repeated OT treatment had lasting positive effects on blood pressure, pain tolerance and corticosterone levels (Petersson et al., 1996; Petersson et al., 1999; Uvnas-Moberg, 1998a; Holst et al., 2002).

The ability of exogenous OT to cause enduring residual changes in behavioral and physiological

traits may reflect the inherent plasticity of OT neural systems. Heightened stimulation of hypothalamic OTRs triggers increased dendritic and peripheral release of OT (e.g., Figure 15.2C) that further stimulates OTRs establishing a positive feedback loop resulting in hypertrophy of the oxytocinergic neurons, decreased astrocytic coverage of neurons, and a subsequent increase in juxtaposition of neurons at the level of somas and dendrites across the entire OT system (for a review, see Theodosis, 2002). Ultimately, this process results in a lasting increase in the productivity and functionality of the OT system, the duration of which can last for months depending on the magnitude and length of the stimulation. Importantly, from a therapeutic perspective, these changes can be induced both *in vivo* and *in vitro* by administration of exogenous OT (Montagnese et al., 1990; Theodosis et al., 1986; Langle et al., 2003).

Results of a recent study from our laboratory provided intriguing preliminary evidence that giving adolescent rats repeated exposure to exogenous OT over a short period in adolescence can have subtle yet significant effects that last into adulthood (Bowen et al., 2011). These behavioral and neural changes included: reduced anxiety, increased sociability, inhibition of the development of alcohol consumption, increased OT in plasma, and upregulated OTR mRNA in the hypothalamus. These findings are of particular interest as they suggest short-term OT treatment is capable of causing lasting behavioral changes that are known to promote long-term abstinence and psychological wellbeing. Furthermore, they suggest that not only might OT have efficacy in treating already developed substance abuse (as previous studies have suggested), but it might also have prophylactic utility in that it appears to be able to cause long-term inhibition of the development of drug self-administration.

These findings warrant further exploration of the motivational and affective processes stimulated by OT that promote cessation of pathological drug consumption, restoration of psychological wellbeing, and maintenance of abstinence. As discussed earlier, it could be the case that OT is shifting people away from the drug of abuse and toward interpersonal relationships, thereby leading to a series of accumulating positive changes that make the individual stronger and less susceptible to drug abuse and the related psychopathologies that are known to promote drug abuse and hinder long-term abstinence.

15.7 Oxytocin as a candidate therapeutic for acute drug withdrawal/detoxification

Clinical approaches to alcohol and drug use disorders face two major challenges: the first is to safely detoxify users and ameliorate craving and emotional distress during acute withdrawal, while the second is to promote long-term abstinence in the months and years following withdrawal. The preclinical results discussed above offer preliminary hope for the second objective, but other intriguing evidence also speaks to the ability of OT to ameliorate physical and behavioral effects associated with acute drug withdrawal.

In seminal preclinical studies, OT was shown to reduce the severity of withdrawal after chronic high-dose administration of opiates (Kovacs et al., 1998), and ethanol (Szabo et al., 1987). A further key study indicated that the withdrawal symptoms arising from sudden precipitated cannabinoid abstinence in rats were reversed by administration of lithium, via an OT-dependent mechanism (Cui et al., 2001). Lithium produces pronounced activation of OT-positive hypothalamic nuclei and the ability of lithium to prevent abstinence was reversed by coadministration of the OT antagonist L-368,899. Moreover, L-368,899 given in the absence of lithium, exacerbated the cannabinoid-withdrawal syndrome. In an analogous fashion, symptoms arising from naloxone-precipitated withdrawal from morphine in mice were attenuated by coadministration of lithium via an OT-dependent mechanism (You et al., 2001).

These findings have prompted preliminary trials of lithium in treating cannabis withdrawal with encouraging results (Winstock et al., 2009) and

such trials could conceivably be extended to future examination of intranasal OT or other OT-releasing compounds as therapeutics. Indeed, a recent randomized, double-blind, placebo controlled clinical trial (Pedersen et al., 2012), in which intranasal OT or placebo was given twice daily for 3 days to alcohol-dependent subjects undergoing medical detoxification, speaks to the potential utility of treating addicted human populations with intranasal OT. The study found that intranasal OT markedly reduced the amount of lorazepam required for detoxification, lowered day 1 and day 2 scores on the Alcohol Withdrawal Symptom Checklist (AWSC) and the Clinical Institute Withdrawal Assessment for Alcohol (CIWA), and significantly reduced self-ratings of alcohol craving and anxiety.

Corticotropin-releasing factor (CRF) is widely considered to be a major determinant of dysphoria during drug withdrawal and CRF antagonists have a potent and well-documented effect in blocking withdrawal-induced anxiety and drug-seeking behavior in rodent models (Heilig and Koob, 2007b; Logrip et al., 2011). OT has marked inhibitory effects on activation of the hypothalamic pituitary adrenal axis (Legros, 2001; Windle et al., 2004; Uvnas-Moberg, 1998b) and strong anxiolytic and antidepressant-like effects, as discussed above. This is clearly worth consideration as the primary mechanism through which OT may ameliorate withdrawal symptoms during detoxification. The ability of OT to perhaps inoculate against future drug use, and rebalance affiliative brain processes, suggests that former drug users might emerge from the clinic following OT-assisted detoxification with an augmented chance of achieving long-term abstinence.

15.8 Conclusions and future directions

The nexus between OT and addiction has reached a most interesting stage of illumination with evidence coming from diverse sources that speaks to the role of the social neuropeptide in addiction-relevant behaviors (Sarnyai, 2011). The early work by Kovacs and colleagues showing that OT can prevent physiological tolerance to opiates, cocaine, and alcohol, was an important departure point, illustrating the ability of the neuropeptide to inhibit addiction-related neuroadaptations in mesolimbic and forebrain sites. A corollary of this is that the endogenous OT system is downregulated as a result of drug abuse and that such adaptations might drive not only tolerance, but also disinhibition of drug-seeking behavior and a loss of interest in social rewards. These predictions have been largely confirmed by the newer preclinical evidence for anticraving and inoculation-like effects of exogenous OT on drug-seeking behavior, although elucidation of specific neuroadpations underlying these effects requires further endeavour. Nonetheless, it has certainly reached the stage where clinical trials of intranasal OT in drug-dependent populations are warranted, and it is to be hoped that the results of such studies will be forthcoming sooner rather than later.

We must, however, face the possibility that the unfavourable physiochemical and pharmacokinetic properties of OT itself will limit its clinical use and that the future of OT-based therapeutics will lie with novel OTR agonists such as the recently developed WAY 267,464 (Ring et al., 2010; Hicks et al., 2012). Even then, there are issues with the ability of OTR agonists to cause what nature always intended OT to do: stimulate uterine contractions and the milk let down reflex. These are complicating side effects of OT agonists that may limit the therapeutic potential of OT, particularly in drug-dependent females of childbearing age. There are also interesting philosophical issues to consider relating to the ability of repeated OT treatment to sculpt personalities and proclivities: do we want to reach a stage where human teenagers are given OT to reduce their future likelihood of drug abuse? What will the world look like if everyone takes OT? One thing seems certain, OT and addiction is a research theme that seems destined to expand, and the future is likely to provide a wealth of exciting findings that will no doubt help to resolve many of the important questions raised in this review.

Acknowledgments

This work was supported by research funding to Iain McGregor from the Australian Research Council and the National Health and Medical Research Council. We are most grateful to our past and current collaborators in this area of research including Dean Carson, Paul Callaghan, Jennifer Cornish, Adam Guastella, Callum Hicks, Glenn Hunt, William Jorgensen, Michael Kassiou, Kirsten Morley, Petra van Nieuwenhuijzen, and Jonathon Arnold. We also acknowledge the generous gift of alcohol-preferring P rats from Professor Andrew Lawrence (Florey Institute, Melbourne).

REFERENCES

Addolorato, G., Leggio, L., Ferrulli, A., Cardone, S., Bedogni, G., Caputo, F., Gasbarrini, G., and Landolfi, R. (2011). Dose-response effect of baclofen in reducing daily alcohol intake in alcohol dependence: secondary analysis of a randomized, double-blind, placebo-controlled trial. *Alcohol and Alcoholism*, 46, 312–317.

Ameisen, O. (2011). High-dose baclofen for suppression of alcohol dependence. *Alcoholism, Clinical and Experimental Research*, 35, 845–846.

Amico, J. A., Vollmer, R. R., Cai, H. M., Miedlar, J. A., and Rinaman, L. (2005). Enhanced initial and sustained intake of sucrose solution in mice with an oxytocin gene deletion. *American Journal of Physiology. Regulatory, Integrative and Comparative Physiology*, 289, R1798–806.

Anacker, A. M. J. and Ryabinin, A. E. (2010). Biological contribution to social influences on alcohol drinking: evidence from animal models. *International Journal of Environmental Research and Public Health*, 7, 473–493.

Baracz, S. J., Rourke, P. I., Pardey, M. C., Hunt, G. E., McGregor, I. S., and Cornish, J. L. (2012). Oxytocin directly administered into the nucleus accumbens core or subthalamic nucleus attenuates methamphetamine-induced conditioend place preference. *Behavioural Brain Research*, 228, 185–193.

Barr, C. S., Dvoskin, R. L., Gupte, M., Sommer, W., Sun, H., Schwandt, M. L., Lindell, S. G., Kasckow, J. W., Suomi, S. J., Goldman, D., Higley, J. D., and Heilig, M. (2009). Functional CRH variation increases stress-induced alcohol consumption in primates. *Proceedings of the National Academy of Sciences*, 106, 14593–14598.

Baskerville, T. A. and Douglas, A. J. (2010). Dopamine and oxytocin interactions underlying behaviors: potential contributions to behavioral disorders. *Cns Neuroscience and Therapeutics*, 16, e92–e123.

Bedi, G., Hyman, D., and De Wit, H. (2010). Is Ecstasy an "empathogen"? Effects of +/− 3,4-methylenedioxymethamphetamine on prosocial feelings and identification of emotional states in others. *Biological Psychiatry*, 68, 1134–1140.

Bedi, G., Phan, K. L., Angstadt, M., and De Wit, H. (2009). Effects of MDMA on sociability and neural response to social threat and social reward. *Psychopharmacology*, 207, 73–83.

Bowen, M. T., Carson, D. S., Spiro, A., Arnold, J. C., and McGregor, I. S. (2011). Adolescent oxytocin exposure causes persistent reductions in anxiety and alcohol consumption and enhances sociability in rats. *PLoS One*, 6, e27237.

Broadbear, J. H., Tunstall, B., and Beringer, K. (2011). Examining the role of oxytocin in the interoceptive effects of 3,4-methylenedioxymethamphetamine (MDMA, 'ecstasy') using a drug discrimination paradigm in the rat. *Addiction Biology*, 16, 202–214.

Budisavljevic, M. N., Stewart, L., Sahn, S. A., and Ploth, D. W. (2003). Hyponatremia associated with 3,4-methylenedioxymethylamphetamine ("Ecstasy") abuse. *American Journal of the Medical Sciences*, 326, 89–93.

Butovsky, E., Juknat, A., Elbaz, J., Shabat-Simon, M., Eilam, R., Zangen, A., Altstein, M., and Vogel, Z. (2006). Chronic exposure to Delta9-tetrahydrocannabinol downregulates oxytocin and oxytocin-associated neurophysin in specific brain areas. *Molecular and Cellular Neurosciences*, 31, 795–804.

Caputo, F. (2011). Gamma-hydroxybutyrate (GHB) for the treatment of alcohol dependence: a call for further understanding. *Alcohol and Alcoholism*, 46, 3.

Carson, D. S., Cornish, J. L., Guastella, A. J., Hunt, G. E., and McGregor, I. S. (2010a). Oxytocin decreases methamphetamine self-administration, methamphetamine hyperactivity, and relapse to methamphetamine-seeking behaviour in rats. *Neuropharmacology*, 58, 38–43.

Carson, D. S., Hunt, G. E., Guastella, A. J., Barber, L., Cornish, J. L., Arnold, J. C., Boucher, A. A., and McGregor, I. S. (2010b). Systemically administered oxytocin decreases methamphetamine activation of the subthalamic nucleus and accumbens core and stimulates

oxytocinergic neurons in the hypothalamus. *Addiction Biology*, 15, 448–463.

Chaviaras, S., Mak, P., Ralph, D., Krishnan, L., and Broadbear, J. H. (2010). Assessing the antidepressant-like effects of carbetocin, an oxytocin agonist, using a modification of the forced swimming test. *Psychopharmacology*, 210, 35–43.

Chen, C., Lin, S., Sham, P., Ball, D., Loh, E., Hsiao, C., Chiang, Y., Ree, S., Lee, C., and Murray, R. (2003). Premorbid characteristics and co-morbidity of methamphetamine users with and without psychosis. *Psychological Medicine*, 33(8), 1407–1414.

Cornish, J. L., Shahnawaz, Z., Thompson, M. R., Wong, S., Morley, K. C., Hunt, G. E., and McGregor, I. S. (2003). Heat increases 3,4-methylenedioxymethamphetamine self-administration and social effects in rats. *European Journal of Pharmacology*, 482, 339–341.

Cui, S. S., Bowen, R. C., Gu, G. B., Hannesson, D. K., Yu, P. H., and Zhang, X. (2001). Prevention of cannabinoid withdrawal syndrome by lithium: involvement of oxytocinergic neuronal activation. *Journal of Neuroscience*, 21, 9867–9876.

Darke, S., Williamson, A., Ross, J., Teesson, M., and Lynskey, M. (2004). Borderline personality disorder, antisocial personality disorder and risk-taking among heroin users: findings from the Australian Treatment Outcome Study (ATOS). *Drug and Alcohol Dependence*, 74, 77–83.

Dawe, S., Davis, P., Lapworth, K., and Mcketin, R. (2009). Mechanisms underlying aggressive and hostile behavior in amphetamine users. *Current Opinion in Psychiatry*, 22, 269–273.

De Souza Villa, P., Menani, J. V., De Arruda Camargo, G. M., De Arruda Camargo, L. A., and Saad, W. A. (2008). Activation of the serotonergic 5-HT1A receptor in the paraventricular nucleus of the hypothalamus inhibits water intake and increases urinary excretion in water-deprived rats. *Regulatory Peptides*, 150, 14–20.

Dumont, G. J., Sweep, F., Van Der Steen, R., Hermsen, R., Donders, A. R. T., Touw, D. J., Van Gerven, J. M. A., Buitelaar, J. K., and Verkes, R. J. (2009). Increased oxytocin concentrations and prosocial feelings in humans after ecstasy (3,4-methylenedioxymethamphetamine) administration. *Social Neuroscience*, 4, 359–366.

Edwards, S. and Self, D. W. (2006). Monogamy: dopamine ties the knot. *Nature Neuroscience*, 9, 7–8.

Emiliano, A. B., Cruz, T., Pannoni, V., and Fudge, J. L. (2006). The interface of oxytocin-labeled cells and serotonin transporter-containing fibers in the primate hypothalamus: A substrate for SSRIs therapeutic effects? *Neuropsychopharmacology*, 32, 977–988.

Fasano, A. and Petrovic, I. (2010). Insights into pathophysiology of punding reveal possible treatment strategies. *Molecular Psychiatry*, 15, 560–573.

Garbutt, J. C., Kampov-Polevoy, A. B., Gallop, R., Kalka-Juhl, L., and Flannery, B. A. (2010). Efficacy and safety of baclofen for alcohol dependence: a randomized, double-blind, placebo-controlled trial. *Alcoholism, Clinical and Experimental Research*, 34, 1849–1857.

Geldenhuys, F. G., Sonnendecker, E. W. W., and De Klerk, M. C. C. (1968). Experience with sodium-gamma-4-hydroxybutyric acid (gamma-OH) in obstetrics. *Journal of Obstetrics and Gynaecology of the British Commonwealth*, 75, 405–413.

Gescheidt, T. and Bares, M. (2011). Impulse control disorders in patients with Parkinson's disease. *Acta Neurologica Belgica*, 111, 3–9.

Gibbens, G. L. and Chard, T. (1976). Observations on maternal oxytocin release during human labor and the effect of intravenous alcohol administration. *American Journal of Obstetrics and Gynecology*, 126, 243–246.

Gimpl, G. and Fahrenholz, F. (2001). The oxytocin receptor system: structure, function, and regulation. *Physiological Reviews*, 81, 629–683.

Goodson, J. L., Schrock, S. E., Klatt, J. D., Kabelik, D., and Kingsbury, M. A. (2009). Mesotocin and nonapeptide receptors promote estrildid flocking behavior. *Science*, 325, 862–866.

Grippo, A. J., Trahanas, D. M., Zimmerman, R. R., Ii, Porges, S. W., and Carter, C. S. (2009). Oxytocin protects against negative behavioral and autonomic consequences of long-term social isolation. *Psychoneuroendocrinology*, 34, 1542–1553.

Hargreaves, G. A., Hunt, G. E., Cornish, J. L., and McGregor, I. S. (2007). High ambient temperature increases 3,4-methylenedioxymethamphetamine-(MDMA, "ecstasy") induced Fos expression in a region-specific manner. *Neuroscience*, 145, 764–774.

Hargreaves, G. A. and McGregor, I. S. (2007). Topiramate moderately reduces the motivation to consume alcohol and has a marked antidepressant effect in rats. *Alcoholism, Clinical and Experimental Research*, 31, 1900–1907.

Heilig, M. and Koob, G. F. (2007a). A key role for corticotropin-releasing factor in alcohol dependence. *Trends in Neurosciences*, 30, 399–406.

Heilig, M. and Koob, G. F. (2007b). A key role for corticotropin-releasing factor in alcohol dependence. *Trends in Neurosciences*, 30, 399–406.

Hicks, C., Jorgensen, W., Brown, C., Fardell, J., Koehbach, J., Gruber, C. W., Kassiou, M., Hunt, G. E., and McGregor, I. S. (2012). The Nonpeptide Oxytocin Receptor Agonist WAY 267,464: Receptor-Binding Profile, Prosocial Effects and Distribution of c-Fos Expression in Adolescent Rats. *Journal of Neuroendocrinology*, 24, 1012–1029.

Hollander, E., Novotny, S., Hanratty, M., Yaffe, R., Decaria, C. M., Aronowitz, B. R., and Mosovich, S. (2003). Oxytocin infusion reduces repetitive behaviors in adults with autistic and Asperger's disorders. *Neuropsychopharmacology*, 28, 193–198.

Holst, S., Uvnas-Moberg, K., and Petersson, M. (2002). Postnatal oxytocin treatment and postnatal stroking of rats reduce blood pressure in adulthood. *Autonomic Neuroscience-Basic and Clinical*, 99, 85–90.

Huber, D., Veinante, P., and Stoop, R. (2005). Vasopressin and oxytocin excite distinct neuronal populations in the central amygdala. *Science*, 308, 245–248.

Hunt, G. E., McGregor, I. S., Cornish, J. L., and Callaghan, P. D. (2011). MDMA-induced c-Fos expression in oxytocin-containing neurons is blocked by pretreatment with the 5-HT-1A receptor antagonist WAY 100635. *Brain Research Bulletin*, 86, 65–73.

Ibragimov, R., Kovacs, G. L., Szabo, G., and Telegdy, G., (1987). Microinjection of oxytocin into limbic-mesolimbic brain structures disrupts heroin self-administration behavior: a receptor-mediated event? *Life Sciences*, 41, 1265–1271.

Jodogne, C., Tirelli, E., Klingbiel, P., and Legros, J. J. (1991). Oxytocin attenuates tolerance not only to the hypothermic but also to the myroelaxant and akinesic effects of ethanol in mice. *Pharmacology, Biochemistry and Behavior*, 40, 261–265.

Kirilly, E., Benko, A., Ferrington, L., Ando, R. D., Kelly, P. A., and Bagdy, G. (2006). Acute and long-term effects of a single dose of MDMA on aggression in Dark Agouti rats. *International Journal of Neuropsychopharmacology*, 9, 63–76.

Kita, I., Yoshida, Y., and Nishino, S. (2006). An activation of parvocellular oxytocinergic neurons in the paraventricular nucleus in oxytocin-induced yawning and penile erection. *Neuroscience Research*, 54, 269–275.

Klavir, O., Flash, S., Winter, C., and Joel, D. (2009). High frequency stimulation and pharmacological inactivation of the subthalamic nucleus reduces 'compulsive' lever-pressing in rats. *Experimental Neurology*, 215, 101–109.

Koerner, B. I. (2010). Secret of AA: After 75 years, we don't know how it works. *Wired*, June 23.

Kovacs, G. L., Sarnyai, Z., Babarczi, E., Szabo, G., and Telegdy, G. (1990). The role of oxytocin dopamine interactions in cocaine-induced locomotor activity. *Neuropharmacology*, 29, 365–368.

Kovacs, G. L., Sarnyai, Z., and Szabo, G. (1998). Oxytocin and addiction: a review. *Psychoneuroendocrinology*, 23, 945–962.

Kovacs, G. L. and Van Ree, J. M. (1985). Behaviorally active oxytocin fragments simultaneously attenuate heroin self-administration and tolerance in rats. *Life Sciences*, 37, 1895–1900.

Kovacs, G.L., Laczi, F., Vecsernyes, M., Hódi, K., Telegdy, G., and László, F. (1987). Limbic oxytocin and arginine 8-vasopressin in morphine tolerance and dependence. *Experimental Brain Research*, 65, 307–311.

Kramer, K. M., Choe, C., Carter, C. S., and Cushing, B. S. (2006). Developmental effects of oxytocin on neural activation and neuropeptide release in response to social stimuli. *Hormones and Behavior*, 49, 206–214.

Krause, E. G., De Kloet, A. D., Flak, J. N., Smeltzer, M. D., Solomon, M. B., Evanson, N. K., Woods, S. C., Sakai, R. R., and Herman, J. P. (2011). Hydration state controls stress responsiveness and social behavior. *Journal of Neuroscience*, 31, 5470–5476.

Langle, S. L., Poulain, D. A., and Theodosis, D. T. (2003). Induction of rapid, activity-dependent neuronal-glial remodelling in the adult rat hypothalamus in vitro. *European Journal of Neuroscience*, 18, 206–214.

Lee, P. R., Brady, D. L., Shapiro, R. A., Dorsa, D. M., and Koenig, J. I. (2005). Social interaction deficits caused by chronic phencyclidine administration are reversed by oxytocin. *Neuropsychopharmacology*, 30, 1883–1894.

Lee, Y., Hamamura, T., Ohashi, K., Fujiwara, Y., and Kuroda, S. (1999). The effect of lithium on methamphetamine-induced regional Fos protein expression in the rat brain. *Neuroreport*, 10, 895–900.

Legros, J.-J. (2001). Inhibitory effect of oxytocin on corticotrope function in humans: are vasopressin and oxytocin ying-yang neurohormones? *Psychoneuroendocrinology*, 26, 649–655.

Leone, M. A., Vigna-Taglianti, F., Avanzi, G., Brambilla, R., and Faggiano, F. (2010). Gamma-hydroxybutyrate (GHB) for treatment of alcohol withdrawal and prevention

of relapses. *Cochrane Database of Systematic Reviews*, CD006266.

Liberzon, I., Trujillo, K. A., Akil, H., and Young, E. A. (1997). Motivational properties of oxytocin in the conditioned place preference paradigm. *Neuropsychopharmacology*, 17, 353–359.

Liu, J. and Wang, L. (2011). Baclofen for alcohol withdrawal. *Cochrane Database of Systematic Reviews*, CD008502.

Logrip, M. L., Koob, G. F., and Zorrilla, E. P. (2011). Role of corticotropin-releasing factor in drug addiction: potential for pharmacological intervention. *CNS Drugs*, 25, 271–287.

Ludwig, M. and Leng, G. (2006). Dendritic peptide release and peptide-dependent behaviours. *Nature Reviews. Neuroscience*, 7, 126–136.

Lukas, M., Bredewold, R., Neumann, I. D., and Veenema, A. H. (2010). Maternal separation interferes with developmental changes in brain vasopressin and oxytocin receptor binding in male rats. *Neuropharmacology*, 58, 78–87.

Lukas, M., Toth, I., Reber, S. O., Slattery, D. A., Veenema, A. H., and Neumann, I. D. (2011). The neuropeptide oxytocin facilitates pro-social behavior and prevents social avoidance in rats and mice. *Neuropsychopharmacology*, DOI: 10.1038/npp.2011.95.

Macdonald, E., Dadds, M. R., Brennan, J. L., Williams, K., Levy, F., and Cauchi, A. J. (2011). A review of safety, side-effects and subjective reactions to intranasal oxytocin in human research. *Psychoneuroendocrinology*, 36, 1114–1126.

Mallet, L., Polosan, M., Jaafari, N., Baup, N., Welter, M. L., Fontaine, D., Du Montcel, S. T., Yelnik, J., Chereau, I., Arbus, C., Raoul, S., Aouizerate, B., Damier, P., Chabardes, S., Czernecki, V., Ardouin, C., Krebs, M. O., Bardinet, E., Chaynes, P., Burbaud, P., Cornu, P., Derost, P., Bougerol, T., Bataille, B., Mattei, V., Dormont, D., Devaux, B., Verin, M., Houeto, J. L., Pollak, P., Benabid, A. L., Agid, Y., Krack, P., Millet, B., and Pelissolo, A. (2008). Subthalamic nucleus stimulation in severe obsessive-compulsive disorder. *New England Journal of Medicine*, 359, 2121–2134.

Marvin, E., Scrogin, K., and Dudas, B. (2010). Morphology and distribution of neurons expressing serotonin 5-HT1A receptors in the rat hypothalamus and the surrounding diencephalic and telencephalic areas. *Journal of Chemical Neuroanatomy*, 39, 235–241.

Matsuoka, T., Sumiyoshi, T., Tanaka, K., Tsunoda, M., Uehara, T., Itoh, H., and Kurachi, M. (2005). NC-1900, an arginine-vasopressin analogue, ameliorates social behavior deficits and hyperlocomotion in MK-801-treated rats: therapeutic implications for schizophrenia. *Brain Research*, 1053, 131–136.

McGregor, I. S., Callaghan, P. D., and Hunt, G. E. (2008). From ultrasocial to antisocial: a role for oxytocin in the acute reinforcing effects and long-term adverse consequences of drug use? *British Journal of Pharmacology*, 154, 358–368.

McGregor, I. S., Gurtman, C. G., Morley, K. C., Clemens, K. J., Blokland, A., Li, K. M., Cornish, J. L., and Hunt, G. E. (2003). Increased anxiety and "depressive" symptoms months after MDMA ("ecstasy") in rats: drug-induced hyperthermia does not predict long-term outcomes. *Psychopharmacology*, 168, 465–474.

McMurray, M. S., Williams, S. K., Jarrett, T. M., Cox, E. T., Fay, E. E., Overstreet, D. H., Walker, C. H., and Johns, J. M. (2008). Gestational ethanol and nicotine exposure: effects on maternal behavior, oxytocin, and offspring ethanol intake in the rat. *Neurotoxicology and Teratology*, 30, 475–486.

Melis, M. R. and Argiolas, A. (2011). Central control of penile erection: a re-visitation of the role of oxytocin and its interaction with dopamine and glutamic acid in male rats. *Neuroscience and Biobehavioral Reviews*, 35, 939–955.

Mithoefer, M. C., Wagner, M. T., Mithoefer, A. T., Jerome, L., and Doblin, R. (2011). The safety and efficacy of +/− 3,4-methylenedioxymethamphetamine-assisted psychotherapy in subjects with chronic, treatment-resistant post-traumatic stress disorder: the first randomized controlled pilot study. *Journal of Psychopharmacology*, 25, 439–452.

Montagnese, C., Poulain, D. A., and Theodosis, D. T. (1990). Influence of ovarian steroids on the ultrastructural plasticity of the adult rat supraoptic nucleus induced by central administration of oxytocin. *Journal of Neuroendocrinology*, 2, 225–231.

Morley, K. C., Gallate, J. E., Hunt, G. E., Mallet, P. E., and McGregor, I. S. (2001). Increased anxiety and impaired memory in rats 3 months after administration of 3,4-methylenedioxymethamphetamine ("ecstasy"). *European Journal of Pharmacology*, 433, 91–99.

Morley, K. C. and McGregor, I. S. (2000). (+/−)-3,4-methylenedioxymethamphetamine (MDMA, 'Ecstasy') increases social interaction in rats. *European Journal of Pharmacology*, 408, 41–49.

Motbey, C. P., Hunt, G. E., Bowen, M. T., Artiss, S., and McGregor, I. S. (2011). Mephedrone

(4-methylmethcathinone, 'meow'): acute behavioural effects and distribution of Fos expression in adolescent rats. *Addiction Biology*, DOI: 10.1111/j.1369-1600.2011.00384.x.

Neumann, I. D. (2007). Stimuli and consequences of dendritic release of oxytocin within the brain. *Biochemical Society Transactions*, 35, 1252–1257.

Neumann, I. D., Torner, L., Toschi, N., and Veenema, A. H. (2006). Oxytocin actions within the supraoptic and paraventricular nuclei: differential effects on peripheral and intranuclear vasopressin release. *American Journal of Physiology. Regulatory, Integrative and Comparative Physiology*, 291, R29–36.

O'Shea, M., McGregor, I. S., and Mallet, P. E. (2006). Repeated cannabinoid exposure during perinatal, adolescent or early adult ages produces similar long-lasting deficits in object recognition and reduced social interaction in rats. *Journal of Psychopharmacology*, 20, 611–621.

O'Shea, M., Singh, M. E., McGregor, I. S., and Mallet, P. E. (2004). Chronic cannabinoid exposure produces lasting memory impairment and increased anxiety in adolescent but not adult rats. *Journal of Psychopharmacology*, 18, 502–508.

Olson, B. R., Drutarosky, M. D., Chow, M.-S., Hruby, V. J., Stricker, E. M., and Verbalis, J. G. (1991). Oxytocin and an oxytocin agonist administered centrally decrease food intake in rats. *Peptides*, 12, 113–118.

Pallanti, S., Bernardi, S., Raglione, L. M., Marini, P., Ammannati, F., Sorbi, S., and Ramat, S. (2010). Complex repetitive behavior: punding after bilateral subthalamic nucleus stimulation in Parkinson's disease. *Parkinsonism and Related Disorders*, 16, 376–380.

Parrott, A. C. (2004). MDMA (3,4-Methylenedioxymethamphetamine) or ecstasy: the neuropsychobiological implications of taking it at dances and raves. *Neuropsychobiology*, 50, 329–335.

Pedersen, C. A., Smedley, K. L., Leserman, J., Jarskog, L. F., Rau, S. W., Kampov-Polevoi, A., Casey, R. L., Fender, T., and Garbutt, J. C. (2012). Intranasal oxytocin blocks alcohol withdrawal in human subjects. *Alcoholism: Clinical and Experimental Research*, DOI: 10.1111/j.1530-0277.2012.01958.x.

Pedraza, C., Davila, G., Martin-Lopez, M., and Navarro, J. F. (2007). Anti-aggressive effects of GHB in OF.1 strain mice: Involvement of dopamine D(2) receptors. *Progress in Neuro-Psychopharmacology and Biological Psychiatry*, 31, 337–342.

Petersson, M., Alster, P., Lundeberg, T., and Uvnäs-Moberg, K. (1996). Oxytocin increases nociceptive thresholds in a long-term perspective in female and male rats. *Neuroscience Letters*, 212, 87–90.

Petersson, M., Hulting, A.-L., and Uvnäs-Moberg, K. (1999). Oxytocin causes a sustained decrease in plasma levels of corticosterone in rats. *Neuroscience Letters*, 264, 41–44.

Popeski, N. and Woodside, B. (2001). Effect of nitric oxide synthase inhibition on fos expression in the hypothalamus of female rats following central oxytocin and systemic urethane administration. *Journal of Neuroendocrinology*, 13, 596–607.

Porges, S. W. (2003). Social engagement and attachment: a phylogenetic perspective. *Annals of the New York Academy of Sciences*, 1008, 31–47.

Qi, J., Yang, J. Y., Song, M., Li, Y., Wang, F., and Wu, C. F. (2008). Inhibition by oxytocin of methamphetamine-induced hyperactivity related to dopamine turnover in the mesolimbic region in mice. *Naunyn-Schmiedeberg's Archives of Pharmacology*, 376, 441–448.

Qi, J., Yang, J. Y., Wang, F., Zhao, Y. N., Song, M., and Wu, C. F. (2009). Effects of oxytocin on methamphetamine-induced conditioned place preference and the possible role of glutamatergic neurotransmission in the medial prefrontal cortex of mice in reinstatement. *Neuropharmacology*, 56, 856–865.

Quinn, H. R., Matsumoto, I., Callaghan, P. D., Long, L. E., Arnold, J. C., Gunasekaran, N., Thompson, M. R., Dawson, B., Mallet, P. E., Kashem, M. A., Matsuda-Matsumoto, H., Iwazaki, T., and McGregor, I. S. (2008). Adolescent rats find repeated Delta(9)-THC less aversive than adult rats but display greater residual cognitive deficits and changes in hippocampal protein expression following exposure. *Neuropsychopharmacology*, 33, 1113–1126.

Ring, R. H., Schechter, L. E., Leonard, S. K., Dwyer, J. M., Platt, B. J., Graf, R., Grauer, S., Pulicicchio, C., Resnick, L., Rahman, Z., Sukoff Rizzo, S. J., Luo, B., Beyer, C. E., Logue, S. F., Marquis, K. L., Hughes, Z. A., and Rosenzweig-Lipson, S. (2010). Receptor and behavioral pharmacology of WAY-267464, a non-peptide oxytocin receptor agonist. *Neuropharmacology*, 58, 69–77.

Rossoni, E., Feng, J., Tirozzi, B., Brown, D., Leng, G., and Moos, F. (2008). Emergent synchronous bursting of oxytocin neuronal network. *PLoS Computational Biology*, 4, e1000123.

Rotzinger, S., Lovejoy, D. A., and Tan, L. A. (2010). Behavioral effects of neuropeptides in rodent models of depression and anxiety. *Peptides*, 31, 736–756.

Russell, J. A., Gosden, R. G., Humphreys, E. M., Cutting, R., Fitzsimons, N., Johnston, V., Liddle, S., Scott, S., and Stirland, J. A. (1989). Interruption of parturition in rats by morphine: a result of inhibition of oxytocin secretion. *Journal of Endocrinology*, 121, 521–536.

Sala, M., Braida, D., Lentini, D., Busnelli, M., Bulgheroni, E., Capurro, V., Finardi, A., Donzelli, A., Pattini, L., Rubino, T., Parolaro, D., Nishimori, K., Parenti, M., and Chini, B. (2011). Pharmacologic rescue of impaired cognitive flexibility, social deficits, increased aggression, and seizure susceptibility in oxytocin receptor null mice: a neurobehavioral model of autism. *Biological Psychiatry*, 69, 875–882.

Sanbe, A., Takagi, N., Fujiwara, Y., Yamauchi, J., Endo, T., Mizutani, R., Takeo, S., Tsujimoto, G., and Tanoue, A. (2008). Alcohol preference in mice lacking the Avpr1a vasopressin receptor. *American Journal of Physiology. Regulatory, Integrative and Comparative Physiology*, 294, R1482–1490.

Sarnyai, Z. (1998). Oxytocin and neuroadptation to cocaine. In: Urban, I., Burbach, J., and Wied, D. (eds.), *Advances in Brain Vasopressin*. Amsterdam, The Netherlands: Elsevier Science.

Sarnyai, Z. (2011). Oxytocin as a potential mediator and modulator of drug addiction. *Addiction Biology*, 16, 199–201.

Sarnyai, Z. and Kovacs, G. L. (1994). Role of oxytocin in the neuroadaptation to drugs of abuse. *Psychoneuroendocrinology*, 19, 85–117.

Scantamburlo, G., Hansenne, M., Fuchs, S., Pitchot, W., Marechal, P., Pequeux, C., Ansseau, M., and Legros, J. J. (2007). Plasma oxytocin levels and anxiety in patients with major depression. *Psychoneuroendocrinology*, 32, 407–410.

Schaller, F., Watrin, F., Sturny, R., Massacrier, A., Szepetowski, P., and Muscatelli, F. (2010). A single postnatal injection of oxytocin rescues the lethal feeding behaviour in mouse newborns deficient for the imprinted Magel2 gene. *Human Molecular Genetics*, 19, 4895–4905.

Schmidt-Mutter, C., Pain, L., Sandner, G., Gobaille, S., and Maitre, M. (1998). The anxiolytic effect of gamma-hydroxybutyrate in the elevated plus maze is reversed by the benzodiazepine receptor antagonist, flumazenil. *European Journal of Pharmacology*, 342, 21–27.

Silva, S. M., Madeira, M. D., Ruela, C., and Paula-Barbosa, M. M. (2002). Prolonged alcohol intake leads to irreversible loss of vasopressin and oxytocin neurons in the paraventricular nucleus of the hypothalamus. *Brain Research*, 925, 76–88.

Sivukhina, E. V., Dolzhikov, A. A., Morozov, I. E., Jirikowski, G. F., and Grinevich, V. (2006). Effects of chronic alcoholic disease on magnocellular and parvocellular hypothalamic neurons in men. *Hormone and Metabolic Research*, 38, 382–390.

Slattery, D. A. and Neumann, I. D. (2008). No stress please! Mechanisms of stress hyporesponsiveness of the maternal brain. *Journal of Physiology*, 586, 377–385.

Slattery, D. A. and Neumann, I. D. (2010). Chronic icv oxytocin attenuates the pathological high anxiety state of selectively bred Wistar rats. *Neuropharmacology*, 58, 56–61.

Subramanian, M. G. (1999). Alcohol inhibits suckling-induced oxytocin release in the lactating rat. *Alcohol*, 19, 51–55.

Sumnall, H. R., Woolfall, K., Edwards, S., Cole, J. C., and Beynon, C. M. (2008). Use, function, and subjective experiences of gamma-hydroxybutyrate (GHB). *Drug and Alcohol Dependence*, 92, 286–290.

Szabo, G., Kovacs, G. L., and Telegdy, G. (1987). Effects of neurohypophyseal peptide hormones on alcohol dependence and withdrawal. *Alcohol and Alcoholism*, 22, 71–74.

Theodosis, D. T. (2002). Oxytocin-secreting neurons: A physiological model of morphological neuronal and glial plasticity in the adult hypothalamus. *Frontiers in Neuroendocrinology*, 23, 101–135.

Theodosis, D. T., Montagnese, C., Rodriguez, F., Vincent, J. D., and Poulain, D. A. (1986). Oxytocin induces morphological plasticity in the adult hypothalamo-neurohypophysial system. *Nature*, 322, 738–740.

Thompson, M. R., Callaghan, P. D., Hunt, G. E., Cornish, J. L., and McGregor, I. S. (2007). A role for oxytocin and 5-HT(1A) receptors in the prosocial effects of 3,4 methylenedioxymethamphetamine ("ecstasy"). *Neuroscience*, 146, 509–514.

Thompson, M. R., Callaghan, P. D., Hunt, G. E., and McGregor, I. S. (2008). Reduced sensitivity to MDMA-induced facilitation of social behaviour in MDMA pre-exposed rats. *Progress in Neuro-Psychopharmacology and Biological Psychiatry*, 32, 1013–1021.

Thompson, M. R., Hunt, G. E., and McGregor, I. S. (2009). Neural correlates of MDMA ("Ecstasy")-induced social interaction in rats. *Social Neuroscience*, 4, 60–72.

Thompson, M. R., Li, K. M., Clemens, K. J., Gurtman, C. G., Hunt, G. E., Cornish, J. L., and McGregor, I. S. (2004).

Chronic fluoxetine treatment partly attenuates the long-term anxiety and depressive symptoms induced by MDMA ('Ecstasy') in rats. *Neuropsychopharmacology*, 29, 694–704.

Tirelli, E., Jodogne, C., and Legros, J. J. (1992). Oxytocin blocks the environmentally conditioned compensatory response present after tolerance to ethanol-induced hypothermia in mice. *Pharmacology, Biochemistry and Behavior*, 43, 1263–1267.

Uvnas-Moberg, K. (1998a). Antistress pattern induced by oxytocin. *News in Physiological Sciences*, 13, 22–25.

Uvnas-Moberg, K. (1998b). Oxytocin may mediate the benefits of positive social interaction and emotions. *Psychoneuroendocrinology*, 23, 819–835.

Uvnas-Moberg, K., Bjokstrand, E., Hillegaart, V., and Ahlenius, S. (1999). Oxytocin as a possible mediator of SSRI-induced antidepressant effects. *Psychopharmacology*, 142, 95–101.

Uvnas-Moberg, K., Bruzelius, G., Alster, P., and Lundeberg, T. (1993). The antinociceptive effect of non-noxious sensory stimulation is mediated partly through oxytocinergic mechanisms. *Acta Physiologica Scandinavica*, 149, 199–204.

Uvnas-Moberg, K., Hillegaart, V., Alster, P., and Ahlenius, S. (1996). Effects of 5-HT agonists, selective for different receptor subtypes, on oxytocin, CCK, gastrin and somatostatin plasma levels in the rat. *Neuropharmacology*, 35, 1635–1640.

Vaccari, C., Lolait, S. J., and Ostrowski, N. L. (1998). Comparative distribution of vasopressin V1b and oxytocin receptor messenger ribonucleic acids in brain. *Endocrinology*, 139, 5015–5033.

Van Nieuwenhuijzen, P., McGregor, I., and Hunt, G. (2009). The distribution of [gamma]-hydroxybutyrate-induced Fos expression in rat brain: Comparison with baclofen. *Neuroscience*, 158, 441–455.

Van Nieuwenhuijzen, P. S., Long, L. E., Hunt, G. E., Arnold, J. C., and McGregor, I. S. (2010). Residual social, memory and oxytocin-related changes in rats following repeated exposure to gamma-hydroxybutyrate (GHB), 3,4-methylenedioxymethamphetamine (MDMA) or their combination. *Psychopharmacology*, 212, 663–674.

Verty, A. N. A., Mcfarlane, J. R., McGregor, I. S., and Mallet, P. E. (2004). Evidence for an interaction between CB1 cannabinoid and oxytocin receptors in food and water intake. *Neuropharmacology*, 47, 593–603.

Volkow, N. D., Baler, R. D., and Goldstein, R. Z. (2011). Addiction: pulling at the neural threads of social behaviors. *Neuron*, 69, 599–602.

Williams, S. K., Cox, E. T., McMurray, M. S., Fay, E. E., Jarrett, T. M., Walker, C. H., Overstreet, D. H., and Johns, J. M. (2009). Simultaneous prenatal ethanol and nicotine exposure affect ethanol consumption, ethanol preference and oxytocin receptor binding in adolescent and adult rats. *Neurotoxicology and Teratology*, 31, 291–302.

Windle, R. J., Kershaw, Y. M., Shanks, N., Wood, S. A., Lightman, S. L., and Ingram, C. D. (2004). Oxytocin attenuates stress-induced c-fos mRNA expression in specific forebrain regions associated with modulation of hypothalamo-pituitary-adrenal activity. *Journal of Neuroscience*, 24, 2974–2982.

Winstock, A. R., Lea, T., and Copeland, J. (2009). Lithium carbonate in the management of cannabis withdrawal in humans: an open-label study. *Journal of Psychopharmacology*, 23, 84–93.

Witt, D. M., Winslow, J. T., and Insel, T. R. (1992). Enhanced social interaction in rats following chronic, centrally infused oxytocin. *Pharmacology, Biochemistry and Behavior*, 43, 855–861.

Wolff, K., Tsapakis, E. M., Winstock, A. R., Hartley, D., Holt, D., Forsling, M. L., and Aitchison, K. J. (2006). Vasopressin and oxytocin secretion in response to the consumption of ecstasy in a clubbing population. *Journal of Psychopharmacology*, 20, 400–410.

Yang, J. Y., Qi, J., Han, W. Y., Wang, F., and Wu, C. F. (2010). Inhibitory role of oxytocin in psychostimulant-induced psychological dependence and its effects on dopaminergic and glutaminergic transmission. *Acta Pharmacologica Sinica*, 31, 1071–1074.

You, Z. D., Li, J. H., Song, C. Y., Lu, C. L., and He, C. (2001). Oxytocin mediates the inhibitory action of acute lithium on the morphine dependence in rats. *Neuroscience Research*, 41, 143–150.

You, Z. D., Li, J. H., Song, C. Y., Wang, C. H., and Lu, C. L. (2000). Chronic morphine treatment inhibits oxytocin synthesis in rats. *Neuroreport*, 11, 3113–3116.

Young, K. A., Gobrogge, K. L., Liu, Y., and Wang, Z. (2011). The neurobiology of pair bonding: insights from a socially monogamous rodent. *Frontiers in Neuroendocrinology*, 32, 53–69.

16

Oxytocin and vasopressin in non-human primates

Benjamin J. Ragen and Karen L. Bales

16.1 Neuroanatomy of OT and AVP in non-human primates

16.1.1 OT and AVP immunoreactive cells and fibers in the brain, spinal cord, and pineal gland

Atunes and Zimmerman (1978) performed one of the first studies using antisera to AVP, OT, and neurophysins to visualize the OT and AVP system in the hypothalamus of non-human primates. They found that there were OT-immunoreactive cells (-ir) and AVP-ir neurons in the paraventricular nucleus (PVN) and the supraoptic nucleus of the hypothalamus (SON). Additionally, cells containing OT and AVP also contained the neurophysin estrogen-stimulated neurophysin (ESN) and nicotine-stimulated neurophysin (NSN), respectively. These two neurophysins are transporter molecules that are important for the trafficking of OT and AVP molecules after translation (de Bree, 2000). OT-ESN cells and AVP-NSN cells were evenly distributed in the PVN while they were localized within the SON.

Subsequent studies examining other non-human primate species found similar results. Currently, OT and AVP neurons have been visualized in three species of macaque (rhesus macaques: *Macaca mulatta*, crab-eating macaques: *Macaca fasciculata*, and Japanese macaques: *Macaca fuscata*) (Atunes and Zimmerman, 1978; Caffé et al., 1989; Ginsberg et al., 1994; Ichimiya et al., 1988; Kawata and Sano, 1982; Sladek and Zimmerman, 1982; Sofroniew et al., 1981), the New World squirrel monkey (*Saimiri sciureus*) (Sofroniew et al., 1981) and the common marmoset (*Callithrix jacchus*) (Wang et al., 1997a; Wang et al., 1997b). While there is great similarity between species in the location of OT and AVP cell bodies and fibers, species differences do appear to exist (Table 16.1 details abbreviations; Tables 16.2 and 16.3). It is sometimes difficult to determine whether these species differences are genuine due to the fact that different studies focus on specific brain areas, particular parts of cells, and use different methodologies. The PVN and SON contain both AVP-ir and OT-ir and/or NSN-ir and ESN-ir cells in all five primate species studied, while the SCN contains only parvocellular AVP-ir neurons (Caffé et al., 1989; Ginsberg et al., 1994; Kawata and Sano, 1982; Sofroniew et al., 1981; Wang et al., 1997a; Wang et al., 1997b). AVP neurons located in the SCN in rhesus macaques are in similar locations to those of the rat and golden hamster (Ueda et al., 1983).

There do appear to be some species differences in OT and AVP cells and fibers in non-human primates. Particularly interesting are comparisons between the polygynous macaques and squirrel monkeys, compared to the socially monogamous marmoset (we once again refer you to Tables 16.2 and 16.3). Only the marmoset had AVP-ir in

Oxytocin, Vasopressin, and Related Peptides in the Regulation of Behavior, ed. E. Choleris, D. W. Pfaff, and M. Kavaliers.

Table 16.1 Detailed abbreviations of brain areas referred to in the text and tables.

Abbreviations
AC = anterior commissure
Acc Hypoth = accessory hypothalamic nuclei
BNST = bed nucleus of the stria terminalis
CeA = central amygdala
CoA = cortical amygdala
DAH = dorsal anterior nucleus
DBB = diagonal band of Broca
DHS = dorsal horn of the spinal cord
DMN = dorsomedial nucleus
Ent = entorhinal cortex
GP = internal segment of the globus pallidus
Hippo = hippocampus
LC = locus coeruleus
LH = lateral hypothalamus
LHB = lateral habenula
LPB = lateral parabrachial nucleus
lPOA = lateral preoptic area
LS = lateral septum
Mamm = mammilary nuclei
MeA = medial amygdala
MPG = medial parabrachial nucleus
mPOA = medial preoptic area
MS = medial septum
NAcc = nucleus accumbens
NST = nucleus of the stria terminalis
NTS = nucleus of the solitary tract
PAG = periaquiductal grey
PEV = periventricular nucleus
PFA = perifornical area
PIC = pyriform cortex
pPOA = periventricular preoptic area
PVN = paraventricular nucleus
SCG = subcallosal cingulate gyrus
SCN = suprachiasmatic nucleus
SN = substantia nigra
SON = supraoptic nucleus
ST = stria terminalis
VHS = ventral horn of the spinal cord
VMN = ventromedial nucleus
VTA = ventral tegmental area

the nucleus accumbens (NAcc) (Wang et al., 1997a), which is a reward area that has been implicated in pair-bonding behavior in rodents (Aragona et al., 2006). In addition, in the marmoset there were sex differences in the number of AVP-ir cells in the stria terminalis, which had not been noted in previous studies of Old World monkeys (Wang et al., 1997a).

OT-ir neurons have also been found in the spinal cord of crab-eating macaques (Swanson and McKellar, 1979). Neurophysin-I and OT fibers show similar staining in the spinal cord with the densest staining between T4-L4 and moderate staining between S1-CO1. This differs from the patchier pattern of neurophysin-I and OT staining in the rat spinal cord. Evidence in rats shows that there is a paraventriculo-spinal tract that stains for neurophysin-I, which would indicate the presence of OT (Swanson, 1977). Squirrel monkeys and rhesus macaques also have fibers staining for neurophysin originating from the PVN and projecting to the brainstem, but these fibers were not examined in the spinal cord (Sofroniew et al., 1981). Further research is needed to elucidate the neuroanatomy of OT and AVP fibers in the spinal cord of non-human primates.

The above-mentioned studies did not report AVP or OT fibers innervating the pineal gland, however, there is evidence that AVP and OT fibers exist in this structure in non-human primates. Ronnekleiv (1988) tracked pineal innervations containing ESN, OT, NSN, and AVP in male and female adult macaques. Fibers containing ESN were found in the stria medullaris, LHB, pineal stalk, and the pineal organ itself. These fibers were located in the perivascular space and between pineal cells. OT fibers were also found in the perivascular space and between pineal cells but did not form the same network seen in ESN stained fibers. Fibers staining for NSN and AVP were similar to those staining for OT, but there were fewer fibers compared to ESN.

16.1.2 OT and AVP receptors in primate brain

OT receptors (OTR) have been difficult to visualize in the primate brain. It has been problematic to find

Table 16.2 Brain areas containing AVP-ir cell bodies, AVP-ir fibers, and AVP mRNA in the rhesus macaque, crab-eating macaque, Japanese macaque, marmoset, squirrel monkey, and rat. We summarize here for the purposes of easy comparison, however, we refer readers to the primary literature for more detail (especially for the rat, for which the literature is extensive).

	AVP					
	Rhesus Macaque	Crab-eating macaque	Japanese macaque	Marmoset	Squirrel Monkey	Rat
PVN	*§	*§†	*§	*§†	*§	*§
SON	*§	*§†	*§	*§†	*§	*§
SCN	*§	*§	*§	*†	*§	*§
Acc	*§	*§†	*§		*§	
Hypoth						
DAH				*†		
DMN		*§				*§
PEV		§	*§	*§		
VMN		§				
LH	*	*		*†		
pPOA				*		
mPOA		§				§
lPOA		§		§		
ST	§			*§	§	§
NST			*§			
BNST	*	*§		*§†		*§
LS	§			§	§	§
MS		§		§		
DBB		*§		§		§
NAcc				§		
AC				§		§
MeA	§	*§			§	*§
CeA	§	§			§	§
CoA		§				§
PFA	*	*	*§			
GP	*	*	*§			
SCG		§				
PIC		§				
Hipp	§	§			§	§
Ent		§				
Mamm		§				
LHB	§	§			§	§
VTA		§				§
SN		§				§
PAG		§				
LC		*§				*§
MPB		§				
LPB		§				
NTS		*§				§
DHS		*§				
VHS						

Key: * = cell bodies, § = fibers, † = mRNA.

References: Atunes and Zimmerman, 1978; Buijs et al., 1978; Caffé et al., 1989; DeVries et al., 1985; Ginsberg et al., 1994; Ichimiya et al., 1988; Kawata and Sano, 1982; Sofroniew et al., 1981; Swaab et al., 1975; Swanson and Sawchenko, 1983; Wang et al., 1997a; Wang et al., 1997b.

Table 16.3 Brain areas containing OT-ir cell bodies and OT-ir fibers in the rhesus macaque, crab-eating macaque, Japanese macaque, marmoset, squirrel monkey, and rat. We summarize here for the purposes of easy comparison, however, we refer readers to the primary literature for more detail (especially for the rat, for which the literature is extensive).

	OT					
	Rhesus Macaque	Crab-eating macaque	Japanese macaque	Marmoset	Squirrel Monkey	Rat
PVN	*§	*§	*§	*§	*§	*§
SON	*§	*§	*§	*§	*§	*§
SCN						
Acc	*§	*§	*§		*§	
Hypoth						
DAH				*		
DMN		§				
PEV		§	*§	*§		
VMN		§		§		
LH	*	*		*		
pPOA				*§		
mPOA		§				
lPOA		§				
ST	§			*§	§	§
NST			*§			
BNST	*	*				
LS						
MS		§				
DBB		§				
NAcc						
AC		§	*§			
MeA		§		*§		
CeA	§	§			§	§
CoA		§		§		
PFA	*	*	*§			
GP	*	*	*§			
SCG						
PIC						
Hipp						
Ent						
Mamm						
LHB		§				
VTA		§				
SN						
PAG		§				
LC		§				§
MPB		§				
LPB		§				
NTS		§				§
DHS		§				
VHS		§				

Key: * = cell bodies, § = fibers, † = mRNA.

References: Atunes and Zimmerman, 1978; Buijs et al., 1978; Caffé et al., 1989; Ginsberg et al., 1994; Ichimiya et al., 1988; Kawata and Sano, 1982; Sofroniew et al., 1981; Swaab et al., 1975; Swanson and Sawchenko, 1983; Wang et al., 1997a; Wang et al., 1997b.

specific antibodies for primate OTR (Toloczko et al., 1997), and the autoradiographic ligands that are used in rodents also bind to AVP receptors (AVPR1a) in primates. One study used the monoclonal antibody, 2F8, to visualize OT receptors in crab-eating macaques (Boccia et al., 2001). The ventral hypothalamic region was examined and OTR were found in the preoptic area and septal nucleus. Specifically, OTR were located on both the cell body and dendrites of the preoptic area but only neuronal fibers in the septal nucleus (Boccia et al., 2001). Another study used the peripherally administered OT antagonist, L368,899, to investigate the regions in which it accumulates in the brain, therefore suggesting the presence of OT receptors (Boccia et al., 2007). After administration, specific limbic regions were analyzed by liquid chromatography/mass spectrometry. L368,899 was found at detectible levels in the hypothalamus, orbitofrontal cortex, amygdala, hippocampus and septum. These results provide indirect evidence that OTR may be located in these brain regions. OTR distribution has also been visualized in the marmoset using ^{125}I-OTA autoradiography, while using a selective AVP agonist to displace binding to AVPR. Receptors were found in the limitans thalamic nucleus, NAcc, motor trigeminal nucleus, substantia gelatinosa of the spinal trigeminal nucleus, DBB, caudate/putamen, superior colliculus, inferior olive, and the dorsal motor nucleus (Schorscher-Petcu et al., 2009).

AVPR1a receptor distribution has been characterized in rhesus macaques (Young et al., 1999b) and common marmosets (Wang et al., 1997a, Schorscher-Petcu et al., 2009). In rhesus macaques, AVPR1a were found in the prefrontal cortex, cingulate cortex, insular cortex, Ent, PIC, LS, presubiculum, subiculum, VMN, Mamm, SON, SCN, BNST, CeA, MeA, DBB, LC, infundibulum, the medial and lateral parabrachial nuclei, principle inferior olive, and inferior olive. In marmosets AVPR1a receptors were located in the olfactory bulb, Islands of Calleja, SCN, BNST, DBB, LS, MeA, NTS, VMN, NAcc, arcuate nucleus, anterodorsal preoptic nucleus, choroid plexus, GP, basal nucleus of Meynert, layer VI in the cortex (Schorscher-Petcu et al., 2009; Wang et al., 1997a), and the prefrontal cortex (Kozorovitskiy et al., 2006). This differing distribution of AVPR1a, which was found mostly in cortical areas, could be due to differences in the promoter region of the *AVPR1a* gene (discussed below).

16.1.3 OT and AVP in cerebrospinal fluid (CSF) and plasma in primates

It is often more difficult to do terminal studies in non-human primates due to their relative scarcity and expensiveness compared to rodents. CSF is often used as a less-invasive alternative to investigate neuropeptide activity within the central nervous system. There is evidence that primates have a circadian rhythm in OT and AVP functioning demonstrated by fluctuating levels of OT and AVP in CSF (Artman et al., 1982; Perlow et al., 1982; Reppert et al., 1983; Reppert et al., 1984). CSF OT levels are three to twelve times higher during light hours compared to dark hours. AVP levels in CSF are also higher during the light hours compared to dark hours, however, the change in levels is not as great as that of OT (Amico et al., 1989; Reppert et al., 1984). The OT rhythm is maintained even when monkeys are housed in constant dark or constant light, and a combination of a 12-hour shift of feeding, care activity, and light are needed to shift this rhythm (Artman et al., 1982). This shift in rhythm takes a few days and the peaks of OT are not as large in comparison to housing under the original lighting schedule (Artman et al., 1982).

Levels of OT in plasma do not follow the same pattern as OT in CSF (Perlow et al., 1982). However, this is not true in rhesus macaques in late pregnancy. Pregnant rhesus experience greater plasma OT during the night compared to day, and it is this increase in OT during the night that is responsible for the increase of uterine contractions (Hirst et al., 1991). ESN follows a similar rhythm as OT, and like OT these fluctuations are only seen in CSF (Perlow et al., 1982). The OT circadian rhythm has been found to be strongest in the lumbar region of the spinal cord (Amico et al., 1989). It is possible that the stronger rhythm is due to a greater number of OT fibers in the

lumbar and thoracic region (Swanson and McKellar, 1979), and local release of OT from these neurons drive the fluctuations in CSF.

Unlike primates, rats do not have a circadian pattern in CSF OT. However, they do have circadian fluctuations in CSF AVP like primates with levels higher during light hours compared to dark hours (Reppert et al., 1983). Unlike the pattern of CSF OT in primates, which has been consistently found, the circadian rhythm of CSF AVP in primates is not as reliable. In one study, the circadian rhythm of CSF AVP was found in less than half of the subjects (Amico et al., 1989). It is difficult to determine the reason for this due to the fact that different macaque species were used, some subjects were ovariectomized, some subjects were ovariectomized and adrenalectomized, and there were only two to five subjects in each group. One factor in this study that differed from others is that these subjects were allowed to move freely, while previous studies had subjects restrained in chairs (Amico et al., 1989; Reppert et al., 1983). It is possible that the stress of restraint may have a role in temporal rhythms of AVP.

The neurobiological regulators of the CSF OT circadian rhythm are currently unknown. Ablations to the SCN, which is a primary component of regulating circadian rhythms (Van Esseveldt et al., 2000), appear to have little or no effect on the rhythmicity of OT levels in CSF (Reppert et al., 1984). The opioid system has been shown to regulate the OT system in non-human primates (Kowalski et al., 1998), however, administration of the opioid antagonist, naloxone, has no effect on the circadian rhythm of OT (Amico et al., 1992). In addition, neither adrenalectomies, castrations, macaque species, nor sex have any effect on the temporal rhythm of CSF OT (Amico et al., 1989). It is possible that the circadian rhythm is driven by the PVN. As discussed before, OT neurons in the spinal cord may originate in the PVN forming part of the paraventriculo-spinal tract (Swanson, 1977). In humans and tree shrews neurophysin projections originating in the PVN have been found to extend to the spinal cord, providing evidence that a paraventriculo-spinal tract containing OT and AVP may exist in non-human primates (Sofroniew et al., 1981). Neuroanatomical studies could demonstrate this, and experiments lesioning the PVN could test whether the CSF OT and AVP circadian rhythms are driven by the PVN instead of the SCN.

16.2 Hormonal modulation of OT and AVP in non-human primates

Neuroanatomical, pharmacological, and observational studies suggest that the OT and AVP systems in non-human primates may be modulated by various hormonal and neurotransmitter systems. Estrogen affects central OT activity in several species (Gimpl and Fahrenholz, 2001; McCarthy, 1995; Witt et al., 1991). Natural fluctuations of estrogen and progesterone are observed in the non-human primate menstrual cycle (Michael et al., 1982). This natural variation in steroid hormones provides an opportunity to study how OT may be correlated with these hormonal systems. Falconer et al. (1980) examined plasma OT during the menstrual cycle of rhesus macaques. Changes in plasma OT existed throughout the menstrual cycle with the highest levels of OT occurring in mid-cycle and OT decreasing as progesterone increased.

Neuroanatomical studies have demonstrated that there are interactions between catecholamines, specifically norepinepherine, and OT and AVP cells in the hypothalamus (Sladek and Zimmermann, 1982; Ginsberg et al., 1994). In rhesus macaques, catecholaminergic varicosities innervate the SON, and these varicosities are colocalized with both AVP and OT cells. The SON colocalizations are more frequent with AVP cells compared to OT cells (Sladek et al., 1982). This study was unfortunately unable to distinguish whether these varicosities were dopaminergic or noradrenergic in nature. Ginsberg et al. (1994) clarified this question by staining for dopamine-β-hydoxylase (DβH), an enzyme that converts dopamine into norepinepherine. In the PVN, DβH-ir varicosities were colocalized with AVP and OT cell bodies and dendrites.

Significantly more DβH-ir varicosities were colocalized with AVP dendrites compared to AVP cell bodies or OT cell bodies and dendrites.

There is evidence that DA may modulate OT and AVP activity (Cameron et al., 1992; Amico et al., 1993), as it does in rats (Melis et al., 1989). Amico et al. (1993) found that apomorphine, a DA agonist for both D1 and D2 receptors, increases both plasma OT (Cameron et al., 1992; Amico et al., 1993) and AVP in rhesus and crab-eating macaques with a greater effect on AVP (Amico et al., 1993). The D2 agonist, LY 163502, also increased plasma OT (Cameron et al., 1992; Amico et al., 1993) and AVP although the D1 agonist, CY 208–243, failed to do so (Cameron et al., 1992; Amico et al., 1993). No sex difference was found in this effect (Amico et al., 1993). Both apomorphine and LY 163502 resulted in an increase in yawning, oral dyskenesia, and hypermobility, and these effects were dose dependent. Additionally, the doses of apomorphine that resulted in greater increases of OT had the most substantial effects on behavior, although behavioral effects were seen without a subsequent increase in OT (Cameron et al., 1992). These findings indicate that it is primarily D2 receptors that influence OT and AVP release and that D1 receptors have little or no effect.

One pharmacological study has found that the opioid system can affect OT in pregnant baboons and can in turn affect uterine contractions (Kowalski et al., 1998). Morphine administration to baboons entering labor results in a decrease of plasma OT and a subsequent decrease in frequency and intensity of uterine contractions. This effect was specifically due to changes in OT release because morphine administration had no effect on OT metabolism (Kowalski et al., 1998). One neuroanatomical study also suggests that there may be an interaction between the opioid system and AVP. Immunocytochemistry visualizing opioid peptide (OP) neurons and AVP containing neurons in juvenile crab-eating macaques found that OP-ir afferents are located on AVP cell bodies (Goldsmith et al., 1991). OP-ir innervations of AVP-ir neurons are located primarily on parvocellular neurons in both the PVN and SON but with more innervations in the PVN.

Research on free-ranging rhesus macaques also demonstrate that genetic variations in the *OPRM1* gene, which codes for the μ opioid receptor, has an effect on OT in the CSF (Hingham et al., 2011). A C77G single nucleotide polymorphism of the *OPRM1* gene exists in rhesus macaques, and individuals containing the G allele have greater binding affinity for β-endorphins (Bond et al., 1998). Interestingly, the effects of having the G allele on CSF OT levels are dependent on the reproductive status of a female. Lactating females with one copy of the G allele have higher levels of CSF OT, while non-pregnant, non-lactating females with one copy of the G allele have lower levels of CSF OT (Hingham et al., 2011).

16.3 OT, AVP, and social behavior in non-human primates

16.3.1 Social behavior and group living in non-human primates

Social touch plays a critical role in non-human primate societies (Dunbar, 2010). Time spent grooming is significantly correlated with group size in various primate taxa (Lehmann et al., 2007). This suggests that grooming is important for maintaining social cohesion in large groups of individuals. Social integration, especially time spent receiving grooming can predict infant survival in baboons (Silk et al., 2003). It is believed that oxytocin is a primary physiological component relating grooming and social cohesion. Touch and warmth have been shown to release OT as well as emulate its effects (Uvnas-Moberg, 1998). It is likely that the OT release during grooming enhances sociability and group cohesion in primate species (Dunbar, 2010).

Species-level differences in CSF OT levels are reflective of differences in social relationships. Bonnet macaques (*Macaca radiata*) are considered more gregarious than pigtail macaques (*Macaca nemestrina*) reflected by more passive contact and

allomaternal behavior (Laudenslager et al., 1990; Rosenblum et al., 1964). Rosenblum et al. (2002) found that the bonnet macaque has significantly higher CSF OT compared to the pigtail macaque. This relationship may not hold true in all cases, particularly within certain species. In free-ranging female rhesus macaques, no correlation between social behaviors and levels of OT or AVP in CSF has been found (Schwandt et al., 2007). However, there is a significant negative correlation between plasma OT and fearful behavior indicated by monkeys low in plasma OT engaging in more fear-grimaces (Schwandt et al., 2007).

A few studies have manipulated the OT and AVP system to see how it affects social behavior in non-human primates. Winslow and Insel (1991b) found that central OT administration in male squirrel monkeys increased aggression and sexual behavior, but this was found only in dominant males. In contrast, OT increased affiliative/associative behaviors and marking behaviors in subordinates. When AVP was administered, there was a decrease in both aggressive and associative behavior in both dominant and subordinate males. The effects of central OT and AVP administration were also examined during separation distress in squirrel monkeys (Winslow and Insel, 1991a). Upon separation, AVP increased autogrooming and scent marking in both dominant and subordinate monkeys. Both OT and AVP decreased isolation peeps and vigilance checking upon separation.

16.3.2 Monogamy and adult attachment

Social monogamy is a mating system characterized by groups consisting of a mated male and female, which display territorial aggression and often biparental care (Kleiman, 1977). OT and AVP have been studied in the context of adult attachment using socially monogamous primates such as tamarins and marmosets (Goldizen, 1990). Snowdon et al. (2010) found correlations between affiliative behavior and urinary oxytocin in cotton-top tamarins (*Sanguinus oedipus*). Pairs that ranked higher in affiliation had higher levels of urinary OT compared to pairs that were ranked as less affiliative. There were no sex differences in levels of OT, but interestingly the type of affiliative behavior that explained the variance in OT was sex specific. For males, variance in OT levels was explained through sexual behavior, while female variance in OT levels was best explained through the duration and frequency of affiliative behaviors such as huddling.

OT and AVP are also affected by separation and reunion with a pair-mate. Common marmoset (*Callithrix jacchus*) males who undergo 48 h of isolation and are later reunited with their attachment figure experience an increase in urinary OT and AVP upon visual contact (Seltzer and Ziegler, 2007). There is no further change when the males are granted physical contact through a mesh screen or when fully reunited with their pair-mate. These results could indicate that visual contact upon reunion increases OT and AVP and reflects social affiliation. However, the increase in OT and AVP was also associated with an increase in cortisol. This may suggest that visual contact without physical contact could be a stressor or result in overall increased feelings of arousal.

OT has also been manipulated in order to examine its effects on pair-bond formation in both male and female black-tufted ear marmosets (*Callithrix penicillata;* Smith et al., 2010). Upon pairing, subjects were given daily administration of intranasal OT or the oral OT antagonist, L368,899. A partner preference test was performed 24 h and 3 weeks after the day of pairing, and affiliative and agonistic behaviors were measured throughout the 3 weeks of cohabitation. During the 3-week cohabitation, OT increased the initiation of huddling, while the OT antagonist decreased proximity and food sharing. OT manipulation had no effect at the 24-h partner preference test, but at the three-week partner preference test chronically OT-treated animals approached their pair-mate first, and the latency to approach their partner compared to a stranger and neutral cage was lower. This provides evidence that OT is important in pair-bond formation in both male and female marmosets.

Manipulation of the AVP system has also been found to affect pair-bonding maintenance in titi

monkeys (Jarcho et al., 2011). The titi monkey (*Callicebus cupreus*) is a socially monogamous New World monkey (Mason, 1966). High doses of intranasal AVP resulted in males approaching their pair-mate more frequently compared to a novel female. AVP also resulted in a decrease in inflammatory gene expression. It is possible that the decrease in inflammatory gene expression was mediated by an increase in cortisol, or alternatively through an anti-inflammatory response secondary to the affiliate behavior.

Whether or not male titi monkeys were pair bonded was predictive of brain glucose metabolism in areas that are likely to contain OT and AVP receptors in that species (Bales et al., 2007). Specifically, lone males differed from males in long-term pair-bonds in the NAcc, MeA, mPOA, LS, and ventral pallidum. They also differed in the SON, although not in the PVN. These findings suggest differential brain activity due to social bonding in many of the same brain areas that have been implicated in rodent studies.

16.3.3 Parental and alloparental behavior in non-human primates

Both AVP and OT have been shown to be involved in parental behavior in rodent models (Bales et al., 2004; Francis and Champagne, 2000; Pedersen, 1997; Wang et al., 1994), as well as in non-human primates. An early experimental manipulation of OT involved central administration to two nulliparous rhesus monkeys (Holman and Goy, 1995). OT administration increased affiliative behavior towards novel infants. Administration of the OT antagonist, L368,899, to a nulliparous female rhesus macaque resulted in a decrease in interest in a novel infant demonstrated by a decrease in lip smacking and looking at an infant (Boccia et al., 2007).

Plasma OT was found to be positively correlated with maternal warmth in free-ranging macaques (Maestripieri et al., 2009), however, levels of CSF OT were not related to maternal behavior (Hingham et al., 2011). This negative finding could be due to behavioral data and CSF being collected at different time points (Hingham et al., 2011). CSF OT in adult female rhesus mothers has been found to be negatively correlated with the age of their infant, therefore mothers with younger infants have greater levels of CSF OT (Parker et al., 2010). It is possible that higher levels of OT are needed in order to provide appropriate care to young infants. However, there appears to be no difference in CSF OT between lactating and non-lactating females. This agrees with previous research that failed to find significant differences in plasma OT between lactating and non-lactating female macaques ($p = 0.07$) (Maestripieri et al., 2009). Lactating and non-lactating rhesus macaques are also similar in that their pulsatile pattern of CSF OT is the same (Challinor et al., 1992). These negative results may be due to the timing of sampling. There are changes in plasma OT in lactating females but this only occurs during nursing bouts. These fluctuations end the day after weaning (Amico et al., 1990).

There is some indirect evidence that exposure to infants can result in a release of OT in non-human primates. Multiparous rhesus macaques exposed to strange infants will adopt them and be affiliative, which is in contrast with nulliparous females who do not associate with novel infants (Holman and Goy, 1980). A subset of the females that expressed this maternal behavior also lactated. Since OT release is important for milk let-down (Gimpl and Fahrenholz, 2001) this provides indirect evidence that multiparous female macaques experience a release of OT upon exposure to infants. Whether OT release is also found centrally is unknown. Interestingly, these multiparous females included intact, ovariectomized, and menopausal females. This suggests that steroid hormones are not necessary for the expression of maternal behavior toward novel infants in rhesus macaques (Holman and Goy, 1980).

A confounding factor of the previous study is that the multiparous females experienced different rearing conditions compared to nulliparous (Holman and Goy, 1980). Multiparous females had experienced caring for younger siblings prior to sexual maturity, while nulliparous females did not have this experience. In a follow-up study captive and wild nulliparous females were used, and the

wild nulliparous populations did express adopting behaviors toward infants. Mammary activity was not measured (Holman and Goy, 1995).

Cooke et al. (1997) examined levels of CSF OT in females before, immediately after, and seven days after parturition. Although there was no change in CSF OT at these different time points, which agrees with previous research, no correlation between OT levels and maternal care was found. This contrasts with findings that plasma OT was positively correlated with maternal warmth (Maestripieri et al., 2009), but agrees with other research (Hingham et al., 2011). One possible explanation is that central and peripheral levels of OT are not correlated and are representing different processes (Amico et al., 1990). Another possibility is that rearing conditions affected outcomes, with one study being done in free-ranging macaques (Maestripieri et al., 2009), and the other in captive macaques (Cooke et al., 1997). Rearing experience has been shown to affect levels of CSF OT with macaques experiencing early adversity showing lower levels of CSF OT compared to those that experienced maternal rearing (Winslow et al., 2003)

In many monogamous species such as marmosets, titi monkeys, and owl monkeys (Genus *Aotus*), males participate in infant care (Goldizen, 1990; Mendoza and Mason, 1986; Wright, 1984). There is evidence that AVP may be involved in paternal behavior in marmosets. Upon fatherhood, marmosets experience an increase in AVPR1a in the prefrontal cortex (Kozorovitskiy et al., 2006). AVPR1a density is also negatively correlated with the age of the youngest offspring, suggesting that AVP is especially important when an offspring is very young.

OT may also play a role in paternal behavior. A study by Saito and Nakamura (2011) tested marmoset fathers with their infants in a food-sharing task and examined how much food the father was willing to share with his infant. Willingness to share food acted as a measure of paternal tolerance. After receiving an intracerebroventricular injection of OT, fathers reduced how frequently they refused to share food with their offspring. This suggests that OTR activation can result in an increase in tolerance toward offspring. Administration of an OTR antagonist had no effect on this behavior. It is possible that, as in male prairie voles (Bales et al., 2004), certain types of male parenting behavior can be facilitated through either the OT or the AVP receptor system, requiring blockade of both types of receptors in order to reduce the occurrence of the behavior. Some socially monogamous primates such as marmosets, tamarins, and titi monkeys are also cooperative breeders with siblings engaging in alloparental care (Goldizen, 1990; Wright, 1984). In titi monkey siblings, plasma AVP at the second month after birth of an infant is negatively correlated with carrying the infant on postnatal weeks two to four (Hostetler et al., 2007). Plasma AVP at the second month after birth as well as the mean plasma OT for six months postbirth is positively correlated with face licking at postnatal weeks two to four (Hostetler et al., 2007). These results indicated that there is a complex interaction between neuropeptide levels, type of alloparental care, and infant development.

16.4 Early experience, social behavior, and the OT/AVP systems in non-human primates

16.4.1 Early experience and OT/AVP in non-human primates

Early experience has an effect on neuropeptide levels as an adult as well as social behavior. Nursery-reared (NR) rhesus macaques have decreased levels of CSF OT as adults compared to mother reared (MR) (Winslow et al., 2003). However, there is no difference in CSF AVP or plasma OT. NR monkeys were also less affiliative and more aggressive. CSF OT was positively correlated with affiliative behaviors, but this was irrespective of rearing history. CSF AVP, however, was negatively correlated with fear behavior specifically fear grimaces.

16.4.2 Aggression and the OT/AVP systems in non-human primates

AVP has been implicated in aggression and territoriality in many species (Bester-Meredith and Marler, 2001; Ferris et al., 1997; Winslow et al., 1993) but there is little evidence investigating the

role of AVP in aggression in non-human primates. One study showing a direct relationship between AVP and aggression found that central administration of AVP in squirrel monkeys decreases aggressive behavior (Winslow and Insel, 1991a). Rank had an effect on the responsiveness to treatment. Dominant males experienced a decrease in aggression with low and high doses, while only high doses were able to decrease aggression in subordinate males. This is in contrast to rodent studies where AVP has been shown to stimulate aggressive behavior (Delville et al., 1996).

A study by Rilling and colleagues (2004) allowed dominant males to engage in sexual behavior with a female and later allowed the male to view the female by herself, or to view a subordinate male engaging in sexual behavior with the same female. This was an attempt to emulate a model of jealousy-induced aggression. Subjects then underwent PET scanning using 18F-FDG to measure glucose uptake in the brain. Relative to watching the female alone, viewing the subordinate male with the female resulted in glucose uptake in the right amygdala, extended amygdala, right temporal pole, and bilateral insular cortex. These data indicate that these brain areas may be involved in anxiety or jealous aggression in male rhesus macaques. AVPR1a receptors are located in some of these same regions such as the amygdala and insular cortex (Young et al., 1999b). There are numerous studies linking aggression and AVP in rodent models (Delville et al., 1996; Ferris et al., 1997), and this remains an open area of research in non-human primates.

16.4.3 Sexual behavior and OT/AVP in non-human primates

OT and AVP are involved in sexual behavior in many species (Carter, 1992; Smock et al., 1998), and have received some attention in primate models. In one study, a multiparous rhesus female given the OT antagonist L368,899 resulted in an elimination of hip-grab behavior, an increased latency for the male to ejaculate, and an increase in the duration of refusing to be mounted and incomplete mounts (Boccia et al., 2007). When a male rhesus macaque observed a female sexual partner (described above) there was an increase in glucose uptake in the MeA demonstrated through PET scanning (Rilling et al., 2004). AVP in the MeA is associated with sexual behavior in many mammalian and non-mammalian species (Smock et al., 1998).

16.4.4 Genetics and heritability of AVP and OT in non-human primates

Recently, there has been intense interest in the *AVPR1a* gene that encodes for the AVPR1a receptor. However, the majority of the research has focused on rodents, specifically those in the genus *Microtus* (Fink et al., 2006; Hammock and Young, 2002; Nair and Young, 2005). Most of the research on the *AVPR1a* gene has focused on variability in the upstream promoter region, which affects social behavior and the distribution of AVPR1a receptors in the brain (Hammock and Young, 2002; Nair and Young, 2005). The monogamous prairie vole has different expression of the AVPR1a receptor in certain brain regions compared to the polygamous montane vole and this has been associated with different lengths of the *AVPR1a* promoter region (Nair and Young, 2005). Viral vectors of the prairie vole *AVPR1a* gene, including its promoter region, into a mouse results in AVPR1a expression in the brain similar to the prairie vole and an increase in sociality (Young et al., 1999a).

The length of the microsatellite region in the promoter appears to dictate AVPR1a expression and behavior in prairie voles, but not perhaps in other species (Hammock and Young, 2005; Fink et al., 2006). Prairie voles containing the long allele in the promoter region have greater AVPR1a receptor expression in the olfactory bulb and LS, two areas that are important for social recognition (Hammock and Young, 2005). Only males with the long allele form a partner preference (under certain test conditions), and pairs containing the long allele exhibit more licking and grooming of pups compared to pairs containing the short allele (Hammock and Young, 2005).

Due to these dramatic effects of promoter length on AVPR1a expression and behavior, this area of the

AVPR1a receptor gene has been studied in primates. There appears to be great variation in the promoter region across various primate taxa, however, a relationship with social behavior is less clear. Donaldson et al. (2008) examined two ~350 bp tandem duplication microsatellite-containing elements in the 5′ flanking region of the *AVPR1a* gene. One duplication (DupA) is located –3730 bp upstream and the second duplication (DupB) is located –3382 bp upstream from the transcription start site. There was a focus on the microsatellite RS3 region located within DupB due to variation in this area in humans and its relationship to social behavior (Walum et al., 2008). A comparison of these two duplications was performed in 13 primate species including six species of apes, four Old World monkey species, two New World monkey species, and one species of strepsirhine, the northern greater galago (*Otolemur garnettii*). Gibbon *AVPR1a* genes contained only DupA, monkeys contained a microsatellite region orthologous to DupB in great apes, and all great apes contained both duplicates, except chimpanzees that were polymorphic for DupB. No pattern was found between microsatellite-containing elements and social organization between species.

Hong et al. (2009) also focused on *AVPR1a* gene variation in 13 primate species including humans, six species of ape, five Old World monkey species, and one New World monkey species. Unlike Donaldson et al. (2008) the RS1 repeat sequence was examined instead of the RS3 sequence. In addition to determining the genotype of various species of primates, they examined whether the length of the RS1 sequence affected gene expression *in vitro* by looking at luciferase expression. In voles, the length of the microsatellite promoter region can affect luciferase expression (Hammock and Young, 2005; Nair and Young, 2005). However, when using the human and chimpanzee regulatory region the length of the promoter region had no effect on luciferase expression. Like the monogamous prairie vole, gibbons and marmosets were expected to have longer alleles due to their monogamous social structure (Kappeler and van Schaik, 2002). Variability was found between species, but the length of this region was not related to social organization.

Rosso et al. (2008) examined two dinucleotide microsatellites located –3956 and –3625 bp upstream from the transcription start site. Species included three monkey species and nine ape species, which also originated from different locations. These microsatellite areas correspond to the DupA and DupB microsatellite-containing elements, respectively (Donaldson et al., 2008). Like previous studies, variation was found between primate species, but this variation did not reflect male mating patterns. These consistent findings suggest that maybe both the non-repetitive and repetitive sequences in the promoter region are important for the social organization in primates.

Although the previous studies included a couple of New World monkey species, the focus was predominantly on apes and cercopithocenes, a group of Old World monkeys. Only the promoter region of the *AVPR1a* gene was examined. In contrast, Babb et al. (2010) investigated individual differences in both the non-coding and coding region of the *AVPR1a* gene in the monogamous owl monkey (*Aotus azarai*) and three other New World monkeys. Species included two monogamous species, titi monkeys (*Callicebus donacophilus*) and saki monkeys (*Pithecia pithecia*), as well as polygamous squirrel monkeys. All New World monkeys showed a GACA repeat pattern in RS1, which is in contrast to the GATA repeat pattern found in Old World monkeys and apes (Donaldson et al., 2008; Rosso et al., 2008). Owl monkeys were unique in that there was conservation in the 5′ upstream region, which is different from prairie voles (Nair and Young, 2005) and the RS3 microsatellite region in humans and chimpanzees (Donaldson et al., 2008). Of particular interest is that there was great intraspecific variation in the coding region of the *AVPR1a* gene in owl monkeys as well as interspecific variation in the coding region within New World monkeys.

These series of studies provide great insight into the *AVPR1a* gene in primates. However, this variability has failed to show any relation to social organization. This is not necessarily surprising, given that monogamy has most likely evolved multiple times (Fraley et al., 2005), and thus probably through at least slightly varying mechanisms.

Although the length of the *AVPR1a* promoter region has differential effects in AVPR1a receptor expression and social behavior in prairie and montane voles, other non-monogamous vole species have similar long promoter regions to that of the prairie vole (Fink et al., 2006).

Although *in vitro* work suggests that the length of the promoter region of the human or chimpanzee *AVPR1a* gene does not affect gene expression (Hong et al., 2009), AVPR1a receptor distribution in the brain in various primate species does suggest a relationship to social organization. There is evidence that marmosets and rhesus macaques have different distributions of AVPR1a receptors (Schorscher-Petcu et al., 2009; Young et al 1999b; Wang et al., 1997a), which is suggestive. This could possibly be due to the differences in their respective promoter regions specifically the GATA repeat found in Old World monkeys and the GACA repeat found in New World monkeys, or due to epigenetic mechanisms driven by early experience.

Intra- and interspecific variation in the coding region of *AVPR1a* of New World monkeys argues for further research in this area (Babb et al., 2010). Specific polymorphisms in the coding region may have the potential to change binding affinity and social behavior. For instance, research into the C77G μ opioid receptor single nucleotide polymorphism in rhesus macaques has discovered effects on attachment behavior (Hingham et al., 2011). Rhesus infants containing the G allele exhibit more physical contact with their mother after repeated separation compared to those individuals without this allele (Barr et al., 2008). It is possible that variation in both the coding and non-coding region is important for influencing social structure in primates.

There have been recent discoveries finding differences in the coding region in the OT gene in primates (this gene codes for the peptide, not the receptor). Lee et al. (2011) compared the OT sequences of one Old World monkey species (rhesus macaque), the northern treeshrew, and five New World monkey species, which included the squirrel monkey, owl monkey, titi monkey, capuchin, and common marmoset. The treeshrew and all of the New World monkey species except for the titi monkey had a single in-frame mutation from thymine to a cytosine, which results in a single amino acid substitution at position 8 of the OT nonapeptide from leucine to a proline, [P8] oxytocin. The mutation has no effect on transcription or translation of the gene. A larger comparison to a wide variety of mammalian species was performed, and only these four New World monkey species and the treeshrew had the [P8] oxytocin. The *AVP* gene was also analyzed in squirrel monkeys but no differences were observed compared to other species. This study might affect interpretation of studies that have used OT administration in New World monkeys to examine behavioral and physiological changes (Smith et al., 2010; Saito and Nakamuro, 2011; Parker et al., 2005), especially since it is unknown how different forms of OT affect binding to OTR. It is possible that changes in response to OT administration in certain New World monkeys may be more robust if [P8] oxytocin is used instead of the more common form.

Due to variations in the *OT* and *AVPR1a* gene in primates, a combination of these systems may be important for the organization of primate social structure since examining only one has not been able to be reflective of specific social structures. In male prairie voles, it appears that AVP plays a greater role in pair bonding and social behavior than OT, however, OT administration in male prairie voles also has the capability to facilitate male social behavior (Cho et al., 1999; Bales et al., 2004). In the owl monkey, for example, the combination of [P8] oxytocin and the variation in the coding region of the *AVPR1a* gene may be important to their monogamous social structure. These new findings argue that these systems should be studied in conjunction and not in isolation.

16.5 OT, AVP, and stress in non-human primates

The relationship between AVP and stress has been relatively well studied in rodents (Liu et al., 2001b; Nishioka et al., 1998; Tanoue et al., 2004) although the relationship between OT and stress is less clear

and may be species and stressor specific (Kalin et al., 1985; Liu et al., 2001b; Nishioka et al., 1998; Parker et al., 2005). OT and AVP often increase after a stressor (Gimpl and Fahrenholz, 2001; Nishioka et al., 1998; Ring, 2005). AVP is released in conjunction with corticotropin-releasing factor to modulate adrenocorticotropin hormone (ACTH) release (Antoni, 1993; Ring, 2005), while it is believed that OT release may be a mechanism to shut down the HPA axis and return the system to homeostasis (Yeğen, 2010). OT, as well as AVP, can serve as a secretagogue of ACTH (Antoni et al., 1983; Schlosser et al., 1994), by acting on V1b receptors.

OT reduced ACTH release in one study of squirrel monkeys. Parker et al. (2005) gave chronic intranasal administration of OT to squirrel monkeys for eight consecutive days. Subjects were isolated after administration on the eighth day. Chronic OT decreased plasma ACTH 90 minutes after the beginning of the isolation stressor. However, there was no effect on cortisol.

The role of OT and stress in non-human primates, however, remains complicated. Rhesus macaques undergoing a daily stressor of 30 min in a confined cage with exposure to loud noise exhibited an increase in ACTH with a decrease in plasma OT at all time points after baseline (Kalin et al., 1985). There was no change in plasma AVP. Rhesus macaques were given dexamethasone (DEX), a synthetic glucocorticoid, for 4 consecutive days. ACTH decreased after treatment, while there was a significant increase in plasma OT on day 4 of DEX challenge. DEX treatment had no effect on plasma AVP (Kalin et al., 1985).

16.6 Conclusions

The study of OT and AVP in non-human primates, and their relationships to behavior, is still underdeveloped when compared to that in rodents. However, with the advent of intranasal neuropeptide administration, and the explosion in genetic techniques and questions, we expect studies in non-human primates to also expand. The development of OT and AVP imaging ligands that cross the blood–brain barrier will also greatly contribute to this research. Non-human primate research has unique contributions to make to our understanding of the role of OT and AVP in human primates.

REFERENCES

Amico, J. A., Challinor, S. M., and Cameron, J. L. (1990). Pattern of oxytocin concentrations in the plasma and cerebrospinal fluid of lactating rhesus monkeys (*Macaca mulatta*): Evidence for functionally independent oxytocinergic pathways in primates. *Journal of Clinical Endocrinology and Metabolism*, 71, 1531–1535.

Amico, J. A., Janosky, J. E., Challinor, S. M., and Cameron, J. L. (1992). Effect of naloxone administration upon the diurnal concentrations of oxytocin in the cerebrospinal fluid of rhesus and cynomolgus monkeys. *Hormone Research*, 38, 171–176.

Amico, J. A., Layden, L. M., Pomerantz, S. M., and Cameron, J. L. (1993). Oxytocin and vasopressin secretion in monkeys administered apomorphine and a D2 receptor agonist. *Life Science*, 52, 1301–1309.

Amico, J. A., Levin, S. C., and Cameron, J. L. (1989). Circadian rhythm of oxytocin in the cerebrospinal fluid of rhesus and cynomolgus monkeys: Effects of castration and adrenalectomy and presence of a caudal-rostral gradient. *Neuroendocrinology*, 50, 624–632.

Antoni, F. A. (1993). Vasopressinergic control of pituitary adrenocorticotropin secretion comes of age. *Frontiers in Neuroendocrinology*, 14, 76–122.

Antoni, F. A., Holmes, M. C., and Jones, M. T. (1983). Oxytocin as well as vasopressin potentiate ovine CRF *in vitro*. *Peptides*, 4, 411–415.

Aragona, B. J., Liu, Y., Yu, Y. J., Curtis, J. T., Detwiler, J.M, Insel, T.R, and Wang, Z. X. (2006). Nucleus accumbens dopamine differentially mediates the formation and maintenance of monogamous pair-bonds. *Nature Neuroscience*, 9, 133–139.

Artman, H. G., Reppert, S. M., Perlow, M. J., et al. (1982). Characterization of the daily oxytocin rhythm in primate cerebrospinal fluid. *Journal of Neuroscience*, 2, 598–603.

Atunes, J. L. and Zimmerman, E. A. (1978). The hypothalamic magnocellular system of the rhesus monkey: An immunocytochemical study. *Journal of Comparative Neurology*, 181, 539–566.

Babb, P. L., Fernandez-Duque, E., and Schurr, T. G. (2010). *AVPR1a* sequence variation in monogamous owl monkey (*Aotus azarai*) and its implications of the evolution of platyrrhine social behavior. *Journal of Molecular Evolution*, 71, 279–297.

Bales, K. L. and Carter, C. S. (2003). Sex differences and developmental effects of oxytocin on aggression and social behavior in prairie voles (*Microtus ochrogaster*). *Hormones and Behavior*, 3, 178–184.

Bales, K. L., Kim. A. J., Lewis-Reese, A. D., and Carter, C. S. (2004). Both oxytocin and vasopressin may influence alloparental behavior in male prairie voles. *Hormones and Behavior*, 5, 354–361.

Bales, K. L., Mason, W. A., Catana, C., Cherry, S. R., and Mendoza, S. P. (2007). Neural correlates of pair-bonding in a monogamous primate. *Brain Research*, 1184, 245–253.

Barr, C. S., Schwandt, M. L., Lindell, S. G., et al. (2008). Variation at the mu-opioid receptor gene (*OPRM1*) influences attachment behavior in infant primates. *Proceedings of the National Academy of Sciences*, 105, 5277–5281.

Bester-Meredith, J. K. and Marler, C. A. (2001). Vasopressin and aggression in cross-fostered California mice (*Peromyscus californicus*) and white-footed mice (*Peromyscus leucopus*). *Hormones and Behavior*, 40, 51–64.

Boccia, M.L, Panicker, A. K., Pedersen, C., and Petrusz, P. (2001). Oxytocin receptors in non-human primate brain visualized with monoclonal antibody. *Neuropharmacology and Neurotoxicology*, 12, 1723–1726.

Boccia, M. L., Goursaud, A. S., Bachevalier, J., Anderson, K. D., and Pedersen, C. A. (2007). Peripherally administered non-peptide oxytocin antagonist, L368,899, accumulates in limbic brain areas: A new pharmacological tool for the study of social motivation in non-human primates. *Hormones and Behavior*, 52, 344–351.

Bond, C., LaForge, K. S., Tian, M., et al. (1998). Single-nucleotide polymorphism in the human mu opioid receptor gene alters β-endorphin binding and activity: Possible implications for opiate addiction. *Proceedings of the National Academy of Sciences, USA*, 95, 9608–9613.

Buijs, R. M., Swaab, D. F., Dogterom, J., and van Leeuwen, F. W. (1978). Intra- and extrahypothalamic vasopressin and oxytocin pathways in the rat. *Cell and Tissue Research*, 186, 423–433.

Caffé, A. R., Van Ryen, P. C., Vand Der Woude, T. P., and Van Leeuwen, F. W. (1989). Vasopressin and oxytocin systems in the brain and upper spinal cord of *Macaca fascicularis*. *Journal of Comparative Neurology*, 287, 302–325.

Cameron, J. L., Pomerantz, S. M., Layden, L. M., and Amicao, J. A. (1992). Dopaminergic stimulation of oxytocin concentrations in the plasma of male and female monkeys by apomorphine and a D2 receptor agonist. *Journal of Clinical Endocrinology and Metabolism*, 75, 855–860.

Carter, C. S. (1992). Oxytocin and sexual behavior. *Neuroscience and Biobehavioral Reviews*, 16, 131–144.

Challinor, S. M., Cameron, J. L., and Amico, J. A. (1992). Pulses of oxytocin in the cerebrospinal fluid of rhesus monkeys. *Hormone Research*, 37, 230–235.

Cho, M. M., DeVries, A. C., Williams, J. R., and Carter, C. S. (1999). The effects of oxytocin and vasopressin on partner preferences in male and female prairie voles (*Microtus ochrogaster*). *Behavioral Neuroscience*, 113, 1071–1079.

Cooke, B., Higley, J. D., Shannon, C., et al. (1997). Rearing history and CSF oxytocin as predictors of maternal competency in rhesus macaques. *Journal of American Primatology*, 42, 102 [Abstract].

De Bree, F. M. (2000). Trafficking of the vasopressin and oxytocin prohormone through the regulated secretory pathway. *Journal of Neuroendocrinology*, 12, 589–594.

Delville, Y., Mansour, K. M., and Ferris, C. F. (1996). Testosterone facilitates aggression by modulating vasopressin receptors in the hypothalamus. *Physiology and Behavior*, 60, 25–29.

DeVries, G. J., Buijs, R. M., Van Leeuwen, F. W., Caffé, A. R., and Swaab, D. F. (1985). The vasopressinergic innervation of the brain in normal and castrated rats. *Journal of Comparative Neurology*, 233, 236–254.

Donaldson, Z. R., Kondrashov, F.A, Putnam, A., et al. (2008). Evolution of a behavior-linked microsatellite-containing element in the 5′ flanking region of the primate *AVPR1A* gene. *BMC Evolutionary Biology*, 8, 180.

Dunbar, R. I. M. (2010). The social role of touch in humans and primates: Behavioural function and neurobiological mechanisms. *Neuroscience and Biobehavioral Reviews*, 34, 260–268.

Ferris, C. F., Melloni, R. H., Koppel, G., et al. (1997). Vasopressin/serotonin interactions in the anterior hypothalamus control aggressive behavior in golden hamsters. *Journal of Neuroscience*, 17, 4331–4340.

Falconer, J., Mitchell, M. D., Mountford, L. A., and Robinson, J. S. (1980). Plasma oxytocin concentrations during the menstrual cycle in the rhesus monkey, *Macaca mulatta*. *Journal of Reproduction and Fertility*, 59, 69–72.

Fink, S., Excoffier, L., and Heckel, G. (2006). Mammalian monogamy is not controlled by a single gene. *Proceedings of the National Academy of Sciences*, 103, 10956–10960.

Fraley, R. C., Brumbaugh, C. C., and Marks, M. J. (2005). The evolution and function of adult attachment: a comparative and phylogenetic analysis. *Journal of Personality and Social Psychology*, 89, 731–746.

Francis, D. D. and Champagne, M. J. (2000). Variations in maternal behavior are associated with differences in oxytocin receptor levels in the rat. *Journal of Neuroendocrinology*, 12, 1145–1148.

Gimpl, G. and Fahrenholz, F. (2001). The oxytocin receptor system: Structure, function and regulation. *Psychological Reviews*, 81, 629–683.

Ginsberg, S. D., Hof, P. R., Young, W. G., and Morrison, J. H. (1994). Noradrenergic innervation of vasopressin and oxytocin-containing neurons in the hypothalamic paraventricular nucleus of the macaque monkey: Quantitative analysis using double-label immunohistochemistry and confocal laser microscopy. *Journal of Comparative Neurology*, 34, 476–491.

Goldizen, A. W. (1990). A comparative perspective on the evolution of tamarin and marmoset social systems. *International Journal of Primatology*, 11, 63–83.

Goldsmith, P. C., Boggan, J. E., and Thind, K. K. (1991). Opioid synapses on vasopressin neurons in the paraventricular and supraoptic nuclei of juvenile monkeys. *Neuroscience*, 45, 709–719.

Hammock, E. A. D. and Young, L. J. (2002). Variation in the vasopressin V1a receptor promoter and expression: Implications for inter- and intraspecific variation in social behaviour. *European Journal of Neuroscience*, 16, 399–402.

Hammock, E. A. D. and Young, L. J. (2005). Microsatellite instability generates diversity in brain and sociobehavioral traits. *Science*, 308, 1630–1634.

Hingham, J. P., Barr, C. S., Hoffman, C. L., et al. (2011). Mu-opioid receptor (OPRM1) variation, oxytocin levels and maternal attachment in free-ranging rhesus macaques *Macaca mulatta. Behavioral Neuroscience*, 125, 131–136.

Hirst, J. J., Haluska, G. J., Cook, M. J., Hess, D. L., and Novy, M. J. (1991). Comparison of plasma oxytocin and catecholamine concentrations with uterine activity in pregnant rhesus monkeys. *Journal of Clinical Endocrinology and Metabolism*, 73, 804–810.

Holman, S. D. and Goy, R. W. (1980). Behavioral and mammary responses of adult female rhesus to strange infants. *Hormones and Behavior*, 14, 348–357.

Holman, S. D. and Goy, R. W. (1995). Experiential and hormonal correlates of care-giving in rhesus macaques. In *Motherhood in Human and Nonhuman Primates: Biosocial Determinants*, Basel: Karger, pp. 87–93.

Hong, K., Matsukawa, R., Hirata, Y., et al. (2009). Allele distribution and effect on reporter gene expression of vasopressin receptor gene (*AVPR1a*)-linked VNTR in primates. *Journal of Neural Transmission*, 116, 535–538.

Hostetler, C. M., Mendoza, S. P., Mason, W. A., and Bales, K. L. (2007). Neuroendocrinology of alloparental care in titi monkeys (*Callicebus cupreus*). *American Journal of Primatology*, 69 (Suppl 1), 58. [Abstract].

Ichimiya, Y., Emson, P. C., and Shaw, F. D. (1988). Localization of vasopressin mRNA-containing neurons in the hypothalamus of the monkey. *Molecular Brain Research*, 4, 81–85.

Insel, T. R., Winslow, J. T., Wang, Z., and Young, L. J. (1998). Oxytocin, vasopressin, and the neuroendocrine basis of pair bond formation. *Advances in Experimental Medicine and Biology*, 449, 215–224.

Jarcho, M. R., Mendoza, S. P., Mason, W. A., Yang, X., and Bales, K. L. (2011). Intranasal vasopressin affects pair bonding and peripheral gene expression in male *Callicebus cupreus. Genes, Brain and Behavior*, 10, 375–383.

Kalin, N. H., Gibbs, D. M., Barksdale, C. M., Shelton, S. E., and Carnes, M. (1985). Behavioral stress decreases plasma oxytocin concentrations in primates. *Life Sciences*, 36, 1275–1280.

Kappeler, P. M. and van Schaik, C. P. (2002). Evolution of primate social systems. *International Journal of Primatology*, 23, 707–740.

Kawata, M. and Sano, Y. (1982). Immunohistochemical identification of the oxytocin and vasopressin neurons in the hypothalamus of the monkey (*Macaca fuscata*). *Anatomy and Embryology*, 165, 151–167.

Kleiman, D. G. (1977). Monogamy in Mammals. *Quarterly Review of Biology*, 52, 39–69.

Kowalski, W. B., Parsons, M. T., Pak, S. C., and Wilson, L. (1998). Morphine inhibits nocturnal oxytocin secretion and uterine contractions in the pregnant baboon. *Biology of Reproduction*, 58, 971–976.

Kozorovitskiy, Y., Hughes, M., Lee, K., and Gould, E. (2006). Fatherhood affects dendritic spines and vasopressin V1a receptors in the primate prefrontal cortex. *Nature Neuroscience*, 9, 1094–1095.

Laudenslager, M. L., Held, P. E., Boccia, M. L., Reite, M. L., and Cohen, J. J. (1990). Behavioral and immunological consequences of brief mother-infant separation: A

species comparison. *Developmental Psychobiology*, 23, 247–264.

Lee, A. G., Cool, D. R., Grunwald, W. C., et al. (2011). A novel form of oxytocin in New World monkeys. *Biology Letters*, 7, 584–587.

Lehmann, J., Korstjens, A. H., and Dunbar, R. I. M. (2007). Group size, grooming and social cohesion in primates. *Animal Behaviour*, 74, 1617–1629.

Liu, Y., Curtis, J. T., and Wang, Z. X. (2001a). Vasopressin in the lateral septum regulates pair bond formation in male prairie voles (*Microtus ochrogaster*). *Behavioral Neuroscience*, 115, 910–919.

Liu, Y., Curtis, J. T., Fowler, C. D., et al. (2001b). Differential expression of vasopressin, oxytocin and corticotrophin-releasing hormone messenger RNA in the paraventricular nucleus of the prairie vole brain following stress. *Journal of Neuroendocrinology*, 13, 1059–1065.

Maestripieri, D., Hoffman, C. L., Anderson, G. M., Carter, C. S., and Higley, J. D. (2009). Mother-infant interactions in free-ranging rhesus macaques: Relationships between physiological and behavioral variables. *Physiology and Behavior*, 96, 613–619.

Mason, W. A. (1966). Social organization of the South American monkey, *Callicebus moloch:* A preliminary report. *Tulane Studies in Zoology*, 13, 23–28.

McCarthy, M. M. (1995). Estrogen modulation of oxytocin and its relation to behavior. *Advances in Experimental Medicine and Biology*, 395, 235–245.

Melis, M. R., Argiolas, A., and Gessa, G. L. (1989). Apomorphine increases plasma oxytocin concentration in male rats. *Neuroscience Letters*, 98, 351–355.

Mendoza, S. P. and Mason, W. A. (1986). Parental division of labour and differentiation of attachments in a monogamous primate (*Callicebus moloch*). *Animal Behaviour*, 34, 1336–1347.

Michael, R. P., Zumpe, D., and Bonsall, R. W. (1982). Behavior of rhesus monkeys during artificial menstrual cycles. *Journal of Comparative and Physiological Psychology*, 96, 875–885.

Nair, H. P. and Young, L. J. (2005). Vasopressin and pair-bond formation: Genes to brain to behavior. *Physiology*, 21, 146–152.

Nishioka, T., Anselmo-Franci, J. A., Li, P., Callahan, M. F., and Morris, M. (1998). Stress increases oxytocin release within the hypothalamic paraventricular nucleus. *Brain Research*, 781, 57–61.

Parker, K. J., Buckmaster, C. L., Schatzberg, A. F., and Lyons, D. M. (2005). Intranasal oxytocin administration attenuates the ACTH stress response in monkeys. *Psychoneuroendocrinology*, 30, 924–929.

Parker, K. J., Hoffman, C. L., Hyde, S. A., Cummings, C. S., and Maestripieri, D. (2010). Effects of age on cerebrospinal fluid oxytocin levels in free-ranging adult female and infant rhesus macaques. *Behavioral Neuroscience*, 124, 428–433.

Pedersen, C. (1997). Oxytocin control of maternal behavior: regulation by sex steroids and offspring stimuli. *Annals of the New York Academy of Sciences*, 807, 126–145.

Perlow, M. J., Reppert, S. M., Artman, H. A., et al. (1982). Oxytocin, vasopressin, and estrogen-stimulated neurophysin: Daily patterns of concentration in cerebrospinal fluid. *Science*, 216, 1416–1418.

Reppert, S. M., Schwartz, W. J., Artman, H. G., and Fisher, D. A. (1983). Comparison of the temporal profiles of vasopressin and oxytocin in the cerebrospinal fluid of the cat, monkey and rat. *Brain Research*, 261, 341–345.

Reppert, S. M., Perlow, M. J., Artman, H. G., et al. (1984). The circadian rhythm of oxytocin in primate cerebrospinal fluid: Effects of destruction of the suprachiasmatic nuclei. *Brain Research*, 307, 384–387.

Rilling, J. K., Winslow, J. T., and Kilts, C. D. (2004). The neural correlates of mate competition in dominant male rhesus macaques. *Biological Psychiatry*, 56, 364–375.

Ring, R. H. (2005). The central vasopressinergic system: Examining the opportunity for psychiatric drug development. *Current Pharmaceutical Design*, 11, 205–225.

Ronnekleiv, O. K. (1988). Distribution in the macaque pineal of nerve fibers containing immunoreactive substance P, vasopressin, oxytocin, and neurophysins. *Journal of Pineal Research*, 1988, 259–271.

Rosenblum, L. A., Kaufman, I. C., and Stynes, A. J. (1964). Individual distance in two species of macaque. *Animal Behaviour*, 12, 338–342.

Rosenblum, L. A., Smith, E. L. P., Altemus, M., et al. (2002). Differing concentrations of corticotropin-releasing factor and oxytocin in the cerebrospinal fluid of bonnet and pigtail macaques. *Psychoneuroendocrinology*, 27, 651–660.

Rosso, L., Keller, L., Kaellmann, H., and Hammond, R. L. (2008). Mating system and *avpr1a* promoter variation in primates. *Biology Letters*, 4, 375–378.

Saito, A. and Nakamura, K. (2011). Oxytocin changes primate paternal tolerance to offspring in food transfer. *Journal of Comparative Physiology A*, 197, 329–337.

Schorscher-Petcu, A., Dupré, A., and Tribollet, E. (2009). Distribution of vasopressin and oxytocin binding sites

in the brain and upper spinal cord of the common marmoset. *Neuroscience Letters*, 461, 217–222.

Schlosser, S. F., Almeida, O. F., Patchev, V. K., Yassouridis, A., and Elands, J. (1994). Oxytocin-stimulated release of adrenocorticotropin from the rat pituitary is mediated by arginine vasopressin receptors of the V1b type. *Endocrinology*, 135, 2058–2063.

Schwandt, M. L., Howell, S., Bales, K., et al. (2007). Associations between the neuropeptides oxytocin and vasopressin and the behavior of free-ranging female rhesus macaques (*Macaca mulatta*). *American Journal of Physical Anthropology Supplement*, 44, 210–211. [Abstract].

Seltzer, L. J. and Ziegler, T. E. (2007). Non-invasive measurement of small peptides in the common marmoset (*Callithrix jacchus*): A radiolabeled clearance study and endogenous excretion under varying social conditions. *Hormones and Behavior*, 51, 436–442.

Silk, J. B., Alberts, S. C., and Altmann, J. (2003). Social bonds of female baboons enhance infant survival. *Science*, 302, 1231–1234.

Sladek, J. R. and Zimmerman, E. A. (1982). Simultaneous monoamine histofluorescence and neuropeptide immunocytochemistry: VI. Catecholamine innervation of vasopressin and oxytocin neurons in the rhesus monkey hypothalamus. *Brain Research Bulletin*, 9, 431–440.

Smith, A. S., Ågmo, A., Birnie, A. K., and French, J. A. (2010). Manipulation of the oxytocin system alters social behavior and attraction in pair-bonding primates, *Callithrix penicillata*. *Hormones and Behavior*, 57, 255–262.

Smock, T., Albeck, D., and Stark, P. (1998). A peptidergic basis for sexual behavior in mammals. *Progress in Brain Research*, 119, 467–481.

Snowdon, C. T., Pieper, B. A., Boe, C. Y., et al. (2010). Variation in oxytocin is related to variation in affiliative behavior in monogamous pairbonded tamarins. *Hormones and Behavior*, 58, 614–618.

Sofroniew, M. V., Weindl. A., Schrell, U., and Wetzstein, R. (1981). Immunohistochemistry of vasopressin, oxytocin and neurophysin in the hypothalamus and extrahypothalamic regions of the human and primate brain. *Acta Histochemica Supplement*, 24, 79–95.

Swaab, D. F., Pool, C. W., and Nijveldt, F. (1975). Immunofluorescence of vasopressin and oxytocin in the rat hypothalamo-neurohypophyseal system. *Journal of Neural Transmission*, 36, 195–215.

Swanson, L. W. (1977). Immunohistochemical evidence for a neurophysin-containing autonomic pathway arising in the paraventricular nucleus of the hypothalamus. *Brain Research*, 128, 346–353.

Swanson, L. W. and McKellar, S. (1979). The distribution of oxytocin- and neurophysin-stained fibers in the spinal cord of the rat and monkey. *Journal of Comparative Neurology*, 188, 87–106.

Swanson, L. W. and Sawchenko, P. E. (1983) Hypothalamic integration: organization of the paraventricular and supraoptic nuclei. *Annual Review of Neuroscience*, 6, 269–324.

Tanoue, A., Ito, S., Honda, K., et al. (2004). The vasopressin V1b receptor critically regulates hypothalamic-pituitary-adrenal axis activity under both stress and resting conditions. *Journal of Clinical Investigation*, 113, 302–309.

Toloczko, D. M., Young, L., and Insel, T. R. (1997). Are there oxytocin receptors in the primate brain? *Annals of the New York Academy of Sciences*, 807, 506–509.

Ueda, S., Kawata, M., and Sano, Y. (1983). Identification of serotonin- and vasopressin immunoreactivities in the suprachiasmatic nucleus of four mammalian species. *Cell Tissue Research*, 234, 237–248.

Uvnas-Moberg, K. (1998). Oxytocin may mediate the benefits of positive social interaction and emotions. *Psychoneuroendocrinology*, 23, 819–835.

Van Esseveldt, L.K.E., Lehman, M. N., and Boer, G. J. (2000). The suprachiasmatic nucleus and circadian time-keeping system revisited. *Brain Research Reviews*, 33, 34–77.

Walum, H., Westberg, L., Henningsson, S., et al. (2008). Genetic variation in the vasopressin receptor 1a gene (*AVPR1a*) associates with pair-bonding behavior in humans. *Proceedings in the National Academy of Sciences*, 105, 14153–14156.

Wang, Z., Ferris, C. F., and DeVries, G. J. (1994). Role of septal vasopressin innervation in paternal behavior in prairie voles (*Microtus ochrogaster*). *Proceedings in the National Academy of Sciences*, 91, 400–404.

Wang, Z., Moody, K., Newman, J. D., and Insel, T. R. (1997a). Vasopressin and oxytocin immunoreactive neurons and fibers in the forebrain of the male and female common marmosets (*Callithrix jacchus*). *Synapse*, 27, 14–25.

Wang, Z., Toloczko, D., Young, L. J., et al. (1997b). Vasopressin in the forebrain of common marmosets (*Callithrix jacchus*): studies with in situ hybridization, immunocytochemistry and receptor autoradiography. *Brain Research*, 768, 147–156.

Williams, J. R., Carter, C. S., and Insel, T. R. (1992). Partner preference development in female prairie voles is facilitated by mating or the central infusion of oxytocin. *Annals of the New York Academy of Sciences*, 652, 487–489.

Winslow, J. T. and Insel, T. R. (1991a). Vasopressin modulates male squirrel monkeys' behavior during social separation. *European Journal of Pharmacology*, 200, 95–101.

Winslow, J. T. and Insel, T. R. (1991b). Social status in pairs of male squirrel monkeys determines the behavioral response to central oxytocin administration. *Journal of Neuroscience*, 11(7), 2032–2038.

Winslow, J. T., Hastings, N., Carter, C. S., Harbaugh, C. R., and Insel, T. R. (1993). A role for central vasopressin in pair bonding in monogamous prairie voles. *Nature*, 365, 545–548.

Winslow, J. T., Noble, P. L. Lyons, C. K., Sterk, S. M., and Insel, T. R. (2003). Rearing effects on cerebrospinal fluid oxytocin concentration and social buffering in rhesus monkeys. *Neuropsychopharmacology*, 28, 910–918.

Witt, D. M., Carter, C. S., and Insel, T. R. (1991). Oxytocin receptor binding in female prairie voles: Endogenous and exogenous oestradiol stimulation. *Journal of Neuroendocrinology*, 3, 155–161.

Wright, P. C. (1984). Biparental care in *Aotus trivirgatus* and *Callicebus moloch*. In *Female primates: Studies by women primatologists*, New York: Alan R. Liss, Inc., pp. 59–75.

Yeğen, B.Ç. (2010). Oxytocin and hypothalamo-pituitary-adrenal axis. *Marmara Pharmaceutical Journal*, 14, 61–66.

Young, L. J., Nilsen, R., Waymire, K. G., Macgregor, G. R., and Insel, T. R. (1999a). Increased affiliative response to vasopressin in mice expressing the vasopressin receptor from a monogamous vole. *Nature*, 400, 766–768.

Young, L. J., Toloczko, D., and Insel, T. R. (1999b). Localization of vasopressin (V1a) receptor binding and mRNA in the rhesus monkey brain. *Journal of Neuroendocrinology*, 11, 291–297.

PART III

Human studies

17

The involvement of oxytocin and vasopressin in fear and anxiety

Animal and human studies

Yoav Litvin and Donald W. Pfaff

17.1 Introduction

The concepts of fear and anxiety have been the focus of much attention due to their fundamental role in the survival of a plethora of organisms and their perturbation in associated human disorders. Oxytocin (OT) and vasopressin (AVP), two of the most important and functionally complex neuropeptides, are intricately involved in fear, anxiety and sociality. OT is involved in prosocial behaviors by ameliorating anxiety, facilitating trust, and promoting female reproductive behavior. AVP is involved in male-typical behaviors, stress and is generally considered anxiogenic. Here, we survey the direct and indirect effects of OT and AVP on fear- and anxiety-related behaviors in both animals and humans, with a particular focus on interactions with systems of stress and sex. We present a novel theory of OT and AVP interaction in the amygdala that may underlie sex differences in anxiety and sociality. Finally, we briefly discuss clinical applications.

17.1.1 Fear and anxiety: Emotions

The neural and behavioral bases of fear and anxiety engulf a range of highly adaptive functions that if perturbed, can lead to extremely debilitating pathologies. In a recent volume dedicated to the progress in the scientific research of fear and anxiety, fear was defined as *"the motivation associated with a number of behaviors that normally occur on exposure to clearly threatening stimuli,"* whereas anxiety as *"the motivation associated with behaviors that occur to potential, signaled or ambiguous threat"* (Blanchard et al., 2008). Hence, fear and anxiety are both subtypes of motivations, commonly referred to as emotions. Here, we define an emotion as any motivation associated with behaviors that occur to changes in homeostasis or reproductive potential.

In light of their primal role in ensuring survival, it is not surprising that both fear and anxiety are emotions that show dramatic analogies within and across species. Darwin (1872) first advocated that emotions: (1) show intraspecies conservation; (2) are innate responses and; (3) evolved from similar expressions in animals, showing interspecies analogies. In support of his view, Darwin pointed out the similarity of facial expressions of emotion in many human and animal societies around the world. Ekman and Friesen (1971) proved Darwin's notion regarding intraspecies conservation of emotional output in their seminal studies of clans from isolated tribes in New Guinea. These tribesmen had no contact with the outside world yet they were able to recognize, differentiate and reproduce facial expressions of emotion, proving that fundamental emotional output is universal.

Oxytocin, Vasopressin, and Related Peptides in the Regulation of Behavior, ed. E. Choleris, D. W. Pfaff, and M. Kavaliers.
Published by Cambridge University Press.

17.1.2 Fear and anxiety; functions

The world is abundant with threats to physical safety and future reproductive ability. In order to meet these challenges, fear and anxiety are emotions that entail a two-pronged adaptive utility. First, they facilitate immediate survival by promoting innate and learned defensive behaviors that are mobilized in response to threats to bodily safety (Blanchard et al., 2009). Second, in many classes of gregarious vertebrates, fear and anxiety promote both survival and reproduction by affecting complex social behaviors; e.g., sexual behaviors (Morgan et al., 2004; Pfaff, 1999), maternal behaviors (Carter and Altemus, 1997), paternal care, courtship, pair bonding (Donaldson and Young, 2008; DeVries et al., 2002; DeVries et al., 1996; Carter et al., 1995), aggression (Veenema and Neumann, 2008), hierarchy formation (Blanchard and Blanchard, 1989), and threat communication (Litvin et al., 2007). In fact, social bonding is considered to result from coupling of neuropeptides involved in the processing of social sensory information and those of reward (Young and Wang, 2004). As such, it stands to reason that the formation and maintenance of social ties are as affected by stress, fear and anxiety, as other reinforcement-dependent behaviors, such as drug addiction (Koob and Kreek, 2007). Lastly, fear and anxiety are dynamic emotions that can be reciprocally affected by the behaviors they facilitate; for example, the loss of a monogamous partner (Bosch et al., 2009) or chronic subordination stress (Litvin et al., 2011) affect behaviors associated with fear and anxiety.

17.1.3 Fear and anxiety: Assessment

The assessment of fear and anxiety in animals entails analysis of the behaviors, which they motivate; that is, the display of innate defensive behaviors toward threatening stimuli, the inhibition of normal non-defensive behaviors (e.g., foraging, grooming, copulation) and the propensity to create conditioned associations between a threat and either the context, environmental cues or both. Researchers investigating the innate aspect of defense have typically focused on exposing animals to predators and associated cues, novel environments and elevated locations (Litvin et al., 2008; Rodgers, 1997), while researchers investigating the learned component have predominantly utilized Pavlovian fear conditioning techniques (Cain and LeDoux, 2008). High throughput assessment of anxiety typically entails quantification of entries into, and times spent in, different quadrants of apparati, such as the open/closed arms of the elevated plus maze (EPM), the light/dark compartments of the light/dark paradigm or around the walls/in the center of an open field.

In humans, assessment of fear and anxiety is limited to non-invasive procedures. These include measures such as heart rate, blood pressure, functional magnetic resonance imaging (fMRI), electroencephalogram (EEG) recordings, and galvanic skin response (GSR). Moreover, alternative methodologies must be employed with respect to administration of compounds in pharmacological clinical trials; whereas in animals it is possible to infuse compounds into both periphery and the brain via a range of invasive routes (e.g., intracranially via site-specific cannulae), practical and ethical considerations render invasive methodologies unfavorable in humans. Thus, significant effort has been directed toward development and usage of alternative routes that circumvent invasive procedures, such as intranasal administration of compounds (e.g., neuropeptides) that do not readily cross the blood–brain barrier (Born et al., 2002). Moreover, intranasal administration can be coupled with fMRI to assess the anatomical correlates of neuropeptide actions.

In contrast to animals, humans can also provide *ex post facto* subjective ratings of experience that are commonly used to assess fear and anxiety. Here, it is important to reflect on the difference between emotion and its mental conscious perception. William James (1884) was the first to differentiate between the initial perception of threat, the ensuing changes in physiology and behavior and the later insight into the experience. In accordance

with James's notion, *ex post facto* subjective ratings of fear and anxiety are the conscious perception of changes in physiology and behavior produced by a clear threat to safety such as activation of the sympathetic nervous system. As stated eloquently by Joseph Ledoux (1998) of the New York University: *"The mental aspect of emotion, the feeling, is a slave to the physiology, not vice versa: we do not tremble because we are afraid or cry because we feel sad; we are afraid because we tremble and sad because we cry."* Thus, subjective descriptions may be useful due to restrictions in human research, but cannot replace physiological and behavioral correlates of fear and anxiety.

17.1.4 Fear and anxiety: Indices

The study and modeling of fear and anxiety in both animals and humans has relied on a thorough analysis of the antecedents and expression of defensive behaviors. Defensive behaviors and their intensity are affected by several factors: (1) *type of threat:* natural animate threats that include predators, associated cues, conspecific attack and non-conspecific competitors evoke dynamic active coping techniques, whereas inanimate dangerous features of the environment are either unavoidable due to their unpredictability (e.g., earthquakes, tsunamis) or are simply evaded (e.g., high cliffs); (Endler, 1986); (2) *context in which the threat is encountered* (see Figure 17.1) – if a plausible escape route is available, an animal will typically flee and/or avoid, yet when trapped it will freeze; (3) *defensive distance* – a key factor differentiating fear from anxiety seems to be the "defensive distance" (McNaughton and Corr, 2004); fear enables an animal to leave a threatening situation (active avoidance), while anxiety facilitates assessment of a dangerous situation (e.g., risk assessment, approach) or withholds entrance (passive avoidance; see Figure 17.1) and; (4) *stimulus intensity,* whereas ambiguous stimuli elicit active risk assessment behaviors and are often associated with a state of anxiety, discrete, present threats elicit flight, avoidance, defensive threat, and attack and are associated with a state of fear. These observations have led to the consensus that defensive behaviors are species typical under parallel circumstances and are similar in form and function (Blanchard et al., 2001). Moreover, studies showing that equivalent brain circuits are activated in several species (in mice: Martinez et al., 2008; in rats: Canteras, 2002; in humans: Mobbs et al., 2007), suggest the existence of a homology of defense across mammals (Blanchard et al., 2001; Perkins and Corr, 2006; Rosen and Schulkin, 1998). An establishment of such a homology enables the foreknowledge of cause–effect relationships in experimental conditions and provides a valid setting for evaluating drug studies, as well as other indices of emotion. In fact, neural structures involved in defense also play a role in associated psychopathologies – panic disorder and phobias are typically associated with fear. Panic involves activation of the periaqueductal gray, phobias of an amygdaloid-hypothalamic centered circuit and anxiety an amygdala-septo-hippocampal system (McNaughton and Corr, 2004)

These behavioral analyses clarify that, although distinct entities, fear and anxiety lie within a continuum and show certain overlap. It is also clear that whereas a behavior may be adaptive in a certain situation, it may indicate pathology in another. Animal models that appropriately manipulate these factors tend to isolate, maximize and potentiate particular defensive behaviors; maladaptive responses in such models include reactivity to a threat that is not present or a mounting of an exaggerated response to an otherwise neutral or even appetitive stimulus. The need to differentiate between fear and anxiety is also evident when assessing the efficacy of various classes of panicolytics/ anxiolytics in both preclinical and clinical trials. Fear is typically associated with fight or flight, and is selectively attenuated by panicolytics, whereas anxiolytics regulate approach withdrawal, that is, risk assessment, which is generally associated with anxiety (Blanchard et al., 2009). Detailed surveys of the differential effects of the major classes of anxiolytics, including benzodiazepines, glutamatergic compounds,

A

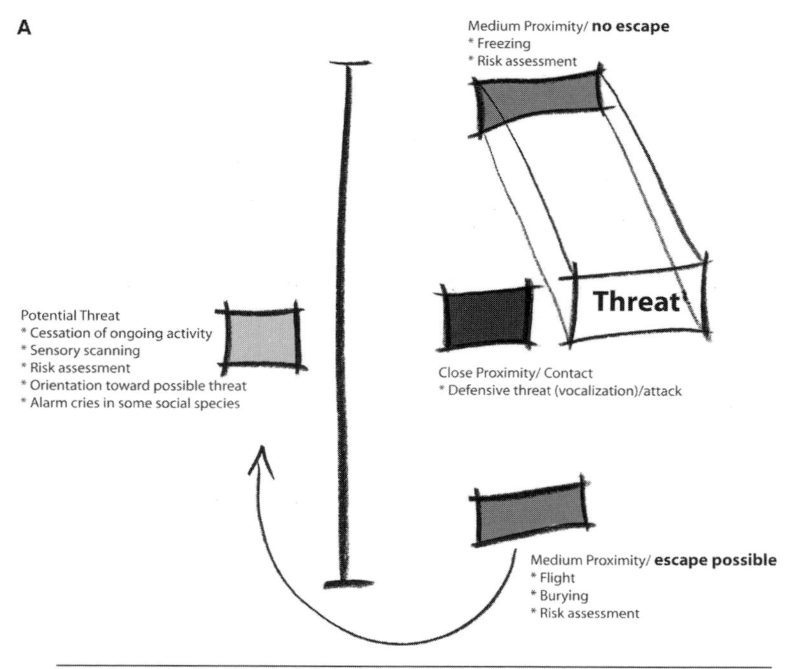

B

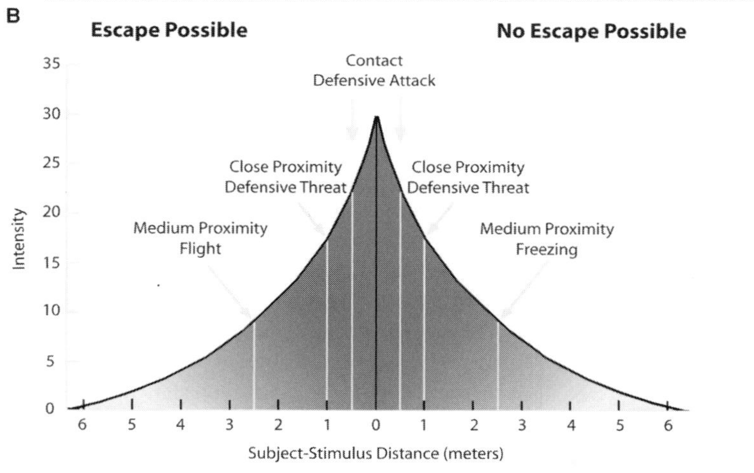

Figure 17.1 Fear- and anxiety-like defensive behaviors are modulated by subject-threat distance and context (e.g., availability of escape). (Graph B adapted from Blanchard et al., 2009.) See color version in plates section.

serotonin-reuptake inhibitors and other peptide receptor ligands are outside the scope of this chapter and have been presented elsewhere (Steckler, 2008; Rowlett, 2008; Guimaraes et al., 2008; Cryan and Dev, 2008; Belzung et al., 2006).

17.1.5 Fear and anxiety: Regulation

Evolutionarily more ancient areas of the brain typically regulate functions necessary for survival and reproduction, yet again signifying the primacy of

fear and anxiety. The hypothalamus coordinates physiological adaptations that facilitate homeostasis and reproduction by integrating and modulating the functions of the autonomic, endocrine, and immune systems (Pfaff et al., 2004). The hypothalamus also facilitates motor function associated with reproduction (Pfaff, 1999) and defense (Canteras, 2002; Markham et al., 2004). These functions are regulated by various subnuclei within the hypothalamus that have afferent and efferent projections to numerous brain structures including the hippocampus, amygdala, septum, mammilary bodies, thalamus, cingulate, prefrontal and orbitofrontal cortices; elements that comprise the limbic system. Hormones that are synthesized within the hypothalamus facilitate communication between these structures. Thus, hormones can modify behavior and physiology that preserves homeostasis and facilitates reproduction (Pfaff et al., 2004).

17.1.6 Regulation of fear and anxiety by oxytocin and arginine vasopressin

Oxytocin (OT) and arginine vasopressin (AVP) are highly conserved and structurally related nonapeptides that function as neurotransmitters and neuromodulators with widespread effects (Landgraf and Neumann, 2004). Research in animals and humans suggests that OT is prosocial, facilitating trust and affiliation and inhibiting systems involved in fear and anxiety, while AVP is involved in male-typical social behaviors (e.g., intermale aggression, scent marking, pair bonding, courtship), stress adaptation, anxiety and aggression (Lim and Young, 2006; Heinrichs and Domes, 2008; Carter, 2007; Volpi et al., 2004; Neumann et al., 2000b; Murgatroyd et al., 2004; van Londen et al., 1997).

OT and AVP are synthesized in the brain and act both peripherally and centrally (Caldwell and Young, 2006):

1. *Peripheral actions* – Oxytocinergic and vasopressinergic magnocellular neurons in the paraventricular (PVN) and supraoptic nuclei (SON) of the hypothalamus project to the neurohypophysis from which they release OT or AVP into the blood stream (Swanson and Sawchenko, 1983; Rhodes et al., 1981). In turn, OT regulates peripheral actions such as lactation (Schafer and Mackenzie, 1911) and parturition (Dale, 1906) via its only known receptor OTR, whereas AVP controls water balance (Bauman, 1965) via its AVPR2 receptor, vasopressor actions (Oliver and Schafer, 1895) via its AVPR1a receptor.
2. *Central actions* – The central actions of OT and AVP are mediated via two routes; the hypothalamic-pituitary adrenal (HPA) stress axis and central synthesis and release.
   a. *The HPA axis* is one of the major neuroendocrine pathways activated in stressful conditions to promote adaptive anxiety-related behaviors (Plotsky, 1991). In concert with the autonomic and immune systems, the HPA axis promotes physiology and behavior that enables adaptive coping and a return to homeostasis (McEwen, 2007). Corticotropin-releasing factor (CRF) and AVP serve as the major secretagogues of the HPA axis in response to a variety of internal and external stimuli (Vale et al., 1981; Kovacs, 1998). AVP synthesized in parvocellular neurons of the PVN travels to the anterior pituitary where it binds to AVPR1b receptors. Here, AVP serves as a major secretagogue of corticotropin (ACTH). AVP also potentiates the effects of corticotropin-releasing factor (CRF) on pituitary corticotrophs (Aguilera et al., 2008; Herman et al., 2008; Sawchenko et al., 1984; Volpi et al., 2004). OT is released intracerebrally in response to stress (Neumann et al., 2000b). The effects of OT on the HPA axis have been more contentious, with some findings indicating an inhibitory effect (Neumann et al., 2000b; Chiodera and Legros, 1981) and some an excitatory one (Gibbs et al., 1984), albeit the latter study investigated the *in-vitro* actions of OT. Regulation of the HPA axis by OT may be at the level of the pituitary and/or the adrenal gland (Legros et al., 1988).
   b. *Central synthesis and release* – OT and AVP are synthesized within – and released from – central nervous system structures to modulate behaviors associated with fear and

anxiety, learning and memory, pain, sociality and feeding via actions on the OT receptor (OTR), and AVP receptors AVPR1a and AVPR1b (Landgraf and Neumann, 2004; Bale et al., 2001; Hernando et al., 2001; Caldwell et al., 2008; Lee et al., 2009). Vasopressinergic and oxytocinergic receptors are present in brain structures implicated in stress, fear, anxiety, and social behaviors (Landgraf and Neumann, 2004; Bale et al., 2001; Hernando et al., 2001). Evidence suggests that the regulation of these receptors within certain brain regions may underlie variations in associated behaviors (Donaldson and Young, 2008). OTR, AVPR1a, and AVPR1b are extensively found in a network of brain structures involved in social behaviors including the medial amygdala (MeA), lateral septum, bed nucleus of the stria terminalis (BNST), medial preoptic area, anterior hypothalamus, ventromedial hypothalamus, tegmentum, and periaqueductal gray (Goodson, 2005; Newman, 1999; De Vries and Buijs, 1983; Landgraf and Neumann, 2004; Hernando et al., 2001; Bale et al., 2001).

17.2 Oxytocin effects on fear and anxiety

Evidence suggests OT attenuates fear and anxiety and increases trust. It is hypothesized that OT may affect fear and anxiety by: (1) modulation of the HPA axis (Neumann, 2002); (2) facilitation of prosocial/affiliative behaviors concomitant with attenuation of neural systems involved in fear and anxiety (Carter, 2007) by binding to medial and central subdivisions of the amygdala, which modulate, respectively, behavioral and autonomic fear reactivity (Huber et al., 2005).

17.2.1 Animal research (see Table 17.1 for examples)

HPA

In response to a variety of stressors, OT is synthesized and released in the PVN, SON, lateral septum, and amygdala (Neumann, 2002). The effects of OT on HPA function depend on stressor type, state and sex of the animal. Though not serving as a major secretagogue of ACTH, OT modulates HPA axis activity via indirect routes. *In vitro* application of OT in combination with CRF potentiates the effects of the latter on release of ACTH from rat hemipituitaries, though OT alone does not cause the release of ACTH (Gibbs et al., 1984). However, the *in vivo* effects of OT on the HPA axis show sexual dimporphism and are sensitive to circulating gonadal steroids. OT has been shown to inhibit both basal and stressor-induced HPA axis markers (ACTH and corticosterone) in males and virgin female rats, an effect that is absent in pregnant and lactating female rats (da Costa et al., 1996; Neumann et al., 1998; Neumann et al., 2000a; Neumann et al., 2000b). Furthermore, evidence suggests that the emotional and physical stressor-induced elevation in OT, corticosterone, CRF and enkephalin are significantly attenuated in pregnant and lactating females (Lightman and Young, 1989; Neumann et al., 1998; Stern et al., 1973). Lactation and parts of the gestation period are characterized by enhanced synthesis (Van Tol et al., 1988; Zingg and Lefebvre, 1988) and release of OT in the brain (Neumann and Landgraf, 1989; Neumann et al., 1993), elevations that likely account for a number of reproductive activities such as maternal behaviors (Insel and Harbaugh, 1989), milk ejection (Freund-Mercier and Richard, 1984), and parturition (Neumann et al., 1996). In contrast, administration of an OT antagonist was shown to be anxiolytic in the EPM in pregnant and lactating rats, but not in males or virgin rats, suggesting a site-specific differential effect of OT on HPA axis and anxiety-related behaviors, depending on hormonal status (Neumann et al., 2000a; Neumann et al., 2000b).

Prosocial behaviors

In gregarious species, social bonds positively affect health and defend against the detrimental effects of stress (Kikusui et al., 2006). One of the prevailing symptoms of various fear and anxiety disorders is disruption in sociality. The lack or loss of an

Table 17.1 Selected examples of studies examining the roles of oxytocin and vasopressin in fear- and anxiety-like behaviors in a variety of popular animal models. Methodologies include knockout, agonist and antagonist administration and small animal imaging.

Animal Models of Fear and Anxiety	Oxytocin	Vasopressin	References
A. Novelty/exploration			
1. Open field	↓	↑	(Uvnas-Moberg et al., 1994; Bielsky et al., 2004; Griebel et al., 2002)
2. Elevated plus-maze	↓	↑	(McCarthy et al., 1996; Bielsky et al., 2004; Griebel et al., 2002)
3. Elevated zero maze	↓	↑	(Ring et al., 2006; Bleickardt et al., 2009)
4. Light/dark test	↓	↑	(Bielsky et al., 2004; Griebel et al., 2002; Blume et al., 2008)
5. Four-plate	↓	↑	(Ring et al., 2006; Serradeil-Le Gal et al., 2005)
B. Social tests			
1. Separation-induced ultrasonic vocalizations	↑	↑	(Bleickardt et al., 2009; Takayanagi et al., 2005)
2. Social defeat-induced anxiety			
a. Acute social defeat	↑	↑	(Griebel et al., 2002; Ebner et al., 2000)
b. Chronic social defeat	–	↑	(Litvin et al., In press)
C. Others			
1. Startle	↓	↓	(Missig et al., 2010; Toufexis et al., 2005)
2. Conditioned fear	↓	↑	(Stoehr et al., 1992)
3. Predator exposure	–	↑	(Ferris et al., 2010)
4. Predator odor	n.s.	–	(Kavaliers et al., 2003)

↑, anxiogenic; ↓, anxiolytic; n.s., nonsignificant role; – has not been assessed. Adapted from Rodgers, 1997 and Litvin et al., 2008.

important social bond has adverse effects on physiology and behavior (DeVries et al., 2003) and may lead to depression (Grippo et al., 2007b; Grippo et al., 2007a; Zisook et al., 1997). The positive effects of strong social bonds are evident when comparing species or strains that differ in their social bonding habits. For example, prairie voles (Microtus ochrogaster), a strain of vole that tend to form life-long monogamous pair bonds, show reduced anxiety in an EPM when compared to their polygamous cousins, meadow voles (Microtus pennsylvanicus) (Stowe et al., 2005). Disruption of a social bond also enhances anxiety; separation from a partner for 4 days increases passive coping behaviors that are equated with depression and tends to increase anxiety-related behaviors in an EPM (Bosch et al., 2009).

An abundance of evidence suggests that OT is involved in social behaviors. Intracerebroventricular (icv) administration of OT potentiates pair bonding in female prairie voles without the necessity of mating (Williams et al., 1994). During mating, concomitant activation of dopaminergic and oxytocinergic systems in a reward center, the nucleus accumbens, reinforces conditioned partner preference (Young and Wang, 2004; Ross et al., 2009). OT levels rise in the PVN of male rats during mating, facilitating risk-taking behaviors and anxiolysis in the EPM for several hours postcopulation (Waldherr and Neumann, 2007), effects blocked by icv administration of an OT antagonist. OT has also been shown to attenuate aggression (Ferris, 2005) and in this manner, enable social investigation and subsequent copulation.

OT has been shown to be involved in maternal behaviors in various mammals (Insel and Harbaugh, 1989; Pedersen and Prange, 1979; Kendrick et al., 1997; Fahrbach et al., 1985). Central

infusion of OT can induce maternal behaviors in virgin females, whereas its receptor blockade inhibits them in dams (Caldwell and Young, 2006). The involvement of OT in maternal behaviors raises the question whether it also independently affects fear and anxiety in pregnant and lactating females. Indeed, this question was examined in a study that compared maternal responses and anxiety-like behaviors in juvenile and adult female prairie voles (Olazabal and Young, 2005). Juveniles showed more affiliation toward novel conspecifics, refrained from negative interactions with the pups (e.g., less attacks) and exhibited less anxiety-like behavior in an open field arena. Interestingly, when compared to females that did not display maternal behaviors and even attacked the pups, "maternal" females were less anxious in an open field, as indicated by elevated number of crossings through the center. These data suggest that OT-mediated anxiolysis may facilitate maternal behaviors.

Several studies support a general anxiolytic effect of OT. In male mice, OT has been shown to be anxiolytic using the elevated zero maze, four-plate, stress-induced hyperthermia tests (Ring et al., 2006) and the open field (Uvnas-Moberg et al., 1994). OTR-deficient male mice pups display a shift from separation-induced ultrasonic vocalizations, to a more active exploratory coping mechanism, in line with a significant role of OTR in modifying social coping (Takayanagi et al., 2005). A study using OT-knockout animals shows that OT has a role in social odor processing, though detection and responsivity to predator odors is unaffected (Kavaliers et al., 2003). The effects of OT have been localized to the PVN, where direct infusion produces anxiolytic effects in the EPM and light/dark tests (Blume et al., 2008). In contrast, a single bout of social defeat has been shown to elevate levels of OT in the lateral septum, though the administration of an OT antagonist does not affect fear and anxiety-like (freezing and exploration, respectively) during the encounter (Ebner et al., 2000). A recent study has shown that OT reduces background anxiety, but not fear conditioning (Missig et al., 2010). The authors suggest that background anxiety may be akin to hypervigilance in post-traumatic stress disorder patients.

Disorders of sociality with abnormal anxiety show clear sexual dimorphism (Choleris et al., 2008). These include major depression, autism spectrum disorder, and schizophrenia. Accordingly, a variety of clinical and preclinical studies show that estrogens and androgens have a profound effect on motivation and emotion (Fink et al., 1996; Pfaff et al., 2004). In fact, the first direct link between biochemical neuronal changes in specific brain regions and a complete mammalian behavior derived from the effects of estrogens in hypothalamic neurons that facilitate lordosis, the primary reproductive behavior of female quadrupeds (Pfaff, 1999). An analysis of sex differences in defensive behaviors show enhanced anxiety in females, especially when involving potential as opposed to discrete threat sources (predator cue versus actual predator), and differential reactivity in such situations to anxiolytics (Blanchard et al., 1991). Furthermore, a recent report shows that females throughout the stages of the estrous cycle display varying intensities of defense when presented with cat odor, with females at estrous stages characterized by higher levels of estrogens showing anxiolyisis, that is, a shift from fear to anxiety-like behaviors (Pentkowski et al., unpublished). In ovariectomized females pretreated with estradiol, peripheral administration of OT exerts anxiolytic effects in the EPM (McCarthy et al., 1996). However, in ovariectomized females that were not treated with estradiol, peripheral OT does not produce anxiolysis. In the same study centrally administered OT even without estrogen replacement was anxiolytic, suggesting that the blood–brain barrier plays a significant role in buffering the effects of peripheral OT, and that estradiol may reduce the threshold of peripheral OT needed to produce anxiolysis.

Various subnuclei of the amygdala have been implicated in the effects of OT. OT acting on the central amygdala has been shown to regulate autonomic components of fear by integrating signals from the basolateral amygdala and cortex and projecting to the hypothalamus and brainstem (Huber

et al., 2005). The MeA is part of a circuit involved in social recognition and anxiety (Spiteri et al., 2010). The MeA is ideally located to mediate social anxiety as it is involved in pheromonal processing (Choleris et al., 2009) through projections to both accessory and main olfactory bulbs (Canteras et al., 1995) and is at the center of a defensive behavioral circuit (Canteras, 2002). Accordingly, activation of the MeA is necessary for social recognition in the mouse (Ferguson et al., 2001); OT knockout mice show marked deficits in social recognition, with icv and intra-MeA OT replacement before the initial encounter rescuing the response (Ferguson et al., 2001).

Estrogens and OT in the MeA are prosocial, attenuating social anxiety and facilitating social recognition, interaction and memory (Choleris et al., 2003; Ferguson et al., 2000; Ferguson et al., 2001; Popik and Van Ree, 1993). Social recognition has been shown to be dependent on synthesis of OT in PVN and binding to OTR in MeA. OTR mRNA levels have been shown to fluctuate in accordance with estrogen levels and throughout the estrus cycle (Bale et al., 1995). Estrogens have been shown to regulate OT synthesis in PVN via effects on estrogen receptor alpha, and OTR expression in the MeA via estrogen receptor beta (Choleris et al., 2003).

17.2.2 Human research and clinical implications

In humans, OT facilitates trust and approach behaviors by reducing behavioral and neuroendocrine responses to social stressors and inhibiting anxiety-related defensive behaviors (Heinrichs and Domes, 2008). OT induces a positive affect in the form of trust in a social setting yet does not induce general calmness and changes in mood in a nonsocial setting (Kosfeld et al., 2005). OT attenuates both HPA output (salivary cortisol) and anxiety in response to the Trier social stress (Heinrichs et al., 2003). Moreover, the combination of OT and social support produces the lowest levels of cortisol and the most pronounced anxiolysis when compared to each treatment alone (Heinrichs et al., 2003). In a study investigating the effects of positive and negative emotional stimuli on OT levels, in response to positive contact (massage) OT levels were found to be increased, whereas they were decreased in response to sadness (Turner et al., 1999). Notably, individuals who did not show a drop in OT levels during times of sadness also showed lower anxiety in close relationships. OT seems to promote other positive emotions; women who were in a relationship had greater increases in OT in response to positive emotion. In addition, OT promotes generosity (Zak et al., 2007). OT has also been shown to improve the ability to infer the mental state of another individual (Domes et al., 2007b). Interestingly, in a more recent study, acute intranasal OT significantly improved fear recognition, but not the recognition of other emotions (Fischer-Shofty et al., 2010). Recently, a study has investigated a naturally occurring genetic polymorphism of OTR (rs53576) and its relationship to empathy and stress reactivity (Rodrigues et al., 2009). Compared to individuals homozygous for the G allele, polymorphic individuals with one or two copies of the A allele exhibited less empathy and displayed higher stress reactivity than GG individuals, as determined by heart rate response during a startle anticipation task and an affective reactivity scale.

Oxytocin, anxiety, and the amygdala

OT has been hypothesized to promote sociality by increasing trust and ameliorating the effects of mild stressors associated with social encounters (McCarthy et al., 1991; McCarthy et al., 1992). The amygdalar complex has been implicated in both fear and social recognition (Adolphs et al., 2005; LeDoux, 2003; Amaral, 2003). When presented with masked pictures of human faces bearing fearful expressions, the amygdala is activated, yet when presented with happy faces, it is inhibited (Whalen et al., 1998). Accordingly, lesions of the amygdala impair responsivity to fearful representations and promote social behavior (Adolphs et al., 2005). Amygdala activation has been associated with untrustworthiness (Winston et al., 2002)

and damage of the amygdala leads to increases in trust (Adolphs et al., 1998). Increases in amygdala activation have been reported in generalized social phobia (Stein et al., 2002) and hypersociability characteristic of patients suffering from Williams–Beuren syndrome has been associated with genetic alterations that affect amygdala activity (Meyer-Lindenberg et al., 2005).

OTR are highly expressed throughout the limbic system, including the amydaloid complex (Landgraf and Neumann, 2004; Huber et al., 2005). As such, the amygdala has been the focus of significant research on the effects of OT on fear, anxiety and sociality. OT reduces activity in the amygdala in concert with attenuating ratings of faces expressing a negative affect (Petrovic et al., 2008). When compared to placebo, intranasal OT has been shown to affect amygdalar responsivity to fearful, angry and happy facial expressions (Domes et al., 2007a). It is suggested that OT reduces uncertainty regarding a social stimulus and by so doing facilitates approach behaviors. Research in adult male humans shows that amygdalar responsivity in response to the presentation of fearful visual stimuli was attenuated in OT-treated versus placebo groups (Kirsch et al., 2005). In this study, OT dramatically attenuated amygdala activation and decoupled it from brainstem regions implicated in autonomic and behavioral components of fear. OT attenuates the amygdala, midbrain regions, and the dorsal striatum, active areas during a perceived breach in trust (Baumgartner et al., 2008). This particular study also found OT-reduced activity in amygdalar efferents related to the expression of fear, the periaqueductal gray and reticular formation, in agreement with preclinical findings (Huber et al., 2005). In a study meant to differentiate between nonsocial and social threats, subjects were presented with pictures of threatening scenes (nonsocial) or pictures of faces (social) and given placebo or intranasal OT (Hariri et al., 2002). Using fMRI, right lateralized activation of the amygdala was present to both types of stimuli. OT inhibited amygdalar activation, with a more pronounced effect on social stimuli activation. In a recent study investigating the effects of OT within a social setting, OT was shown to be involved in intergroup conflict management by facilitating in-group trust, cooperation and defense but not offense against out groups, a behavioral pattern referred to as parochial altruism (De Dreu et al., 2010). Collectively, these studies show that OT is prosocial, attenuating fear and anxiety and facilitating approach by regulating activity in the amygdala.

17.3 Vasopressin effects on fear and anxiety

AVP is involved in male-typical social behaviors (e.g., intermale aggression, scent marking, pair bonding, courtship), stress adaptation, anxiety, and aggression (Murgatroyd et al., 2004; Ring, 2005; Griebel et al., 2003; Donaldson et al., 2010; Carter, 2007). It is hypothesized that AVP may affect fear and anxiety by: (1) modulation of the HPA axis and; (2) central synthesis and release in the PVN, SON, the MeA and the BNST (De Vries and Buijs, 1983; Caffe et al., 1987; Landgraf and Neumann, 2004).

17.3.1 Animal research (see Table 17.1 for examples)

HPA

AVP synthesized in the parvocellular neurons of the PVN is released in the external zone of the median eminence and transported via the portal blood circulation to the anterior pituitary. There, AVP binds to AVPR1b and stimulates ACTH release by potentiating the actions of CRF (Aguilera et al., 2008). Chronic stress has been shown to increase AVP content in the PVN and AVPR1b in the anterior pituitary (Volpi et al., 2004), suggesting a potential shift from CRF to AVP control of the HPA axis and implicating AVPR1b as a potential therapeutic target for associated disorders. Indeed, affective disorders such as major depression, anxious-retarded depression and obsessive-compulsive disorder are characterized by atypical AVP levels or receptor activity (Purba et al., 1996; Zhou et al., 2001; van Londen et al., 1997). Furthermore, the AVPR1b

selective antagonist SSR149415 was found to be anxiolytic in a variety of tests and antidepressant in the forced swim test (Griebel et al., 2002), though these effects were present in both intact and hypophysectomized rats, suggesting AVP effects on behavior were at AVPR1b outside of the pituitary.

Central synthesis and release

AVP is centrally synthesized and released in the PVN, SON, the MeA and the BNST (De Vries and Buijs, 1983; Caffe et al., 1987; Landgraf and Neumann, 2004) to modulate a variety of behaviors. Vasopressinergic cells and fibers show interspecies conservation and sexual dimorphism, with markedly higher levels in males likely attributable to differential expression of gonadal steroids, such as testosterone and estrogens (De Vries, 2008). Testosterone affects AVP production via androgen and estrogen signaling and these effects can be localized to the BNST and MeA, regions abundant in vasopressinergic neurons that coexpress estrogen and androgen receptors. Gonadal steroids in the BNST and the MeA induce AVP expression, which is released in the lateral septum and ventral pallidum (De Vries, 2008). Testosterone may interact with AVP to attenuate acoustic startle in the presence of bright ambient light, a task that is dependent on the BNST (Toufexis et al., 2005). In contrast, estradiol was shown to enhance anxiogenesis in an EPM when administered concomitantly with AVP (McCarthy et al., 1996).

Some reports show contradictory evidence regarding the role of central AVP in anxiety; AVP administered intraseptum as well as peripherally (i.p.) was shown to be anxiolytic (Appenrodt et al., 1998), whereas application of an AVP antagonist intraseptum is anxiolyic as well (Liebsch et al., 1996). A study supports the latter finding; AVPR1 antisense oligodeoxynucleotide administration into the lateral septum is anxiolytic, as indicated by increased entries and percent of time spent in the open arms of an EPM (Landgraf et al., 1995). In addition, polymorphisms in the AVPR1a gene (Hammock et al., 2005) correlate with anxiety. AVPR1a may be involved in predator-induced fear and its conditioning (Ferris et al., 2010), though the use of an appetitive-conditioned stimulus (sucrose) in this study complicates the interpretation of these findings.

Selective breeding is commonly used to model extreme phenotypes within a normal population. In rats bred for extremes in anxiety-related behavior on the EPM (Liebsch et al., 1998), hyperanxiety correlates with high levels of synthesis and release of AVP from both magnocellular and parvocellular neurons of the PVN (Wigger et al., 2004). In line with these findings, intra-PVN application of an AVP1 receptor antagonist results in a decrease in anxiety-like behavior in rats bred for high anxiety, implicating AVP in PVN in trait anxiety (Wigger et al., 2004). In addition, the high anxiety phenotype shows single nucleotide polymorphisms in the AVP gene promoter that result in increased AVP expression, which may lead to anxiogenesis (Murgatroyd et al., 2004).

In an effort to discern AVP actions on AVPR1a or AVPR1b to affect anxiety, pharmacological as well as gene knockdown/out methodologies have been employed. AVPR1a receptor knockouts show a reduction in anxiety-like behaviors in the EPM, light/dark and open field paradigms (Bielsky et al., 2004). JNJ-17308616, a novel AVPR1a antagonist has been shown to be anxiolytic in the EPM, elevated zero maze, isolation induced ultrasonic vocalization and mouse marble-burying tests (Bleickardt et al., 2009). AVPR1b is also significantly involved in fear and anxiety-related behaviors (Griebel et al., 2003; Caldwell and Young, 2006). Although the AVPR1b knockout does not show changes in anxiety-like behaviors in the EPM, open field or light/dark apparati (Egashira et al., 2005), SSR149415, a selective nonpeptide AVPR1b antagonist produces anxiolysis in the light/dark, EPM, four-plate and punished drinking tests, (Griebel et al., 2002; Serradeil-Le Gal et al., 2005) as well as antidepressant effects that are localized to the lateral septum (Stemmelin et al., 2005). SSR149415 does not affect measures of risk assessment, which are used as indices of generalized anxiety disorder (GAD), yet affected defensive aggression in a defense test battery and social defeat

stress-induced anxiety (Griebel et al., 2002). These results suggest that the AVPR1b receptor is involved in stress disorders facilitated by a traumatic event (e.g., post-traumatic stress disorder). Recently, our laboratory assessed the effects of acute SSR149415 on chronic social defeat-induced social anxiety in male mice (Litvin et al., 2011). SSR149415 attenuated freezing and risk assessment in response to a novel male conspecific in chronically defeated animals, yet did not significantly affect avoidance. Collectively, these results suggest a novel therapeutic value for AVPR1b antagonists in the treatment of anxiety disorders (with a social component) that arise from both acute and chronic trauma.

17.3.2 Human research

Affective disorders such as major depression are characterized by an overactive HPA axis.HPA hyperactivity may be due to CRF overexpression (Licinio and Gold, 1991) and/or adrenal sensitization to ACTH (Charlton and Ferrier, 1989). Gold et al. (1978) were the first to postulate a role for AVP in affective disease due to its effects on sleep, pain, and memory. High cerebrospinal fluid concentration of AVP and CRF were found in depressed patients, with fluoxetine treatment attenuating both neuropeptides (De Bellis et al., 1993). High plasma AVP values have also been associated with a family history of depression and mixed anxiety and retardation (Goekoop et al., 2006). Small nucleotide polymorphism (SNP) haplotypes in the AVPR1b gene have been shown to protect against recurrent major depression, demonstrating increased susceptibility to affective disorders in patients with alternative polymorphisms (van West et al., 2004).

Intranasal AVP administration differentially affects men and women; in men, AVP reduces the perceived friendliness of same-sex conspecifics, whereas in women AVP has the opposite effect. AVP reduces heart-rate decelerations in both sexes and is anxiogenic in response to angry faces, as assessed by the State/Trait Anxiety Inventory (Thompson et al., 2006). These results are in accordance with the "tend and befriend" hypothesis by Taylor et al. (2000), which states that in contrast to unisex typical flight or fight responses in the face of danger, in females social anxiety promotes affiliative behaviors that serve to reduce stress of both self and offspring.

The amygdala has been implicated in the effects of AVP on mood. Amygdala reactivity to fearful faces has been shown to be associated with genetic variations in AVPR1a (Meyer-Lindenberg et al., 2009). Using a negative emotional processing paradigm, a recent study provides evidence for vasopressinergic modulation of an amygdala-medial prefrontal cortex circuit (Zink et al., 2010). AVP maintained activity in the subgenual cingulate and modified the interaction between subgenual and supragenual cingulate cortices. These changes may reflect disinhibition of the amygdala and consequent fear, as the medial prefrontal cortex has been shown to regulate amygdalar activity by means of negative feedback (Quirk et al., 2003).

17.4 Oxytocin and vasopressin interactions

The findings presented so far clarify that both OT and AVP are involved in fear and anxiety, at times in opposing fashions (Legros, 2001; Viviani and Stoop, 2008). Though divergent in their peripheral actions, OT and AVP affect similar systems within the central nervous system. OT and AVP show strong linkage to gonadal steroid expression and reproductive stage; evidence suggests OT is largely prosocial and may be particularly involved in female reproductive behaviors, while AVP is predominantly expressed in males, where it facilitates male-typical social behaviors. Together with their structural similarities and cross reactivity, these findings suggest that OT and AVP may interact on a behavioral level. Indeed, OT and AVP have been shown to have opposite effects on consolidation and expression of conditioned freezing to a context associated with shock; OT attenuates freezing when given prior to testing (expression) or immediately after training (consolidation), while AVP significantly enhances freezing when given before testing and tends to increase

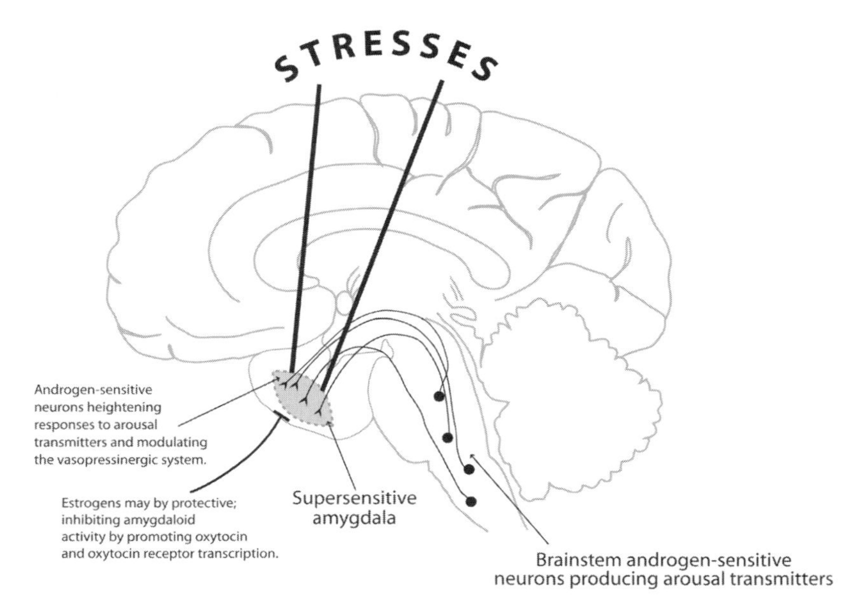

Figure 17.2 Schematic medial view of the human right cerebral hemisphere, cerebellum, and brainstem. The axons of brainstem androgen-sensitive neurons secrete arousal-related neurotransmitters such as norepinephrine (NE), dopamine (DA), serotonin (5 HydroxyTryptamine, 5HT), and others. Their widely activating effects on the cortex and subcortical nuclei, notably the amygdala, heighten alertness. In the amygdala, axonal inputs from these androgen-sensitive brainstem neurons increase sensitivity to fear-producing situations. Androgens may promote activation of systems associated with anxiety and male-typical behaviors by modulating vasopressinergic expression in the amygdala. Estrogens may serve a protective function by facilitating transcription of oxytocin and the oxytocin receptor. From uterine life on, boys are exposed to higher levels of androgens than girls; consequently they tend to react more strongly to various exogenous stresses (adapted from Pfaff et al., 2011).

freezing when administered after training (Stoehr et al., 1992). AVP and OT injected into the ventral hippocampus show opposing effects on the expression of active avoidance and its extinction; OT attenuates avoidance and catalyzes extinction, while AVP promotes avoidance and inhibits extinction (Ibragimov, 1990). OT and AVP may also interact at the pituitary level to modulate the HPA axis. OT has been shown to inhibit AVP-induced ACTH levels (Legros et al., 1982), while AVP promotes them.

As demonstrated in both the animal and human literature, the amygdala seems to regulate OT and AVP effects on mood. Viviani and Stoop (2008) have developed a cellular model that may underlie the opposing effects of AVP and OT on autonomic expression of fear and anxiety. Simply stated, AVP excites neurons in the medial division of the central amygdala to promote anxiogenesis. OT excites GABAergic neurons in the lateral part of the central amygdala, which in turn project to and inhibit the medial division, thus inhibiting the actions of AVP.

The relationships between OT, estrogen, AVP, testosterone, anxiety and sociality have inspired a theory regarding the sexual dimorphic nature of the social anxiety-related symptoms in autism spectrum disorders (ASD); that is, higher prevalence of autism in males (Pfaff et al., 2011). In males, brain arousal is enhanced by testosterone

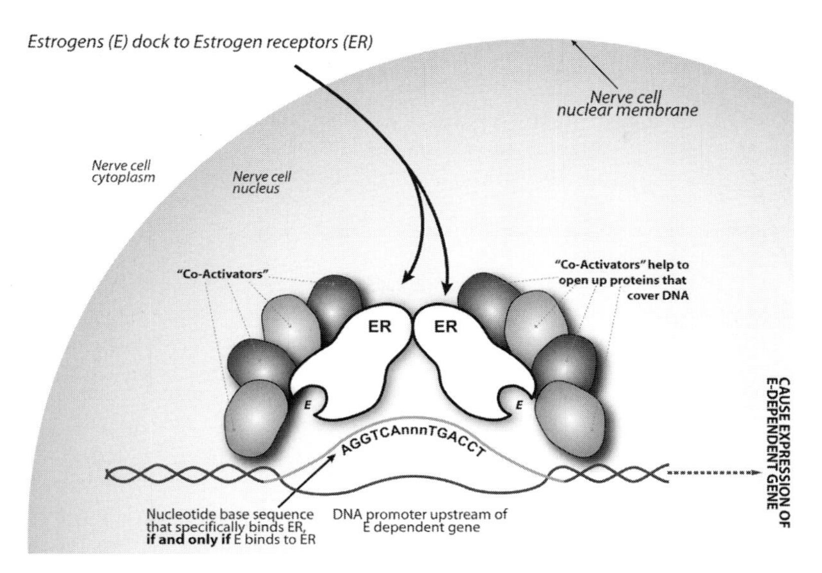

Figure 17.3 Schematic diagram of the role of Estrogens (E) on DNA of neurons in the brainstem and amygdala (see Figure 17.2), not to mention other neurons and cells in somatic tissues. E enter the nucleus from the neuronal cytoplasm after binding to Estrogen Receptors (ER). Only when E are bound to ER do the coactivator proteins that cover the coiled DNA open up the chromatin, uncovering and uncoiling the DNA to allow gene transcription. Uncoiling enables the E-carrying ER to interact with the DNA nucleotide base sequence of the promoter that commands the downstream expression of E-dependent genes (downstream is off the right side of the diagram). Downstream effects may include activation of oxytocinergic neuropeptide and/or receptor expression (adapted from Pfaff et al., 2011).

binding to androgen receptors in brainstem neurons (Pfaff, 2006). Arousal-related brainstem neurons activate the amygdala resulting in fear, anxiety and social avoidance typical in ASD. In addition to higher basal androgens in males, elevated prenatal androgenic exposure may induce hyperarousal and consequent social anxiety by modulating brainstem efferents, amygdaloid receptors and the vasopressinergic system (see Figure 17.2). In females, in addition to lower expression of androgens limiting hyperexcitability, estrogenic effects on the oxytocinergic system would protect against the deleterious effects of prenatal stress. On the cellular level, gonadal steroids may bind to associated receptors and facilitate gene transcription, which may lead to downstream effects such as expression of OT, AVP and/or their receptors (see Figure 17.3). Furthermore, lower levels of OT in concert with higher levels of AVP in brain regions associated with anxiety (e.g., BNST and MeA; De Vries, 2008), would predispose males to some psychopathologies associated with both anxiety and sociality (e.g., ASD).

17.5 Clinical implications

The combination of animal and human studies are beginning to delineate the involvement of OT and AVP in human disorders of affect; fear, anxiety, and sociality. Novel techniques such as analysis of SNPs or intranasal neuropeptide administration coupled with fMRI are unveiling the involvement of specific targets for clinical intervention. A thorough elucidation of the oxytocinergic and vasopressignergic systems and their interactions with hormonal systems related to both sex and

stress may shed light on the antecedents of affective and social disorders such as anxiety, depression, autism, and schizophrenia.

REFERENCES

Adolphs, R., Gosselin, F., Buchanan, T. W., Tranel, D., Schyns, P., and Damasio, A. R. (2005). A mechanism for impaired fear recognition after amygdala damage. *Nature*, 433, 68–72.

Adolphs, R., Tranel, D., and Damasio, A. R. (1998). The human amygdala in social judgment. *Nature*, 393, 470–474.

Aguilera, G., Subburaju, S., Young, S., and Chen, J. (2008). The parvocellular vasopressinergic system and responsiveness of the hypothalamic pituitary adrenal axis during chronic stress. *Prog Brain Res*, 170, 29–39.

Amaral, D. G. (2003). The amygdala, social behavior, and danger detection. *Ann N Y Acad Sci*, 1000, 337–347.

Appenrodt, E., Schnabel, R., and Schwarzberg, H. (1998). Vasopressin administration modulates anxiety-related behavior in rats. *Physiol Behav*, 64, 543–547.

Bale, T. L., Davis, A. M., Auger, A. P., Dorsa, D. M., and Mccarthy, M. M. (2001). CNS region-specific oxytocin receptor expression: importance in regulation of anxiety and sex behavior. *J Neurosci*, 21, 2546–2552.

Bale, T. L., Dorsa, D. M., and Johnston, C. A. (1995). Oxytocin receptor mRNA expression in the ventromedial hypothalamus during the estrous cycle. *J Neurosci*, 15, 5058–5064.

Bauman, J. W., Jr. (1965). Effect of hypophysectomy on the renal concentrating ability of the rat. *Endocrinology*, 77, 496–500.

Baumgartner, T., Heinrichs, M., Vonlanthen, A., Fischbacher, U., and Fehr, E. (2008). Oxytocin shapes the neural circuitry of trust and trust adaptation in humans. *Neuron*, 58, 639–650.

Belzung, C., Yalcin, I., Griebel, G., Surget, A., and Leman, S. (2006). Neuropeptides in psychiatric diseases: an overview with a particular focus on depression and anxiety disorders. *CNS Neurol Disord Drug Targets*, 5, 135–145.

Bielsky, I. F., Hu, S. B., Szegda, K. L., Westphal, H., and Young, L. J. (2004). Profound impairment in social recognition and reduction in anxiety-like behavior in vasopressin V1a receptor knockout mice. *Neuropsychopharmacology*, 29, 483–493.

Blanchard, D. C., Hynd, A. L., Minke, K. A., Minemoto, T., and Blanchard, R. J. (2001). Human defensive behaviors to threat scenarios show parallels to fear- and anxiety-related defense patterns of non-human mammals. *Neurosci Biobehav Rev*, 25, 761–770.

Blanchard, D. C., Litvin, Y., Pentkowski, N. S., and Blanchard, R. J. (2009). Defense and aggression. In Berntson, G. G., and Cacioppo, J. T. (Eds.) *Handbook of Neuroscience for the Behavioral Sciences*. Hoboken, New Jersey, John Wiley and Sons.

Blanchard, D. C., Shepherd, J. K., De Padua Carobrez, A., and Blanchard, R. J. (1991). Sex effects in defensive behavior: baseline differences and drug interactions. *Neurosci Biobehav Rev*, 15, 461–468.

Blanchard, R. J. and Blanchard, D. C. (1989). Antipredator defensive behaviors in a visible burrow system. *J Comp Psychol*, 103, 70–82.

Blanchard, R. J., Blanchard, D. C., Griebel, G., and Nutt, D. J. (2008). Introduction to the handbook on fear and anxiety. In Blanchard, R. J., Blanchard, D. C., Griebel, G., and Nutt, D. J. (Eds.) *Handbook of Anxiety and Fear*. Amsterdam, Elsevier Academic Press.

Bleickardt, C. J., Mullins, D. E., Macsweeney, C. P., Werner, B. J., Pond, A. J., Guzzi, M. F., Martin, F. D., Varty, G. B., and Hodgson, R. A. (2009). Characterization of the V1a antagonist, JNJ-17308616, in rodent models of anxiety-like behavior. *Psychopharmacology (Berl)*, 202, 711–718.

Blume, A., Bosch, O. J., Miklos, S., Torner, L., Wales, L., Waldherr, M., and Neumann, I. D. (2008). Oxytocin reduces anxiety via ERK1/2 activation: local effect within the rat hypothalamic paraventricular nucleus. *Eur J Neurosci*, 27, 1947–1956.

Born, J., Lange, T., Kern, W., Mcgregor, G. P., Bickel, U., and Fehm, H. L. (2002). Sniffing neuropeptides: a transnasal approach to the human brain. *Nat Neurosci*, 5, 514–516.

Bosch, O. J., Nair, H. P., Ahern, T. H., Neumann, I. D., and Young, L. J. (2009). The CRF system mediates increased passive stress-coping behavior following the loss of a bonded partner in a monogamous rodent. *Neuropsychopharmacology*, 34, 1406–1415.

Caffe, A. R., Van Leeuwen, F. W., and Luiten, P. G. (1987). Vasopressin cells in the medial amygdala of the rat project to the lateral septum and ventral hippocampus. *J Comp Neurol*, 261, 237–252.

Cain, C. K. and Ledoux, J. E. (2008). Brain mechanisms of Pavlovian and instrumental aversive conditioning. In Blanchard, R. J., Blanchard, D. C.,

Griebel, G., and Nutt, D. J. (Eds.) *Handbook of Anxiety and Fear*. Amsterdam, Elsevier Academic Press.

Caldwell, H. K., Lee, H. J., Macbeth, A. H., and Young, W. S., III. (2008). Vasopressin: behavioral roles of an "original" neuropeptide. *Prog Neurobiol*, 84, 1–24.

Caldwell, H. K. and Young, W. S. I. (2006). Oxytocin and Vasopressin: Genetics and Behavioral Implications. IN Lajtha, A. (Ed.) *Handbook on Neurochemistry and Molecular Neurobiology*. New York, Springer.

Canteras, N. S. (2002). The medial hypothalamic defensive system: hodological organization and functional implications. *Pharmacol Biochem Behav*, 71, 481–491.

Canteras, N. S., Simerly, R. B., and Swanson, L. W. (1995). Organization of projections from the medial nucleus of the amygdala: a PHAL study in the rat. *J Comp Neurol*, 360, 213–245.

Carter, C. S. (2007). Sex differences in oxytocin and vasopressin: implications for autism spectrum disorders? *Behav Brain Res*, 176, 170–186.

Carter, C. S. and Altemus, M. (1997). Integrative functions of lactational hormones in social behavior and stress management. *Ann NY Acad Sci*, 807, 164–174.

Carter, C. S., Devries, A. C., and Getz, L. L. (1995). Physiological substrates of mammalian monogamy: the prairie vole model. *Neurosci Biobehav Rev*, 19, 303–314.

Charlton, B. G. and Ferrier, I. N. (1989). Hypothalamo-pituitary-adrenal axis abnormalities in depression: a review and a model. *Psychol Med*, 19, 331–336.

Chiodera, P. and Legros, J. J. (1981). [Intravenous injection of synthetic oxytocin induces a decrease of cortisol plasma level in normal man]. *C R Seances Soc Biol Fil*, 175, 546–549.

Choleris, E., Clipperton-Allen, A. E., Plhan, A., and Kavaliers, M. (2009). Neuroendocrinology of social information processing in rats and mice. *Front Neuroendocrinol*, 30, 442–459.

Choleris, E., Devidze, N., Kavaliers, M., and Pfaff, D. W. (2008). Steroidal/neuropeptide interactions in hypothalamus and amygdala related to social anxiety. *Prog Brain Res*, 170, 291–303.

Choleris, E., Gustafsson, J. A., Korach, K. S., Muglia, L. J., Pfaff, D. W., and Ogawa, S. (2003). An estrogen-dependent four-gene micronet regulating social recognition: a study with oxytocin and estrogen receptor-alpha and -beta knockout mice. *Proc Natl Acad Sci USA*, 100, 6192–6197.

Cryan, J. F. and Dev, K. K. (2008). The glutamatergic system as a potential therapeutic target for the treatment of anxiety disorders. In Blanchard, R. J., Blanchard, D. C., Griebel, G., and Nutt, D. J. (Eds.) *Handbook of Anxiety and Fear*. Amsterdam, Elsevier Academic Press.

Da Costa, A. P., Wood, S., Ingram, C. D., and Lightman, S. L. (1996). Region-specific reduction in stress-induced c-fos mRNA expression during pregnancy and lactation. *Brain Res*, 742, 177–184.

Dale, H. H. (1906). On some physiological actions of ergot. *J Physiol*, 34, 163–206.

Darwin, C. (1872). *The Expression of the Emotions in Man and Animals*, Chicago, University of Chicago Press, 1872/1965.

De Bellis, M. D., Gold, P. W., Geracioti, T. D., Jr., Listwak, S. J., and Kling, M. A. (1993). Association of fluoxetine treatment with reductions in CSF concentrations of corticotropin-releasing hormone and arginine vasopressin in patients with major depression. *Am J Psychiatry*, 150, 656–657.

De Dreu, C. K., Greer, L. L., Handgraaf, M. J., Shalvi, S., Van Kleef, G. A., Baas, M., Ten Velden, F. S., Van Dijk, E., and Feith, S. W. (2010). The neuropeptide oxytocin regulates parochial altruism in intergroup conflict among humans. *Science*, 328, 1408–1411.

De Vries, G. J. (2008). Sex differences in vasopressin and oxytocin innervation of the brain. *Prog Brain Res*, 170, 17–27.

De Vries, G. J. and Buijs, R. M. (1983). The origin of the vasopressinergic and oxytocinergic innervation of the rat brain with special reference to the lateral septum. *Brain Res*, 273, 307–317.

Devries, A. C., Devries, M. B., Taymans, S. E., and Carter, C. S. (1996). The effects of stress on social preferences are sexually dimorphic in prairie voles. *Proc Natl Acad Sci USA*, 93, 11980–11984.

Devries, A. C., Glasper, E. R., and Detillion, C. E. (2003). Social modulation of stress responses. *Physiol Behav*, 79, 399–407.

Devries, A. C., Guptaa, T., Cardillo, S., Cho, M., and Carter, C. S. (2002). Corticotropin-releasing factor induces social preferences in male prairie voles. *Psychoneuroendocrinology*, 27, 705–714.

Domes, G., Heinrichs, M., Glascher, J., Buchel, C., Braus, D. F., and Herpertz, S. C. (2007a). Oxytocin attenuates amygdala responses to emotional faces regardless of valence. *Biol Psychiatry*, 62, 1187–1190.

Domes, G., Heinrichs, M., Michel, A., Berger, C., and Herpertz, S. C. (2007b). Oxytocin improves "mind-reading" in humans. *Biol Psychiatry*, 61, 731–733.

Donaldson, Z. R., Spiegel, L., and Young, L. J. (2010). Central vasopressin V1a receptor activation is independently necessary for both partner preference formation and expression in socially monogamous male prairie voles. *Behav Neurosci*, 124, 159–163.

Donaldson, Z. R. and Young, L. J. (2008). Oxytocin, vasopressin, and the neurogenetics of sociality. *Science*, 322, 900–904.

Ebner, K., Wotjak, C. T., Landgraf, R., and Engelmann, M. (2000). A single social defeat experience selectively stimulates the release of oxytocin, but not vasopressin, within the septal brain area of male rats. *Brain Res*, 872, 87–92.

Egashira, N., Tanoue, A., Higashihara, F., Fuchigami, H., Sano, K., Mishima, K., Fukue, Y., Nagai, H., Takano, Y., Tsujimoto, G., Stemmelin, J., Griebel, G., Iwasaki, K., Ikeda, T., Nishimura, R., and Fujiwara, M. (2005). Disruption of the prepulse inhibition of the startle reflex in vasopressin V1b receptor knockout mice: reversal by antipsychotic drugs. *Neuropsychopharmacology*, 30, 1996–2005.

Ekman, P. and Friesen, W. V. (1971). Constants across cultures in the face and emotion. *J Pers Soc Psychol*, 17, 124–129.

Endler, J. A. (1986). *Defense Against Predators*, Chicago, University of Chicago Press.

Fahrbach, S. E., Morrell, J. I., and Pfaff, D. W. (1985). Possible role for endogenous oxytocin in estrogen-facilitated maternal behavior in rats. *Neuroendocrinology*, 40, 526–532.

Ferguson, J. N., Aldag, J. M., Insel, T. R., and Young, L. J. (2001). Oxytocin in the medial amygdala is essential for social recognition in the mouse. *J Neurosci*, 21, 8278–8285.

Ferguson, J. N., Young, L. J., Hearn, E. F., Matzuk, M. M., Insel, T. R., and Winslow, J. T. (2000). Social amnesia in mice lacking the oxytocin gene. *Nat Genet*, 25, 284–288.

Ferris, C. F. (2005). Vasopressin/oxytocin and aggression. *Novartis Found Symp*, 268, 190–198; discussion 198–200, 242–253.

Ferris, C. F., Smerkers, B., Simon, N. G., Lu, S.-F., Heindel, N., Guillon, C., Fabio, K., Lacey, C. J., Garippa, C., and Brownstein, M. J. (2010). Imaging of conditioned predatory fear: Modulation by V1a receptors. *Society for Neuroscience*, 48, 1396–1405.

Fink, G., Sumner, B. E., Rosie, R., Grace, O., and Quinn, J. P. (1996). Estrogen control of central neurotransmission: effect on mood, mental state, and memory. *Cell Mol Neurobiol*, 16, 325–344.

Fischer-Shofty, M., Shamay-Tsoory, S. G., Harari, H., and Levkovitz, Y. (2010). The effect of intranasal administration of oxytocin on fear recognition. *Neuropsychologia*, 48, 179–184.

Freund-Mercier, M. J. and Richard, P. (1984). Electrophysiological evidence for facilitatory control of oxytocin neurones by oxytocin during suckling in the rat. *J Physiol*, 352, 447–466.

Gibbs, D. M., Vale, W., Rivier, J., and Yen, S. S. (1984). Oxytocin potentiates the ACTH-releasing activity of CRF(41) but not vasopressin. *Life Sci*, 34, 2245–2249.

Goekoop, J. G., De Winter, R. P., De Rijk, R., Zwinderman, K. H., Frankhuijzen-Sierevogel, A., and Wiegant, V. M. (2006). Depression with above-normal plasma vasopressin: validation by relations with family history of depression and mixed anxiety and retardation. *Psychiatry Res*, 141, 201–211.

Gold, P. W., Goodwin, F. K., and Reus, V. I. (1978). Vasopressin in affective illness. *Lancet*, 1, 1233–1236.

Goodson, J. L. (2005). The vertebrate social behavior network: evolutionary themes and variations. *Horm Behav*, 48, 11–22.

Griebel, G., Simiand, J., Serradeil-Le Gal, C., Wagnon, J., Pascal, M., Scatton, B., Maffrand, J. P., and Soubrie, P. (2002). Anxiolytic- and antidepressant-like effects of the non-peptide vasopressin V1b receptor antagonist, SSR149415, suggest an innovative approach for the treatment of stress-related disorders. *Proc Natl Acad Sci USA*, 99, 6370–6375.

Griebel, G., Simiand, J., Stemmelin, J., Gal, C. S., and Steinberg, R. (2003). The vasopressin V1b receptor as a therapeutic target in stress-related disorders. *Curr Drug Targets CNS Neurol Disord*, 2, 191–200.

Grippo, A. J., Cushing, B. S., and Carter, C. S. (2007a). Depression-like behavior and stressor-induced neuroendocrine activation in female prairie voles exposed to chronic social isolation. *Psychosom Med*, 69, 149–157.

Grippo, A. J., Gerena, D., Huang, J., Kumar, N., Shah, M., Ughreja, R., and Carter, C. S. (2007b). Social isolation induces behavioral and neuroendocrine disturbances relevant to depression in female and male prairie voles. *Psychoneuroendocrinology*, 32, 966–980.

Guimaraes, F. S., Carobrez, A. P., and Graeff, F. G. (2008). Modulation of anxiety behaviors by 5-HT-interacting drugs. In Blanchard, R. J., Blanchard, D. C., Griebel, G., and Nutt, D. J. (Eds.) *Handbook of Anxiety and Fear*. Amsterdam, Elsevier Academic Press.

Hammock, E. A., Lim, M. M., Nair, H. P., and Young, L. J. (2005). Association of vasopressin 1a receptor levels with a regulatory microsatellite and behavior. *Genes Brain Behav*, 4, 289–301.

Hariri, A. R., Tessitore, A., Mattay, V. S., Fera, F., and Weinberger, D. R. (2002). The amygdala response to emotional stimuli: a comparison of faces and scenes. *Neuroimage*, 17, 317–323.

Heinrichs, M., Baumgartner, T., Kirschbaum, C., and Ehlert, U. (2003). Social support and oxytocin interact to suppress cortisol and subjective responses to psychosocial stress. *Biol Psychiatry*, 54, 1389–1398.

Heinrichs, M. and Domes, G. (2008). Neuropeptides and social behaviour: effects of oxytocin and vasopressin in humans. *Prog Brain Res*, 170, 337–350.

Herman, J. P., Flak, J., and Jankord, R. (2008). Chronic stress plasticity in the hypothalamic paraventricular nucleus. *Prog Brain Res*, 170, 353–364.

Hernando, F., Schoots, O., Lolait, S. J., and Burbach, J. P. (2001). Immunohistochemical localization of the vasopressin V1b receptor in the rat brain and pituitary gland: anatomical support for its involvement in the central effects of vasopressin. *Endocrinology*, 142, 1659–1668.

Huber, D., Veinante, P., and Stoop, R. (2005). Vasopressin and oxytocin excite distinct neuronal populations in the central amygdala. *Science*, 308, 245–248.

Ibragimov, R. (1990). Influence of neurohypophyseal peptides on the formation of active avoidance conditioned reflex behavior. *Neurosci Behav Physiol*, 20, 189–193.

Insel, T. R. and Harbaugh, C. R. (1989). Lesions of the hypothalamic paraventricular nucleus disrupt the initiation of maternal behavior. *Physiol Behav*, 45, 1033–1041.

James, W. (1884). What is an emotion? *Mind*, 9, 188–205.

Kavaliers, M., Colwell, D. D., Choleris, E., Agmo, A., Muglia, L. J., Ogawa, S., and Pfaff, D. W. (2003). Impaired discrimination of and aversion to parasitized male odors by female oxytocin knockout mice. *Genes Brain Behav*, 2, 220–230.

Kendrick, K. M., Da Costa, A. P., Broad, K. D., Ohkura, S., Guevara, R., Levy, F., and Keverne, E. B. (1997). Neural control of maternal behaviour and olfactory recognition of offspring. *Brain Res Bull*, 44, 383–395.

Kikusui, T., Winslow, J. T., and Mori, Y. (2006). Social buffering: relief from stress and anxiety. *Philos Trans R Soc Lond B Biol Sci*, 361, 2215–2228.

Kirsch, P., Esslinger, C., Chen, Q., Mier, D., Lis, S., Siddhanti, S., Gruppe, H., Mattay, V. S., Gallhofer, B., and Meyer-LINDENBERG, A. (2005). Oxytocin modulates neural circuitry for social cognition and fear in humans. *J Neurosci*, 25, 11489–11493.

Koob, G. and Kreek, M. J. (2007). Stress, dysregulation of drug reward pathways, and the transition to drug dependence. *Am J Psychiatry*, 164, 1149–1159.

Kosfeld, M., Heinrichs, M., Zak, P. J., Fischbacher, U., and Fehr, E. (2005). Oxytocin increases trust in humans. *Nature*, 435, 673–676.

Kovacs, K. J. (1998). Functional neuroanatomy of the parvocellular vasopressinergic system: transcriptional responses to stress and glucocorticoid feedback. *Prog Brain Res*, 119, 31–43.

Landgraf, R., Gerstberger, R., Montkowski, A., Probst, J. C., Wotjak, C. T., Holsboer, F., and Engelmann, M. (1995). V1 vasopressin receptor antisense oligodeoxynucleotide into septum reduces vasopressin binding, social discrimination abilities, and anxiety-related behavior in rats. *J Neurosci*, 15, 4250–4258.

Landgraf, R. and Neumann, I. D. (2004). Vasopressin and oxytocin release within the brain: a dynamic concept of multiple and variable modes of neuropeptide communication. *Front Neuroendocrinol*, 25, 150–176.

Ledoux, J. (2003). The emotional brain, fear, and the amygdala. *Cellular and Molecular Neurobiology*, 23, 727–738.

Ledoux, J. E. (1998). *The Emotional Brain: the Mysterious Underpinnings of Emotional Life*, New York, Simon and Schuster.

Lee, H. J., Macbeth, A. H., Pagani, J. H., and Young, W. S., III. (2009). Oxytocin: the great facilitator of life. *Prog Neurobiol*, 88, 127–151.

Legros, J. J. (2001). Inhibitory effect of oxytocin on corticotrope function in humans: are vasopressin and oxytocin ying-yang neurohormones? *Psychoneuroendocrinology*, 26, 649–655.

Legros, J. J., Chiodera, P., and Demey-PONSART, E. (1982). Inhibitory influence of exogenous oxytocin on adrenocorticotropin secretion in normal human subjects. *J Clin Endocrinol Metab*, 55, 1035–1039.

Legros, J. J., Chiodera, P., and Geenen, V. (1988). Inhibitory action of exogenous oxytocin on plasma cortisol in normal human subjects: evidence of action at the adrenal level. *Neuroendocrinology*, 48, 204–206.

Licinio, J. and Gold, P. W. (1991). Role of corticotrophin releasing hormone 41 in depressive illness. *Baillieres Clin Endocrinol Metab*, 5, 51–58.

Liebsch, G., Montkowski, A., Holsboer, F., and Landgraf, R. (1998). Behavioural profiles of two Wistar rat lines selectively bred for high or low anxiety-related behaviour. *Behav Brain Res*, 94, 301–310.

Liebsch, G., Wotjak, C. T., Landgraf, R., and Engelmann, M. (1996). Septal vasopressin modulates anxiety-related behaviour in rats. *Neurosci Lett*, 217, 101–104.

Lightman, S. L. and Young, W. S., III. (1989). Lactation inhibits stress-mediated secretion of corticosterone and oxytocin and hypothalamic accumulation of corticotropin-releasing factor and enkephalin messenger ribonucleic acids. *Endocrinology*, 124, 2358–2364.

Lim, M. M. and Young, L. J. (2006). Neuropeptidergic regulation of affiliative behavior and social bonding in animals. *Horm Behav*, 50, 506–517.

Litvin, Y., Blanchard, D. C., and Blanchard, R. J. (2007). Rat 22kHz ultrasonic vocalizations as alarm cries. *Behav Brain Res*, 182, 166–172.

Litvin, Y., Murakami, G., and Pfaff, D. W. (2011). Effects of chronic social defeat on behavioral and neural correlates of sociality: Vasopressin, oxytocin and the vasopressinergic V1b receptor. *Physiol Behav*, 103, 393–403.

Litvin, Y., Pentkowski, N. S., Pobbe, R. L., Blanchard, D. C., and Blanchard, R. J. (2008). Unconditioned models of fear and anxiety. In Blanchard, R. J., Blanchard, D. C., Griebel, G., and Nutt, D. J. (Eds.) *Handbook of Anxiety and Fear*. Amsterdam, Elsevier Academic Press.

Markham, C. M., Blanchard, D. C., Canteras, N. S., Cuyno, C. D., and Blanchard, R. J. (2004). Modulation of predatory odor processing following lesions to the dorsal premammillary nucleus. *Neurosci Lett*, 372, 22–26.

Martinez, R. C., Carvalho-Netto, E. F., Amaral, V. C., Nunes-DE-SOUZA, R. L., and Canteras, N. S. (2008). Investigation of the hypothalamic defensive system in the mouse. *Behav Brain Res*, 192, 185–190.

McCarthy, M. M., Chung, S. K., Ogawa, S., Kow, L. M., and Pfaff, D. W. (1991). Behavioral effects of oxytocin: is there a unifying principle. In Jard, S., and Jamison, R. (Eds.) *Vasopressin, Colloque INSERM*. John Libbey Eurotext.

McCarthy, M. M., Kow, L. M., and Pfaff, D. W. (1992). Speculations concerning the physiological significance of central oxytocin in maternal behavior. In Pedersen, C. A., Jirikowski, G. F., Caldwell, J. D., and Insel, T. R. (Eds.) *Oxytocin in Maternal, Sexual and Social Behaviors*. New York, Ann. NY Acad. Sci.

McCarthy, M. M., Mcdonald, C. H., Brooks, P. J., and Goldman, D. (1996). An anxiolytic action of oxytocin is enhanced by estrogen in the mouse. *Physiol Behav*, 60, 1209–1215.

McEwen, B. S. (2007). Physiology and neurobiology of stress and adaptation: central role of the brain. *Physiol Rev*, 87, 873–904.

Mcnaughton, N. and Corr, P. J. (2004). A two-dimensional neuropsychology of defense: fear/anxiety and defensive distance. *Neurosci Biobehav Rev*, 28, 285–305.

Meyer-Lindenberg, A., Hariri, A. R., Munoz, K. E., Mervis, C. B., Mattay, V. S., Morris, C. A., and Berman, K. F. (2005). Neural correlates of genetically abnormal social cognition in Williams syndrome. *Nat Neurosci*, 8, 991–993.

Meyer-Lindenberg, A., Kolachana, B., Gold, B., Olsh, A., Nicodemus, K. K., Mattay, V., Dean, M., and Weinberger, D. R. (2009). Genetic variants in AVPR1A linked to autism predict amygdala activation and personality traits in healthy humans. *Mol Psychiatry*, 14, 968–975.

Missig, G., Ayers, L. W., Schulkin, J., and Rosen, J. B. (2010). Oxytocin reduces background anxiety in a fear-potentiated startle paradigm. *Neuropsychopharmacology*, 35, 2607–2616.

Mobbs, D., Petrovic, P., Marchant, J. L., Hassabis, D., Weiskopf, N., Seymour, B., Dolan, R. J., and Frith, C. D. (2007). When Fear is Near: Threat Imminence Elicits Prefrontal-Periaqueductal Gray Shifts in Humans. *Science*, 317, 1079–1083.

Morgan, M. A., Schulkin, J., and Pfaff, D. W. (2004). Estrogens and non-reproductive behaviors related to activity and fear. *Neurosci Biobehav Rev*, 28, 55–63.

Murgatroyd, C., Wigger, A., Frank, E., Singewald, N., Bunck, M., Holsboer, F., Landgraf, R., and Spengler, D. (2004). Impaired repression at a vasopressin promoter polymorphism underlies overexpression of vasopressin in a rat model of trait anxiety. *J Neurosci*, 24, 7762–7770.

Neumann, I., Douglas, A. J., Pittman, Q. J., Russell, J. A., and Landgraf, R. (1996). Oxytocin released within the supraoptic nucleus of the rat brain by positive feedback action is involved in parturition-related events. *J Neuroendocrinol*, 8, 227–233.

Neumann, I. and Landgraf, R. (1989). Septal and Hippocampal Release of Oxytocin, but not Vasopressin, in the Conscious Lactating Rat During Suckling. *J Neuroendocrinol*, 1, 305–308.

Neumann, I., Russell, J. A., and Landgraf, R. (1993). Oxytocin and vasopressin release within the supraoptic and paraventricular nuclei of pregnant, parturient and lactating rats: a microdialysis study. *Neuroscience*, 53, 65–75.

Neumann, I. D. (2002). Involvement of the brain oxytocin system in stress coping: interactions with the hypothalamo-pituitary-adrenal axis. *Prog Brain Res*, 139, 147–162.

Neumann, I. D., Johnstone, H. A., Hatzinger, M., Liebsch, G., Shipston, M., Russell, J. A., Landgraf, R., and Douglas, A. J. (1998). Attenuated neuroendocrine responses to emotional and physical stressors in pregnant rats involve adenohypophysial changes. *J Physiol*, 508 (Pt 1), 289–300.

Neumann, I. D., Torner, L., and Wigger, A. (2000a). Brain oxytocin: differential inhibition of neuroendocrine stress responses and anxiety-related behaviour in virgin, pregnant and lactating rats. *Neuroscience*, 95, 567–575.

Neumann, I. D., Wigger, A., Torner, L., Holsboer, F., and Landgraf, R. (2000b). Brain oxytocin inhibits basal and stress-induced activity of the hypothalamo-pituitary-adrenal axis in male and female rats: partial action within the paraventricular nucleus. *J Neuroendocrinol*, 12, 235–243.

Newman, S. W. (1999). The medial extended amygdala in male reproductive behavior. A node in the mammalian social behavior network. *Ann NY Acad Sci*, 877, 242–257.

Olazabal, D. E. and Young, L. J. (2005). Variability in "spontaneous" maternal behavior is associated with anxiety-like behavior and affiliation in naive juvenile and adult female prairie voles (Microtus ochrogaster). *Dev Psychobiol*, 47, 166–178.

Oliver, G. and Schafer, E. A. (1895). On the Physiological Action of Extracts of Pituitary Body and certain other Glandular Organs: Preliminary Communication. *J Physiol*, 18, 277–279.

Pedersen, C. A. and Prange, A. J., Jr. (1979). Induction of maternal behavior in virgin rats after intracerebroventricular administration of oxytocin. *Proc Natl Acad Sci USA*, 76, 6661–6665.

Pentkowski, N. S., Litvin, Y., Blanchard, D. C., Vasconcellos, A., King, L. B., and Blanchard, R. J. (Unpublished). Effects of Estrous Cycle Stage on Unconditioned and Conditioned Defensive Behavior in Female Long-Evans Hooded Rats., University of Hawaii at Manoa.

Perkins, A. M. and Corr, P. J. (2006). Reactions to threat and personality: psychometric differentiation of intensity and direction dimensions of human defensive behaviour. *Behav Brain Res*, 169, 21–28.

Petrovic, P., Kalisch, R., Singer, T., and Dolan, R. J. (2008). Oxytocin attenuates affective evaluations of conditioned faces and amygdala activity. *J Neurosci*, 28, 6607–6615.

Pfaff, D. W. (1999). *Drive: Neurobiological and Molecular Mechanisms of Sexual Motivation*, Cambridge, Massachusetts, MIT Press.

Pfaff, D. W. (2006). *Brain Arousal and Information Theory: Neural and Genetic Mechanisms*, Cambridge, Massachusetts and London, England, Harvard University Press.

Pfaff, D. W., Phillips, M. I., and Rubin, R. T. (2004). *Principles of Hormone/Behavior Relations*, Amsterdam, Elsevier Academic Press.

Pfaff, D. W., Rapin, I., and Goldman, S. (2011). Male predominance in autism: neuroendocrine influences on arousal and social anxiety. *Autism Res*, 4, 163–176.

Plotsky, P. M. (1991). Pathways to the secretion of adrenocorticotropin: a view from the portal. *J Neuroendocrinol*, 3, 1–9.

Popik, P. and Van Ree, J. M. (1993). Social transmission of flavored tea preferences: facilitation by a vasopressin analog and oxytocin. *Behav Neural Biol*, 59, 63–68.

Purba, J. S., Hoogendijk, W. J., Hofman, M. A., and Swaab, D. F. (1996). Increased number of vasopressin- and oxytocin-expressing neurons in the paraventricular nucleus of the hypothalamus in depression. *Arch Gen Psychiatry*, 53, 137–143.

Quirk, G. J., Likhtik, E., Pelletier, J. G., and Pare, D. (2003). Stimulation of medial prefrontal cortex decreases the responsiveness of central amygdala output neurons. *J Neurosci*, 23, 8800–8807.

Rhodes, C. H., Morrell, J. I., and Pfaff, D. W. (1981). Immunohistochemical analysis of magnocellular elements in rat hypothalamus: distribution and numbers of cells containing neurophysin, oxytocin, and vasopressin. *J Comp Neurol*, 198, 45–64.

Ring, R. H. (2005). The central vasopressinergic system: examining the opportunities for psychiatric drug development. *Curr Pharm Des*, 11, 205–225.

Ring, R. H., Malberg, J. E., Potestio, L., Ping, J., Boikess, S., Luo, B., Schechter, L. E., Rizzo, S., Rahman, Z., and Rosenzweig-Lipson, S. (2006). Anxiolytic-like activity of oxytocin in male mice: behavioral and autonomic evidence, therapeutic implications. *Psychopharmacology (Berl)*, 185, 218–225.

Rodgers, R. J. (1997). Animal models of 'anxiety': where next? *Behav Pharmacol*, 8, 477–496; discussion 497–504.

Rodrigues, S. M., Saslow, L. R., Garcia, N., John, O. P., and Keltner, D. (2009). Oxytocin receptor genetic variation relates to empathy and stress reactivity in humans. *Proc Natl Acad Sci USA*, 106, 21437–21441.

Rosen, J. B. and Schulkin, J. (1998). From normal fear to pathological anxiety. *Psychol Rev*, 105, 325–350.

Ross, H. E., Cole, C. D., Smith, Y., Neumann, I. D., Landgraf, R., Murphy, A. Z., and Young, L. J. (2009). Characterization of the oxytocin system regulating affiliative behavior in female prairie voles. *Neuroscience*, 162, 892–903.

Rowlett, J. K. (2008). Subtype-selective GABAA/benzodiazepine receptor ligands for the treatment of anxiety disorders. In Blanchard, R. J., Blanchard, D. C., Griebel, G., and Nutt, D. J. (Eds.) *Handbook of Anxiety and Fear*. Amsterdam, Elsevier Academic Press.

Sawchenko, P. E., Swanson, L. W., and Vale, W. W. (1984). Corticotropin-releasing factor: co-expression within distinct subsets of oxytocin-, vasopressin-, and neurotensin-immunoreactive neurons in the hypothalamus of the male rat. *J Neurosci*, 4, 1118–1129.

Schafer, E. A. and Mackenzie, K. (1911). The action of animal extracts on milk secretion. *Proceedings of the Royal Society of London Series B-Containing Papers of a Biological Character*, 84, 16–22.

Serradeil-Le Gal, C., Wagnon, J., III, Tonnerre, B., Roux, R., Garcia, G., Griebel, G., and Aulombard, A. (2005). An overview of SSR149415, a selective nonpeptide vasopressin V(1b) receptor antagonist for the treatment of stress-related disorders. *CNS Drug Rev*, 11, 53–68.

Spiteri, T., Musatov, S., Ogawa, S., Ribeiro, A., Pfaff, D. W., and Agmo, A. (2010). The role of the estrogen receptor alpha in the medial amygdala and ventromedial nucleus of the hypothalamus in social recognition, anxiety and aggression. *Behav Brain Res*, 210, 211–220.

Steckler, T. (2008). Peptide receptor ligands to treat anxiety disorders. In Blanchard, R. J., Blanchard, D. C., Griebel, G., and Nutt, D. J. (Eds.) *Handbook of Anxiety and Fear*. Amsterdam, Elsevier Academic Press.

Stein, M. B., Goldin, P. R., Sareen, J., Zorrilla, L. T., and Brown, G. G. (2002). Increased amygdala activation to angry and contemptuous faces in generalized social phobia. *Arch Gen Psychiatry*, 59, 1027–1034.

Stemmelin, J., Lukovic, L., Salome, N., and Griebel, G. (2005). Evidence that the lateral septum is involved in the antidepressant-like effects of the vasopressin V1b receptor antagonist, SSR149415. *Neuropsychopharmacology*, 30, 35–42.

Stern, J. M., Goldman, L., and Levine, S. (1973). Pituitary-adrenal responsiveness during lactation in rats. *Neuroendocrinology*, 12, 179–191.

Stoehr, J. D., Cramer, C. P., and North, W. G. (1992). Oxytocin and vasopressin hexapeptide fragments have opposing influences on conditioned freezing behavior. *Psychoneuroendocrinology*, 17, 267–271.

Stowe, J. R., Liu, Y., Curtis, J. T., Freeman, M. E., and Wang, Z. (2005). Species differences in anxiety-related responses in male prairie and meadow voles: the effects of social isolation. *Physiol Behav*, 86, 369–378.

Swanson, L. W., and Sawchenko, P. E. (1983). Hypothalamic integration: organization of the paraventricular and supraoptic nuclei. *Annu Rev Neurosci*, 6, 269–324.

Takayanagi, Y., Yoshida, M., Bielsky, I. F., Ross, H. E., Kawamata, M., Onaka, T., Yanagisawa, T., Kimura, T., Matzuk, M. M., Young, L. J., and Nishimori, K. (2005). Pervasive social deficits, but normal parturition, in oxytocin receptor-deficient mice. *Proc Natl Acad Sci USA*, 102, 16096–16101.

Taylor, S. E., Klein, L. C., Lewis, B. P., Gruenewald, T. L., Gurung, R. A., and Updegraff, J. A. (2000). Biobehavioral responses to stress in females: tend-and-befriend, not fight-or-flight. *Psychol Rev*, 107, 411–429.

Thompson, R. R., George, K., Walton, J. C., Orr, S. P., and Benson, J. (2006). Sex-specific influences of vasopressin on human social communication. *Proc Natl Acad Sci USA*, 103, 7889–7894.

Toufexis, D., Davis, C., Hammond, A., and Davis, M. (2005). Sex differences in hormonal modulation of anxiety measured with light-enhanced startle: possible role for arginine vasopressin in the male. *J Neurosci*, 25, 9010–9016.

Turner, R. A., Altemus, M., Enos, T., Cooper, B., and Mcguinness, T. (1999). Preliminary research on plasma oxytocin in normal cycling women: investigating emotion and interpersonal distress. *Psychiatry*, 62, 97–113.

Uvnas-Moberg, K., Ahlenius, S., Hillegaart, V., and Alster, P. (1994). High doses of oxytocin cause sedation and low doses cause an anxiolytic-like effect in male rats. *Pharmacol Biochem Behav*, 49, 101–106.

Vale, W., Spiess, J., Rivier, C., and Rivier, J. (1981). Characterization of a 41-residue ovine hypothalamic peptide that stimulates secretion of corticotropin and beta-endorphin. *Science*, 213, 1394–1397.

Van Londen, L., Goekoop, J. G., Van Kempen, G. M., Frankhuijzen-Sierevogel, A. C., Wiegant, V. M., Van Der Velde, E. A., and De Wied, D. (1997). Plasma levels of arginine vasopressin elevated in patients with major depression. *Neuropsychopharmacology*, 17, 284–292.

Van Tol, H. H., Bolwerk, E. L., Liu, B., and Burbach, J. P. (1988). Oxytocin and vasopressin gene expression in the hypothalamo-neurohypophyseal system of the rat during the estrous cycle, pregnancy, and lactation. *Endocrinology*, 122, 945–951.

Van West, D., Del-Favero, J., Aulchenko, Y., Oswald, P., Souery, D., Forsgren, T., Sluijs, S., Bel-KACEM, S., Adolfsson, R., Mendlewicz, J., Van DUIJN, C., Deboutte, D., Van Broeckhoven, C., and Claes, S. (2004). A major SNP haplotype of the arginine vasopressin 1B receptor protects against recurrent major depression. *Mol Psychiatry*, 9, 287–292.

Veenema, A. H. and Neumann, I. D. (2008). Central vasopressin and oxytocin release: regulation of complex social behaviours. *Prog Brain Res*, 170, 261–276.

Viviani, D. and Stoop, R. (2008). Opposite effects of oxytocin and vasopressin on the emotional expression of the fear response. *Prog Brain Res*, 170, 207–218.

Volpi, S., Rabadan-Diehl, C., and Aguilera, G. (2004). Vasopressinergic regulation of the hypothalamic pituitary adrenal axis and stress adaptation. *Stress*, 7, 75–83.

Waldherr, M. and Neumann, I. D. (2007). Centrally released oxytocin mediates mating-induced anxiolysis in male rats. *Proc Natl Acad Sci USA*, 104, 16681–16684.

Whalen, P. J., Rauch, S. L., Etcoff, N. L., Mcinerney, S. C., Lee, M. B., and Jenike, M. A. (1998). Masked presentations of emotional facial expressions modulate amygdala activity without explicit knowledge. *J Neurosci*, 18, 411–418.

Wigger, A., Sanchez, M. M., Mathys, K. C., Ebner, K., Frank, E., Liu, D., Kresse, A., Neumann, I. D., Holsboer, F., Plotsky, P. M., and Landgraf, R. (2004). Alterations in central neuropeptide expression, release, and receptor binding in rats bred for high anxiety: critical role of vasopressin. *Neuropsychopharmacology*, 29, 1–14.

Williams, J. R., Insel, T. R., Harbaugh, C. R., and Carter, C. S. (1994). Oxytocin administered centrally facilitates formation of a partner preference in female prairie voles (Microtus ochrogaster). *J Neuroendocrinol*, 6, 247–250.

Winston, J. S., Strange, B. A., O'Doherty, J., and Dolan, R. J. (2002). Automatic and intentional brain responses during evaluation of trustworthiness of faces. *Nat Neurosci*, 5, 277–283.

Young, L. J. and Wang, Z. (2004). The neurobiology of pair bonding. *Nat Neurosci*, 7, 1048–1054.

Zak, P. J., Stanton, A. A., and Ahmadi, S. (2007). Oxytocin increases generosity in humans. *PLoS One*, 2, e1128.

Zhou, J. N., Riemersma, R. F., Unmehopa, U. A., Hoogendijk, W. J., Van Heerikhuize, J. J., Hofman, M. A., and Swaab, D. F. (2001). Alterations in arginine vasopressin neurons in the suprachiasmatic nucleus in depression. *Arch Gen Psychiatry*, 58, 655–662.

Zingg, H. H. and Lefebvre, D. L. (1988). Oxytocin and vasopressin gene expression during gestation and lactation. *Brain Res*, 464, 1–6.

Zink, C. F., Stein, J. L., Kempf, L., Hakimi, S., and Meyer-Lindenberg, A. (2010). Vasopressin modulates medial prefrontal cortex-amygdala circuitry during emotion processing in humans. *J Neurosci*, 30, 7017–7022.

Zisook, S., Paulus, M., Shuchter, S. R., and Judd, L. L. (1997). The many faces of depression following spousal bereavement. *J Affect Disord*, 45, 85–94; discussion 94–95.

18

Oxytocin instantiates empathy and produces prosocial behaviors

Jorge A. Barraza and Paul J. Zak

18.1 Introduction

In January of 2007, Wesley Autrey was waiting for the New York City subway with his two young daughters when a young man nearby starting seizing and fell onto the tracks. With a train coming, Mr. Autrey leapt toward the tracks and moved the man, Cameron Hollopeter, between two tracks as the subway raced toward them. Autrey lay on top of Hollopeter to hold him down as the subway passed over them. Both lived.

Autrey's actions are an extreme form of altruistic behavior; that is, a behavior that is individually costly but benefits another (West, Griffin, and Gardner, 2006). This chapter focuses on prosocial behaviors, by which we mean intentional altruistic behaviors. Indeed, prosocial behaviors are so quotidian that we often fail to notice the countless ways people assist others, often with no ability or expectation of reciprocation. Did human evolution really select for survival of the nicest? (Bowles, 2009) If so, there must be one or more neural substrates promoting prosociality.

The neuropeptide oxytocin (OT) has been implicated in a number of prosocial behaviors (Zak, 2011; also see Campbell, 2010; Meyer-Lindenberg, 2008; Tost et al., 2010). We divide the research on prosocial behaviors into four domains: reciprocity, trust, generosity, and altruistic helping. Although each behavior is likely to have its own distinct neural processes that produce it, we have found that they all depend on OT. Our lab has taken a comprehensive approach to the study of prosocial behaviors and OT, developing both correlational and causal evidence. Our approach first designs tasks that induce OT release. Much of the early OT research relied on basal OT levels that are both low and highly variable and are unpredictive of most prosocial behaviors in humans. It is the release of OT, not its basal level, that motivates social engagement. Once we have confirmed that a prosocial behavior induces OT release, we then establish the causal relationship by infusing OT, showing that we can directly manipulate the behavior being studied.

This chapter reviews and puts into context recent findings on the role OT in producing prosocial behaviors in humans. We also connect this physiologic mechanism to psychological states that produce these behaviors, particularly empathy. Empathy causes one to attend to another's plight, and can be a motivation to invest resources to assist someone in distress. An OT-empathy relationship has been suggested based on similar behavior effects, but several lines of emerging research appear to show that OT does instantiate the experience of empathy.

18.2 Prosocial behavior

Prosocial behavior can be a single event (e.g., donating money to a beggar) or can extend over a long

Oxytocin, Vasopressin, and Related Peptides in the Regulation of Behavior, ed. E. Choleris, D. W. Pfaff, and M. Kavaliers.
Published by Cambridge University Press.

period of time (e.g., joining a volunteer organization). For this review, we focus on prosocial behaviors that can be studied in the lab. These include reciprocity, trust, generosity, and altruistic helping. Reciprocity and generosity, as operationalized in the proceeding studies, can be classified as sharing tasks in which benefits for another reduce one's own resources. Trust can be classified as potentially cooperative, where people can obtain a mutually beneficial outcome. Finally, we discuss our recent research done on a distinct form of prosociality, altruistic helping, in which assistance occurs at a distance and many external incentives to benefit another are absent.

Prior to reviewing the literature, we note that there are a number of benefits to using economic games to study human prosociality. The first is that the measure of prosociality – money transferred to another person – is active, objective, and scales to measure degrees of prosociality. This means that there is meaning behind an incremental change, unlike differences across Likert-type 7-point scales. Secondly, behavior is not gratuitous; one cannot simply claim to be trusting or generous, a tangible action must take place in order to classify it as such. For instance, to gauge trust, money must be sent to another person without a guarantee of return. Finally, measuring prosociality using money motivates participants in experiments to attend to the task by adding value to the decision.

18.2.1 Reciprocity

A key ingredient to cooperation is reciprocity. Reciprocity is an important aspect in many evolutionarily based explanations for the existence of prosocial behaviors (e.g., Gintis, 2000; Nowak and Sigmund, 2006; Trivers, 1971). Reciprocation can be either positive or negative, that is, responding in-kind to how one is treated. This reflects a tit-for-tat strategy that is among the best strategies in prisoner's dilemma tournaments (Axelrod, 1984) and is part of the evolutionary model of strong reciprocity (Gintis, 2000).

Positive reciprocity. Reciprocity that comes at an individual cost may make sense when extended cooperation is possible. However, humans appear to return favors even knowing that the initial giver will not know that the favor was reciprocated (Burger, Sanchez, Imberi, and Grande, 2009). A turning point in the study of reciprocity was the discovery of the role of OT (Zak et al., 2004; 2005; Zak, 2007; Zak, 2008). The impact of OT on reciprocity (and trust, discussed below) has been largely studied using a sequential cooperative dilemma known as the trust game (Berg, Dickhaut, and McCabe, 1995).

In the trust game (TG), participants are placed into dyads and randomly assigned to the roles of decision-maker 1 (DM1) and decision-maker 2 (DM2). Both DMs are allocated a fixed dollar amount, often $10, as an endowment from the experimenters. After instruction, DM1 can choose to send any integer amount (including zero) of his or her $10 to the DM2 in his or her dyad. Participants know that whatever is sent comes out of DM1's account and is tripled in DM2's account. DM2 is then prompted to send some integer amount back to DM1 (including zero). Transfers from DM2 to DM1 are not multiplied and constitute a dollar-for-dollar allocation out of DM2's account and into DM1's account. The typical version of the trust game used in OT studies has participants making a single decision as either DM1 or DM2 so that the effects of reputation are minimized. Moreover, decisions are made via computer so that matched participants are not identifiable to each other or to the experimenters. This minimizes extraneous factors that might influence the decision (e.g., partner demographics).

The consensus view in experimental economics is that the amount that DM1 sends to DM2 is a measure of *trust* (Smith, 1998). The more money DM1 sends to DM2, the greater degree of trust since more money is at risk if DM2 defects. Similarly, the money sent by DM2 to DM1 is an index of the former's *reciprocity* or *trustworthiness*, that is, the amount DM2 reciprocates given a signal of trust from DM1. In the TG, 98% of DM2s who are sent money return at least some (Smith, 1998; Zak, 2008). Remember

that participants typically play a single round of the TG, so there is no chance of future cooperation as a rationale for reciprocity. There are therefore strong incentives for DM2s to defect. Moreover, the interaction is computer mediated and anonymous, so defectors can escape identification. If there are no tangible incentives to reciprocate, then what motivates or signals reciprocity?

Using the TG, we found that peripheral OT for DM2s who received an intentional signal of trust was an average of 41% higher compared to DM2 controls receiving the same average amount of money determined by a random draw and therefore not denoting trust ($N = 67$, $p = 0.05$; Zak et al., 2004; 2005). We have also found that the within-subjects change in OT is proportional to the money received (Morhenn et al., 2008). The TG was the first non-reproductive stimulus shown in humans to our knowledge that stimulated OT release. Among nine other hormones assayed, including arginine vasopressin, testosterone, dihydrotestosterone, adrenocorticotropic hormone (ACTH), cortisol, prolactin, estradiol, and progesterone; none had direct or indirect effects on OT or reciprocity.

In a related study, we sought to test if raising OT would subsequently increase reciprocity. Based on the animal literature showing that belly stroking sometimes raises OT, we used licensed massage therapists to give participants 15-min moderate pressure back massages prior to making decisions in the TG (Morhenn et al., 2008). A control group rested quietly for 15 min on different days and was therefore unaware that others received massages. Blood draws preceded the massages and followed the TG decision. We found that massage alone did not raise OT ($N = 24$, $p = 0.62$). However, in DM2s massage appeared to prime OT release when combined with the receipt of a monetary transfer denoting trust. The change in OT for DM2s who were massaged and trusted was 16% higher than DM2 controls who received the same average monetary transfer denoting trust ($N = 32$, $p = 0.006$). The most intriguing finding was that reciprocation was 243% higher for DM2s in the massage-primed group relative to DM2 controls. The increase in OT for this group predicted the amount of money reciprocated by DM2s ($r = 0.43$, $p = 0.03$). As in our previous studies, no direct or indirect effects of hormones that interact with OT were found.

Even though the majority of participants in these studies tend to follow the reciprocity norm, about five percent of DM2s in the TG are unconditional non-reciprocators. These participants return zero or near zero, no matter how much money they receive from DM1s. OT levels for non-reciprocating participants were more than one standard deviation above the average DM2 OT level (>470 pg/ml), indicating a possible OT dysregulation. Nonreciprocators also have psychological traits similar to psychopaths (Zak, et al., 2005. A study of those with social anxiety disorder ($N = 24$) found a similar OT dysregulation relative to controls ($N = 20$) in which greater symptom severity was associated with higher basal OT ($p = 0.04$; Hoge et al., 2008).

Negative reciprocity. Negative reciprocity, also termed altruistic punishment or moralistic punishment, is an intentional action in which one punishes someone who has harmed oneself or another individual (e.g., Ostrom, Gardner, and Walker, 1994). Many experiments show that people are willing to incur a cost in order to "punish" others who violate reciprocity norms (e.g., Kurzban et al., 2007; Camerer, 2003). The use of resources to punish a non-reciprocator in a one-shot setting is prosocial because it provides a disincentive for defection that can motivate the defector to cooperate with others in future interactions.

One way to measure negative reciprocity in the lab is to have participants make decisions in the ultimatum game (UG; Güth, Schmittberger, and Schwarze, 1982). In the UG, participants are randomly assigned the roles of DM1 and DM2 and placed into dyads and decisions are made in sequence and anonymously. After instruction, DM1 offers a split of an endowment (e.g., \$10) s/he received from the experimenters to DM2 who has no endowment. DM2 then decides whether to accept or reject the offer from DM1. If DM2 rejects, both DMs receive nothing. If DM2 accepts, the funds are distributed according to the offer made by DM1. A

rejection by DM2 is costly punishment since DM2 is incurring a cost (ending up with zero) to punish DM1 for acting in a selfish manner. Note that the UG is a zero-sum game producing winners and losers; in contrast, the TG is a positive-sum game and allows for win-win outcomes.

OT infusion does not affect the punishment threshold (the minimum amount of money DM2 is willing to accept from DM1; Zak, Stanton, and Ahmadi, 2007); and plasma OT is uncorrelated with DM2 rejections (Barraza and Zak, 2009). Yet, testosterone (T), which inhibits OT binding (Insel, Young, Witt, and Crews, 1993; Arsenijevic and Tribollet, 1998), appears to promote punishment. Men with high endogenous T are more likely to reject low offers in the UG as DM2s (Burnham, 2007). Moreover, Zak, Borja, et al. (2005) find that the biologically active metabolite of T, dihydrotestosterone (DHT), increases for DM2 men in the TG who receive low trust signals (i.e., small transfers from DM1) and predicts nonreciprocity. This effect was not found for women.

Stronger evidence comes from a study that administered 10 g of transdermal T to men using a double-blind within-subjects design with before and after blood draws to measure the change in T levels (Zak et al., 2009). Men with artificially elevated T, which was roughly doubled over baseline (average increases: total T 60%; free T 97%, DHT 128%, all $p < 0.05$), did not set an overall higher punishment threshold as DM2s in the UG. However, the punishment threshold increased linearly for all three measures of T ($N = 200$, r range 0.15 to 0.23, p range $=$ 0.001 to 0.03). In other words, these artificial "alpha males" set the bar higher for what they deemed as a fair distribution and were willing to punish others for violating this sharing norm at a cost to themselves.

Although negative reciprocity can produce a prosocial outcome, it is unlikely that punishment of non-cooperators is motivated by prosocial considerations. Several studies have shown that punishing those who show low levels of reciprocity is rewarding, particularly for men (e.g., Singer et al., 2006; De Quervain, 2004). In the UG, DM2s report rejecting offers because they feel anger toward DM1s (Pillutla and Murnighan, 1996). DM2s who receive stingy offers also have greater activity in the anterior insula, a brain region associated with visceral disgust (Sanfey et al., 2003). Moreover, punishment may serve reputational advantages (Gintis, Smith and Bowles, 2001), as it is enhanced by the presence of others (Kurzban et al., 2007).

18.2.2 Trust

Trust is an essential component of cooperation and other forms of collective action (Zak, 2012). Trust occurs when the other party is deemed to be trustworthy. The subgame perfect Nash equilibrium prediction of a one-shot trust game is for DM1 to send zero and for DM2 to return zero, since there is no requirement for DM2 to reciprocate and thus no reason for DM1 to trust. However, less than 10% of DM1s chose this strategy (Smith, 1998; Zak, 2008). So why would anyone trust a stranger when there are no repercussions if the stranger keeps the money? Moreover, what helps people determine when and whom to trust?

We have used the TG to examine if OT administration would impact the decision to trust a stranger. After a one-hour loading period, a 24IU dose of OT given intranasally more than doubled the number of DM1s who sent all their money to the DM2s in their dyads (45% vs. 22% for those on placebo; Kosfeld et al., 2005). In this study, OT had no effect on an objective risk-taking task where participants made choices with a computer rather than a human. Participants in the OT condition did not have cognition or mood changes mediating these effects. There was no effect on DM2s in this experiment, mostly likely due to the high degree of trust and the subsequent action of endogenous OT release. Others have found that OT infusion increases evaluations of trustworthiness of strangers in healthy adults using the TG (Theodoriou, Rowe, Penton-Voak, and Rogers, 2009) and for people with Asperger's syndrome using a cooperative task similar to the trust game (Andari et al., 2010).

Brain-imaging experiments have discovered that OT facilitates trust primarily by reducing amygdala activation, as well as in the anterior cingulate, dorsal striatum and midbrain regions (Baumgartner, et al., 2008; Zak et al., 2006). Baumgartner and colleagues (2008) had participants play multiple rounds of the TG as DM1, with feedback that their trust had been "betrayed" (non-reciprocal DM2s). OT participants were more likely to continue to trust after betrayal relative to placebo participants.

These findings open the issue of whether OT leads to indiscriminant trust. A recent study finds that additional information on TG partners may inhibit OT's effect on trust. In order to manipulate the trustworthiness of a person (i.e., DM2), Mikolajczak and colleagues (2010a), provided participants playing the TG as DM1s with information on their DM2 dyadic partner. DM2s were either presented as reliable or unreliable based on information on the academic field (e.g., philosophy vs. marketing) and leisure activities (practicing first aid vs. playing violent games). DM1s in the OT condition (32IU) were more likely to trust dyadic partners who were presented as reliable compared DM1s receiving a placebo. An OT-induced increase in trust by DM1s was not found when dyadic partners were presented as untrustworthy. A within-subjects study using participants with Asperger's syndrome also finds a contingent OT-trust effect (Andari et al., 2010). Participants played a computerized ball-tossing game with three fictitious players that, unbeknownst to the participants, reciprocated a ball toss at varying rates. An allocation of money was made each time the ball was "tossed" to a person, so there was incentive to throw to those likely to reciprocate. Participants on 24IU OT were more likely to toss the ball to the player that exhibited a greater probability of reciprocation, as compared to themselves on placebo. Moreover, participants were also more likely to express feelings of trust toward the cooperative player when on OT than on placebo. In other words, OT participants showed a stronger preference to trust those who were worthy of trust.

There are now examples of other forms of trust being enhanced by OT in addition to the TG. Mikolajczak and colleagues (2010b) demonstrated that participants receiving 32IU OT trusted strangers more with personal information. In their study, participants were asked to complete a questionnaire where they reported intimate sexual practices (e.g., use of toys, sado-masochism practices). After a 45-min drug load period, participants were then instructed to place the questionnaire in an envelope that they were able to seal and tape before handing to the experimenter, if desired. Participants in the OT condition showed more "trust" in handing the experimenter an unsealed envelope (60% vs. 3.3% for those on placebo). This indicates that the OT-trust relationship appears to extend beyond behaviors where there is possible self-gain.

18.2.3 Generosity

We have used the UG to investigate generosity. Generosity is defined as giving more than is needed in order to satisfy expectations in an exchange (Zak et al., 2007). By giving DM2s the option to reject an offer, the UG requires that DM1s consider how the DM2 in the dyad would react to an offer. That is, an effective choice in the UG requires that DM1 take the perspective of DM2. A purely resource-driven model of human behavior predicts that DM2 should accept any positive offer from DM1 (Zak, 2011). However, experimental studies find that most offers smaller than 30% are nearly always rejected by DM2s in Western countries (Camerer, 2003).

Plasma OT release has been found to correlate indirectly with DM1 transfers in the UG (Barraza and Zak, 2009). Participants in this study viewed an emotionally charged video prior to playing the UG. Although the change in OT did not predict generosity, OT did predict the experience of empathy in response to the video. Those who were more empathically engaged made more generous offers in the UG ($N = 56$, $r = 0.24$, $p = 0.05$).

OT infusion has been used in the UG to establish a causal relationship with generosity (Zak et al., 2007). In this experiment, each participant made decisions as both DM1 and DM2 with actual

assignment of roles determined by random draw after choices were made. Generosity was operationalized as the amount DM1 offers exceeding one's minimal acceptable offer as DM2. Infusing 40IU of OT increased generosity by 80% relative to those who received a placebo ($N = 34$, $p = 0.005$).

We have also found that testosterone administration reduces generosity (Zak et al., 2009). Men whose testosterone was artificially raised, compared to themselves on placebo, were 27% less generous in the UG ($N = 200$, $p = 0.04$). The reduction in generosity declined linearly as levels of total-, free-, and dihydrotestosterone rose (r range: –0.19 to –0.31, $p = 0.01$). For example, generosity by participants in the lowest decile of DHT averaged 85% higher ($3.65 out of $10) compared to generosity by those in the highest decile of DHT ($0.55 out of $10).

18.2.4 Altruistic helping

Behavior in the TG and UG can be classified as jointly cooperative, that is, the individual may benefit tangibly and directly by acting prosocially. Altruistic helping, such as anonymous unilateral transfers, lacks this self-benefiting motive. In laboratory studies, we remove external incentives that may drive altruistic helping, for instance, increasing one's social status or building a positive reputation. In experimental economics the dictator game (DG) has been used to measure altruism, removing the extrinsic and self-serving benefits that are typically involved in such acts (e.g., Camerer, 2003). The DG is similar to the UG in that DM1 begins with an allocated sum (e.g., $10) and DM2 has nothing. DM1 is asked to choose some amount to send to DM2 (including zero), but unlike the UG, DM2 has no choice but to accept the offer. As a result, DM1 has no incentive to consider DM2's perspective and likely response to a transfer. Indeed, without any need to consider DM2's response, offers tend to be much lower in the DG than in the UG (e.g., Zak et al., 2007).

An intranasal infusion of 40IU of OT did not affect DM1 transfers in the DG ($3.77 vs. $3.58 placebo, $p = 0.51$; Zak et al., 2007). The reason for this appears to be that, unlike the UG, emotional engagement (via perspective taking) is absent in the DG. This is consistent with the lack of an effect from 10g of transdermal T administration on choices in the DG. Men on T, compared to themselves, did not offer less in the DG ($N = 200$; T: $3.34, placebo: $3.56, $p = 0.86$).

Genetic research indicates that OT and related peptides may be involved in altruistic helping in the DG. Variations in both the arginine vasopressin 1a (AVPR1a) RS3 promote repeat region (Knafo et al., 2008) and single-nucleotide polymorphisms (SNPs) in the oxytocin receptor gene (OXTR rs1042778; Israel et al., 2009) show significant associations with DM1 transfers in the DG. Yet, a subsequent study by another research team was unable to replicate the association between 9 SNPs of OXTR (including rs1042778) and DG allocations (Apicella et al., 2010).

Charitable giving. Our lab has recently examined whether OT affects another form of altruistic helping, charitable giving. Charitable giving is an indirect form of altruistic helping done through institutions. Charitiable donations are typically made without any direct exposure to the beneficiary or direct knowledge of how the money will be used. Like the DG, charitable giving is a unilateral transfer of money to strangers. However, charitable giving provides a motive to transfer money because of the perceived need of eventual recipients; perceived need tends to increase the expression of empathy (Batson, 1991; Davis, 1996; de Waal, 2008).

In a study investigating emotionally induced OT release (Barraza and Zak, 2009), we found no correlation between basal or emotionally reactive OT and charitable donations ($p = 0.45$). However, we found a positive correlation between the amount donated to charity and DM1 transfers in the UG ($r = 0.36$, $p = 0.004$). Since OT has a direct effect on DM1 transfers in the UG, it is possible that OT infusion could affect charitable donations.

Barraza and colleagues (2011) examined whether 40IU of OT would increase the likelihood and size of donations. Participants were first allowed to earn varying amounts of money by making monetary

decisions in the lab since out-of-the-lab donations are made from earned income, not windfalls. Next, participants were presented with an option to donate to one of two charities, the American Red Cross or the Palestinian Red Crescent Society. OT did not significantly affect the decision to make a donation to charity (40% vs. 32% placebo, $N = 132$, $p = 0.15$), but for people who decided to donate, those on oxytocin gave 48% more money than those on placebo (OT: \$4.76; placebo: \$3.22, $p = 0.03$). This difference was largely driven by donations to the American Red Cross (OT: \$5.12, placebo: \$3.09, $p = 0.04$); OT infusion had no significant effect on donations to the Red Crescent Society relative to placebo ($p = 0.35$). This result is consistent with the in-group prosocial preference of OT infusion found by other researchers (e.g., de Dreu et al., 2010).

A related study had participants view public service announcements (PSA; Lin, Sparks, Morin, and Zak, 2011). After receiving 40IU of OT ($N = 40$; $OT = 20$), participants watched 16 PSAs that featured social and health related issues. To make donations salient, participants earned \$5 for correctly answering one question regarding content following each PSA. Next, they were given an opportunity to donate some of the money they earned to the charitable causes in the ads. Those who received OT donated to 33% of the causes compared to participants receiving the placebo who donated to 21% ($N = 538$; $p = 0.001$). OT also increased the average donation by 56% compared to controls (OT = \$0.84; placebo= \$0.54, $p < 0.001$).

18.3 Relating physiological and psychological mechanisms

The review above has shown that OT promotes a variety of prosocial behaviors in human beings. People in these experiments have difficulty during debriefing explaining why they willingly shared resources in one-shot anonymous settings (Zak, 2011). This is consistent with the highest densities of OT receptors being found mostly in subcortical brains regions (Barberis and Tribollet, 1996). Although multiple motives may drive prosocial behavior, empathy is a likely proximal mechanism for other-regarding behavior. The role of empathy in motivating prosocial behaviors has been proposed by philosophers like Adam Smith (1759) and in evolutionary models of reciprocal altruism (e.g., Trivers, 1971). Since 2007, a handful of studies have been published examining the relationship between empathy and OT directly.

18.3.1 Empathy

Here, we discuss three distinct forms of empathy found in the literature and their relation to OT; two forms of emotional empathy – empathic distress and empathic concern (compassion), and a cognitive form of empathy, perspective taking (the process of inferring the mental states of others). The former two are more associated with affective states while the latter is believed to be a primarily cognitive process.

Empathic distress. Empathy conceptualized as empathic distress, or personal distress, is an aversive state brought on by witnessing physical or emotion pain in another (Batson, 1991; Davis, 1983; 1996). Empathic distress is characterized by reactive and aversive feelings (e.g., worry, anxiety, discomfort) that are focused on the self (e.g., Batson, 1991; Davis, 1996). Brain-imaging studies have examined empathic distress by having participants view another person receiving a painful stimulus, or view the facial expression of someone in pain (i.e., an empathy-for-pain paradigm, see Singer and Lamm, 2009). These studies find shared activation for pain in self and pain in others in the anterior insula, a brain region associated with the affective experience of pain. Singer and colleagues (2008) tested the effects of OT on the experience of empathy using the empathy-for-pain paradigm and subsequent behavior in the trust game. Participants received either 24IU of OT or a placebo intranasally prior to the pain procedure and trust game. The authors found that OT did not affect brain activation in regions previously found to be associated with empathy (e.g., anterior insula) for self-experienced

pain or for other-witnessed pain. In addition, OT did not impact decisions in the trust game. The authors concluded that OT does not promote empathy; however, this result only applies to a particular kind of empathy, empathic distress.

While empathic distress can bring awareness to the suffering of another, it appears to do so via sympathetic arousal rather than through a desire to engage with another. The lack of an effect of OT on empathic distress is not surprising since OT functions as an anxiolytic for moderate stress (Bartz and Hollander, 2006; Heinrichs, Baumgartner, Kirschbaum, and Ehlert, 2003). By reducing the degree of vicarious arousal, OT may reduce distress but also allow for other-focused empathic states to occur.

Empathic concern. A second form of empathy is described as an other-focused emotion that is ultimately felt *for* another person (Batson, 2009; Batson et al., 2009). Deriving from the "parental instinct," empathic concern generates an impulse to protect others and is perhaps the "root of all altruism" (McDougall, 1926). Many refer to this affective state as compassion (also "pity," "sympathy," and "empathy," see Batson, 2009), which is classified as one of the human virtues (Peterson and Seligman, 2004) and resides in the family of moral emotions (Haidt, 2003). In a series of studies, Decety and colleagues (see Decety and Lamm, 2009) have shown that empathic distress and empathic concern are separable phenomena in terms of brain function. Moreover, those who become physiologically overaroused (elevated heart rate and skin conductance) experience distress, and become motivated to address egoistic concerns (Eisenberg and Fabes, 1990; Eisenberg et al., 2004; Hoffman, 1981). Alternatively, those who are aware of distress in others and are able to regulate the arousal that arises from it are more likely to experience empathic concern (i.e., sympathy; Eisenberg and Fabes, 1990).

We find that empathic concern, and not empathic distress, is associated with endogenous OT release (Barraza and Zak, 2009). Using a 100-s video of a 2-year-old boy who has terminal brain cancer narrated by his father, we asked viewers to rate a series of adjectives relating to their affective states after viewing the video. We found a 47% increase in OT immediately after viewing this video relative to baseline ($N = 23$, $p = 0.004$). This increase in OT was positively correlated with self-reported empathic concern ($r = 0.20$, $p = 0.01$) after controlling for self-reported empathic distress. We also found a positive correlation between self-reported concern and DM1 generosity in the UG ($r = 0.24$, $p = 0.05$). The analyses for self-reported empathic distress yielded negative or null correlations. These null findings for empathic distress parallel those from Singer and colleagues (2008) investigating the effects of OT infusion on empathy using the empathy-for-pain paradigm.

We also tested whether trait empathy was correlated with emotionally reactive OT. Participants were measured on dispositional empathic distress, concern, and perspective taking using the Interpersonal Reactivity Index (Davis, 1983). The increase in OT after viewing the emotional video was significantly and positively correlated with scores on dispositional empathic concern, but not with empathic distress or perspective-taking trait measures. This study was the first to provide direct evidence that OT is associated with empathy. It is the other-focused nature of empathic concern that appears to induce OT release and to promote prosocial behaviors.

Perspective taking. Empathy has been viewed as a cognitive function where one is able to imagine the feeling state of another without sharing the particular state themselves or feeling for the person's plight (e.g., Batson, 2009; Davis, 2005). Perspective taking has been found to increase empathic concern for people belonging to different stigmatized social groups like the homeless, persons with AIDS, and drug users (e.g., Batson, Early, and Salvarani, 1997; Davis, 2005; Davis et al., 2004; Lamm, Batson, and Decety, 2007). Moreover, perspective-taking instructions decrease brain activity in pain regions when witnessing pain in others (Lamm, Batson, and Decety, 2007), possibly enhancing the likelihood of experiencing empathic concern.

Research in humans has found that OT modulates the ability to infer the emotions of others. Domes and colleagues (2007) tested if OT affected

performance on the "Reading the Mind in the Eyes" (RMET), a task that measures the ability to read emotional states in others. When given 24IU OT intranasally, participants were able to accurately identify more emotional faces in the RMET than those given placebo (72% vs 69% placebo, $p = 0.02$). OT also increased the ability to correctly assess emotions that were difficult to identify by those on placebo ($p < 0.006$). A separate study finds that an oxytocin receptor (OTR) polymorphism (rs53576) is also related to performance in RMET, with homozygous GG allele performing better than AA/AG participants (Rodrigues et al., 2009). These researchers also found that the same allelic variations related to self-reported trait empathy scores (as measured by composite of the empathic concern, perspective taking, and fantasy subscales of the Interpersonal Reactivity Index; Davis, 1983), with GG reporting higher trait empathy than AA/AG participants.

This research suggests that OT allows humans to infer the emotional states of others. However, the evidence appears to be stronger for a link between OT and emotional rather than cognitive forms of empathy. For instance, a recent study by Hurlemann et al. (2010) found that while OT infusion enhanced "emotional empathy" (how much they *feel* for a target in an image), there was no difference for "cognitive empathy" (ability to accurately infer the emotional state of a target in an image) relative to placebo. Although presented separately, it is likely that these different empathic states co-occur or initiate the experience of one another. Perspective taking may make someone aware of the plight of others, thereby facilitating the experience of empathic concern. In turn, feeling empathy for another may motivate greater interest in the emotional and psychological state of another. Through the release of OT, these empathic states can lead to prosocial behaviors.

18.4 Conclusion

Human beings feel empathy for individuals with whom they are unacquainted, as evidenced by prosocial assistance in and outside the lab. The review of findings herein has shown that OT release in the brain is one physiologic factor that changes the self–other relationship by causing human beings to experience empathy, and empathic concern in particular. The OT infusion studies from our lab and others establish the causal relationship between prosocial behaviors and OT. Taken together, this research shows that OT is part of the neurophysiology of human prosociality, even producing prosocial behaviors where these is little incentive to help others. Future research is likely to find many other examples of prosocial behaviors associated with OT. For instance, unpublished data from our lab suggests that in both a college-aged and general adult sample, those who regularly volunteer have higher basal OT and a larger change in OT after a stimulus than non-volunteers. Engagement in habituated prosocial behaviors, like volunteering, may result in a positive feedback loop tuning the oxytocinergic system to be more responsive to social stimuli.

We have also found that women who release more OT after a trust stimulus, compared to those who release less report greater happiness with their lives ($r = 0.31$, $p = 0.05$; Grosberg, Merlin and Zak, 2012). These "super-releasers" had fewer depressive symptoms ($r = -0.35$, $p = 0.05$), and had more sex ($r = 0.29$, $p = 0.05$) with fewer partners ($r = -0.33$, $p = 0.022$). High OT releasers appear to be happier because of the rich social milieu that they inhabit, with higher quality romantic relationships ($r = .42$, $p = 0.01$) and more friends ($r = 0.27$, $p = 0.06$). As gregariously social creatures, most human beings crave social connections. Indeed, those with larger social networks report better health (e.g., Myers, 2001). Social relationships are sustained by prosocial behaviors, and OT appears to be a critical component of human prosociality.

REFERENCES

Andari, E., Duhamel, J. R., Zalla, T., Herbrecht, E., Leboyer, M., and Sirigu, A. (2010). Promoting social behavior with oxytocin in high-functioning autism spectrum disorders.

Proceedings of the National Academy of Sciences of the United States of America, 107, 4389–4394.

Arsenijevic Y. and Tribollet, E. (1998). Region-specific effect of testosterone on oxytocin receptor binding in the brain of the aged rat. *Brain Research*, 785, 167–170.

Apicella, C. L., Cesarini, D., Johannesson, M., Dawes, C. T., Lichtenstein, P., Wallace, B., et al. (2010). No Association between oxytocin receptor (OXTR) gene polymorphisms and experimentally elicited social preferences. *PLoS One*, 5, e11153.

Axelrod, R. (1984). *The Evolution of Cooperation*. New York: Basic Books.

Barberis, C. and Tribollet, E. (1996). Vasopressin and oxytocin receptors in the central nervous system. *Critical Reviews in Neurobiology*, 10, 119–154.

Barraza, J. A., McCullough, M. E., Ahmadi S., and Zak, P. J. (2011). Oxytocin infusion increases charitable donations regardless of monetary resources. *Hormones and Behavior*, 60, 148–151.

Barraza, J. A. and Zak, P. J. (2009). Empathy toward strangers triggers oxytocin release and subsequent generosity. *Annals of the New York Academy of Sciences*, 1167, 182–189.

Bartz, J. A. and Hollander, E. (2006). The neuroscience of affiliation: Forging links between basic and clinical research on neuropeptides and social behavior. *Hormones and Behavior*, 50, 518–528.

Batson, C. D. (1991). *The altruism question: Toward a Social-Psychological Answer*. Hillsdale: Lawrence Erlbaum.

Batson, C. D. (2009). These things called empathy: Eight related but distinct phenomena. In J. Decety and W. Ickes (Eds.), *The Social Neuroscience of Empathy*, Cambridge: MIT Press, pp. 3–16.

Batson, C. D., Early, S., and Salvarani, G. (1997). Perspective taking: Imagining how another feels versus imagining how you would feel. *Personality and Social Psychology Bulletin*, 23, 751–758.

Batson, C. D., Ahmad, N., and Lishner, D. A. (2009). Empathy and altruism. In S. J. Lopez and C. R. Snyder (Eds.), *Oxford Handbook of Positive Psychology*, 2nd edn, New York: Oxford University Press, pp. 417–426.

Baumgartner, T., Heinrichs, M., Vonlanthen, A., Fischbacher, U., and Fehr, E. (2008). Oxytocin shapes the neural circuitry of trust and trust adaptation in humans. *Neuron*, 58, 639–650.

Berg, J., Dickhaut, J., and McCabe, K. (1995). Trust, reciprocity, and social history. *Games and Economic Behavior*, 10, 122–142.

Bowles, S. (2009). Did warfare among ancestral hunter-gatherer groups affect the evolution of human social behaviors? *Science*, 324, 1293–1298.

Burger, J. M., Sanchez, J., Imberi, J. E., and Grande, L. R. (2009). The norm of reciprocity as an internalized social norm: Returning favors even when no one finds out. *Social Influence*, 4, 11–17.

Burnham, T. C. (2007). High-testosterone men reject low ultimatum game offers. *Proceedings of the Royal Society B*, 274, 2327–2330.

Camerer, C. (2003). *Behavioral Game Theory Experiments in Strategic Interaction*. New York: Russell Sage Foundation.

Campbell, A. (2010). Oxytocin and human social behavior. *Personality and Social Psychology Review*, 14, 281–295.

Davis, M. H. (1983). Measuring individual differences in empathy: Evidence for a multidimensional approach. *Journal of Personality and Social Psychology*, 44, 113–126.

Davis, M. H. (1996). *Empathy: A Social Psychological Approach*. Boulder: Westview Press.

Davis, M. H. (2005). A 'constituent' approach to the study of perspective taking: What are its fundamental elements? In Malle, Bertram F., and Hodges, Sarah D. (Eds.). *Other Minds: How Humans Bridge the Divide Between Self and Others*, New York: Guilford Press, pp. 44–55.

Davis, M. H., Soderlund, T., Cole, J., Gadol, E., Kute, M., Myers, M., et al. (2004). Cognitions associated with attempts to empathize: How do we imagine the perspective of another? *Personality and Social Psychology Bulletin*, 30, 1625–1635.

Decety, J. and Lamm, C. (2009). Empathy versus personal distress: Recent evidence from social neuroscience. In J. Decety and W. Ickes (Eds.), *The Social Neuroscience of Empathy*, Cambridge: MIT press, pp. 199–213.

De Dreu, C. K. W., Greer, L. L., Handgraaf, M. J. J., Shalvi, S., Van Kleef, G. A., Baas, M., et al. (2010). The neuropeptide oxytocin regulates parochial altruism in intergroup conflict among humans. *Science*, 328, 1408–1258.

de Quervain, D. J. F., Fischbacher, U., Treyer, V., Schellhammer, M., Schnyder, U., Buck, A., et al. (2004). The neural basis of altruistic punishment. *Science*, 305, 1254–1258.

de Waal, F. B. M. (2008). Putting the altruism back into altruism: The evolution of empathy. *Annual Review of Psychology*, 59, 279–300.

Domes, G., Heinrichs, M., Michel, A., Berger, C., and Herpertz, S. C. (2007). Oxytocin improves 'mind-reading' in humans. *Biological Psychiatry*, 61, 731–733.

Eisenberg, N. and Fabes, R. A. (1990). Empathy: Conceptualization, measurement, and relation to prosocial behavior. *Motivation and Emotion*, 14, 131–149.

Eisenberg, N., Valiente, C., Champion, C., and Miller, A. G. (2004). Empathy-related responding: Moral, social, and socialization correlates. In *The Social Psychology of Good And Evil*, New York, NY: Guilford Press, pp. 386–415.

Gintis, H. (2000). Strong reciprocity and human sociality. *Journal of Theoretical Biology*, 206,169–179.

Gintis, H., Smith, E. A., and Bowles, S. (2001). Costly signaling and cooperation. *Journal of Theoretical Biology*, 213, 103–119.

Grosberg, D., Merlin, R., and Zak, P. J. (2011). What makes women happy: biological correlates with life satisfaction in women, in review.

Güth, W., Schmittberger, R., and Schwarze, B. (1982). An experimental analysis of ultimatum bargaining. *Journal of Economic Behavior and Organization*, 3, 367–388.

Haidt, J. (2003). The moral emotions. In R. J. Davidson, K. R. Scherer, and H. H. Goldsmith (Eds.), *Handbook of Affective Science*, Oxford: Oxford University Press, pp. 852–870.

Heinrichs, M., Baumgartner, T., Kirschbaum, C., and Ehlert, U. (2003). Social support and oxytocin interact to suppress cortisol and subjective responses to psychosocial stress. *Biological Psychiatry*, 54, 1389–1398.

Hoffman, M. L. (1981). Is altruism part of human nature? *Journal of Personality and Social Psychology*, 40, 121–137.

Hoge, E. A., Pollack, M. H., Kaufman, R. E., Zak, P. J., and Simon, N. M. (2008). Oxytocin levels in social anxiety disorder. *CNS Neuroscience and Therapeutics*, 14, 165–170.

Hurlemann, R., Patin, A., Onur, O. A., Cohen, M. X., Baumgartner, T., Metzler, S., et al. (2010). Oxytocin enhances amygdala-dependent, socially reinforced learning and emotional empathy in humans. *Journal of Neuroscience*, 30, 4999–5007.

Insel, T. R., Young, L., Witt, D. M., and Crews, D. (1993). Gonadal steroids have paradoxical effects on brain oxytocin receptors. *Journal of Neuroendocrinology*, 5, 619–628.

Israel, S., Lerer, E., Shalev, I., Uzefovsky, F., Riebold, M., Laiba, E., et al. (2009). The oxytocin receptor (OXTR) contributes to prosocial fund allocations in the dictator game and the social value orientations task. *PLoS One*, 4, e5535.

Kosfeld, M., Heinrichs, M., Zak, P. J., Fischbacher, U., and Fehr, E. (2005). Oxytocin increases trust in humans. *Nature*, 435, 673–676.

Knafo A., Israel, S., Darvasi, A., Bachner-Melman, R., Uzefovsky, F., Cohen, L., Feldman, E., et al. (2008). Individual differences in allocation of funds in the dictator game associated with length of the arginine vasopressin 1a receptor RS3 promoter region and correlation between RS3 length and hippocampal RNA. *Genes, Brain, and Behavior*, 7, 266–275.

Kurzban, R., DeScioli, P., and O'Brien, E. (2007). Audience effects on moralistic punishment. *Evolution and Human Behavior*, 28, 75–84.

Lamm, C., Batson, C. D., and Decety, J. (2007). The neural basis of human empathy: Effects of perspective-taking and cognitive appraisal: An event-related fMRI study. *Journal of Cognitive Neuroscience*, 19, 42–58.

Lin, P-Y., Sparks, N., Morin, C., Johnson, W. D., and Zak, P. J. (2013). Oxytocin increases the influence of advertising, PLoS ONE 8(2): e56934. doi:10.1371/journal.pone.0056934.

McDougall, W. (1926). *An Introduction to Social Psychology*. London: Methuen.

Meyer-Lindenberg, A. (2008). Impact of prosocial neuropeptides on human brain function. *Progress in Brain Research*, 170, 463–470.

Myers, D. G. (2001). Close relationships and quality of life. In *Well-Being: The Foundations of Hedonic Psychology*. New York: Russell Sage Foundation.

Mikolajczak, M., Gross, J. J., Lane, A., Corneille, O., de Timary, P., and Luminet, O. (2010a). Oxytocin makes people trusting, not gullible. *Psychological Science*, 21, 1072–1074.

Mikolajczak, M., Pinon, N., Lane, A., de Timary, P., and Luminet, O. (2010b). Oxytocin not only increases trust when money is at stake, but also when confidential information is in the balance. *Biological Psychology*, 85, 182–184.

Morhenn, V. B., Park, J. W., Piper, E., and Zak, P. J. (2008). Monetary sacrifice among strangers is mediated by endogenous oxytocin release after physical contact. *Evolution and Human Behavior*, 29, 375–383.

Nowak, M. A. and Sigmund, D. (2005). Evolution of indirect reciprocity. *Nature*, 437, 1291–1298.

Ostrom, E., Gardner, J., and Walker, R. (1994). *Rules, Games, and Common-Pool Resources*. Univ. of Michigan Press, Ann Arbor.

Peterson, C. and Seligman, M. E. P. (2004). *Character Strengths and Virtues: A Handbook and Classification*. Oxford University Press, Oxford.

Pillutla, M. M. and Murnighan, J. K. (1996). Unfairness, anger, and spite: Emotional rejections in ultimatum

offers. *Organizational Behavior and Human Decision Processes*, 68, 208–224.

Rodrigues, S. M., Saslow, L. R., Garcia, N., John, O. P., and Keltner, D. (2009). Oxytocin receptor genetic variation relates to empathy and stress reactivity in humans. *Proceedings of the National Academy of Sciences of the United States of America*, 106, 21437–21441.

Sanfey, A. G., Rilling, J. K., Aronson, J. A., Nystrom, L. E., and Cohen, J. D. (2003). The neural basis of economic decision-making in the ultimatum game. *Science*, 300, 1755–1758.

Singer, T., Seymour, B., O'Doherty, J. P., Stephan, K. E., Dolan, R. J., and Frith, C. D. (2006). Empathic neural responses are modulated by the perceived fairness of others. *Nature*, 439, 466–469.

Singer, T. and Lamm, C. (2009). The social neuroscience of empathy. In *Year in Cognitive Neuroscience 2009* Vol. 1156, Oxford: Blackwell Publishing, pp. 81–96.

Singer, T., Snozzi, R., Bird, G., Petrovic, P., Silani, G., Heinrichs, M., et al. (2008). Effects of oxytocin and prosocial behavior on brain responses to direct and vicariously experienced pain. *Emotion*, 8, 781–791.

Smith, A. (1759). *The theory of moral sentiments.* Online Library of Liberty, http://oll.libertyfund.org/accessed 3/13/09.

Smith, V. L. (1998). The two faces of Adam Smith. *Southern Economic Journal*, 65, 1–19.

Theodoridou, A., Rowe, A. C., Penton-Voak, I. S., and Rogers, P. J. (2009). Oxytocin and social perception: Oxytocin increases perceived facial trustworthiness and attractiveness. *Hormones and Behavior*, 56, 128–132.

Tost, H., Kolachana, B., Hakimi, S., Lemaitre, H., Verchinski, B. A., Mattay, V. S., et al. (2010). A common allele in the oxytocin receptor gene (OXTR) impacts prosocial temperament and human hypothalamic-limbic structure and function. *Proceedings of the National Academy of Sciences of the United States of America*, 107, 13936–13941.

Trivers, R. L. (1971). Evolution of reciprocal altruism. *Quarterly Review of Biology*, 46, 35–57.

West, S. A., Griffin, A. S., and Gardner, A. (2006). Social semantics: Altruism, cooperation, mutualism, strong reciprocity, and group selection. *Journal of Evolutionary Biology*, 20, 415–432.

Zak, P. J. (2011). The physiology of moral sentiments. *Journal of Economic Behavior and Organization*, 77, 53–65.

Zak, P. J. (2007). The neuroeconomics of trust. In *Renaissance in Behavioral Economics.* R. Frantz, Ed. New York: Routledge, pp. 17–33.

Zak, P. J. (2008). The neurobiology of trust. *Scientific American*, June, pp. 88–95.

Zak, P. J. (2012). *The Moral Molecule: The Source of Love and Prosperity.* New York: Dutton.

Zak, P. J., Borja, K., Matzner, W. T., and Kurzban, R. (2005). The neuroeconomics of distrust: Sex differences in behavior and physiology. *American Economic Review Papers and Proceedings*, 95, 360–363.

Zak, P. J., Kurzban, R., Ahmadi, S., Swerdloff, R. S., Park, J., Efremidze, L., et al. (2009). Testosterone administration decreases generosity in the ultimatum game. *PLoS One*, 4, e8330.

Zak, P. J., Kurzban, R., and Matzner, W. T. (2004). The neurobiology of trust. *Annals of the New York Academy of Sciences*, 1032, 224–227.

Zak, P. J., Kurzban, R., and Matzner, W. T. (2005). Oxytocin is associated with human trustworthiness. *Hormones And Behavior*, 48, 522–527.

Zak, P. J., Park, J., Ween, J., and Graham, S. (2006). An fMRI study of interpersonal trust with exogenous oxytocin infusion. Poster presented at Society for Neuroscience Program No. 2006-A-130719-SfN.

Zak, P. J., Stanton, A. A., and Ahmadi, S. (2007). Oxytocin increases generosity in humans. *PLoS One*, 2, e1128.

19

Oxytocin and vasopressin in human sociality and social psychopathologies

Richard P. Ebstein, Idan Shalev, Salomon Israel, Florina Uzefovsky, Reut Avinun, Ariel Knafo, Nurit Yirmiya, and David Mankuta

19.1 Introduction

Gene Robinson has coined the term "sociogenomics" to describe the genetic basis for social life (Robinson et al., 2005) across diverse animal taxa. The main aim of sociogenomics is to understand the complex pathways by which genes contribute to the wiring of social behavior. Since many animals display varying degrees of social behavior, the evolution and selection of genes that foster such behavior is of considerable interest and the identification of such genes has become a fascinating area of research leveraging on the increasing power of cutting-edge genomic tools. Interestingly, the evolution of large primate brain size has been attributed to the need for "Machiavellian intelligence" to enable individuals to successfully manipulate and engage in group living. It appears that the technology of social networking evidenced today in human society by Facebook and Twitter had its origins early on in the more ancient carbon-based neocortex.

Although it is clear that genes encode many aspects of social behavior also in humans (Ebstein et al., 2010), how complex human social behaviors may drive gene evolution is less well understood. A good example of this concept is provided by the domestication of cattle in the Neolithic and how this change in life style impacted human gene selection by extending beyond weaning lactose tolerance in some European populations (Burger et al., 2007). Lactase (the enzyme that breaks down lactose) persistence is a dominant Mendelian trait and the absence of this mutation in early European farmers argues for the "culture-historical hypothesis," whereby lactase persistence alleles were rare until the advent of dairying early in the Neolithic but then rose rapidly in frequency under natural selection. One wonders if the advent of the age of social networking that appears to be a major preoccupation of a large proportion of the population in developed countries, might not have an evolutionary edge by selecting for some kinds of "twitter" genes.

Core social behaviors are common to most animals whether they dwell alone or in groups. All animals need to find food, create a nest or shelter to rear young and defend it from predators, acquire a mate to reproduce and successfully parent their offspring. Animals that live in groups need of course to cooperate and coordinate many of those activities that reflect the social requirements of living together. Sociality is mediated by multiple systems including sensory, autonomic, emotional, and motor mechanisms that permit or prevent at the simplest level approach or withdrawal behaviors. The endocrine system and the plethora of signaling molecules that have developed over the course of vertebrate evolution figure prominently in mediating social behaviors from fish to humans. Of

Oxytocin, Vasopressin, and Related Peptides in the Regulation of Behavior, ed. E. Choleris, D. W. Pfaff, and M. Kavaliers.
Published by Cambridge University Press.

particular interest in this regard are two nonapeptide hormones, oxytocin (OT), and arginine vasopressin (AVP) that in mammals underpin a number of social and affiliative behaviors.

19.2 OT

Beyond the long-known peripheral effects of OT on parturition and lactation, a wealth of animal studies have elaborated the role of OT, or their analogs such as isotocin and vasotocin (Goodson, 2008), in molding social behavior across the vertebrates (Insel, 2010). In the past few years the role of OT has also been examined in our own species, and similar to what has been learned from animal studies, it appears that this nonapeptide also influence social behaviors in humans (Heinrichs et al., 2009; Donaldson and Young, 2008). Indeed, OT has been suggested as the "great facilitator of life" in a recent review (Lee et al., 2009).

In humans, intranasal administration of OT has been shown to increase trust (Kosfeld et al., 2005), facilitate mind reading (Domes et al., 2007), enhance human memory for social identity (Savaskan et al., 2008), increase positive communication between couples (Ditzen et al., 2008), increase gaze to the eye region (Guastella et al., 2007) and increase generosity (Zak et al., 2007). Intriguingly, OT plasma levels have been linked to individual patterns of maternal–fetal attachment (Levine et al., 2007) and salivary OT levels were associated with bonding to own parents and inversely related to psychological distress, particularly depressive symptoms (Gordon et al., 2008). Social anxiety symptom severity, adjusted for age and gender in a healthy group of subjects, was associated with higher plasma oxytocin levels (Hoge et al., 2008). Imaging studies reinforce the role of OT in influencing human social behavior with evidence demonstrating that OT modulates the amygdala and other brain regions (Baumgartner et al., 2008).

19.1.2 OT and mirror neurons

Neural activity reflected in EEG oscillations has been shown to reliably manifest perception as well as execution of biological motion. In particular, mu rhythms, measured between 8–12 Hz over somato-motor regions, are desynchronized and their power attenuated when engaging in motor activity (Gastaut et al., 1952), and also while observing actions executed by someone else (e.g., Muthukumaraswamy and Johnson, 2004; Muthukumaraswamy et al., 2004). These characteristics led authors to tentatively link the suppression of mu rhythms with a human mirror neuron system (for a review see (Pineda, 2005)). The mirror neuron system (MNS), originally discovered in the monkey (Rizzolatti et al., 1996), is thought to have evolved in humans into a wider neural system, enabeling simulation (and from it understanding) of other's intentions, thoughts and feelings (Gallese et al., 2004). In the last few years, several studies of typical participants linked EEG mu suppression to higher social information processing including social skills (Oberman et al., 2007; Oberman and Ramachandran, 2007), theory of mind (Pineda and Hecht, 2009; Perry et al., 2010b) and empathy (Cheng et al., 2008a; Cheng et al., 2008b). Using similar paradigms, other studies found deactivation also in a higher frequency range (beta range, 15–25 Hz, (e.g., Muthukumaraswamy and Singh, 2008). Supporting a link between social cognition and this EEG manifestation several studies of autistic spectrum disorders (ASD) found abnormal mu suppression when ASD individuals viewed actions performed by others despite normal suppression when they actually performed the same actions (Martineau et al., 2004; Oberman et al., 2008) but see (Raymaekers et al., 2009).

Little is known regarding the underlying neurotransmitter pathways mediating mirror neuron systems and we recently addressed this question using a pharmacological strategy. In our study (Perry et al., 2010a), we investigated the effect of intranasally administered OT on the suppression of EEG rhythms in the low and high alpha/mu range, and in the beta band (15–25 Hz) using the same point-light stimuli and design as used by Perry and colleagues (Perry et al., 2010b). We hypothesized that OT, a neuropeptide that partially shapes human social cognition, would enhance alpha/mu and beta

suppression. To our knowledge this is the first study (Perry et al., 2010a) to link mirror neuron systems, hypothesized to play a role in social cognition, and OT, a neuropeptide that in many mammals including man has striking effects on affiliative and social behaviors.

In the 8–10 Hz, a significant main effect of treatment showed that suppression was significantly enhanced in the OT versus the placebo conditions, see Figure 19.1A. There was no main effect of region or laterality; the OT-enhanced suppression was widespread across the scalp (occipital, central and frontal regions). Since the suppression index reflects the ratio between the power during the experimental (biological motion) conditions and the power during the baseline (non-biological motion) condition, the difference between OT and placebo could be caused by either enhanced suppression in the biological motion conditions following the OT treatment, or by a reduced suppression in the non-biological motion condition following this treatment. In order to differentiate between these two alternatives, we conducted a second ANOVA in which the factors were treatment and motion condition (biological, non-biological) and the dependent variable was the log of the power in each condition (across regions and sites). This ANOVA revealed no main effect for treatment, an expected effect of motion ($p = 0.001$) and, notably, an interaction between treatment and motion ($p < 0.05$), revealing a binary effect of OT, in opposite directions: enhancing suppression for biological motion and reducing suppression for non-biological motion, thus generating the significant interaction Figure 19.1B.

This study is a first step in investigating the effects of OT on EEG oscillations in the human brain and demonstrates that this hormone modulates EEG rhythms in the alpha/mu and beta ranges differentially in tasks of biological motion and non-biological motion. Importantly, by linking mirror neuron activity to oxytocin we provide the first evidence that this nonapeptide, which accumulating evidence suggests is a pre-eminent social hormone in humans as well as in lower mammals, modulates an EEG oscillation that has been hypothesized to partially mediate higher social functions such as theory of mind and empathy. Indeed, by showing that OT enhances mu suppression the Perry et al. study (Perry et al., 2010a) reveals a common groundwork that appears to link two previously separate focuses of social cognition in humans – social hormones and mirror neurons.

19.2 AVP

The actions of OT and AVP are frequently in opposing partnership (Carter et al., 2008). OT shows anxiolytic and antistress effects, both in females and in males, effects that were localized within the central amygdala and the hypothalamic paraventricular nucleus (Neumann, 2008). On the other hand, AVP is associated with arousal and vigilance, and in defensive behaviors, such territoriality and mate guarding (Carter, 1998; Landgraf and Neumann, 2004).

Most of the knowledge on the effects of AVP in modulating social behavior comes from animal models, in which it has been primarily linked to a diverse variety of social behaviors especially in males. The hormone is expressed in greater quantities in the brains of male versus female voles (Bamshad et al., 1993) and even humans (Ishunina and Swaab, 1999). Mating induces secretion of AVP in male prairie voles but not in females (Wang et al., 1994), this further implicates AVP in mating-induced changes in male voles' social behaviors. As part of the postmating male social behavior, AVP causes aggression upon intrusions to one's territory. Moreover, AVP antagonists injected centrally inhibit creation of mate preference and aggressive behaviors towards intruders, while infusion of AVP induces partner preference behavior even without mating (Winslow et al., 1993). AVP's role as the initiator of male social behavior is further supported by the finding that androgens control AVP synthesis in several brain areas (De Vries et al., 1994).

Despite these advances in animal models, much less is known about the role of AVP in human social behavior. Our initial studies of AVP and human social cognition have been powered by a neurogenetic strategy and we generated evidence

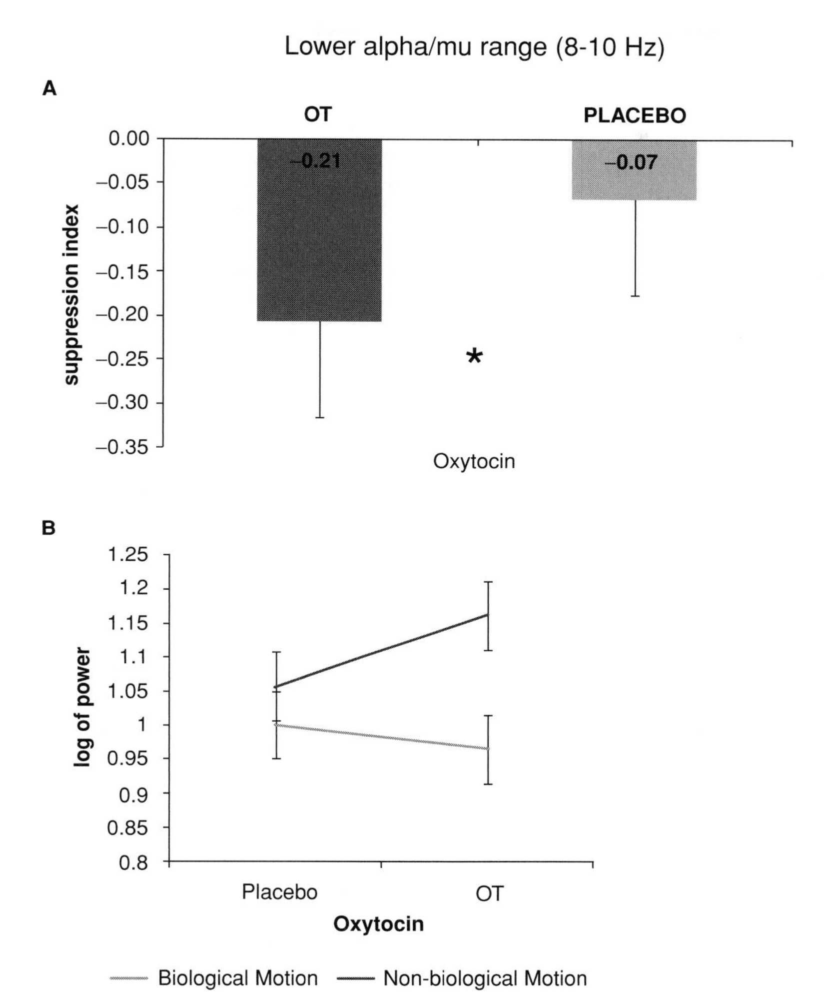

Figure 19.1 (A) Suppression in the 8–10 Hz range, OT versus placebo. Both bars show suppression for the biological motion conditions compared to the non-biological condition, but this suppression is enhanced significantly by OT. Error bars represent standard error (SE). (B) 8–10 Hz interaction between treatment x motion. OT had an opposite effect on EEG for perception of biological versus non-biological stimuli.

for a role of the arginine vasopressin 1a receptor gene (AVPR1a) in a number of social behaviors in non-clinical subjects (reviewed in (Ebstein et al., 2010)).

We next employed a complementary pharmacological strategy by examining specific behaviors following intranasal administration of AVP. Studies using intranasal administration to investigate the role of AVP in human social cognition are scarce, but the work done so far suggests a similar effect to the one found in rodents, that is, AVP has a sex-specific effect on social interactions, specifically promoting aggression in males. Notably, this correlation was stronger in males than in females. Thompson and colleagues (Thompson et al., 2006; Thompson et al., 2004) showed that intranasal

administration of AVP stimulates agonistic facial motor patterns in response to the faces of unfamiliar men and decreases perceptions of the friendliness of those faces. In contrast, in women, AVP had an opposite effect of increasing perceptions of the friendliness of those faces. Taken together, these findings point to a major role of AVP in modulating social communication.

Given that AVP seems to have a more prominent role in male-typical behavior as discussed above, our group examined in two separate investigations the effect of intranasal AVP on two central facets of human social behavior solely on male subjects, social stress (Shalev et al., 2011) and empathy (Uzefovsky et al., 2011).

19.2.2 AVP and social stress

Social behavior requires the drive to approach others and the diminution of stress and fear that is naturally elicited by proximity to others. The nonapeptide arginine vasopressin (AVP), which has been related to social behavior and hypothalamus–pituitary–adrenal (HPA) axis regulation, seems to be a prime candidate for modulating stress in social situations (Goodson, 2008).

To shed light on the neurobiological substrate of AVP in modulating social behavior in the context of social stress, we employed the Trier Social Stress Test (TSST) (Kirschbaum et al., 1993), a paradigm that has proven particularly effective in evaluating psychosocial stress under controlled laboratory conditions. The TSST employs brief public speaking and a mental arithmetic paradigm to evoke a robust salivary cortisol response in subjects. Gender (Uhart et al., 2006; Kirschbaum et al., 1995), genetics (Kumsta et al., 2007; Shalev et al., 2009), and environmental stressors (Macmillan et al., 2009) among other factors (Kudielka et al., 2009) influence individual's responses during the TSST. In this context, HPA axis reactivity is indexed by measuring salivary cortisol while central nervous system reactivity is measured by monitoring blood pressure and pulse rate. Given AVP's established dual roles in modulating both the HPA axis and social signaling, we hypothesized that AVP prior to the TSST would lead to an interaction with the stress response and heightened cortisol reactivity when compared to placebo. Furthermore, given the long evolutionary history of AVP as a social hormone (Goodson, 2008), we hypothesized that the effects of AVP on the stress response could be specifically attributed to the social evaluative elements of the TSST. We therefore designed a set of studies to investigate the influence of AVP on HPA reactivity under a set of conditions varying in social evaluation and exposure to stress. We began by investigating the effect of AVP on HPA axis reactivity in the full TSST and hypothesized that intranasal AVP would enhance salivary cortisol output.

Despite indirect evidence for its contribution to social signaling (Goodson, 2008), little is known regarding the role of AVP in the context of human social stress. For example, it could be the case that intranasal AVP, even in the absence of stressful cues, would directly activate the HPA axis, resulting in increased cortisol levels. Furthermore, it may also be the case that stressors in general, even those in the absence of the social evaluative threat produced in the TSST may interact with AVP to trigger a rise in cortisol. Hence, we tested the hypothesis that the effect of intranasal AVP on the salivary cortisol response is contingent upon social contexts. To address this issue, we implemented three additional experimental conditions.

The first experiment was titled the "no task" group and controlled for direct physiological influences of AVP administration on HPA reactivity under a no stress condition. In this experiment, subjects were simply administered intranasal AVP or placebo while sitting by themselves in a controlled environment, in the absence of stressful stimuli. The second experiment was titled the "no audience," in which participants engaged in a modified TSST, in the absence of audience and cameras, and consequently in the absence of social evaluative threats yet still retaining enough stressors to trigger a cortisol response. The third experiment employed an exercise bike ("bike ergometry"), also in the absence of audience and cameras, which was designed to

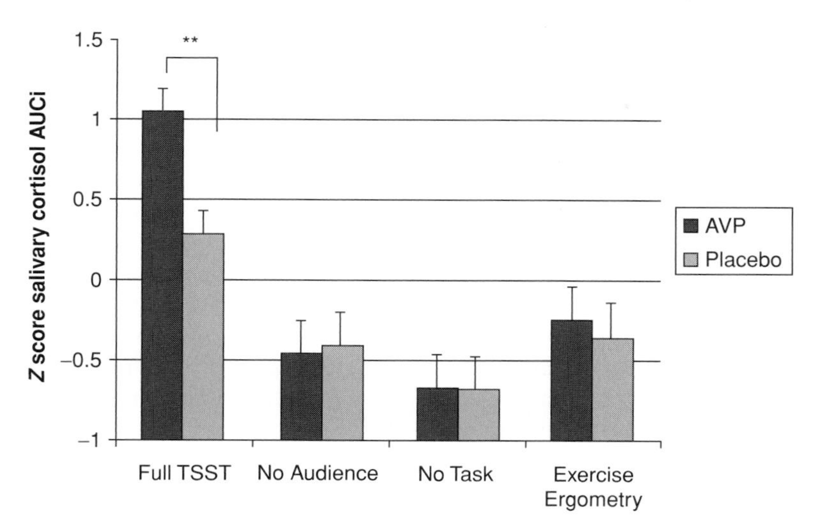

Figure 19.2 Area under the curve with respect to increase of salivary cortsiol (AUCi, Z-score) (+ SEM) grouped by AVP and placebo in all conditions (full TSST, "no audience," "no task" and "bike ergometry"). $^{**}p < 0.005$.

evoke physiological stress (cortisol, blood pressure and heart rate) but not a social stress response. The purpose of these three additional experiments was to isolate the social evaluative threat and determine the specificity of AVP effect on this component of the TSST procedure.

If intranasal AVP resulted in a direct physiological response then we would expect to observe increases in cortisol levels in the AVP group in all four conditions. Alternatively, if intranasal AVP was sensitive to stress responses in general, then we would expect to observe AVP's effects on cortisol reactivity for the full TSST, the "no audience" and the "bike ergometry" conditions. However, if AVP's effects were specific to contexts that contained social evaluative threats, then we would expect to observe an effect of AVP on cortisol response in the full TSST condition only.

We examined each treatment group in separate general linear model repeated measures tests and by univariate analysis comparing area under the curve with respect to increase AUCi (Figure 19.2). AVP augments the response to social stress solely in the experimental condition of full TSST. No AVP-related augmentation was seen in the three control conditions. Similarly, when comparing AUCi (Figure 19.3), there was a highly significant effect of AVP on AUCi in the full TSST condition ($p = 0.001$) but not in the other control conditions.

Despite the intense interest generated by the role of neuropeptides as social hormones that modulate dyadic and group interactions with conspecifics across vertebrates (Ebstein et al., 2009), our study is the only one of a very few investigations (Thompson et al., 2006; Pietrowsky et al., 1996; Born et al., 2002; Zink et al., 2010) to use intranasal AVP toward unraveling the mode of AVP action on the human social brain and particularly its role in social stress. AVP enhanced the impact of the audience on the TSST task and appears to increase the subjects' sensitivity to the social milieu and specifically the presence of observers. As argued by Dickerson and Kemeny (Dickerson and Kemeny, 2004), status in humans is conferred through hedonic processes that relate to respect, self-esteem, acceptance, and positive social attention.

19.2.3 AVP and empathy

At the core of human social behavior is the ability to empathize, that is, the ability to accurately

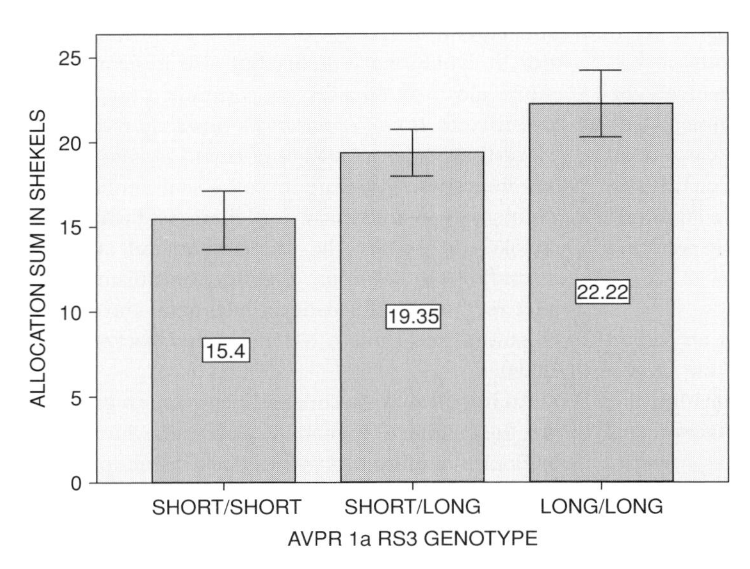

Figure 19.3 Allocation amount (continuous variable) grouped by AVPR1a RS3 long vs. short genotype. Error bars are SEM. One-way ANOVA: $n = 203$, $F = 3.456$, $P = 0.033$. SPSS post hoc analysis using the Tukey's HSD test showed a significant difference between short/short vs. long/long ($P = 0.025$).

perceive another's emotions and to respond to them in an appropriate way. Empathy, which molds the fabric of social interactions and underlies all forms of social communication, has been hypothesized to drive prevalent social processes such as altruism and prosocial behavior (Batson et al., 1991). Importantly, empathy is a multifaceted concept that consists of at least two processes: cognitive empathy (CE) – the ability to perceive what the other is feeling, and the focus of this report – and emotional empathy – matching emotional response to the feelings of the other (Davis, 1980). Facial expressions are salient representatives of our emotions and an important focus of social communication between conspecifics. Interpreting facial expression appears to be a robust indicator of CE (Baron-Cohen et al., 2001). Accordingly, the "reading the mind in the eyes test" (RMET), employed in our investigation, is a widely used instrument for evaluating CE in both non-clinical and clinical settings (Baron-Cohen et al., 1999; Baron-Cohen et al., 2001).

We first examined the overall effect of treatment (AVP versus placebo) on the ability of male participants to correctly recognize the emotions exhibited in the eye pictures (Uzefovsky et al., 2011). We analyzed the data employing univariate ANOVA with treatment as the sole variable and found, as predicted, that the AVP group made significantly more errors than the placebo group ($t(37) = 2.199$, $p = 0.034$, $D = 0.72$). Given the known gender-specific effect of AVP (Thompson et al., 2006; Thompson et al., 2004) we hypothesized that the perceived gender of the individuals photographed might be a factor in how AVP affects emotion recognition. The RMET includes photographs of both men and women and when scores were stratified by the sex of the observed pictures, the effect of AVP was not found when males observed female photographs. In contrast, in gender-matched photos; that is, when males observed male photographs, a large ($D = 0.91$) effect was found for the administration of AVP ($t(37) = -2.77$, $p = 0.009$).

We further hypothesized that the valence of the emotion exhibited would play a role in how AVP affects emotion recognition, since AVP was shown to induce aggressive emotions in men (Thompson

et al., 2004). Additional evidence for a role of AVP in human aggression is suggested by a study of cerebrospinal fluid levels of AVP that were positively correlated with aggressive behaviors in subjects suffering from personality disorders (Coccaro et al., 1998), We therefore sorted the RMET pictures by those displaying negative versus positive emotions. The effect of AVP on reducing empathy was only observed for negative emotions ($t\,(37) = -2.38$, $p = 0.023$) and not for positive emotions

Our study (Uzefovsky et al., 2011) is an important step towards understanding the role of AVP in modulating human social communication and demonstrates that AVP has a specific effect on cognitive empathy in males. AVP diminishes cognitive empathy in males, but only toward others males. Importantly, this finding echoes previous knowledge derived from animal models showing a role for AVP in aggression related male social behavior. For example, in male prairie voles, mating-mediated AVP release induces aggression toward other males and strange females (Winslow and Insel, 1993). In voles, aggression was directed toward both sexes, while in this report the effect was limited to photos of males. In humans, physical aggressive behavior by males is more common than by females (Archer and Coyne, 2005), and it is more commonly directed towards other males. Notably, feelings of aggression, which increase the propensity for a physical attack, are accompanied by diminished ability to empathize with the possible victim (Preston and de Waal, 2002). Therefore, AVP would be hypothesized to diminish cognitive empathy of males toward other males, and not towards females, as is reflected in our current findings (Uzefovsky et al., 2011).

19.3 Nonapeptides and other-regarding behaviors (altruism)

A prominent trait that characterizes our species in particular is altruism, providing benefit for others at the cost of one's own fitness. The evolution of altruism presents a conundrum to evolutionary theory, viz. how can natural selection prefer individuals displaying prosocial traits, over those that carry selfish ones? Not only is altruism a paradox in evolutionary theory but also in economics since altruism appears to contradict the prime directive of *Homo economicus* to maximize profit. Nevertheless, by all accounts *Homo sapiens* shows an inordinate measure of prosocial tendencies. Numerous evolutionary explanations have been invoked to explain the phenomenon of altruism including kin selection, reciprocal altruism, indirect reciprocity, altruistic punishment and others (Sigmund and Hauert, 2002; Fehr and Rockenbach, 2004).

An important experimental approach to unraveling the enigma of human altruistic behavior and its biological underpinnings has been the adoption of paradigms from the field of behavioral economics. In particular, the use of incentivized "games" based on the principle of "put your money where your mouth is" has provided deep insights into the neurogenetics, neuroendocrinology and neural correlates of prosocial human behavior. The first study using a molecular genetic approach combined with a classic behavioral economic paradigm, the Dictator game, was carried out by our own group (Knafo et al., 2008). In the Dictator game the first mover is endowed with a fixed sum of money and "dictates" how much to allocate to the second player who is a passive recipient. We employed a candidate gene approach and based on translational evidence from the extensive studies of the vole (Young et al., 2008; Hammock, 2007; Hammock et al., 2005; Hammock and Young, 2005) and hypothesized that the length of the arginine vasopressin 1a receptor promoter (*AVPR1a*) RS3 microsatellite would predict allocations of funds in the anonymous Dictator game just as the length of the equivalent promoter region in the vole determines parenting and affiliative behaviors. Indeed, as shown in Figure 19.3 increasing lengths of the RS3 microsatellite are associated with increasing giving behavior. We also demonstrated for the first time in humans that increasing length of this microsatellite was characterized by increasing AVPR1a mRNA levels in postmortem hippocampal specimens. It is encouraging to note that the Dublin group (Tansey et al., 2011) has recently also reported that shorter repeat

alleles of RS1 and RS3 decreased relative promoter activity in the human neuroblastoma cell line SH-SY5Y. Furthermore, Meyer-Lindenberg et al. (Meyer-Lindenberg et al., 2008) showed that the long *AVPR1a* alleles predicted greater amygdala activation during functional imaging employing an emotional face-matching paradigm. Further evidence of the functionality of the promoter region microsatellite length was a study by our own group of prepulse inhibition (Levin et al., 2009). Altogether, it appears that in humans, similar to the vole, the repeat regions of the promoter region of the *AVPR1a* gene are functionally significant and influence transcription of this important gene.

We extended our studies of human altruism (Israel et al., 2009) and examined association with the oxytocin receptor (OXTR). The OT receptor gene is present in single copy in the human genome and was mapped to the gene locus 3p25–3p26.2 (Gimpl and Fahrenholz, 2001). The gene spans 17 kb and contains 3 introns and 4 exons. Exons 1 and 2 correspond to the 5′ non-coding region. Exons 3 and 4 encode the amino acids of the OT receptor. Intron 3, which is the largest at 12 kb, separates the coding region immediately after the putative transmembrane domain 6. Exon 4 contains the sequence encoding the seventh transmembrane domain, the COOH terminus, and the entire 3-non-coding region, including the polyadenylation signals.

Association (101 male and 102 female students) using a robust family-based test between 15 single-tagging SNPs (htSNPs) across the OXTR was demonstrated with both the Dictator game and social values orientation paradigm (Van Lange, 1999). Three htSNPs across the gene region showed significant association with both of the two games. The most significant association was observed with rs1042778 ($p = 0.001$). Overall, our results demonstrate that genetic polymorphisms for the *OXTR* gene are associated with human prosocial decision. This investigation converges with a large body of animal research showing that OT is an important social hormone across vertebrates including *Homo sapiens*. Individual differences in prosocial behavior have been shown by twin studies to have a substantial genetic basis (Rushton, 2004; Knafo and Plomin, 2006) and the current investigation demonstrates that common variants in the oxytocin receptor gene, an important element of mammalian social circuitry, partially underlie such individual differences.

A note of caution is needed regarding the role of *OXTR* SNPs in contributing to Dictator game allocations since a study by a Swedish group (Apicella et al., 2010) failed to replicate our findings. They do not find any strong evidence for a role for the *OXTR* SNPs typed in our study (Israel et al., 2009) as a source of individual differences in Dictator game giving, trust or trustworthiness in either an additive model or a dominance model. However, a SNP of considerable interest, rs53576, was associated with prosocial behavior in men, albeit the association did not survive correction for multiple testing. Notably, this SNP has been associated initially in a Chinese study with autism (Wu et al., 2005), with ADHD (Park et al., 2010), with neural correlates and temperament traits (Tost et al., 2010), with depression and separation anxiety along with SNP rs2254298 (Costa et al., 2009), in the generation of affectivity, emotional loneliness and IQ (Lucht et al., 2009), with empathy scores employing the reading of the mind in the eyes test (Rodrigues et al., 2009), parenting style (Bakermans-Kranenburg and van Ijzendoorn, 2008), but not with attachment insecurities (Gillath et al., 2008).

All of the studies so far discussed of OT and AVP were focused on aspects of normal behavior and normal sociality. However, accumulating evidence suggests that OT and AVP also contribute to psychopathology. In particular, these two nonapeptides have been linked to autism spectrum disorders (ASD), a condition characterized by deficits in social cognition. In the next section we discuss the role of these neuropeptides in ASD.

19.4 OT and AVP in psychopathology

The first suggestions that OT and AVP might contribute to ASD appeared more than a decade ago (Freeman, 1997; Waterhouse et al., 1996; Panksepp, 1993; Modahl et al., 1992; Insel et al., 1999). The

connection between these hormones and ASD is a good example of translational medicine and the worthwhileness of extrapolating results from animal experiments to a human disease. As first proposed by Kanner (Kanner, 1943) "autistic children have come into the world with an innate inability to form the usual biologically provided affective contact with people." The identification of AVP and OT as pre-eminent social hormones that mediate affiliative and attachment behaviors in lower mammals suggested the reasonable notion that deficits in AVP and OT neuropathways could also contribute to ASD etiology. It remained to be demonstrated that these two neuropeptides are also social hormones in *H. sapiens* but in the past decade a wealth of evidence has greatly strengthened the idea that also in people AVP and OT also modulate workings of the social brain.

The first direct evidence for a connection between ASD and autism was the study by Modahl (Modahl et al., 1998) who measured plasma OT levels in autism subjects. Despite individual variability and overlapping group distributions, the autistic group had significantly lower plasma OT levels than the normal group. OT increased with age in the normal but not the autistic children. Elevated OT was associated with higher scores on social and developmental measures for the normal children, but was associated with lower scores for the autistic children. These relationships were strongest in a subset of autistic children identified as "aloof." Although making inferences to central OT functioning from peripheral measurement is difficult, the data suggest that OT abnormalities may exist in autism, and it was suggested that more direct investigation of central nervous system OT function was warranted.

Direct experimental evidence suggesting a role of AVP and OT in ASD was provided by association studies linking *AVPR1a* with idiopathic autism (Yirmiya et al., 2006; Wassink et al., 2004; Kim et al., 2002) and discussed in the following paragraph. The human AVP V1a (*AVPR1a*) receptor gene is relatively simple, 2 exons and 1 intron, located at 12q14–15 with 3 polymorphisms located in the 5′ flanking region and one in the intron (Thibonnier, 2004; Thibonnier et al., 1994). The 5′ flanking region microsatellites, RS1 and RS3, have received the most attention, with links to a diverse set of interpersonal skills from sibling relationship to musical ability to economic decision making (Israel et al., 2008).

Kim and colleagues (Kim et al., 2002) genotyped two microsatellite polymorphisms (RS3 and RS1) from the 5' flanking region of AVPR1A for 115 autism trios and found nominally significant transmission disequilibrium between autism and RS1 by a family-based association test that was not significant after Bonferroni correction. In a second study by Wassink and colleagues (Wassink et al., 2004) association was observed with both promoter region markers but only when the patients were stratified by their language characteristics; significance was observed in the normal language families. The third consecutive study linking *AVPR1a* and ASD was carried out by our own group (Yirmiya et al., 2006). In our study, we failed to observe significant association between RS1 and RS3 and autism but observed significant transmission disequilibrium between an intronic microsatellite (AVR), which is in moderate linkage disequilibrium with the promoter region microsatellites. In our study, haplotype analysis showed significant transmission disequilibrium in autism families when all three microsatellites (two promoter region markers and the intronic microsatellite) were analyzed. Notably, we also observed association between *AVPR1a* and daily living and communication skills in these subjects using the Vineland Adaptive Behavioral Scales (VABS) and ADOS-G. These results prompted us to suggest the notion that the association between this gene and social deficits provides a likely molecular genetic and behavioral mechanism by which this receptor contributes risk for autism, similar to its role in determining social communications in the vole. The importance of the vasopressin system in processing social information in mammals is further underscored in a study by Tobin et al., 2010 who demonstrated that the rat olfactory bulb contains a large population of interneurons that express vasopressin, that blocking the actions of vasopressin in the olfactory bulb impairs the

social recognition abilities of rats and that vasopressin agonists and antagonists can modulate the processing of information by olfactory bulb neurons. To summarize, both translational research and more direct molecular genetic studies indicate a provisional role for the *AVPR1a* receptor in contributing to the etiology of ASD.

An interesting study that strengthens the connection between *AVPR1a* and human social relationships that very much resonates with the vole saga is from Sweden by Walum and colleagues (Walum et al., 2008). They report an association between one of the human *AVPR1A* repeat polymorphisms (RS3), the second most common repeat allele 334 base pairs, and traits reflecting pair-bonding behavior in men, including partner bonding, perceived marital problems, and marital status, and show that the RS3 genotype of the males also affects marital quality as perceived by their spouses. These results suggest an association between a single gene and pair-bonding behavior in humans, and indicate that the well-characterized influence of AVP on pair bonding in voles may be of relevance also for humans.

Intriguingly, the 334 bp risk allele that doubles the risk of marital crisis, is the identical allele overtransmitted in ASD (Kim et al., 2002). Prompted by these observations in the Walum et al. (Walum et al., 2008) and Kim et al. (Kim et al., 2002) studies regarding the RS3 second most common repeat (334 or 327 bp depending on the PCR amplification methods used in each study), we thought it worthwhile to take a closer look at this allele in our study of autism (Yirmiya et al., 2006). Notably, our data also shows overtransmission of this second most common allele (RS3 allele 5): 29.9% (ASD) versus 22.1% (see table 1 in (Yirmiya et al., 2006)) which is borderline significant, and uncorrected for testing multiple alleles (p=0.06) (see table 3 (Yirmiya et al., 2006)). Nevertheless, the direction is similar, strengthening the notion that the second most common RS3 allele contributes risk for dysfunctional social behaviors. Furthermore, a recent imaging study by Meyer-Lindenberg's group (Meyer-Lindenberg et al., 2008) showed that the second most common 334 bp risk allele of RS3 (present in 21.3% of subjects) showed differential overactivation of the left and right amygdala.

Additional evidence that the *AVPR1a* RS3 334/327 bp allele characterizes in carriers less than optimal social relationships and behavior is a study carried out by our Jerusalem group (Avinun et al., 2011). Avinun et al. (Avinun et al., 2011) examined preschoolers (3.5-year-old twins) for altruistic behavior employing the Dictator game and the resource allocation game (Fehr et al., 2008) adapted for this age group. The results showed for the first time a differential effect of a genetic variant (RS3 334/327 bp) on altruism, which is well in line with evolutionary predictions: Increasing altruism towards kin and decreasing altruism towards non-kin. Population-based analysis was used to examine the association between *AVPR1A* and pure altruism as measured in the Dictator game modified for toddlers. The RS3 risk allele was examined against all others (comparing 36 carriers of at least one copy with 62 non-carriers). Results for the Dictator game were significant (Kendall's tau-c = –0.27, p =0.004), with carriers of the 327 bp allele being 4 times less likely to donate more than one sticker chart to a stranger. In contrast, when altruism in the context of kin selection was examined opposite results were observed. These results by Avinun et al. underscore the complexity of examining and interpreting allelic effects in human social relationships. Gene effects play out across a multidimensional space and over the life course and it is important to capture the richness of such behaviors with studies from early development to adulthood as well as across diverse laboratory-based models. Gene effects that impinge on other directed preferences are context dependent and hence it is not surprising that allelic differences are observed on the background of kin versus non-kin altruism that is an important distinction reflecting deep evolutionary roots (Hamilton, 1963).

19.5 Endophenotypes

The path from a genetic polymorphism to risk for autism (or for that matter any complex

disorder) is non-linear and reflects the wide gap separating DNA that encodes elements of synaptic transmission from behavioral phenotypes. Moreover, the syndrome of autism itself is complex, reflecting considerable heterogeneity at both the behavioral and molecular genetic level. Indeed, we have argued (Levy and Ebstein, 2008) that the time is ripe to loosen the ties between biological research and clinical syndromes, moving on to allow across-syndrome cohort selection, including non-clinical subjects, based on trait homogeneity, which can then guide the search for biological endophenotypes of behavior. Furthermore, given the inherent complexity of mapping genes to behavior and considering the non-linearity of such mapping, the involvement of epigenetic processes and the interplay of environmental factors in development, a necessary intermediate step in understanding the links between genetics and behavior is to decompose the syndromic disorders into intermediate or endophenotypes (Gottesman and Gould, 2003). For example, a very promising approach is mapping brain endophenotypes (Meyer-Lindenberg and Weinberger, 2006). The fact that clinical syndromes capture real-life phenomena that have medical, educational, economical and social-emotional implications provides a powerful motivation for studying syndromes. Yet, when it comes to biological research, especially molecular genetic studies, none of these motivations seem compelling and relevant classifications must be based on a different set of considerations.

We believe that the current review of OT and AVP research in both non-clinical and clinical syndrome strengthens the argument made by Yonata and Ebstein (Levy and Ebstein, 2008) that studying endophenotypes of complex behavioral disorders such as autism in non-syndromic subject cohorts is highly informative and at least for ASD, has greatly extended our understanding of the role of these two neuropeptides in contributing to pathophysiolgy. Understanding some of the deficits in the social brain in ASD is first rooted in translational biology, viz. the animal model illustrated by the vole. This translational approach led to an initial series of molecular genetic studies linking *AVPR1a* and *OXTR* polymorphisms to the syndrome. Importantly, studies in non-clinical cohort revealed the rich contours of OT and AVP neurobiology in molding social and affiliative behaviors in humans with profound implications for understanding deficits in social cognition in syndromic disorders such as autism. Most importantly, we believe these investigations of non-clinical cohorts have outlined some of the salient biological underpinnings, as well as provided the scientific basis, for clinical trials now in progress in several centers that are testing OT as a novel therapeutic agent in ASD.

In the next section of this review we discuss a new player in the AVP-OT story, CD38, a story that again illustrates the importance of animal research in generating candidates for contributing to human diseases and, importantly, how quick recognition of the translational importance of early findings can be immediately applied in human studies towards understanding complex behavioral disorders such as autism.

19.6 CD38 and autism spectrum disorders

A seminal paper by Higashida and his group led to the discovery that OT release in the brain is mediated by ADP-ribosyl cyclase and /or CD38 (Jin et al., 2007). They used CD38 gene knockout mice (*Cd38–/–*), and discovered that CD38-dependent cyclic ADP ribose (cADPR)- and NAADP-sensitive intracellular Ca^{2+} mobilization plays a key role in OT release from soma and axon terminals of hypothalamic neurons, with marked effects on social behavior. In particular, maternal behavior was dependent on OT, and social amnesia in males was evident in the absence of this hormone. CD38 cleaves NAD and NADP, generating cADPR, NAADP, and ADPR (Higashida et al., 2001). cADPR mobilizes Ca^{2+} from ryanodine-sensitive intracellular Ca^{2+} stores in the endoplasmic reticulum and NAADP liberates it from other pools located in lysosomes or secretory granules. The two molecules act as second messengers independent of inositol 1,4,5-trisphosphate (IP3)

(Higashida et al., 2001). To summarize, CD38 is a multifunctional molecule (ecto-enzyme) combining enzymatic and receptor properties and playing a key role in various physiological processes in the tissues (proliferation, differentiation, migration, adhesion, and secretion). In the brain, CD38 is found in neurons and glial cells, shows intracellular or plasma membrane location, and is enriched in neuronal perikarya and dendrites (Ceni et al., 2006; Mizuguchi et al., 1995). CD38 is critical for OT but not AVP release (Jin et al., 2007).

The accumulating evidence discussed above, showing that OT plays a central role in both normal as well as dysfunctional social relationships / cognition (Ebstein et al., 2009; Israel et al., 2008), *ipso facto* targets CD38, a key mediator of OT brain release, as a potential focus of interest in normal human social behaviors as well as disorders of social cognition, especially autism (Bartz and McInnes, 2007; Young, 2007). In the past year, two research groups have independently addressed the role of CD38 in autism in human subjects. Higashida and his colleagues (Munesue et al., 2010) analyzed 10 single-nucleotide polymorphisms (SNPs) and mutations of CD38 by resequencing DNAs mainly from a case-control study in Japan, and Caucasian cases mainly recruited to the Autism Genetic Resource Exchange (AGRE). CD38 SNPs, rs6449197 and rs3796863 showed significant associations with a subset of ASD subjects (IQ $>$ 70; designated as high functioning autism / HFA) in 104 AGRE family trios, but not with Japanese 188 HFA subjects. Interestingly, a mutation / rare polymorphism that caused tryptophan to replace arginine at amino acid residue 140 (R140W; (rs1800561, 4693C $>$ T)) was found in 0.6–4.6% of the Japanese population and was associated with ASD in the smaller case-control study. The SNP was clustered in pedigrees in which the fathers and brothers of T-allele-carrier probands had ASD or ASD traits. In this cohort OT plasma levels were lower in subjects with the T allele than in those without.

In our first study of CD38 (Lerer et al., 2010), we examined all tagging SNPs across the CD38 gene region in 170 subjects diagnosed with ASD from 149 families (see (Lerer et al., 2008) for description of the subjects). Individual SNPs and haplotypes were tested for association with ASD. Additionally, the relationship between diabetes, autism and CD38 (Atladottir et al., 2009), as well as the use of CD38 as a disease marker (Malavasi et al., 2008), suggests that it would also be worthwhile to explore CD38 expression in immune cell lines derived from ASD patients. These considerations prompted us to measure CD38 gene expression in lymphoblastoid cell lines (LBC) derived from both ASD subjects and unaffected parents. We also include in the gene expression and family-based association analysis the SNP (rs3796863), which proved significantly associated with ASD in the Munesue et al., 2010 study, see Figure 19.4.

We first examined association between CD38 tagging SNPs and DSM IV ASD. ASD subjects were evenly grouped into high and low functioning based on an IQ cutoff of 70. This subject stratification was aimed at reducing phenotypic heterogeneity in the autism sample. Significant association was observed between low functioning ASD and three to seven haplotypes. The results are significant ($p < 0.05$) following permutation testing. Importantly, the SNP (and the "C" allele) identified in the Munesue et al., 2010 study (rs3796863), which they found significantly associated with ASD, is located in all except one of the significant haplotypes in our study (Lerer et al., 2010).

We then asked the question whether CD38 expression in peripheral cells a biomarker for ASD? CD38 mRNA levels in LBC derived from subjects with autism and unaffected parents were examined. A highly significant reduction (SPSS ANOVA-Affected status: $F = 14.72$, $p = 0.0002$, df = 1; Sex: $F = 4.680$, $p = 0.033$, df = 1; Interaction: affected x sex, $F = 2.304$, $p = 0.132$, df = 1) in CD38 expression was observed in cells from the DSM IV ASD subjects ($N = 44$) compared to "unaffected" parents ($N = 40$). The main effects are observed for diagnosis and sex. These first results (Lerer et al., 2010) have now been partially replicated in a new study from our laboratory (Riebold et al., 2011).

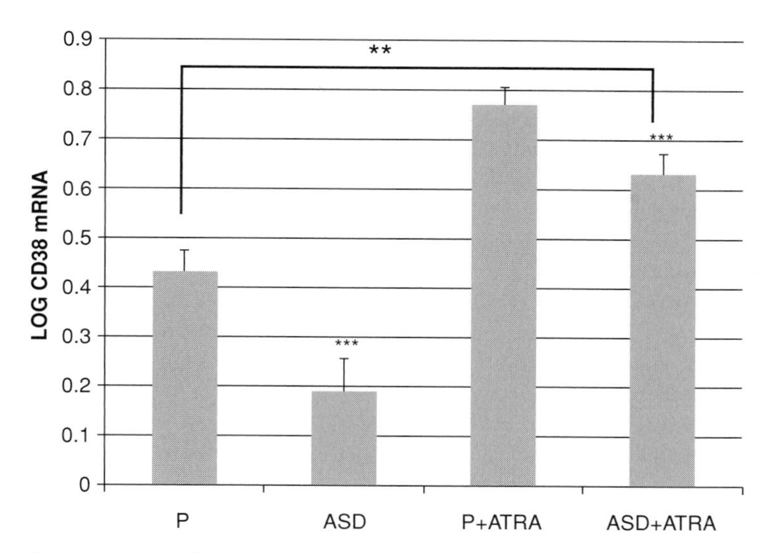

Figure 19.4 The effect of 48 hours 0.1 μm ATRA treatment on CD38 mRNA levels in LBC lines. **, independent samples t-test, $t = -3.199$ $p = 0.002$; prolonged ATRA treatment elevates reduced CD38 mRNA levels in LBC lines from ASD patients ($N = 42$) above parental (P) basal expression ($N = 78$). Also, basal and induced CD38 mRNA levels are significantly reduced in ASD cell lines compared to parental (P) cell lines (***, independent t-test; $p < 0.001$).

In our subsequent investigation (Riebold et al., 2011), we have reanalyzed the EBV lines described in the first report, significantly adding to the sample with 38 new cell lines so that in the second investigation for each proband both of their parents were now included in the analysis. Cells in culture, or frozen lines were first thawed, and then cultured, and their CD38 mRNA levels measured. It was important to determine whether expression of CD38 is stable and is maintained despite repeated cycles of freezing and thawing. The new results (Figure 19.4) confirm that CD38 expression in ASD patient lines is substantially lower than in those derived from the patients' parents. Although these results are not a fully independent replication, we believe they nevertheless considerably strengthen our first findings that reduced CD38 transcription is a characteristic of peripheral lymphocyte cells derived from ASD subjects (Lerer et al., 2010).

Retinoids are a class of compounds consisting of retinol (vitamin A) and its derivatives and synthetic analogs (Alvarez et al., 2011; Theodosiou et al., 2010). Natural retinoids are fundamental for many physiologic processes, such as reproduction, growth, and cellular differentiation. These effects are mediated by binding to, and activating two different types of nuclear receptors, the retinoic acid (RA) receptors (RARs) and retinoid X receptors (RXRs).

All-trans retinoic acid (ATRA) is a potent inducer of CD38 (Ferrero and Malavasi, 2002) suggesting the possibility that this compound can be used to "rescue" cells exhibiting low CD38 synthesis and hence might be a novel therapeutic strategy in treatment of autism. We wanted to determine whether the diminished expression of CD38 in ASD could be reversed through simple treatment with ATRA. Such a demonstration would provide *in vitro* "proof of principle" that retinoids could play a role in the clinical treatment of ASD.

Following 48 h of ATRA treatment, the results indicate that the CD38 gene in the EBV lines obtained from the ASD probands conserves its ability to respond with a significant induction of CD38 mRNA

(Figure 19.4). The parental lines display the same ability, although to a lesser extent (paired t test $t=-13.26$ $p<0.001 \pm$ ATRA). These results, demonstrating that ATRA can elevate CD38 levels in cells obtained from ASD subjects who show impaired CD38 transcription, strengthen the notion that vitamin A and related retinoids are potential therapeutic agents in the treatment of ASD. Indeed, retinoids are widely used as treatment modules in a spectrum of diseases including acne and psoriasis (Trapasso et al., 2009), cancer (Siddikuzzaman et al., 2010), and as a dietary supplement reduces child mortality between 6 months to five years in low and middle income countries (Imdad et al., 2010). We put forward the prospect that retinoids are well worth examining as therapeutic agents in autism and possibly other disorders that are characterized by dysfunctional social cognition / relationships especially where OT has been suggested to play a role.

We also examined the role of genotype in mediating the response to ATRA. Cell lines were genotyped for the rs6449182 SNP, which leads to a C→G variation. This SNP is located in intron 1 of the regulatory region of human CD38, proximal to RARE. The presence of the allele G is reported as being paralleled by increased binding of the transcription factor E2A(Saborit-Villarroya et al., 2011). Furthermore, the G allele marks an increased risk in CLL patients of transformation into Richter's syndrome (Aydin et al., 2008).

The results from our second study (Riebold et al., 2011), indicate that the presence of the G allele is paralleled by i) reduced transcriptional levels of CD38 mRNA, a characteristic shared by ASD with those of the parental lines although in LBC lines obtained from ASD probands the difference does not attain statistical significance. Furthermore, ii) the G allele is accompanied by reduced sensitivity to ATRA treatment (+ ATRA treatment in parental lines $CC = 0.82 \pm 0.04$; $CG = 0.69 \pm 0.04$ $p = 0.04$).

Our results showing that CD38 expression is reduced in ASD prompted us to examine whether its expression levels might also reflect phenotypical characteristics of ASD further enhancing the value of this ectoenzyme as a potential biomarker. We looked at social functioning measures that were available for these probands since such deficits are a core clinical characteristic of autism. The results obtained clearly show a significant correlation between transcriptional levels of CD38 mRNA and IQ and Vineland Adaptive Behavioral Scores (VABS) scores (Sparrow et al., 1984), except for VABS socialization. Nonetheless, the correlation with the VABS total scores does prove significant ($r = 0.431$, $p = 0.008$ $N = 42$). See Figure 19.5.

Our studies of CD38 add to the growing list of potential biomarkers in ASD. Importantly, a correlation was observed between expression levels of CD38 in LBC derived from these subjects and social and communication skills that are core deficits in this disorder. Notably, the potential of CD38 expression as a diagnostic indicator for ASD was a hypothesis-driven idea catalyzed by the seminal study of Higashida and his colleagues (Jin et al., 2007) in the CD38 knockout mouse and reinforced by two independent molecular genetic studies showing association between SNPs in the CD38 gene and ASD (Munesue et al., 2010; Lerer et al., 2010).

In addition to the potential of CD38 as a biomarker that may prove useful in early diagnosis of illness, the study of CD38 expression in peripheral lymphocytes has allowed us to model the potential of retinoic acids as a therapeutic agent in ASD. Indeed, we have shown that LBC derived from ASD subjects and characterized by reduced CD38 transcription can be "rescued" by simple treatment with all-trans retinoic acid. We believe these results provide the first "proof of principle" for a novel therapeutic strategy in treatment of ASD by enhancing OT secretion in the brain indirectly by ATRA induction of CD38 followed by mobilization of ryanodine-sensitive intracellular Ca^{2+} stores from the endoplasmic reticulum that in turn release OT.

19.7 Future directions

A decade ago little was known about the neurobiology of social and affiliative behaviors in humans

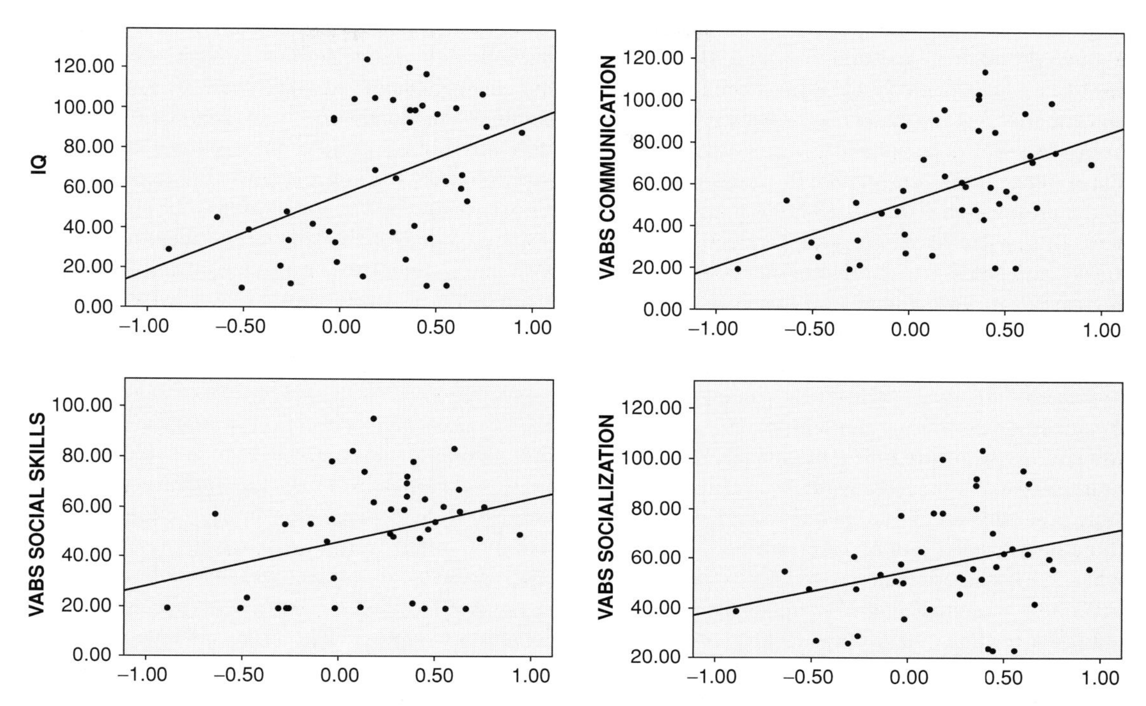

Figure 19.5 Correlation between CD38 mRNA expression and IQ and VABS subscores in lymphoblast lines derived from ASD subjects. Pearson correlations (IQ $r = 0.431$, $p = 0.004$ $N = 42$; VABS communication $r = 0.487$, $p = 0.001$ $N = 42$; VABS social skills $r = 0.329$, $p = 0.034$ $N = 42$; VABS socialization $r = 0.294$, $p = 0.059$ $N = 42$).

despite the fact that we are an intensely social species. Adolphs in a 2003 review (Adolphs, 2003) noted "... that our social nature defines what makes us human, what makes us conscious or what gave us our large brains." Knowledge regarding the social brain in humans has advanced by leveraging neuroimaging, neuroendocrinology and more recently neurogenetics. Neuroimaging studies have delineated the neural correlates of social cognition including perception of social signals, motivational evaluation and representation, somatic emotional response and modulation of cognition, and the representation of emotional response and social reasoning (Adolphs, 2003).

The best known sensory system in many animals for processing and transmitting social information is pheromones (Dulac and Torello, 2003). However, in humans processing of social information is mainly visual and face perception has been the primary focus for much of human social neuroscience during the past few years, growing out of earlier neurophysiologic and recent neuroimaging studies that demonstrated cells or fields in the monkey temporal cortex respond to faces (Tsao et al., 2003). Future studies would benefit by employing a neurogenetic strategy and stratifying responses to intranasal OT by genotype. Additionally, both the *OXTR* (Gregory et al., 2009; Kusui et al., 2001; Mizumoto et al., 1997) and *AVPR1a* genes are characterized by CpG islands and hence combining both sequence as well as epigenetic markings along these two genes offers the possibility of informing individual differences in response to nonapeptides by factoring in environmental variables using methylation of CpG islands as a proxy.

The discovery of the primate mirror neurons system has had a dramatic impact on social neuroscience by providing a brain mechanism for social learning. Intriguingly, a recent study by our group, discussed above, suggests that OT modulates the brain's mirror neuron system (Perry et al., 2010a). This report is the first hint linking a novel primate brain pathway suggested to mediate social skills with a social hormone. As noted by Iacoboni (Iacoboni, 2009) social psychology studies have demonstrated that imitation and mimicry are pervasive, automatic, and facilitate empathy. The demonstration that OT modulates mu suppression may provide an important neurochemical underpinning to the mechanism by which this nonapeptide enhances empathy and other regarding behaviors such as altruism and trust in humans. Indeed, mirror neurons embody the overlap between perception and action predicted by the ideomotor framework by discharging both during action execution and during action observation. By firing during actions of the self and of other individuals, mirror neurons may provide a remarkably simple neural mechanism for recognizing the actions of others. So-called broadly congruent mirror neuron cells provide a flexible coding of actions of self and others. It has been suggested that broadly congruent mirror neurons are ideal cells to support cooperative behavior among people (Newman-Norlund et al., 2007). Future research needs to further examine the role of OT in modulating mirror neuron function towards a deeper understanding of how these nonapeptides impact such a wide range of social and affiliative behaviors in humans. Such studies carried out on sufficient numbers of subjects would profitably employ a neurogenetic strategy and stratify individual differences in OT modulation of mu suppression by using the subject's genotype information.

The recent study by our own group on how intranasal AVP modulates cognitive empathy (Uzefovsky et al., 2011) as well as the previous work of Thompson and colleagues (Thompson et al., 2006; Thompson et al., 2004) provide evidence for a role of AVP in processing facial information in humans contingent on gender. Furthermore, an imaging genomic's strategy carried out by Meyer-Lindenberg and colleagues showed that human amygdala function is strongly associated with genetic variation in *AVPR1a*. Another focus of AVP action on the social brain is our demonstration that intranasal administration of this nonapeptide enhances HPA axis reactivity but only in social contexts (Shalev et al., 2011). We note that studies employing intranasal administration of AVP have lagged considerably behind OT – a gap that can perhaps be attributed to the commercial availability of oxytocin as a nasal spray. Future research should aim at closing this gap and more studies need to be carried out using intranasal AVP towards understanding the role of this hormone, coupled with OT, in shaping the social brain in humans.

Another area of considerable interest is the role of other elements in OT neurotransmission pathways, especially CD38 and retinoids, in affecting the social brain in non-clinical subjects. Our own studies suggest a provisional role of CD38 in conferring vulnerability to ASD (Riebold et al., 2011, Lerer et al., 2010; Ebstein et al., 2009), investigations that need to be extended to non-clinical subjects. Furthermore, the induction by all-trans retinoic acid of CD38, and our demonstration that cell lines showing decreased CD38 transcription can be rescued in vitro by ATRA treatment (Riebold et al., 2011, Lerer et al., 2010), suggests the possibility that vitamin A may have a modulatory role on the normal social brain. Indeed, intranasal administration of vitamin A drops might be predicted to mimic OT treatment perhaps using paradigms such as the trust game.

Finally, deficits in social functioning characterize a number of psychopathologies including such disorders as autism, schizophrenia, personality disorders, and social phobias. Understanding the molecular architecture of the social brain from neural correlates to neurogenetics in non-clinical subjects will help unravel the complexity of psychopathologies characterized by deficits in social cognition. Furthermore, such an understanding is a two-way street and deficits in the social brain characteristic of several psychopathologies will also

enable a deeper appreciation of empathy and mentalization processes in non-clinical subjects.

Acknowledgments

Financial support from the National University of Singapore (Decision Making Under Urbanization: A Neurobiological and Experimental Economics Approach), the Ministry of Education at Singapore (Biological Economics and Decision Making), the AXA research foundation (Biology of Decision Making under Risk) and the Templeton Foundation (Genes, God and Generosity: The Yin Yang of DNA and Culture), are gratefully acknowledged.

REFERENCES

Adolphs, R. (2003). Cognitive neuroscience of human social behaviour. *Nat Rev Neurosci*, 4, 165–178.

Alvarez, S., Bourguet, W., Gronemeyer, H., and De Lera, A. R. (2011). Retinoic acid receptor modulators: a perspective on recent advances and promises. *Expert Opin Ther Pat*, 21, 55–63.

Apicella, C. L., Cesarini, D., Johannesson, M., Dawes, C. T., Lichtenstein, P., Wallace, B., Beauchamp, J., and Westberg, L. (2010). No association between oxytocin receptor (OXTR) gene polymorphisms and experimentally elicited social preferences. *PLoS One*, 5, e11153.

Archer, J. and Coyne, S. M. (2005). An integrated review of indirect, relational, and social aggression. *Pers Soc Psychol Rev*, 9, 212–230.

Atladottir, H. O., Pedersen, M. G., Thorsen, P., Mortensen, P. B., Deleuran, B., Eaton, W. W., and Parner, E. T. (2009). Association of family history of autoimmune diseases and autism spectrum disorders. *Pediatrics*, 124, 687–694.

Avinun, R., Israel, S., Shalev, I., Gritsenko, I., Bornstein, G., Ebstein, R.P., and Knafo, A. (2011). AVPR1A variant associated with preschoolers' lower altruistic behavior. *PLoS one*, 6.

Aydin, S., Rossi, D., Bergui, L., D'Arena, G., Ferrero, E., Bonello, L., Omede, P., Novero, D., Morabito, F., Carbone, A., Gaidano, G., Malavasi, F., and Deaglio, S. (2008). CD38 gene polymorphism and chronic lymphocytic leukemia: a role in transformation to Richter syndrome? *Blood*, 111, 5646–5653.

Bakermans-Kranenburg, M. J. and Van Ijzendoorn, M. H. (2008). Oxytocin receptor (OXTR) and serotonin transporter (5-HTT) genes associated with observed parenting. *Soc Cogn Affect Neurosci*, 3, 128–134.

Bamshad, M., Novak, M. A., and De Vries, G. J. (1993). Sex and species differences in the vasopressin innervation of sexually naive and parental prairie voles, Microtus ochrogaster and meadow voles, Microtus pennsylvanicus. *J Neuroendocrinol*, 5, 247–255.

Baron-Cohen, S., Wheelwright, S., Hill, J., Raste, Y., and Plumb, I. (2001). The "Reading the Mind in the Eyes" Test revised version: a study with normal adults, and adults with Asperger syndrome or high-functioning autism. *J Child Psychol Psychiatry*, 42, 241–251.

Bartz, J. A. and McInnes, L. A. (2007). CD38 regulates oxytocin secretion and complex social behavior. *Bioessays*, 29, 837–841.

Batson, C. D., Batson, J. G., Slingsby, J. K., Harrell, K. L., Peekna, H. M., and Todd, R. M. (1991). Empathic joy and the empathy-altruism hypothesis. *J Pers Soc Psychol*, 61, 413–426.

Baumgartner, T., Heinrichs, M., Vonlanthen, A., Fischbacher, U., and Fehr, E. (2008). Oxytocin shapes the neural circuitry of trust and trust adaptation in humans. *Neuron*, 58, 639–650.

Born, J., Lange, T., Kern, W., McGregor, G. P., Bickel, U., and Fehm, H. L. (2002). Sniffing neuropeptides: a transnasal approach to the human brain. *Nat Neurosci*, 5, 514–516.

Burger, J., Kirchner, M., Bramanti, B., Haak, W., and Thomas, M. G. (2007). Absence of the lactase-persistence-associated allele in early Neolithic Europeans. *Proc Natl Acad Sci USA*, 104, 3736–3741.

Carter, C. S. (1998). Neuroendocrine perspectives on social attachment and love. *Psychoneuroendocrinology*, 23, 779–818.

Carter, C. S., Grippo, A. J., Pournajafi-Nazarloo, H., Ruscio, M. G., and Porges, S. W. (2008). Oxytocin, vasopressin, and sociality. *Prog Brain Res*, 170, 331–336.

Ceni, C., Pochon, N., Villaz, M., Muller-Steffner, H., Schuber, F., Baratier, J., De Waard, M., Ronjat, M., and Moutin, M. J. (2006). The CD38-independent ADP-ribosyl cyclase from mouse brain synaptosomes: a comparative study of neonate and adult brain. *Biochem J*, 395, 417–426.

Cheng, Y., Lee, P. L., Yang, C. Y., Lin, C. P., Hung, D., and Decety, J. (2008a). Gender differences in the mu rhythm of the human mirror-neuron system. *PLoS ONE*, 3, e2113.

Cheng, Y., Yang, C. Y., Lin, C. P., Lee, P. L., and Decety, J. (2008b). The perception of pain in others suppresses somatosensory oscillations: a magnetoencephalography study. *Neuroimage*, 40, 1833–1840.

Coccaro, E. F., Kavoussi, R. J., Hauger, R. L., Cooper, T. B., and Ferris, C. F. (1998). Cerebrospinal fluid vasopressin levels: correlates with aggression and serotonin function in personality-disordered subjects. *Arch Gen Psychiatry*, 55, 708–714.

Costa, B., Pini, S., Gabelloni, P., Abelli, M., Lari, L., Cardini, A., Muti, M., Gesi, C., Landi, S., Galderisi, S., Mucci, A., Lucacchini, A., Cassano, G. B., and Martini, C. (2009). Oxytocin receptor polymorphisms and adult attachment style in patients with depression. *Psychoneuroendocrinology*, 34, 1506–1514.

Davis, M. (1980). A multidimensional approach to individual differences in empathy. *JSAS Catalog of Selected Documents in Psychology*, 10, 85.

De Vries, G. J., Wang, Z., Bullock, N. A., and Numan, S. (1994). Sex differences in the effects of testosterone and its metabolites on vasopressin messenger RNA levels in the bed nucleus of the stria terminalis of rats. *J Neurosci*, 14, 1789–1794.

Dickerson, S. S. and Kemeny, M. E. (2004). Acute stressors and cortisol responses: a theoretical integration and synthesis of laboratory research. *Psychol Bull*, 130, 355–391.

Ditzen, B., Schaer, M., Gabriel, B., Bodenmann, G., Ehlert, U., and Heinrichs, M. (2008). Intranasal Oxytocin Increases Positive Communication and Reduces Cortisol Levels During Couple Conflict. *Biol Psychiatry*.

Domes, G., Heinrichs, M., Michel, A., Berger, C., and Herpertz, S. C. (2007). Oxytocin improves "mind-reading" in humans. *Biol Psychiatry*, 61, 731–733.

Donaldson, Z. R. and Young, L. J. (2008). Oxytocin, vasopressin, and the neurogenetics of sociality. *Science*, 322, 900–904.

Dulac, C. and Torello, A. T. (2003). Molecular detection of pheromone signals in mammals: from genes to behaviour. *Nat Rev Neurosci*, 4, 551–562.

Ebstein, R. P., Israel, S., Chew, S. H., Zhong, S., and Knafo, A. (2010). Genetics of human social behavior. *Neuron*, 65, 831–844.

Ebstein, R. P., Israel, S., Lerer, E., Uzefovsky, F., Shalev, I., Gritsenko, I., Riebold, M., Salomon, S., and Yirmiya, N. (2009). Arginine vasopressin and oxytocin modulate human social behavior. *Ann NY Acad Sci*, 1167, 87–102.

Fehr, E., Bernhard, H., and Rockenbach, B. (2008). Egalitarianism in young children. *Nature*, 454, 1079–1083.

Fehr, E. and Rockenbach, B. (2004). Human altruism: economic, neural, and evolutionary perspectives. *Curr Opin Neurobiol*, 14, 784–790.

Ferrero, E. and Malavasi, F. (2002). *A Natural History of the Human CD38 Gene*. Kluwer Academic Publishers, Dordrecht, 81–99.

Freeman, W. J. (1997). Neurohumoral brain dynamics of social group formation. Implications for autism. *Ann NY Acad Sci*, 807, 501–503.

Gallese, V., Keysers, C., and Rizzolatti, G. (2004). A unifying view of the basis of social cognition. *Trends Cogn Sci*, 8, 396–403.

Gastaut, H., Terzian, H., and Gastaut, Y. (1952). [Study of a little electroencephalographic activity: rolandic arched rhythm]. *Mars Med*, 89, 296–310.

Gillath, O., Shaver, P. R., Baek, J. M., and Chun, D. S. (2008). Genetic Correlates of Adult Attachment Style. *Pers Soc Psychol Bull*, 34, 1396–1405.

Gimpl, G. and Fahrenholz, F. (2001). The oxytocin receptor system: structure, function, and regulation. *Physiol Rev*, 81, 629–683.

Goodson, J. L. (2008). Nonapeptides and the evolutionary patterning of sociality. *Prog Brain Res*, 170, 3–15.

Gordon, I., Zagoory-Sharon, O., Schneiderman, I., Leckman, J. F., Weller, A., and Feldman, R. (2008). Oxytocin and cortisol in romantically unattached young adults: Associations with bonding and psychological distress. *Psychophysiology*.

Gottesman, I. I. and Gould, T. D. (2003). The endophenotype concept in psychiatry: etymology and strategic intentions. *Am J Psychiatry*, 160, 636–645.

Gregory, S. G., Connelly, J. J., Towers, A. J., Johnson, J., Biscocho, D., Markunas, C. A., Lintas, C., Abramson, R. K., Wright, H. H., Ellis, P., Langford, C. F., Worley, G., Delong, G. R., Murphy, S. K., Cuccaro, M. L., Persico, A., and Pericak-Vance, M. A. (2009). Genomic and epigenetic evidence for oxytocin receptor deficiency in autism. *BMC Med*, 7, 62.

Guastella, A. J., Mitchell, P. B., and Dadds, M. R. (2007). Oxytocin increases gaze to the eye region of human faces. *Biol Psychiatry*.

Hamilton, W. D. (1963). The evolution of altruistic behavior. *American naturalist*, 354–356.

Hammock, E. A. (2007). Gene regulation as a modulator of social preference in voles. *Adv Genet*, 59, 107–127.

Hammock, E. A., Lim, M. M., Nair, H. P., and Young, L. J. (2005). Association of vasopressin 1a receptor levels with a regulatory microsatellite and behavior. *Genes Brain Behav*, 4, 289–301.

Hammock, E. A. and Young, L. J. (2005). Microsatellite instability generates diversity in brain and sociobehavioral traits. *Science*, 308, 1630–1634.

Heinrichs, M., Von Dawans, B., and Domes, G. (2009). Oxytocin, vasopressin, and human social behavior. *Front Neuroendocrinol*, 30, 548–557.

Higashida, H., Hashii, M., Yokoyama, S., Hoshi, N., Chen, X. L., Egorova, A., Noda, M., and Zhang, J. S. (2001). Cyclic ADP-ribose as a second messenger revisited from a new aspect of signal transduction from receptors to ADP-ribosyl cyclase. *Pharmacol Ther*, 90, 283–296.

Hoge, E. A., Pollack, M. H., Kaufman, R. E., Zak, P. J., and Simon, N. M. (2008). Oxytocin levels in social anxiety disorder. *CNS Neurosci Ther*, 14, 165–170.

Iacoboni, M. (2009). Imitation, Empathy, and Mirror Neurons. *Annu Rev Psychol*, 60, 653–670.

Imdad, A., Herzer, K., Mayo-Wilson, E., Yakoob, M. Y., and Bhutta, Z. A. (2010). Vitamin A supplementation for preventing morbidity and mortality in children from 6 months to 5 years of age. *Cochrane Database Syst Rev*, 12, CD008524.

Insel, T. R. (2010). The challenge of translation in social neuroscience: a review of oxytocin, vasopressin, and affiliative behavior. *Neuron*, 65, 768–779.

Insel, T. R., O'Brien, D. J., and Leckman, J. F. (1999). Oxytocin, vasopressin, and autism: is there a connection? *Biol Psychiatry*, 45, 145–157.

Ishunina, T. A. and Swaab, D. F. (1999). Vasopressin and oxytocin neurons of the human supraoptic and paraventricular nucleus: size changes in relation to age and sex. *J Clin Endocrinol Metab*, 84, 4637–4644.

Israel, S., Lerer, E., Shalev, I., Uzefovsky, F., Reibold, M., Bachner-Melman, R., Granot, R., Bornstein, G., Knafo, A., Yirmiya, N., and Ebstein, R. P. (2008). Molecular genetic studies of the arginine vasopressin 1a receptor (AVPR1a) and the oxytocin receptor (OXTR) in human behaviour: from autism to altruism with some notes in between. *Prog Brain Res*, 170, 435–449.

Israel, S., Lerer, E., Shalev, I., Uzefovsky, F., Riebold, M., Laiba, E., Bachner-Melman, R., Maril, A., Bornstein, G., Knafo, A., and Ebstein, R. P. (2009). The oxytocin receptor (OXTR) contributes to prosocial fund allocations in the dictator game and the social value orientations task. *PLoS ONE*, 4, e5535.

Jin, D., Liu, H. X., Hirai, H., Torashima, T., Nagai, T., Lopatina, O., Shnayder, N. A., Yamada, K., Noda, M., Seike, T., Fujita, K., Takasawa, S., Yokoyama, S., Koizumi, K., Shiraishi, Y., Tanaka, S., Hashii, M., Yoshihara, T., Higashida, K., Islam, M. S., Yamada, N., Hayashi, K., Noguchi, N., Kato, I., Okamoto, H., Matsushima, A., Salmina, A., Munesue, T., Shimizu, N., Mochida, S., Asano, M., and Higashida, H. (2007). CD38 is critical for social behaviour by regulating oxytocin secretion. *Nature*, 446, 41–45.

Kanner, L. (1943). Autistic disturbances of affective contact. *Nerv. Child*, 2, 217–250.

Kim, S. J., Young, L. J., Gonen, D., Veenstra-Vanderweele, J., Courchesne, R., Courchesne, E., Lord, C., Leventhal, B. L., Cook, E. H., Jr., and Insel, T. R. (2002). Transmission disequilibrium testing of arginine vasopressin receptor 1A (AVPR1A) polymorphisms in autism. *Mol Psychiatry*, 7, 503–507.

Kirschbaum, C., Klauer, T., Filipp, S. H., and Hellhammer, D. H. (1995). Sex-specific effects of social support on cortisol and subjective responses to acute psychological stress. *Psychosom Med*, 57, 23–31.

Kirschbaum, C., Pirke, K. M., and Hellhammer, D. H. (1993). The 'Trier Social Stress Test' – a tool for investigating psychobiological stress responses in a laboratory setting. *Neuropsychobiology*, 28, 76–81.

Knafo, A., Israel, S., Darvasi, A., Bachner-Melman, R., Uzefovsky, F., Cohen, L., Feldman, E., Lerer, E., Laiba, E., Raz, Y., Nemanov, L., Gritsenko, I., Dina, C., Agam, G., Dean, B., Bornstein, G., and Ebstein, R. P. (2008). Individual differences in allocation of funds in the dictator game associated with length of the arginine vasopressin 1a receptor RS3 promoter region and correlation between RS3 length and hippocampal mRNA. *Genes Brain Behav*, 7, 266–275.

Knafo, A. and Plomin, R. (2006). Prosocial behavior from early to middle childhood: genetic and environmental influences on stability and change. *Dev Psychol*, 42, 771–786.

Kosfeld, M., Heinrichs, M., Zak, P. J., Fischbacher, U., and Fehr, E. (2005). Oxytocin increases trust in humans. *Nature*, 435, 673–676.

Kudielka, B. M., Hellhammer, D. H., and Wust, S. (2009). Why do we respond so differently? Reviewing determinants of human salivary cortisol responses to challenge. *Psychoneuroendocrinology*, 34, 2–18.

Kumsta, R., Entringer, S., Koper, J. W., Van Rossum, E. F., Hellhammer, D. H., and Wust, S. (2007). Sex specific associations between common glucocorticoid receptor gene variants and hypothalamus-pituitary-adrenal axis responses to psychosocial stress. *Biol Psychiatry*, 62, 863–869.

Kusui, C., Kimura, T., Ogita, K., Nakamura, H., Matsumura, Y., Koyama, M., Azuma, C., and Murata, Y. (2001).

DNA methylation of the human oxytocin receptor gene promoter regulates tissue-specific gene suppression. *Biochem Biophys Res Commun*, 289, 681–686.

Landgraf, R. and Neumann, I. D. (2004). Vasopressin and oxytocin release within the brain: a dynamic concept of multiple and variable modes of neuropeptide communication. *Front Neuroendocrinol*, 25, 150–176.

Lee, H. J., Macbeth, A. H., Pagani, J. H., and Young, W. S., III. (2009). Oxytocin: the great facilitator of life. *Prog Neurobiol*, 88, 127–151.

Lerer, E., Levi, S., Israel, S., Yaari, M., Nemanov, L., Mankuta, D., Nurit, Y., and Ebstein, R. P. (2010). Low CD38 expression in lymphoblastoid cells and haplotypes are both associated with autism in a family-based study. *Autism Res*, 3, 293–302.

Lerer, E., Levi, S., Salomon, S., Darvasi, A., Yirmiya, N., and Ebstein, R. P. (2008). Association between the oxytocin receptor (OXTR) gene and autism: relationship to Vineland Adaptive Behavior Scales and cognition. *Mol Psychiatry*, 13, 980–988.

Levin, R., Heresco-Levy, U., Bachner-Melman, R., Israel, S., Shalev, I., and Ebstein, R. P. (2009). Association between arginine vasopressin 1a receptor (AVPR1a) promoter region polymorphisms and prepulse inhibition. *Psychoneuroendocrinology*, 34, 901–908.

Levine, A., Zagoory-Sharon, O., Feldman, R., and Weller, A. (2007). Oxytocin during pregnancy and early postpartum: individual patterns and maternal-fetal attachment. *Peptides*, 28, 1162–1169.

Levy, Y., and Ebstein, R. P. (2009). Research review: crossing syndrome boundaries in the search for brain endophenotypes. *Journal of Child Psychology and Psychiatry*, 50(6), 657–668.

Lucht, M. J., Barnow, S., Sonnenfeld, C., Rosenberger, A., Grabe, H. J., Schroeder, W., Volzke, H., Freyberger, H. J., Herrmann, F. H., Kroemer, H., and Rosskopf, D. (2009). Associations between the oxytocin receptor gene (OXTR) and affect, loneliness and intelligence in normal subjects. *Prog Neuropsychopharmacol Biol Psychiatry*, 33, 860–866.

Macmillan, H. L., Georgiades, K., Duku, E. K., Shea, A., Steiner, M., Niec, A., Tanaka, M., Gensey, S., Spree, S., Vella, E., Walsh, C. A., Bellis, M. D., Meulen, J. V., Boyle, M. H., and Schmidt, L. A. (2009). Cortisol response to stress in female youths exposed to childhood maltreatment: Results of the youth mood project. *Biol Psychiatry*.

Malavasi, F., Deaglio, S., Funaro, A., Ferrero, E., Horenstein, A. L., Ortolan, E., Vaisitti, T., and Aydin, S. (2008). Evolution and function of the ADP ribosyl cyclase/CD38 gene family in physiology and pathology. *Physiol Rev*, 88, 841–886.

Martineau, J., Schmitz, C., Assaiante, C., Blanc, R., and Barthelemy, C. (2004). Impairment of a cortical event-related desynchronisation during a bimanual load-lifting task in children with autistic disorder. *Neurosci Lett*, 367, 298–303.

Meyer-Lindenberg, A., Kolachana, B., Gold, B., Olsh, A., Nicodemus, K. K., Mattay, V., Dean, M., and Weinberger, D. R. (2008). Genetic variants in AVPR1A linked to autism predict amygdala activation and personality traits in healthy humans. *Mol Psychiatry*.

Meyer-Lindenberg, A. and Weinberger, D. R. (2006). Intermediate phenotypes and genetic mechanisms of psychiatric disorders. *Nat Rev Neurosci*, 7, 818–827.

Mizuguchi, M., Otsuka, N., Sato, M., Ishii, Y., Kon, S., Yamada, M., Nishina, H., Katada, T., and Ikeda, K. (1995). Neuronal localization of CD38 antigen in the human brain. *Brain Res*, 697, 235–240.

Mizumoto, Y., Kimura, T., and Ivell, R. (1997). A genomic element within the third intron of the human oxytocin receptor gene may be involved in transcriptional suppression. *Mol Cell Endocrinol*, 135, 129–138.

Modahl, C., Fein, D., Waterhouse, L., and Newton, N. (1992). Does oxytocin deficiency mediate social deficits in autism? *J Autism Dev Disord*, 22, 449–451.

Modahl, C., Green, L., Fein, D., Morris, M., Waterhouse, L., Feinstein, C., and Levin, H. (1998). Plasma oxytocin levels in autistic children. *Biol Psychiatry*, 43, 270–277.

Munesue, T., Yokoyama, S., Nakamura, K., Anitha, A., Yamada, K., Hayashi, K., Asaka, T., Liu, H. X., Jin, D., Koizumi, K., Islam, M. S., Huang, J. J., Ma, W. J., Kim, U. H., Kim, S. J., Park, K., Kim, D., Kikuchi, M., Ono, Y., Nakatani, H., Suda, S., Miyachi, T., Hirai, H., Salmina, A., Pichugina, Y. A., Soumarokov, A. A., Takei, N., Mori, N., Tsujii, M., Sugiyama, T., Yagi, K., Yamagishi, M., Sasaki, T., Yamasue, H., Kato, N., Hashimoto, R., Taniike, M., Hayashi, Y., Hamada, J., Suzuki, S., Ooi, A., Noda, M., Kamiyama, Y., Kido, M. A., Lopatina, O., Hashii, M., Amina, S., Malavasi, F., Huang, E. J., Zhang, J., Shimizu, N., Yoshikawa, T., Matsushima, A., Minabe, Y., and Higashida, H. (2010). Two genetic variants of CD38 in subjects with autism spectrum disorder and controls. *Neurosci Res*, 67, 181–191.

Muthukumaraswamy, S. D. and Johnson, B. W. (2004). Changes in rolandic mu rhythm during observation of a precision grip. *Psychophysiology*, 41, 152–156.

Muthukumaraswamy, S. D., Johnson, B. W., and McNair, N. A. (2004). Mu rhythm modulation during observation

of an object-directed grasp. *Brain Res Cogn Brain Res*, 19, 195–201.

Muthukumaraswamy, S. D. and Singh, K. D. (2008). Modulation of the human mirror neuron system during cognitive activity. *Psychophysiology*, 45, 896–905.

Neumann, I. D. (2008). Brain oxytocin: a key regulator of emotional and social behaviours in both females and males. *J Neuroendocrinol*, 20, 858–865.

Newman-Norlund, R. D., Van Schie, H. T., Van Zuijlen, A. M., and Bekkering, H. (2007). The mirror neuron system is more active during complementary compared with imitative action. *Nat Neurosci*, 10, 817–818.

Oberman, L. M., Pineda, J. A., and Ramachandran, V. S. (2007). The human mirror neuron system: A link between action observation and social skills. *Soc Cogn Affect Neurosci*, 2, 62–66.

Oberman, L. M. and Ramachandran, V. S. (2007). The simulating social mind: the role of the mirror neuron system and simulation in the social and communicative deficits of autism spectrum disorders. *Psychol Bull*, 133, 310–327.

Oberman, L. M., Ramachandran, V. S., and Pineda, J. A. (2008). Modulation of mu suppression in children with autism spectrum disorders in response to familiar or unfamiliar stimuli: the mirror neuron hypothesis. *Neuropsychologia*, 46, 1558–1565.

Panksepp, J. (1993). Commentary on the possible role of oxytocin in autism. *J Autism Dev Disord*, 23, 567–569.

Park, J., Willmott, M., Vetuz, G., Toye, C., Kirley, A., Hawi, Z., Brookes, K. J., Gill, M., and Kent, L. (2010). Evidence that genetic variation in the oxytocin receptor (OXTR) gene influences social cognition in ADHD. *Prog Neuropsychopharmacol Biol Psychiatry*, 34(4), 697–702.

Perry, A., Bentin, S., Shalev, I., Israel, S., Uzefovsky, F., Bar-On, D., and Ebstein, R. P. (2010a). Intranasal oxytocin modulates EEG mu/alpha and beta rhythms during perception of biological motion. *Psychoneuroendocrinology*, 34, 1446–1453.

Perry, A., Troje, N. F., and Bentin, S. (2010b). Exploring motor system contributions to the perception of social information: Evidence from EEG activity in the mu/alpha frequency range. *Soc Neurosci*, 5, 272–284.

Pietrowsky, R., Struben, C., Molle, M., Fehm, H. L., and Born, J. (1996). Brain potential changes after intranasal vs. intravenous administration of vasopressin: evidence for a direct nose-brain pathway for peptide effects in humans. *Biol Psychiatry*, 39, 332–340.

Pineda, J. A. (2005). The functional significance of mu rhythms: translating “seeing” and “hearing” into “doing.” *Brain Res Brain Res Rev*, 50, 57–68.

Pineda, J. A. and Hecht, E. (2009). Mirroring and mu rhythm involvement in social cognition: are there dissociable subcomponents of theory of mind? *Biol Psychol*, 80, 306–314.

Preston, S. D. and De Waal, F. B. (2002). Empathy: Its ultimate and proximate bases. *Behav Brain Sci*, 25, 1–20; discussion 20–71.

Raymaekers, R., Wiersema, J. R., and Roeyers, H. (2009). EEG study of the mirror neuron system in children with high functioning autism. *Brain Res*, 1304, 113–121.

Riebold, M., Mankuta, D., Lerer, B., Israel, S., Zhong, S., Nemanov, L., Monakhov, M., Levi, S., Yirmiya, N., Yaari, M., Malavasi, F., and Ebstein, R. P. (2011). All-trans-Retinoic-Acid (ATRA) upregulates reduced CD38 transcription in lymphoblastoid cell lines from autism spectrum disorder *Molecular Medicine*, 17(7–8), 799–800.

Rizzolatti, G., Fadiga, L., Gallese, V., and Fogassi, L. (1996). Premotor cortex and the recognition of motor actions. *Brain Res Cogn Brain Res*, 3, 131–141.

Robinson, G. E., Grozinger, C. M., and Whitfield, C. W. (2005). Sociogenomics: social life in molecular terms. *Nat Rev Genet*, 6, 257–270.

Rodrigues, S. M., Saslow, L. R., Garcia, N., John, O. P., and Keltner, D. (2009). Oxytocin receptor genetic variation relates to empathy and stress reactivity in humans. *Proc Natl Acad Sci USA*, 106(50), 21437–21441.

Rushton, J. P. (2004). Genetic and environmental contributions to pro-social attitudes: a twin study of social responsibility. *Proc Biol Sci*, 271, 2583–2585.

Saborit-Villarroya, I., Vaisitti, T., Rossi, D., D'Arena, G., Gaidano, G., Malavasi, F., and Deaglio, S. (2011). E2A is a transcriptional regulator of CD38 expression in chronic lymphocytic leukemia. *Leukemia*, 25, 479–488.

Savaskan, E., Ehrhardt, R., Schulz, A., Walter, M., and Schachinger, H. (2008). Post-learning intranasal oxytocin modulates human memory for facial identity. *Psychoneuroendocrinology*, 33(3), 368–374.

Shalev, I., Israel, S., Uzefovsky, F., Gritsenko, I., Kaitz, M., and Ebstein, R. P. (2011). Vasopressin needs an audience: Neuropeptide elicited stress responses are contingent upon perceived social evaluative threats. *Hormones and behavior*, 60(1), 121–127. doi: 10.1016/j.yhbeh.2011.04.005

Shalev, I., Lerer, E., Israel, S., Uzefovsky, F., Gritsenko, I., Mankuta, D., Ebstein, R. P., and Kaitz, M. (2009). BDNF Val66Met polymorphism is associated with HPA axis reactivity to psychological stress characterized by genotype and gender interactions. *Psychoneuroendocrinology*, 34, 382–388.

Siddikuzzaman, Guruvayoorappan, C., and Berlin Grace, V. M. (2010). All Trans Retinoic Acid and Cancer. *Immunopharmacol Immunotoxicol*, 33, 241–249.

Sigmund, K. and Hauert, C. (2002). Primer: altruism. *Curr Biol*, 12, R270–272.

Sparrow, S. S., Balla, D. A., and Cicchetti, D. V. (1984). *Vineland Adaptive Behavior Scales*, Minneapolis, Minnesota: American Guidance Services.

Tansey, K. E., Hill, M. J., Cochrane, L. E., Gill, M., Anney, R. J., and Gallagher, L. (2011). Functionality of promoter microsatellites of arginine vasopressin receptor 1A (AVPR1A): implications for autism. *Mol Autism*, 2, 3.

Theodosiou, M., Laudet, V., and Schubert, M. (2010). From carrot to clinic: an overview of the retinoic acid signaling pathway. *Cell Mol Life Sci*, 67, 1423–1445.

Thibonnier, M. (2004). Genetics of vasopressin receptors. *Curr Hypertens Rep*, 6, 21–26.

Thibonnier, M., Auzan, C., Madhun, Z., Wilkins, P., Berti-Mattera, L., and Clauser, E. (1994). Molecular cloning, sequencing, and functional expression of a cDNA encoding the human V1a vasopressin receptor. *J Biol Chem*, 269, 3304–3310.

Thompson, R., Gupta, S., Miller, K., Mills, S., and Orr, S. (2004). The effects of vasopressin on human facial responses related to social communication. *Psychoneuroendocrinology*, 29, 35–48.

Thompson, R. R., George, K., Walton, J. C., Orr, S. P., and Benson, J. (2006). Sex-specific influences of vasopressin on human social communication. *Proc Natl Acad Sci USA*, 103, 7889–7894.

Tobin, V. A., Hashimoto, H., Wacker, D. W., Takayanagi, Y., Langnaese, K., Caquineau, C., Noack, J., Landgraf, R., Onaka, T., Leng, G., Meddle, S. L., Engelmann, M., and Ludwig, M. (2010). An intrinsic vasopressin system in the olfactory bulb is involved in social recognition. *Nature*, 464, 413–417.

Tost, H., Kolachana, B., Hakimi, S., Lemaitre, H., Verchinski, B. A., Mattay, V. S., Weinberger, D. R., and Meyer-Lindenberg, A. (2010). A common allele in the oxytocin receptor gene (OXTR) impacts prosocial temperament and human hypothalamic-limbic structure and function. *Proc Natl Acad Sci USA*, 107, 13936–13941.

Trapasso, E., Cosco, D., Celia, C., Fresta, M., and Paolino, D. (2009). Retinoids: new use by innovative drug-delivery systems. *Expert Opin Drug Deliv*, 6, 465–483.

Tsao, D. Y., Freiwald, W. A., Knutsen, T. A., Mandeville, J. B., and Tootell, R. B. (2003). Faces and objects in macaque cerebral cortex. *Nat Neurosci*, 6, 989–995.

Uhart, M., Chong, R. Y., Oswald, L., Lin, P. I., and Wand, G. S. (2006). Gender differences in hypothalamic-pituitary-adrenal (HPA) axis reactivity. *Psychoneuroendocrinology*, 31, 642–652.

Uzefovsky, F., Shalev, I., Israel, S., Knafo, A., and Ebstein, R. P. (2012). Vasopressin selectively impairs emotion recognition in men. *Psychoneuroendocrinology*, 37(4), 576–580. doi: 10.1016/j.psyneuen.2011.07.018

Van Lange, P. A. M. (1999). The pursuit of joint outcomes and equality in outcomes: An integrative model of social value orientation. *Journal of Personality and Social Psychology*, 77, 337–349.

Walum, H., Westberg, L., Henningsson, S., Neiderhiser, J. M., Reiss, D., Igl, W., Ganiban, J. M., Spotts, E. L., Pedersen, N. L., Eriksson, E., and Lichtenstein, P. (2008). Genetic variation in the vasopressin receptor 1a gene (AVPR1A) associates with pair-bonding behavior in humans. *Proc Natl Acad Sci USA*, 105, 14153–14156.

Wang, Z., Smith, W., Major, D. E., and De Vries, G. J. (1994). Sex and species differences in the effects of cohabitation on vasopressin messenger RNA expression in the bed nucleus of the stria terminalis in prairie voles (Microtus ochrogaster) and meadow voles (Microtus pennsylvanicus). *Brain Res*, 650, 212–218.

Wassink, T. H., Piven, J., Vieland, V. J., Pietila, J., Goedken, R. J., Folstein, S. E., and Sheffield, V. C. (2004). Examination of AVPR1a as an autism susceptibility gene. *Mol Psychiatry*, 9, 968–972.

Waterhouse, L., Fein, D., and Modahl, C. (1996). Neurofunctional mechanisms in autism. *Psychol Rev*, 103, 457–489.

Winslow, J. T., Hastings, N., Carter, C. S., Harbaugh, C. R., and Insel, T. R. (1993). A role for central vasopressin in pair bonding in monogamous prairie voles. *Nature*, 365, 545–548.

Winslow, J. T. and Insel, T. R. (1993). Effects of central vasopressin administration to infant rats. *Eur J Pharmacol*, 233, 101–107.

Wu, S., Jia, M., Ruan, Y., Liu, J., Guo, Y., Shuang, M., Gong, X., Zhang, Y., Yang, X., and Zhang, D. (2005). Positive association of the oxytocin receptor gene (OXTR) with autism in the Chinese Han population. *Biol Psychiatry*, 58, 74–77.

Yirmiya, N., Rosenberg, C., Levi, S., Salomon, S., Shulman, C., Nemanov, L., Dina, C., and Ebstein, R. P. (2006). Association between the arginine vasopressin 1a receptor (AVPR1a) gene and autism in a family-based study: mediation by socialization skills. *Mol Psychiatry*, 11, 488–494.

Young, K. A., Liu, Y., and Wang, Z. (2008). The neurobiology of social attachment: A comparative approach to behavioral, neuroanatomical, and neurochemical studies. *Comp Biochem Physiol C Toxicol Pharmacol*, 148, 401–410.

Young, L. J. (2007). Regulating the social brain: a new role for CD38. *Neuron*, 54, 353–356.

Zak, P. J., Stanton, A. A., and Ahmadi, S. (2007). Oxytocin increases generosity in humans. *PLoS ONE*, 2, e1128.

Zink, C. F., Stein, J. L., Kempf, L., Hakimi, S., and Meyer-Lindenberg, A. (2010). Vasopressin modulates medial prefrontal cortex-amygdala circuitry during emotion processing in humans. *J Neurosci*, 30, 7017–7022.

20

Oxytocin and autism

Joshua J. Green, Bonnie Taylor, and Eric Hollander

20.1 Introduction

Over the past three decades, growing research has implicated an association between the peptide hormone oxytocin (OT) and autism spectrum disorders (ASD) due to the increasingly documented involvement of OT in the regulation of social and repetitive behaviors (Modahl et al., 1992; Panksepp, 1992; Waterhouse et al., 1996; McCarthy and Altemus, 1997; Insel et al., 1999; Hollander et al., 2003; Lim et al., 2005; Hollander and Bartz, 2006; Carter, 2007; Bartz and Hollander, 2008; Guastella et al., 2008; Gregory et al., 2009; Guastella et al., 2010). In this chapter, we review research to date that has investigated the relationship between OT and the presentation or regulation of several key deficits associated with ASD. First, we provide an overview of ASD and of OT, followed by a brief review of the current theories surrounding the etiology of the disorder. Next, we review genetic research of the oxytocin receptor gene (OTR) followed by an overview of experimental therapeutic research involving the administration of OT in individuals with ASD. Finally, suggestions for directions that future research should address are discussed. Of note, only minimal animal and translational research is reviewed, as it can be found discussed at length elsewhere in the present volume.

20.2 Autism spectrum disorder

ASD refers to a group of neurodevelopmental disorders (autistic disorder, Asperger Syndrome and pervasive developmental disorder not otherwise specified) characterized by deficits within three core symptom domains: social interaction, speech and communication, and repetitive or compulsive behaviors with restricted interests. Formal DSM-IV diagnostic criteria for autistic disorder can be found in Textbox 20.1.

The social impairment commonly seen in individuals with ASD often manifests as a lack of eye contact while conversing, difficulty maintaining a conversation, low social and emotional reciprocity or mutual enjoyment in social activities, a diminished ability of imitation or mimicry (Dapretto et al., 2006), and impaired daily interaction skills (APA, 1994; Bartz and Hollander, 2008). In addition, it is thought that individuals with ASD have marked difficulty with empathy and in understanding the thoughts or intentions of others – an idea increasingly referred to as "theory of mind." (Baron-Cohen et al.,1985; Boria et al., 2009). Broadly speaking, individuals with ASD have difficulty engaging in two-way or reciprocal interactions, often leaving others with the impression that they are socially awkward or strangely detached from those around them.

Oxytocin, Vasopressin, and Related Peptides in the Regulation of Behavior, ed. E. Choleris, D. W. Pfaff, and M. Kavaliers.
Published by Cambridge University Press.

Textbox 20.1 DSM-IV diagnostic criteria for autistic disorder

I. A total of six (or more) items from (A), (B), and (C), with at least two from (A), and one each from (B) and (C)

(A) Qualitative impairment in social interaction, as manifested by at least two of the following:

1. Marked impairments in the use of multiple non-verbal behaviors such as eye-to-eye gaze, facial expression, body posture, and gestures to regulate social interaction.
2. Failure to develop peer relationships appropriate to developmental level.
3. A lack of spontaneous seeking to share enjoyment, interests, or achievements with other people, (e.g., by a lack of showing, bringing, or pointing out objects of interest to other people).
4. Lack of social or emotional reciprocity (note: in the description, it gives the following as examples: not actively participating in simple social play or games, preferring solitary activities, or involving others in activities only as tools or "mechanical" aids).

(B) Qualitative impairments in communication as manifested by at least one of the following:

1. Delay in, or total lack of, the development of spoken language (not accompanied by an attempt to compensate through alternative modes of communication such as gesture or mime).
2. In individuals with adequate speech, marked impairment in the ability to initiate or sustain a conversation with others.
3. Stereotyped and repetitive use of language or idiosyncratic language.
4. Lack of varied, spontaneous make-believe play or social imitative play appropriate to developmental level.

(C) Restricted repetitive and stereotyped patterns of behavior, interests and activities, as manifested by at least two of the following:

1. Encompassing preoccupation with one or more stereotyped and restricted patterns of interest that is abnormal either in intensity or focus.
2. Apparently inflexible adherence to specific, nonfunctional routines or rituals.
3. Stereotyped and repetitive motor mannerisms (e.g hand or finger flapping or twisting, or complex whole-body movements).
4. Persistent preoccupation with parts of objects.

II. Delays or abnormal functioning in at least one of the following areas, with onset prior to age 3 years:

(A) social interaction;

(B) language as used in social communication;

(C) symbolic or imaginative play

III. The disturbance is not better accounted for by Rett's Disorder or childhood disintegrative disorder.

Because individuals with ASD tend to think in more concrete terms, social activities like "small talk" and understanding abstract language, such as jokes and metaphors, are difficult. Furthermore, difficulties in face recognition, (Szatmari et al., 1990; Davies et al., 1994; Barton, 2003), and the interpretation or processing of the affective states of others through both facial expressions (Hobson et al., 1988; Tantam et al., 1989) and tone of voice, also appear to be impaired (Hobson et al., 1988; Rutherford et al., 2002).

In addition to social impairment, repetitive behaviors and/or restricted interests are a domain of concern. These symptoms may present as simple, but highly repetitive motor behaviors called

stereotypies or "self-stimulation" behaviors, such as rocking, spinning, hand-flapping, or clapping. In contrast, some individuals may present with more complex repetitive behaviors that manifest as rigid routines involving a specific interest or preoccupation.

The need for the development of novel therapeutic strategies targeting ASD has increased considerably as the prevalence of autism is similarly increasing due to a variety of reasons such as the broadening of the diagnostic criteria, increased public and professional awareness, earlier diagnosis, and the availability of services.

20.3 Etiology of autism

Attempts to understand the etiology of ASD have been approached from myriad perspectives. Research has focused on the contribution of genetic causes, environmental influences, viral or bacterial infection, immune dysfunction, mirror neuron deficiency, hyperfunctioning of microcircuits, amygdala dysfunction, abnormalities in the locus coeruleus-noradrenergic system, serotonin dysfunction, and the involvement of oxytocin. Indeed, while this review will focus on OT, human social behavior is mediated through multiple systems including the serotonergic and dopaminergic reward systems. It may be that many different combinations of genetic deviations exist, whether genetic or epigenetically caused, that give rise to various phenotypes of ASD. This may partially explain the heterogeneity seen in the disorder. Ultimately, understanding the core pathogenesis of ASD will result in the development of more targeted treatment alternatives, which, as we will discuss in this chapter, is progressing in the case of oxytocin.

Some theories, such as a dysfunctional mirror neuron system (MNS) (Dapretto et al., 2006; Wan et al., 2010;), a dysgenesis or agenesis of the olfactory bulbs (Brang and Ramachandran, 2010), a hyperfunctioning of local neural microcircuits dubbed the *Intense World Theory* (Markram and Markram, 2010), and a dysfunction with the amygdala, (Baron-Cohen et al., 2000) support the involvement of OT and AVP in the etiology of autism. For example, the olfactory bulbs contain a high density of OT and AVP receptors that project to many of the regions (such as the amygdala) implicated in the MNS and the *Intense World Theory*, as well as many of the cognitive faculties (i.e., empathy) that comprise the social deficits seen in ASD (Gallese et al., 2004). Relatedly, fMRI research has demonstrated that OT exerts effects on brain regions in the midbrain and amygdala, whereby it affects social behaviors involved in trust and fear (Kosfeld et al., 2005; Kirsch et al., 2005) Furthermore, music making and music-based therapies known to stimulate many regions of the MNS, including those that are implicated in emotional recognition, language and empathy, have demonstrated temporal ameliorations of many of the symptoms associated with ASD (Boso et al., 2007, Kim et al., 2008, Koelsch, 2009; Finnigan and Starr, 2010; Wan et al., 2010). Likewise, at least one study has suggested that soothing music may increase OT levels in the brain (Nilsson, 2009). As such, the association with these peptides and ASD continues to garner the interest of researchers across scientific disciplines.

20.4 Oxytocin and vasopressin

As explained elsewhere in this volume, oxytocin (OT) is a nonapeptide (i.e., it has nine amino acids) that is synthesized in magnocellular neurons in both the paraventricular and supraoptic nuclei of the hypothalamus. It is released into the bloodstream by way of axon terminals in the posterior pituitary (Burbach et al., 2006). OT exerts effects both peripherally and centrally. Peripherally, it is involved in milk ejection, orgasm, and the facilitation of uterine contractions. Important to autism-related symptomatology, centrally located OT plays a neuromodulatory role in aspects of social affiliation and social cognition (Burbach et al., 2006; Hammock and Young, 2006).

Less attention has focused on the possible role of AVP in the pathophysiology of autism. Instead, a

contributing role of AVP has been hypothesized for mood and anxiety disorders since AVP plays a role in both endocrine and neural stress responses (Surget and Belzung, 2008); AVP receptor antagonism appears to decrease anxiety in receptor ligand studies in animal models (Griebel et al., 2002; Salome et al., 2006). In this respect, it is of interest that effective treatment with a selective serotonin reuptake inhibitor to reduce cognitively mediated repetitive behaviors, such as an insistence on sameness and on following routines and rituals, is hypothesized to be associated with a reduction in anxiety in autistic patients. Thus, future studies aimed at exploring the relationship between AVP receptor antagonism, repetitive behaviors, and anxiety may be warranted.

Since the role of OT, in particular, has been empirically studied in relation to the symptom manifestation and treatment of the repetitive behaviors and social deficits seen in ASD, the present review will focus exclusively on the research pertaining to OT.

20.5 Genetic research on the oxytocin receptor gene

As mentioned previously, significant research has been dedicated to determining the genetic factors that contribute to the etiology of ASD. Because of an increasingly established link between OT and autism, some genetic research has investigated the oxytocin receptor gene (OTR). To date, several studies have found associations with various single nucleotide polymorphisms (SNPs). For example, Wu et al. (2005) found a significant association between two SNPs in the *OTR* and ASD (rs2254298 and rs53576) in a group of 195 Han Chinese parent–offspring autism trios. A second study involving a European-origin population (recruited and conducted in the US), also found a significant association at rs2254298, but not rs53576 (Jacob et al., 2007). For the rs2254298 SNP, they recorded an overtransmission of the G allele to probands with autistic disorder, whereas overtransmission of the A allele was found in the Chinese Han population. In both cases, however, the G allele was more frequent than the A allele. A third study (Lerer et al., 2008) of 133 families conducted in Jerusalem corroborated the association between ASD and the SNP rs2254298: the association corresponded with lower scores on the daily living skills and communication domains of the Vineland Adaptive Behavior scale. Additionally, they reported haplotype associations of rs237897 and rs13316193. In accordance with the findings of Jacob et al., they did not find significance of rs53576. A fourth investigation (Yrigollent et al., 2008) did not analyze any of the same SNPs, but did find that rs2254298 is in the same linkage disequilibrium block as rs2268493 – the SNP that showed associations with both multivariate and univariate phenotypes in their analysis. The results of these studies provide some evidence of an allelic association between certain genes that may control affiliative behaviors in ASD. Of note, polymorphisms in the AVP receptor gene (Avpr) have also been associated with autism, highlighting the partnership that these two peptides seem to share (Kim et al., 2002; Wassink et al., 2004; Yirmiya et al., 2006).

In contrast to the above, at least two studies of the OTR report conflicting results. Wermter et al. (2009) reported that one SNP (rs2270465) of the OTR was associated with ASD, but that it did not remain significant after correction for multiple testing. For all other 21 SNPs analyzed, they found no associations. In addition, Tansey et al. (2010) analyzed 18 SNPs at the OTR for association in three independent autism samples from Ireland, Portugal, and the United Kingdom and concluded that common genetic variation in OTR does not appear to be associated with the etiology of ASD, at least in Caucasian samples. To account for this discrepancy, they cite ethnic differences as a possible confound and reference the complexity of the OTR. Curiously however, Tansey et al., reported that the SNPs rs237897 and rs13316193, the same with which Lerer et al. (2008) found associations, are involved in transcriptional regulation of OTR. This led the authors to suggest that OTR expression may be influenced by temporal or contextual factors, such as an epigenetic modification of expression during specific development periods (Tansey et al., 2010).

A recent investigation addressed possible epigenetic factors. Gregory et al. (2009) analyzed the promoter region of the OTR (from both peripheral blood mononuclear cells and cells in the temporal cortex) and found this region to be hypermethylated in independent cohorts with autism compared to controls. In addition, their analyses indicated that cells from the temporal cortex showed decreased levels of expression (20% less in autistic males compared to controls). Overall, this suggests that increased methylation of the promoter region in the OTR correlates with decreased expression of the gene in this tissue type. They also identified a deletion in the OTR in an autism proband and in both his mother with OCD and an affected sibling who had epigenetic misregulation of this gene through aberrant gene silencing by DNA methylation. This increased methylation was associated with decreased OTR mRNA expression in temporal cortex, implying that epigenetic regulation of OTR may be involved in the early development and etiology of autism. In relation to epigenetic variables, Page et al. (2009) conducted research in which they describe the role that two genes, "Pten" and "SLC6A4" (serotonin transporter), might play in ASD. Their research suggests that these two genes, which act on their respective pathways – the PI3K and serotonin pathways – possibly confer susceptibility to ASD. Specifically, they found that Pten haploinsufficent mice produced macrocephaly across genders as well as deficits in social approach behavior in females. Macrocephaly is common in individuals with ASD, especially during early development, and as such their results suggest that these two genes may influence each other to produce this phenotype often seen in ASD. Since ASD is a developmental disorder necessarily diagnosed before the age of 3, exploring the discussed genetic and epigenetic influences is both necessary an timely.

20.6 Limitations of genetic research

One important factor that confounds all research in ASD is the inherent heterogeneity of autistic symptomatology, which ranges in both type and severity. Undoubtedly, the word "spectrum" in autism spectrum disorders refers to this. The problem of heterogeneity seems to be a particularly important hurdle to overcome for genetic research since there is also variability when working at the level of genes, and since different expressions of ASD phenotypes may be related to specific genes. Seen in this light, it becomes clear that the genetic factors associated with ASD are highly complex both because of the lack of homogeneity in ASD, but also because even within a well-defined group there exists significant individual variability. Indeed, less than 10% of ASD cases are attributable to single-gene disorders (Gregory et al., 2009). Overall, this heterogeneity, the multiple contributing loci and the inevitable epigenetic and gene–environment interactions, contribute to a complicated genetic picture for ASD.

20.7 Clinical trials: Experimental therapeutics and related research

In addition to genetic research, other animal and experimental research over the last three decades has increasingly substantiated the role of OT in human social behaviors. As such, a thorough understanding of the biology of ASD may involve a more accurate understanding of the neurobiological roles of OT and related peptides. We are only at the inception of the development of such a picture. Despite this, the existing preliminary research is decidedly noteworthy.

Some of the first studies investigating the relationship between OT and ASD measured the blood plasma levels of OT in both healthy individuals and those with ASD. For example, one study found decreased levels of OT in blood plasma of prepubescent autistic individuals as well as increased OT precursor levels compared to age-matched controls (Modahl et al., 1998). In addition, their follow-up study corroborated this finding (Green et al., 2001). Although not the primary focus of the investigation, Andari et al. (2010) also reported

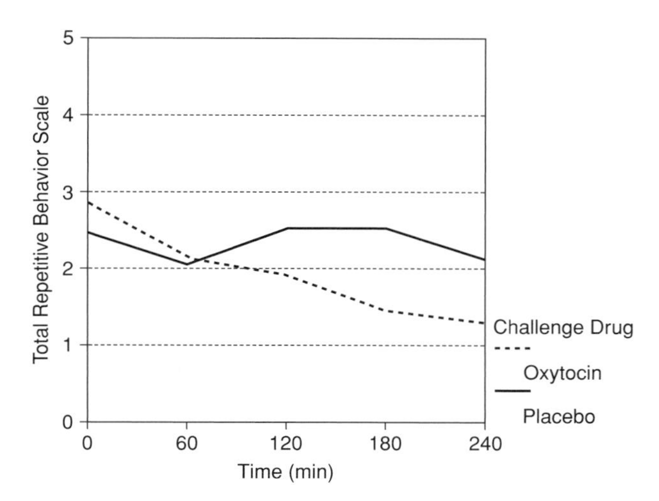

Figure 20.1 Effects of oxytocin vs. placebo infusion on repetitive behaviors in autism spectrum disorder patients over time. Mean scores were significantly lower over time following oxytocin *vs.* placebo ($F = 3.487$, df = 4, 52, $p = 0.027$).

significantly decreased blood plasma OT concentration in adults with ASD compared to healthy controls. In contrast, Jansen et al. (2006) report *increased* OT blood plasma levels in adults with ASD. This discrepancy highlights the complexity of drawing any conclusions about the relationship between peripheral blood plasma levels of OT and ASD. As Jansen et al. (2006) suggest, the differing patient populations among the studies (children vs. adults), as well as the known heterogeneity found in ASD, may partially account for these conflicting results. Further, the relation between OT and the autistic brain may be related to synthesis and/or function, rather than simply a dysfunction of quantity (Green et al., 2001). Moreover, blood plasma levels of OT do not necessarily correlate with OT levels in the brain because it is not known how much OT crosses the blood–brain barrier, making data from peripheral readings difficult to interpret.

Thanks to decades of animal research implicating OT in repetitive behaviors and social affiliation, research investigating the effects of OT on the symptoms of ASD has increased substantially since the turn of the century. Generally speaking, research has focused on treating the social deficits associated with ASD, and to date, at least one study has investigated the use of OT on repetitive behaviors.

20.7.1 Oxytocin as a treatment for social deficits

Much experimental therapeutic research of OT in individuals with ASD has focused on the social domain. Hollander et al. (2007) investigated the ability of individuals with ASD to assign an emotional significance to certain intonations in speech. This is an important skill that contributes to some of the social and language difficulties seen in ASD (Gervais et al., 2004; Kuhl et al., 2005). In this double-blind placebo-controlled crossover study, 15 adults diagnosed with either autism or Asperger's Disorder underwent two challenge days in which they received either a continuous infusion of intravenous synthetic OT (Pitocin) or placebo. During the challenges, participants were presented with four versions of the same neutral sentence that varied in emotional intonation, such that the sentence was prerecorded with either a happy, indifferent, angry, or sad emotional intonation. The participant then had to point to a word that they felt best reflected the emotional tone of the sentence. A mixed regression analysis indicated significant pretest to post-test improvement for all conditions except for participants who received placebo second. This result is likely due to the finding that those individuals who received OT infusion on the first visit showed retention of the ability to comprehend affective speech on the second visit two weeks later. That is, they did not show a tendency to revert to baseline when retested after a 2-week delay, suggesting a role of OT in the development or maintenance of social memories. By comparison, those individuals who received the placebo first did show a tendency to revert to baseline (see Figure 20.1).

OT has also been associated with an increase in gazing to the eye region of the face. Guastella et al. (2008) conducted a double-blind, randomized,

placebo-controlled, between-subject design trial of intranasal OT in 52 healthy male volunteers by presenting them with 24 neutral human faces and tracking their eye movements. Their results showed that those on OT significantly fixated more on the eye region of the faces. This is a relevant finding since individuals with ASD often avoid eye contact, instead focusing mostly on the mouth (Baron-Cohen et al., 2001; Dalton et al., 2005). They also tend to have difficulty recognizing emotions in the face or facial expressions (Domes et al., 2007). One question is whether those with ASD simply find the eye region uninteresting or otherwise no more important than other parts of the face. And, if so, does an administration of OT selectively alter the experience of these individuals such that the eye region becomes subjectively more important and/or interesting? Andari et al. (2010) presented neutral pictures of human faces to healthy controls and to subjects with high-functioning ASD and had them determine the gender and gaze direction of each face. While subjects looked at the pictures, the total fixation time for various regions of the face (i.e., eyes, nose, mouth, forehead, cheeks, and outside of facial contour) and the number of eye saccades were measured. Compared to controls, individuals with ASD underexplored the face, and in particular, avoided the eye region, and did so with a higher saccade rate than healthy controls. Analysis of OT compared to placebo revealed that in subjects with high-functioning ASD, OT increased the amount of fixation time on the eye region selectively, (other regions were not significantly different across both conditions) and significantly reduced saccade frequency (e.g., increased duration of fixation). Although promising, the authors note that the gaze time on the face and eye region under the OT condition still nevertheless remained significantly lower than that of healthy controls. It is of interest to note that the authors suggest OT may reduce the fear or anxiety that is associated with gazing at the eye region, and thus increase the gaze time and decreased saccade frequency. This seems partially supported by fMRI research that has demonstrated fear-reducing effects of OT (Kirsch et al., 2005).

In addition to gaze direction, the study by Andari et al. (2010) also utilized a type of social outcome measure (in the form of a ball-tossing computer game) that simulates social exchanges to assess the effect of OT on socially relevant cues. By analyzing the individual's ball-toss choices, they determined that while on placebo, high-functioning individuals with ASD demonstrated no preference or discrimination between three fictitious interaction partners, comprised of a "good" partner (i.e., would throw the ball back to the user most often), a "neutral" partner (no preference), and a "bad" partner (rarely return the ball). In comparison, healthy subjects predictably threw the ball to the good player significantly more than the other two. After OT administration, ASD subjects approached the same pattern as their non-ASD counterparts, such that they engaged more often with the good player. Since the game involved a process of determining the reciprocity of each player, it may be viewed as a social learning assessment, implying that OT facilitated social learning. If this is the case, OT may act as a way to "open the door" to social learning, though more research is certainly needed to bolster this hypothesis.

Additional research has widened the putative role of OT in human social behavior. Improvements in both emotional recognition (Guastella et al., 2010) and "empathic accuracy" (Bartz et al., 2010) have been associated with intranasal OT administration. In the first study, Guastella et al. highlight the difficulty that individuals with ASD have in recognizing emotional facial expressions as evaluated by the reading the mind in the eyes test (RMET). In this test, a participant is presented with a picture of the eye region of various faces and asked to choose an adjective out of a list of four that best describes the emotional expression. Sixteen males between the ages of 12 and 19 diagnosed with DSM-IV criteria for autistic disorder completed the RMET 45 min after an intranasal administration of OT. Interestingly, when the tasks of the RMET outcome measure were separated out into "easy" and "hard" items, only the scores on the easy items were significant compared to placebo. In a healthy, non-ASD male population,

other research has found stronger effects of OT for hard items of social cognition rather than easier items (Domes et al., 2007). Guastella et al. suggest that the positive effect of OT may be most prominent for items that are of moderate difficulty for participants, and less so when the tests lead to floor or ceiling effects. In other words, OT may only raise the bar to a certain threshold, beyond which OT exerts little effect.

The second study by Bartz et al. (2010) assessed the effects of intranasal OT on an empathic accuracy task in 27 healthy, non-ASD males and correlated this data with baseline scores on the autism spectrum quotient (AQ), a measure in which lower scores indicate greater social-cognitive proficiency. Although no participants were diagnosed with ASD, results indicated that participants with low AQ scores performed well in both placebo and OT conditions on the empathic accuracy task. In contrast participants with high AQ scores (i.e., more ASD-like traits) performed poorly on placebo, but equally well on OT. The authors suggest that OT selectively improved empathetic accuracy for those with social cognitive deficits (high AQ), but it had no effect on individuals who were more socially proficient on the AQ. This is in line with the suggestion by Guastella et al. (2010) that OT may improve social-cognitive skills (such as empathic accuracy and emotional recognition) only to a point (ceiling effect), after which there may be no improvement. It may be that ASD individuals fail to develop the social skills associated with empathy as fast as their healthy counterparts. As such, an OT challenge may to allow these individuals to temporarily reach this level, while healthy controls or individuals with otherwise normal social-cognitive development do not show improvement because their base level is already at or around the threshold that OT can help one attain.

Finally, a pilot study (in review at the time of the writing of this chapter) assessed the longer-term efficacy of intranasal OT versus placebo in 19 adults with ASD over a period of six weeks. After six weeks on intranasal OT, improvements on various outcome measures that are associated with aspects of social cognition (RMET), repetitive behaviors, and emotional well-being, were observed. Minimal side effects were noted. (Anagnostou et al., 2010 presented at IMFAR). Indeed, there is a need for larger, and longer investigations to reveal the full potential of OT on individuals with ASD.

20.7.2 Oxytocin as a treatment for repetitive behaviors

Hollander et al. (2003) conducted a double-blind placebo-controlled crossover trial of OT in 15 adults with either autism or Asperger's Disorder as confirmed by both the autism diagnostic interview-revised (ADI-R) and DSM-IV criteria. Patients served as their own controls and as such underwent two challenge days in which they received either a continuous infusion of intravenous synthetic OT (Pitocin) or placebo over a period of four hours, titrated up every 15 min. A repeated measures analysis of variance (ANOVA) demonstrated that on OT, both the severity of the repetitive behaviors and the number of different types of these behaviors significantly decreased over time (see Figure 20.2).

The results of this study suggest that OT may mediate the severity and frequency of repetitive behaviors seen in ASD. Curiously, both animal research and human research in OCD have demonstrated that OT tends to increase or induce repetitive behaviors rather than abate them (Bartz and Hollander, 2008). This contrast highlights both the complexity of the role of OT in human behavior, but also implies important differences between OCD and ASD. In sum, although the exact role that OT plays in repetitive behaviors is unclear, it remains nevertheless an important area for future research.

20.8 Limitations of experimental research

While the research to date is exciting and promising, several limitations must be highlighted. First, because of the relative heterogeneity seen in ASD,

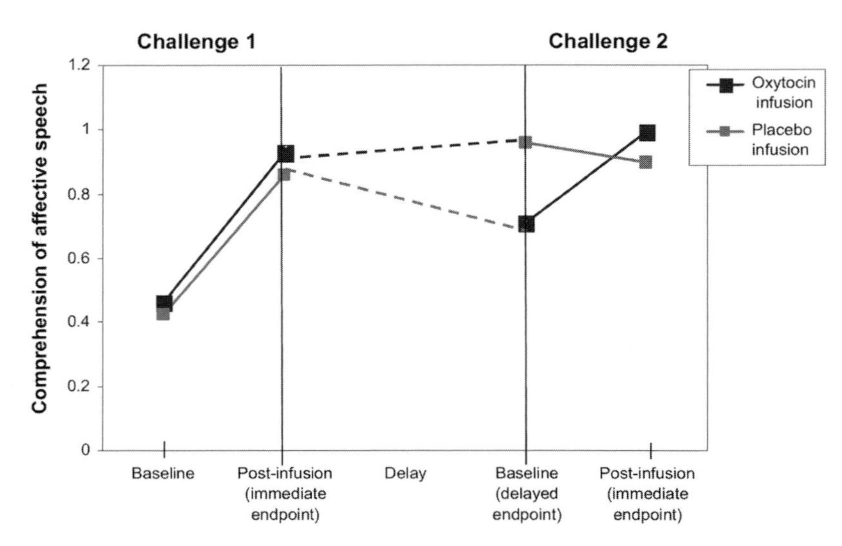

Figure 20.2 Estimated linear trends (on the basis of mixed regression analysis) across time in the dichotomized scores of the affective speech comprehension as a function of conditions (oxytocin vs. placebo) and order of administration (oxytocin first vs. placebo first). Dotted line represents interval between the first and second challenge procedures (days: mean = 16.07; SD = 14.26). See color version in plates section.

more accurate and sensitive measures need to be developed to establish more homogenous types or subgroups. As a result, current research often reveals somewhat foggy results as a consequence of significant individual variability. Second, earlier research appears to demonstrate that OT and AVP have limited access to the brain because of the blood–brain barrier, such that it is unknown what amount reaches the brain (Meisenberg and Simmons, 1983). One consequence of this limited access is that peripheral measures of OT are not necessarily very meaningful because they do not correlate with the levels that reach the brain. Nor is it an ethical option to measure OT centrally because of the invasive nature of such a procedure. Additionally, OT is known to interact with dopaminergic and serotonergic systems in the brain, and therefore its action in the brain is both more complex, and probably more widespread than currently known (McDougle et al., 1999), especially considering the recent suggestions of its role in the olfactory system and the MNS. Third, outcome measures for the sociocognitive symptom domain are in their infancy compared to those that seek to assess repetitive behaviors. Further work is needed to develop outcome measures that are acceptable for registration trials in autism. Accordingly, the repetitive behavior domain is a good target to focus on because more valid outcome measures allow a smoother path for the development of OT as a novel treatment strategy for ASD. Finally, the lack of PET ligands for the OT receptor in humans, and the lack of available small molecule agonists for use in humans, both represent additional hurdles to overcome for future research as they prevent us from determining the distribution of OT (and AVP) receptors in the human brain and the central effects of the peripherally administered peptide.

20.9 Future directions

In order for a full understanding of ASD, we will need to find solutions to address the limitations discussed above. Future research should attempt to improve our methods defining more

homogeneous subgroups of ASD, investigate more thoroughly the OT and AVP systems, particularly during developmental years, and seek to bridge differing approaches to solving common dilemmas. That is, a combination of genetic, epigenetic, experimental, neuroimaging, and therapy-based approaches will help better elucidate the role of the OT system in mediating specific domains in autism, and in developing effective therapeutic agents. In regard to OT and AVP, more research is needed to determine the safety and efficacy of such agents in various patient populations, and within these populations, which target symptoms are most consistently influenced by various therapeutic agents.

Acknowledgments

The authors acknowledge funding by the National Alliance for Research on Schizophrenia and Depression (NARSAD), the Seaver Foundation, and the National Institute of Mental Health (NIH-NIMH) Studies to Advance Autism Research and Treatment (STAART) centers. Dr. Eric Hollander is named as an inventor on a patent application for oxytocin in autism.

Financial disclosures

Dr. Eric Hollander has applied for a patent for oxytocin in autism. Dr. Hollander has consulted in the past to Neuropharm and Nastech.

Related funding: Dr. Hollander has received funding to do a clinical trial along with functional imaging with intranasal oxytocin in high functioning ASD adults from a NARSAD Distinguished Investigator Award, and at the time of this study was funded by the NIH STAART Program 1 U54 MH66673–01.

REFERENCES

American Psychiatric Association. (1994). Task Force on DSM-IV. *Diagnostic and statistical manual of mental disorders (DSM-IV)*. 4th edn. Washington, DC: American Psychiatric Association.

Andari, E., Duhamel, J. R., Zalla, T., Herbrecht, E., Leboyer, M., and Sirigu, A. (2010). Promoting social behavior with oxytocin in high-functioning autism spectrum disorders. *Proc Natl Acad Sci USA*, 107, 4389–4394.

Baron-Cohen, S., Leslie, A. M., and Frith, U. (1985). Does the autistic child have a "theory of mind"? *Cognition*, 21, 37–46.

Baron-Cohen, S., Ring, H. A., Bullmore, E. T., et al. (2000). The amygdala theory of autism. *Neurosci Biobehav Rev*, 24, 355–364.

Baron-Cohen, S., Wheelwright, S., Hill, J., Raste, Y., and Plumb, I. (2001). The "Reading the Mind in the Eyes" Test, revised version: A study with normal adults, and adults with Asperger syndrome or high-functioning autism. *J Child Psychol Psychiatry*, 42, 241–251.

Barton, J. J. (2003). Disorders of face perception and recognition. *Neurol Clin*, 21: 521–548.

Bartz, J. and Hollander, E. (2008). Oxytocin and experimental therapeutics in autism spectrum disorders. *Prog Brain Res*, 170, 451–462.

Bartz, J. A., Zaki, J., Bolger, N., et al. (2010). Oxytocin selectively improves empathic accuracy. *Psychological Science*, 21, 1426–1428.

Boso, M., Emanuele, E., Minazzi, V., Abbamonte, M., and Politi, P. (2007). Effect of long-term interactive music therapy on behavior profile and musical skills in young adults with severe autism. *J Altern Complement Med*, 13, 709–712.

Brang, D. and Ramachandran, V. S. (2010). Olfactory bulb dysgensis, mirror neuron system dysfunction, and autonomic dysregulation as the neural basis for autism. *Med Hypotheses*, 74, 919–921.

Burbach, J. P. H., Young, L. J., and Russell, J. A. (2006). Oxytocin: synthesis, secretion, and reproductive functions. In Neill, J. D., Ed. *Knobil and Neill's Physiology of Reproduction*. 3rd edn. Amsterdam: Elsevier, pp. 3055–3128.

Carter, C. S. (2007). Sex differences in oxytocin and vasopressin: implications for autism spectrum disorders? *Behav Brain Res*, 176, 170–186.

Dalton, K. M., Nacewicz, B. M., Johnstone, T., et al. (2005). Gaze fixation and the neural circuitry of face processing in autism. *Nature Neurosc*, 8, 519–526.

Dapretto, M., Davies, M. S., Pfeifer, J. H., et al. (2006). Understanding emotions in others: mirror neuron dysfunction in children with autism spectrum disorders. *Nat Neurosci*, 9, 28–30.

Davies, S., Bishop, D., Manstead, A. S., and Tantam, D. (1994). Face perception in children with autism and Asperger's syndrome. *J Child Psychol Psychiatry*, 35, 1033–1057.

Domes, G., Heinrichs, M., Michel, A., Berger, C., and Herpertz, S. (2007). Oxytocin improves "mind-reading" in humans. *Biol Psychiatry*, 61, 731–733.

Finnigan, E. and Starr, E. (2010). Increasing social responsiveness in a child with autism. A comparison of music and non-music interventions. *Autism*, 14, 321–348.

Gallese, V., Keysers, C., and Rizzolatti, G. (2004). A unifying view of the basis of social cognition. *Trends Cogn Sci*, 8, 396–403.

Gervais, H., Belin, P., Boddaert, N., et al. (2004). Abnormal cortical voice processing in autism. *Nat Neurosci*, 8, 801–802.

Gregory, S. G., Connelly, J. J., Towers, A., et al. (2009). Genomic and epigenetic evidence for oxytocin receptor deficiency in autism. *BMC Med*, 7.

Green, L., Fein, D., Modahl, C., et al. (2001). Oxytocin and autistic disorder: alterations in peptide forms. *Biol Psychiatry*, 50, 609–613.

Griebel, G., Simiand, J., Serradeil-le Gal, C., et al. (2002). Anxiolytic- and antidepressant-like effects of the non-peptide vasopressin V1b receptor antagonist, SSR149415, suggest an innovative approach for the treatment of stress-related disorders. *Proc Natl Acad Sci USA*, 99, 6370–6375.

Guastella, A. J., Einfeld, S. L., Gray, K. M., et al. (2010). Intranasal oxytocin improves emotion recognition for youth with autism spectrum disorders. *Biol Psychiatry*, 67, 692– 694.

Guastella, A. J., Mitchell, P. B., and Dadds, M. R. (2008). Oxytocin increases gaze to the eye region of human faces. *Biol Psychiatry*, 1, 3–5.

Hammock, E. A. and Young, L. J. (2006). Oxytocin, vasopressin, and pair bonding: implications for autism. *Philos Trans R Soc Lond B Biol Sci*, 361, 2187–2198.

Hobson, R. P., Ouston, J., and Lee, A. (1988). Emotion recognition in autism: coordinating faces and voices. *Psychol Med*, 18, 911–923.

Hollander, E. and Bartz, J. A. (2006). The neuroscience of affiliation: forging links between basic and clinical research on neuropeptides and social behavior. *Horm Behav*, 50, 518–528.

Hollander, E., Bartz, J., Chaplin, W., et al. (2007). Oxytocin increases retention in social cognition in autism. *Biol Psychiatry*, 61, 498–450.

Hollander, E., Novotny, S., Hanratty, M., et al. (2003). Oxytocin infusion reduces repetitive behaviors in adults with autistic and Asperger's disorders. *Neuropsychopharmacology*, 28, 193–198.

Insel, T. R., O'Brien, D. J., and Leckman, J. F. (1999). Oxytocin, vasopressin, and autism: Is there a connection? *Biol Psychiatry*, 45, 145–157.

Jacob, S., Brune, CW., Carter, C. S., et al. (2007). Association of the oxytocin receptor gene (*OXTR*) in Caucasian children and adolescents with autism. *Neurosci Lett*, 417, 6–9.

Jansen, L. M., Gispen-De Wied, C. C., Wiegant, V. M., et al. (2006). Autonomic and neuroendocrine responses to a psychosocial stressor in adults with autistic spectrum disorder. *J Autism Dev Disord*, 36, 891–899.

Kim, J., Wigram, T., and Gold, C. (2008). The effects of improvisational music therapy on joint attention behaviors in autistic children: a randomized controlled study. *J Autism Dev Disord*, 38, 1758–1766.

Kim, S. J., Young, L. J., Gonen, D., et al. (2002). Transmission disequilibrium testing of arginine vasopressin receptor 1A (AVPR1A) polymorphisms in autism. *Mol Psychiatry*, 7, 503–507.

Kirsch, P., Esslinger, C., Chen, Q., et al. (2005). Oxytocin modulates neural circuitry for social cognition and fear in humans. *J Neurosci*, 25, 11489–11493.

Kosfeld, M., Heinrichs, M., Zak, P. J., Fischbacher, U., and Fehr, E. (2005). Oxytocin increases trust in humans. *Nature*, 435, 673–676.

Kuhl, PK., Coffey-Corina, S., Padden, D., and Dawson, G. (2005). Links between social and linguistic processing of speech in preschool children with autism: behavioral and electrophysiological measures. *Dev Sci*, 8, F1–F12.

Lerer, E., Levi, S., Salomon, S., et al. (2008). Association between the oxytocin receptor (*OXTR*) gene and autism: relationship to Vineland Adaptive Behavior Scales and cognition. *Mol Psychiatry*, 13, 980–988.

Lim, M. M., Bielsky, I. F., and L. J. Young (2005). Neuropeptides and the social brain: potential rodent models of autism. *Int J Dev Neurosci*, 23, 235–243.

Markram, K. and Markram, H. (2010). The intense world theory – a unifying theory of the neurobiology of autism. *Front Hum Neurosci*, 21, 224.

McCarthy, M. M. and Altemus, M. (1997). Central nervous system actions of oxytocin and modulation of behavior in humans. *Mol Med Today*, 3, 269–275.

McDougle, C. J., Barr, L. C., Goodman, W. K., and Price, L. H. (1999). Possible role of neuropeptides in obsessive compulsive disorder. *Psychoneuroendocrinology*, 24, 1–24.

Meisenberg, G. and Simmons, W. H. (1983). Minireview. Peptides and the blood-brain barrier. *Life Sci*, 32, 2611–2623.

Modahl, C., Fein, D., Waterhouse, L., and Newton, N. (1992). Does oxytocin deficiency mediate social deficits in autism? *J Autism Dev Disord*, 22, 449–451.

Modahl, C., Green, L., Fein, D., et al. (1998). Plasma oxytocin levels in autistic children. *Biol Psychiatry*, 43, 270–277.

Nilsson, U. (2009). Soothing music can increase oxytocin levels during bed rest after open-heart surgery: a randomised control trial. *J Clin Nurs*, 18, 2153–2161.

Page, D. T., Kuti, O. J., Prestia, C., and Sur, M. (2009). Haploinsufficiency for *Pten* and *Serotonin transporter* cooperatively influences brain size and social behavior. *Proc Natl Acad Sci USA*, 106, 1989–1994.

Panksepp, J. (1992). Oxytocin effects on emotional processes: separation distress, social bonding, and relationships to psychiatric disorders. *Ann NY Acad Sci*, 652, 243–252.

Rutherford, M. D., Baron-Cohen, S., and Wheelwright, S. (2002). Reading the mind in the voice: a study with normal adults and adults with Asperger syndrome and high functioning autism. *J Autism Dev Disord*, 32, 189–194.

Salome, N., Stemmelin, J., Cohen, C., and Griebel, G. (2006). Differential roles of amygdaloid nuclei in the anxiolytic- and antidepressant-like effects of the V1b receptor antagonist, SSR149415, in rats. *Psychopharmacology (Berl)*, 187, 237–244.

Surget, A. and Belzung, C. (2008). Involvement of vasopressin in affective disorders. *Eur J Pharmacol*, 583, 340–349.

Szatmari, P., Tuff, L., Finlayson, M. A., and Bartolucci, G. (1990). Asperger's syndrome and autism: neurocognitive aspects. *J Am Acad Child Adolesc Psychiatry*, 29, 130–136.

Tansey, K. E., Keeley, J., Brookes, B., et al. (2010). Oxytocin receptor (OXTR) does not play a major role in the aetiology of autism: Genetic and molecular studies. *Neuroscience Letters*, 474, 163–167.

Tantam, D., Monaghan, L., Nicholson, H., and Stirling, J. (1989). Autistic children's ability to interpret faces: a research note. *J Child Psychol Psychiatry*, 30, 623–630.

Wan, C. Y., Demaine, K., Zipse, L., Norton, A., and Schlaug, G. (2010). From music making to speaking: Engaging the mirror neuron system in autism. *Brain Research Bulletin*, 82, 161–168.

Wassink, T. H., Piven, J., Vieland, V. J., et al. (2004). Examination of AVPR1a as an autism susceptibility gene. *Mol. Psychiatry*, 9, 968–972.

Waterhouse, L., Fein, D., and Modahl, C. (1996). Neurofunctional mechanisms in autism. *Psychol Rev*, 103, 457–489.

Wermter, A. K., Kamp-Becker, I., Hesse, P., et al. (2010). Evidence for the involvement of genetic variation in the oxytocin receptor gene (*OXTR*) in the etiology of autistic disorders on high-functioning level. *Am J Med Genet B Neuropsychiatr Genet*, 153B, 629–639.

Wu, S., Jia, M., Ruan, Y., et al. (2005). Positive association of the oxytocin receptor gene (*OXTR*) with autism in the Chinese Han population. *Biol Psychiatry*, 58, 74–77.

Yirmiya, N., Rosenberg, C., Levi, S., et al. (2006) Association between the arginine vasopressin 1a receptor (AVPR1a) gene and autism in a family-based study: mediation by socialization skills. *Mol Psychiatry*, 11, 488–494.

Yrigollent, C. M., Han, S. S., Kochetkova, A., et al. (2008) Genes Controlling Affiliative Behavior as Candidate Genes for Autism. *Biol Psychiatry*, 63, 911–916.

21

Oxytocin, vasopressin, and related peptides in the regulation of behavior

Where next?

Elena Choleris, Donald W. Pfaff, and Martin Kavaliers

We editors believe that by dividing this text into three parts, (i) Anatomy, Function, and Development of the Oxytocin (OT) and Vasopressin (AVP) systems, (ii) Comparative Approach in Behavioral Studies, and (iii) Human Studies, we have made clear that each of these three sections has its own room for further development using three different sets of methodological approaches. In addition, and importantly, we think the interactions among these three approaches to studies of the OT and AVP systems have the potential to lead to the development of entirely new lines of research as well as clinical applications. For example, fruitful information for human applications may come from a better understanding of evolutionary precursors of OT and AVP, their original primary function, and interactions with other neurochemical systems.

21.1 Anatomy, function, and development

While our understanding of the anatomy, function and development of the oxytocin and vasopressin system has made very important progress (Chapters 1–4), there is ample room for further research at this level. The regulation of the OT and AVP system is through a complicated series of different molecular and neurohormonal mechanisms (e.g., CD38 and vitamin A, Chapters 1, 3, and 19), several of which are still not well understood. For example, it is known that OT, AVP, and their receptors are regulated by the sex hormones. Indeed, the theme of estrogenic and androgenic regulation of OT and AVP is highly recurrent in this book. However, to what extent the various hormone receptors and their genomic, non-genomic, long-term, and rapid mechanisms of action affect OT, AVP and related peptides, is still unclear. Also, the implications of this for our understanding of evolution, behavior, and applicability to human, especially in regards to age and sex-related differences in action are important factors that are too often neglected in investigations.

In addition, the mechanisms of action of OT and AVP in various brain areas are not fully understood (Chapter 2). In particular, with the exception of a few very promising investigations with adaptive behavioral responses (Chapter 2), we need to expand our knowledge of the behavioral implications of these various brain mechanisms and whether they are related to early-life organizational or later activational effects (Chapter 4). Also, which of these mechanisms represent ancestral adaptations and which have secondarily evolved for species-specific behavioral and physiologic adaptations, is certainly worthy of further investigations.

Finally, whether and how other systems interplay with OT and AVP requires further research effort. In particular, the interaction of OT, AVP and related nonapeptides with other neurotransmitter

Oxytocin, Vasopressin, and Related Peptides in the Regulation of Behavior, ed. E. Choleris, D. W. Pfaff, and M. Kavaliers.
Published by Cambridge University Press.

systems that are known to affect multiple behaviors is little understood. Initial studies in this direction show interesting and promising results. For example, dopamine is important for nonapeptide mediation of social bonds (Chapter 8), gamma-aminobutyric acid, norepinephrine and the hypothalamic–pituitary–adrenal axis may be involved in parental behavior, social bonds, anxiety, and stress responses (Chapters 9, 10, 17, and 19), norepinephrine, serotonin and several other neurotransmitters in territory marking and aggression (Chapter 12). Our understanding of the exact nature of these interactions is often incomplete and offers ground for future research. A better knowledge of the functional integration of a relatively recent mammalian system (OT and AVP) with evolutionary older and highly conserved systems (e.g., dopamine, serotonin) would not only further our understanding of these systems and their evolution, but it could also lead to the development of novel combined drug targets for clinical applications.

21.2 Comparative approach in behavioral studies

The comparative approach in behavioral studies on OT, AVP, and related peptides such as isotocin (IT), mesotocin (MT), and arginine vasotocin (AVT) is a leading theme of this book and represents a crucial area of research that has highlighted the phylogenetic evolution of these systems and their physiologic and behavioral functions. Thanks to these studies we know that the nonapeptide AVT and its localization in the preoptic/hypothalamic area, with projections to the pituitary and mid/hindbrain is the ancestral system from which AVP, IT, MT, OT, and related peptides have evolved. In addition, projections of these neurons throughout the brain have been described in all groups, including fish and amphibians (Chapters 5 and 6) suggesting that brain/behavior effects of these systems are conserved traits within the vertebrates. Finally, that these nonapeptides are involved in social, sexual, and reproductive behaviors in a manner that is interlinked with the action of sex and stress hormones, also appears a shared, possibly early evolved, trait (Chapters 5 and 6). However, the behavioral effects on these nonapeptides and their brain mechanisms of action across the vertebrates are an area that requires further investigations. Research within each of the classes of vertebrates covered in this book, fish, amphibian, bird, and mammals, still has several gaps, which have been extensively highlighted in the various chapters and we refer the readers to those. We should point out here, however, the striking fact that among the several classes studied, that of the reptiles, has been largely neglected and clearly needs investigations. Similarly, the promising research on the system's precursors in invertebrates deserves development. In particular, investigations in the behavioral implications of these systems are very limited (see Preface). It would be especially interesting to further investigate the behavioral functions of these peptides in invertebrate species with high and complex levels of social behavior such as bees, ants, and termites.

The predominance of chapters focusing on mammalian species in this book (Chapters 8–20) reflects the high prevalence of investigations on the involvement of OT and AVP in the mediation of mammalian behavior. In mammals, comparative studies have pointed at evolutionary paths leading to OT and AVP involvement in social behavior. AVPs role in social behavior may have stemmed from its involvement in blood pressure and renal regulation, which, via the implications of these functions for urine concentration, may have lead to AVP mediation of territory marking (Chapter 12), aggression (Chapters 8, 11, 12, and 16), social recognition (Chapters 13 and 14), and male social bonds (Chapters 8 and 16). OT's role in mammalian social behavior instead may have originated from OT's function in parturition and milk letdown, which has lead to OT's involvement in maternal and paternal behavior in various species (Chapters 8, 9, 10, and 16), social recognition (Chapters 8, 13, and 14), and social bonds (Chapters 8, 9, 10, and 16). While the studies with

mammals are the most extensive among the vertebrate species, even here there is ample room for further research. For example, the role of AVP and OT in mammalian social, sexual and maternal behavior shows species differences (Chapters 8 and 9) that point at a key involvement of OT and AVP in the evolution of different types of mammalian social systems (monogamous, polygamous, eusocial, various degrees of territoriality) and parental responses (Chapter 9). However, these comparative studies have been limited to a few species within a small number of rodent genuses (e.g., *Microtus, Peromyscus, Heterocephalus*), and to an even smaller number of primate species (Chapters 8 and 16). A clear understanding of the evolution of the social and non-social functions of the OT and AVP systems in mammals needs further comparative studies within a larger number of mammalian taxonomic groups before a true phylogenetic analysis is possible.

Following a phylogenetic leading theme, the last part of the book completes our coverage of the nonapeptides in the vertebrates, from fish to humans, and describes recent developments in translational research that are leading to the use of OT and AVP in medical applications. These chapters span from the involvement of OT and AVP in drug addiction (Chapter 15), stress, fear, anxiety (Chapters 16, 18, and 19), and social behavior (Chapters 18 and 19) to social psychopathologies such as autism spectrum disorder (Chapters 19 and 20). While showing important research data these chapters also open several questions that require further investigation. For example, animal research points at a potential use of OT or OT analogs to treat or prevent drug addiction and drug-seeking behavior (Chapter 15). Similarly, there seems to be potential for the use of OT and/or AVP antagonists for reducing stress and anxiety (Chapter 16). Clinical trials assessing this potential use in human populations are yet to be run. Conversely, the literature on OT prosocial effects (Chapter 18) has led to ongoing clinical trials where intranasal OT is used to treat autism. Initial results (Chapter 20) are very promising and further development of these lines of research is certainly warranted. For example, they may be expanded to other social pathologies such as schizophrenia, social phobias, and various personality disorders. However, the mechanisms of action of the OT and AVP intranasal administrations are little understood. In particular, given the extremely low blood–brain barrier penetration of OT and AVP how much of the peptides reach relevant brain areas is currently unclear (Chapter 20). Hence, whether the behavioral effects of intranasal OT and AVP are mediated by central mechanisms is still an open question that needs investigating. In addition, given that disorders of sociality, such as autism, are related to altered development, a better understanding of early organizational and activational effects of OT and AVP is needed. This is particularly important if we want to understand the ontogenetic origins of social disorders as well as the potential for interventions later in life.

Finally, we should emphasize the issue of technical limitations that have been important restraining factors to the advancement of the field. To date, the lack of highly specific drugs and radioligands for non-mammalian species (Chapters 8 and 9), imaging ligands that can cross the blood–brain barrier (Chapters 16 and 20), small-molecule peptides that cross the blood–brain barrier (Chapter 20), and of selective antibodies for the OT receptor (Chapter 9) has created a serious obstacle to the development of investigations on the OT, AVP and related systems, that requires further research and technical advancement.

The field is fast moving and exciting. We look forward to the next book where the numerous still open questions described above will have been tackled.

Index